The Impact of Stereochemistry on Drug Development and Use

CHEMICAL ANALYSIS

A SERIES OF MONOGRAPHS ON
ANALYTICAL CHEMISTRY AND ITS APPLICATIONS

Editor
J. D. WINEFORDNER

VOLUME 142

A WILEY INTERSCIENCE PUBLICATION

JOHN WILEY & SONS, INC.

New York / Chichester / Weinheim / Brisbane / Singapore / Toronto

The Impact of Stereochemistry on Drug Development and Use

HASSAN Y. ABOUL-ENEIN

King Faisal Specialist Hospital and Research Centre
Saudi Arabia

and

IRVING W. WAINER

McGill University Montreal, Canada

A WILEY-INTERSCIENCE PUBLICATION

JOHN WILEY & SONS, INC.

New York / Chichester / Weinheim / Brisbane / Singapore / Toronto

This text is printed on acid-free paper.

Library of Congress Cataloging in Publication Data:

The impact of stereochemistry on drug development and use / [edited
by] Hassan Y. Aboul-Enein and Irving W. Wainer.
p. cm.—(Chemical analysis; v. 142)
Includes bibliographical references and index.
ISBN 0-471-59644-2 (alk. paper)
1. Chiral drugs. I. Aboul-Enein, Hassan Y. II. Wainer, Irving
W. III. Series.
RS429. I48 1997
615′.7—dc20 96-33491

Printed in the United States of America

10 9 8 7 6 5 4 3 2 1

To my wife Nagla and my sons Youssef, Faisal, and Basil,
who have filled my life with love, joy, and pride.

Hassan Y. Aboul-Enein

CONTENTS

CONTRIBUTORS

Laila I. Abou Basha, Bioanalytical and Drug Development Laboratory, Biological and Medical Research (mBC-03), King Faisal Specialist Hospital and Research Centre, P.O. Box 3354, Riyadh 11211, Saudi Arabia.

Hassan Y. Aboul-Enein, Bioanalytical and Drug Development Laboratory, Biological and Medical Research (mBC-03), King Faisal Specialist Hospital and Research Centre, P.O. Box 3354, Riyadh 11211, Saudi Arabia.

Sut Ahuja, Ahuja Consulting, 27 Monsey Heights, Monsey, New York, 10952.

Stig G. Allenmark, Department of Organic Chemistry, University of Göteburg, S-41296 Göteburg, Sweden.

Gordon L. Amidon, College of Pharmacy, The University of Michigan, 428 Church Street, Ann Arbor, Michigan 48109-1065.

Thomas A. Baillie, Department of Drug Metabolism, Merck Research Laboratories, WP26A-2044, West Point, Pennsylvania 19486.

Leslie Z. Benet, Department of Pharmacy, University of California San Francisco, San Francisco, California 94143-0446.

Carlo Bertucci, Centro Studio CNR Macromolecole Stereordinate Otticamente Attive, Dipartmento di Chimica e Chimica Industriale, Università di Pisa, via Risorgimento, 35, 56126 Pisa, Italy.

Gottfried Blaschke, Institute of Pharmaceutical Chemistry, University of Münster, Hittorfstr. 58-62, D-48419, Münster, Germany.

Bernard Blessington, Pharmaceutical Chemistry, Bradford University, Bradford BD7 IDP, United Kingdom

Jacek Bojarski, Department of Organic Chemistry, College of Medicine, Jagiellonian University, Medyczna 9, 30-688 Krakow, Poland.

Phyllis R. Brown, Department of Chemistry, University of Rhode Island, Kingston, Rhode Island 02881.

Ralf Büschges, Boehringer Ingelheim KG, Pharmaceutics Department, D-55216 Ingelheim, Germany.

Enrico Domenici, Department of Microbiology, Glaxo Wellcome SpA, Medicine Research Centre, 37135 Verona, Italy.

Markus Fluck, Institut für Organische Chemie der Universität, Auf der Morgenstelle 18, D-72076, Tübingen, Germany.

Eric Francotte, Pharmaceuticals Division, Research Department, K-122. P. 25, CIBA-GEIGY Limited, CH-4002 Basel, Switzerland.

Alexandra Glausch, Institut für Organische Chemie der Universität, Auf der Morgenstelle 18, D-72076, Tübingen, Germany.

Michael Gross, Chiros International, Box 193, HoHoKus, New Jersey 07423.

Andrew J. Hutt, Department of Pharmacy, King's College London, Mahresa Road, London, SW3 6LX, United Kingdom.

Seigo Iwakawa, Department of Hospital Pharmacy, School of Medicine, Kobe University, Chuou-Ku, Kobe 650c, Japan.

Hansjörg Jakubetz, Institut für Organische Chemie der Universität, Auf der Morgenstelle 18, D-72026 Tübingen, Germany.

Martin Jung, Institut für Organische Chemie der Universität, Tübingen, Auf der Morgenstelle 18, D-72076 Germany.

Sylvie Laganière, Biopharmaceutics and Pharmacodynamics Division, Bureau of Drug Research, Health Protection Branch, 2-West, Banting Building, Tunney's Pasture, Ottawa, Ontario, Canada KIA 0L2.

Peter Langguth, Astra Hässle AB, Kärragatan 5, S-43183 Mölndal, Sweden.

Weiyong Li, The R. W. Johnson Pharmaceutical Research Institute, Spring House, Pennsylvania 19477.

Elke Lipka, TSRL Inc., 540 Avis Drive, Ann Arbor, Michigan, 48108.

Eric Martin, Department of Pharmacology, Johann Wolfgang Goethe-University, Marie-Curie Str. 9, Geb. N 260, D-60053 Frankfurt, Germany.

Jonathan P. Mason, Department of Pharmacy, King's College London, Manresa Road, London, SW3 6LX, United Kingdom.

Sabine Mayer, Institut für Organische Chemie der Universität, Auf der Morgenstelle 18, D-72076 Tübingen, Germany.

Werner Möhrke, Procter & Gamble Pharmaceuticals, D-64331 Weiterstadt, Germany.

Simona Negura, Institut für Organische Chemie der Universität, Auf der Morgenstelle 18, D-720726 Tübingen, Germany.

Yoshio Okamoto, Department of Applied Chemistry, School of Engineering, Nagoya University, Chikusa-ku, Nagoya 464-01, Japan.

Carlo Rosini, Dipartimento di Chimica, Università della Basilicata a Potenza, via Nazario Sauro, 85, 85100 Potenza, Italy.

Thomas M. Rossi, The R. W. Johnson Pharmaceutical Research Institute, Spring House, Pennsylvania 19477.

Piero Salvadori, Centro Studio CNR Macromolecole Stereordinate Otticamente Attive, Dipartmento di Chimica e Chimica Industriale, Universià di Pisa, via Risergimento, 35, 56126 Pisa, Italy.

Kathleen M. Schultz, Department of Medicinal Chemistry, School of Pharmacy, University of Washington, Seattle, Washington 98195.

Volker Schurig, Institut für Organische Chemie der Universität, Auf der Murgenstelle 18, D-72076 Tübingen, Germany.

Hildegard Spahn-Langguth, Department of Pharmacy, Martin-Luther-University Halle-Wittenberg, Wolfgang-Langenbeck-Strasse, D-06120 Halle/Saale, Germany.

Michael E. Swartz, Waters Corporation, Department TB, 34 Maple Street, Milford, Massachusetts 01757.

Timothy J. Ward, Department of Chemistry, Millsaps College, Jackson, Mississippi 39210.

Karen D. Ward, Department of Chemistry, Millsaps College, Jackson, Mississippi 39210.

Dorothee Wistuba, Institut fur Organische Chemie der Universität, Auf der Mongenstelle 18, 72076 Tübingen, Germany.

Eiji Yashima, Department of Applied Chemistry, School of Engineering, Nagoya University, Chikusa-ku, Nagoya 464-01, Japan.

Parnian Zia-Amirhosseini, Department of Pharmacy, University of California San Francisco, San Francisco, California 94143-0446.

PREFACE

With the separation of the enantiomorphic crystals of ammonium sodium tartrate, Louis Pasteur uncovered one of the wonders of nature—the interrelationship of symmetry and asymmetry (1). His observations led him to formulate a proposal which is the foundation of molecular stereochemistry: "The optical activity of organic solutions is determined by molecular asymmetry, which produces nonsuperimposable mirror image structures" (2).

This proposition initiated an intense study of the physicochemical properties and theoretical nature of stereochemistry. During the remainder of the nineteenth century, the work of Pasteur, Van't Hoff, Le Bel, and Wislicenus expanded and clarified the concept of chiral molecules (3–5). During this period, the biological and pharmacological implications of stereochemistry were largely ignored, even though nature was the source of chiral chemicals. To some extent this was necessary since a chemical foundation was required to build a biological understanding.

The understanding and appreciation of the role stereochemistry plays in pharmacology also stems from the work of Pasteur. In 1858, Pasteur reported that the *dextro* form of ammonium tartrate was more rapidly destroyed by the mold *Penicillium glaucum* than the *levo* isomer (6). During the next 50 years, numerous similar examples were reported in diverse biological fields. Then, in 1908, Abderhalde and Müller reported the differential effects of (–)- and (+)-epinephrine on blood pressure (pressor effects) (7) and chirality entered mainstream pharmacological research. By the 1930s, Cushny (8) and Easson and Stedman (9) had laid the basis for the initial theoretical understanding of stereochemical differences in pharmacological activity.

As a result of the observations by Cushny, Easson, and Stedman, stereochemistry became an integral part of medicinal chemistry; in particular, a core element in the study of quantitative structure–activity relationships (QSAR). However, stereochemistry essentially remained buried and chiral drugs continued to be developed as racemates. Single isomer drugs were considered only if there were readily available single isomer starting materials (e.g., steroids, antibiotics) or inescapable pharmacological consequences associated with one of the isomers (e.g., *dextro*-methorphan was developed as an over-the-counter antitussive since *levo*-methorphan is a registered narcotic).

To a great extent, stereochemistry could be relegated to a secondary consideration because the pharmaceutical industry and the regulatory agencies lacked adequate analytical techniques. What you could not measure you could not require. This situation has dramatically changed over the past 15 years with the development of analytical methods capable of the rapid separation and accurate measurement of enantiomeric composition.

The breakthrough in this area came with the development of a commercially available HPLC chiral stationary phase (CSP) by W. H. Pirkle (10). This revolutionized the analytical and preparative separations of enantiomeric compounds and initiated a rapid increase in commercially available HPLC and GC CSPs. These technological advances have, in turn, produced an increased interest in the *in vivo* pharmacological fate of the separate enantiomers of chiral substances; particularly from the drug regulatory agencies. The response has been a sustained rise in the number of studies concerned with the pharmacokinetic and metabolic disposition of enantiomeric drugs. At the present time, these studies are routine procedures in the development and testing of new drugs, for both racemic and single-isomers formulations.

Perhaps the most cogent statement of the implications of stereochemical differences in pharmacokinetics and pharmacodynamics was presented by Ariëns (11):

> Often only one isomer is therapeutically active, but this does not mean that the other is really inactive. It may very well contribute to the side effects. The therapeutically nonactive isomer in a racemate should be regarded as an impurity (50% or more). It is emphasized how in clinical pharmacology, and particularly in pharmacokinetics, neglect of stereoselectivity in action leads to the performance of expensive, highly sophisticated scientific nonsense.

The phrase "highly sophisticated scientific nonsense" has been widely quoted and perhaps has been the single most important ideological statement in the current growth of stereochemical awareness in the pharmaceutical industry.

The direct connection between the advancement of enantioselective technology and the discovery, development, and marketing of chiral drugs has resulted in a rapid maturing of this field of research. It is safe to say that if a chiral substance exhibits pharmacological activity, the properties of the separate enantiomers of that substance will be defined, measured, and evaluated in relationship to the development of racemic or single-isomer therapeutic agent.

This is possible because adequate quantities of the separate isomers can be easily prepared and their pharmacological properties and fate determined alone and in the presence of the opposite enantiomer. These are accepted and, in fact, required experiments which are now routinely published in a variety of scientific journals and reported at international meetings. This was not always the case.

As with many new technologies, enantioselective analytical and pharmacological studies were often difficult to report in established journals. Indeed, new journals such as *Chirality* and *Tetrahedron Asymmetry* were created to give this new technology a voice. Chiral symposia, workshops, and meetings were formed to facilitate the dissemination of the scientific advancements and to educate colleagues. As the theory and practice of stereochemistry in pharmaceutical development grew, the symposia, workshops, and meetings spawned a number of books detailing these advancements.

These publications and advancements are the foundation for the present volume. This collection contains a series of articles that describe a mature field of work, one that has been reintegrated into the mainstream of the analytical and pharmacological sciences.

The initial chapters of this work address the pharmacological consequences of stereochemistry. The fact that enantiomers have different fates and effects is not surprising since living organisms contain numerous chiral biopolymers such as proteins, enzymes, cellular surfaces, etc. In fact, it is safe to say that any active pharmacological process has the possibility of being enantioselective or enantiospecific, and it probably is. Thus, protein binding, biotransformation, receptor binding, active transport into and out of cells, intracellular sequestration, DNA binding, etc. have been shown to discriminate between drug enantiomers. This situation has been generalized in "Pfeiffer's rule," which states that the more potent a drug, the more likely it is to show stereoselectivity due to a greater steric demand for tight receptor binding (12). It should be noted that since the physicochemical properties of enantiomers are equivalent, passive pharmacological processes, for example absorption, will be the same for both isomers.

Drug metabolism is a key aspect in the pharmacological fate and effect of a therapeutic agent. The enantioselectivities of the Phase I and Phase II microsomal transformations have been extensively studied and are a rich source of biological information. In addition, nonmicrosomal transformations have also shown stereochemical preferences. For examples, the aldoketo reductase mediated reduction of the prochiral drug metyrapone to the chiral metyrapol is not enantioselective in men, but has a pronounced enantioselectivity in women with (+)-metyrapol favored over the (–)-enantiomer by a factor of 1.6:1 (13).

However, these investigations have inherent problems stemming from the duality of enantiomers, which are at the same time chemically identical and spatially different molecules. This situation is illustrated by the results of a study of a metabolic enantiomeric interaction involving the competitive inhibition of (S)-warfarin-7-hydroxylase by (R)-warfarin (14). The authors of this study concluded that:

1. The kinetic parameters defining the interactions of two enantiomers of a racemic drug with the cytochrome P-450s or other macromolecular systems in the living organism can only be properly defined from experiments with the pure enantiomers;
2. an enantiomer of a racemic drug may contribute significantly to biological effect not by its inherent activity but by altering the pharmacokinetics of the eutomer;
3. enantiomeric interactions are not easily detected unless directly sought and may be relatively common.

Three chapters are devoted to the stereochemical aspects of drug metabolism and will give the reader an excellent overview of this area. Three additional chapters address particular aspects of chirality and drug activity. One presents the toxicological consequences of the stereoselectivity in the xenobiotic metabolism of alkenes and a second describes the synthesis, chromatographic resolution, and biological activity of chiral barbiturates. The remaining chapter presents the case of ethambitol, where the lack of a clear understanding of the stereochemical composition of the drug has hindered its effective therapeutic application.

The latter chapter also reflects another by-product of the current preoccupation with drug stereochemistry, the "racemic switch." In the strategy, currently racemic drugs are reevaluated with the intention of developing a single-isomer formulation. The goal is a new chirally pure drug with a better therapeutic index than the racemate and, perhaps, new clinical applications. These possibilities are illustrated by the use of verapamil (VER) in the treatment of adriamycin-resistant tumors.

VER is a chiral calcium channel blocking drug widely used in the therapy of hypertension, supraventricular arrhythmias, and angina pectora (15). The enantiomers of VER have different pharmacodynamic and pharmacokinetic properties; for example, S-(–)-VER is 10 to 20 times more potent than R-(+)-VER [16]. In addition to its cardiovascular activities, VER has another possible clinical application in cancer chemotherapy as a modifier of multidrug resistance (MDR). Initial *in vitro* experiments have demonstrated that the presence of VER in the incubation media increased the cytotoxicity of vinca alkaloid and anthracycline derivatives in several resistant tumor cell lines; in particular adriamycin resistant cell lines (17). One proposed source of MDR is a decrease in the accumulation of intracellular concentrations of the antineoplastic agents due to increased expression of a glycoprotein which acts as an efflux pump for cyctostatic drugs (18). The effect of VER on MDR is due to the inhibition of this efflux pump (18).

Based upon the *in vitro* experiments, several clinical Phase I trials were carried out using VER in combination with adriamycin (19) and vinblastine (20).

However, these trials were not successful. Plasma levels which were comparable to the effective *in vitro* concentrations could not be achieved due to the cardiotoxicity of VER. The results of one study involving the treatment of MDR ovarian cancer patients with VER and adriamycin were summarized as follows:

> However, the high infusion rates of verapamil (9 μg/kg/min) required to achieve these plasma levels produced an unacceptable degree of cardiac toxicity. Two patients developed transient atropine-responsive complete heart block and four patients developed transient congestive heart failure with increases in pulmonary capillary wedge pressure... Future studies should use less cardiotoxic calcium channel blockers that can be safely administered to produce the plasma levels required for *in vitro* sensitization of drug resistant cells (19).

One less cardiotoxic compound is the "inactive" isomer of VER, (R)-(+)-VER. While this isomer of VER has on 1/10 to 1/20 of the negative dromotropic, inotropic, and vasodilating activity of (S)-(–)-VER, it has equivalent activity in the modification of MDR (17, 21). Clinical Phase I trials of (R)-(+)-VER are currently underway (21).

The next group of chapters in this volume addresses the preparation of chirally pure compounds and the analytical determination of stereochemical composition. The predominate theme is chromatographic enantioselective resolution on chiral stationary phases. This is the correct emphasis since, as stated above, the development of commercially available CSPs was the major technological advance which triggered the current chiral explosion. These chapters are presented by experts in the field and are recommended to the reader.

Chromatographic resolutions on CSPs are not the only approach to the determination of optical purity. Indirect determination based upon diastereomeric derivatizations is still a popular and effective strategy and is discussed in one of the chapters. In addition, the use of circular dichroism (CD) spectroscopy in stereochemical and analytical determinations is also discussed. Of further interest is the contribution describing the application of CD spectroscopy to the study of drug-protein binding. This is an interesting and powerful technique which often provides pharmacological information unattainable by other means.

The last section of the volume presents the current status of the regulatory–pharmaceutical industry debate concerning guidelines for the development and approval of stereoisomeric drugs. These chapters do not present the regulatory guidelines in their final form, especially since the international harmonization of drug regulations is an on-going process. However, they present the historical evolution of the debate as well as an outline of the final product.

No matter what form the final regulatory guidelines take, it is clear that drug development can no longer occur without consideration of drug stereochemistry. This is quite a change from the situation in the early 1980s. It is our hope that this volume will give the reader some sense of the magnitude of the stereochemical revolution that has occurred in the past 15 years and an understanding of the maturity of this area of research. We have in one sense come full circle and stand alongside Pasteur in amazement of nature's duality, symmetry, and dissymmetry, and its chemical and pharmacological consequences. It will be fascinating to see what the next 15 years will bring.

Montreal, Quebec, Canada
January 1996
IRVING W. WAINER

Riyadh, Saudi Arabia
January 1996
HASSAN Y. ABOUL-ENEIN

REFERENCES

1. Pasteur, L. (1848) *Ann. Chim. Phys.*, **24**, 442.
2. Pasteur, L. (1901) in *The Foundation of Stereochemistry: Memoirs by Pasteur, Van't Hoff, Le Bel and Wislicenus*, (G. M. Richardson, trans. and ed.), American Book Co., New York, pp. 1–33.
3. For a historical perspective of the development of stereochemistry, see D. E. Drayer in *Drug Stereochemistry: Analytical Methods and Pharmacology, Second Edition* (Wainer, I. W., ed) Marcel Dekker, New York, 1993, pp. 1–24.
4. Partington, J. R. (1964) *A History of Chemistry, Vol. 4*, MacMillan and Co., Ltd., London, pp. 749–764.
5. Weyer, J. (1974) *Angew. Chemie. Internat. Ed.*, **13**, 591.
6. Pasteur, L. (1858) *Compt. Rend.*, **46**, 615.
7. Abderhalden, E. and Müller, F. (1908) *Z. Physiol. Chem.*, **58**, 185.
8. Cushny, A. R. (1926) *Biological Relations of Optically Isomeric Substances*, Ballière, Tindall and Cox, London.
9. Easson, L. H. and Stedman, E. (1933) *Biochem. J.*, **27**, 1257.
10. Pirkle, W. H., Finn, J. M., Schriner, J. L. and Hamper, B. C. (1981) *J. Am. Chem. Soc.*, **103**, 3964.
11. Ariëns, E. J. (1984) *Eur. J. Clin. Pharmacol.*, **26**, 663.
12. Pfeiffer, C. C. (1956) *Science*, **124** , 29.
13. Chiarotto, J. A. and Wainer, I. W. (1995) *Pharm. Sci.*, **1**, 79.
14. Kunze, K. L., Eddy, A. C., Gibaldi, M., and Trager, W. F. (1991) *Chirality*, **3**, 24.
15. Ellrodt, G., Chew, C. Y. C. and Sing, B. N. (1980) *Circulation*, **62**, 669.
16. Echizen, H., Manz, M. and Eichelbaum, M. (1988) *J. Cardiovasc. Pharmacol.*, **12**, 543.

17. Plumb, J. A., Milroy, R. and Kaye, S. B. (1990) *Biochem. Pharmacol.*, **39**, 787.
18. I. Pastan I, and Gottesman, M. M. (1987) *N. Engl. J. Med.*, **316**, 1388.
19. Ozols, R. F., Cunnion, R. E., Klecker, R. W., Hamilton, T. C., Ostchega, Y., Parrillo, J. E., and Yound, R. C. (1987) *J. Clin. Oncol.*, **5**, 641.
20. Cairo, M. S., Siegel, S., Anas, N., and Sender, L. (1989) *Cancer Res.*, **49**, 1063.
21. Haussermann, K., Benz, B., Gekeler, V., Schumacher, K., and Eichelbaum, M. (1991) *Biochem. Pharmacol.*, **40**, 53.

CHEMICAL ANALYSIS

A SERIES OF MONOGRAPHS ON ANALYTICAL CHEMISTRY AND ITS APPLICATIONS

J. D. Winefordner, *Series Editor*

Vol. 1. **The Analytical Chemistry of Industrial Poisons, Hazards, and Solvents,** *Second Edition.* By the late Morris B. Jacobs

Vol. 2. **Chromatographic Adsorption Analysis.** By Harold H. Strain (*out of print*)

Vol. 3. **Photomeric Determination of Traces of Metals.** *Fourth Edition*
Part I: General Aspects. By E. B. Sandell and Hiroshi Onishi
Part IIA: Individual Metals, Aluminum to Lithium. By Hiroshi Onishi
Part IIB: Individual Metals, Magnesium to Zirconium. By Hiroshi Onishi

Vol. 4. **Organic Reagents Used in Gravimetric and Volumetric Analysis.** By John F. Flagg (*out of print*)

Vol. 5. **Aquametry: A Treatise on Methods for the Determination of Water.** *Second Edition* (*in three parts*). By John Mitchell, Jr. and Donald Milton Smith

Vol. 6. **Analysis of Insecticides and Acaricides.** By Francis A. Gunther and Roger C. Blinn (*out of print*)

Vol. 7. **Chemical Analysis of Industrial Solvents.** By the late Morris B. Jacobs and Leopold Schetlan

Vol. 8. **Colorimetric Determination of Nonmentals.** *Second Edition.* Edited by the late David F. Boltz and James A. Howell

Vol. 9. **Analytical Chemistry of Titanium Metals and Compounds**. By Maurice Codell

Vol. 10. **The Chemical Analysis of Air Pollutants.** By the late Morris B. Jacobs

Vol. 11. **X-Ray Spectrochemical Analysis.** *Second Edition.* By L. S. Birks

Vol. 12. **Systematic Analysis of Surface-Active Agents.** *Second Edition.* By Milton J. Rosen and Henry A. Goldsmith

Vol. 13. **Alternating Current Polarography and Tensammetry**. By B. Breyer and H. H. Bauer

Vol. 14. **Flame Photometry.** By R. Herrmann and J. Alkemade

Vol. 15. **The Titration of Organic Compounds** (*in two parts*). By M. R. F. Ashworth

Vol. 16. **Complexation in Analytical Chemistry: A Guide for the Critical Selection of Analytical Methods Based on Complexation Reactions.** By the late Anders Ringbom

Vol. 17. **Electron Probe Microanalysis.** *Second Edition.* By L. S. Briks

Vol. 18. **Organic Complexing Reagents: Structure, Behavior, and Application to Inorganic Analysis.** By D. D. Perrin

Vol. 19. **Thermal Analysis.** *Third Edition.* By Wesley Wm. Wendlandt

Vol. 20. **Amperometric Titrations.** By John T. Stock

Vol. 21. **Reflectance Spectroscopy**. By Wesley Wm. Wnndlandt and Harry G. Hecht

Vol. 22. **The Analytical Toxicology of Industrial Inorganic Poisons.** By the late Morris B. Jacobs

Vol. 23. **The Formation and Properties of Precipitates.** By Alan G. Walton

Vol. 24. **Kinetics in Analytical Chemistry.** By Harry B. Mark, Jr. and Garry A. Rechnitz

Vol. 25. **Atomic Absorption Spectroscopy.** *Second Edition* By Morris Slavin

Vol. 26. **Characterization of Organometallic Compounds** (*in two parts*). Edited by Minoru Tsutsui

Vol. 27. **Rock and Mineral Analysis.** *Second Edition.* By Wesley M. Johnson and John A. Maxwell

Vol. 28. **The Analytical Chemistry of Nitrogen and Its Compounds** (*in two parts*). Edited by C. A. Streuli and Philip R. Averell

Vol. 29. **The Analytical Chemistry of Sulfur and Its Compounds** (*in three parts*). By J. H. Karchmer

Vol. 30. **Ultramicro Elemental Analysis**. By Günther Tölg

Vol. 31. **Photometric Organic Analysis** (*in two parts*). By Eugene Sawicki

Vol. 32. **Determination of Organic Compounds: Methods and Procedures.** By Frederick T. Weiss

Vol. 33. **Masking and Demasking of Chemical Reactions.** By D. D. Perrin

Vol. 34. **Neutron Activation Analysis.** By D. De Soete, R. Gijbels, and J. Hoste

Vol. 35. **Laser Raman Spectroscopy.** By Marvin C. Tobin

Vol. 36. **Emission Spectrochemical Analysis.** By Morris Slavin

Vol. 37. **Analytical Chemistry of Phosphorus Compounds.** Edited by M. Halmann

Vol. 38. **Luminescence Spectroscopy in Analytical Chemistry.** By J. D. Winefordner, S. G. Schulman, and T. C. O'Haver

Vol. 39. **Activation Analysis with Neutron Generators.** By Sam S. Nargolwalla and Edwin P. Przybylowicz

Vol. 40. **Determination of Gaseous Elements in Metals.** Edited by Lynn L. Lewis, Laben, M. Melnick, and Ben D. Holt

Vol. 41. **Analysis of Silicons.** Edited by A. Lee Smith

Vol. 42. **Foundations of Ultracentrifugal Analysis.** By H. Fujita

Vol. 43. **Chemical Infrared Fourier Transform Spectroscopy.** By Peter R. Griffiths

Vol. 44. **Microscale Manipulations in Chemistry.** By T. S. Ma and V. Horak

Vol. 45. **Thermometric Titrations.** By J. Barthel

Vol. 46. **Trace Analysis: Spectroscopic Methods for Elements.** Edited by J. D. Winefordner

Vol. 47. **Contamination Control in Trace Element Analysis.** By Morris Zief and James W. Mitchell

Vol. 48. **Analytical Applications of NMR.** By D. E. Leyden and R. H. Cox

Vol. 49. **Measurement of Dissolved Oxygen.** By Michael L. Hitchman

Vol. 50. **Analytical Laser Spectroscopy.** Edited by Nicolo Omenetto

Vol. 51. **Trace Element Analysis of Geological Materials.** By Roger D. Reeves and Robert R. Brooks

Vol. 52. **Chemical Analysis by Microwave Rotational Spectroscopy.** By Ravi Varma and Lawrence W. Hrubesh

Vol. 53. **Information Theory As Applied to Chemical Analysis.** By Karel Eckschlager and Vladimir Stepanek

Vol. 54. **Applied Infrared Spectroscopy: Fundamentals, Techniques, and Analytical Problem-Solving.** By A. Lee Smith

Vol. 55. **Archaeological Chemistry.** By Zvi Goffer

Vol. 56. **Immobilized Enzymes in Analytical and Clinical Chemistry.** By P. W. Carr and L. D. Bowers

Vol. 57. **Photoacoustics and Photoacoustic Spectroscopy**. By Allan Rosencwaig

Vol. 58. **Analysis of Pesticide Residues.** Edited by H. Anson Moye

Vol. 59. **Affinity Chromatography.** By William H. Scouten

Vol. 60. **Quality Control in Analytical Chemistry.** *Second Edition.* By G. Kateman and L. Buydens

Vol. 61. **Direct Characterization of Fineparticles.** By Brian H. Kaye

Vol. 62. **Flow Injection Analysis.** By J. Ruzicka and E. H. Hansen

Vol. 63. **Applied Electron Spectroscopy for Chemical Analysis.** Edited by Hassan Windawi and Floyd Ho

Vol. 64. **Analytical Aspects of Environmental Chemistry.** Edited by David F. S. Natusch and Philip K. Hopke

Vol. 65. **The Interpretation of Analytical Chemical Data by the Use of Cluster Analysis.** By D. Luc Massart and Leonard Kaufman

Vol. 66. **Solid Phase Biochemistry: Analytical and Synthetic Aspects.** Edited by William H. Scouten

Vol. 67. **An Introduction to Photoelectron Spectroscopy.** By Pradip K. Ghosh

Vol. 68. **Room Temperature Phosphorimetry for Chemical Analysis.** By Tuan Vo-Dinh

Vol. 69. **Potentiometry and Potentiometric Titrations.** By E. P. Serjeant

Vol. 70. **Design and Application of Process Analyzer Systems.** By Paul E. Mix

Vol. 71. **Analysis of Organic and Biological Surfaces.** Edited by Patrick Echlin

Vol. 72. **Small Bore Liquid Chromatography Columns: Their Properties and Uses.** Edited by Raymond P. W. Scott

Vol. 73. **Modern Methods of Particle Size Analysis.** Edited by Howard G. Barth

Vol. 74. **Auger Electron Spectroscopy.** By Michael Thompson, M. D. Baker, Alec Christie, and J. F. Tyson

Vol. 75. **Spot Test Analysis: Clinical, Environmental, Forensic and Geochemical Applications.** By Ervin Jungreis

Vol. 76. **Receptor Modeling in Environmental Chemistry.** By Philip K. Hopke

Vol. 77. **Molecular Luminescence Spectroscopy: Methods and Applications** (*in three parts*). Edited by Stephen G. Schulman

Vol. 78. **Inorganic Chromatographic Analysis.** By John C. McDonald

Vol. 79. **Analytical Solution Calorimetry.** Edited by J. K. Grime

Vol. 80. **Selected Methods of Trace Metal Analysis: Biological and Environmental Samples.** By Jon C. VanLoon

Vol. 81. **The Analysis of Extraterrestrial Materials.** By Isidore Adler

Vol. 82. **Chemometrics.** By Muhammad A. Sharaf, Deborah L. Illman, and Bruce R. Kowalski

Vol. 83. **Fourier Transform Infrared Spectrometry.** By Peter R. Griffiths and James A. de Haseth

Vol. 84. **Trace Analysis: Spectroscopic Methods for Molecules.** Edited by Gary Christian and James B. Callis

Vol. 85. **Ultratrace Analysis of Pharmaceuticals and Other Compounds of Interest.** Edited by S. Ahuja

Vol. 86. **Secondary Ion Mass Spectrometry: Basic Concepts, Instrumental Aspects, Applications and Trends.** By A. Benninghoven, F. G. Rüdenauer, and H. W. Werner

Vol. 87. **Analytical Applications of Lasers.** Edited by Edward H. Piepmeier

Vol. 88. **Applied Geochemical Analysis.** By C. O. Ingamells and F. F. Pitard

Vol. 89. **Detectors for Liquid Chromatography.** Edited by Edward S. Yeung

Vol. 90. **Inductively Coupled Plasma Emission Spectroscopy: Part I: Methodology, Instrumentation, and Performance; Part II: Applications and Fundamentals.** Edited by J. M. Boumans

Vol. 91. **Applications of New Mass Spectrometry Techniques in Pesticide Chemistry.** Edited by Joseph Rosen

Vol. 92. **X-Ray Absorption: Principles, Applications, Techniques of EXAFS, SEXAFS, and XANES.** Edited by D. C. Konnigsberger

Vol. 93. **Quantitative Structure-Chromatographic Retention Relationships.** By Roman Kaliszan

Vol. 94. **Laser Remote Chemical Analysis.** Edited by Raymond M. Measures

Vol. 95. **Inorganic Mass Spectrometry.** Edited by F. Adams, R. Gijbels, and R. Van Grieken

Vol. 96. **Kinetic Aspects of Analytical Chemistry.** By Horacio A. Mottola

Vol. 97. **Two-Dimensional NMR Spectroscopy.** By Jan Schraml and Jon M. Bellama

Vol. 98. **High Performance Liquid Chromatography.** Edited by Phyllis R. Brown and Richard A. Hartwick

Vol. 99. **X-Ray Fluorescence Spectrometry.** By Ron Jenkins

Vol. 100. **Analytical Aspects of Drug Testing.** Edited by Dale G. Deutsch

Vol. 101. **Chemical Analysis of Polycyclic Aromatic Compounds.** Edited by Tuan Vo-Dinh

Vol. 102. **Quadrupole Storage Mass Spectrometry.** By Raymond E. March and Richard J. Hughes

Vol. 103. **Determination of Molecular Weight.** Edited by Anthony R. Cooper

Vol. 104. **Selectivity and Detectability Optimizations in HPLC.** By Satinder Ahuja

Vol. 105. **Laser Microanalysis.** By Lieselotte Moenke-Blankenburg

Vol. 106. **Clinical Chemistry.** Edited by E. Howard Taylor

Vol. 107. **Multielement Detection Systems for Spectrochemical Analysis.** By Kenneth W. Busch and Marianna A. Busch

Vol. 108. **Planar Chromatography in the Life Sciences.** Edited by Joseph C. Touchstone

Vol. 109. **Fluorometric Analysis in Biomedical Chemistry: Trends and Techniques Including HPLC Applications.** By Norio Ichinose, George Schwedt, Frank Michael Schnepel, and Kyoko Adochi

Vol. 110. **An Introduction to Laboratory Automation.** By Victor Cerda and Guillermo Ramis

Vol. 111. **Gas Chromatography: Biochemical, Biomedical, and Clinical Applications.** Edited by Ray E. Clement

The Impact of Stereochemistry on Drug Development and Use

CHAPTER

1

CHIRALITY AND DRUG HAZARDS

HASSAN Y. ABOUL-ENEIN and LAILA I. ABOU BASHA

Bioanalytical and Drug Development Laboratory
Biological and Medical Research (MBC-03)
King Faisal Specialist Hospital and Research Centre
Riyadh 11211
Saudi Arabia

Symmetry or asymmetry is one of the interesting features of geometric figures with two or more dimensions. One of the most common asymmetric molecules is a tetravalent carbon atom with four different ligands attached to it (Figure 1.1).

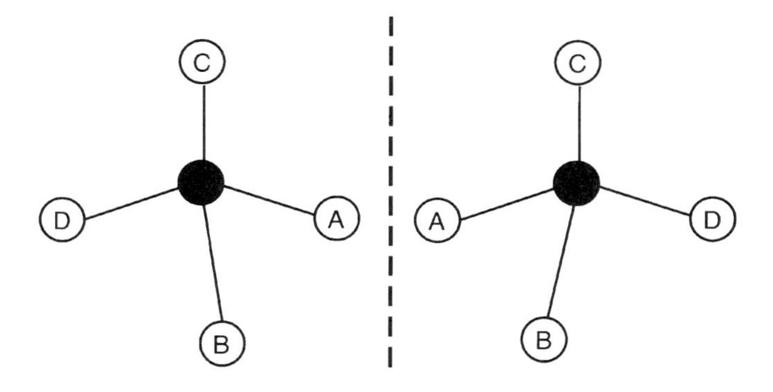

Figure 1.1. An asymmetric tetravalent carbon atom and its mirror image.

1.1. DEFINITION

When molecules composed of the same constituents have the same structural formulas (are like one another with respect to which atoms are joined) but

The Impact of Stereochemistry on Drug Development and Use, Edited by Hassan Y. Aboul-Enein and Irving W. Wainer. Chemical Analysis Series, Vol. 142.
ISBN 0-471-59644-2

differ from each other only in the way the atoms or groups are oriented in space, they are defined as stereoisomers (1). Stereoisomers can be optical or geometrical isomers. Optical isomers are members of a set of stereoisomers, at least two of which are optically active or chiral; geometrical isomers are members of a set of stereoisomers that contains no optically active members.

When such optical isomers are nonsuperimpossible mirror images, they are enantiomeric and called antipodes. The property of nonsuperimpossibility is termed chirality and the substructural feature that gives rise to this asymmetry is called the chiral center. When stereoisomers are not mirror images, they are termed diastereomeric and can exist when the molecules contain more than one chiral center (2). In diastereomers, stereoisomers are classified as meso, *cis*, *trans*, Z, and E members.

The particular arrangement of atoms in space that distinguishes one stereoisomer from another is termed a configuration.

Carbon is not the only atom that can act as asymmetric center. Phosphorus, sulfur, and nitrogen are among other atoms that form chiral molecules (1).

There are three types of stereoisomers.

(*1*) *Enantiomers.* Enantiomeric compounds in which the asymmetric center is a tetravalent carbon, as in Figure 1.1, represent the largest class of chiral molecules. The tetrahedral orientation of the bonds to a tetravalent carbon is such that when four nonidentical ligands are present, the mirror image of the molecules is nonsuperimpossible, and the molecule is enantiomeric and chiral. When two of the ligands are identical, the mirror image is superimpossible, and the molecule possesses a plane of symmetry and is achiral. Enantiomers can be distinguished empirically in terms of their ability to rotate polarized light: to the right (+) as a dextrorotatory isomer and to the left (–) as a levorotatory isomer. Enantiomers are identical in chemical and physical properties except for the sign of rotation of plane-polarized light.

(*2*) *Diastereomers.* Diastereomers are optical isomers that are not related as an object and its mirror image. Unlike enantiomers, the physical and chemical properties of diastereomers can differ and it is not unusual for them to have different melting and boiling points, refractive indices, solubilities, etc. Their optical rotation can differ in both sign and magnitude.

A common diastereomeric molecule is one that contains two asymmetric carbons. The compounds ephedrine and pseudoephedrine (Figure 1.2) illustrate this situation. In these molecules, the asymmetric carbon 2 and 2′ are identical, whereas carbons 1 and 1′ are mirror images. The diastereomer (ephedrine and pseudoephedrine) exists as a member of an enantiomeric pairs, that is, (+)- and (–)-ephedrine and (+)- and (–)-pseudoephedrine,

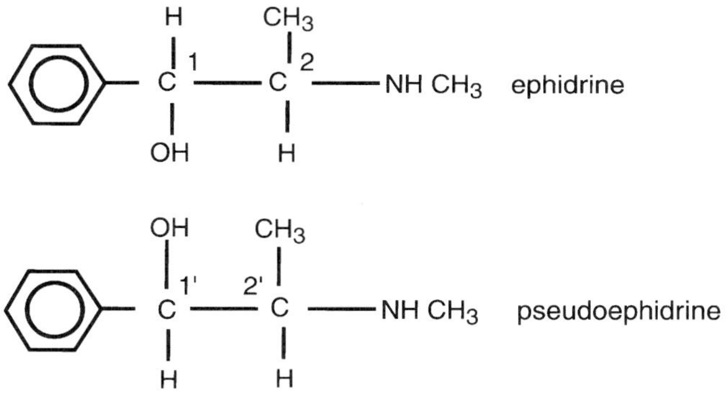

Figure 1.2. The structure of diastereomeric molecules ephidrine and pseudoephidrine.

respectively. Thus, a diastereomeric molecule with two asymmetric centers is most often represented by four stereoisomers.

(*3*) *Geometrical Isomers.* Molecules that contain a carbon–carbon double bond, alkenes, and similar double-bonded systems, C=N, for example, can exist as stereoisomers. This situation is illustrated by 2-butene (Figure 1.3). These sets of isomers are not optically active.

In this molecule, the two methyl groups can be found on the same or opposite side of the double bond. The molecule is defined as the *cis*- or Z-isomers when these groups are on the same side, and *trans* or E when they are on the opposite side. The Z comes from the German *zusammen* that means together; E comes from the German *entgegen* that means opposite.

The configuration around a chiral center can be described in terms of the two conventions that are next described.

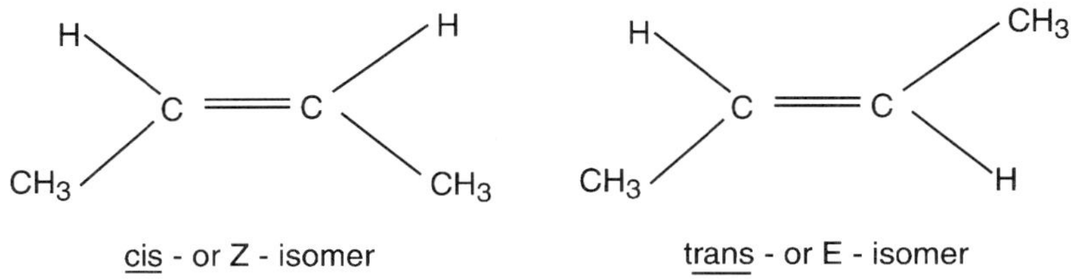

Figure 1.3. The isomers of 2-butene.

1.1.1. The Fisher Convention

To assign a configuration, the molecule under investigation must be chemically converted to glyceraldehyde or to another molecule of known configuration. Once that is accomplished, a *R* or *S* configuration is assigned accordingly to the D or L configuration (3). The sign of rotation cannot be used as a priority in assigning a configuration, because it does not always correspond. For example, L-alanine has a (+) sign of rotation, whereas the sign of rotation for L-glycereldehyde is (–).

The D, L convention is used today mainly for defining the absolute configuration of α animo acids and sugars. Because of this, the Fisher convention has been almost replaced by the Cahn–Ingold–Prelog convention.

1.1.2. The Cahn–Ingold–Prelog Convention

This convention was designated by its originators as the *sequence rule* since it designates the sequence of substituents around the asymmetric center (4). To apply this method, the chiral center is aligned in space so that the smallest group is oriented away from the viewer. The priority order is established by rules based on atomic number. Once aligned in space, the arrangement of the remaining three groups is defined as either a *R* or *S* configuration, depending on whether the decreasing priority order is clockwise (*R*, rectus) or counter-clockwise (*S*, sinister).

This convention can be used to specify the configuration of a chiral center. Also, it is extremely useful for describing diastereomers. In this case, each chiral center is designated independently and the configuration of the whole can be easily assigned. For example, instead of (+)- and (–)-pseudoephedrine, the assigned configuration is (*R*, *S*)- and (*S*, *R*)-ephedrine and (*R*, *R*)- and (*S*, *S*)-pseudoephedrine. The lowercase letters d and l are no longer used.

1.2. CHIRALITY AND BIOLOGICAL ACTIVITY

Chirality is a prominent feature of most biological processes, and the enantiomers of a bioactive molecule often possesses different biological effects. The phenomenon of enantioselectivity in biological action is not restricted to a pharmaceutical context but is characteristic of all biologically active agents, including insecticides herbicides, flavors and fragrances, and food additives. For example, it is well known that the herbicidal activity of 2-phenoxy-proprionic acid groups of agrochemicals resides predominantly in the *R*-enantiomer. Consequently, there is increasing environmental pressure to market these compounds as single enantiomers, thereby reducing their ecological impact (5).

Drugs that are derived from natural products are usually obtained in the optically active or pure form of a single isomer. However, drugs that are produced by chemical synthesis are usually a mixture of equal parts of two, four, or more isomers, depending on the number of asymmetric centers.

The concept of stereoselectivity in drug action, transport, and elimination has been well characterized for a select number of investigated endogenous and exogenous chemicals (6–8). When these pharmacological or physiological processes have been investigated, stereoselectivity has usually been found (9, 10).

The concept of stereoselective binding of a drug to a receptor evolved through historical development. Pasteur was the first to notice the close juxtaposition of biological catalysis and biological activity. In his studies of the underlying causes of both fermentation and infectious diseases, he became convinced that microbes, although capable of mediating a useful biotransformation, could also bring death and destruction to mankind. Thus, at the molecular level biocatalysis and biological activity are two sides of the same coin; the former involves the reaction of a substrate with a polymer (enzyme) and the latter a biopolymer (enzyme reception site) with a bioactive molecule. The common denominator is chirality (11).

In his studies of sugars, Emil Fischer (12) observed that the enzyme emulsion catalyzes the hydrolysis of β-methyl-D-glucoside as a substrate. This observation led Fischer to propose his famous *lock-and-key concept* of enzyme specificity. But this simple explanation led to the popular *one-enzyme one-substrate concept*, a misconception formula prevalent among chemists.

Most drugs are specific and their action is usually explained on the basis of receptor theory. The basic idea of receptor sites can be traced back to Paul Ehrlich who proposed the term chemoreceptors to describe these binding sites (13).

Receptor molecules in the body are proteins that exhibit high affinities for binding ligands of a certain molecular structure; this is completely analogous to enzyme substrate binding. The binding of a substrate to a receptor triggers a mechanism (e.g., the modification of enzyme activity, transport of ions, etc.) that manifests itself in a biological response.

Whatever their physiological function, receptors have one thing in common. They are themselves chiral molecules and can, therefore, be expected to be enantioselective in their binding to messenger molecules.

The idea of enantioselectivity in drug–receptor interaction dates back to 1926 when Cushny (14) proposed that different bioactivities of two enantiomers arise from binding to sites of the same chirality. This was further

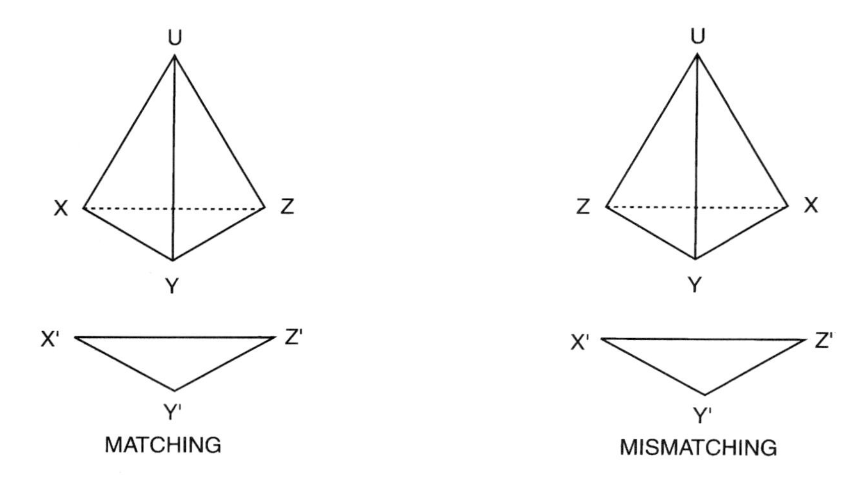

Figure 1.4. The Easson–Stedman three-point contact model.

elaborated by Easson and Stedman (15) in 1933 to become the widely accepted *three-point contact* model (Figure 1.4). They postulated that the (more) active enantiomer binds more tightly because the sequence of the three groups around the asymmetric carbon atom, XYZ, forms a triangular face of tetrahedron that matches a complementary triad of the chiral binding site ′X′Y′Z of the receptor protein. The less active enantiomer binds ineffectively since it has a mirror-image sequence of the three groups, XYZ, that leads to a mismatch with the receptor binding sites. The Easson–Stedman hypothesis provides a sound basis for understanding stereoselectivity in drug–receptor interactions, for example, those between adrenergic receptors and adrenergic blockers or stimulants (9). When the body has a sudden need for energy (e.g., in a fight or flight situation), the adrenergic hormone, noradnenaline, is released. This neurotransmitter, or chemical messenger, subsequently triggers a system that, in turn, gives rise to an increase in heart rate and force (i.e., surge of energy). The three-point contact model (Figure 1.4) for pharmacological enantioselectivity became widely accepted and was later applied to understanding enzyme specificity following its discovery by Ogston (16) in 1948. Furthermore, the model has formed a useful basis for understanding the chromatographic separation of enantiomers on chiral columns.

A further consideration of the enantioselectivity of drug–receptor interactions led Pfeiffer (17) in 1956 to postulate that "the lower the effective dose of a drug, the greater the difference in pharmacological effect of the optical isomers." This simple statement became known as *Pfeiffer's rule*. It is logical as a receptor–drug interaction involves the lock-and-key fit of the molecule with

Table 1.1. Nomenclature for Stereospecificity in Biological Activity

Eutomer: isomer with higher affinity (aff_{eu})
Distomer: isomer with lower affinity (aff_{dis})
Eudismic ratio (ER): aff_{eu}/aff_{dis}
Eudismic index (EI): log aff_{eu}-log aff_{dis}
ER and EI are measures of the sterospecificity of the substance

the right configuration at the receptor site; the better the fit, the better the drug. If both enantiomers can fit into the active site, it is unlikely to be a good fit. This can be compared to a hand and a glove; if both a right and left hand can fit equally well into a glove, the glove is unlikely to be well-fitting.

Lehmann et al. (18) introduced the term *eudismic ratio* to describe Pfeiffer's rule, presented in Table 1.1. The eudismic ratio is defined as the ratio of the activity of an active enantiomer (eutomer) to that of the less active enantiomer (the distomer) and the logarithm of a eudismic ratio in the eudismic index.

A significant correlation is often present when the eudismic index is plotted against the affinity of the eutomer, producing straight lines. The slope of these lines is a quantitative measure of the stereoselectivity of the system and is known as the eudismic-affinity quotient.

Eudismic analysis has led to the quantification of data from many series of compounds and provides a validation of Pfeiffer's rule.

The interaction of chiral along with a generalized receptor has been extensively reviewed by Crossley (19). In this case, the drug is assumed to have sites on a b c d capable of specific complimentary sites on receptor A B C D. Thus, Aa will be a significant interaction, where Ab, Ac, and Ad are of minimal interaction and possibly repulsive. Considering all possible interaction in Figure 1.5 by keeping each group a, b, c, and d fixed in turn, we find that the *R*-enantiomer will have one four-way and eight one-way significant interactions. Similarly, the *S*-enantiomer will have six two-way significant interactions. So, the *S*-enantiomer has some affinity for the receptor, and the magnitude of this will depend on the relative importance of the individual interactions and also cooperation between them. In the enantiomer, such cooperation is high and there is a high degree of molecular fit at the receptor; then the *S*-enantiomer will be essentially inactive at this particular receptor. Whether the *S*-enantiomer produces a response depends on the efficacy at the receptor produced by the two-way significant interactions.

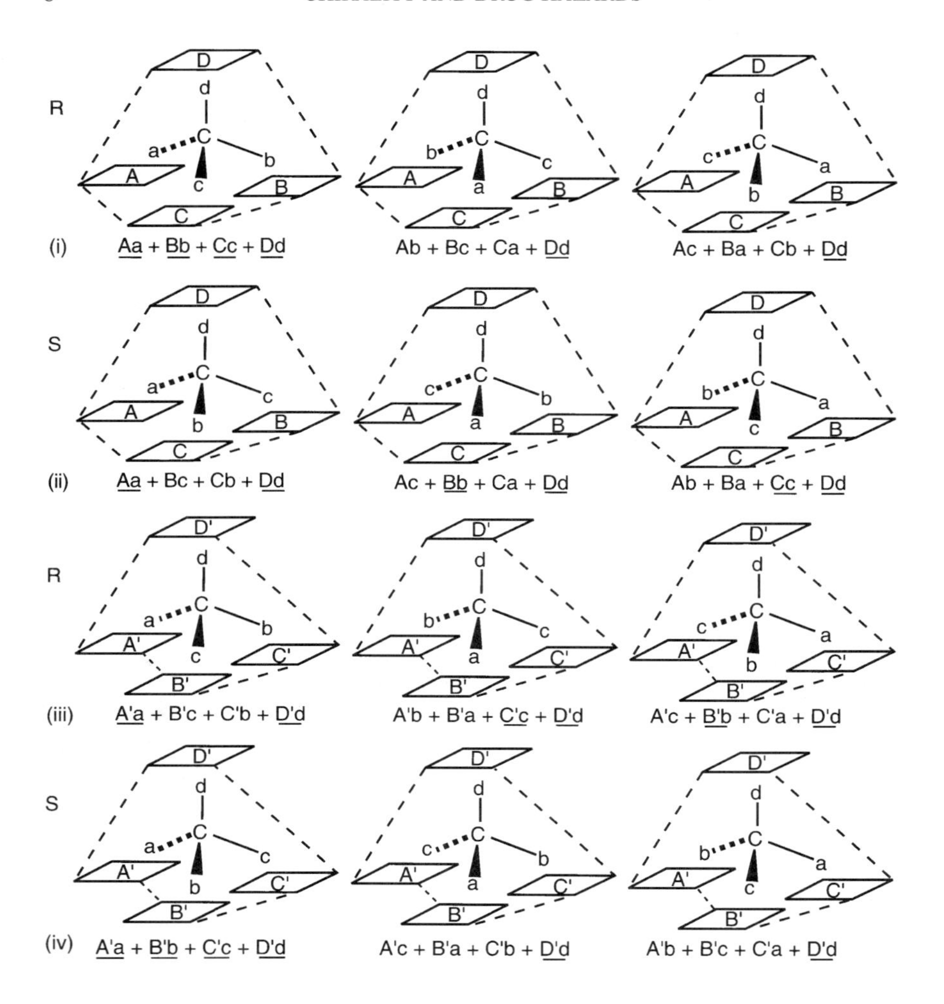

Figure 1.5. Chiral interactions with a receptor. Major interactions are indicated as Aa. (i) With d fixed, the *R*-enantiomer can produce one four-way and two one-way major interactions and, if we allow each group to be fixed in turn, there are a total of eight one-way interactions. (ii) The *S*-enantiomer produces three two-way major interactions with d fixed. If we follow a conformational change in the receptor, (iii) the *R*-enantiomer now produces three two-way major interactions and (iv) the *S*-isomer has one four-way and two one-way major interactions, again with d fixed (19).

When the activity of the *R*-enantiomer is maximized, the *S*-enantiomer still has the possibility of a significant number of interactions with the receptor. So, its activity may not, therefore, be zero and it will be able to antagonize the effect of the *R* form, at least to some extent (19).

The previous treatment assumes that there is no conformational change in the receptor, but this is not necessarily the case, for all conformational changes can produce more than one state in which a drug may interact, including inactive states.

There is a direct relationship between the eudismic ratio and the effectiveness of the drug; that is, the higher the eudismic ratio, the more effective the drug or the lower its effective dosage. The development of this concept has shown that enantioselectivity in biological action is the rule and not the exception.

The stereoselective binding of drugs to receptors led us to the conclusion that specific drugs, those that act at very low concentration, exert their effects by interacting with a specific receptor site in the living cell. The primary mechanism so far is the formation of a reversible drug–receptor complex. This triggers a secondary mechanism such as the opening of anion channels, or catalyzes the formation of a "second messenger" often cyclic AMP (cAMP). Other kinases are then activated. This chain of reactions finally results in physiological change attributed to the drug (20). The same mechanisms also operate with endogenous agents such as hormones and neurotransmitters.

The enormous complexity of living systems and the remoteness of cause from effect (i.e., drug administration and pharmacological action) introduce many complications into the study of such relationships.

Molecular pharmacologists and physical scientists have sought to simplify the experimental system as much as possible by omitting irrelevant factors like drug transport and metabolism and putting it on a level accessible to molecular manipulations and precise physico-chemical methods. The recent development of the methodology of quantitative binding experiments on membrane preparation and later on isolated receptors has become more sophisticated, precise, and simple, and has led to the increasing realization of drug–receptor interaction at the molecular level. This has allowed direct experimental access to receptor binding sites and the recent development of several complementary receptor models and the characterization of molecular properties of drug receptors. Ariëns et al. (21), Burt (22), Hollenberg (23), among others, and their many co-workers are the molecular pharmacologists in the forefront of the spectacular and explosive progress of "receptonology" since the 1970s.

1.3. THE MAIN PHASES IN BIOLOGICAL ACTION

For an understanding of the biological effect of drugs (24), it is important to distinguish three main phases in their action.

1.3.1. Initial Exposure Phase

This is governed by the activity of the drug (affinity for receptors, agonist or antagonist activities, and the activity of metabolites) and also by tissue specificity due to receptor differentiation and distribution for the parent compound and its metabolites. This phase is followed by the pharmacokinetic phase.

1.3.2. Pharmacokinetic Phase

This phase involves absorption, distribution (e.g., plasma binding), metabolism (routes and rates), and excretion (e.g., clearance rates and routes).

The exposure and pharmacodynamic phases together determine the bioavailability of a drug.

1.3.3. Pharmacodynamic Phase

It involves the interaction of the bioactive agent with the molecular site of action (receptors, enzymes, etc.) in the target tissue, leading to the observed effect.

It is important to realize that enantioselectivity plays an important role in not only the pharmacodynamic phase but also the pharmacokinetic phase.

The pharmacological effect of a drug is influenced by both its pharmacodynamic and pharmacokinetic phases and enantioselectivity plays an important role in both. The therapeutic inactive isomer (distomer) of a racemic drug should be regarded as an impurity that possesses a different pharmacological entity.

1.4. CHIRALITY AND PHARMACOLOGICAL ACTION

Different pharmacodynamics and pharmacokinetics of the eutomer and a drug can lead to a variety of effects attributable to a distomer in racemic inactive isomers. The following examples will illustrate the different situations that are encountered with regard to pharmacological activity.

(*1*) *All pharmacological activity may reside in one enantiomer.* An example is α-methyldopa, where all the antihypertensive activity is in the *S*-isomer. The *R*-isomer in this case might be regarded as an isomeric impurity. If we consider the *R*-isomer inactive, the definition of inactivity is limited to those pharmacologic tests that have been used. But such is not the case as the antihypertensive agent α-methyldopa is one of the first and classic examples of

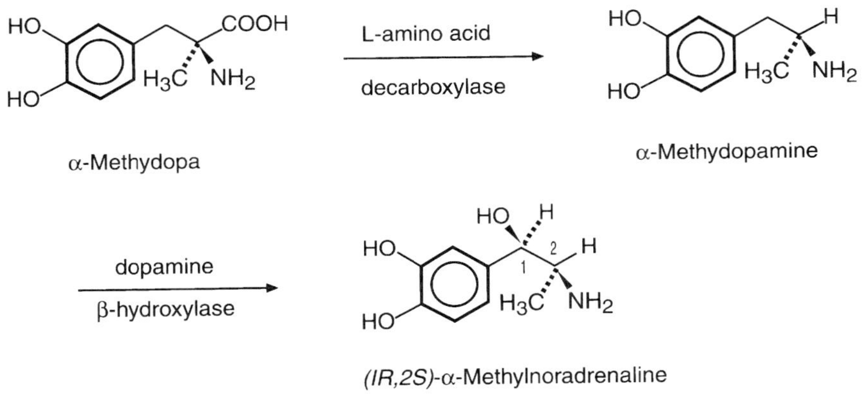

L-amino acid decarboxylase

α-Methydopa

α-Methydopamine

dopamine β-hydroxylase

(IR,2S)-α-Methylnoradrenaline

Figure 1.6. In vivo biotransformation of α-methyldopa.

a synthetic chiral drug that was developed as a single enantiomer (25). α-Methyldopa suppresses the formation of nor-adrenaline by inhibiting the enzymatic decarboxylation of L-dopa. In doing so, α-methyladrenaline (Figure 1.6) later also contributes to the overall antihypertensive effect, presumably by blocking β-adrenergic receptors. It is only the *S*-enantiomer of α-methyldopa that acts as a substrate for the aromatic amino acid decarboxylase and hence produces an antihypertensive effect.

(*2*) *Enantiomers may have identical qualitative and quantitative pharmacological activity*. The isomers of promethazine have nearly equivalent properties with regard to antihistaminic properties and toxicity. In practical terms, the two previously mentioned possibilities are the least complicated, in that the isomers act with the same pharmacological characteristics regardless of the type of isomers.

(*3*) *Enantiomers have qualitatively similar pharmacological activity, but have different quantitative potencies:*

(a) β-Adrenergic blocking agents, for example, propranolol
(b) Verapamil

1.4.1. (a) Propranolol

The structures of enantiomers of β-blockers are presented in Figure 1.7. The side chain with an isopropyl or bulkier substituent on the amine appears to favor interaction with β-receptors. The nature of the substituent on the aromatic ring determines whether the effect will be predominantly activation

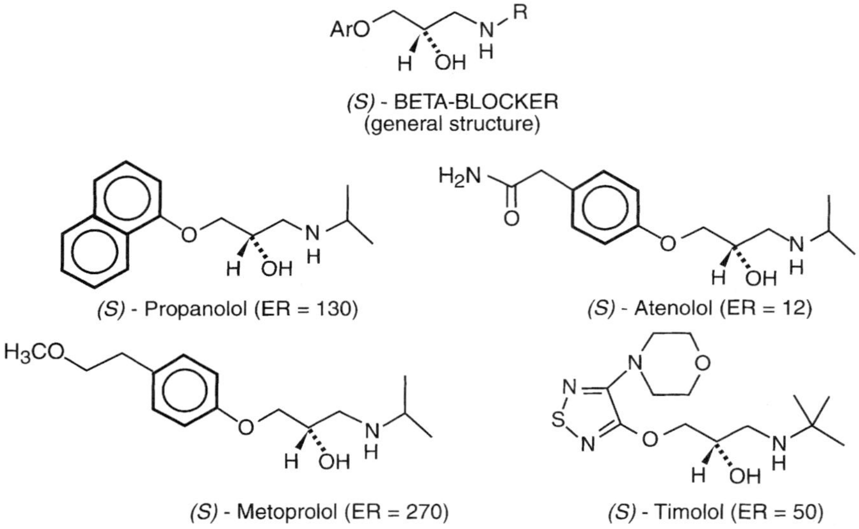

(S) - BETA-BLOCKER
(general structure)

(S) - Propanolol (ER = 130)

(S) - Atenolol (ER = 12)

(S) - Metoprolol (ER = 270)

(S) - Timolol (ER = 50)

Figure 1.7. Structures of the eutomers of β-blockers (ER eudismic ratio).

or blockage. These substituents also affect cardio-selectivity. The alphatic hydroxyl group appears to be essential for activity, and the *S*(–)-enantiomers of the β-adrenergic antagonist are much more potent than the *R*(+)-enantiomers.

Propranolol, the first distributed β-adrenergic blocking drug, is a nonselective $\beta_{1,2}$-adrenoceptor blocking agent with a membrane-stabilizing effect, but intrinsic sympathomimetic effect or combined α-adrenergic blockage. Propranolol like most β-adrenoceptor blocking drugs, with the exception of timolol and penbutolol (which are marketed as *S*-isomers), is commercially available as a racemate. In animals, *S*(–)-propranolol, the levorotary isomer, is about 100 times more potent than *R*(+)-propranolol in blocking ionotropic and chronotropic responses to isoproterenol (β_1, β_2) (26, 27). Both isomers possess a similar potency for the membrane-stabilizing action that is characteristic of class I antiarrhythmic drugs and is also referred to as the quinidine-like effect. Similar differences in isomer potency have been demonstrated in humans (28).

If β-blockers are used as antiarrythmic drugs exclusive of β-blocking properties, then separate preclinical and clinical studies would have to be conducted for the relatively inactive β-blocking isomers.

In hypertensive patients, the same daily doses of racemic propranolol and the non-β blocking *R*(+)-isomer inhibited thrombin and arachidonic acid-induced platelet aggregation and thromboxane synthesis (29). Only the β-blocking *S*(−)-isomer is effective against angina.

For some psychiatric conditions and in migraine headache treatment efficacy is not stereoselective. Stereoselectivity is not a characteristic of propranolol inhibition of insulin release, elevation of plasma triglycerides, or inhibition of thyroxine metabolism.

The disposition of propranolol is stereoselective, *S*(−)-propranolol is bound to a greater extent in plasma than the *R*(+)-isomer (free fraction 10.9% versus 12.2%, respectively). This is explained by stereoselective binding to α_1 acid glycoprotein (30). In humans given racemic propranolol, the S(−)-/R(+)-propranolol plasma concentration ratio is dependent on the route of administration [1.16:1, intravenous (31); 2.45:1 oral (32)].

1.4.2. (b) Verapamil

Verapamil is one of the recently available calcium channel blocking drugs that are used in anginal therapy (33). The single chiral center in the chemical structure of verapamil results in two stereoisomers: *S*(−)-verapamil and *R*(+)-verapamil (Figure 1.8). Verapamil products are marketed as a racemic mixture of *R*- and *S*-enantiomers (34, 35). The total pharmacological activity of a verapamil dose is determined by the combined bioavailability of both individual enantiomers. The *S*-enantiomer is preferentially metabolized after oral dosing, resulting in the *R*-enantiomer being more prevalent in the systemic circulation (35). The actual value of the enantiomeric ratio (*R*:*S*) in the plasma, however, is determined by a combination of factors.

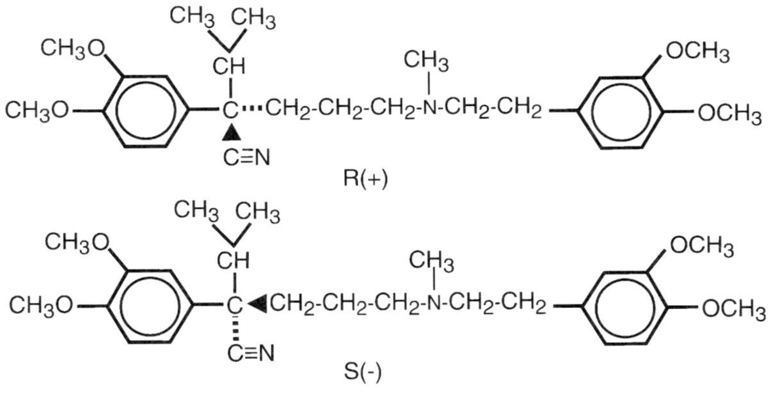

Figure 1.8. The absolute configuration of verapamil enantiomers.

Pharmacokinetic differences between the two enantiomers (e.g., volume of distribution, protein binding, clearances) (36, 37) are the source of differences, but external factors such as the route of administration (e.g., oral or intravenous) and the rate of absorption are also critical (38). The type of drug formulation (e.g., tablet or capsule) is a factor influencing the enantiomer ratio observed in the systemic circulation. The verapamil modulation of calcium influx slows atrioventricular conduction, reduces myocardial contractility and systemic vascular resistance, and results in coronary and peripheral vasodilatation. Verapamil is indicated for controlling hypertension, paroxysmal supraventricular tachycardia, and rapid ventricular atrial flutter or fibrillation (39, 40).

The *R*(+)-enantiomer of verapamil is potent in increasing coronary blood flow; *S*(–)-verapamil has a much less significant cardiodepressant effect (41). *R*(+)-verapamil is an effective antianginal drug with less toxicity. It is the drug of choice in subgroups of patients with vasospastic angina, where a decrease in coronary flow is the principle mechanism.

(*4*) *Both isomers have independent therapeutic potencies.* A chiral drug has a desirable, but different therapeutic effect for both isomers. It can occur in natural products (e.g., quinidine antiarrythmic and quinine antimalarial drug) and also with synthetic drugs. An example is propoxyphene (Figure 1.9). Dextropropoxyphene is an analgesic agent, whereas levopropoxyphene is devoid of analgesic properties but is an effective antitussive (42).

(*5*) *The distomer exhibits undesirable side-effects.* In some cases, the distomer may exhibit undesirable (toxic) side-effects that are not characteristic of a eutomer. The well-known and tragic example is thalidomide, which was marketed in the 1960s as a sedative and administered as a racemate. At that time, it was not known that the *R*-enantiomer is an effective sedative, whereas the *S*-enantiomer is teratogenic and causes fetal anomalies (43).

Ketamine is another example. *S*(+)-ketamine is an active anesthetic and analgesic, and undesirable side-effects including hallucination and agitation, are associated with the *R*(–)-distomer (38). Other examples of this category are shown in Figure 1.10.

(*6*) *Some enantiomers exhibit therapeutic advantages.* Indacrinone (diuretic) is a good example of enantiomers exhibiting therapeutic advantages (Figure 1.11). The *R*(+)-enantiomer is the active diuretic; in common with some other diuretics, the *R*-isomer possesses the undesireable side-effect of uric acid retention. The *S*-enantiomer acts as an uricosuric, promoting uric acid secretion and, therefore, antagonizing the undesired side-effect of the *R*-isomer. It provides a good argument for marketing indacrinone as a racemate. Studies have shown that a *R*:*S* mixture of 9:1 affords an optimal therapeutic effect (44).

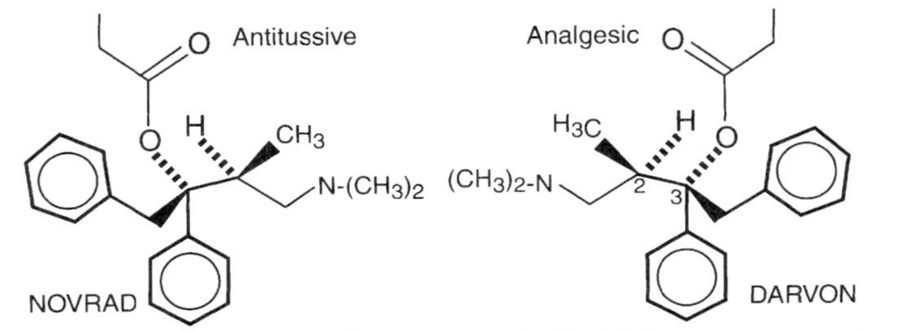

Antitussive — NOVRAD — (2S,3R)-(-)-Levopropoxyphene

Analgesic — DARVON — (2S,3S)-(+)-Dextropropoxyphene

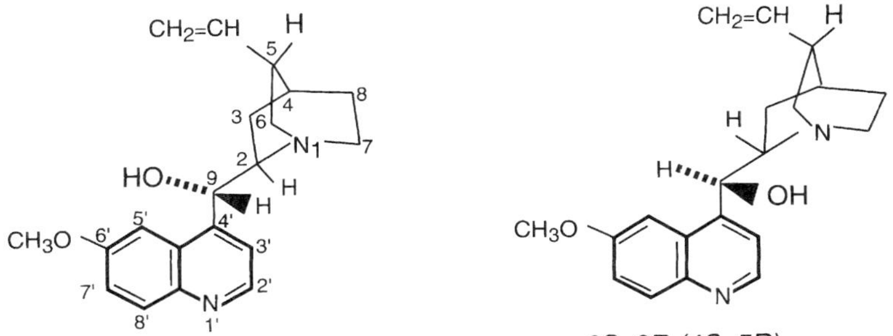

2R, 9S (4S, 5R) QUINIDINE

2S, 9R (4S, 5R) QUININE

Figure 1.9. Enantiomers exibiting different therapeutic effect.

From the previous examples, it is evident that for chiral drugs the use of non-isomer-specific drug assays for the assessment of drug potency or bioavailability may provide an incomplete or invalid conclusion.

Differences in the potential activities (site of action, *in vitro* and *in vivo* metabolism, pharmacokinetics) of the separate enantiomeric forms of racemic drugs have necessitated the inclusion of stereochemistry as an essential component of the description of a drug containing one or more chiral centers.

The stereoisomeric composition of drug substances is rapidly becoming an important issue in the development, approval, and clinical use of pharmaceutical preparations. Accordingly, drug regulatory authorities currently adopt a strict policy to develop drugs only in their single stereoisomer form if the data justify marketing a single enantiomer to ensure optimal efficacy and avoid unwanted side-effects (45, 46).

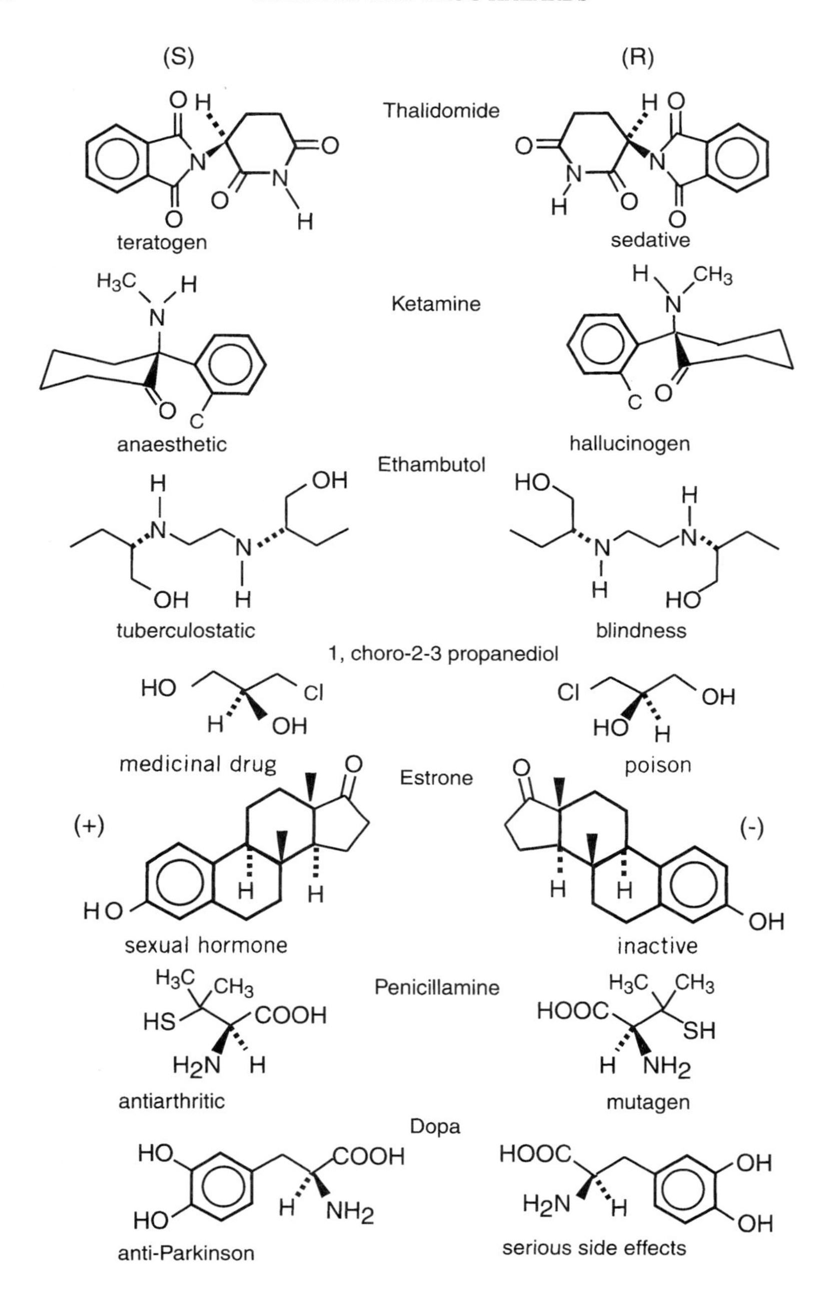

Figure 1.10. Examples of chiral drugs, where the distomer exhibits undesirable or toxic side-effects.

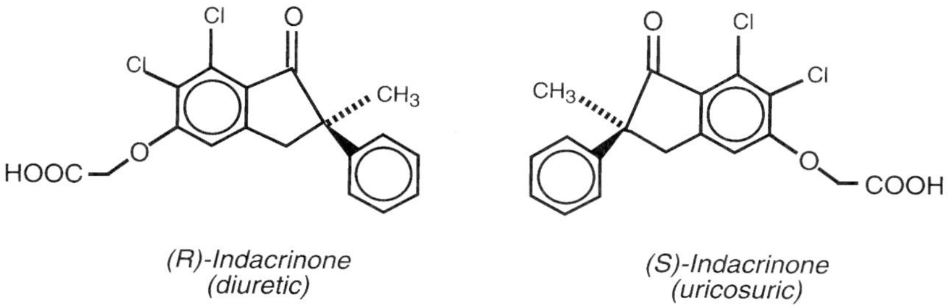

(R)-Indacrinone
(diuretic)

(S)-Indacrinone
(uricosuric)

Figure 1.11. Indacrinone enantiomers.

1.5. CONCLUSION

The recent development of the methodology of quantitative binding experiments of isolated receptors has led to the increasing realization and characterization of drug–receptor interaction at the molecular level. Also, these studies lead to the observation of interactions between enantiomers for receptor recognition processes.

Many drug enantiomers have different pharmacological activities in man, and this seems to be the result of intersubject variation in the ratios of the concentration of individual drug enantiomers in human plasma. Such has already been demonstrated for propranolol and verapamil. Stereoselectivity in drug absorption, distribution, metabolism, and elimination produces differences in isomer activity, ranging from unwanted toxicity to no significance to enhanced activity. A greater understanding of drugs and receptors will follow when testing single enantiomers becomes routine.

REFERENCES

1. Wainer, I. W. and Marcotte, A. L. (1988). In *Drug Stereochemistry, Analytical Methods and Pharmacology*, I. W. Wainer and D. E. Drayer (eds), New York: Marcel Dekker, p. 31.
2. Mislow, K. (1965). *Introduction to Stereochemistry*, New York: W. A. Benjamin, p. 78.
3. Fischer, E. (1919). *Chem. Ber.*, **524**, 129.
4. Khan, R. S., Ingold, C. K., and Prelog, V. (1966). Specification of molecular chirality. *Agrew. Chem. Int. Ed.*, **5**, 385.
5. Sheldon, R. (1990). *Chem. & Ind.*, **2**, 212.
6. Portoghese, P. S. (1970). *Ann. Rev. Pharmacol*, **10**, 51.

7. Lehmann, F. P. A. (1978). In *Receptors and Recognition*, Vol. 5, P. Curtrecasas and M. F. Greaves (eds.), London: Chapman and Hall, p. 1.
8. William, K. and Lee, E. (1985). *Drugs*, **30**, 333.
9. Ariëns, E. J. (1984). *Eur. J. Clin. Pharmacol.*, **26**, 663.
10. Simonyi, M. (1984). *Med. Res. Rev.*, **4**, 359.
11. Wynberg, H. (1986). *Topic Stereochem.*, **16**, 87.
12. Fischer, E. (1894). *Ber. Dtsch. Chem. Ges.*, **27**, 2895.
13. Ehrlich, J. N. (1913). *Br. Med. J.*, **11**, 353.
14. Cushny, A. R. (1926). *Biological Relationships of Optically Active Substances*. London: Bailliene, Tindall and Cox, p. 53.
15. Easson, L. H. and Stedman, E. (1933). *Biochem. J.*, **27**, 1257.
16. Ogston, A. G. (1948). *Nature*, **162**, 963.
17. Pfeiffer, C. C. (1956). *Science*, **124**, 29.
18. Lehmann, P. A., Rodrigues de Miranda, J. F., and Ariëns., E. J. (1976). *Progr. Drug Res.*, **20**, 707.
19. Crossley, R. (1992). *Tetrahedron*, **48**, 38, 8155.
20. Nogrady, T. (1988). In *Medicinal Chemistry*, 2nd ed., Oxford, England: Oxford University Press, p. 58.
21. Ariëns, E. J., Beld, A. J., Rodrigues de Miranda, J. F., Simonis, A. M. (1979). In The Receptors, R. D. O'Brien (ed.), New York: Plenum Press, p. 33.
22. Burt, D. R. (1985). In *Neurotransmitter Receptor Binding*, 2nd ed. H. J. Yamamura, S. J. Enna, and M. J. Kuhar (eds.), New York: Raven Press, p. 41.
23. Hollenberg, M. D. (1985). In *Neurotransmitter Receptor Binding.*, 2nd ed., H. J. Yamamura, S. J. Enna, and M. J. Kuhar (eds.), New York: Raven Press, p. 1.
24. Ariëns, E. J. (1989). In *Chiral Separation by HPLC*, A. M. Krstulovic (ed.), Chichester: Ellis Horwood, p. 31.
25. Baldwin, J. J., and Abrams, W. B. (1988). In *Drug Stereochemistry, Analytical Methods and Pharmacology*, I. W. Wainer, and D. E. Drayer (eds.), New York: Marcel Dekker, p. 311.
26. Barnett, A. M. and Cullum, V. A. (1968). *Br. J. Pharmacol.*, **34**, 43.
27. Howe, R. and Rao, B. S. (1968). *J. Med. Chem.*, **11**, 1118.
28. Bennet, D., Balcon, R., Hoy, J., and Sowton, E. (1970). *Thorax*, **25**, 86.
29. Campbell, R. W. F., Murray, A., and Julian, D. G. (1981). *Br. Heart J.*, **46**, 351.
30. Albani, F., Riva, R., Contin, M., and Baruzzi, R. (1984). *Br. J. Clin. Pharmacol.*, **18**, 244.
31. Olanoff, L. S., Walle, T., Walle, U. K., Cowart, T. D., and Gaffney, T. E. (1984). *Clin. Pharmacol. Ther.*, **35**, 755.
32. Silber, B., Holford, N. H. G., and Riegelman (1982). *J. Pharm. Sci.*, **71**, 699.
33. Antman, E. M., Stone, P. H., Muller, J. E., and Braunwald, E. (1980). *Ann. Int. Med.*, **93**, 875.

34. Einchelbaum, M., Mikus, G., and Vogelgesang, B. (1984). *Br. J. Clin. Pharmacol.*, **17**, 453.
35. Mikus, G., Einchelbaum, M., Fisher, C., Crumulka, S., Klotz, U., and Kroemer, H. K. (1990). *J. Pharmacol. Exp. Ther.*, **253**, 1042.
36. Gross, A. S., Heuer, B., and Eincelbaum, M. (1988). *Biochem. Pharmacol.*, **37**, 4623.
37. Vogelgesang, B., Echizen, H., Schmidt, E., and Einchelbaum, M. (1984). *Br. J. Clin. Pharmacol.*, **18**, 733.
38. Harder, S., Thurman, P., Siewert, M., Blume, H., Tuber, T., and Rietbrock, N. (1991). *J. Cardiovasc. Pharmacol.*, **17**, 207.
39. Chatterjee, K., Rouleau, J. L., and Parmley, W. W. (1984). *J. Am. Med. Assoc.*, **252**, 1170.
40. Dustant, H. P. (1989). *Hypertension*, **13** (Suppl. I), 1137.
41. Satoh, K., Yanagisawa, T., and Taira, N. J. (1980). *Cardiovasc. Pharmacol.*, **2**, 309.
42. Hyneck, M., Deut, J., and Hook, J. B. (1990). In *Chirality in Drug Design and Synthesis*, C. Brown (ed.), New York: Academic Press, p. 1.
43. De Camp, W. H. (1989). *Chirality*, **1**, 2.
44. Tobert, J. A., Cirillo, V. J., Hitzenberger, G., James, I., Pryor, J., Cook, T., Buntinx, A., Holmes, J. B., and Lutterbech, P. M. (1981). *Clin. Pharmacol. Ther.*, **29**, 344.
45. FDA's Policy Statement for the Development of New Stereoisomeric Drugs (1992). *Chirality*, **4**, 338.
46. CPMP Note for Guidance: Investigation of Chiral Active Substances, Commission of the European Communities (1993).

CHAPTER

2

STEREOCHEMISTRY IN THE DRUG DEVELOPMENT PROCESS: ROLE OF CHIRALITY AS A DETERMINANT OF DRUG ACTION, METABOLISM, AND TOXICITY

THOMAS A. BAILLIE

Merck Research Laboratories
Department of Drug Metabolism
West Point, Pennsylvania 19486

KATHLEEN M. SCHULTZ

Department of Medicinal Chemistry
School of Pharmacy
University of Washington
Seattle, Washington 98195

2.1. INTRODUCTION

The intrinsic asymmetry of receptors, enzymes, and other endogenous macromolecules represents the basis for biological discrimination between the stereoisomeric forms of chiral drugs and other xenobiotics. Such discrimination leads to the well-known phenomenon of "chiral recognition" (Testa, 1989), in which the enantiomers of a chiral drug may exhibit differences in biological activity (pharmacological or toxicological) or in one or more of the processes of absorption, distribution, metabolism, and excretion (Drayer, 1986; Borgström et al., 1990; Lee and Williams, 1990; Levy and Boddy, 1991; Williams, 1991; Wainer, 1993). Although these concepts have been recognized for decades, stereochemistry in drug development received little attention prior to the early 1980s, and new therapeutic agents containing one or more center of asymmetry generally were developed as mixtures of the isomeric forms for reasons of synthetic accessibility and economic feasibility. Around this period, however, a number of factors brought about a change in the *status quo* in the development of chiral drugs, the most significant of which were the

The Impact of Stereochemistry on Drug Development and Use, Edited by Hassan Y. Aboul-Enein and Irving W. Wainer. Chemical Analysis Series, Vol. 142.
ISBN 0-471-59644-2

advent of efficient methods for large-scale stereoselective organic synthesis (and the commercial availability of the necessary chiral reagents) and the introduction of chromatographic systems for the stereoselective analysis of drugs and their metabolites in biological fluids. In parallel with these advances in synthetic and analytical chemistry, which were quickly implemented by the pharmaceutical industry, the techniques of molecular biology were being adopted by pharmacologists to clone and express human enzymes, receptors, and other drug targets for the development of rapid *in vitro* screens for biological activity. As a consequence, drug discovery programs could generate, in a relatively short period of time, many active compounds that were both highly potent and selective in their interaction with biological targets. Indeed, both potency and selectivity often was conferred by the introduction of one or more chiral centers into the candidate drug molecule.

The above trend for the development of potent drugs with at least one center of asymmetry has interesting biological consequences. For molecules with a single asymmetric center (which will form the focus of this chapter), Pfeiffer's rule states that an inverse relationship exists between the effective dose of the drug and the ratio of activity of its enantiomers (Pfeiffer, 1956). In other words, the more potent the compound, the larger will be the difference in biological activity between its enantiomers. It follows, therefore, that the new generation of extremely potent chiral drug candidates typically will have one highly active enantiomer and a relatively inert, or possibly even toxic, antipode.

In light of these considerations, and in view of a proliferation of reports on differences in the biological activity and metabolic fate of individual enantiomers of racemic drugs, the question, "Why should one continue to develop racemic mixtures for therapeutic use rather than the pure active enantiomers?" attracted serious attention for the first time. Indeed, this issue was a popular topic for discussion in the scientific literature of the 1980s, and became a somewhat contentious one for the pharmaceutical industry that was being subject to increasing pressure to develop optically pure therapeutic agents. Following a series of timely, if rather provocative, articles by Ariëns on the development of racemic drugs (Ariëns, 1984; 1986; 1993), the subject of stereochemistry in pharmaceutical research and development was highlighted at various national and international symposia. These gave rise to a series of proceedings and thoughtful commentaries on the subject of chirality in drug development (Caldwell, 1989; De Camp, 1989; Holmes et al., 1990; Testa and Trager, 1990; Cayen, 1991; Hutt, 1991; Wilson and Walker, 1991; Gross et al., 1993). Subsequently, several regulatory agencies, including the U.S. Food and Drug Administration (FDA), issued guidelines on the development of stereoisomers (FDA, 1992), and current efforts to harmonize drug development internationally undoubtedly will incorporate elements of these (Shindo and Caldwell, 1991; Rauws and Groen, 1994). The general consensus to emerge from these debates was that when a practical and economically viable route to the enantiomer with the desired activity can be established on a production scale, this should be the preferred form of the drug for development. Use of the racemate, it was argued, ought to be viewed as polytherapy (with "fixed ratio combinations") and treated accordingly by regulatory agencies.

Today, the proportion of optically pure drugs in development relative to racemic mixtures is markedly higher than was the case even five years ago, and this trend is likely to continue in the future (Stinson, 1994). However, despite the current bias toward optically pure drugs, there remain situations where the decision to develop one enantiomer over the racemate may not be clear-cut, but must be based on a consideration of the physicochemical behavior, metabolic fate, and toxicological potential of both optical isomers. Clearly, this is not a trivial undertaking for the sponsor, in that it usually requires the investment of significant effort and resources. Although such considerations add further to the argument in favor of optically pure drugs, the purpose of this chapter is to discuss examples where a dogmatic approach to the selection of the more active enantiomer for marketing might not be the most appropriate course of action.

2.2. INTERCONVERSION OF STEREOISOMERS

In a document published by the Pharmaceutical Manufacturers' Association ad hoc Committee on Racemic Mixtures (Holmes et al., 1990), it was pointed out that one situation in which a decision likely would be made to develop the racemate as opposed to a pure enantiomer would be where "the enantiomers are rapidly interconverted *in vitro* and/or *in vivo* on a time scale such that administering a single enantiomer offers no advantage." Clearly, the stability of the chiral center will be of key importance in deciding whether to develop a pure active enantiomer or the racemic mixture. As discussed by Testa and coworkers (Testa et al., 1993), there are two types of configurational stability: namely, "pharmaceutical" and "pharmacological." Whereas the former refers to the stability of the chiral center during the manufacturing process and shelf-life, the latter (which will be the focus in this chapter) pertains to stability under physiological conditions during the life-time of the drug *in vivo*.

A number of examples have been reported of chiral drugs that exhibit a lack of pharmacological configurational stability, including the sedative-hypnotic benzodiazepine oxazepam and simple derivatives thereof (Figure 2.1). Oxazepam racemizes spontaneously in solution *in vitro* (Yang and Lu, 1992), and kinetic studies have indicated that the half-life of racemization in an aqueous medium at 37 °C and pH 7.4 is in the order of 1–4 minutes (Aso

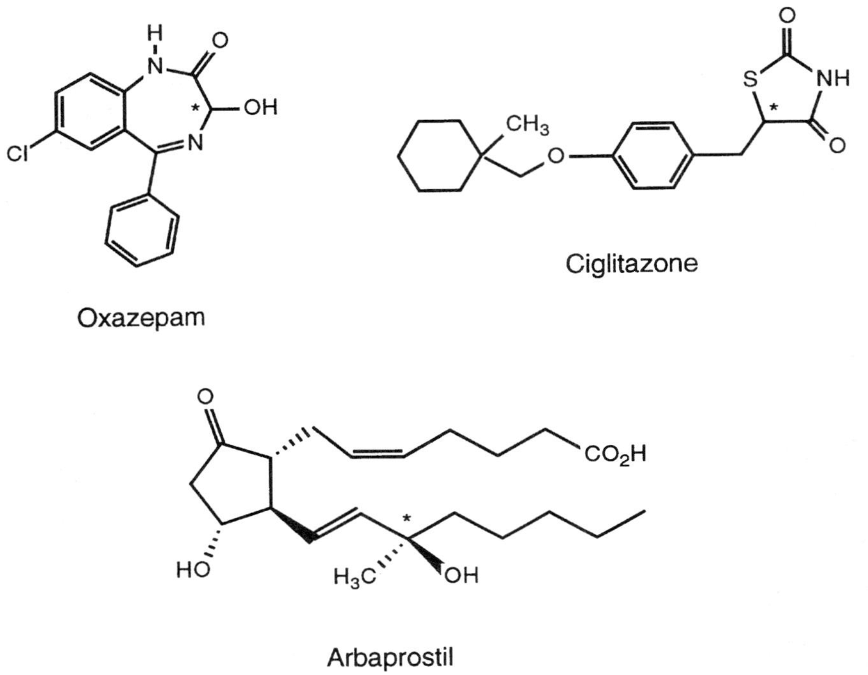

Oxazepam

Ciglitazone

Arbaprostil

Figure 2.1. Examples of compounds that exhibit configurational instability under physiological conditions. The location of the chiral center involved is indicated in each case with an asterisk.

et al., 1988; Yang and Lu, 1989). Since the elimination half-life of the drug in humans may be as long as 21 hours, oxazepam should be viewed as a rapidly equilibrating mixture of two enantiomeric forms, and it is highly unlikely that any therapeutic advantage would be realized by developing only one enantiomer. Interestingly, whereas the mechanism of interconversion of *R*- and *S*-oxazepam originally was believed to involve reversible opening of the 7-membered ring, more recent studies have implicated both base- and acid-catalyzed processes that do not involve a ring-opened intermediate (Yang, 1994; Yang and Bao, 1994). It should be noted that although such *in vitro* experiments may be valuable in predicting the behavior of the chiral center *in vivo*, they should be complemented by appropriate *in vivo* studies in which configurational stability in animals or humans is assessed by means of a stereoselective assay.

5-Substituted thiazolidinedione derivatives represent a relatively new structural class of antidiabetic agents that are under clinical evaluation (Hofmann and Colca, 1992). These compounds possess a chiral center at C-5 on the thiazolidinedione ring system, and in one study on the prototype of this family (ciglitazone; Figure 2.1), the two enantiomers of the drug were resolved by crystallization of their (+)- and (–)-1-phenylethylamine salts and evaluated for biological activity *in vivo* and configurational stability *in vitro* (Sohda et al., 1984). The results of this study indicated that the optical isomers displayed essentially the same antidiabetic and hypolipidemic activities in mice, and that they underwent spontaneous racemization in solution. The authors concluded, therefore, that the absence of a pharmacological difference between (+) and (–)-ciglitazone probably was a consequence of the lability of the proton at the chiral center of the thiazolidinedione ring under physiological conditions. Thus, if configurational instability on a timescale shorter than the *in vivo* half-life is common to other members of this structural class, there would seem to be little rationale for considering a single enantiomer for drug development. Similarly, a number of chiral drugs in which the asymmetric center is benzylic have been shown to undergo rapid racemization *in vitro* (Pepper et al., 1994), and semiempirical rules to predict the configurational stability of a chiral center of the type R″R′RC–H have been proposed (Testa et al., 1993).

Chiral allylic alcohols represent an interesting class of compounds from a biological point of view, since they are prone to acid-catalyzed racemization *in vitro* and therefore may be expected to be configurationally unstable in gastric acid following oral dosing. Indeed, arbaprostil (Figure 2.1), a synthetic analog of prostaglandin E_2 that was investigated by Upjohn as a potential antiulcer drug, was found to undergo epimerization of the allylic alcohol at C-15 upon exposure to mineral acid (Merritt and Bronson, 1978). Since it was known that the pharmacological activity of arbaprostil resided exclusively in the 15(*S*) epimer, Upjohn proposed to develop the inactive 15(*R*) form for therapeutic use on the premise that conversion of this prodrug to the active form in the stomach would occur to a significant degree only under conditions of high acidity. Thus, arbaprostil would be "activated" by the very condition that the drug was designed to treat, namely, excess gastric acid secretion. Under normal levels of stomach acidity, it was argued, racemization would be low and systemic exposure to the pharmacologically potent 15(*S*) isomer would be minimized. This example illustrates the importance of understanding the chemistry of the drug molecule prior to making a decision on the appropriate isomer for development.

A second example of configurational instability involving a chiral allylic alcohol is the new antiepileptic agent stiripentol (Figure 2.2), the two enantiomers of which appear to have similar anticonvulsant activities in rats (Shen et al., 1992). As in the case of arbaprostil, stiripentol undergoes acid-catalyzed racemization *in vitro* in a time- and pH-dependent manner (Tang et al., 1994). Therefore, following oral administration of either enantiomer to animals, one

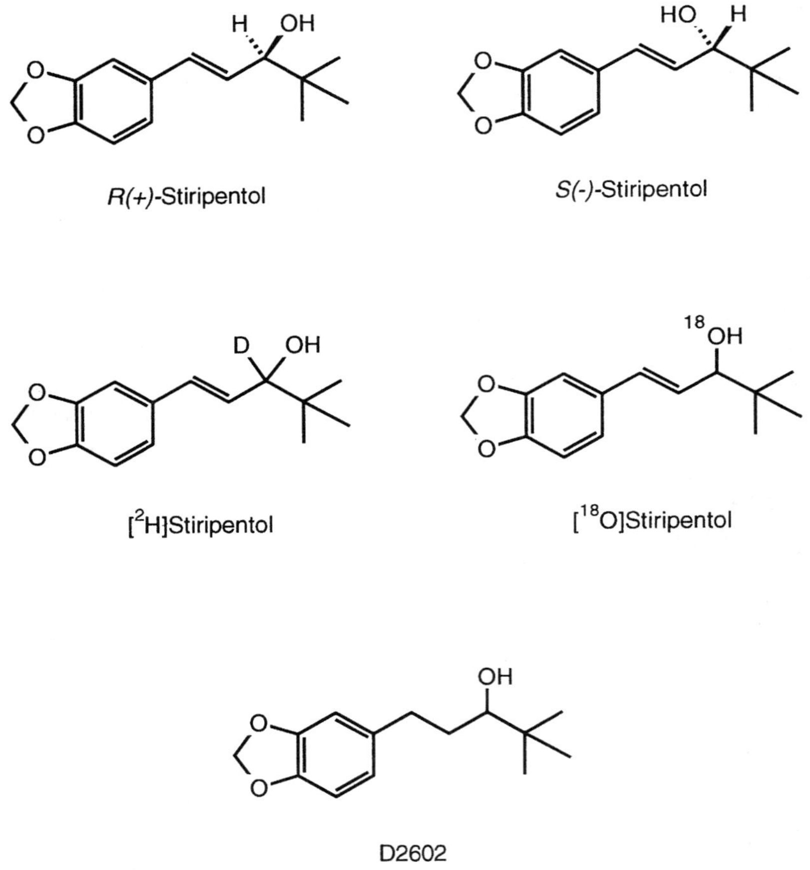

H OH
R(+)-Stiripentol

HO H
S(-)-Stiripentol

D OH
[^{2}H]Stiripentol

^{18}OH
[^{18}O]Stiripentol

OH
D2602

Figure 2.2. Structures of *R*- and *S*-stiripentol, [^{2}H]stiripentol, [^{18}O]stiripentol, and D2602.

might expect to observe significant racemization of the drug in gastric acid and, as a result, that blood would contain a mixture of the two optical isomers. However, when rats were dosed orally with *R*- or *S*-stiripentol and the configuration of the drug in the systemic circulation was analyzed by a stereoselective HPLC method, it was found that whereas *R*-stiripentol had been transformed extensively to its antipode, little inversion was evident following dosing with *S*-stiripentol (Figure 2.3). In order to explain this unexpected result, which seemed to indicate that only one enantiomer of stiripentol undergoes chiral inversion *in vivo*, a series of mechanistic studies were per-

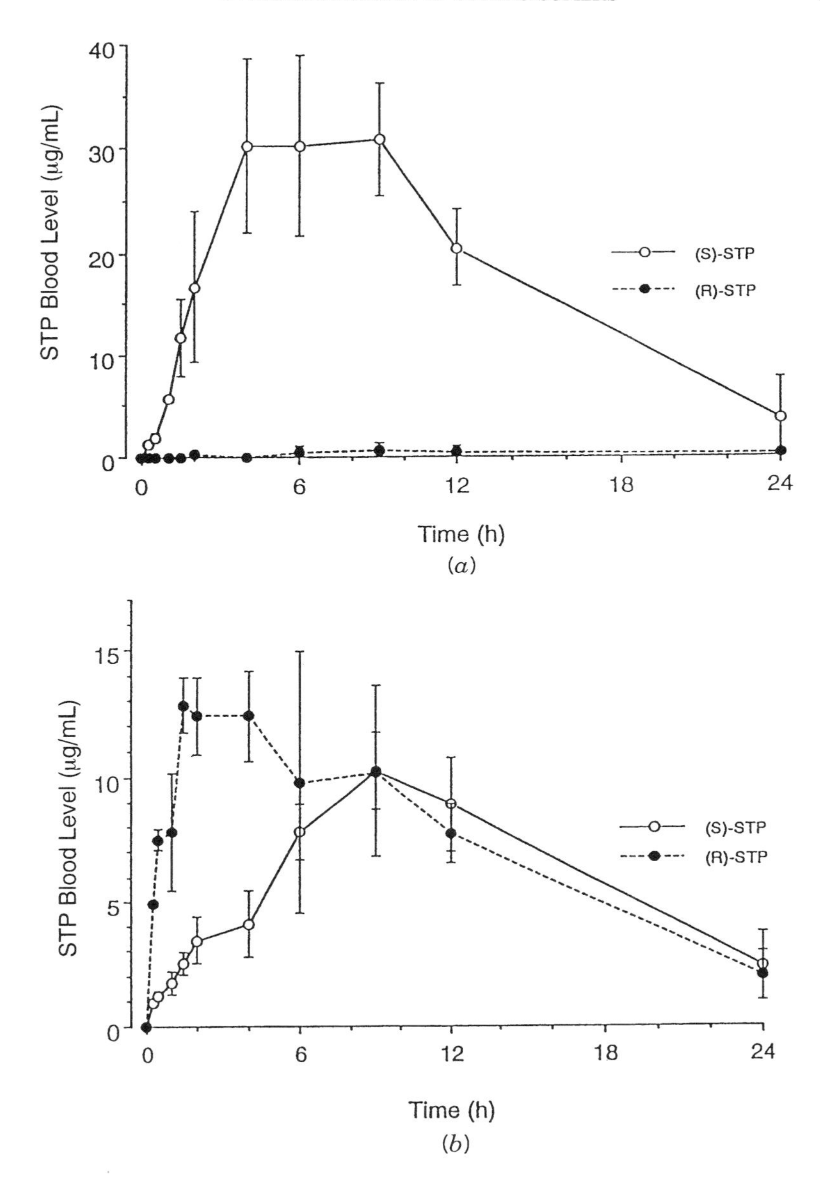

Figure 2.3. Blood concentration versus time plots for the individual enantiomers of stiripentol (STP) in rats. Groups of rats ($N \geq 3$) received a single oral dose (300 mg kg^{-1}) of (*a*) *S*-STP, or (*b*) *R*-STP, and the concentrations of each enantiomer of the drug were measured as a function of time by means of a stereoselective HPLC assay. (Adapted from Zhang et al., 1994, with permission.)

formed using the rat as an animal model (Zhang et al., 1994). It was found that the transformation of *R*-stiripentol to the *S* form was dependent on the presence of the side-chain double bond because the enantiomers of the corresponding saturated alcohol (D2602; Figure 2.2) did not interconvert *in vivo*. Experiments with analogs of stiripentol labeled with deuterium or oxygen-18 at the chiral center (Figure 2.2) showed that, whereas deuterium was retained *in vivo*, partial loss of the ^{18}O occurred from both enantiomers of the drug. Hence, the mechanism for the *R* to *S* conversion could not involve a stereoselective alcohol oxidation–reduction sequence at the chiral center, but rather proceeded by way of C—O bond cleavage. Interestingly, pretreatment of rats with pentachlorophenol, an inhibitor of sulfation and glucuronidation, led to a marked decrease in the rate of conversion of *R*-stiripentol to its antipode, suggesting that the unusual stereoselective pharmacokinetics of this drug were mediated, at least in part, by an enantioselective conjugation process (Zhang et al., 1994). Further experiments showed that the inversion phenomenon was observed only after oral dosing, and not following intravenous or intraperitoneal administration. Moreover, when stiripentol was given orally, it was found that the drug present at various points along the gastrointestinal tract became progressively enriched in molecules of *R* configuration, such that the free drug in cecum, large intestine, and feces consisted largely of the *R*-enantiomer, regardless of the configuration of the administered drug (Tang et al., 1994). Finally, a key finding was the observation that stiripentol undergoes metabolism in the rat to a glucuronide conjugate which is excreted largely in bile, and that this conjugation process is highly stereoselective for the *R*-enantiomer (Tang et al., 1995).

Collectively, the above studies led to the construction of a scheme (Figure 2.4) that accounts satisfactorily for all the experimental observations on the apparent chiral inversion of stiripentol *in vivo*. Thus, following oral administration of either *R*- or *S*-stiripentol, both enantiomers are subject to partial racemization under the low pH of gastric acid. The resulting mixture of isomers leaves the stomach and enters the small intestine from which they are effectively absorbed and passed to the liver via the portal circulation. Upon reaching the liver, the drug is subject to enantioselective glucuronidation, which serves to preferentially extract molecules of *R* configuration. Indeed, the capacity of the liver for glucuronidation of the *R*-enantiomer is such that when *S*-stiripentol is given orally, molecules of *R* configuration (formed under the action of gastric acid) are essentially completely removed by glucuronidation. As a consequence, analysis of the stereochemistry of the unconjugated drug circulating in peripheral blood fails to reveal the existence of the *S* to *R* conversion. Conversely, when the *R*-enantiomer is administered, the concentration of molecules of *R* stereochemistry leaving the stomach and absorbed from the gastrointestinal tract is so much higher than that of *S*-stiripentol that

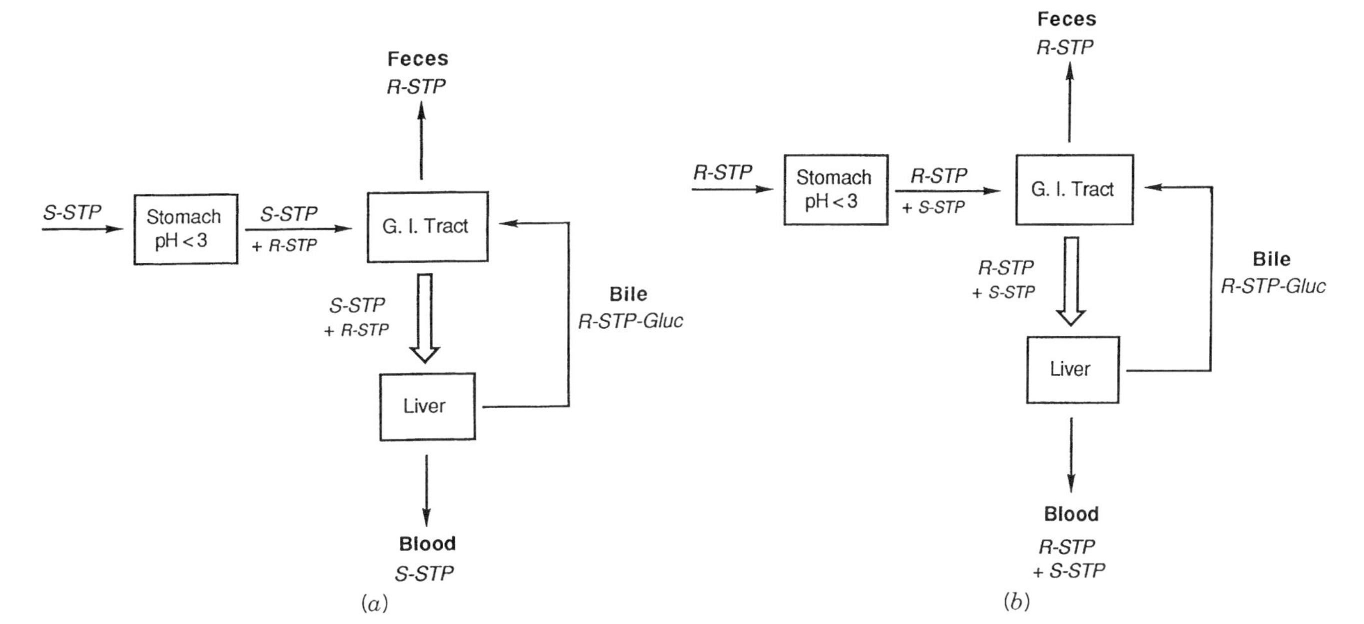

Figure 2.4. Schematic representation of the proposed role of different organ systems in the apparent metabolic chiral inversion of stiripentol in the rat. (*a*) depicts the fate of orally administered *S*-stiripentol (*S*-STP), whereas (*b*) depicts that of *R*-stiripentol (*R*-STP). In the acidic environment of the stomach, both enantiomers undergo partial chemical racemization such that a mixture of enantiomers is passed to the gastrointestinal (G. I.) tract for absorption. However, as a result of saturable, enantioselective first-pass glucuronidation of the *R*-isomer, this species appears in blood only when its concentration in the portal system exceeds a certain threshold. This explains why *R*-stiripentol is not observed in the systemic circulation when the *S*-enantiomer is given, but why *S*-stiripentol is observed in the blood of rats dosed with the *R*-enantiomer. This scheme also accounts for the appearance of *R*-stiripentol glucuronide in bile and of the corresponding deconjugated *R*-isomer in feces. (Adapted from Tang et al., 1994, with permission.)

hepatic conjugation is saturated, and molecules of both configurations now appear in the systemic circulation. The glucuronide conjugate eliminated via the bile (which consists almost exclusively of the glucuronide of *R*-stiripentol) is hydrolyzed back to free drug by bacterial β-glucuronidase in the gut, and this explains the progressive enrichment of stiripentol in descending portions of the gastrointestinal tract in molecules of *R* stereochemistry. Pentachlorophenol inhibits glucuronidation *in vivo*, and this accounts for the ability of this compound to suppress the apparent chiral inversion of *R*-stiripentol. The requirement for the side-chain double bond is linked to the propensity of the allylic alcohol moiety to undergo acid-catalyzed racemization in the stomach, which proceeds via a resonance-stabilized carbonium ion intermediate. Clearly, racemization proceeding according to this mechanism would not be observed with the saturated alcohol (D2602; Figure 2.2), but would be consistent with the partial loss of the isotopic label from [^{18}O]stiripentol observed for both enantiomers of the drug present in the systemic circulation (Zhang et al., 1994). Therefore, the apparent unidirectional chiral inversion of stiripentol *in vivo* results from a combination of three factors: (1) acid-catalyzed partial chemical racemization of both enantiomers in gastric acid, (2) enantioselective, saturable glucuronidation of *R*-stiripentol in the liver, leading to preferential entry of the *S*-enantiomer into the systemic circulation, and (3) excretion of *R*-stiripentol glucuronide into bile, followed by hydrolysis of the conjugate in the gut to regenerate free *R*-stiripentol that appears in feces. This example illustrates the point that, in view of a complex interplay between chemical and metabolic processes, the disposition of this chiral drug *in vivo* could not have been explained on the basis of blood, urine, and fecal data alone. However, by gaining a detailed mechanistic insight into the basis for the unusual chiral inversion phenomenon, it was possible not only to rationalize the enantioselective pharmacokinetics of stiripentol in the rat, but also to predict that there would be little advantage to developing this new drug as a pure enantiomer in light of its propensity to racemize in gastric acid. Preliminary clinical studies with the enantiomers of stiripentol have lent support to this conclusion (Tang et al., 1995).

A different type of configurational instability is observed with certain 2-arylpropionic acid derivatives ("profens"), an important class of nonsteroidal antiinflammatory drugs with analgesic and antipyretic activities (Nicholson, 1982). By virtue of their 2-substituted propionic acid structure, the profens contain a chiral center, and it has been demonstrated *in vitro* that the pharmacological activity of these agents resides predominantly (and in some cases almost exclusively) in the *S*-enantiomers (reviewed in Williams, 1990). Surprisingly, however, the individual optical isomers of several profens, including ibuprofen (Figure 2.5), the prototype of the series, were found to exhibit closely similar antiinflammatory activities *in vivo* (Hutt and Caldwell,

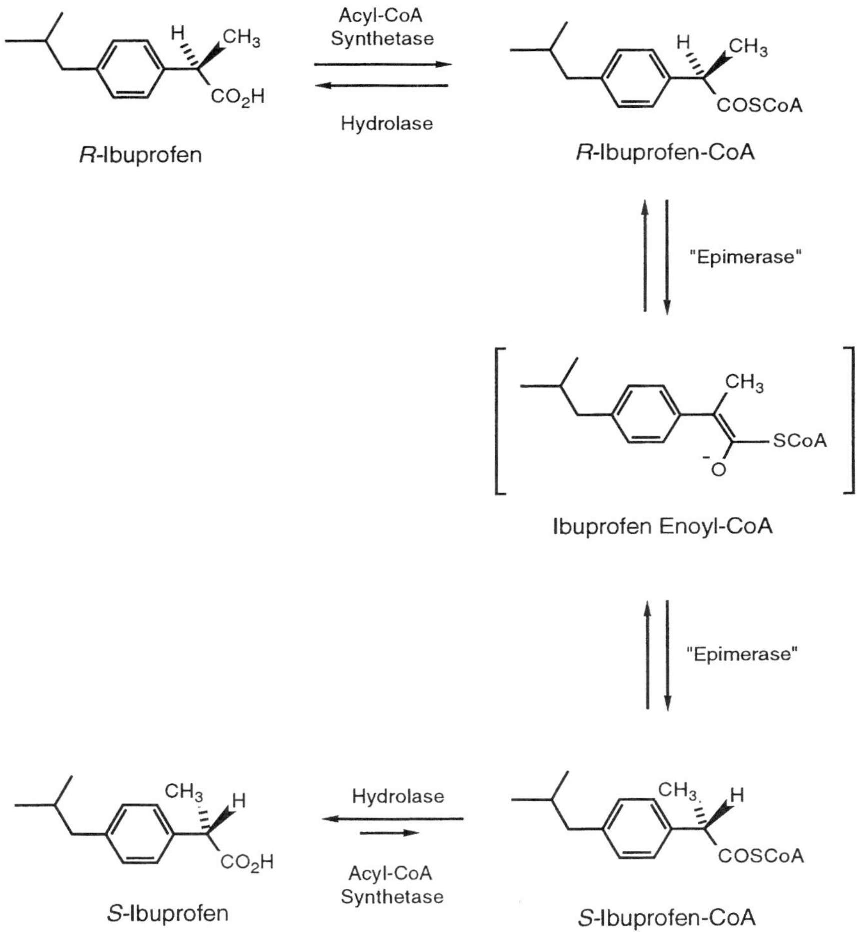

H CH3
CO2H
R-Ibuprofen

Acyl-CoA Synthetase
Hydrolase

H CH3
COSCoA
R-Ibuprofen-CoA

"Epimerase"

CH3
SCoA
-O
Ibuprofen Enoyl-CoA

"Epimerase"

CH3 H
CO2H
S-Ibuprofen

Hydrolase
Acyl-CoA Synthetase

CH3 H
COSCoA
S-Ibuprofen-CoA

Figure 2.5. Proposed scheme for the metabolic chiral inversion of *R*-ibuprofen, illustrating the pivotal role of the coenzyme A thioester intermediate in the process. Since *S*-ibuprofen is inefficiently converted to its coenzyme A conjugate, little inversion of *S*-ibuprofen is observed, and consequently the process appears to be unidirectional (*R* to *S*, but not vice versa). (Adapted from Sanins et al., 1991, with permission.)

1984; Williams, 1990). Largely as a result of pioneering studies on ibuprofen by Adams and co-workers (Adams et al., 1967; 1976), it was established that the reason for this paradoxical observation was that the pharmacologically less active *R*-enantiomers were metabolically transformed to their more active

S antipodes in a process that was essentially unidirectional in nature. Thus, when administered *in vivo*, the *R*-enantiomers of the profens, in effect, serve as prodrugs of their active *S* counterparts (Hutt and Caldwell, 1983, 1984; Caldwell et al., 1988). For many years, the mechanism of this unusual process remained obscure, and it was unclear why some profens underwent chiral inversion, whereas others apparently did not.

As a result of a number of *in vitro* and *in vivo* investigations on the metabolic fate of the profens, the majority of which have adopted ibuprofen as a model compound, the biochemical mechanism of the chiral inversion process has now become reasonably well defined (Figure 2.5). In the first step, which is critical to the outcome of the overall reaction, the profen is converted to the corresponding coenzyme A (CoA) thioester derivative in a process that exhibits a marked stereochemical preference for the *R*-enantiomer (Knadler and Hall, 1990). This key reaction, which is catalyzed primarily by microsomal long-chain acyl-CoA ligase (Knights et al., 1992), affords the intermediate that is subject to epimerization by an enzyme termed "2-arylpropionyl-CoA epimerase" (Shieh and Chen, 1993). Actually, the latter enzyme catalyzes the interconversion of the *R* and *S* forms of the profen-CoA, probably via a symmetrical enolate anion, as was first demonstrated with the aid of a specifically deuterated variant of *R*-ibuprofen (Sanins et al., 1991). When the resulting *S*-profen-CoA thioester is hydrolyzed to yield the free acid, the consequence is a net *R* to *S* conversion. The reverse reaction (*S* to *R*) is disfavored because of the steric preference of the acyl-CoA ligase for substrates of *R* configuration, although it should be pointed out that the ligase is not completely stereospecific and some examples of modest *S* to *R* inversion have been reported (reviewed in Shirley et al., 1994).

The above scheme, which accounts satisfactorily for the observed metabolic chiral inversion of *R*-ibuprofen and several of its congeners, also is consistent with the failure of other profens, for example, flurbiprofen, to exhibit this phenomenon. Thus, if neither the *R*- nor *S*-enantiomers of a certain profen act as substrates for the acyl-CoA ligase, the drug will not form a CoA thioester intermediate, will fail to gain access to the epimerase enzyme, and therefore will not undergo chiral inversion. This hypothesis was tested in the case of flurbiprofen (Figure 2.6), when the behavior of the enantiomers of the native drug and their corresponding synthetic CoA thioesters was examined *in vitro* in rat liver mitochondria (Porubek et al., 1991). The results showed that whereas the free acid forms of *R*- and *S*-flurbiprofen did not undergo detectable inversion in this system, the CoA thioesters epimerized rapidly. As a consequence of these mechanistic insights, one may predict that any chiral 2-substituted propionic acid which can be converted to the corresponding CoA thioester has the potential to exhibit metabolic chiral inversion and, indeed, examples of the phenomenon have been reported for xenobiotics

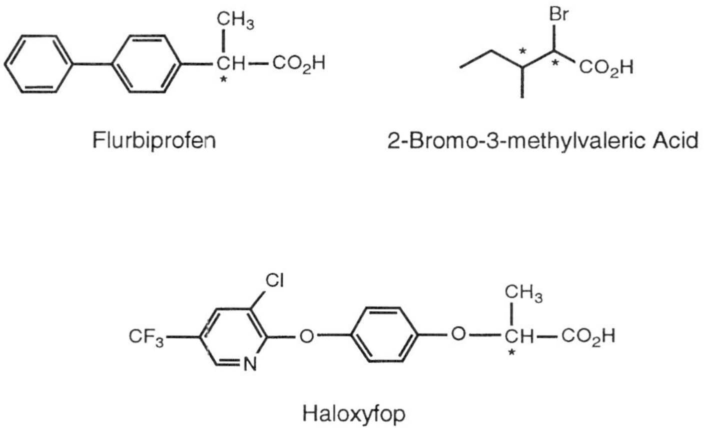

Flurbiprofen

2-Bromo-3-methylvaleric Acid

Haloxyfop

Figure 2.6. Structures of flurbiprofen, 2-bromo-3-methylvaleric acid, and haloxyfop. The location of the chiral center(s) is indicated in each case with an asterisk.

other than members of the profen family, for example, 2-bromo-3-methylvaleric acid (Polhuijs et al., 1992) and haloxyfop (Bartels and Smith, 1989) (Figure 2.6).

From the standpoint of new drug development, it may be argued that racemic profens are less desirable than the pure *S*-isomers as therapeutic agents since in those cases where chiral inversion is observed, the *R*-enantiomers may be considered, at best, as prodrugs of their active *S* forms. In addition, however, it should be borne in mind that the *R*-enantiomers may elicit adverse biological effects through the entry of their CoA derivatives into pathways of lipid metabolism, and studies are underway in several laboratories to investigate this possibility. In the case of those profens that do not undergo chiral inversion, the *R*-enantiomer may be viewed as an undesirable "isomeric ballast" (Ariëns, 1986). In either situation, these considerations have led several pharmaceutical manufacturers to propose "enantiomeric switches," whereby a drug marketed previously as a racemate is reintroduced as a pure enantiomer for reasons of enhanced efficacy and/or safety (Stinson, 1992). Ibuprofen is a case in point, the *S*-enantiomer of which currently is being developed by Merck as an over-the-counter analgesic with pharmacokinetic and therapeutic characteristics superior to those of generic racemic ibuprofen (Carey, 1993). A number of other marketed racemic drugs also are being considered for "enantiomeric switches" (Carey, 1993).

Examples of functional groups other than those cited above that exhibit configurational instability as a result of metabolic transformations have been discussed in a recent commentary by Testa et al. (1993). For instance, the oxidation of a chiral secondary alcohol, followed by stereoselective reduction of the ketone thus formed, can lead to epimerization at the chiral center. Chiral *N*-oxides may be reduced to tertiary amines, reoxidation of which can occur to yield a mixture of the two enantiomers in a different proportion to that present in the starting material. Similarly, metabolic reduction of a chiral sulfoxide to the corresponding achiral sulfide, followed by stereoselective reoxidation, can lead to net interconversion of the sulfoxide enantiomers (Kashiyama et al., 1994). It should not be assumed, therefore, that drugs containing chiral centers of the above types necessarily will retain their stereochemical identities when administered *in vivo*, but this should be established experimentally through the application of stereoselective assays of the drug isomers present in the systemic circulation.

2.3. STEREOSELECTIVE TOXICITY

Just as the two enantiomers of a chiral drug may possess discrete pharmacological activities, they also can differ in toxic potential as a consequence of different interactions with critical biological targets, and several examples of stereoselective toxicity have been described in the literature (see, e.g., FDA, 1992; Vermeulen and te Koppele, 1993). In some situations, therefore, the choice between developing one enantiomer over its antipode may be based on toxicological considerations. In cases where both enantiomers are pharmacologically active, yet only one proves to be toxic, the decision to develop the nontoxic species would be straightforward. However, situations may arise where some degree of toxicity resides in both enantiomers, and selection of the more appropriate candidate for development must be based on a consideration of the relative pharmacological and toxicological activities of both optical isomers, that is, on their respective therapeutic indices. An interesting example of the role of toxicology in the selection of a drug candidate may be found in studies on MK-0571 (Figure 2.7), an orally active racemic leukotriene antagonist that was evaluated as a potential drug for the therapy of bronchial asthma. In the course of safety assessment studies in mice, which were conducted on both the racemate and individual enantiomers of MK-0571, it was found that the drug caused peroxisome proliferation, as manifest by increased liver weights, increased peroxisomal volume density, and a pleiotropic induction of enzyme activities associated with peroxisomal proliferation (Grossman et al., 1992). However, these toxic effects were markedly enantioselective, since virtually all the peroxisomal proliferating activity proved to be associated with the *S*-enantiomer. Based on this finding, the decision was made to focus on the *R*-enantiomer for development, even

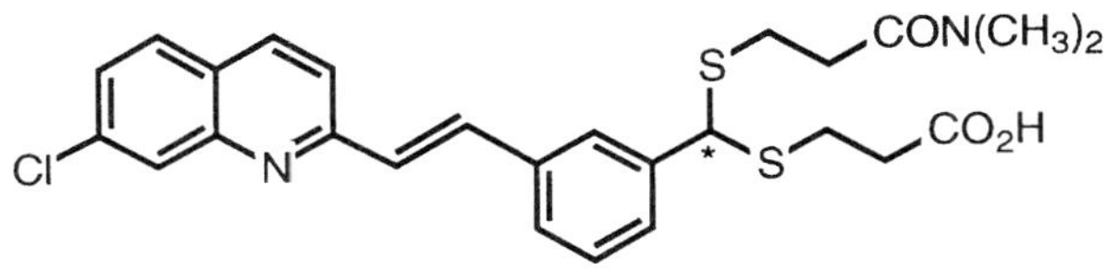

MK-0571

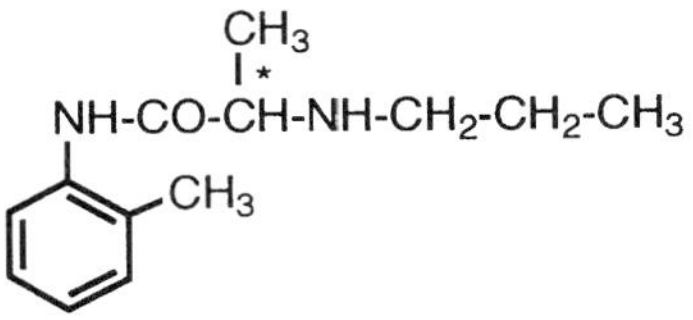

Prilocaine

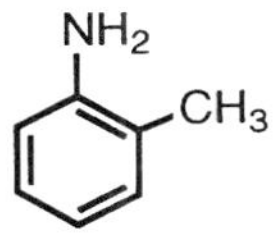

o-Toluidine

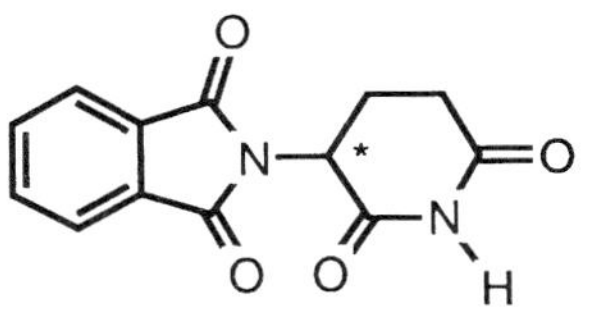

Thalidomide

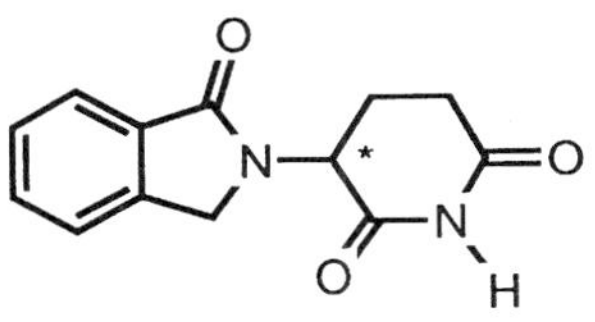

EM 12

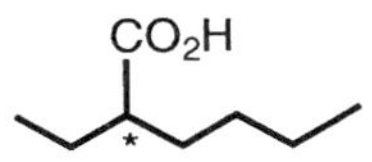

2-Ethylhexanoic Acid

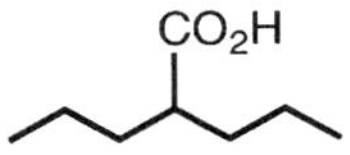

Valproic Acid

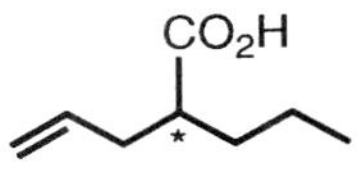

2-*n*-Propyl-4-pentenoic Acid

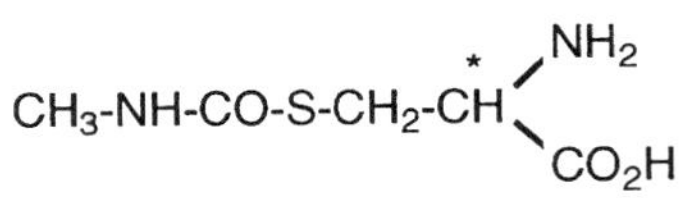

S-(N-methylcarbamoyl)cysteine (SMC)

Figure 2.7. Examples of compounds that have been implicated in enantioselective toxicities. The location of the chiral center is indicated with an asterisk in each case.

though this isomer was less potent than its antipode with respect to leukotriene antagonist activity. Interestingly, peroxisomal proliferation with MK-0571 was not observed in the rat, apparently as a result of differences in the pharmacokinetics of the drug in rats versus mice that led to a lower exposure of liver tissue to the more toxic enantiomer in the rat relative to the mouse (Grossman et al., 1992). This example illustrates the importance of evaluating the pharmacological, toxicological, and pharmacokinetic properties of each enantiomer of a chiral drug in more than one animal species in order to make an informed decision on the appropriate stereoisomer of a chiral drug for clinical development.

It is now widely recognized that the toxic effects of a drug or other xenobiotic may be mediated, not by the parent compound, but by one or more of its metabolites, and therefore it is imperative to establish the identities and biological activities of the major circulating metabolites of a candidate drug at an early stage of development. One of the first examples of enantioselective metabolism leading to enantioselective toxicity was the case of prilocaine (Figure 2.7), a local anesthetic drug that undergoes hydrolysis *in vivo* to *o*-toluidine that, in turn, causes methemoglobinemia (Åkerman and Ross, 1970). Intravenous administration of the (–)-enantiomer of prilocaine to cats caused more rapid formation of methemoglobin than was observed after dosing with the (+)-enantiomer, and this difference was found to be due to more rapid hydrolysis of (–)-prilocaine to its toxic (achiral) metabolite. However, despite these differences in the rates of methemoglobin formation, measurements in blood drawn over the 6-hour postdose period demonstrated that, overall, the total amounts of methemoglobin produced by the two enantiomers of the drug and by the racemic mixture were similar, and it was concluded that there would be no advantage in developing one pure enantiomer of prilocaine over the racemic mixture.

Perhaps the most widely publicized claim of enantioselectivity in drug-mediated toxicity has been with thalidomide (Figure 2.7), a sedative that was introduced into clinical practice in Europe and proved, tragically, to be a human teratogen. Although thalidomide was marketed as the racemic mixture, studies on the efficacy and safety of the individual enantiomers (conducted after the drug had been withdrawn from use) appeared to indicate that the teratogenic properties of thalidomide resided in the *S*-enantiomer (Blaschke et al., 1979). It was speculated, therefore, that the drug-induced birth defects could have been avoided if thalidomide had been developed as the pure *R*-enantiomer. However, as discussed in a critical review of the available data (De Camp, 1989), the evidence that *S*-thalidomide is indeed the teratogenic species is ambiguous at best, being based on experiments that failed to take into account the fact that the drug racemizes rapidly *in vivo* (Knoche and Blaschke, 1994). Similar problems with configurational stability have confounded studies with the enantiomers of the teratogenic thalidomide analog, EM12 (Figure 2.7), that racemize both *in vitro* and *in vivo* (Schmahl et al., 1988; 1989). Clearly, based on these observations, it is difficult to interpret the results of studies on the teratogenicity of thalidomide and related phthalimide derivatives *in vivo* in terms of enantioselective toxicity. In contrast, there is good evidence that the enantiomers of 2-ethylhexanoic acid (Figure 2.7), a metabolite of the commonly used phthalate plasticizer di-(2-ethylhexyl)phthalate, differ markedly in their teratogenic potential (Hauck et al., 1990). Thus, the *S*-enantiomer failed to yield any teratogenic or embryotoxic response in mice, whereas its antipode was highly teratogenic and embryotoxic in this animal model; the racemic mixture gave a response that was intermediate between those of the individual isomers. Also, the structurally related carboxylic acid 2-*n*-propyl-4-pentenoic acid, a chiral metabolite of the achiral antiepileptic drug valproic acid (Figure 2.7), has been found to exhibit enantioselectivity in its embryotoxic effects, inasmuch as the *S* form was appreciably more potent in inducing exencephaly in mice than its antipode (Hauck and Nau, 1989). Interestingly, 2-*n*-propyl-4-pentenoic acid has been shown to be generated from valproate by an enantioselective cytochrome P-450-dependent desaturation reaction that, in the case of isolated rat hepatocytes, led to a mixture of *R*- and *S*-isomers in a ratio of 83:17 (Porubek et al., 1989). Several other metabolites of valproate, generated by mitochondrial β-oxidation, also have been shown to be formed from the parent drug with marked enantiotopic differentiation (Shirley et al., 1993). Thus, valproate represents a valuable model compound for mechanistic investigations of the adverse effects of aliphatic carboxylic acids, particularly with regard to the role of stereoselective metabolism in the generation of toxic intermediates (Baillie and Sheffels, 1995).

It should be noted that, in comparison with other disciplines, for example pharmacology (where considerable interest has centered on the role of stereochemistry as a mediator of biological response), relatively little attention appears to have been devoted to stereochemical issues in the field of toxicology, at least judging from material published to date in the open literature. However, investigations on stereoselective aspects of the toxicity of foreign compounds are likely to yield important information on mechanisms of toxicity at the molecular level. For example, recent studies with the enantiomers of *S*-(*N*-methylcarbamoyl)cysteine (SMC; Figure 2.7), a carbamoylating metabolite of the toxin methyl isocyanate, demonstrated that whereas both L- and D-SMC (corresponding to *R* and *S* stereochemistry, respectively) inhibited glutathione reductase in rat hepatocytes, only the D-isomer produced a sustained depletion of glutathione in these cells (Kassahun et al., 1994). These findings indicate that while both isomers were able to carbamoylate, and thereby inhibit, glutathione reductase, the by-product released from L-SMC, namely, the endogenous amino acid L-cysteine, could be utilized by the liver cells for *de novo* glutathione biosynthesis. In contrast, the by-product (D-cysteine) released from D-SMC could not be utilized in this way, and thus replenishment of glutathione via biosynthesis was not possible due to the unnatural configuration of this amino acid. Hence, from the viewpoint of developing a compound such as SMC for clinical use (e.g., as a potential radiosensitizer for use in cancer chemotherapy), an understanding of the role of stereochemistry in its mechanism of action clearly would be essential for the rational selection of one isomer over the other. Similar considerations apply to the use of *N*-acetylcysteine as a chemoprotective agent, since only the L-isomer has been found to be effective as an antidote for acetaminophen poisoning in mice (Corcoran and Wong, 1986), whereas both enantiomers protect against oxygen-induced lung injury in rats (Särnstrand et al., 1995). It is to be hoped that, with the heightened awareness of the key role of stereochemistry in other areas of biological science, increasing attention will be paid to stereochemical issues in future toxicological research.

2.4. CONCLUSIONS

The examples discussed in this chapter illustrate the importance of taking into consideration stereochemical factors in assessing the overall pharmacological, toxicological, and metabolic characteristics of candidate drugs that are pertinent to the selection of a particular new chemical entity for development. Clearly, many additional factors enter into the decision-making process, such as economic feasibility, patent protection, and, most important of all, projected clinical utility. In addition, stereoselective drug interactions, many of which result from isozyme-selective enzyme inhibition, are beginning to attract more attention now that the experimental tools are available with which to study such phenomena (Gibaldi, 1993). It has been argued that regulatory authorities should not impose on the pharmaceutical industry rigid requirements for the development of pure stereoisomers (Holmes et al., 1990; Testa and Trager, 1990; Cayen, 1991). Rather, each submission should be considered on a case-by-case basis, where the decision to market a pure enantiomer or the racemic mixture is based on a sound scientific rationale where maximum benefit to the patient is the ultimate objective.

ACKNOWLEDGMENTS

We would like to thank Drs. Bogdan Matuszewski and Chiu Kwan (Department of Drug Metabolism, Merck Research Laboratories) for critically reviewing this chapter.

REFERENCES

Adams, S. S., Cliffe, E. E., Lessel, B., and Nicholson, J. S. (1967). Some biological properties of 2-(4-isobutylphenyl)-propionic acid. *J. Pharm. Pharmacol.*, **56**, 1686.

Adams, S. S., Bresloff, P., and Mason, C. G. (1976). Pharmacological differences between the optical isomers of ibuprofen: evidence for metabolic inversion of the (–) isomer, *J. Pharm. Pharmacol.*, **28**, 256–257.

Åkerman, B. and Ross, S. (1970). Stereospecificity of the enzymatic biotransformation of the enantiomers of prilocaine (Citanest). *Acta Pharmacol. Toxicol.*, **28**, 445–453.

Ariëns, E. J. (1984). Stereochemistry, a basis for sophisticated nonsense in pharmacokinetics and clinical pharmacology. *Eur. J. Clin. Pharmacol.*, **26**, 663–668.

Ariëns, E. J. (1986). Stereochemistry: A source of problems in medicinal chemistry. *Med. Res. Rev.*, **6**, 451–466.

Ariëns, E. J. (1993). Nonchiral, homochiral and composite chiral drugs. *Trends Pharmacol. Sci.*, **14**, 68–75.

Aso, Y., Yoshioka, S., Shibazaki, T., and Uchiyama, M. (1988). The kinetics of the racemization of oxazepam in aqueous solution. *Chem. Pharm. Bull.* (*Tokyo*), **36**, 1834–1840.

Baillie, T. A. and Sheffels, P. R. (1995). Valproic acid. Chemistry and biotransformation. In *Antiepileptic Drugs*, 4th ed., R. H. Levy, R. H. Mattson, B. S. Meldrum (eds.), New York: Raven Press, pp. 589–604.

Bartels, M. J. and Smith, F. A. (1989). Stereochemical inversion of haloxyfop in the Fischer 344 rat. *Drug Metab. Disp.*, **17**, 286–291.

Blaschke, G., Kraft, H. P., Fickentscher K., and Köhler, F. (1979). Chromatographische racemattrennung von thalidomide und teratogene Wirkung der enantiomere. *Arzneim.-Forsch.* (*Drug Res.*), **29**, 1640–1642.

Borgström, L., Kennedy, B.-M., Nilsson, B., and Angelin, B. (1990). Relative absorption of the two enantiomers of terbutaline after duodenal administration. *Eur. J. Clin. Pharmacol.*, **38**, 621–623.

Caldwell, J. (1989). St. Mary's Discussion Forum: Racemates and enantiomers: scientific and regulatory aspects. *Chirality*, **1**, 249–250.

Caldwell, J., Hutt, A. J., and Fournel-Gigleux, S. (1988). The metabolic chiral inversion and dispositional enantioselectivity of the 2-arylpropionic acids and their biological consequences. *Biochem. Pharmacol.*, **37**, 105–114.

Carey, J. (1993). New resolutions in drug design, *Chem. Brit.*, **29**, 1053–1056.

Cayen, M. N. (1991). Racemic mixtures and single stereoisomers: Industrial concerns and issues in drug development. *Chirality*, **3**, 94–98.

Corcoran, G. B. and Wong, B. K. (1986). Role of glutathione in prevention of acetaminophen-induced hepatotoxicity by *N*-acetyl-L-cysteine in vivo: Studies with *N*-acetyl-D-cysteine in mice. *J. Pharmacol. Exp. Ther.*, **253**, 54–61.

De Camp, W. H. (1989). The FDA perspective on the development of stereoisomers. *Chirality*, 1, 2–6.

Drayer, D. E. (1986). Pharmacodynamic and pharmacokinetic differences between drug enantiomers in humans: An overview. *Clin. Pharmacol. Ther.*, **40**, 125–133.

FDA (1992). FDA's policy statement for the development of new stereoisomeric drugs. *Chirality*, **4**, 338–340.

Gibaldi, M. (1993). Stereoselective and isozyme-selective drug interactions. *Chirality*, **5**, 407–413.

Gross, M., Cartwright, A., Campbell, B., Bolton, R., Holmes, K., Kirkland, K., Salmonson, T., and Robert, J.-L. (1993). Regulatory requirements for chiral drugs. *Drug Inf. J.*, **27**, 453–457.

Grossman, S. J., DeLuca, J. G., Zamboni, R. J., Keenan, K. P., Patrick, D. H., Herold, E. G., van Zwieten, M. J., and Zacchei, A. G. (1992). Enantioselective induction of peroxisomal proliferation in CD-1 mice by leukotriene antagonists. *Toxicol. Appl. Pharmacol.*, **116**, 217–224.

Hauck, R.-S. and Nau, H. (1989). Asymmetric synthesis and enantioselective teratogenicity of 2-*n*-propyl-4-pentenoic acid (4-en-VPA), an active metabolite of the anticonvulsant drug, valproic acid. *Toxicol. Lett.*, **49**, 41–48.

Hauck, R.-S., Wegner, C., Blumtritt, P., Fuhrhop, J.-H., and Nau, H. (1990). Asymmetric synthesis and teratogenic activity of (*R*)- and (*S*)-2-ethylhexanoic acid, a metabolite of the plasticizer di-(2-ethylhexyl)phthalate. *Life Sci.*, **46**, 513–518.

Hofmann, C. A. and Colca, J. R. (1992). New oral thiazolidinedione antidiabetic agents as insulin sensitizers. *Diab. Care*, **15**, 1075–1078.

Holmes, K. D., Baum, R. G., Brenner, G. S., Eaton, C. R., Gross, M., Grundfest, C. C., Margerison, R. B., Morton, D. R., Murphy, P. J., Palling, D., Repic, O., Simon, R., and Stoll, R. E. (1990). Comments on enantiomerism in the drug development process. *Pharmaceut. Technol.*, **14**, 46–52.

Hutt, A. J. (1991). Drug chirality: Impact on pharmaceutical regulation. *Chirality*, **3**, 161–164.

Hutt, A. J. and Caldwell, J. (1983). The metabolic chiral inversion of 2-arylpropionic acids—a novel route with pharmacological consequences. *J. Pharm. Pharmacol.*, **35**, 693–704.

Hutt, A. J. and Caldwell, J. (1984). The importance of stereochemistry in the clinical pharmacokinetics of the 2-arylpropionic acid non-steroidal anti-inflammatory drugs. *Clin. Pharmacokinet.*, **9**, 371–373.

Kashiyama, E., Todaka, T., Odomi, M., Tanokura, Y., Johnson, D. B., Yokoi, T., Kamataki, T., and Shimizu, T. (1994). Stereoselective pharmacokinetics and interconversions of flosequinan enantiomers containing chiral sulphoxide in rat. *Xenobiotica*, **24**, 369–377.

Kassahun, K., Jochheim, C., and Baillie, T. A. (1994). Effect of carbamate thioester derivatives of methyl- and 2-chloroethyl isocyanate on glutathione levels and glutathione reductase activity in isolated rat hepatocytes. *Biochem. Pharmacol.*, **48**, 587–594.

Knadler, M. P. and Hall, S. D. (1990). Stereoselective arylpropionyl-CoA thioester formation *in vitro*. *Chirality*, **2**, 67–73.

Knights, K. M., Talbot, U. M., and Baillie, T. A. (1992). Evidence of multiple forms of rat liver microsomal coenzyme a ligase catalysing the formation of 2-arylpropionyl-coenzyme a thioesters. *Biochem. Pharmacol.*, **44**, 2415–2417.

Knoche, B. and Blaschke, G. (1994). Stereoselectivity of the *in vitro* metabolism of thalidomide. *Chirality*, **6**, 221–224.

Lee, E. J. D. and Williams, K. M. (1990). Chirality. Clinical pharmacokinetic and pharmacodynamic considerations. *Clin. Pharmacokinet.*, **18**, 339–345.

Levy, R. H. and Boddy, A. V. (1991). Stereoselectivity in pharmacokinetics: A general theory. *Pharmaceut. Res.*, **8**, 551–556.

Merritt, M. V. and Bronson, G. E. (1978). Kinetics of epimerization of 15*(R)*-methyl-prostaglandin E_2 and of 15*(S)*-methylprostaglandin E_2 as a function of pH and temperature in aqueous solution. *J. Am. Chem. Soc.*, **100**, 1891–1895.

Nicholson, J. S. (1982). Ibuprofen. In *Chronicles of Drug Discovery*, Vol. 1, J. S. Bindra, and D. Lednicer (eds.), New York: Wiley, pp. 149–172.

Pepper, C., Smith, H. J., Barrell, K. J., Nicholls, P. J., and Hewlins, M. J. E. (1994). Racemization of drug enantiomers by benzylic proton abstraction at physiological pH. *Chirality*, **6**, 400–404.

Pfeiffer, C. C. (1956). Optical isomerism and pharmacological action, a generalization. *Science*, **124**, 29–31.

Polhuijs, M., Tergau, A. C., and Mulder, G. J. (1992). Chiral inversion and stereoselective glutathione conjugation of the four 2-bromo-3-methylvaleric acid stereoisomers in the rat *in vivo* and *in vitro*. *J. Pharmacol. Exp. Ther.*, **260**, 1349–1354.

Porubek, D. J., Barnes, H., Meier, G. P., Theodore, L. J., and Baillie, T. A. (1989). Enantiotopic differentiation during the biotransformation of valproic acid to the hepatotoxic olefin 2-*n*-propyl-4-pentenoic acid. *Chem. Res. Toxicol.*, **2**, 35–40.

Porubek, D. J., Sanins, S. M., Stephens, J. R., Grillo, M. P., Kaiser, D. G., Halstead, G. W., Adams, W. J., and Baillie, T. A. (1991). Metabolic chiral inversion of flurbiprofen-CoA *in vitro*. *Biochem. Pharmacol.*, **42**, R1–R4.

Rauws, A. G. and Groen, K. (1994). Current regulatory (draft) guidance on chiral medicinal products: Canada, EEC, Japan, United States. *Chirality*, **6**, 72–75.

Sanins, S. M., Adams, W. J., Kaiser, D. G., Halstead, G. W., Hosley, J., Barnes, H., and Baillie, T. A. (1991). Mechanistic studies on the metabolic chiral inversion of *R*-ibuprofen in the rat. *Drug Metab. Dispos.*, **19**, 405–410.

Särnstrand, B., Tunek, A., Sjödin, K., and Hallberg, A. (1995). Effects of *N*-acetylcysteine stereoisomers on oxygen-induced lung injury in rats. *Chem.-Biol. Inter.*, **94**, 157–164.

Schmahl, H.-J., Heger, W., and Nau, H. (1989). The enantiomers of the teratogenic thalidomide analogue EM12. *Toxicol. Lett.*, **45**, 23–33.

Schmahl, H.-J., Nau, H., and Neubert, D. (1988). The enantiomers of the teratogenic thalidomide analogue EM12: 1. Chiral inversion and plasma pharmacokinetics in the marmoset monkey. *Arch. Toxicol.*, **62**, 200–204.

Shen, D. D., Levy, R. H., Savitch, J. L., Boddy, A. V., Lepage, F., and Tombret, F.

(1992). Comparative anticonvulsant potency and pharmacokinetics of (+)- and (–)-enantiomers of stiripentol. *Epilepsy Res.*, **12**, 29–36.

Shieh, W.-R. and Chen, C.-S. (1993). Purification and characterization of novel "2-arylpropionyl-CoA epimerases" from rat liver cytosol and mitochondria. *J. Biol. Chem.*, **268**, 3487–3493.

Shindo, H. and Caldwell, J. (1991). Regulatory aspects of the development of chiral drugs in Japan: A status report. *Chirality*, **3**, 91–93.

Shirley, M. A., Hu, P., and Baillie, T. A. (1993). Stereochemical studies on the β-oxidation of valproic acid in isolated rat hepatocytes. *Drug Metab. Disp.*, **21**, 580–586.

Shirley, M. A., Guan, X., Kaiser, D. G., Halstead, G. W., and Baillie, T. A. (1994). Taurine conjugation of ibuprofen in humans and in rat liver *in vitro*. Relationship to metabolic chiral inversion. *J. Pharmacol. Exp. Ther.*, **269**, 1166–1175.

Sohda, T., Mizuno, K., and Kawamatsu, Y. (1984). Studies on antidiabetic agents. VI. Asymmetric transformation of (±)-5-[4-(1-methylcyclohexylmethoxy)benzyl]-2,4-thiazolidinedione (ciglitazone) with optically active 1-phenylethylamines. *Chem. Pharm. Bull.* (*Tokyo*), **32**, 4460–4465.

Stinson, S. C. (1992). Chiral drugs. *Chem. Brit*, **28**, 46–79.

Stinson, S. C. (1994). Chiral drugs. New single-isomer products on the chiral drug market create demand for enantiomeric intermediates and enantioseletive technologies. *Chem. Eng. News*, **72** (38), 38–72.

Tang, C., Lepage, F., and Baillie, T. A. (1995). Unpublished results.

Tang, C., Zhang, K., Lepage, F., Levy, R. H., and Baillie, T. A. (1994). Metabolic chiral inversion of stiripentol in the rat. II. Influence of route of administration. *Drug Metab. Disp.*, **22**, 554–560.

Testa, B. (1989). Mechanisms of chiral recognition in xenobiotic metabolism and drug–receptor interactions. *Chirality*, **1**, 7–9.

Testa, B. and Trager, W. F. (1990). Racemates versus enantiomers in drug development: Dogmatism or pragmatism? *Chirality*, **2**, 129–133.

Testa, B., Carrupt, P.-A., and Gal, J. (1993). The so-called "interconversion" of stereoisomeric drugs: An attempt at clarification, *Chirality*, **5**, 105–111.

Vermeulen, N. P. E. and te Koppele, J. M. (1993). Stereoselective biotransformation. Toxicological consequences and implications. In *Drug Stereochemistry. Analytical Methods and Pharmacology*, 2nd ed. I. W. Wainer (ed.), New York: Marcel Dekker, pp. 245–280.

Wainer, I. W. (ed.) (1993). *Drug Stereochemistry. Analytical Methods and Pharmacology*, 2nd. ed. New York: Marcel Dekker.

Williams, K. M. (1990). Enantiomers in arthritic disorders. *Pharmacol. Ther.*, **46**, 273–295.

Williams, K. M. (1991). Molecular asymmetry and its pharmacological consequences. *Adv. Pharmacol.*, **22**, 57–135.

Wilson, K. and Walker, J. (eds.) (1991). Chirality and its importance in drug development. *Biochem. Soc. Trans.*, **19**, 443–475.

Yang, S. K. (1994). Acid-catalyzed stereoselective heteronucleophilic substitution and racemization of 3-*O*-methyloxazepam and 3-*O*-ethyloxazepam. *Chirality*, **6**, 175–184.

Yang, S. K. and Bao, Z. (1994). Base-catalyzed racemization of 3-*O*-acyloxazepam. *Chirality*, **6**, 321–328.

Yang, S. K. and Lu, X.-L. (1989). Racemization kinetics of enantiomeric oxazepams and stereoselective hydrolysis of enantiomeric oxazepam 3-acetates in rat liver microsomes and brain homogenate. *J. Pharmaceut. Sci.*, **78**, 789–795.

Yang, S. K. and Lu, X.-L. (1992). Resolution and stability of oxazepam enantiomers. *Chirality*, **4**, 443–446.

Zhang, K., Tang, C., Rashed, M., Cui, D., Tombret, F., Botte, H., Lepage, F., Levy, R. H., and Baillie, T. A. (1994). Metabolic chiral inversion of stiripentol in the rat. I. Mechanistic studies. *Drug Metab. Disp.*, **22**, 544–553.

CHAPTER

3

STEREOCHEMICAL ASPECTS OF DRUG METABOLISM

JONATHAN P. MASON and ANDREW J. HUTT

Department of Pharmacy
King's College London,
Manresa Road, London, SW3 6LX,
United Kingdom.

"Oh, life is a crooked thing."[1]

3.1. INTRODUCTION

Life does indeed seem to be a crooked thing, particularly at a molecular level. Historically, the connection between molecular asymmetry and the fundamental processes of life was made early in the nineteenth century. By the end of the century, as a result of the pioneering work of such giants of the scientific community as van't Hoff, Le Bel, and Pasteur, the three-dimensional theory of molecular structure had become established (Drayer, 1993) and by the middle of the present century, it had become generally accepted that interactions between small molecules (drugs, amino acids, sugars, etc.) and biological macromolecules, for example, enzymes, receptors, etc., occur at restricted portions of the macromolecules, that is, active and/or binding sites, through which they mediate their biological functions (Cornforth, 1974; 1984; Lehman et al., 1976; Testa and Mayer, 1988). Since molecular asymmetry is common in biological systems, it should come as no surprise that when chiral molecules are involved, the processes of pharmacodynamics and pharmacokinetics, that is, absorption, distribution, metabolism, and excretion, generally exhibit profound stereoselectivity (Testa and Mayer, 1988; Campbell, 1990).

[1]This deliberate misquotation (substituting *life* for *love*) of the Irish poet, W. B. Yeats, was made by Sir John Cornforth in a review in which he considered the stereochemistry of life (Cornforth, 1984).

The Impact of Stereochemistry on Drug Development and Use, Edited by Hassan Y. Aboul-Enein and Irving W. Wainer. Chemical Analysis Series, Vol. 142.
ISBN 0-471-59644-2

Drug metabolism studies have made and continue to make fundamental contributions to the drug discovery and development process, to the elucidation of the mechanisms of action of both new and existing drugs, and also to the understanding of the molecular mechanisms of drug and foreign compound toxicity. Metabolic studies have an essential role in the safety evaluation of a new chemical entity (NCE) and the regulatory guidelines for safety evaluation all make reference to metabolic and pharmacokinetic data. Although, at present, there is no absolute requirement from the major regulatory bodies for the marketing of single stereoisomers, there is a requirement for companies to specify the nature of the material used in preclinical studies. In addition, any potential problems associated with the use of stereoisomeric mixtures of NCEs, particularly with regard to toxicology, pharmacology, and pharmacokinetics, including metabolism, should be addressed (Cartwright, 1991; Rauws and Groen, 1994).

The present chapter is confined primarily to stereoselectivity as it applies to the processes of biotransformation; however, reference will occasionally be made to the stereoselectivity of other pharmacokinetic processes. This chapter will review the underlying mechanisms of enantioselectivity in drug and foreign compound metabolism and examine the role of inter- and intraspecies differences in expression of the enzymes involved in xenobiotic biotransformation.

Historically, the reactions of drug metabolism have been divided into two groups (Williams, 1959; Conti and Bickel, 1977). The first of these groups are termed the *phase I* reactions, where the xenobiotic is transformed by a functionalization reaction, for example, oxidation, reduction, or hydrolysis, resulting in the introduction, or unmasking of a functional group in the xenobiotic. The *phase II* reactions involve conjugation of the xenobiotic, or a phase I metabolite of the xenobiotic, with an endogenous moiety, such as glucuronic acid, sulfate, glutathione, or an amino acid, for example, glycine (Table 3.1). The majority of xenobiotics are eliminated as one or more metabolites, produced as a result of one, or both of these phases of metabolism. Since most xenobiotics are relatively lipophilic, the primary purpose of xenobiotic metabolism is to increase the aqueous solubility of the compound. As a result of the increased aqueous solubility, most metabolites are excreted in the urine, although a number, particularly glucuronides, are excreted in bile. In general, phase II metabolites account for the bulk of the inactive, excreted material.

As pointed out above, stereoselectivity may be observed during drug absorption, particularly where the compound resembles a nutrient for which active transport systems are available; examples of interest include L-dopa, L-methotrexate, and L-cephalexin. Stereoselectivity may also be observed in distribution, as a consequence of the differential binding of drug enantiomers

Table 3.1. Principle Reactions Involved in Phase I and Phase II Xenobiotic Biotransformation

Phase I

Reaction Type	Pathway
Oxidation	Aliphatic or aromatic hydroxylation; epoxidation; heteroatom dealkylation (N—, S—, or O—); heteroatom oxidation (N—, S—, etc); oxidative dehalogenation
Reduction	Nitro reduction to amine/hydroxylamine; carbonyl reduction to alcohol; reductive dehalogenation
Hydrolysis	Esters; amides; hydrazides

Phase II

1. Activated Conjugating Agent

Reaction Type	Coenzyme	Enzyme	Functional Group
Glucuronidation	UDPGA[a]	UDPGT[b]	—OH, —COOH, —NH_2, —NR_2, —SH, —C—H (acidic)
Sulfation	PAPS[c]	Sulfotransferase	—OH, —NH_2, —SO_2NH_2
Methylation	S-adenosylmethionine	Methyltransferase	—OH, —NH_2
Acetylation	Acetylcoenzyme A	Acetyltransferase	—OH, —NH_2
Hydration		Epoxide hydrolase	Epoxides

2. Activated Xenobiotic

Reaction Type	Endogenous Conjugation Agent	Activated Intermediate
Amino acid conjugation	Amino acid, for example, glycine, glutamine, taurine	Coenzyme A thioesters of xenobiotic carboxylic acids
Glutathione conjugation	γ-glutamylcysteinylglycine (glutathione, or GSH)	Epoxides; arene oxides; haloalkanes; nitroalkanes; alkenes; aromatic halo-and nitro-compounds

[a]UDPGA=Uridine diphosphate glucuronic acid.
[b]UDPGT=UDP-glucuronosyl-transferase.
[c]PAPS=3′-phosphoadenosine-5-phosphosulphate.

to plasma proteins, particularly albumin and α_1-acid glycoprotein (AGP) and tissue proteins. In addition, stereoselectivity may be observed in renal excretion as a result of active secretion or reabsorption, as has been seen with metoprolol, disopyramide, pindolol, and chloroquine. In general, such differences tend to be modest, enantiomeric ratios for such processes being of the order of two or less; however, exceptions to this generality do exist (Campbell, 1990).

Probably the most important contribution to differences in the pharmacokinetic properties of a mixture of isomers occurs as a result of stereoselective drug metabolism. Stereoselectivity in drug metabolism appears to be the rule, rather than the exception, which perhaps should not be surprising, as the reactions involve a direct interaction between the substrate and a chiral enzyme. Stereochemical effects in drug metabolism have only relatively recently become the focus of attention. The reasons for this are two-fold. First, investigations of the metabolism and pharmacokinetics of individual enantiomers, following the administration of racemates, have been hampered in the past by a lack of sufficiently sensitive analytical techniques capable of separating individual enantiomers following their isolation from biological fluids (Caldwell et al., 1988a). Second, meaningful *in vitro* studies into the underlying mechanisms of drug metabolism only became possible in the late 1950s and early 1960s, following the characterization of the microsomal metabolizing systems and, more recently, with the isolation and cloning of the enzymes involved and their expression in various vector systems, for example, bacteria and yeast (Guengerich et al., 1993; Guengerich, 1994).

3.2. STEREOCHEMICAL ASPECTS OF DRUG METABOLISM

The stereoselectivity of the reactions of drug metabolism may be categorized into three reaction types in terms of their selectivity with respect to the substrate and/or product (Jenner and Testa, 1973). Thus, there may be *substrate selectivity*, where one isomer is metabolized more rapidly than the other; *product stereoselectivity*, where one particular stereoisomer is produced preferentially; or a combination of the two, that is, *substrate-product stereoselectivity*, where one enantiomer is preferentially metabolized to yield a particular product (Low and Castagnoli, 1978; Testa, 1989). Alternatively, the stereoselectivity of drug metabolism may be described in terms of the stereochemical consequences of the reactions (Caldwell et al., 1988a). The following events have been defined:

1. *Prochiral to chiral* transformations, where a prochiral center of a symmetric molecule is preferentially metabolized to form one of two possible

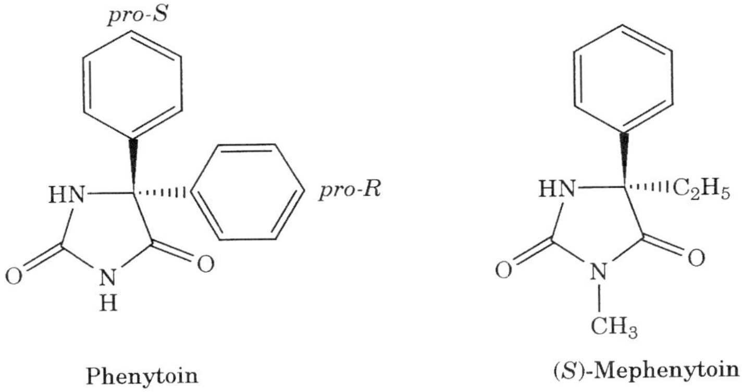

Figure 3.1. Structures of phenytoin and (*S*)-mephenytoin.

enantiomers. Examples of this reaction type include the formation of (*R*)-2-*n*-propyl-4-pentenoic acid from valproic acid (Porubek et al., 1988) and the stereoselective *para*-hydroxylation of phenytoin (Figure 3.1), resulting in the formation of a greater than 90% enantiomeric excess of (*S*)-4′-hydroxyphenytoin in man (Poupaert et al., 1975). In the dog, however, phenytoin undergoes oxidation to yield the *R*-isomer of the *meta*-hydroxylated product (Maguire et al., 1978). Thus, the oxidation of phenytoin exhibits species-dependent regio- and stereoselectivity. Conjugation may also result in the formation of a chiral metabolite from a prochiral drug, as demonstrated by the N-glucosylation of amobarbital (Figure 3.2). This reaction results in the introduction of a chiral center in the substrate at C5 of the ring system (Soine et al., 1986).

2. *Chiral to chiral* transformations, in which the two enantiomers of a compound are differentially metabolized at a site remote from the chiral center, for example, (*S*)-warfarin (Figure 3.3) is selectively 7-hydroxylated in man (Lewis et al., 1974), whereas in the rat, the *R*-isomer undergoes this biotransformation (Pohl et al., 1976a; 1976b).

3. *Chiral to diastereoisomer* transformations, where a second chiral center is introduced into the drug by either a phase I reaction at a prochiral center, (examples include side-chain oxidation of pentobarbitone to yield the corresponding alcohol derivatives and keto-reduction of warfarin to yield a pair of diastereoisomeric alcohols), or the substrate undergoes conjugation with a chiral conjugating agent, for example, the stereoselective glucuronidation of oxazepam (Sisenwine et al., 1982).

4. *Chiral to achiral* transformations, where the substrate undergoes transformation at the chiral center, resulting in a loss of asymmetry, for example,

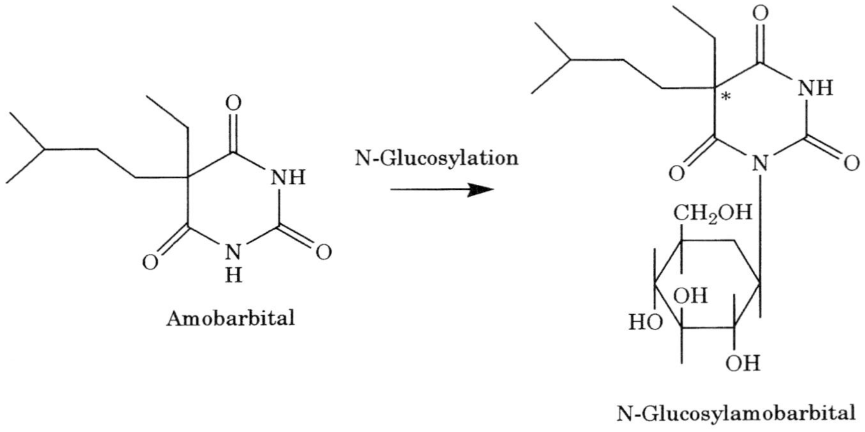

O
NH
N
H
O
O
Amobarbital

N-Glucosylation

O
*
NH
O
N
O
CH_2OH
OH
HO
OH
N-Glucosylamobarbital

Figure 3.2. The conjugation of amobarbital with glucose to yield diastereoisomeric N-glucosides.

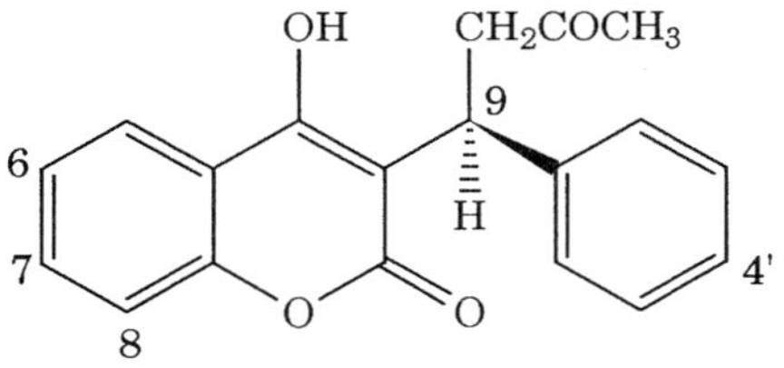

OH
CH_2COCH_3
9
6
7
8
H
O
O
4'

Figure 3.3. Structure of (*S*)-warfarin.

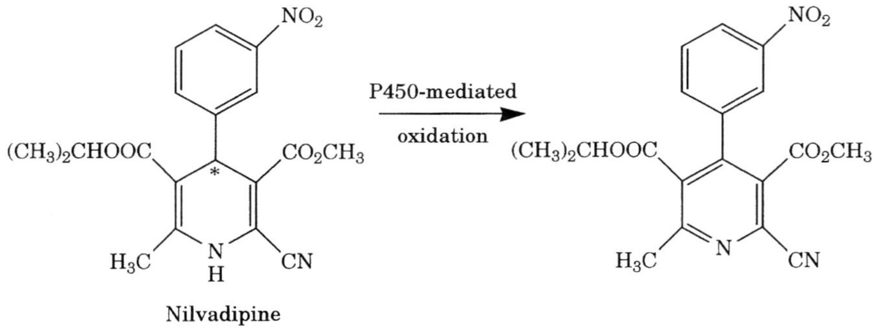

NO_2
$(CH_3)_2CHOOC$
*
CO_2CH_3
H_3C
N
H
CN
Nilvadipine

P450-mediated
oxidation

NO_2
$(CH_3)_2CHOOC$
CO_2CH_3
H_3C
N
CN

Figure 3.4. The oxidation of nilvadipine resulting in loss of chirality.

the aromatization of the dihydropyridine calcium channel blocking agents, for example, nilvadipine (Figure 3.4), which undergoes cytochrome P450-mediated oxidation to yield the corresponding achiral pyridine analogue. This reaction exhibits species-dependent stereoselectivity. In rats, (+)-nilvadipine undergoes preferential aromatization, whereas in man and dogs, the (–)-isomer is the preferred substrate (Niwa et al., 1989; Tokuma et al., 1989).

5. *Chiral inversion*, where one enantiomer of the substrate is biochemically converted into its antipode, with no other change to the structure, for example, the 2-arylpropionic acid nonsteroidal antiinflammatory drugs (NSAIDs), collectively known as the "profens." With these drugs, examples of which include ibuprofen and ketoprofen, the pharmacologically inactive *R*-enantiomers are selectively inverted to form their active *S*-antipodes (Hutt and Caldwell, 1983; Caldwell et al., 1988a).

One of the most important aspects of xenobiotic metabolism, particularly in relation to preclinical safety evaluation, is the differential metabolism of substrates by different species, examples of which have been indicated above. Species differences in drug metabolism may be either *quantitative*, where the same metabolic pathways are utilized to different extents, or *qualitative*, where different pathways are utilized. Such species differences arise as a result of differences in the expression of the enzymes involved. Species differences may present a significant problem in the evaluation of new chemical entities as there is no guarantee that humans will metabolize the drug in the same manner and to a similar extent as the test species used. With respect to stereoisomers, the problem of species variability is compounded as the products of metabolism may be "the same," but the variation may occur in the stereoselectivity of the metabolic pathway. Thus, without taking stereochemistry into account, information obtained from animal studies may be misleading and this may have considerable significance for the human situation.

3.2.1. Phase I Enzyme Systems

The major phase I oxidative enzyme systems are the cytochrome P450-dependent microsomal mixed-function oxidase (MMFO) system and the microsomal flavin-containing monooxygenases (FMOs), although there are other oxidative enzyme systems, such as monoamine oxidase (MAO) and various alcohol and aldehyde dehydrogenases, which are relatively less important with respect to xenobiotic metabolism. Other phase I enzyme systems include hydrolytic enzymes, such as esterases, and a variety of reductases. The MMFO system and the FMOs are termed monooxygenases, that is, they activate molecular dioxygen and oxidize substrates by insertion of a single oxygen atom. Both enzymes require NADPH, molecular oxygen (O_2), and magnesium ions (Mg^{2+}) and catalyze essentially the same reaction, resulting in the formation of an oxidized product.

3.2.1.1. Cytochrome P450

The cytochrome P450 enzymes (P450s) are a family of iron-containing hemoproteins. These enzymes are involved in the metabolism of both endogenous and foreign compounds and can oxidize and, under certain conditions, reduce a wide diversity of substrates (Guengerich, 1992a; 1992b; 1994). During the last two decades, several P450 isoenzymes have been isolated and classified according to their protein and gene sequences (Nelson et al., 1993). The purification and sequencing of multiple P450s in different laboratories have resulted in numerous trivial names for the same P450 isoenzyme (Cheng and Schenkman, 1982; Guengerich et al., 1982; Ryan et al., 1982; 1984). In 1987, a standardized nomenclature for the P450s was proposed, based on similarities between amino acid coding sequences, which was updated in 1991[2] (Nebert et al., 1987; 1991). By 1993, the P450 superfamily consisted of 221 different genes and 12 putative pseudogenes (Nelson et al., 1993). The term "P450" refers to the cytochrome P450 gene superfamily, if the isoenzymes differ by more than 60% in amino acid sequence, then they are consigned to different families and subfamilies. Mammalian sequences within the same subfamily are always greater than 55% identical (Nelson et al., 1993).

Since the 1970s, considerable effort has been directed toward the isolation and characterization of the P450s. This work has led to an examination of the relationships between structure and function. The mammalian P450s can be divided into those P450 proteins mainly involved in xenobiotic metabolism and those involved in physiological metabolic reactions (Guengerich and Shimada, 1991). Those involved in steroidogenesis, namely, the CYP11, CYP17, CYP19, CYP21, and CYP27 subfamilies, and in the metabolism of cholesterol and bile acids, the CYP7 and CYP51 subfamilies, exhibit a high degree of regio- and stereospecificity (Nebert et al., 1991; Coon et al., 1992). Coincidentally, in evolutionary terms, these are also the oldest mammalian P450s.

In contrast to the aforementioned isoenzymes, those P450s in the CYP1, CYP2, and CYP3 subfamilies are multifunctional and exhibit a relatively low degree of substrate specificity. These P450s are involved in the metabolism of a plethora of chemically diverse xenobiotic substrates, and their catalytic

[2]It was recommended that workers use the italicized root symbol *CYP* (*Cyp* for murine genes), an arabic number to designate the family, a letter to indicate the subfamily, and a second arabic numeral to represent the individual gene, whereas for the gene product, that is, mRNA and protein, the nonitalicized root CYP should be used (Nebert et al., 1991). These conventions will be used throughout this chapter.

activities can be markedly enhanced by the administration of various compounds known to either induce their *de novo* synthesis, for example, phenobarbitone, rifampicin, etc., or to inhibit their catalytic activity. As a result, these P450s are of particular interest in pharmacological and toxicological studies (Guengerich, 1994). However, although in many cases these P450s are relatively nonselective, this is not always the case and there are numerous examples of isoenzymes in the CYP1, CYP2, and CYP3 families demonstrating substrate selectivity, resulting in regio- and/or stereoselectivity (Table 3.2).

One of the best examples of such selectivity is the P450-mediated aliphatic hydroxylation of steroid hormones, which is generally highly stereoselective, if not stereospecific. The isoenzyme-specific regioselectivity of the oxidation of testosterone has been studied using purified P450s, isolated from various animal species, including man. These data are summarized in Table 3.3. The oxidation of both testosterone and progesterone exhibits regio- and stereospecificity. It has been suggested that the regio- and stereoselective hydroxylation of testosterone could be useful both as a probe for the functional characterization of the multiplicity of hepatic P450s and to gain information as to the catalytic mechanism of certain P450s (Swinney et al., 1987; Waxman, 1988; Vermeulen and te Koppele, 1993). For example, although the 6β-hydroxylation of testosterone in the rat is mediated by CYP2C13, this same isoenzyme catalyzes both the 6β- and 16α-hydroxylation of progesterone (Swinney et al., 1987). In addition, stereospecificity may depend on the substrate, for example, rat CYP2A2 mediates the hydroxylation of the 15-position of both testosterone and progesterone. However, whereas for testosterone the ratio of 15α- to 15β-hydroxylated products is greater than 13, for progesterone the situation is reversed, the ratio being 0.39:1 (Jansson et al., 1985).

The oxidation of steroids also highlights species differences in isoenzyme selectivity. Thus, although the 6β-hydroxylation of testosterone in the rat is catalyzed primarily by CYP2C13 (Cheng and Schenkman, 1983; Ryan et al., 1984; Matsumoto et al., 1986; Funae and Imaoka, 1987), in man and monkey this reaction is catalyzed by isoenzymes in the CYP3A subfamily (Kawano et al., 1987; Komori et al., 1988; Wrighton et al., 1989; Dalet-Beluche et al., 1992). In addition to species differences, there are also sex-dependent differences in the metabolism of steroids. For example, in the male rat, the 6β-hydroxylation of testosterone, progesterone, and androstenedione is catalysed by CYP2A1, while CYP1A1 catalyzes both the 2α-hydroxylation of testosterone and the 16α-hydroxylation of androstenedione. Yet in the female rat, the major metabolic transformation of these three steroids is 15β-hydroxylation catalyzed by CYP1A2 (Waxman, 1988).

The 16α-hydroxylation of progesterone is mediated in the rabbit by CYP2C3 and in man by CYP2C9. In the rabbit, there is a variant enzyme,

Table 3.2. Substrate-Product Enantioselectivity of Oxidation of Various Substrates by Selected Human Cytochrome P450 Isoenzymes

Isoenzyme	Substrates	Stereochemistry of Metabolites	References
1A2	(*R*)-warfarin	(*R*)-6-hydroxywarfarin	(Rettie et al., 1992)
		(*R*)-7-hydroxywarfarin	
		(*R*)-8-hydroxywarfarin	
2C9	(*S*)-warfarin	(*S*)-7-hydroxywarfarin	(Rettie et al., 1992)
		(*S*)-6-hydroxywarfarin	
		(*S*)-4′-hydroxywarfarin	
	(*S*)-mephenytoin	(*S*)-4′-hydroxymephenytoin	(Ged et al., 1988)
	(*S*)-hexobarbitone	(*S*)-3′α-hydroxyhexobarbitone	(Kato et al., 1992)
	(*R*)-hexobarbitone	(*R*)-3′β-hydroxyhexobarbitone	(Kato et al., 1992)
2D6	Debrisoquine	(*S*)-4-hydroxydebrisoquine	
	(*S*)-propafenone	(*S*)-5-hydroxypropafenone	(Kroemer et al., 1991)
	(*R*)-metoprolol	(*R*)-O-desmethylmetoprolol	(Kim et al., 1993; Mautz et al., 1995)
	(*S*)-metoprolol	(*S*)-α-hydroxymetoprolol	
	(1′*R*)-bufuralol	(1′*R*,1″*S*)-hydroxybufuralol	(Mautz et al., 1995)
	(1′*S*)-bufuralol	(1′*S*,1″*R*)-hydroxybufuralol	
3A4	(*R*)-warfarin	(*R*)-10-hydroxywarfarin	(Rettie et al., 1992)
		(*R*)-4′-hydroxywarfarin	
	(*S*)-warfarin	(*S*)-10-hydroxywarfarin	(Rettie et al., 1992)
		(*S*)-6-hydroxywarfarin	
		(*S*)-4′-hydroxywarfarin	
		(*S*)-7′-hydroxywarfarin	
		(*S*)-8′-hydroxywarfarin	

Table 3.3. Regioselective Hydroxylation of Testosterone, Progesterone, and Androstenedione by Rat Hepatic Cytochrome P450 Isoenzymes

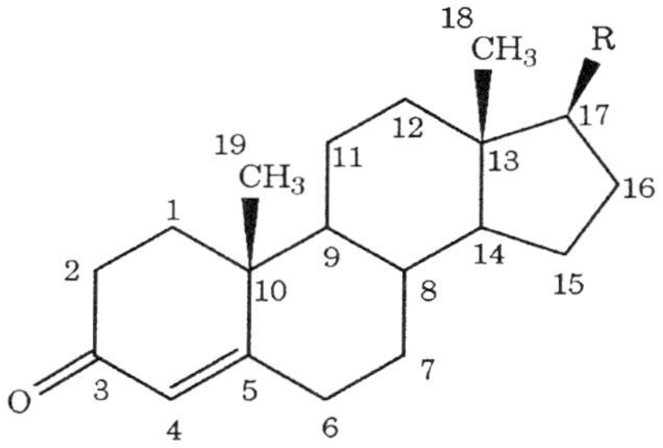

Progesterone R = $-COCH_3$
Testosterone R = $-OH$
Androstenedione R = $=O$

	Substrate		
P450 Isoenzyme	Progesterone Reactions(s) (Ratio)	Testosterone Reaction(s) (Ratio)	Androstenedione Reaction(s) (Ratio)
CYP1A1	$6\beta/16\alpha$ (2:1)	6β	6β
CYP1A2	$6\beta/16\alpha$ (4:1)	6β	6β
CYP2A1	$7\alpha/6\alpha$ (14:1)	$7\alpha/6\alpha$ (21:1)	$7\alpha/6\beta$ (30:1)
CYP2B1	16α	$17/16\alpha/16\beta$ (1.5:1.3:1)	$16\beta/16\alpha$ (13:1)
CYP2B2	16α	$17/16\alpha/16\beta$ (1.6:1.1:1)	$16\beta/16\alpha$ (12:1)
CYP2C6	$21/16\alpha$ (1:1)		
CYP2C7	$16\alpha/2\alpha$ (2:1)	16α	
CYP2C11	$16\alpha/15\alpha/6\beta/2\alpha/$ (102:1:9:132)	$16\alpha/6\beta/2\alpha$ (40:1:37)	$16\alpha/6\beta/2\alpha$ (50:4:1)
CYP2C12	15α	$15\alpha/1\alpha$ (1:1)	
CYP2C13	$16\alpha/15\alpha/6\beta$ (20:1:17)	$16\alpha/15\alpha/6\beta$ (1:3:13)	$16\alpha/6\beta$ (1:7)
CYP3A1		6β	
CYP3A2		6β	

Data derived from Swinney et al. (1987) and Ogri et al. (1994).

CYP2C3v, that differs from CYP2C3 by one amino acid residue, a serine residue in CYP2C3 being replaced by threonine in CYP2C3v. This variation changes the regio- and stereoselectivity of progesterone oxidation from 16α-hydroxylation to 6β-hydroxylation (Hsu et al., 1993), suggesting an alternative binding mode of the substrate within the active site.

Molecular modeling studies indicate that when progesterone is orientated for 16α-hydroxylation within the active sites of CYP2C3 and CYP2C9, the A-ring is in close proximity to the serine residue, allowing hydrogen bonding between the serine hydroxyl group and the progesterone A-ring carbonyl oxygen. When serine is changed to threonine, the *β*-methyl group of threonine sterically hinders hydrogen bond formation. However, when progesterone is reorientated within the putative active site to the optimal position for 6*β*-hydroxylation, the D-ring carbonyl group can form a hydrogen bond with the threonine hydroxyl group (Lewis, 1995).

CYP2C9 is also responsible for the oxidation of (*S*)-mephenytoin and tolbutamide, and similar molecular modeling studies have shown that both of these compounds fit into the putative active site and that the correct orientation of both compounds gives rise to hydrogen bond possibilities at the same serine residue. Interestingly, the distance between the serine residue and the hæm oxene atom is 9.4 Å, which is close to the distance between the site of metabolism and the relevant hydrogen bonding function in each substrate. This distance is 7.9 Å for tolbutamide and (*S*)-mephenytoin and 10.7 Å in progesterone (Lewis, 1995). It would appear in the case of these substrates at least, and probably as a general hypothesis, that the stereoselectivity of P450-mediated oxidation is primarily a function of the enzyme structure and the chirality of the enzyme active site.

Species differences in both regio- and stereoselectivity are also seen in the metabolism of the coumarin anticoagulant, warfarin (Figure 3.3). In both rat and man, (*S*)-warfarin is cleared at a faster rate than its *R*-antipode, 4 times faster in rat and 10 times faster in man (Thijssen and Baars, 1987). Warfarin can undergo a number of metabolic transformations, including hydroxylation at the 4′-, 6-, 7-, 8-, or 10-positions, and ketone reduction in the side chain, to form the diastereoisomeric "warfarin alcohols" (Kaminsky et al., 1984; Hermans and Thijssen, 1989; 1991). The relative extents of these various pathways vary considerably between species (Caldwell et al., 1988a). In man, *in vivo*, the principal metabolite is 7-hydroxywarfarin, which is stereoselectively formed from the pharmacologically more potent *S*-enantiomer, with small amounts of (*S*)-6- and (*S*)-4′-hydroxywarfarin and (*S*, *S*)-warfarin alcohol. In contrast, (*R*)-warfarin is subject to both oxidation and reduction, resulting mainly in (*R*)-6- and some (*R*)-7-hydroxywarfarin as well as (*R*, *S*)-warfarin alcohol (Gibaldi, 1993). In human hepatic microsomes, (*R*)-warfarin is oxidised to form (*R*)-6-, (*R*)-8-, and (*R*)-10-hydroxywarfarins, while the *S*-isomer is oxidized to form (*S*)-7- and (*S*)-4′-hydroxywarfarins (Kaminsky et al., 1984). In the rat, however, although 7-hydroxywarfarin is also the major metabolite, the *R*-isomer is the preferred substrate for this transformation (Pohl et al., 1976a; 1976b).

The 6- and 7-hydroxylations of warfarin are P450-mediated oxidations and since these positions in the coumarin ring are adjacent, it is possible that both metabolites may be formed from a single epoxide intermediate. The different stereochemical composition of these products suggests that either they arise through the action of different isoenzymes, or the epoxide undergoes stereoselective rearrangement (Trager and Jones, 1987). It has recently been shown that in man the 7-hydroxylation of warfarin is specifically catalyzed by CYP2C9. This isoform is highly stereoselective for (*S*)-warfarin and generates both (*S*)-7-hydroxy- and (*S*)-6-hydroxywarfarin in a ratio of 3.5:1 (Rettie et al., 1992). Other human P450 isoforms have been shown to be both regio- and stereoselective with respect to warfarin oxidation. Thus, CYP1A2 is stereoselective for (*R*)-warfarin and regioselective for the 6-position, whereas CYP3A4 is selective for the formation of (*R*)-10-hydroxywarfarin (Rettie et al., 1992). Since warfarin undergoes such a diversity of metabolic biotransformations, it may prove to be a useful probe drug for investigations of human hepatic P450 profiles, either *in vivo* or *in vitro* (Kaminsky et al., 1984).

It was noted earlier that between species there may be differences in both the rate and route of metabolism of xenobiotics and such differences have been highlighted in the examples reviewed above. Quantitative differences reflect variations in the affinity of homologous enzymes for the same substrate, whereas qualitative differences reflect metabolic defects arising as the species under examination lacks the appropriate enzyme. It is equally true that such differences may also exist within species, that is, metabolic transformations may show strain differences. For many drugs, their rate of metabolism exhibits a normal distribution within the population. In some cases, however, there occur large intraspecies variations in xenobiotic metabolism and discrete subpopulations are apparent within a population, the so-called genetic polymorphisms.

In man, one of the best examples of such a genetic polymorphism is the so-called debrisoquine 4-hydroxylation phenotype (Idle and Smith, 1979). Debrisoquine, a prochiral substrate, undergoes P450-mediated metabolism, catalyzed in man by the specific P450 isoenzyme CYP2D6, to yield the chiral metabolite 4-hydroxydebrisoquine (Figure 3.5). In addition, debrisoquine

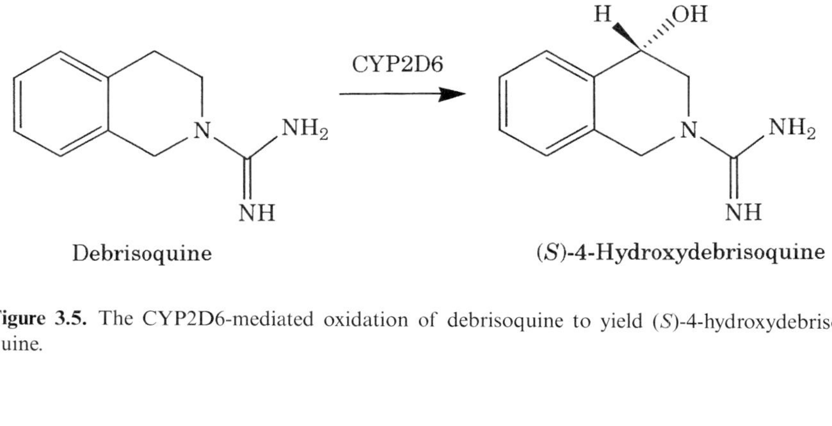

Figure 3.5. The CYP2D6-mediated oxidation of debrisoquine to yield (*S*)-4-hydroxydebrisoquine.

undergoes P450-mediated aromatic hydroxylation at the 5-, 6-, 7-, and 8-positions, yielding various phenolic metabolites (Idle and Smith, 1979). Within the human population, there are two distinct phenotypes, namely, the so-called extensive metabolizers (EMs) of debrisoquine, who express CYP2D6 and can carry out the 4-hydroxylation reaction, and poor metabolizers (PMs), who exhibit defective expression of the required isoenzyme and excrete only relatively minor quantities of 4-hydroxydebrisoquine. It has been estimated that some 5–10% of the Caucasian population do not express this isoenzyme. This polymorphism is associated with the compromised metabolism of more than 30 drugs. The known substrates for CYP2D6 are structurally diverse and include debrisoquine, propafenone, the *β*-adrenergic receptor antagonists, propranolol, metoprolol, and bufuralol, the tricyclic antidepressants, desipramine and nortriptyline, and other drugs such as codeine, dextromethorphan, and methoxyamphetamine (Islam et al., 1991). Many of these agents are chiral and the reactions show stereoselectivity with respect to the substrate, or alternatively, if prochiral, to the product.

In the case of the 4-hydroxylation of debrisoquine, in EMs a high degree of stereoselectivity, favoring the formation of the *S*-enantiomer of the product, has been demonstrated (Eichelbaum, 1988; Eichelbaum et al., 1988), the enantiomeric excess of (*S*)-4-hydroxydebrisoquine being greater than 97%. In contrast, in PMs there was not only decreased formation of 4-hydroxydebrisoquine, but also a loss of stereoselectivity. In these subjects, the enantiomeric excess with respect to (*S*)-4-hydroxydebrisoquine ranged from 90–28%, and a negative correlation between urinary metabolic ratio (the ratio of 4-hydroxydebrisoquine to debrisoquine excreted in urine) and enantiomeric excess was found, that is, the less 4-hydroxydebrisoquine formed, the lower the product stereoselectivity of the reaction (Eichelbaum et al., 1988; Meese and Eichelbaum, 1989).

It has recently been observed that the 5-hydroxylation of the antiarrythmic drug, propafenone (Figure 3.6), cosegregates with the debrisoquine 4-hydroxylation polymorphism (Siddoway et al., 1987; Kroemer et al., 1989). Both enantiomers of propafenone are equipotent as sodium channel blockers, however, the *S*-enantiomer also exhibits *β*-adrenergic receptor antagonist activity. When the individual enantiomers are administered, (*S*)-propafenone is cleared more rapidly than its *R*-antipode. However, when the racemate is used, the converse is true and the *R*-enantiomer is cleared at a much greater rate. It has been shown that both enantiomers are potent inhibitors of the CYP2D6-mediated 5-hydroxylation, but (*R*)-propafenone is a more potent inhibitor of the 5-hydroxylation of its *S*-antipode than the *S*-enantiomer is of the *R*-isomer. Thus, *in vivo*, when the racemate is administered, (*R*)-propafenone inhibits the clearance of (*S*)-propafenone, leading to accumulation of the *S*-enantiomer (Kroemer et al., 1991). This may have therapeutic

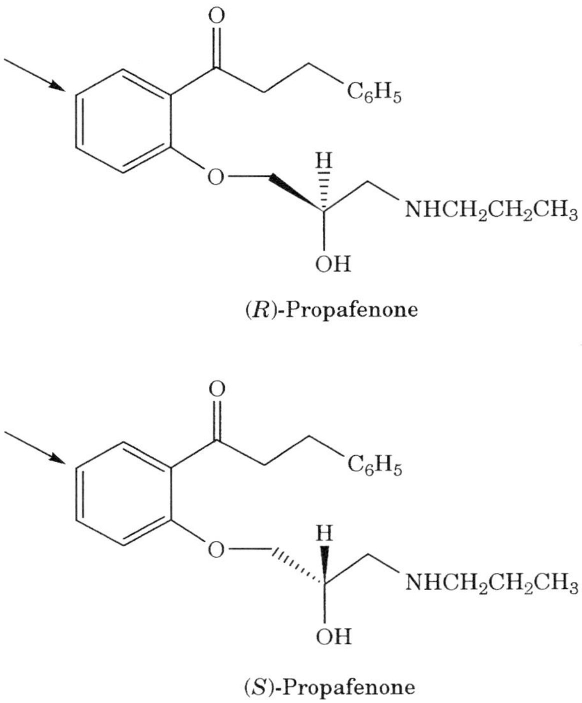

(*R*)-Propafenone

(*S*)-Propafenone

Figure 3.6. Structures of (*R*)-and (*S*)-propafenone. Arrows indicate the position of hydroxylation.

consequences in terms of the debrisoquine polymorphism, since PMs may achieve concentrations of (*S*)-propafenone sufficient to elicit β-blockade. This could explain the observation that 67% of PMs reported CNS side-effects, compared with only 14% of EMs who suffered similar problems (Meese and Eichelbaum, 1989).

As mentioned above, CYP2D6 is recognized as a principal enzyme in the oxidative metabolism of a number of β-adrenoreceptor antagonists, such as propranolol, metoprolol, and bufuralol (Lennard et al., 1986). Both metoprolol and bufuralol (Figure 3.7) exhibit regio- and stereoselectivity in their metabolism by CYP2D6. In the case of metoprolol, the *R*-enantiomer is selectively O-demethylated, whereas for its *S*-antipode there is a preference for α-hydroxylation (Kim et al., 1993; Mautz et al., 1995). Bufuralol also undergoes regio- and enantioselective CYP2D6-mediated benzylic hydroxylation to form diastereoisomeric 1″-hydroxybufuralols (Weerawarna et al., 1991). The benzylic hydroxylation of (1′*R*)-bufuralol yields an enantiomeric excess of

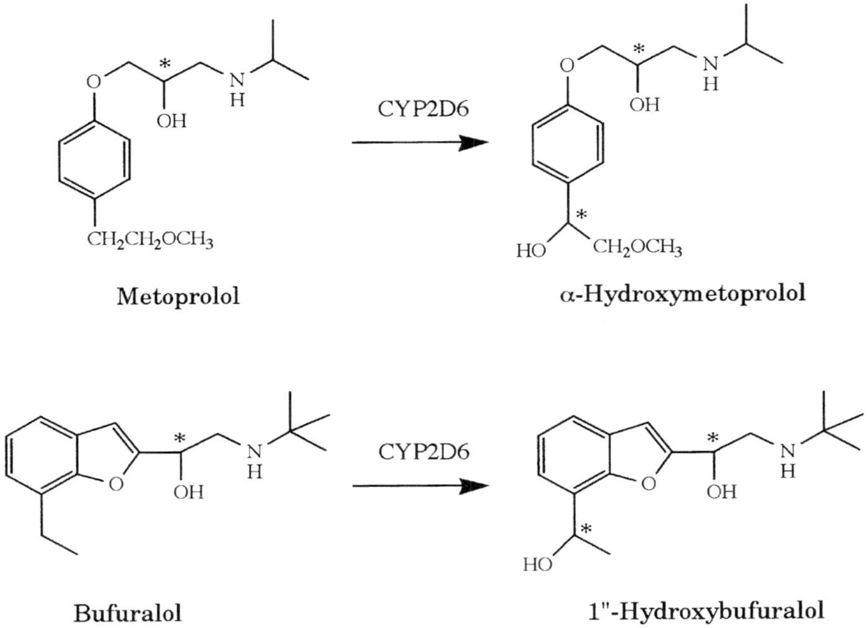

Figure 3.7. The CYP2D6-mediated oxidation of metoprolol and bufuralol.

(1′*R*, 1″*S*)-hydroxybufuralol, while the 1′*S*-enantiomer undergoes benzylic hydroxylation to form a slight excess of the (1′*S*, 1″*R*)-hydroxylated product (Mautz et al., 1995).

A significant illustration of the importance of chirality in drug disposition with respect to the human genetic polymorphisms is seen with the racemic antiepileptic agent mephenytoin (Figure 3.1), and its N-demethylated metabolite, nirvanol (Kupfer et al., 1981; Kupfer and Preisig, 1984). The 4′-hydroxylation of both mephenytoin and nirvanol in man is catalyzed, stereoselectively, by CYP2C9 (Umbenhauer et al., 1987; Ged et al., 1988; Relling et al., 1990). Similar to the case of the debrisoquine oxidation phenotype, the stereoselective hydroxylation of mephenytoin is under genetic control, with some 2–5% of Caucasians and 20% of Japanese populations being deficient in this pathway (Nakamura et al., 1985). In EMs, (*S*)-mephenytoin undergoes rapid 4′-hydroxylation, whereas its *R*-antipode undergoes N-demethylation to form (*R*)-nirvanol, but at a much slower rate, resulting in a 200-fold difference in the oral clearance of the enantiomers of mephenytoin (Wedlund et al., 1985). In PMs, however, (*S*)-mephenytoin undergoes negligible 4′-hydroxylation, the majority of the dose being N-demethylated to yield (*S*)-nirvanol at a rate similar to that of the *R*-enantiomer.

3.2.1.2. Flavin-Containing Monooxygenase

The flavin-containing monooxygenase (FMO), or microsomal FAD-containing enzyme, is a polymeric protein that contains one mole of FAD as the prosthetic group per protein monomer. The enzyme is localized in the endoplasmic reticulum and was first purified from pig liver microsomes by Ziegler and Mitchell (1972). FMO is found in several species, including man, rat, pig, rabbit, etc., and in various tissues, including liver, lung, kidney, and skin (Dannan and Guengerich, 1982). The sites of oxidative attack by the FMOs are nucleophilic centers, such as nitrogen, sulfur, phosphorus, and selenium, in a multitude of structurally diverse compounds. The overall stoichiometry of the reaction is essentially the same as that of the P450s, but the catalytic mechanism is different. The catalytic cycle of FMO involves flavin reduction, binding of molecular oxygen, internal electron transfer to oxygen to form the hydroperoxyflavin complex (E-FAD-OOH), substrate binding, release of product, and dissociation of $NADP^+$, yielding the oxidized enzyme (Ziegler and Poulsen, 1978; Ziegler, 1988; 1993). The porcine hepatic FMO purified by Ziegler and Mitchell (1972) was originally assumed to be a single enzyme. However, recent investigations into the molecular biology of the enzyme system have demonstrated the existence of at least five isoforms[3] (Lawton et al., 1994). The five isoforms have all been identified in human tissues and their cDNAs have been cloned.

There appear to be age- and tissue-dependent differences in expression of the individual isoforms. Thus, FMO1 was cloned from human fetal liver cDNA (Dolphin et al., 1991) and this isoform is expressed in the human fetus in several tissues, but in the adult human the mRNA is found mainly in the kidney. This suggests that during development, expression of FMO1 is "switched off" in all tissues except the kidney. This contrasts with both the pig and rabbit, in which FMO1 is the major isoform in adult liver (Phillips et al., 1995). FMO2 was cloned from adult human lung cDNA and the *FMO2* gene appears to be expressed predominantly in this tissue. FMO3 was cloned from adult human liver cDNA and the *FMO3* gene is expressed mainly in this tissue. The FMO3 mRNA is found in low abundance in human fetal liver and lung and in human adult kidney and lung, suggesting that the expression of

[3] As with the nomenclature system described for the P450 gene family, the symbols for the genes and cDNAs are italicized and given an arabic numeral, for example, *FMO1*, whereas the gene products, that is, the mRNAs and proteins, are nonitalicized, for example, FM01 (Lawton et al., 1994).

FMO3 is increased during development. FMO4 was cloned from human hybridizing clones and the mRNA is found in low abundance in several human fetal and adult tissues. The *FMO4* gene appears to be constitutively expressed. FMO5 was cloned from human adult liver cDNA and the *FMO5* gene is expressed in this tissue (Phillips et al., 1995).

The expression of FMO isoforms also appears to be sex-dependent. In studies using the pregnant rabbit, it was found that expression of FMO2 in the lung was induced by both progesterone and glucocorticoids, but the enhanced expression during pregnancy correlated only to progesterone levels. In the rabbit kidney, FMO2 expression correlated to plasma glucocorticoid levels, whereas in the liver, FMO1 levels increased during pregnancy and it was found that this increase was influenced by both progesterone and cortisol (Lee et al., 1995).

In common with the P450s, the FMOs frequently exhibit stereoselectivity in oxidation. If the substrate is chiral, then there may be substrate stereoselectivity, as has been observed in the oxidation of the enantiomers of the tobacco alkaloid, nicotine. Alternatively, N-oxidation of prochiral tertiary amines, which yield chiral N-oxides as metabolites, may exhibit product stereoselectivity, if not total stereospecificity. In addition, chiral N-oxides may undergo stereoselective enzymatic reduction. Thus, the enantiomeric composition of the N-oxide measured as an end product of metabolism may be the result of two independent metabolic processes.

(*S*)-Nicotine is the natural isomer found in tobacco, although up to 10% (*R*)-nicotine may be formed during combustion (Klus and Kuhn, 1977; Crooks et al., 1992), and it is, therefore, important to examine the metabolic fate of both enantiomers of nicotine. As the pyrrolidine nitrogen in nicotine is prochiral, the N′-oxidation of racemic nicotine can potentially result in the formation of four stereoisomers (Figure 3.8). The N′-oxidation of nicotine is mediated mainly by FMO (Williams et al., 1990), although there is evidence for the involvement of P450. For example, it has been shown that rat liver CYP2B1, mouse liver CYP2B10, and rabbit lung CYP4A4 catalyze formation of (*S*)-nicotine-N′-oxide, with a *trans*:*cis* ratio of approximately 4:1 (Cashman et al., 1995).

In rat, mouse, hamster, and guinea pig hepatic preparations, (*S*)-nicotine is oxidized to form both N′-oxide diastereoisomers, namely *cis*-(1′*R*, 2′*S*)- and *trans*-(1′*S*, 2′*S*)-nicotine-N′-oxides, the reaction being product stereoselective for the *cis*-diastereoisomer. In contrast, the N′-oxidation of (*R*)-nicotine is stereoselective for the *trans*-(1′*R*, 2′*R*)-nicotine-N′-oxide. When the N′-oxidation of both enantiomers is considered, product formation leads primarily to the 1′*R*-configured N′-oxide (Crooks, 1993). In studies using purified pig hepatic FMO (FMO1), (*R*)-nicotine exhibited stereospecific N′-oxidation to yield the *trans*-(1′*R*, 2′*R*)-N′-oxide, whereas (*S*)-nicotine gave rise to both

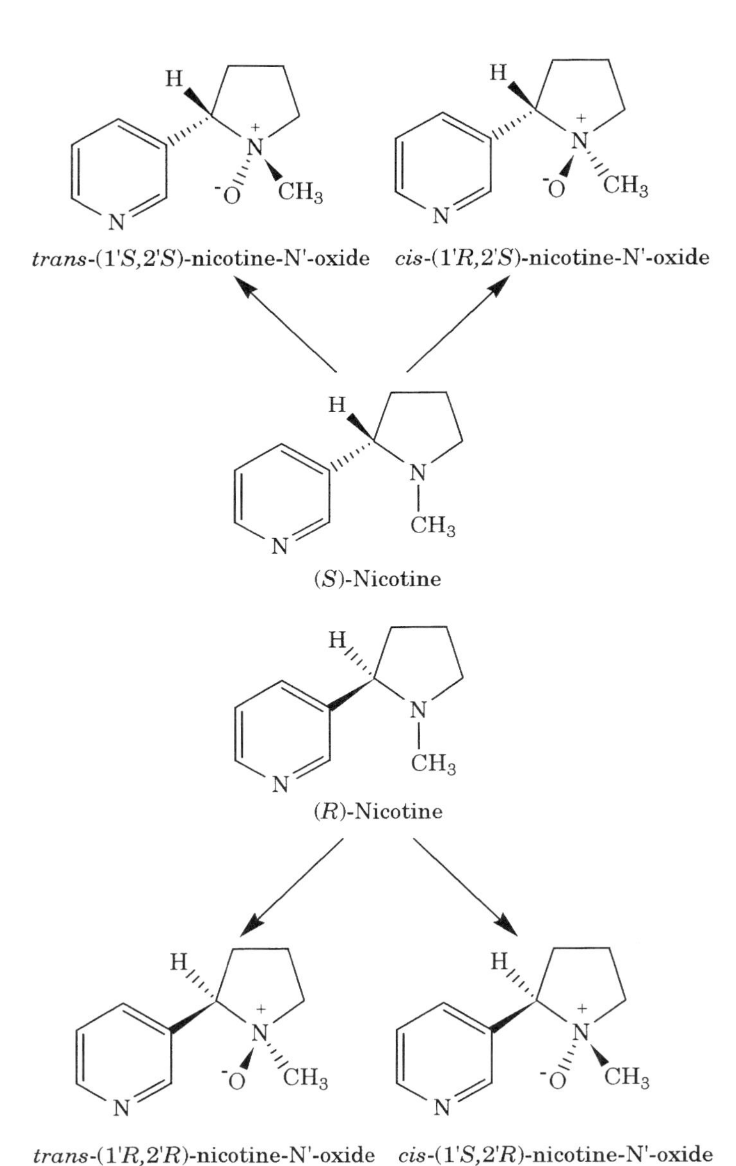

Figure 3.8. The N-oxidation of the enantiomers of nicotine to yield diastereoisomeric N′-oxides. *Cis*- and *trans*- refer to the position of the methyl group relative to the pyridine ring.

diastereoisomers in equal quantities. It was also found that (*R*)-nicotine was a better substrate for the purified hog enzyme than its *S*-antipode (Damani et al., 1988). In contrast to studies using animal models, studies using human hepatic preparations and cDNA-expressed human hepatic FMO3 (Cashman et al., 1995) suggest that FMO3 stereospecifically catalyzes the N′-oxidation of (*S*)-nicotine to form *trans*-(1′*S* 2′*S*)-N′-oxide. It has been suggested that (*S*)-nicotine N′-oxidation may prove to be useful as a selective probe of human FMO3 activity (Cashman et al., 1995).

The nicotine N′-oxides also undergo reduction back to nicotine, both *in vivo* and *in vitro* and such reactions may also exhibit stereoselectivity. Thus, *cis*-(1′*R*, 2′*S*)-nicotine-N′-oxide is reduced more rapidly than its *trans*-1′*S*, 2′*S*-diastereoisomer *in vitro* by hepatic microsomal and intestinal microfloral reductases. Such reversible transformations, together with their corresponding stereoselectivities, obviously complicate the interpretation of the total stereochemical composition of nicotine N′-oxides with respect to their use as probe compounds for FMO activity (Crooks, 1993).

The stereoselectivity of N-oxidation of achiral compounds containing prochiral nitrogen centers has, until recently, received scant attention. One compound that has been studied *in vitro* is N-ethyl-N-methylaniline (EMA) and its oxidation to the corresponding chiral N-oxide (Figure 3.9). Incubation of EMA with rat, rabbit, mouse, and guinea pig hepatic microsomal preparations resulted in N-oxidation, EMA-N-oxide accounting for approximately 10% of the total metabolic products formed after a 60-minute incubation period. This N-oxidation was shown to be product-stereoselective for the *S*-enantiomer; the enantiomeric excess of (*S*)-EMA-N-oxide ranged from approximately 40% with the guinea pig preparation, to some 50% in the rat and mouse preparations and 68% with the rabbit (Hadley et al., 1993; 1994a). In studies using purified porcine hepatic FMO (FMO1), N-oxidation was the only metabolic transformation and, again, the formation of EMA-N-oxide was stereoselective for the *S*-isomer, with a mean ratio of *S*:*R* of 4:1 (Hadley et al., 1994b). Neither inhibition, nor activation of FMO affected product stereoselectivity. Determination of the kinetics of the reaction demonstrated that the K_m's for the formation of both enantiomers of the N-oxide were identical, but the V_{max} values differed, indicating that a single oxygenating species was responsible for the N-oxidation, but with a degree of flexibility of the substrate at the active site, allowing two productive binding modes to occur.

Stereoselectivity in the formation of EMA-N-oxide has also been examined using human FMO1 and FMO3 expressed in both bacterial and insect cells. Both systems catalyzed the N-oxidation of EMA, the activity of the eukaryotic system being greater than that of the prokaryotic system. As with the studies using rodent hepatic microsomes and purified porcine FMO, both systems exhibited stereoselectivity toward the formation of (*S*)-EMA-N-

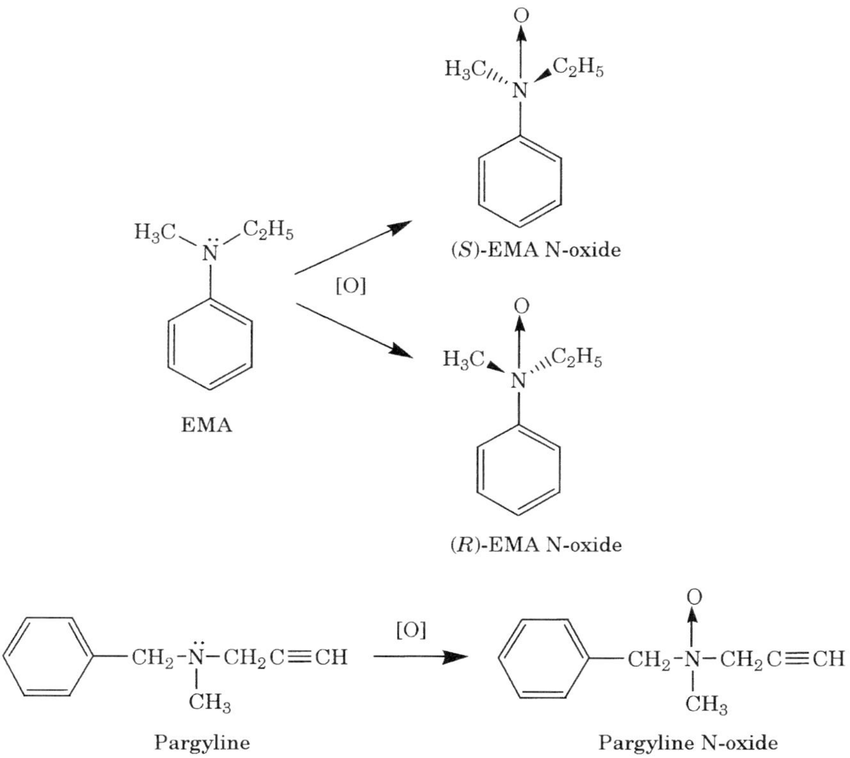

EMA

[O]

(*S*)-EMA N-oxide

(*R*)-EMA N-oxide

Pargyline

[O]

Pargyline N-oxide

Figure 3.9. FMO-mediated N-oxidation of N-ethyl-N-methylaniline (EMA) and pargyline.

oxide. FMO1 resulted in an 85% enantiomeric excess of the *S*-enantiomer, whereas FMO3 gave an enantiomeric excess of some 60%, irrespective of the expression vector used (Phillips et al., 1995).

Stereoselectivity in FMO-catalyzed N-oxidation has also been observed with the monoamine oxidase (MAO) inhibitor, pargyline (N-benzyl-N-methyl-2-propynylamine). The *in vitro* N-oxidation of pargyline (Figure 3.9) is highly enantioselective for the formation of the (+)-enantiomer. In rabbit and guinea pig hepatic microsomes, there was approximately an 80% enantiomeric excess of (+)-pargyline-N-oxide, in rat preparations the enantiomeric excess was 86%, whereas in mouse hepatic microsomes the enantiomeric excess was 90%. When purified porcine hepatic FMO (FMO1) was used, the N-oxidation of pargyline was found to be stereospecific (Hadley et al., 1994c). This latter finding suggests that the stereoselectivity observed using microsomal preparations may be an indication of the involvement of other isoforms of FMO, or other enzymes catalyzing the N-oxidation. Initial kinetic studies suggest that the N-oxidation of pargyline may indeed be mediated by more than one isoform/enzyme (Hadley et al., 1994c). This is supported by the results of studies using human FMO1 and FMO3 expressed in insect cells. FMO1 was stereospecific toward formation of the (+)-enantiomer of the N-oxide, while FMO3 produced mainly the (–)-enantiomer (Phillips et al., 1995).

In addition to stereoselectivity in N-oxidation, the FMO-catalyzed sulfoxidation of prochiral thioethers may also exhibit stereoselectivity. Sulfoxides exist as tetrahedral structures, the sulfur atom bearing three different ligands and a single electron pair. The configurational stability of such compounds leads to the production of two enantiomers following oxidation of achiral sulfur-containing compounds, for example, the histamine H_2-antagonist, cimetidine (Holland, 1988; Boyd et al., 1989).

In man cimetidine sulfoxide (Figure 3.10) is quantitatively the most important oxidation product of cimetidine (Mitchell et al., 1982a). Similarly to N-oxides, sulfoxides may subsequently undergo reduction back to the parent thioether. Indeed, this is the case with cimetidine, since following administration of the sulfoxide to both man and guinea pig, the major route of metabolism was reduction to cimetidine (Mitchell et al., 1982b; Mitchell and Waring, 1989). Thus, the enantiomeric composition of material excreted in urine may arise as a result of two independent pathways. The enantiomeric composition of cimetidine sulfoxide in urine, following the administration of

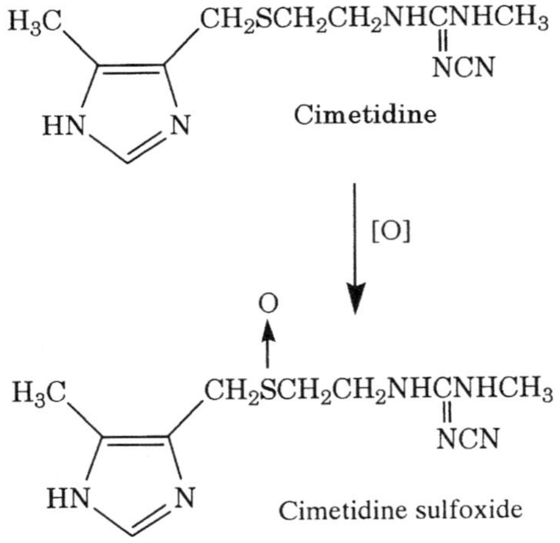

H3C CH2SCH2CH2NHCNHCH3
NCN
HN N
Cimetidine

[O]

O
H3C CH2SCH2CH2NHCNHCH3
NCN
HN N
Cimetidine sulfoxide

Figure 3.10. The S-oxidation of cimetidine.

cimetidine to man, has been investigated by two groups. The (+)-enantiomer of the sulfoxide was present in excess, with approximately 70% of the material in this form, the ratio of (+):(–) being 71:29 (Kuzel et al., 1994) and 75:25 (Cashman et al., 1993). In rat urine, the ratio of (+):(–) was found to be 57:43, that is, the sulfoxidation of cimetidine was selective with respect to the (+)-enantiomer in both species (Kuzel et al., 1994). The absolute configuration of the two enantiomers has yet to be determined, but Cashman and co-workers (1995) have suggested that (+)-cimetidine-S-oxide has the *S* configuration and the (–)-isomer has the *R* configuration.

When cimetidine was incubated with human liver microsomes, S-oxidation was stereoselective for the (+)-enantiomer, resulting in a (+/–) ratio of 84:16, an enantiomeric composition in good agreement with that observed *in vivo*. In contrast, when pig liver microsomes and purified pig hepatic FMO1 were used, S-oxidation was enantioselective for the (–)-isomer, with a (–/+) ratio of 57:43 (Cashman et al., 1993; Cashman, 1995). Since FMO3 is the major FMO isoform in human adult liver, these data suggest that FMO3 and FMO1 exhibit opposite product enantioselectivities for cimetidine sulfoxidation, and it has been suggested that cimetidine sulfoxidation could be a useful bioindicator for the presence of FMO1 and FMO3 *in vivo*. Thus, a species which exhibits a predominance of cimetidine sulfoxidation to produce the (+)-S-oxide possesses mainly FMO3 activity, whereas a species with a predominance of FMO1 activity will excrete mainly (–)-cimetidine-S-oxide (Cashman, 1995; Cashman et al., 1995).

One of the most widely studied groups of sulfur-containing model compounds are the alkyl *p*-tolyl sulfides (Figure 3.11). The sulfoxidation of these compounds has been studied using various preparations and it has been shown that the stereochemistry of the product sulfoxide depends on the enzyme effecting the reaction. For example, the FMO-mediated sulfoxidation of 4-tolylethyl sulfide yields predominantly (>95%) the *R*-enantiomer of the sulfoxide, whereas purified P450s favor (>80%) the production of the *S*-enantiomer (Light et al. 1982; Waxman et al., 1982). Similar differences in the stereoselectivity in P450- and FMO-mediated sulfoxidation have also been reported with 2-aryl substituted-1,3-oxathiolanes, 2-methyl-1,3-benzodithiole (Lomri et al., 1993) and the organophosphate insecticide, phorate (Hodgson and Levi, 1992).

Rettie et al. (1990) have carried out a systematic evaluation of the stereoselectivity of the sulfoxidation of a series of prochiral alkyl *p*-tolyl sulfides. Purified rabbit lung FMO (FMO2) stereoselectively catalyzed the sulfoxidation of methyl-, ethyl-, *n*-propyl-, and *iso*-propyl-*p*-tolyl sulfides, in each case forming an enantiomeric excess of the *R*-configured sulfoxide. The stereoselectivity of the oxidation decreased with increasing alkyl chain length, that is, the increasing steric bulk of the alkyl group decreases stereoselectivity (Rettie

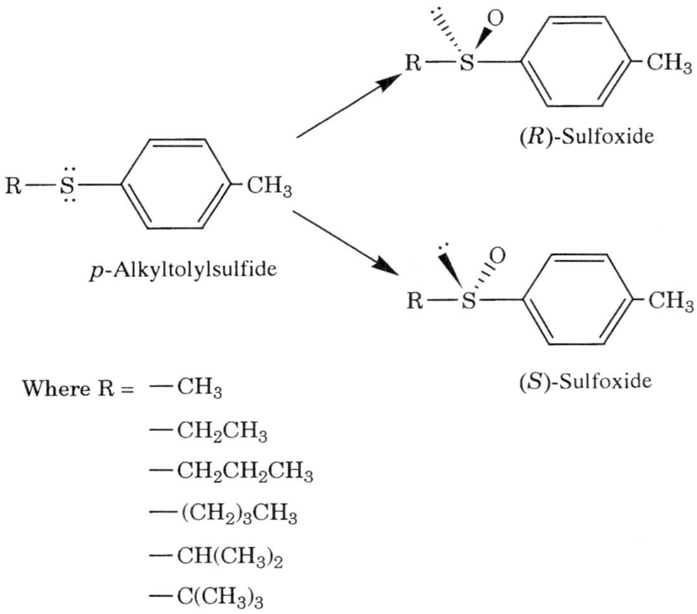

Figure 3.11. The S-oxidation of *p*-alkyltolysulfides.

et al., 1990; 1995). Thus, methyl-*p*-tolyl sulfide was essentially stereospecifically sulfoxidized (> 99% *R*), whereas with the *iso*-propyl analogue sulfoxidation resulted in only 63% *R*-sulfoxide. When purified minipig liver FMO (FMO1) was used, a similar trend was observed, although in this case the stereoselectivity in sulfoxidation was reduced compared with rabbit lung FMO. Thus, oxidation of the methyl analogue resulted in 91% *R*-sulfoxide, while oxidation of the *iso*-propyl analogue showed a slight preference for the *S*-configured sulfoxide (59% *S*). These data contrast with those obtained using purified rabbit P450 LM2 (CYP2B4), which is stereoselective toward formation of the *S*-configured sulfoxides. There was a greater than 80% enantiomeric excess of the *S*-sulfoxide with all four analogues (Rettie et al., 1995).

It was initially suggested that stereoselective reduction, by P450, aldehyde oxidase, or cytosolic NADPH-dependent N-oxide reductase, of the products of FMO-mediated oxidation may account for the resultant stereochemical composition of the products. This hypothesis was discounted due to either cell location or P450-inhibition (Rettie et al., 1990). In a later study, it was found that mouse, rat, rabbit, hamster, and minipig liver microsomal FMO (FMO1) catalyzed the sulfoxidation of methyl-*p*-tolyl sulfide to form greater than 76%

enantiomeric excess of the *R*-configured sulfoxide, whereas in human adult liver microsomes, there was little or no stereoselectivity toward the *R*-enantiomer (0–40% enantiomeric excess). In both human adult kidney and fetal liver microsomal preparations, there was an 86–97% enantiomeric excess of *R*-sulfoxide (Sadeque et al., 1992). These data suggest that the human adult FMO is markedly different from that found in rodent livers, which is consistent with FMO3 being the major isoform in human adult liver, whereas FMO1 is predominant in rodent, porcine, and human fetal liver and human adult kidney (Phillips et al., 1995).

The sulfoxidation of alkyl-*p*-tolyl sulfides has also been studied using rabbit cDNA expressed as FMO1, FMO2, FMO3, and FMO5 in *Escherichia coli* (Rettie et al., 1994) and purified rabbit liver FMO1 (Rettie et al., 1995). Both FMO1 and FMO2 exhibited high affinity toward methyl-*p*-tolyl sulfide (K_m<10 μM), whereas FMO3 exhibited low affinity (K_m>100 μM), and FMO5 did not catalyze sulfoxidation of any of the compounds. FMO1 was almost stereospecific toward formation of (*R*)-methyl-*p*-tolyl sulfoxide, with greater than 90% *R*-configured sulfoxide being formed from ethyl-, *n*-propyl-, and *n*-butyl-analogues, whereas FMO3 showed little stereoselectivity, ranging from 50% *R*-sulfoxide for the methyl-analogue to 88% *R*-sulfoxide with the *n*-butyl analogue (Rettie et al., 1995).

In rabbit hepatic microsomes, sulfoxidation of methyl-*p*-tolyl sulfide was best described by a two-enzyme model, with the enantiomeric excess of (*R*)-methyl-*p*-tolyl sulfoxide decreasing with increasing substrate concentration, suggesting that both FMO1 and FMO3 contribute to the sulfoxidation of this agent in rabbit liver (Rettie et al., 1994). Since the kinetic and stereochemical properties of hepatic microsomes were similar to those of FMO1, methyl-*p*-tolyl sulfide may prove to be a useful agent for discriminating between FMO1 and FMO3 catalysis in rabbit hepatic preparations. In addition, since the *n*-alkyl-*p*-tolyl sulfides can differentiate between multiple FMO isoforms in various tissues, they may prove particularly useful for the functional characterization of FMO isoforms present in primate tissues (Rettie et al., 1995).

Stereoselective sulfoxidation has been shown to occur with the endogenous substrate, methionine. In studies using rabbit cDNA-expressed FMO1, FMO2, and FMO3 in *E. coli* (Duescher et al., 1994; Elfarra, 1995), (+)-L-methionine sulfoxide was the predominant metabolite. With FMO3 the ratio of (+):(–) was 8:1, and with FMO1 the ratio was 1.5:1, whereas with FMO2 the ratio was 0.7:1. The K_m value obtained with FMO3 was much lower than that obtained with either FMO1 or FMO2 (Elfarra, 1995). Kinetic studies using rabbit liver and kidney microsomal preparations gave K_m values comparable to that found with FMO3, which suggests that FMO3 is the major isoform involved in methionine sulfoxidation in rabbit liver and kidney. Further support for this suggestion comes from the finding that in rabbit liver and kidney microsomes, ratios of (+):(–) sulfoxide were 8:1 and 6:1, respectively (Duescher et al., 1994).

3.2.1.3. Reductive Drug Metabolism

The reduction of a functional group in a xenobiotic may be the consequence of the actions of a redox system, for example, the reduction of aldehydes and ketones to alcohols is mediated by the same enzyme (alcohol dehydrogenase) that catalyzes the oxidation of alcohols to ketones and aldehydes. Other reductions may be catalyzed by microsomal enzymes, such as P450 and NADPH-P450-reductase, or by bacterial reductases in the G.I. tract. Chiral molecules may be stereoselectively metabolized via reductive pathways. Such metabolism may occur at a site distant from the chiral center, possibly resulting in the introduction of a new chiral center, that is, chiral to diastereoisomer transformation, as seen with the keto-reduction of warfarin (Lewis et al., 1974). Alternatively, a prochiral molecule may undergo reductive metabolism resulting in the formation of a chiral product, that is, prochiral to chiral transformation. A good example of the latter is seen with the selective 5-HT_2 receptor antagonist, ketanserin, that undergoes reductive metabolism of its carbonyl function. Ketanserin exhibits product stereoselectivity to form a greater than 95% enantiomeric excess of the *R*-carbinol product (Meese and Eichelbaum, 1989).

As mentioned above, in addition to a number of oxidative metabolic transformations, warfarin also undergoes reductive metabolism. The keto-reduction of warfarin exhibits species differences in substrate stereoselectivity. In man, *in vivo*, the *R*-enantiomer is preferentially reduced to form (*R*, *S*)-warfarin alcohol, whereas in the rat this reaction is stereoselective for the *S*-enantiomer, forming the *S*,*S*-alcohol (Lewis et al., 1974; Pohl et al., 1976a; Hermans and Thijssen, 1989). *In vitro*, however, both rat and human systems exhibit similar stereoselectivities, (*R*)-warfarin being the preferred substrate, resulting in the formation of the *R*,*S*-alcohol (Hermans and Thijssen, 1989). Other oral anticoagulants also undergo stereoselective metabolism, for example, in man 4′-nitrocoumarol (acenocoumarol, or nicoumalone) undergoes metabolism by reduction of both the keto- and aromatic nitro-groups, as well as 6- and 7-hydroxylation, with the *S*-isomer being cleared at a faster rate than its *R*-antipode, although it is unknown which reaction is stereoselective (Dieterle et al., 1977). *In vitro*, in rat, human, porcine, equine, and rabbit cytosolic systems, ketone reduction of 4′-nitrocoumarol, 4′-chlorowarfarin, and warfarin is selective for conversion of the *R*-enantiomers to their respective *R*, *S*-alcohols. The stereoselectivity of keto-reduction of 4′-nitrocoumarol is greater than 4′-chlorowarfarin that is, in turn, greater than warfarin (Hermans and Thijssen, 1989).

3.2.1.4. Hydrolysis

There are a number of hydrolytic enzymes widely distributed in the body. Esters can be hydrolyzed in the plasma, by nonspecific cholinesterases and other esterases, or in the liver, by esterases that are specific for groups of compounds. The plasma esterases are nonspecific and may also hydrolyze amides, although at a slower rate than the corresponding esters. However, amides are more likely to be hydrolyzed by more specific liver amidases. There are several examples of the enantioselective hydrolysis of chiral esters and amides.

Enantioselective ester hydrolysis is seen with oxazepam-3-acetate. In mouse and human liver preparations, the *R*-enantiomer was preferentially hydrolyzed by a microsomal esterase. However, in mouse and human brain preparations, hydrolysis catalyzed by a mitochondrial esterase was selective toward the *S*-enantiomer. In rat intestinal preparations, microsomal esterases selectively hydrolyzed the *R*-enantiomer, whereas cytosolic esterases were selective for its *S*-antipode (Yang et al., 1993).

Enantioselectivity in the hydrolysis of amides is observed with the local anaesthetic, prilocaine. High doses of prilocaine result in the development of ferrihemoglobinemia, a toxic reaction resulting from the oxidation of the ferrous iron of hemoglobin. The hydrolysis of prilocaine to *o*-toluidine, the cause of ferrihemoglobin formation, is enantioselective for the *R*-enantiomer. Thus as a consequence of selectivity in hydrolysis, the rate but not the extent of ferrihemoglobin formation, is greater following administration of (*R*)- than (*S*)-prilocaine (Akerman and Ross, 1970).

3.2.2. Phase II enzyme systems

3.2.2.1. Glucuronidation

Glucuronidation is perhaps the most common of all xenobiotic conjugation reactions. This is most likely due to the fact that the conjugating agent, uridine diphosphate glucuronic acid (UDPGA), the enzymes, and other cofactors are involved in normal mammalian carbohydrate biosynthesis. The enzymes involved in glucuronidation, the *UDP-glucuronosyltransferases* (UDPGTs), are microsomal and membrane-dependent and found in most mammalian tissues. The mechanism of reaction of UDPGA with a substrate involves nucleophilic substitution and is associated with an inversion of configuration from the α-glucuronic acid in UDPGA to form a β-glucuronide (Figure 3.12).

In recent years, a number of UDPGT isoenzymes have been isolated, purified, and characterized (Tephly, 1992). The UDPGTs appear to have marked similarities and are highly conserved. The major differences in their functional properties depend primarily on differences in N-terminal amino acid

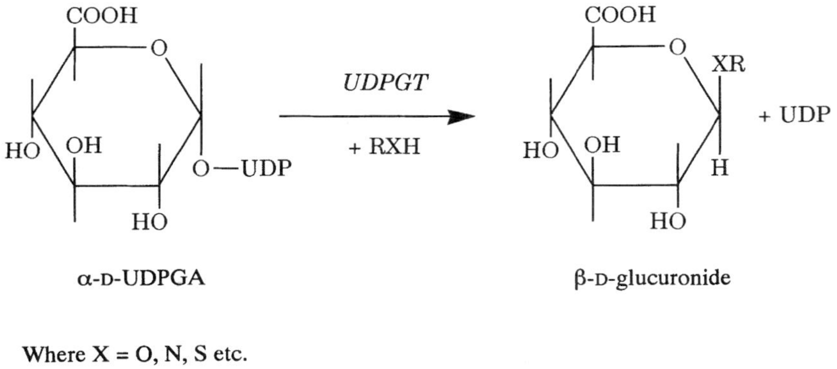

UDPGT

+ RXH

+ UDP

α-D-UDPGA

β-D-glucuronide

Where X = O, N, S etc.

Figure 3.12. General mechanism of glucuronidation, showing the inversion of configuration from α-D-UDPGA to form a β-D-glucuronide.

sequences. Human liver microsomes possess UDPGTs that are qualitatively different from other animals, and there are some UDPGTs in human liver that are not present in lower animals (Tephly, 1992).

There are numerous examples of enantioselectivity in glucuronidation, for example, (*S*)-propranolol undergoes glucuronidation at a faster rate than its *R*-antipode (Walle et al., 1983). Another example of stereoselective glucuronidation is seen with the benzodiazepine anxiolytic, oxazepam, that exhibits species-dependent stereoselective glucuronidation. In man, rabbit, and dog *in vivo*, (*S*)-oxazepam is more extensively glucuronidated than its *R*-antipode, with *S*:*R* ratios ranging from 2:1 to 3.4:1, whereas in the rhesus monkey, the *R*-enantiomer is more extensively conjugated, with an *S*:*R* ratio of 0.55:1. In *in vitro* studies, using dog, pig, rabbit, and rat hepatic preparations, (*S*)-oxazepam is more rapidly glucuronidated than (*R*)-oxazepam, whereas in monkey liver preparations, it is the *R*-enantiomer that is more rapidly conjugated (Sisenwine et al., 1982). The stereoselectivity of oxazepam glucuronidation has also been used as a probe for the induction of UDPGTs in the rabbit. Pretreatment of the animals with inducers of the GT_1 enzyme, for example, 3-methylcholanthrene and β-naphthoflavone, resulted in changes in the *S*:*R* ratio compared with control, but the total rate of formation of the glucuronides was unaltered. It was suggested that the decrease in the rate of formation of the *S*-glucuronide was due to induction of the isoenzyme(s) responsible for the conjugation of the *R*-enantiomer. Inducers of GT_2, for example, phenobarbitone, resulted in an increase in enzyme activity with no effect on the stereochemical composition of the product (Yost and Finley, 1985).

A similar situation is observed with the antidepressant, oxaprotiline. In humans, *in vivo*, glucuronidation is the major route of metabolism of oxaprotiline, with (*S*)-oxaprotiline-O-glucuronide predominating (Dieterle et al., 1984). It is interesting to note that the glucuronides of both oxazepam and oxaprotiline exhibit stereoselectivity in their enzyme-mediated hydrolysis. In both cases, the *S*-glucuronides are more easily cleaved by β-glucuronidase than the *R*-glucuronides (Ruelius et al., 1979; Dieterle et al., 1984). In the case of oxazepam, β-glucuronidase from *E.coli* hydrolyzes (*S*)-oxazepam-O-glucuronide at a rate some 400 times faster than that of the *R*-glucuronide, whereas bovine β-glucuronidase exhibits less stereoselectivity and molluscan β-glucuronidase (from *Helix pomatia*) exhibits little or no stereoselectivity (Ruelius et al., 1979).

Another example of stereoselectivity in glucuronidation is seen with the opiate analgesic, morphine. In the rat the natural compound, (–)-morphine, is metabolized via both 3-O-glucuronidation and N-demethylation. For the synthetic compound, (+)-morphine, the major route of metabolism is 6-O-glucuronidation. This enantiomer also undergoes 3-O-glucuronidation, albeit at a rate some eight times slower than the (–)-isomer, as well as N-demethylation. Overall, (+)-morphine is glucuronidated at half the rate of its (–)-antipode (Rane et al., 1985).

Enantioselectivity in glucuronidation is also observed in the metabolism of the quinolone antibiotic, ofloxacin. Following administration to the rat, the major route of metabolism of ofloxacin is glucuronidation, whereas in dogs, monkeys, and humans, glucuronidation is only a minor route of metabolism. In the rat, *in vivo*, the area under the plasma concentration-time curve (AUC) of (*R*)-ofloxacin is some 1.6 times that of its *S*-antipode. The elimination half-life of the *S*-enantiomer is 1.8 hours, whereas that of the *R*-isomer is 3 hours (Okazaki et al., 1989). These pharmacokinetic differences arise as a result of the stereoselective glucuronidation of (*S*)-ofloxacin. In addition, *in vitro* studies, using rat hepatic microsomal preparations, have indicated relatively minor differences in values for the apparent K_m for the glucuronidation of the two enantiomers, but a 6.5-fold difference in V_{max} in favor of the *S*-enantiomer. Further studies indicate that (*R*)-ofloxacin is a competitive inhibitor of the glucuronidation of its *S*-antipode, with a K_i value of 2.92 mM. As a result of this enantiomeric interaction in glucuronidation, the serum concentrations of (*S*)-ofloxacin are markedly increased, following administration of the racemate, resulting in a 1.7-fold increase in AUC compared with that observed following an equivalent dose of the single enantiomer (Okazaki et al., 1989; 1991). This example illustrates the potential problems that may arise when comparing dispositional studies using racemates and individual enantiomers.

Enantio- and organ-specificity of glucuronidation, is seen with one of the metabolites of nortryptiline, *E*-10-hydroxynortryptiline. In human liver microsomal preparations, (+)-*E*-10-hydroxynortryptiline is stereospecifically glucuronidated (Dumont et al., 1987), whereas in duodenal microsomal preparations, glucuronidation is confined to the (–)-enantiomer (Dahl-Puustinen et al., 1989).

3.2.2.2. Sulfation

Sulfation is a major metabolic pathway for a number of phenols, as well as some amines, alcohols, and thiols. Sulfation involves the transfer of sulfate, from the activated endogenous sulfate donor, 3′-phosphoadenosine-5′-phosphosulfate (PAPS), to a suitable nucleophilic group, for example, hydroxyl, amino, etc., within the substrate. Most drugs and their oxidative metabolites that are substrates for glucuronidation are also candidates for sulfation. The principal factor governing whether sulfation or glucuronidation occurs is concentration. Thus at low substrate concentrations sulfation generally predominates, whereas at high substrate concentrations, glucuronidation is preferred.

This metabolic reaction is catalyzed by a group of enzymes known collectively as the *sulfotransferases* (STs). These soluble enzymes are found in the cytosol of many tissues, including the liver, kidney, and G.I. tract, and in many species. A number of isoforms of STs have been isolated and characterized, and these isoforms are named according to their preferred substrates. Thus, there are phenol STs (PSTs), alcohol STs, arylamine STs, and steroid STs. The former three STs are relatively nonspecific and will catalyze the sulfation of numerous xenobiotics, whereas the steroid STs are specific either for a single steroid, or for steroids of a particular type. Several STs have been isolated, purified, and characterized from various species (Falany, 1991). Thus far, three human liver STs have been identified, namely, dehydroepiandrosterone-ST (DHEA-ST), the monoamine form of PST (M-PST), and the phenol form of PST (P-PST). In human liver P-PST predominates (Campbell et al., 1987), whereas in human jejunum there are only low levels of P-PST and M-PST is the predominant ST (Sundaram et al., 1989).

There are a few reports of enantioselectivity in sulfation. In man, (+)-4′-hydroxypropranolol (4′-HOP) is stereoselectively sulfated; thus, the urinary ratio of (+):(–) 4′-HOP-sulfate is 4.6 (Walle et al., 1984). The sulfation of 4′-HOP also exhibits species-dependent stereoselectivity (Christ and Walle, 1985; Walle and Walle, 1991). In both hamster and dog cytosolic preparations, (+)-4′-HOP was sulfated more rapidly than its (–)-antipode, with (+):(–) ratios of 1.6 and 1.4, respectively (Christ and Walle, 1985). In rat cytosol, the (–)-enantiomer was preferred, with a (+):(–) ratio of 0.72 (Walle and Walle, 1991).

In human liver cytosol, the sulfation of 4′-HOP exhibited both high- and low-affinity components. The high-affinity component was highly stereoselective toward the (+)-enantiomer and the V_{max} for (+)-4′-HOP sulfation was some 4.6 times greater than that for the (–)-isomer, whereas the low-affinity component showed no stereoselectivity. Thus, overall sulfation was selective for the (+)-isomer (Walle and Walle, 1991). In further studies using purified human cytosolic PSTs, it was demonstrated that the high-affinity component of 4′-HOP sulfation was catalyzed by M-PST, which exhibited a V_{max} (+):(–) ratio of 4.7, whereas P-PST exhibited no stereoselectivity (Walle and Walle, 1992).

A number of β-adrenergic receptor agonists, including terbutaline, prenalterol, isoproterenol, and salbutamol, exhibit stereoselectivity in their PST-catalyzed sulphation (Morgan, 1990). In human liver cytosol, the (+)-enantiomers of both terbutaline and prenalterol are preferentially sulfated (Walle and Walle, 1990). The sulfation of terbutaline was catalyzed by M-PST with a V_{max} (+):(–) ratio of 1.8, whereas P-PST was unreactive toward terbutaline (Walle and Walle, 1992). In rat liver cytosol, the (+):(–) terbutaline sulfate ratio was 7.3 and the V_{max} for the sulfation of (+)-terbutaline was some eight-fold greater than for the (–)-enantiomer (Walle and Walle, 1989).

In human hepatic cytosol, isoproterenol was sulfated by both high- and low-affinity reactions, catalyzed by M-PST and P-PST, respectively. Both enzymes were selective for (+)-isoproterenol with five-fold greater efficiency than for the (–)-isomer. In human jejunal cytosolic preparations and human platelets, both of which exhibited only M-PST activity, (+)-isoproterenol was sulfated with six-fold higher efficiency than its (–)-antipode (Pesola and Walle, 1993). The enantioselectivity of sulfation toward (+)-isoproterenol appears to be due mainly to differences in K_m, whereas for 4′-HOP and terbutaline, enantioselective sulfation is the result of differences in V_{max}. It is interesting that both M-PST and P-PST were enantioselective toward (+)-isoproterenol, since in the case of both 4′-HOP and terbutaline, only M-PST was enantioselective (Pesola and Walle, 1993). In contrast to these studies, with both human liver cytosol and purified PSTs, (–)-salbutamol is sulfated some eight times more efficiently than its (+)-antipode (Walle et al., 1993).

3.2.2.3. Glutathione Conjugation

Glutathione (GSH), the sulfur-containing tripeptide, γ-glutamylcysteinylglycine, is the most abundant nonprotein thiol naturally occurring in mammalian cells and is found in the majority of tissues at concentrations in the millimolar range. Most glutathione is present in the reduced thiol form (GSH), although a number of mixed disulfides are formed, including glutathione disulfide (GSSG), glutathione-protein disulfides, and thioethers. The reduced thiol form is maintained via the enzyme *glutathione reductase*. Within the cell, the bulk of GSH is found in the cytosol, with some in the mitochondria and some in various extracellular fluids (Holmgrën, 1985).

Glutathione is involved in detoxication acting either as a nucleophile, conjugation occurring via the thiol group, or as a reductant in the removal of reactive oxygen species. When functioning as a nucleophile, GSH forms conjugates with electrophilic species, either through nucleophilic substitution (S_N2-type reactions), or through addition reactions, for example, addition to double bonds and epoxide ring systems. Glutathione conjugation is usually mediated via a family of *glutathione S-transferases* (GSTs), although if the xenobiotic is sufficiently electrophilic, spontaneous reaction may occur. The GSTs catalyze the biotransformation of a broad range of substrates and are located in both the microsomal and cytosolic fractions. The cytosolic enzymes are the most abundant and have been most extensively studied. A number of human, rat, and mouse cytosolic GSTs have been isolated, purified, and characterized (Mannervik and Danielson, 1988; Hiratsuka et al., 1990). These isoenzymes are dimeric proteins, which may be either homo- or heterodimers, and belong to different multigene families. They have been classified according to their isoelectric points as either acidic (pi-family), basic (alpha-family), or neutral (mu-family) and details of their nomenclature are summarized in Table 3.4 (Ketterer and Mulder, 1990).

Once formed, GSH conjugates are transported into the circulation, or if they are formed in the liver, they may be excreted in the bile. Glutathione conjugates in the circulation undergo a series of hydrolytic reactions, mediated by *γ-glutamyltranspeptidase* and *aminopeptidase* to yield cysteine conjugates (*premercapturic acids*), which are subsequently N-acetylated, resulting in the formation of *mercapturates* (N-acetylcysteine conjugates). Cysteine conjugates arising from the above pathway may be cleaved by a family of enzymes, the *cysteine conjugate β-lyases* (C-S lyases), to produce thiolated metabolites

Table 3.4. Nomenclature and Classification of Rat, Human, and Mouse Cytosolic Glutathione S-Transferase (GST) Isoenzymes

Species	Class		
	Alpha (Basic)[a]	Mu (Neutral)[a]	Pi (Acidic)[a]
Rat	1–1	3–3	7–7
	1–2	3–4	
	2–2	4–4	
		6–6	
Human	α–ε	μ	π
Mouse	MI	MIII	MII

[a]Acidic, neutral, and basic refer to the isoelectric point.
From Ketterer and Mulder (1990) and te Koppele and Mulder (1991).

(Tateishi, 1983; Bakke and Gustafsson, 1984). The C-S lyases are cytosolic enzymes found in kidney, liver, and intestinal microflora, particularly in the caecum. At present, little is known concerning the stereoselectivity of this reaction.

The GSTs catalyze the nucleophilic attack of GSH at numerous substrates, including epoxides, and arylalkyl- and alkyl-halides. This substitution involves the inversion of configuration at the attacked carbon atom (S_N2) with the substrate undergoing a Walden inversion (Figure 3.13). There are numerous examples of stereoselectivity in GSH conjugation, for example, the conjugation of the 1-halo-1-phenylethanes and 2-halooctanes. The conjugation of 1-chloro- and 1-bromo-1-phenylethane has been shown to occur using rat liver cytosolic preparations. Using purified isoenzymes of the alpha- and mu-families, the reaction was substrate-selective (four- to five-fold) for the *S*-enantiomers in each case, with the product, as a result of the Walden inversion, having the *R*-configuration (Mangold and Abdel-Monem, 1983). In contrast to the above, no stereoselectivity was observed in the conjugation of the enantiomers of 2-iodo- and 2-bromooctane. Inversion of configuration of the products was found, with the reaction rate of the iodo compound being 10-fold greater than that of the bromo derivative, whereas the chloro derivative did not react (Ridgewell and Abdel-Monem, 1987).

Stereoselectivity in glutathione conjugation is also seen with a number of epoxide metabolites of aromatic hydrocarbons and alkenes. Epoxides appear to be good substrates for mu-family GSTs and their conjugation with GSH frequently exhibits both regio- and stereoselectivity. For example, the conjugation of benzo[a]pyrene-4,5-oxide (BPO) with GSH can occur at both epoxide carbon atoms. However, in rat hepatic cytosolic preparations, the reaction preferentially occurs at the *R*-configured carbon. Thus, conjugation of (+)-(4*S*, 5*R*)-BPO preferentially yields the (5*S*)-glutathionyl-(4*S*)-hydroxy conjugate, whereas (–)-(4*R*, 5*S*)-BPO yields predominantly the (4*S*)-glutathionyl-(5*S*)-hydroxy stereoisomer (Hernandez et al., 1980; Armstrong et al., 1981a).

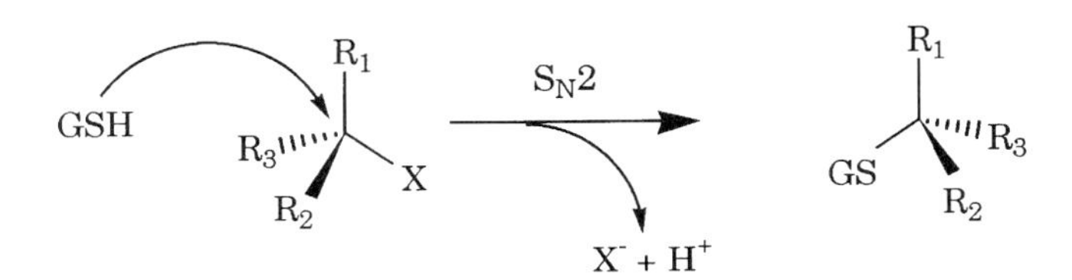

Figure 3.13. General mechanism of inversion of chirality of halides resulting from glutathione conjugation.

Studies with purified GSTs from rat and other species suggest that attack at *R*-configured carbons of epoxides with two benzylic carbon atoms is the main determining factor in both the regio- and stereoselectivity of glutathione conjugation of benzylic epoxide carbons. Studies using BPO, pyrene-4,5-oxide, and *cis*-stilbene oxide (SBO) have shown that mu-family GSTs preferentially catalyze conjugation with *R*-configured carbon atoms (>95%). Alpha-family GSTs are also preferential for *R*-carbons (70%), whereas pi-family GSTs show greater than 95% preference for *S*-configured carbons (Watabe et al., 1985; Dostal et al., 1986; 1988; de Smidt et al., 1987; te Koppele and Mulder, 1991). Purified rat GST 4-4 exhibits greater than 98% stereoselectivity for the *R*-configured carbon of *cis*-(1*R*, 2*S*)-SBO, resulting in the formation of (1*S*, 2*S*)-glutathionyl-SBO (Figure 3.14). Preference for the *R*-configured carbon is also seen with *trans*-(1*R*, 2*R*)-SBO. With (4*R*, 5*S*)-pyrene-4,5-oxide, human

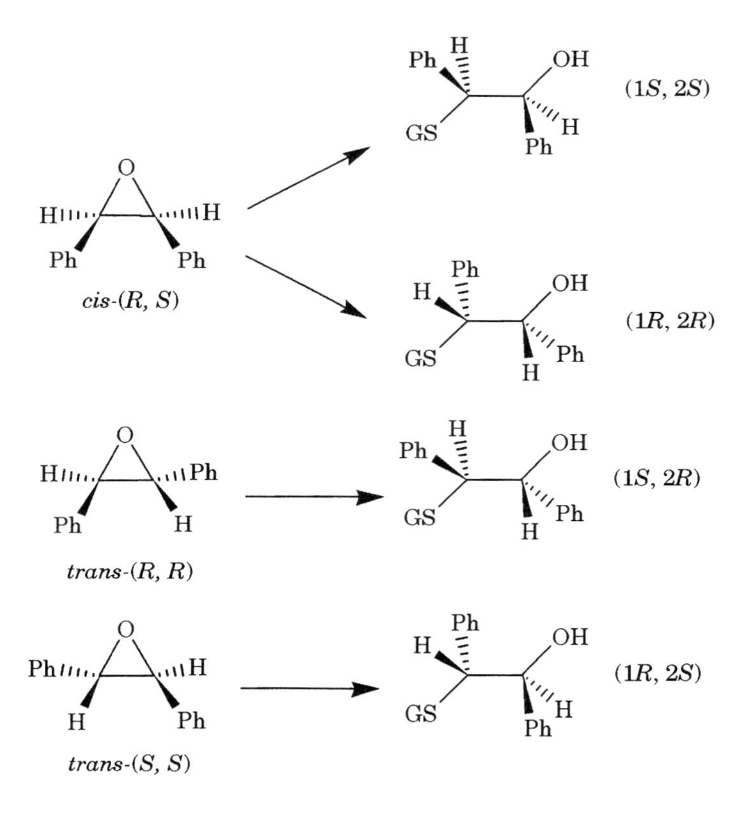

Figure 3.14. The possible stereochemical consequences of the glutathione conjugation of stilbene oxide stereoisomers.

GST-μ exhibits greater than 99% preference for the *R*-configured carbon (Dostal et al., 1988). Overall, it was shown that the mu-family GSTs exhibited the highest catalytic efficiencies and degree of stereoselectivity toward *R*-configured carbons. Although the alpha-family GSTs also preferentially attacked *R*-configured carbons, their catalytic efficiencies were one to two orders of magnitude lower than those of mu-family GSTs. In contrast, pi-family GSTs exhibited stereoselectivity toward *S*-configured carbons, with intermediate catalytic efficiency (de Smidt et al., 1987; Dostal et al., 1988; te Koppele and Mulder, 1991).

When a compound forms an epoxide possessing a single benzylic carbon atom, GSTs preferentially attack the benzylic carbon, due to its higher electrophilicity, irrespective of its configuration. Examples of such compounds include 1,2,3,4-tetrahydronaphthalene-1,2-oxide, benzo[a]pyrene-7,8-dihydrodiol-9,10-oxide (7,8,9,10-BPDO), and *trans*-*β*-methylstyrene oxide; in all these cases, glutathione conjugation occurs at the benzylic carbon, irrespective of its configuration (Watabe et al., 1983; 1984; 1985). For example, the 7*R*, 8*S*, 9*S*, 10*R* stereoisomer of BPDO [(+)-*anti*-BPDO], the most carcinogenic isomer of this metabolite of benzo[a]pyrene is preferentially conjugated with GSH at the benzylic 10-position. This reaction has been shown to be catalyzed by several GSTs, including rat subunits 4 and 7, and human GST-*μ* (Jernström et al., 1985; Robertson et al., 1986a; 1986b). In contrast, its (–)-7*S*, 8*R*, 9*R*, 10*S*-antipode undergoes limited conjugation at this position, the selectivity of conjugation (+):(–) being 49:1 (Robertson and Jernström, 1986).

From the above, it is obvious that the majority of substrates for GSH conjugation are electrophilic species and therefore potentially highly toxic. As a result, relatively few studies have investigated the stereoselectivity of the reaction *in vivo*. This problem has been addressed in recent years by the use of the now obsolete hypnotic drug α-bromoisovalerylurea (BIU) (te Koppele and Mulder, 1991). In rats, the majority of the dose (70–75%) of BIU is excreted as a variety of mercapturates (te Koppele et al., 1986; Polhuijs et al., 1991). It was shown that (*R*)-BIU had a shorter terminal half-life than its *S*-antipode, suggesting that there is stereoselectivity in the glutathione conjugation of BIU (te Koppele et al., 1986; Mulder et al., 1991). In *in vitro* studies, using rat liver microsomes, isolated hepatocytes and perfused rat liver, the rate of glutathione conjugation of (*R*)-BIU, which is preferentially catalyzed by mu-family GSTs to form (*S*)-glutathionylisovalerylurea, is higher than for its *S*-antipode, which forms (*R*)-glutathionylisovalerylurea. The conjugation of the *S*-enantiomer is preferentially catalyzed by alpha-family GSTs. In contrast, with the corresponding α-bromoisovaleric acid, the stereoselectivity of glutathione conjugation favors the *S*-enantiomer, the reaction being essentially stereospecific. Since the stereoselective GSH conjugation of BIU is catalyzed by the alpha- and mu-family GSTs, the rate of excretion of

BIU-mercapturates offers a noninvasive *in vivo* probe for the assessment of isoenzyme-selective GSH conjugation in experimental animals and possibly humans (Vermeulen, 1989).

3.2.2.4. Epoxide Hydrolaze

Epoxides are chemically reactive due to ring strain and polarization. Most epoxides to which mammals are exposed arise as the result of P450-mediated oxidation of aromatic compounds and alkenes. Most epoxides are short-lived due to their high chemical reactivity and spontaneously rearrange to form phenols. Epoxides with longer half-lives may be subject to metabolism via the enzyme *epoxide hydrolaze* (EH). This microsomal, membrane-bound enzyme, found in close association with P450 in most mammalian tissues (Lu and Miwa, 1980), transforms reactive electrophilic epoxides to unreactive vicinal diols. In common with many of the enzymes involved in xenobiotic metabolism, the EHs may be induced by a variety of compounds, including polycyclic aromatic hydrocarbons, halogenated compounds, for example, perfluorodecanoic acid (PFDA), and phenolic antioxidants. The mechanism of the reaction catalyzed by EH involves nucleophilic attack by H_2O from the side of the molecule opposite to the epoxide ring, resulting in the formation of *trans*-dihydrodiols, which may be subject to further oxidative metabolism, resulting in the formation of dihydrodiolepoxides. In contrast to the initial epoxide metabolites, these dihydrodiolepoxides are generally poor substrates for EH (Oesch, 1980; Timms et al., 1987).

A number of examples of stereoselective EH-mediated hydration have been reported in the literature. For example, cyclohexene oxide is hydrated to form *trans*-(1*R*, 2*R*)-dihydroxycyclohexane (Jerina et al., 1970). Jerina and coworkers (1970) also demonstrated that, in an ^{18}O-enriched medium ($H_2{}^{18}O$), naphthalene oxide was hydrated (30%), with the ^{18}O regiospecifically introduced in the 2-position to form *trans*-(1*R*, 2*R*)-naphthalene-1,2-dihydrodiol. It has since been shown that EH regiospecifically adds H_2O at the allylic position of the arene oxide and the stereochemistry at the reacting carbon center is reversed in the product diol compared to the epoxide substrate (Armstrong et al., 1981b).

Examination of the stereoselectivity of EH based on the stereochemical composition of the resulting diol metabolite is complicated by the potential contribution of the P450 isoenzyme mediating the initial epoxidation. This situation may be illustrated by a consideration of the metabolism of benzo[a]pyrene, which undergoes regioselective CYP1A1-mediated oxidation to yield predominantly the (7*R*, 8*S*)-benzo[a]pyrene-7,8-oxide. The hydration of this by EH results in the formation of *trans*-(7*R*, 8*R*)-benzo[a]pyrene-7,8-dihydrodiol (Figure 3.15). However, the apparently selective formation

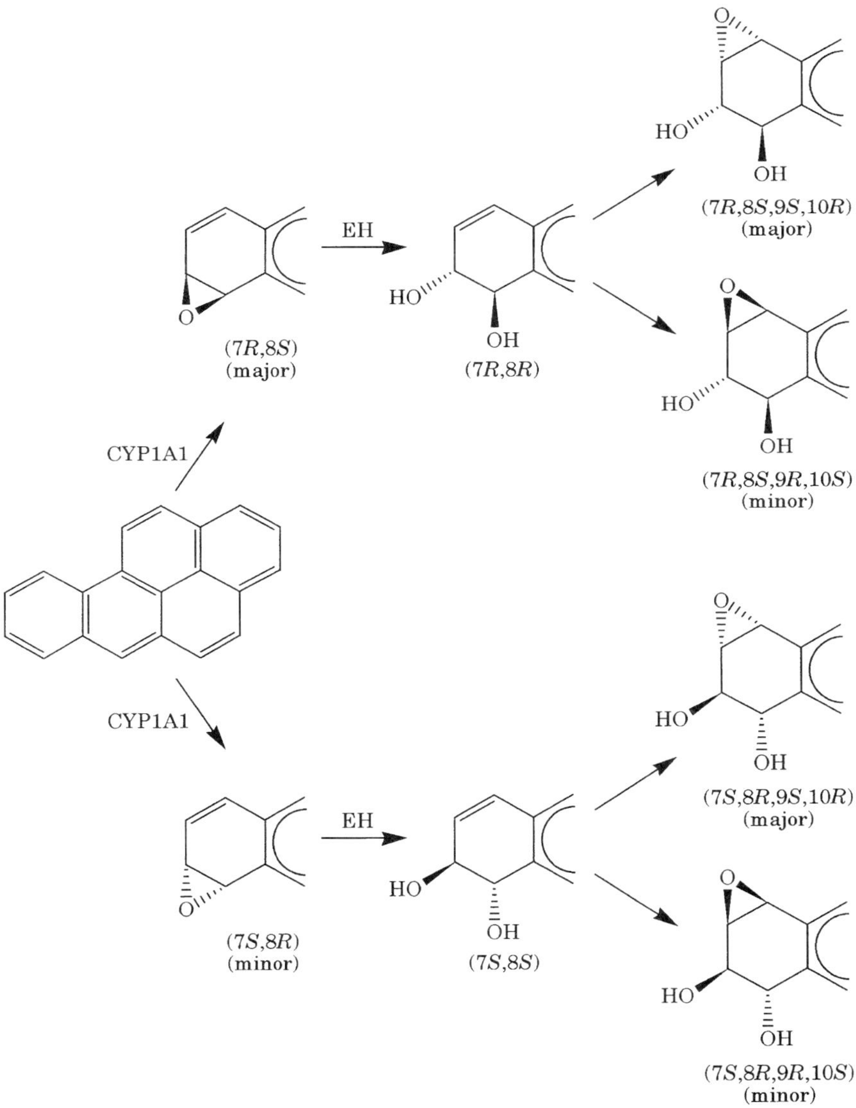

CYP1A1

(7*R*,8*S*)
(major)

EH

HO OH
(7*R*,8*R*)

HO OH
(7*R*,8*S*,9*S*,10*R*)
(major)

HO OH
(7*R*,8*S*,9*R*,10*S*)
(minor)

CYP1A1

(7*S*,8*R*)
(minor)

EH

HO OH
(7*S*,8*S*)

HO OH
(7*S*,8*R*,9*S*,10*R*)
(major)

HO OH
(7*S*,8*R*,9*R*,10*S*)
(minor)

Figure 3.15. Stereoselectivity of the CYP1A1 and epoxide hydrolaze-mediated metabolism of benzo[a]pyrene to yield stereoisomeric dihydrodiolepoxide metabolites. The relative extents of the reactions are indicated by the "major" and "minor" notations under the appropriate structures.

of this particular stereoisomer is mainly due to the selectivity of the initial oxidation, as it has been demonstrated that the (7*S*, 8*R*)-oxide is the preferred substrate for EH (Levin et al., 1980). Both the (7*R*, 8*R*)- and (7*S*, 8*S*)-dihydrodiols of benzo[a]pyrene undergo further P450-mediated oxidation to yield predominantly the (7*R*, 8*S*, 9*S*, 10*R*)- and (7*S*, 8*R*, 9*S*, 10*R*)-7,8-dihydrodiol-9,10-epoxides, respectively. It is interesting to note that CYP1A1 and EH conspire to yield predominantly the (7*R*, 8*S*, 9*S*, 10*R*)-7,8-dihydrodiol-9,10-epoxide, which is the most potent carcinogenic dihydrodiol epoxide of benzo[a]pyrene (Levin et al., 1980). The CYP1A1 and EH metabolism of other aromatic compounds, including benzene, 4-chlorobenzene, naphthalene, chrysene, anthracene, and benz[a]anthracene, predominantly results in the formation of *R*,*R*-dihydrodiols (Jerina et al., 1970; Armstrong et al., 1981b; van Bladeren et al., 1985).

Phenyloxirane, the major mutagenic metabolite of styrene, exhibits both regio- and stereoselectivity in its hydrolysis by EH. In rabbit liver microsomes, in an ^{18}O-enriched medium ($H_2{}^{18}O$), *rac*-phenyloxirane was hydrated to form phenylethanediol, with more than 90% of the ^{18}O regiospecifically attached at the nonbenzylic carbon (Jerina et al., 1970). When *rac*-phenyloxirane was incubated with rat hepatic microsomes, it was demonstrated that (*R*)-phenyloxirane was preferentially hydrated compared with its *S*-antipode and that the *S*-enantiomer appeared to be a poor substrate for EH. However, when the individual enantiomers were used, it was found that the *S*-enantiomer was a better substrate for EH than its *R*-antipode, suggesting that (*R*)-phenyloxirane acts as a competitive inhibitor of the EH-catalyzed hydration of the *S*-enantiomer (Watabe et al., 1983).

3.2.2.5. Methylation

The methylation of xenobiotics is catalyzed mainly by nonspecific methyltransferases found in the lung and by physiological methyltransferases, including catechol O-methyltransferase (COMT), imidazole N-methyltransferase, and thiol S-methyltransferase, in other tissues. The methylation pathway involves the conjugation of nucleophilic groups, such as hydroxyls, amines, and thiols, with the methyl group of the cofactor, S-adenosylmethionine (SAM). In contrast to other conjugation reactions, methylation involves the addition of a lipophilic moiety to the substrate and thus frequently leads to formation of a less polar product, thereby hindering excretion of the compound.

Few examples of stereoselectivity in methylation have been reported. The best example is probably that of nicotine. In guinea pigs, both *in vivo* and *in vitro*, using various tissue preparations, (*R*)-nicotine undergoes essentially stereospecific N-methylation, that is, at the pyridyl nitrogen, to form the (*R*)-N-methylnicotinium ion (Figure 3.16). This reaction is catalyzed by

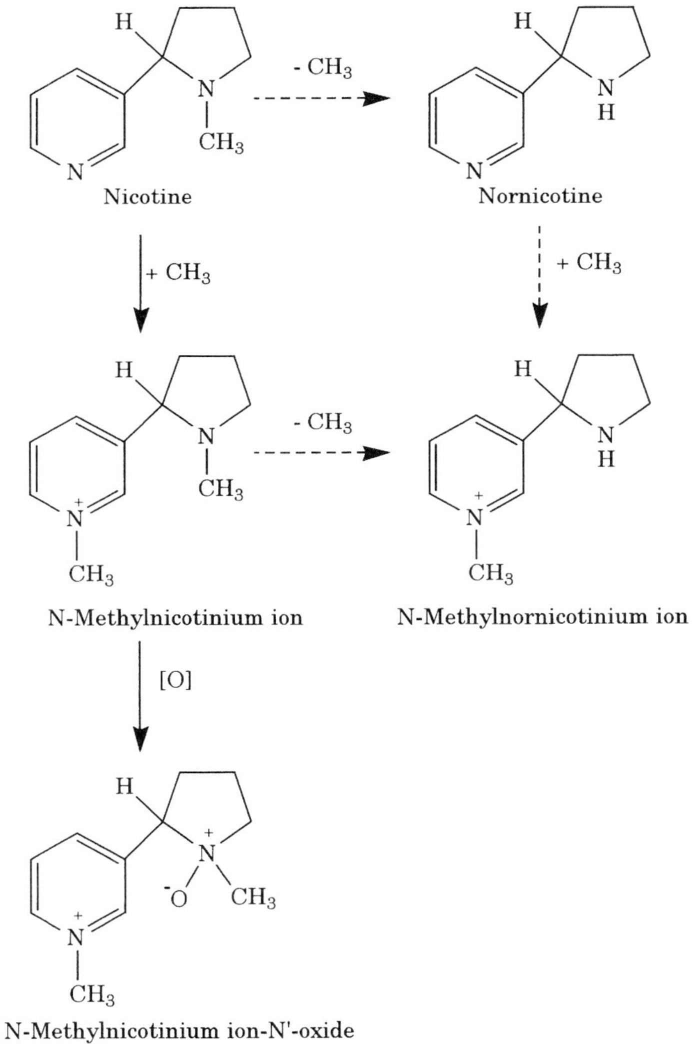

Nicotine

$- CH_3$

Nornicotine

$+ CH_3$

$+ CH_3$

N-Methylnicotinium ion

$- CH_3$

N-Methylnornicotinium ion

[O]

N-Methylnicotinium ion-N'-oxide

Figure 3.16. The N-methylation of nicotine and certain of its metabolites.

guinea pig lung *aromatic azaheterocycle N-methyltransferase* (Cundy et al., 1985a; 1985b; 1985c; Crooks, 1993). In contrast, (*S*)-nicotine is not methylated and, in fact, acts as a competitive inhibitor of the N-methylation of the *R*-enantiomer (Cundy et al., 1985a). No N′-methylated metabolites of either enantiomer were detected. Aromatic azaheterocycle N-methyltransferase activity has been shown to be widely distributed throughout guinea pig tissues, including lung, liver, spleen and brain (Cundy et al., 1985b). Isolated guinea pig alveolar macrophages, which possess a number of methyltransferase activities, avidly N-methylated (*R*)-nicotine, but not the *S*-enantiomer (Gairola et al., 1988). In common with other guinea pig preparations, (*S*)-nicotine inhibited the methylation of its *R*-antipode in this system (Crooks, 1993).

The *in vitro* N-methylation of nicotine has also been demonstrated using rabbit liver cytosolic preparations (Damani et al., 1986). Two amine N-methyltransferases, designated A and B, were isolated from rabbit liver cytosol. These two enzymes were shown to exhibit broad and overlapping substrate selectivities, for example, both will catalyze N-methylation of aromatic heterocycles to form quaternary ammonium metabolites (Crooks et al., 1988a). When nicotine enantiomers were incubated with the isolated enzymes, both A and B catalyzed the N-methylation of (*R*)-nicotine, but not that of its *S*-antipode. In contrast, both enantiomers of cotinine were substrates for both enzymes (Crooks et al., 1988a). However, when a crude preparation was used, containing both N-methyltransferases, (*R*)-nicotine was methylated at either nitrogen atom to form the (*R*)-N-methylnicotinium and (*R*)-N′-methylnicotinium ions, while the *S*-enantiomer was methylated solely at the pyridyl nitrogen (Damani et al., 1986). Human liver cytosolic preparations were able to N-methylate both nicotine enantiomers with stereoselectivity toward (*R*)-nicotine (Crooks and Godin, 1987).

The oxidative metabolites of nicotine, which all retain the basic pyridine nitrogen atom in their structure, may also be methylated. For example, the N-methylcotinium ion has been detected as a human urinary metabolite of (*S*)-nicotine (McKennis et al., 1963) and as a metabolite of (*R*)-nicotine, but not (*S*)-nicotine, in the guinea pig (Cundy et al., 1984). In addition, the N-methylnornicotinium ion (Figure 3.16) has been isolated from guinea pig urine following the administration of (*R*)-nicotine, but not (*S*)-nicotine. Detection of this metabolite provides a possibly unique example of both N-demethylation and N-methylation occurring within the same molecule (Sato and Crooks, 1985). The N-methylnicotinium ion N′-oxide (Figure 3.16) has also been isolated and characterized as a urinary metabolite of (*R*)-nicotine in the guinea pig (Pool et al., 1986; Crooks et al., 1988b). This metabolite was excreted as a mixture of both *cis*-(1′*S*, 2′*R*)- and *trans*-(1′*R*, 2′*R*)-diastereoisomers, formed in a ratio of 1.6:1. In contrast, (*S*)-nicotine did not afford this metabolite. It appears that the formation of N-methylnicotinium ion N′-oxide probably results from the N′-oxidation of N-methylnicotine (Pool et al., 1986).

3.2.2.6. Chiral Inversion

The 2-arylpropionic acid derivatives, the so-called "profens," for example, ibuprofen, flurbiprofen, naproxen, etc. (Figure 3.17), are a major group of nonsteroidal antiinflammatory drugs (NSAIDs) widely used in clinical prac-

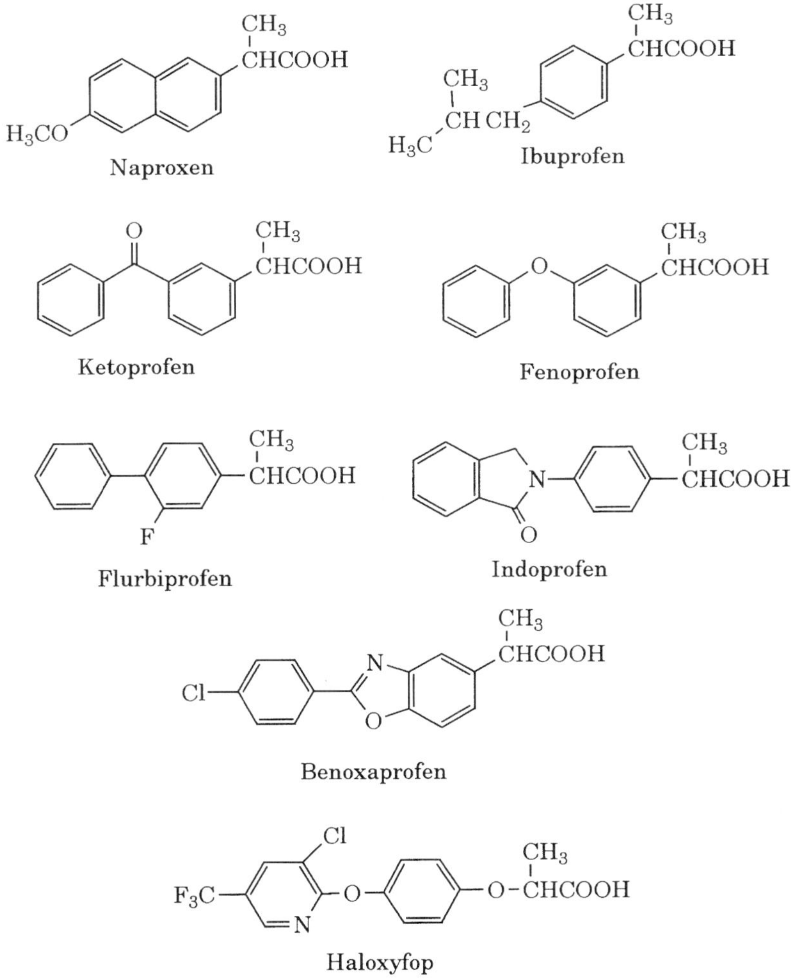

CH_3
CHCOOH
H_3CO
Naproxen

CH_3
CHCOOH
CH_3
CH CH_2
H_3C
Ibuprofen

O
CH_3
CHCOOH
Ketoprofen

CH_3
O
CHCOOH
Fenoprofen

CH_3
CHCOOH
F
Flurbiprofen

CH_3
N
CHCOOH
O
Indoprofen

CH_3
N
CHCOOH
Cl
O
Benoxaprofen

Cl
CH_3
F_3C
O
O-CHCOOH
N
Haloxyfop

Figure 3.17. The structures of various "profen" NSAIDs and "oxyprofen" herbicides.

tice (Caldwell et al., 1988a; Kean et al., 1991). The antiinflammatory actions of the profens occur as a result of their ability to inhibit one or more of the enzymes involved in the conversion of arachidonic acid to prostaglandins, leukotrienes, and other inflammatory mediators. For most of these drugs, their *in vitro* pharmacological activity resides in the *S*-enantiomers, with their *R*-antipodes being virtually inactive. For many of the profens, however, when they are evaluated *in vivo*, the eudismic ratios are greatly reduced (see Caldwell et al., 1988b and the references cited there). The majority of the profens are marketed and administered as the racemates, the exceptions being (*S*)-naproxen, (*S*)-flunoxaprofen, (*S*)-ibuprofen, in Austria and (*S*)-ketoprofen in Spain (Harrison et al., 1970; Roszkowski et al., 1971; Kean et al., 1989; Lennard et al., 1990).

The large discrepancy between the *in vitro* and *in vivo* potencies prompted studies into the metabolism of the enantiomers. In addition to various oxidative and conjugative metabolic transformations, the profens undergo a unidirectional metabolic inversion of chirality of the *R*-acid to its *S*-antipode, with no other change to the molecule (Hutt and Caldwell, 1983). It appears that this reaction is a general feature of the disposition of the profens, although there can be substantial differences in both the rate and extent of inversion, which appears to vary with both the substrate and the species under investigation (Hutt and Caldwell, 1983; 1984; Caldwell et al., 1988a; 1988b). Thus, in man the degree to which inversion occurs may be negligible, as is the case with indoprofen (Tamassia et al., 1984), intermediate, for example, ibuprofen (Lee et al., 1984), or essentially complete, as with fenoprofen (Rubin et al., 1985). The clinical implications of chiral inversion are unclear, but it may assume considerable biological significance in light of the fact that pharmacological activity resides in the *S*-enantiomers. Thus, when significant inversion occurs, the *R*-enantiomers may be considered to act as pro-drugs for the active agent, for example, fenoprofen and ibuprofen.

Chiral inversion was first demonstrated with ibuprofen (Adams et al., 1976; Kaiser et al., 1976). In *in vivo* studies, high plasma concentrations of (*S*)-ibuprofen were seen following administration of the *R*-enantiomer, but no (*R*)-ibuprofen was seen after administration of the *S*-isomer. The total inversion of ibuprofen in man has been estimated to range between 57–71% of the dose (Lee et al., 1985). In addition to chiral inversion, ibuprofen undergoes other metabolic reactions, including aliphatic oxidation, amino acid conjugation, and glucuronidation (Kaiser et al., 1976; Lee et al., 1985; El Mouelhi et al., 1987; Shirley et al., 1994).

Considerable attention has been focused on the mechanism of the inversion reaction. The majority of mechanistic studies have employed ibuprofen, the prototype compound in this therapeutic class, as a model compound. The original hypothesis was that the inversion reaction involves stereospecific formation of the *R*-profenyl-coenzyme A (CoA) thioesters (Wechter et al., 1974). The resultant (*R*)-2-arylpropionyl-CoA thioester is then epimerized, possibly via the actions of *methylmalonyl-CoA racemase*, or a related enzyme (Williams and Day, 1985; Tracy and Hall, 1992). The epimerized CoA thioesters are then cleaved to yield both *R*- and *S*-enantiomers. Alternative mechanisms have been proposed, for example, the formation of oxidative products, such as, 2,3-dehydroibuprofen, or nonenzymatic dissociation of the α-methine proton (Lan et al., 1976; Mayer et al., 1988). These hypotheses have been discounted in light of the findings of a series of *in vivo* and *in vitro* investigations by a number of workers (Williams et al., 1986; Chen et al., 1991; Sanins et al., 1991; Tracy and Hall, 1991; 1992; Tracy et al., 1993). The proposed mechanism of the chiral inversion reaction is shown in Figure 3.18 (Knihinicki et al., 1991; Sanins et al., 1991).

The enzymes involved in the formation of the CoA thioesters have not been identified. However, it has been demonstrated, *in vitro*, that both palmitic and octanoic acids are capable of inhibiting the formation of ibuprofenyl-CoA. This suggests that microsomal and mitochondrial medium- or long-chain fatty acyl-CoA synthetases are involved (Knights et al., 1989; Knadler and Hall, 1990; Tracy et al., 1993). Interestingly, propionic acid had no effect on (*R*)-ibuprofenyl-CoA formation, suggesting that short-chain fatty acyl-CoA synthetase is not involved (Tracy et al., 1993).

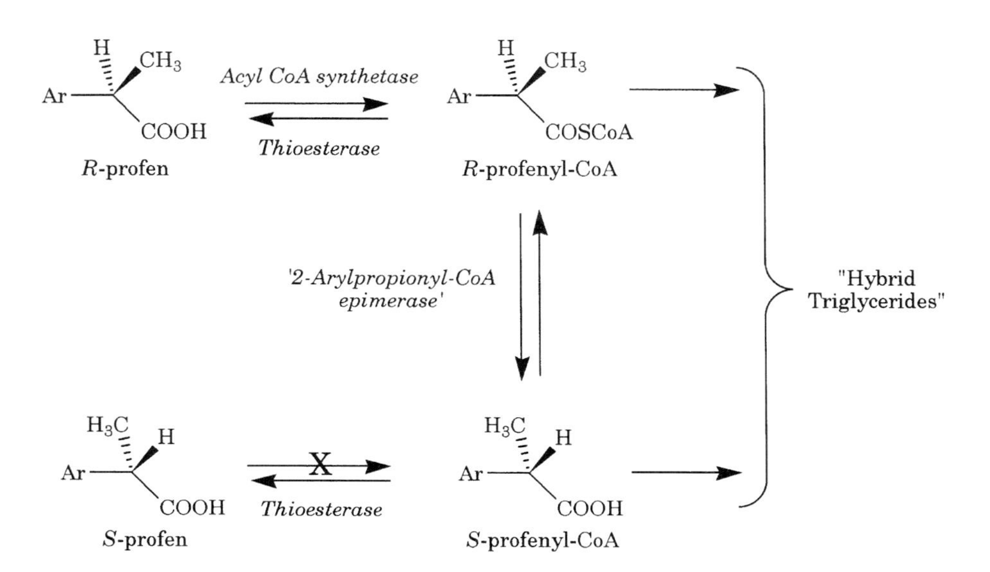

Figure 3.18. The mechanism of the metabolic chiral inversion of the 2-arylpropionic acids.

Further evidence for the involvement of CoA and long-chain fatty acyl-CoA ligase comes from studies of the inversion of ibuprofen in rats that had been chronically treated with clofibrate, a hypolipidemic agent known to both increase hepatic CoA levels (Skrede and Halvorsen, 1979) and induce long-chain fatty acyl-CoA ligase (Knights et al., 1991). Chronic treatment of rats with clofibrate resulted in enhanced chiral inversion in liver homogenates (Knights et al., 1991). These findings are further supported by studies showing that both enantiomers of ibuprofen, fenoprofen, ketoprofen, and naproxen inhibit the formation of palmitoyl-CoA in rat liver microsomes (Knights and Jones, 1992). In addition, both the hypolipidemic agent, bezafibrate, and (*S*)-ibuprofen inhibit the formation of ibuprofenyl-CoA *in vitro* (Tracy et al., 1993). The ability of (*S*)-ibuprofen to inhibit the (*R*)-ibuprofenyl-CoA thioester formation is consistent with the findings of *in vivo* studies in which, following the administration of racemic ibuprofen to dogs, the AUC of (*R*)-ibuprofen was increased when compared with the AUC when (*R*)-ibuprofen was given alone (Ahn et al., 1991).

So far as the epimerization step is concerned, a "*2-arylpropionyl-CoA epimerase*," which is located in both mitochondria and cytosol, has been purified and found to be different from the methylmalonyl-CoA epimerase, a mitochondrial enzyme (Shieh and Chen, 1993). Neither the epimerization nor the hydrolysis is stereoselective and the epimerization is bidirectional (Chen et al., 1991; Knihinicki et al. 1991; Tracy and Hall, 1992; Shieh and Chen, 1993). It has been shown that the epimerization reaction is more efficient than the hydrolysis of the thioesters (Tracy and Hall, 1991). The hypothesis that the stereoselective formation of (*R*)-ibuprofenyl-CoA is the first step in the inversion reaction has recently been disputed by Menzel and co-workers (1994), who have suggested that the stereoselectivity of inversion is determined by the formation of (*R*)-ibuprofenyl-adenylate. Obviously, more work is required before the mechanism of the inversion reaction is unequivocally determined.

The formation of CoA thioesters may account for some of the toxicity associated with the profens. For example, both (*R*)-ibuprofen and (*R*)-fenoprofen are stereoselectively incorporated into adipose tissue in the form of "hybrid" triglycerides (Sallustio et al., 1988; Knihinicki et al., 1989). The reaction involves the acyl transfer of the profenyl moiety from the 2-arylpropionyl-CoA thioesters to glycerol (Williams et al., 1986). These hybrid triglycerides may accumulate, leaving long-term residues that interfere with normal lipid metabolism and membrane function and, since they are lipophilic, may cross the blood-brain barrier, leading to CNS disturbances (Fears et al., 1978; Crayford and Hutson, 1980; Caldwell and Marsh, 1983; Williams et al., 1986). Since the putative epimerase enzyme inverts the *R*-isomer CoA thioesters to the *S*-isomer CoA thioesters, then both are capable of forming hybrid triglycerides (Williams et al., 1986). Since the *S*-profens do not form CoA thioesters and, therefore, cannot form hybrid triglycerides, and the *S*-isomers are pharmacologically active, there is a strong argument for administering the *S*-enantiomers alone.

There are considerable species differences in both the rate and extent of inversion. The most marked species difference is observed with benoxaprofen (Opren). The inversion half-life of benoxaprofen is some 40 times longer in man than in rat. Thus, the toxicity testing of benoxaprofen may have been inadequate, because rats were exposed to considerably less *R*-isomer than humans (Simmonds et al., 1980).

The related 2-aryloxypropionate herbicides, for example, haloxyfop (Figure 3.17), also undergo chiral inversion in mammals (Bartels and Smith, 1989). In the case of these compounds, the *S*-enantiomers are inverted to their *R*-antipodes. This reversal in the direction of inversion only arises due to the fact that the stereochemical designation of the "oxyprofens" according to the Cahn–Ingold–Prelog sequence rules is reversed compared with the "profen" NSAIDs, that is, the *R*-enantiomers of the oxyprofens correspond to the *S*-enantiomers of the profens.

3.3. CONCLUDING REMARKS

In recent years, we have come to appreciate the fact that stereochemical factors play a significant role in drug disposition and, therefore, contribute to our understanding of the related fields of clinical pharmacology and toxicology. Since the processes of biotransformation involve an interaction between a xenobiotic and a chiral bioenvironment, it is probably the case that a pair of enantiomers will never exhibit identical metabolic behaviors. An examination of both the regio- and stereoselectivity of metabolism may yield valuable information concerning the nature of the drug-enzyme interaction.

A potentially useful approach in drug metabolism studies is the application of the techniques of molecular biology in combination with those of the stereospecific analytical methodologies. Such an approach using isoenzyme expression systems, together with either prochiral substrates that undergo transformation to yield chiral metabolites, or enantiomeric mixtures of substrates, may allow the drug metabolism scientist to characterize the drug metabolizing enzymes in terms of their regio- and stereoselectivity toward selected substrates (Oguri et al., 1994; Pelkonnen and Breimer, 1994). A combination of the above approach together with molecular modeling studies, as exemplified by the examples of progesterone and (*S*)-mephenytoin outlined above, allow the drug metabolism scientist to "probe" enzyme active site geometry and rationalize the differences observed between enzymes and isoenzymes. Such data will allow the characterization of human and animal tissue samples with respect to isoenzyme composition and, ultimately, if appropriate probe molecules can be developed for use in man, facilitate the determination of the relative levels of specific isoenzymes *in vivo*. Such investigations will make a considerable contribution to our understanding of the clinical and toxicological aspects of drug use in man and facilitate the extrapolation of data obtained from animal models and *in vitro* studies to man.

REFERENCES

Adams, S. S., Bresloff, P., and Mason, C. G. (1976). Pharmacological differences between the optical isomers of ibuprofen: Evidence for metabolic inversion of the (–)-isomer. *J. Pharm. Pharmacol.*, **28**, 256–257.

Ahn, H.-Y., Amidon, G. L., and Smith, D. E. (1991). Stereoselective systemic disposition of ibuprofen enantiomers in the dog. *Pharm. Res.*, **8**, 1186–1190.

Akerman, B. and Ross, S. (1970). Stereospecificity of the enzymatic biotransformation of the enantiomers of prilocaine (Citanest). *Acta. Pharmacol. Toxicol.*, **28**, 445–453.

Armstrong, R. N., Levin, W., Ryan, D. E., Thomas, P. E., Mah, H. D., and Jerina, D. M. (1981a). Stereoselectivity of rat liver cytochrome $P450_c$ on formation of benzo[a]pyrene-4,5-oxide. *Biochem. Biophys. Res. Comm.*, **100**, 1077–1084.

Armstrong, R. N., Kedzierski, B., Levin, W., and Jerina, D. M., (1981b). Enantioselectivity of microsomal epoxide hydrolase toward arene oxide substrates. *J. Biol. Chem.*, **256**, 4726–4733.

Bakke, J. and Gustafsson, J.-A. (1984). Mercapturic acid pathway metabolites of xenobiotics: Generation of potentially toxic metabolites during enterohepatic circulation. *Trend. Pharmacol. Sci.*, **5**, 517–521.

Bartels, M. J. and Smith, F. A. (1989). Stereochemical inversion of haloxyfop in the Fischer 344 rat. *Drug Metab. Disp.*, **17**, 286–291.

Boyd, D. R., Walsh, C. T., and Jack Chen, Y. C., (1989). S-Oxygenase. II. Chirality of sulphoxidation reactions. In *Sulphur-Containing Drugs and Related Organic Compounds*, Vol. 2, Part A, L. A. Damani (ed.), New York: Wiley, pp. 66–99.

Caldwell, J. and Marsh, M. V., (1983). Inter-relationship between xenobiotic metabolism and lipid biosynthesis. *Biochem. Pharmacol.*, **32**, 1667–1672.

Caldwell, J., Winter, S. M., and Hutt, A. J. (1988a). The pharmacological and toxicological significance of the stereochemistry of drug disposition. *Xenobiotica*, **18**, 59–70.

Caldwell, J., Hutt, A. J., and Fournel-Gigleux, S. (1988b). The metabolic chiral inversion and dispositional enantioselectivity of the 2-arylpropionic acids and their biological consequences. *Biochem. Pharmacol.*, **37**, 105–114.

Campbell, D. B. (1990). Stereoselectivity in clinical pharmacokinetics and drug development. *Eur. J. Drug Metab. Pharmacokin.*, **15**, 109–125.

Campbell, N. R. C., van Loon, J. A., and Weinshilboum, R. M. (1987). Human liver phenol sulfotransferase: Assay conditions, biochemical properties and partial purification of isoenzymes of the thermostable form. *Biochem. Pharmacol.*, **36**, 1435–1446.

Cartwright, A. C. (1991). Regulatory implications: An authority's perspective. *Biochem. Soc. Trans.*, **19**, 465–467.

Cashman, J. R. (1995). Structural and catalytic properties of the mammalian flavin-containing monooxygenase. *Chem. Res. Toxicol.*, **8**, 165–181.

Cashman, J. R., Park, S. B., Yang, Z.-C., Washington, C. B., Gomez, D. Y., Giacomini, K. M., and Brett, C. M., (1993). Chemical, enzymatic and human enantioselective S-oxygenation of cimetidine. *Drug Metab. Disp.*, **21**, 587–597.

Cashman, J. R., Park, S. B., Berkman, C. E., and Cashman, L. E. (1995). Role of hepatic flavin-containing monooxygenase 3 in drug and chemical metabolism in adult humans. *Chem.-Biol. Interact.*, **96**, 33–46.

Chen, C.-S., Shieh, W.-R., Lu, P.-H., Harriman, S., and Chen, C.-Y. (1991). Metabolic stereoisomeric inversion of ibuprofen in mammals. *Biochem. Biophys. Acta*, **1078**, 411–417.

Cheng, K.-C. and Schenkman, J. B. (1982). Purification and characterisation of two constitutive forms of rat liver microsomal cytochrome P450. *J. Biol. Chem.*, **257**, 2378–2385.

Cheng, K.-L. and Schenkman, J. B. (1983). Testosterone metabolism by cytochrome P450 isoenzymes RLM_3 and RLM_5 and by microsomes. *J. Biol. Chem.*, **258**, 11738–11744.

Christ, D. D. and Walle, T. (1985). Stereoselective sulphate conjugation of 4-hydroxypropranolol *in vitro* by different species. *Drug Metab. Disp.*, **13**, 380–381.

Conti, A. and Bickel, M. (1977). History of drug metabolism: Discoveries of the major pathways in the 19th century. *Drug. Metab. Rev.*, **6**, 1–50.

Coon, M. J., Ding, X., Pernecky, S. J., and Vaz, D. N. (1992). Cytochrome P450: Progress and predictions. *FASEB J.*, **6**, 669–673.

Cornforth, J. W. (1974). Enzymes and stereochemistry. *Tetrahedron*, **30**, 1515–1524.

Cornforth, J. W. (1984). Stereochemistry of life. *Interdisc. Sci. Rev.*, **9**, 107–112.

Crayford, J. V. and Hutson, D. H., (1980). Endobiotic triglyceride formation. *Endobiotics*, **10**, 349–354.

Crooks, P. A. (1993). N-Oxidation, N-methylation and N-conjugation reactions of nicotine. In *Nicotine and Related Alkaloids. Absorption, Distribution, Metabolism and Excretion*, J. W. Gorrod, and J. Wahren (eds.), London: Chapman and Hall, pp. 81–109.

Crooks, P. A. and Godin, C. S. (1987). N-Methylation of nicotine enantiomers by human liver cytosol. *J. Pharm. Pharmacol.*, **40**, 153–154.

Crooks, P. A., Godin, C. S., Damani, L. A., Ansher, S. S., and Jakoby, W. A. (1988a). Formation of quaternary amines by N-methylation of azaheterocycles with homogeneous amine N-methyltransferases. *Biochem. Pharmacol.*, **37**, 1673–1677.

Crooks, P. A., Pool, W. F., and Damani, L. A. (1988b). N-Methyl-N′-oxonicotinium ion: Synthesis and stereochemistry. *Chem. Indust.* (*London*), **37**, 95–96.

Crooks, P. A., Godin, C. S., and Pool, W. F. (1992). Enantiomeric purity of nicotine in tobacco smoke condensate. *Med. Sci. Res.*, **20**, 879–880.

Cundy, K. C., Godin, C. S., and Crooks, P. A. (1984). Evidence of stereospecificity in the *in vivo* methylation of [^{14}C]-(±)-nicotine in the guinea pig. *Drug Metab. Disp.*, **12**, 755–759.

Cundy, K. C., Crooks, P. A., and Godin, C. S. (1985a). Remarkable substrate-inhibitor properties of nicotine enantiomers towards a guinea pig lung aromatic azaheterocycle N-methyltransferase. *Biochem. Biophys. Res. Comm.*, **128**, 312–316.

Cundy, K. C., Godin, C. S., and Crooks, P. A. (1985b). Stereospecific *in vitro* N-methylation of nicotine in guinea pig tissues by an S-adenosylmethionine-dependent N-methyltransferase. *Biochem. Pharmacol.*, **34**, 281–284.

Cundy, K. C., Sato, M., and Crooks, P. A. (1985c). Stereospecific *in vivo* N-methylation of nicotine in the guinea pig. *Drug Metab. Disp.*, **13**, 175–185.

Dahl-Puustinen, M. L., Dumont, E., and Bertilsson, L. (1989). Glucuronidation of *E*-10-hydroxynortryptiline in human liver, kidney and intestine: Organ-specific differences in enantioselectivity. *Drug Metab. Disp.*, **17**, 433–436.

Dalet-Beluche, I., Boulenc, X., Fabre, G., Maurel, P., and Bonfils, C. (1992). Purification of two cytochrome P450 isozymes related to CYP2A and CYP3A gene families from monkey (baboon, *Papio papio*) liver microsomes. Cross reactivity with human forms. *Eur. J. Biochem.*, **204**, 641–648.

Damani, L. A., Shaker, M. S., Godin, C. S., Crooks, P. A., Ansher, S. S., and Jakoby, W. B. (1986). The ability of amine N-methyltransferases from rabbit liver to N-methylate azaheterocycles. *J. Pharm. Pharmacol.*, **38**, 547–550.

Damani, L. A., Pool, W. F., Crooks, P. A., Kaderlik, R. K., and Ziegler, D. M. (1988). Stereoselectivity in the N′-oxidation of nicotine isomers by flavin-containing monooxygenases. *Mol. Pharmacol.*, **33**, 702–705.

Dannan, G. A. and Guengerich, F. P. (1982). Immunochemical comparison and quantitation of microsomal flavin-containing monooxygenases in various hog, mouse, rat, rabbit, dog and human tissues. *Mol. Pharmacol.*, **22**, 787–794.

de Smidt, P. C., McCarrick, M. A., Darnow, J. N., Mervic, M., and Armstrong R. N. (1987). Stereoselectivity and enantioselectivity of glutathione S-transferase toward stilbene oxide substrates. *Biochem. Int.*, **14**, 401–408.

Dieterle, W., Faigle, J. W., Montigel, C., Sulc, M., and Theobald, W. (1977). Biotransformation and pharmacokinetics of acenocoumarol (Sintrom) in man. *Eur. J. Clin. Pharmacol.*, **11**, 367–375.

Dieterle, W., Faigle, J. W., Kriemler, H. P., and Winkler, T. (1984), The metabolic fate of [^{14}C]-oxaprotiline. HCL in man. II. Isolation and identification of metabolites. *Xenobiotica*, **14**, 311–319.

Dolphin, C. T., Shephard, E. A., Povey, S., Palmer, C. N. A., Ziegler, D. M., Ayesh, R., Smith, R. L., and Phillips, I. R. (1991). Cloning, primary sequence and chromosomal mapping of a human flavin-containing monooxygenase. *J. Biol. Chem.*, **266**, 12379–12385.

Dostal, L. A., Aitio, A., Harris, C., Bhatia, A. V., Hernandez, O., and Bend, J. R. (1986). Cytosolic glutathione S-transferases in various rat tissues differ in stereoselectivity with polycyclic arene and alkene oxide substrates. *Drug Metab. Disp.*, **14**, 303–309.

Dostal, L. A., Guthenberg, C., Mannervik, B., and Bend, J. R. (1988). Stereoselectivity and regioselectivity of purified human glutathione transferases π, α-ε and μ with alkene and polycyclic arene oxide substrates. *Drug Metab.Disp.*, **16**, 420–424.

Drayer, D. E. (1993). The early history of stereochemistry. From the discovery of molecular asymmetry and the first resolution of a racemate by Pasteur to the asymmetrical chiral carbon of van't Hoff and Le Bel. In *Drug Stereochemistry. Analytical Methods and Pharmacology*, 2nd ed., I. W. Wainer, (ed.), New York: Marcel Dekker, pp. 1–24.

Duescher, R. J., Lawton, M. P., Philpot, R. M., and Elfarra, A. A. (1994). Flavin-containing monooxygenase (FMO)-dependent metabolism of methionine and evidence for FMO3 being the major FMO involved in methionine sulfoxidation in rabbit liver and kidney microsomes. *J. Biol. Chem.*, **269**, 17525–17530.

Dumont, E., von Bahr, C., Perry, Jr., T. L., and Bertilsson, L. (1987). Glucuronidation of the enantiomers of *E*-10-hydroxynortryptiline in human and rat liver microsomes. *Pharmacol. Toxicol.*, **61**, 335–341.

Eichelbaum, M. (1988). Pharmacokinetic and pharmacodynamic consequences of stereoselective drug metabolism in man. *Biochem. Pharmacol.*, **37**, 93–96.

Eichelbaum, M., Bertilsson, L., Kupfer, A., Steiner, E., and Meese, C. O. (1988). Enantioselectivity of 4-hydroxylation in extensive and poor metabolizers of debrisoquine. *Br. J. Clin. Pharmacol.*, **25**, 505–508.

Elfarra, A. A. (1995). Potential role of the flavin-containing monooxygenases in the metabolism of endogenous compounds. *Chem.-Biol. Interact.*, **96**, 47–55.

El Mouelhi, M., Ruelius, H. W., Fenselau, C., and Dulik, D. M. (1987). Species-dependent enantioselective glucuronidation of three 2-arylpropionic acids. *Drug Metab. Disp.*, **15**, 767–772.

Falany, C. N. (1991). Molecular enzymology of human liver cytosolic sulfotransferases. *Trend. Pharmacol. Sci.*, **12**, 255–259.

Fears, R., Baggaley, K. H., Alexander, R., Morgan, B., and Hindley, R. M. (1978). The participation of ethyl 4-benzyloxybenzonate (BRL 10894) and other aryl-substituted acids in glycerolipid metabolism. *J. Lipid Res.*, **19**, 3–11.

Funae, Y. and Imaoka, S. (1987). Purification and characterisation of liver microsomal cytochrome P450 from untreated male rats. *Biochim. Biophys. Acta*, **926**, 349–358.

Gairola, C., Godin, C. S., Houdi, A. A., and Crooks, P. A. (1988). Inhibition of histamine N-methyltransferase activity in guinea pig pulmonary alveolar macrophages by nicotine. *J. Pharm. Pharmacol.*, **40**, 724–726.

Ged, C., Umbenhauer, D. R., Beelew, T. M., Bork, R. W., Sirvastava, P. K., Shinriki, N., Lloyd, R. S., and Guengerich, F. P. (1988). Characterisation of cDNAs, mRNAs and proteins related to human liver microsomal cytochrome P450 (S)-mephenytoin 4′-hydroxylase. *Biochemistry.*, **27**, 6929–6940.

Gibaldi, M. (1993). Stereoselective and isozyme-selective drug interactions. *Chirality*, **5**, 407–413.

Guengerich, F. P. (1992a). Characterisation of human cytochrome P450 enzymes. *FASEB J.*, **6**, 745–748.

Guengerich, F. P. (1992b). Human cytochrome P450 enzymes. *Life Sci.*, **50**, 1471–1478.

Guengerich, F. P. (1994). Catalytic selectivity of human cytochrome P450 enzymes: Relevance to drug metabolism and toxicity. *Toxicol. Lett.*, **70**, 133–138.

Guengerich, F. P., and Shimada, T. (1991). Oxidation of toxic and carcinogenic chemicals by human cytochrome P450 enzymes. *Chem. Res. Toxicol.*, **4**, 391–407.

Guengerich, F. P., Dannan, G. A., Wright, S. T., Martin, M. V., and Kaminsky, L. S. (1982). Purification and characterisation of liver microsomal cytochromes P450: Electrophoretic, spectral, catalytic and immunochemical properties and inducibility of eight isozymes isolated from rats treated with phenobarbital, or β-naphthoflavone. *Biochemistry*, **21**, 6019–6030.

Guengerich, F. P., Gillam, E. M., Ohmori, S., Sandhu, P., Brian, W. R., Sari, M. A., and Iwasaki, M. (1993). Expression of human cytochrome P450 enzymes in yeast and bacteria and relevance to studies on catalytic specificity. *Toxicology.*, **82**, 21–37.

Hadley, M. R., Oldham, H. G., Camilleri, P., Murphy, J., Hutt, A. J., and Damani, L. A. (1993). Stereoselective microsomal N-oxidation of N-ethyl-N-methylaniline. *Biochem. Pharmacol.*, **45**, 1739–1742.

Hadley, M. R., Oldham, H. G., Damani, L. A., and Hutt, A. J. (1994a). Species variability in the stereoselective N-oxidation of N-ethyl-N-methylaniline. *Br. J. Pharmacol.*, **111**, 326P.

Hadley, M. R., Oldham, H. G., Damani, L. A., and Hutt, A. J. (1994b). Asymmetric metabolic N-oxidation of N-ethyl-N-methylaniline by purified flavin-containing monooxygenase. *Chirality*, **6**, 98–104.

Hadley, M. R., Svajdlanka, E., Damani, L. A., Oldham, H. G., Tribe, J., Camilleri, P., and Hutt, A. J. (1994c). Species variability in the stereoselective N-oxidation of pargyline. *Chirality*, **6**, 91–97.

Harrison, I. T., Lewis, B., Nelson, P., Rooks, W., Roszkowski, A., Tomolonis, A., and Fried, J. H. (1970). Non-steroidal anti-inflammatory agents I: 6-substituted 2-naphthylacetic acids. *J. Med. Chem.*, **13**, 203–205.

Hermans, J. J. R., and Thijssen, H. H. W. (1989). The *in vitro* ketone reduction of warfarin and analogues. Substrate stereoselectivity, product stereoselectivity and species differences. *Biochem. Pharmacol.*, **38**, 3365–3370.

Hermans, J. J. R. and Thijssen, H. H. W. (1991). Comparison of the rat liver microsomal metabolism of the enantiomers of warfarin and 4′-nitrowarfarin (Acenocoumarol). *Xenobiotica*, **21**, 295–307.

Hernandez, O., Walker, M., Cox, R. H., Faureman, G. L., Smith, B.R. and Bend, J. R. (1980). Regiospecificity and stereospecificity in the enzymatic conjugation of glutathione with (+)-benzo[a]pyrene 4,5-oxide. *Biochem. Biophys. Res. Comm.*, **96**, 1494–1501.

Hiratsuka, A., Sebata, N., Kawashima, K., Okuda, H., Ogura, K., Watabe, T., Satoh, K., Hatayama, I., Tsuchida, S., Ishikawa, T., and Sato, K. (1990). A new class of rat glutathione S-transferase Yrs-Yrs inactivating reactive sulphate esters as metabolites of carcinogenic arylmethanols. *J. Biol. Chem.*, **265**, 11973–11981.

Hodgson, E. and Levi, P. E. (1992). The role of the flavin-containing monooxygenase (EC 1.14.13.8) in the metabolism and mode of action of agricultural chemicals. *Xenobiotica*, **22**, 1175–1183.

Holland, H. L. (1988). Chiral sulfoxidation by biotransformation of organic sulfides. *Chem. Rev.*, **88**, 473–485.

Holmgrën, A. (1985). Thioredoxin. *Ann. Rev. Biochem.*, **54**, 237–271.

Hsu, M. H., Griffin, K. J., Wang, Y., Kemper, B., and Johnson, E. F. (1993). A single amino acid substitution confers progesterone 6β-hydroxylase activity to rabbit cytochrome P4502C3. *J. Biol.Chem.*, **268**, 6939–6944.

Hutt, A. J., and Caldwell, J. (1983). The metabolic chiral inversion of 2-arylpropionic acids—a novel route with pharmacological consequences. *J. Pharm. Pharmacol.*, **35**, 693–704.

Hutt, A. J. and Caldwell, J. (1984). The importance of stereochemistry in the clinical pharmacokinetics of the 2-arylpropionic acid non-steroidal anti-inflammatory drugs. *Clin. Pharmacokin.*, **9**, 371–373.

Idle, J. R. and Smith, R. L. (1979). Polymorphisms of oxidation at carbon centres of drugs and their clinical significance. *Drug Metab. Rev.*, **9**, 301–317.

Islam, S. A., Wolf, C. R., Lennard, M. S., and Sternberg, M. J. E. (1991). A three-dimensional molecular template for substrates of human cytochrome P450 involved in debrisoquine 4-hydroxylation. *Carcinogenesis.*, **12**, 2211–2219.

Jansson, I., Mole, J., and Schenkman, J. B. (1985). Purification and characterisation of a new form (RLM2) of liver microsomal P450 from untreated rat. *J. Biol. Chem.*, **260**, 7084–7093.

Jenner, P. and Testa, B. (1973). The influence of stereochemical factors on drug disposition. *Drug Metab. Rev.*, **2**, 117–184.

Jerina, D. M., Ziffer, H., and Daly, J. W. (1970). The role of the arene-oxepin system in the metabolism of aromatic substrates. IV. Stereochemical considerations of dihydrodiol formation and dehydrogenation. *J. Am. Chem. Soc.*, **92**, 1056–1061.

Jernström, B., Martinez, M., Meyer, D. J., and Ketterer, B. (1985). Glutathione conjugation of the carcinogenic and mutagenic electrophile ($\pm$)-7β,8α-dihydroxy-9α,10α-oxy-7,8,9,10-tetrahydrobenzo[a]pyrene catalyzed by purified rat liver glutathione transferases. *Carcinogenesis*, **6**, 85–90.

Kaiser, D. G., van Giessen, G. J., Reischer, R. J., and Wechter, W. J. (1976). Isomeric inversion of ibuprofen (*R*)-enantiomer in humans. *J. Pharm. Sci.*, **65**, 269–273.

Kaminsky, L. S., Dunbar, D. A., Wang, P. P., Beaune, P., Larrey, D., Guengerich, F. P., Schnellmann, R. G., and Sipes, I. G. (1984). Human hepatic cytochrome P450 composition as probed by *in vitro* microsomal metabolism of warfarin. *Drug Metab.Disp.*, **12**, 470–476.

Kato, R., Yamazoe, Y., and Yasumori, T. (1992). Polymorphism in stereoselective hydroxylations of mephenytoin and hexobarbital by Japanese liver samples in relation to cytochrome P450 human-2 (IIC9). *Xenobiotica*, **22**, 1083–1092.

Kawano, S., Kamataki, T., Yasumori, T., Yamazoe, Y., and Kato, R. (1987). Purification of human liver cytochrome P450 catalysing testosterone 6β-hydroxylation. *J. Biochem.*, **102**, 493–501.

Kean, W. F., Lock, C. J. L., Rishcke, J., Butt, R., Buchanan, W. W., and Howard-Lock, H. E. (1989). Effect of *R* and *S* enantiomers of naproxen on aggregation and thromboxane production in human platelets. *J. Pharm. Sci.*, **78**, 324–327.

Kean, W. F., Lock, C. J. L., and Howard-Lock, H. E. (1991). Chirality in anti-rheumatic drugs. *Lancet*, **338**, 1565–1568.

Ketterer, B. and Mulder, G. J. (1990). Glutathione conjugation. In *Conjugation Reactions in Drug Metabolism*, G. J. Mulder, (ed.), London: Taylor and Francis Ltd., pp. 307–364.

Kim, M., Shen, D. D., Eddy, A. C., and Nelson, W. L. (1993). Inhibition of the enantioselective oxidative metabolism of metoprolol by verapamil in human liver microsomes. *Drug Metab. Disp.*, **21**, 309–317.

Klus, H. and Kuhn, H. (1977). A study of the optical activity of smoke nicotines. *Fachliche Mitt. Oesterr. Tabokregie*, **17**, 331–336.

Knadler, M.P,. and Hall, S.D. (1990) Stereoselective arylpropionyl-CoA thioester formation, *in vitro*. *Chirality*, **2**, 67–73.

Knights, K. M. and Jones, M. E. (1992). Inhibition kinetics of hepatic microsomal long-chain fatty acid-CoA ligase by 2-arylpropionic acid non-steroidal anti-inflammatory drugs. *Biochem. Pharmacol.* **43**, 1465–1471.

Knights, K. M., Drew, R., and Meffin, P. J. (1989). Enantiospecific formation of fenoprofen coenzyme A thioester *in vitro*. *Biochem. Pharmacol.*, **37**, 3539–3542.

Knights, K. M., Addinall, T. F., and Roberts, B. J. (1991). Enhanced chiral inversion of *R*-ibuprofen in liver from rats treated with clofibric acid. *Biochem. Pharmacol.*, **41**, 1775–1777.

Knihinicki, R. D., Williams, K. M., and Day, R. O. (1989). Chiral inversion of 2-arylpropionic acid non-steroidal anti-inflammatory drugs. I: *In vitro* studies of ibuprofen and flurbiprofen. *Biochem. Pharmacol.*, **38**, 4389–4395.

Knihinicki, R. D., Day, R. O., and Williams, K. M. (1991). Chiral inversion of 2-arylpropionic acid non-steroidal anti-inflammatory drugs. II: Racemisation and hydrolysis of (*R*)- and (*S*)-ibuprofen-CoA thioesters. *Biochem.Pharmacol.*, **42**, 1905–1911.

Komori, M., Hashizume, T., Ohi, H., Miura, T., Kitada, M., Nagashima, K., and Kamataki, T. (1988). Cytochrome P450 in human liver microsomes: High-performance liquid chromatographic isolation of three forms and their characterisation. *J. Biochem.*, **104**, 912–916.

Kroemer, H. K., Mikus, G., Kronbach, T., Mayer, U. A., and Eichelbaum, M. (1989). *In vitro* characterisation of the human cytochrome P450 involved in polymorphic oxidation of propafenone. *Clin. Pharmacol. Ther.*, **45**, 28–33.

Kroemer, H. K., Fischer, C., Meese, C. O., and Eichelbaum, M. (1991). Enantiomer/enantiomer interaction of (*S*)- and (*R*)-propafenone for cytochrome P450IID6-catalysed 5-hydroxylation: *In vitro* evaluation of the mechanism. *Mol. Pharmacol.*, **40**, 135–142.

Kupfer, A. and Preisig, R. (1984). Pharmacogenetics of mephenytoin: A new drug hydroxylation polymorphism in man. *Eur. J. Clin. Pharmacol.*, **26**, 753–759.

Kupfer, A., Roberts, R. K., Schenker, S., and Branch, R. A. (1981). Stereoselective metabolism of mephenytoin in man. *J. Pharmacol. Exp. Ther.*, **218**, 193–199.

Kuzel, R. A., Bhasir, S. K., Oldham, H. G., Damani, L. A., Murphy, J., Camilleri, P., and Hutt, A. J. (1994). Investigations into the chirality of the metabolic sulfoxidation of cimetidine. *Chirality*, **6**, 607–614.

Lan, S. J., Kripalani, K. J., Dean, A. V., Egli, P., Difazio, L. T., and Schreiber, E. C. (1976). Inversion of optical configuration of α-methylfluorene-2-acetic acid (cicloprofen) in rats and monkeys. *Drug Metab.Disp.*, **4**, 330–339.

Lawton, M. P., Cashman, J. R., Cresteil, T., Dolphin, C. T., Elfarra, A. A., Hines, R. N., Hodgson, E., Kimura, T., Ozols, J., Phillips, I. R., Philpot, R. M., Poulsen, L. L., Rettie, A. E., Shephard, E. A., Williams, D. E., and Ziegler, D. M. (1994). A nomenclature system for the mammalian flavin-containing monooxygenase gene family based on amino acid sequence identities. *Arch.Biochem.Biophys.*, **308**, 254–257.

Lee, E. J. D., Williams, K. M., Graham, G. G., Day, R. O., and Champion, G. D., (1984) Liquid chromatographic determination and plasma concentration profile of optical isomers of ibuprofen in humans. *J. Pharm. Sci.*, **73**, 1542–1544.

Lee, E. J. D., Williams, K. M., Day, R., Graham, G., and Champion, D. (1985). Stereoselective disposition of ibuprofen enantiomers in man. *Br. J. Clin. Pharmacol.*, **19**, 669–674.

Lee, M.-Y., Smiley, S., Kadkhodayan, S., Hines, R. N., and Williams, D. E. (1995). Developmental regulation of flavin-containing monooxygenase (FMO) isoforms 1 and 2 in pregnant rabbit. *Chem.-Biol. Interact.*, **96**, 75–85.

Lehmann, P. A., Rodrigues de Miranda, J. F., and Ariëns, E. J. (1976). Stereoselectivity and affinity in molecular pharmacology. *Prog. Drug Res.*, **20**, 101–142.

Lennard, M. S., Crewe, H. K., Tucker, G. T., and Woods, H. F. (1986). Metoprolol oxidation by rat liver microsomes. Inhibition by debrisoquine and other drugs. *Biochem. Pharmacol.*, **35**, 2757–2761.

Lennard, M. S., Tucker, G. T., and Woods, H. F. (1990). Stereoselectivity in pharmacokinetics and drug metabolism. In *Comprehensive Medicinal Chemistry*. Vol. 5: *Biopharmaceutics*, J. B. Taylor. (ed.), Oxford: Pergamon Press, pp. 187–204.

Levin, W., Buening, M. K., Wood, A. W., Chang, R. L., Kedzierski, B., Thakker, D. R., Boyd, D. R., Gadginanath, G. S., Armstrong, R. N., Yagi, H., Karle, J. M., Slaga, T. J., Jerina, D. M., and Conney, A. H. (1980). An enantiomeric interaction in the metabolism and tumorigenicity of (+)- and (–)-benzo[a]pyrene-7,8-oxide. *J. Biol. Chem.*, **255**, 9067–9074.

Lewis, D. F. V. (1995). Three-dimensional models of human and other mammalian microsomal P450s constructed from an alignment with P450102 ($P450_{bm3}$). *Xenobiotica*, **25**, 333–366.

Lewis, R. J., Trager, W. F., Chan, K. K., Breckenridge, A. M., Orme, M.l'E., Rowland, M., and Shary, W. (1974). Warfarin. Stereochemical aspects of its metabolism and the interaction with phenylbutazone. *J. Clin. Invest.*, **53**, 1607–1617.

Light, D. R., Waxman, D. J., and Walsh, C. (1982). Studies on the chirality of sulfoxidation catalysed by bacterial cyclohexane monooxygenase and hog liver flavin adenine dinucleotide containing monooxygenase. *Biochemistry.*, **21**, 2490–2498.

Lomri, N., Yang, Z. and Cashman, J. R. (1993). Regio- and stereoselective oxygenations by adult human liver flavin-containing monooxygenase 3. Comparison with forms 1 and 2. *Chem. Res. Toxicol.*, **6**, 800–807.

Low, L. K. and Castagnoli, N. (1978). Enantioselectivity in drug metabolism. In *Annual Reports in Medicinal Chemistry*. Vol. 13, F. H. Clarke (ed.), New York: Academic Press, pp. 304–315.

Lu, A. Y. H. and Miwa, G. T. (1980). Molecular properties and biological functions of microsomal epoxide hydrase. *Ann. Rev. Pharmacol. Toxicol.*, **20**, 513–531.

Maguire, J. H., Butler, T. C., and Dudley, K. H. (1978). Absolute configuration of (+)-*S*-(3-hydroxyphenyl)-5-phenylhydantoin, the major metabolite of 5,5-diphenylhydantoin in the dog. *J. Med. Chem.*, **21**, 1294–1297.

Mangold, J. B. and Abdel-Monem, M. M. (1983). Stereochemical aspects of conjugation reactions catalysed by rat liver glutathione S-transferase isozymes. *J. Med. Chem.*, **26**, 66–71.

Mannervik, B. and Danielson, H. (1988). Glutathione transferases—structure and catalytic activity. *CRC Crit. Rev. Biochem.*, **23**, 281–334.

Matsumoto, T., Emi, Y., Kawabata, S., and Omura, T. (1986). Purification and characterisation of three male-specific and one female-specific forms of cytochrome P450 from rat liver microsomes. *J. Biochem.*, **100**, 1359–1371.

Mautz, D. S., Nelson, W. L., and Shen, D. D. (1995). Regioselective and stereoselective oxidation of metoprolol and bufuralol catalysed by microsomes containing cDNA-expressed human P4502D6. *Drug Metab. Disp.*, **23**, 513–517.

Mayer, J. M., Bartolucci, C., Maitre, J.-M., and Testa, B. (1988). Metabolic chiral inversion of anti-inflammatory 2-arylpropionates: Lack of reaction in liver homogenates and study of methine proton acidity. *Xenobiotica*, **18**, 533–543.

McKennis, Jr., H., Turnbull, L. B., and Bowman, E. R. (1963). N-Methylation of nicotine and cotinine *in vivo*. *J. Biol. Chem.*, **238**, 719–723.

Meese, C. O., and Eichelbaum, M. (1989). Stereoselective drug metabolism and pharmacodynamic consequences. In *Xenobiotic Metabolism and Disposition*, R. Kato, R. W. Estabrook, and M. N. Cayen (eds.), London: Taylor and Francis, pp.179–184.

Menzel, S., Waibel, R., Brune, K., and Geisslinger G. (1994). Is the formation of *R*-ibuprofenyl-adenylate the first stereoselective step of chiral inversion? *Biochem. Pharmacol.*, **48**, 1056–1058.

Mitchell, S. C., and Waring, R. H. (1989). The fate of cimetidine sulphoxide in the guinea-pig. *Xenobiotica*, **19**, 179–188.

Mitchell, S. C., Idle, J. R., and Smith, R. L. (1982a). The metabolism of [^{14}C]-cimetidine in man. *Xenobiotica*, **12**, 283–292.

Mitchell, S. C., Idle, J. R., and Smith, R. L. (1982b). Reductive metabolism of cimetidine sulphoxide in man. *Drug Metab. Disp.*, **10**, 289–290.

Morgan, D. J. (1990). Clinical pharmacokinetics of β-agonists. *Clin. Pharmacokin.*, **18**, 270–294.

Mulder, G. J, te Koppele, J. M., Schipper, C. G. M., Snel, W. H. M., Pang, K. S., and Polhuijs, M. (1991). Stereoselectivity of glutathione conjugation *in vivo*, in the perfused liver and in isolated hepatocytes. *Drug Metab. Rev.*, **23**, 311–330.

Nakamura, K., Goto, F., Ray, W. A., McAllister, C. B., Jacqz, E., Wilkinson, G. R., and Branch, R.A. (1985). Interethnic differences in genetic polymorphism of debrisoquine and mephenytoin hydroxylation between Japanese and Caucasian populations. *Clin. Pharmacol. Ther.*, **38**, 402–408.

Nebert, D. W., Adesnik, M., Coon, M. J., Estabrook, R. W., Gonzalez, F. J., Guengerich, F. P., Gunsalus, I. C., Johnson, E. F., Kemper, B., Levin, W., Phillips, I.R., Sato, R., and Waterman, M. R. (1987). The P450 gene superfamily: recommended nomenclature. *DNA*, **6**, 1–11.

Nebert, D. W., Nelson, D. R., Coon, M. J., Estabrook, R. W., Feyereisen, R., Fujii-Kuriyama, Y., Gonzalez, F. J., Guengerich, F. P., Gunsalus, I. C., Johnson, E. F., Loper, J. C., Sato, R., Waterman, M. R., and Waxman, D.J. (1991). The P450 gene superfamily: Update on new sequences, gene mapping, and recommended nomenclature. *DNA Cell Biol.*, **10**, 1–14.

Nelson, D. R., Kamataki, T., Waxman, D. J., Guengerich, F. P., Estabrook, R.W., Feyereisen, R., Gonzalez, F. J., Coon, M. J., Gunsalus, I. C., Gotoh, O., Okuda, K., and Nebert, D. W. (1993). The P450 superfamily: Update on new sequences, gene mapping, accession numbers, early trivial names of enzymes and nomenclature. *DNA Cell Biol.*, **12**, 1–51.

Niwa, T., Tokuma, Y., Nakagawa, K., and Noguchi, H. (1989). Stereoselective oxidation of nilvadipine, a new dihydropyridine calcium antagonist, in rat and dog liver. *Drug Metab. Disp.*, **17**, 64–68.

Oesch, F. (1980). Microsomal epoxide hydrolase. In E*nzymatic Basis of Detoxication*, Vol. II, W. B. Jakoby (ed.), New York: Academic Press, pp. 277–290.

Oguri, K., Yamada, H., and Yoshimura, H. (1994). Regiochemistry of cytochrome P450 isozymes. *Ann. Rev. Pharmacol. Toxicol.*, **34**, 251–279.

Okazaki, O., Kurata, T., and Tachizawa, H. (1989). Stereoselective metabolic disposition of enantiomers of ofloxacin in rat. *Xenobiotica*, **19**, 419–429.

Okazaki, O., Kurata, T., Hakwani, H., and Tachizawa, H. (1991). Stereoselective glucuronidation of ofloxacin in rat liver microsomes. *Drug Metab.Disp.*, **19**, 376–380.

Pelkonnen, O. and Breimer, D. D. (1994). Role of environmental factors in the pharmacokinetics of drugs: Considerations with respect to animal models, P450 enzymes and probe drugs. In *Handbook of Experimental Pharmacology*. Vol. 110: *Pharmacokinetics of Drugs*, P. G. Welling, and L. P. Balant (eds.), Berlin: Springer-Verlag, pp.289–332.

Pesola, G. R. and Walle, T. (1993), Stereoselective sulfate conjugation of isoproterenol in humans: Comparison of hepatic, intestinal and platelet activity. Chirality, **5**, 602–609.

Phillips, I. R., Dolphin, C. T., Clair, P., Hadley, M. R., Hutt, A. J., McCombie, R.R., Smith, R. L., and Shephard, E. A. (1995). The molecular biology of the flavin-containing monooxygenases of man. *Chem.Biol. Interact.*, **96**, 17–32.

Pohl, L. R., Bales, R., and Trager, W. F. (1976a). Warfarin: Stereochemical aspects of its metabolism *in vivo* in the rat. *Res.Comm. Chem. Path. Pharmacol.*, **15**, 233–256.

Pohl, L. R., Nelson, S. D., Porter, W. R., Trager, W. F., Fasco, M., Baker, F. D., and Fenton, J. W. (1976b). Warfarin—stereochemical aspects of its metabolism by rat liver microsomes. *Biochem. Pharmacol.*, **25**, 2153–2162.

Polhuijs, M., Meijer, D. K. F., and Mulder, G. J. (1991). The fate of diastereomeric glutathione conjugates of 2-bromoisovalerylurea in blood in the rat *in vivo* and in the perfused liver. Stereoselectivity in biliary and urinary excretion. *J. Pharmacol. Exp. Ther.*, **256**, 458–461.

Pool, W. F., Houdi, A. A., Damani, L. A., Layton, W. J., and Crooks, P. A. (1986). Isolation and characterisation of N-methyl-N′-oxonicotinium ion, a new urinary metabolite of *R*-(+)-nicotine in the guinea pig. *Drug Metab. Disp.*, **14**, 574–579.

Porubek, D. J., Barnes, H., Theodore, L. J., and Baillie, T. A. (1988). Enantioselective synthesis and preliminary metabolic studies of the optical isomers of 2-*n*-propyl-4-pentenoic acid, a hepatotoxic metabolite of valproic acid. *Chem. Res. Toxicol.*, **1**, 343–348.

Poupaert, J. H., Cavalier, R., Claesen, M. H., and Dumont, P. A. (1975). Absolute configuration of the major metabolite of 5,5-diphenylhydantoin, 5-(4′-hydroxyphenyl)-5-phenylhydantoin. *J. Med. Chem.*, **18**, 1268–1271.

Rane, A., Gawronska-Szklarz, B., and Svensson, J.-O. (1985). Natural (–)- and unnatural (+)-enantiomers of morphine: Comparative metabolism and effect of morphine and phenobarbital treatment. *J. Pharmacol. Exp. Ther.*, **234**, 761–765.

Rauws, A. G. and Groen, K. (1994). Current regulatory (draft) guidance on chiral medicinal products: Canada, EEC, Japan, United States. *Chirality*, **6**, 72–75.

Relling, M. V., Aoyama, T., Gonzalez, F. J., and Meyer, U. A. (1990). Tolbutamide and mephenytoin hydroxylation by human cytochrome P450s in the CYP2C subfamily. *J. Pharmacol. Exp. Ther.*, **252**, 442–447.

Rettie, A. E., Bogucki, B. D., Lim, I., and Meier, G. P. (1990). Stereoselective sulfoxidation of a series of alkyl *p*-tolyl sulfides by microsomal and purified flavin-containing monooxygenases. *Mol. Pharmacol.*, **37**, 643–651.

Rettie, A. E., Korzekwa, K. R., Kunze, K. L., Lawrence, R. F., Eddy, A. C., Aoyama, T., Gelboin, H. V., Gonzalez, F. J., and Trager, W. F. (1992). Hydroxylation of warfarin by human cDNA-expressed cytochrome P450: A role for P4502C9 in the aetiology of (*S*)-warfarin-drug interactions. *Chem. Res. Toxicol.*, **5**, 54–59.

Rettie, A. E., Lawton, M. P., Jafar, A., Sadeque, M., Meier, G. P., and Philpot, R. M. (1994). Prochiral sulfoxidation as a probe for multiple forms of the microsomal flavin-containing monooxygenase: Studies with rabbit FMO1, FMO2, FMO3 and FMO5 expressed in *Escherichia coli*. *Arch. Biochem. Biophys.*, **311**, 369–377.

Rettie, A. E., Meier, G. P., and Sadeque A. J., (1995). Prochiral sulfides as *in vitro* probes for multiple forms of the flavin-containing monooxygenase. *Chem. Biol. Interact*, **96**, 3–15.

Ridgewell, R. E. and Abdel-Monem, M.M. (1987). Stereochemical aspects of the glutathione S-transferase-catalysed conjugations of alkyl halides. *Drug Metab. Disp.*, **15**, 82–90.

Robertson, I. G. and Jernström, B. (1986). The enzymatic conjugation of glutathione with bay-region diol-epoxides of benzo[a]pyrene, benz[a]anthracene and chrysene. *Carcinogenesis.*, **7**, 1633–1636.

Robertson, I. G., Guthenberg, C., Mannervik, B., and Jernström, B. (1986a). Differences in stereoselectivity and catalytic efficiency of three human glutathione transferases in the conjugation of glutathione with 7β,8α-dihydroxy-9α,10α-oxy-7,8,9,10-tetrahydrobenzo[a]pyrene. *Cancer Res.*, **46**, 2220–2224.

Robertson, I. G., Jensson, H., Mannervik, B., and Jernström, B. (1986b). Glutathione transferases in rat lung: The presence of transferase 7-7, highly efficient in the conjugation of glutathione with the carcinogenic (+)-7β,8α-dihydroxy-9α,10α-oxy-7,8,9,10-tetrahydrobenzo[a]pyrene. *Carcinogenesis.*, **7**, 295–299.

Roszkowski, A. P., Rooks, W. H., Tomolonie, A. J., and Miller, L.M. (1971). Anti-inflammatory and analgesic properties of *d*-2-(6′-methoxy-2′-naphthyl)propionic acid (Naproxen). *J. Pharmacol. Exp. Ther.*, **179**, 114–123.

Rubin, A., Knadler, M. P., Ho, P. P. K., Bechtol, L. D., and Wolen, R. L. (1985). Stereoselective inversion of (*R*)-fenoprofen to (*S*)-fenoprofen in humans. *J. Pharm. Sci.*, **74**, 82–84.

Ruelius, H. W., Tio, C. O., Knowles, J. A., McHugh, S. L., Schillings, R. T., and Sisenwine, S. F. (1979). Diastereoisomeric glucuronides of oxazepam. Isolation and stereoselective enzymic hydrolysis. *Drug Metab. Disp.*, **7**, 40–43.

Ryan, D. E., Thomas, P. E., Reik, L. M., and Levin, W. (1982). Purification, characterisation and regulation of five rat hepatic microsomal cytochrome P450 isoenzymes. *Xenobiotica*, **12**, 727–744.

Ryan, D. E., Iida, S., Wood, A. W., Thomas, P. E., Lieber, C. S., and Levin, W. (1984). Characterisation of three highly purified cytochromes P450 from hepatic microsomes of adult male rats. *J. Biol. Chem.*, **259**, 1239–1250.

Sadeque, A. J. M., Eddy, A. C., Meier, G. P., and Rettie, A. E. (1992). Stereoselective sulfoxidation by human flavin-containing monooxygenase. Evidence for catalytic diversity between hepatic, renal and fetal forms. *Drug Metab. Disp.*, **20**, 832–839.

Sallustio, B. C., Meffin, P. J., and Knights, K. M. (1988). The stereospecific incorporation of fenoprofen into rat hepatocyte and adipocyte triacylglycerols. *Biochem. Pharmacol.*, **37**, 1919–1923.

Sanins, S. M., Adams, W. J., Kaiser, D. G., Halstead, G. W., Hosley, J., Barnes, H., and Baillie, T. A. (1991). Mechanistic studies on the metabolic chiral inversion of (*R*)-ibuprofen in the rat. *Drug Metab. Disp.*, **19**, 405–410.

Sato, M., and Crooks, P. A. (1985). N-Methylnornicotinium ion, a new *in vivo* metabolite of *R*-(+)-nicotine. *Drug Metab. Dispos.*, **13**, 348–353.

Shieh, W.-R. and Chen, C.-S. (1993). Purification and characterisation of novel "2-arylpropionyl-CoA epimerases" from rat liver cytosol and mitochondria. *J. Biol.Chem.*, **268**, 3487–3493.

Shirley, M. A., Guan, X., Kaiser, D. G., Halstead, G. W., and Baillie, T. A. (1994).

Taurine conjugation of ibuprofen in humans and in rat liver *in vitro*. Relationship to metabolic chiral inversion. *J. Pharmacol. Exp. Ther.*, **269**, 1166–1175.

Siddoway, L. A., Thompson, K. A., McAllister, C.B., Wang, T., Wilkinson, G. R., Roden, D. M., and Woosley, R. L. (1987). Polymorphisms of propafenone metabolism and disposition in man: Clinical and pharmacokinetic consequences, *Circulation*, **75**, 785–791.

Simmonds, R. G., Woodage, T. J., Duff, S. M., and Green, J. N., (1980). Stereospecific inversion of *R*-(–)-benoxaprofen in rat and man. *Eur. J. Drug Metab. Pharmacokin.*, **5**, 169–172.

Sisenwine, S. F., Tio, C. O., Hadley, F. V., Liu, A.-L., Kimmel, H. B., and Ruelius, H.W. (1982). Species-related differences in the stereoselective glucuronidation of oxazepam. *Drug Metab. Disp.*, **10**, 605–608.

Skrede, S. and Halvorsen, O. (1979). Increased biosynthesis of CoA in the liver of rats treated with clofibrate. *Eur. J. Biochem.*, **98**, 223–229.

Soine, W. H., Soine, P. J., Overton, B. W., and Garrettson, L. K. (1986), Product enantioselectivity in the N-glucosylation of amobarbital. *Drug Metab. Disp.*, **14**, 619–621

Sundaram, R. S., Szumlanski, C., Otterness, D., van Loon, J. A., and Weinshilboum, R.M. (1989). Human intestinal phenol sulfotransferase: Assay conditions, activity levels and partial purification of the thermolabile form. *Drug Metab. Disp.*, **17**, 255–264.

Swinney, D. C., Ryan, D. E., Thomas, P. E., and Levin, W. (1987). Regioselective progesterone hydroxylation catalysed by eleven rat hepatic cytochrome P450 isozymes, *Biochemistry*. **26**, 7073–7083.

Tamassia, V., Jannuzzo, M. G., Moro, E., Stegnjaich, S., Groppi, W., and Nicolis, F. B. (1984). Pharmacokinetics of the enantiomers of indoprofen in man. *Int. J. Clin. Pharmacol. Res.*, **4**, 223–230.

Tateishi, M. (1983). Methylthiolated metabolites. *Drug Metab. Rev.*, **14**, 1207–1234.

te Koppele, J. M. and Mulder, G. J. (1991). Stereoselective glutathione conjugation by subcellular fractions and purified glutathione S-transferases. *Drug Metab. Rev.*, **23**, 331–354.

te Koppele, J. M., Dogterom, P., Vermeulen, N. P. E., Meijer, D. K. F., van der Gen, A., and Mulder, G. J. (1986). α-Bromoisovalerylurea as a model substrate for studies on pharmacokinetics of glutathione conjugation in the rat. II. Pharmacokinetics and stereoselectivity of metabolism and excretion *in vivo* and in the perfused liver. *J. Pharmacol. Exp. Ther.*, **239**, 905–914.

Tephly, T. R. (1992). Isolation and purification of UDP-glucuronosyltransferases. *Chem. Res. Toxicol.*, **3**, 509–516.

Testa, B. (1989). Conceptual and mechanistic overview of stereoselective drug metabolism. In *Xenobiotic Metabolism and Disposition*, R. Kato, R.W. Estabrook, and M. N., Cayen (eds.) London: Taylor and Francis, pp.153–160.

Testa, B., and Mayer, J. M. (1988). Stereoselective drug metabolism and its significance in drug research. In *Progress in Drug Research*, Vol. 32, E. Jucker (ed.), Basel: Birkhäuser Verlag, pp. 249–303.

Thijssen, H. H. W., and Baars, L. G. M. (1987). The biliary excretion of acenocoumarol in the rat: Stereochemical aspects: *J. Pharm. Pharmacol.*, **39**, 655–657.

Timms, C., Schladt, L., Robertson, L., Rauch, P., Schramm, H., and Oesch, F. (1987). The regulation of rat liver epoxide hydrolases in relation to that of other drug-metabolising enzymes. In *Drug Metabolism—from Molecules to Man*, D. J. Benford, J. W. Bridges, and G. G. Gibson (eds.), London: Taylor and Francis, pp. 55–69.

Tokuma, Y., Fujiwara, T., Niwa, T., Hashimoto, T., and Naguchi, H. (1989). Stereoselective disposition of nilvadipine, a new dihydropyridine calcium antagonist in the rat and dog. *Res. Comm. Chem. Path. Pharmacol.*, **63**, 249–262.

Tracy, T. S. and Hall, S. D. (1991). Determination of the epimeric composition of ibuprofenyl-CoA. *Anal. Biochem.*, **195**, 24–29.

Tracy, T. S. and Hall, S. D. (1992). Metabolic inversion of (*R*)-ibuprofen: Epimerisation and hydrolysis of ibuprofenyl-coenzyme A. *Drug Metab. Disp.*, **20**, 322–327.

Tracy, T. S., Wirthwein, D. P., and Hall, S. D. (1993). Metabolic inversion of (*R*)-ibuprofen: Formation of ibuprofenyl-coenzyme A. *Drug Metab. Disp.*, **21**, 114–120.

Trager, W. F. and Jones, J. P. (1987). Stereochemical considerations in drug metabolism. In *Progress in Drug Metabolism*, Vol. 10, J. W. Bridges, L. F. Chasseaud, and G. G. Gibson (eds.), London: Taylor and Francis, pp. 55–83.

Umbenhauer, D. R., Martin, M. V., Lloyd, R. S., and Guengerich, F. P. (1987). Cloning and sequence determination of a complementary DNA related to human liver microsomal cytochrome P450 *S*-mephenytoin 4-hydroxylase. *Biochemistry*, **26**, 1094–1099.

van Bladeren, P. J., Sayer, J. M., Ryan, D. E., Thomas, P. E., Levin, W., and Jerina, D.M. (1985). Differential stereoselectivity of cytochromes P450b and P450c in the formation of naphthalene and anthracene 1,2-oxides. The role of epoxide hydrolase in determining the enantiomer composition of the 1,2-dihydrodiols formed. *J. Biol. Chem.*, **260**, 10226–10235.

Vermeulen, N. P. E. (1989). Analysis of mercapturic acids as a tool in biotransformation, biomonitoring and toxicological studies. *Trend. Pharmacol. Sci.*, **10**, 177–181.

Vermeulen, N. P. E., and te Koppele, J. M. (1993). Stereoselective biotransformation. Toxicological consequences and implications. In *Drug Stereochemistry. Analytical Methods and Pharmacology*, 2nd ed., I. W. Wainer (ed.), New York: Marcel Dekker, pp. 245–280.

Walle, T. and Walle, U. K. (1990). Stereoselective sulphate conjugation of racemic terbutaline by human liver cytosol. *Br. J. Clin. Pharmacol.*, **30**, 127–133.

Walle, T. and Walle, U. K. (1991). Stereoselective sulfate conjugation of racemic 4-hydroxypropranolol by human and rat liver cytosol. *Drug Metab. Disp.*, **19**, 448–453.

Walle, T. and Walle, U. K. (1992). Stereoselective sulfate conjugation of 4-hydroxypropranolol and terbutaline by the human liver phenol-sulfotransferases. *Drug Metab. Disp.*, **20**, 333–336.

Walle, T., Wilson, M. J., Walle, U. K., and Bai, S. A. (1983). Stereochemical composition of propranolol metabolites in the dog using stable isotope-labelled pseudo-racemates. *Drug Metab. Disp.*, **11**, 544–549.

Walle, T, Walle, U. K., Wilson, M. J., Fagan, T. C., and Gaffney, T. E. (1984). Stereoselective ring oxidation of propranolol in man. *Br. J. Clin. Pharmacol.*, **18**, 741–747.

Walle, T., Walle, U. K., Thornburg, K. R., and Schey, K. L. (1993). Stereoselective sulfation of albuterol in humans. Biosynthesis of the sulfate conjugate by HEP G2 cells. *Drug Metab. Disp.*, **21**, 76–80.

Walle, U. K. and Walle, T. (1989). Stereoselective sulfation of terbutaline by the rat liver cytosol: Evaluation of experimental approaches. *Chirality*, **1**, 121–126.

Watabe, T., Ozawa, N., and Hiratsuka, A. (1983). Studies on metabolism and toxicity of styrene. VI. Regioselectivity in glutathione S-conjugation and hydrolysis of racemic, *R*- and *S*-phenyloxiranes in rat liver. *Biochem. Pharmacol.*, **32**, 777–785.

Watabe, T., Hiratsuka, A., and Tsurumori, T. (1984). Regiospecific and diastereoselective inactivation of mutagenic 9,10-dihydrobenz[a]pyrene 7,8-oxide by hepatic cytosolic glutathione S-transferase. *Biochem. Pharmacol.*, **33**, 4051–4056.

Watabe, T., Hiratsuka, A., and Tsurumori, T. (1985). Enantioselectivity in glutathione conjugation of 1,2-epoxy-1,2,3,4-tetrahydronaphthalene by hepatic glutathione S-transferase. *Biochem. Biophys. Res. Comm.*, **130**, 65–70.

Waxman, D. J. (1988). Interactions of hepatic cytochromes P450 with steroid hormones. Regioselectivity and stereospecificity of steroid metabolism and hormonal regulation of rat P450 enzyme expression. *Biochem. Pharmacol.*, **37**, 71–84.

Waxman, D. J., Light, D. R., and Walsh, C. (1982). Chiral sulfoxidations catalysed by rat liver cytochromes P450. *Biochemistry.*, **21**, 2499–2507.

Wechter, W. J., Loughhead, D. G., Reischer, R. J., Van Giessen, G. J., and Kaiser, D. G. (1974). Enzymatic inversion at saturated carbon: Nature and mechanism of the inversion of (*R*)-(–)-*p*-isobutylhydratropic acid. *Biochem. Biophys. Res. Comm.*, **61**, 833–837.

Wedlund, P. J. Aslanian, W. S., Jacqz, E., McAllister, C.B., Branch, R. A., and Wilkinson, G. R. (1985). Phenotypic differences in mephenytoin pharmacokinetics in normal subjects. *J. Pharmacol. Exp. Ther.*, **234**, 662–669.

Weerawarna, S. A., Geisshüssler, S. M., Murthy, S. S., and Nelson, W. L. (1991). Enantioselective and diastereoselective hydroxylation of bufuralol. Absolute configurations of the 7-(1-hydroxyethyl)-2-[1-hydroxy-2-(*tert*-butylamino)ethyl]benzofurans, the benzylic hydroxylation metabolites. *J. Med. Chem.*, **34**, 3091–3097.

Williams, D. E., Shiganaga, M. K., and Castagnoli Jr., N. (1990). The role of cytochromes P-450 and flavin-containing monooxygenase in the metabolism of (*S*)-nicotine by rabbit lung. *Drug Metab. Disp.*, **18**, 418–428.

Williams, K. M., and Day, R. O. (1985). Stereoselective disposition—Basis for variability in response to NSAID's. *Agents and Actions*, **17**, 119–126.

Williams, K., Day, R., Knihinicki, R., and Duffield, A. (1986). The stereoselective uptake of ibuprofen enantiomers into adipose tissue. *Biochem. Pharmacol.*, **35**, 3403–3405.

Williams, R. T. (1959). *Detoxication Mechanisms. The Metabolism and Disposition of Drugs, Toxic Substances and Other Organic Compounds*, London: Chapman and Hall.

Wrighton, S. A., Ring, B. J., Watkins, P. B., and VandenBranden, M. (1989). Identification of a polymorphically expressed member of the human cytochrome P450III family. *Mol. Pharmacol.*, **36**, 97–105.

Yang, S. K., Huang, A., and Huang, J. D. (1993). Enantioselectivity of esterases in human brain. *Chirality*, **5**, 565–568.

Yost, G. S., and Finley, B. L. (1985). Stereoselective glucuronidation as a probe of induced forms of UDP-glucuronyltransferase in rabbits. *Drug Metab. Disp.*, **13**, 5–8.

Ziegler, D. M. (1988). Flavin-containing monooxygenases: catalytic mechanism and substrate specificities. *Drug Metab. Rev.*, **19**, 1–32.

Ziegler, D. M. (1993). Recent studies on the structure and function of multisubstrate flavin-containing monooxygenases, *Ann. Rev. Pharmacol. Toxicol.*, **33**, 179–199.

Ziegler, D. M. and Mitchell, C. H. (1972). Microsomal oxidase IV: Properties of a mixed-function amine oxidase isolated from pig liver microsomes. *Arch. Biochem. Biophys.*, **150**, 116–125.

Ziegler, D. M., and Poulsen, L. L. (1978). Hepatic microsomal mixed-function amine oxidase. In *Methods in Enzymology*, Vol. 52, S. Fleischer and L. Packer (eds.), New York: Academic Press, pp.142–151.

CHAPTER

4

SOME EXAMPLES FOR STEREOSELECTIVE BIOTRANSFORMATION OF DRUGS

GOTTFRIED BLASCHKE

Institute of Pharmaceutical Chemistry
University of Münster
D-48419, Münster, Germany

4.1. INTRODUCTION

According to Professor E. J. Ariëns, who had recognized very early the importance of stereoselectivity in pharmacology, it is "scientific nonsense" to investigate the metabolism of chiral drugs without consideration of stereoselective aspects. Indeed, the metabolism of numerous drugs is highly stereoselective, depending on the type of metabolic conversion. In this chapter, some methods used in this research are demonstrated with examples investigated in our group.

4.2. METHODS FOR THE INVESTIGATION OF STEREOSELECTIVE BIOTRANSFORMATIONS

Two different methods can be used in principle. The simplest procedure is to investigate the metabolism of both optically pure enantiomers in separate experiments. Provided that no racemization or interconversion of enantiomers occurs, the amounts of unchanged drug and its metabolites can be analyzed directly for each enantiomer by chromatography on achiral phases, avoiding expensive chiral stationary phases. But experiments with three samples —both enantiomers and for comparison with the racemate—under exactly the same conditions are necessary in order to be able to evaluate the data.

The more common procedure is the application of the racemic drug and subsequent chiral analysis of nonmetabolized drug and its metabolites. This

The Impact of Stereochemistry on Drug Development and Use, Edited by Hassan Y. Aboul-Enein and Irving W. Wainer. Chemical Analysis Series, Vol. 142.
ISBN 0-471-59644-2

assay can be done after derivatization of the enantiomers with optically active reagents to diastereomers that are separable on achiral stationary phases. The most popular assay, however, is the direct chiral separation of the drug and its metabolites on one of the numerous chiral stationary phases (CSPs) by gas chromatography or HPLC. Recently, capillary electrophoresis (CE) in the presence of chiral additives also proved to be an efficient method for direct enantioselective assays in metabolism studies.

4.3. INVESTIGATIONS WITH ENANTIOMERS: PRAZIQUANTEL AND MEFLOQUINE

Praziquantel (the structure in Figure 4.1, H instead of OH), the most important antischistosomal drug, is used as the racemate, although the *R*(–)-enantiomer proved to be at least 10^5 times more effective than the respective *S*(+)-enantiomer (1). Pure enantiomers of praziquantel were obtained easily from the racemate on a preparative scale by low-pressure liquid chromatography on microcrystalline cellulose triacetate as the chiral stationary phase (2).

Both enantiomers and the racemate were applied to rats. Serum samples were analyzed for unmetabolized praziquantel and its main metabolites in rats, *cis*- and *trans*-4-hydroxypraziquantel, by HPLC on an ODS column (3).

In these chromatograms, 30 minutes after the oral application of *S*(+)- and R(–)-praziquantel, the peaks for nonmetabolized praziquantel (PZQ), *trans*-4-hydroxy praziquantel (trans) and *cis*-4-hydroxy praziquantel (cis), were quantified. As an Internal Standard (IS), a praziquantel analogue with cylopentyl instead of the cyclohexyl substituent was used.

R(–)-praziquantel is much faster metabolized. Correspondingly, much more *cis*- and *trans*-4-hydroxy praziquantel is formed from *R*(–)-praziquantel. Additionally, the ratio between the *cis* and *trans* hydroxy isomer is different in both experiments. After application of *R*(–)-praziquantel, this ratio is almost 1:1, whereas from *S*(+)-praziquantel, much less *trans*-isomer is formed. Furthermore, this ratio of *trans*/*cis* is time-dependent, increasing with the time interval after application of the enantiomers. As expected, the racemate is metabolized to 4-hydroxy praziquantel with a *trans*/*cis* ratio between the figures of both enantiomers. In these experiments, the praziquantel enantiomers did not racemize or were interconverted.

The antimalarial drug mefloquine (Figure 4.2) is also used as a racemic drug. The biotransformation *in vitro* was investigated using rat liver microsomes (4).

Incubation of both enantiomers and also as a control of the racemate with rat microsomes revealed again the enantioselective formation of the main metabolite, a carboxylic acid (Figure 4.2). (–)-Mefloquine is the better substrate for liver cell fractions compared to the (+)-enantiomer. From the race-

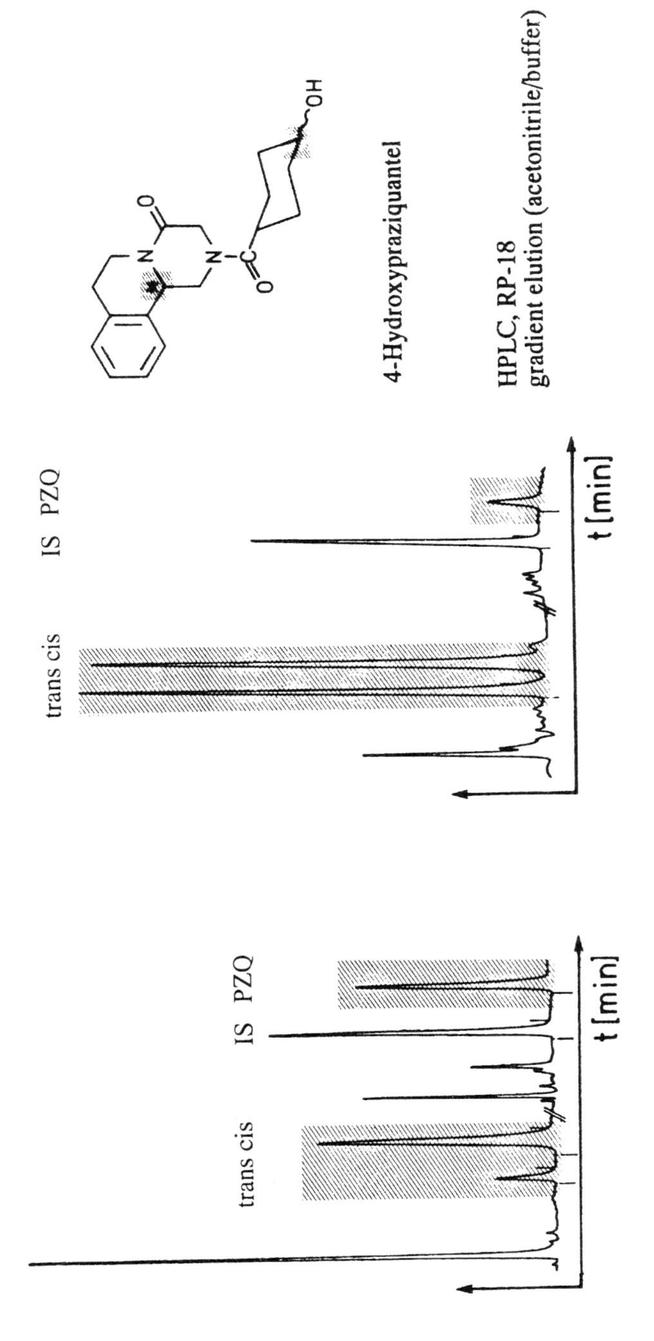

Figure 4.1. HPLC chromatograms of rat serum samples, 30 minutes after the oral application of *S*(+)- and R(−)-praziquantel.

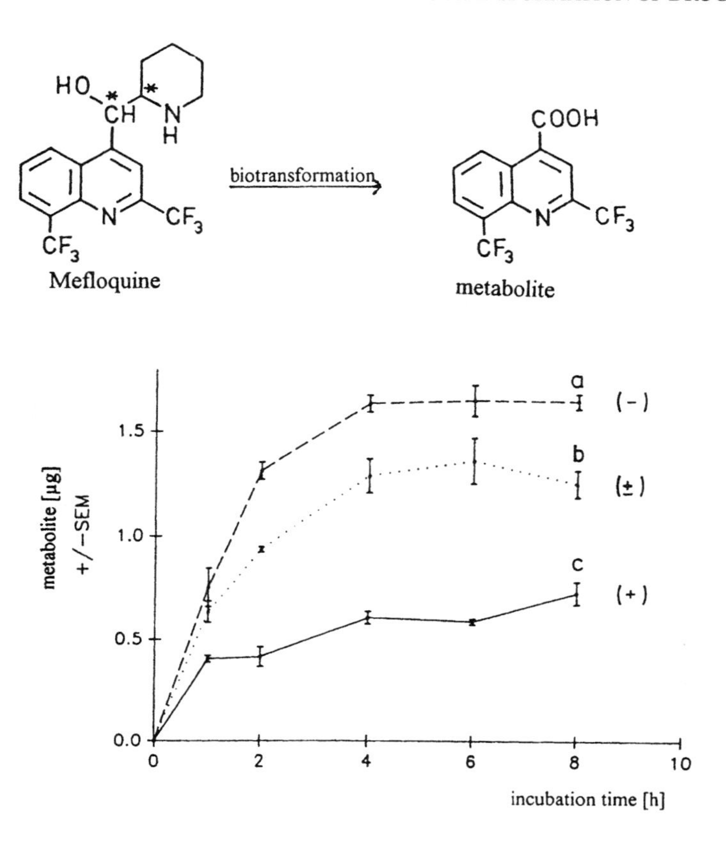

Figure 4.2. Formation of the achiral main metabolite of mefloquine.Incubation of (*a*) (−)-, (*b*) (±)-, and (*c*) (+)-mefloquine with rat liver microsomes. HPLC, RP-18 silica.

mate, the amounts of the metabolites are, as expected, again between both values for the enantiomers. The assay was performed by HPLC on achiral ODS silica. The configurational stability of both mefloquine enantiomers during the incubation experiment was secured in additional investigations by chiral analyses.

4.4. CHIRAL ASSAYS AFTER CHIRAL DERIVATIZATION: MEFLOQUINE AND CHLORPHENOXAMINE

In order to analyze the enantiomers of mefloquine in plasma after application of the racemic drug to human volunteers, another approach was used. The secondary amino group of mefloquine was reacted with optically active naphthyl ethyl isocyanate to the diastereomeric urea derivatives that are separable by HPLC on silica gel. Using a larger excess of reagent resulted in the formation of two additional peaks in the chromatograms, probably due to a second derivatization step at the hydroxy group. This assay was adapted for the chiral assay of mefloquine in human plasma (4). From 27 human plasma samples after oral application of the racemic drug, almost all samples showed a much higher amount of (–)-mefloquine, the ratios of (–):(+)-mefloquine being in the range of 1.7:11. Only in the serum samples of one patient was almost racemic mefloquine analyzed. Typical chromatograms are shown in Figure 4.3. In the chromatograms of blank serum, no interfering peaks were observed.

A direct separation of mefloquine on protein columns as CSPs by HPLC was also possible, but very broad peaks were observed that did not allow the chiral assay in plasma samples. A second example for chiral derivatizations is the antihistamine chlorphenoxamine (Figure 4.4). The enantiomers were

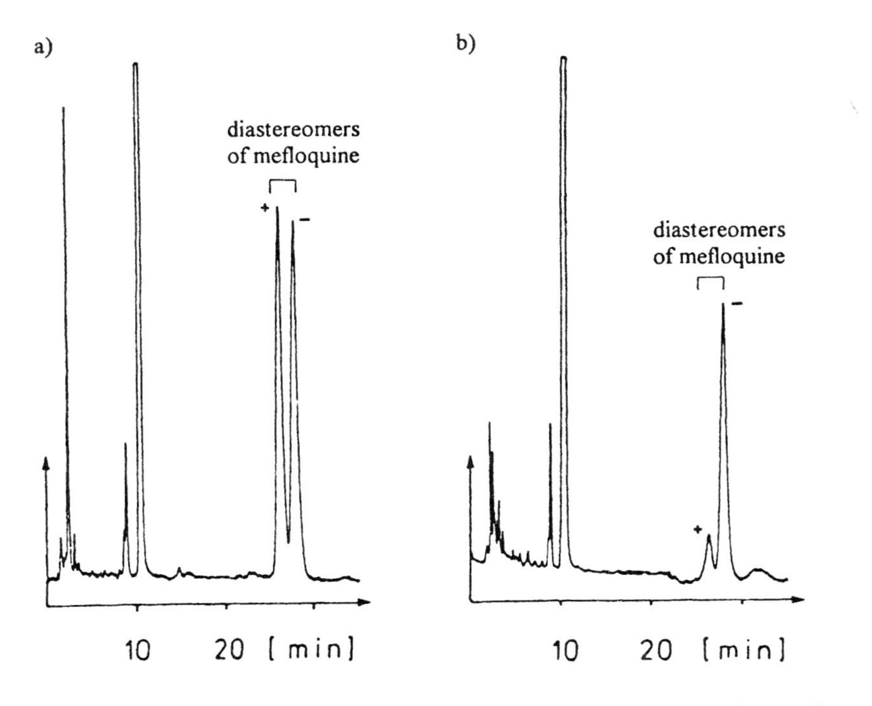

Figure 4.3. (*a*) Blank plasma, spiked with racemic mefloquine. (*b*) Plasma sample of one patient after the oral application of mefloquine. Each sample was derivatized with *S*(+)-1-(1-naphthyl)-ethyl isocyanate. HPLC chromatograms, separations on silica gel (*n*-hexane/ethyl acetate 3:1).

obtained by fractional crystallization salts of the racemic base with (+)- and (–)-dibenzoyltartaric acid. All attempts to determine the enantiomeric purity by HPLC on different chiral stationary phases failed because no analytical separation was obtained. Therefore, chiral derivatization was used again: as the first step, the tertiary amine chlorphenoxamine was N-demethylated by reaction with a chloroformate. The resulting secondary amine (the structure in Figure 4.4, $NHCH_3$ instead of $N(CH_3)_2$) was derivatized with optically active naphthyl ethyl isocyanate, forming two diastereomeric urea derivatives

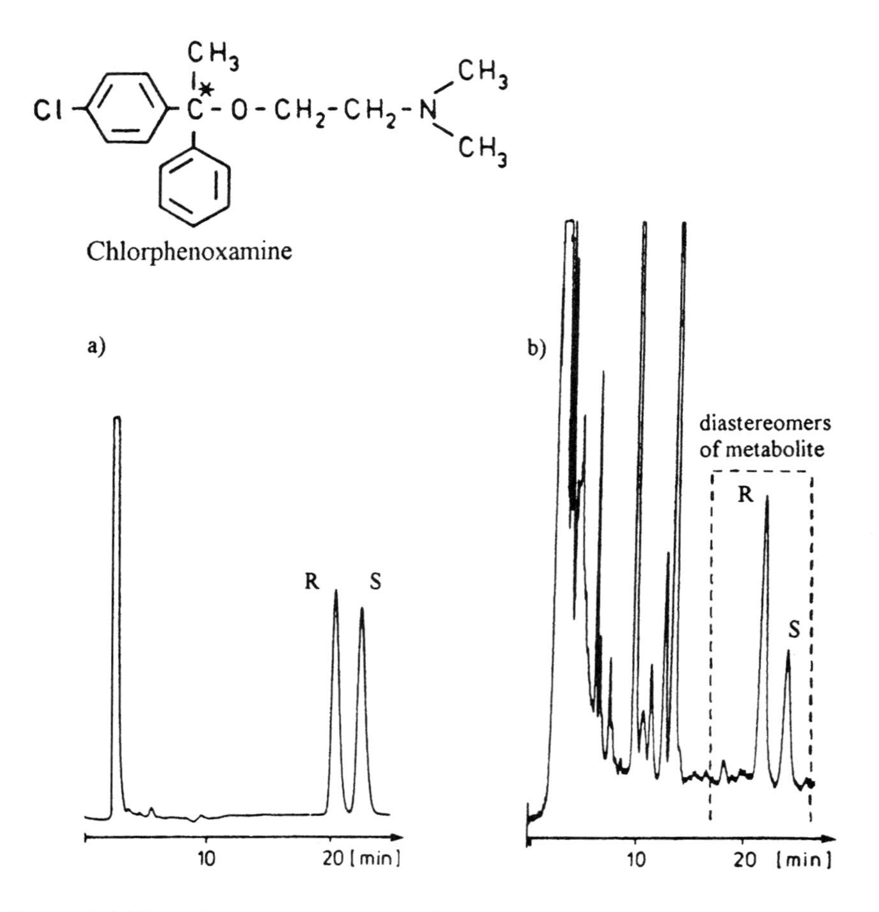

Figure 4.4. (*a*) Racemic chlorphenoxamine. (*b*) Urine sample 2–4 hours after the oral application of racemic chlorphenoxamine and extraction of the metabolite N-demethyl chlorphenoxamine [$NH(CH_3)$ instead of $N(CH_3)_2$]. Both samples were derivatized with R(–)-1-(1-naphthyl)ethyl isocyanate and separated by HPLC on silica gel.

that were separated by HPLC on silica gel. In Figure 4.4*a*, the chromatogram of N-demethylated and derivatized racemic chlorphenoxamine is shown. The two diastereomers are baseline separated. Both enantiomers of chlorphenoxamine showed after demethylation and derivatization only one of the corresponding peaks indicating enantiomeric purity.

N-demethyl chlorphenoxamine also is the main chiral metabolite of chlorphenoxamine. For the chiral assay in urine samples after oral application of the racemic drug to human volunteers, urine samples were collected, extracted, and derivatized directly with *R*-(–)-1-naphthyl ethyl isocyanate. In human urine, higher amounts of *R*-N-demethyl chlorphenoxamine were always found, indicating a stereoselective metabolism (Figure 4.4*b*). The enantiomeric ratio of the metabolite ratio again is time-dependent. The ratios R/S are much higher in the first time intervals after application than in the later intervals (5).

Chiral derivatizations are used if a direct chiral resolution on CSPs is not possible on chiral columns or a chiral column is not available in the laboratory. An advantage can sometimes be the introduction of a chromophor like the naphthyl group with a high absorption coefficient and/or fluorescent properties, enabling a more sensitive and selective detection. The most serious disadvantages of the derivatization method are time-consuming derivatization steps and side reactions during derivatization. Furthermore, the accuracy of the assay depends on the enantiomeric purity of the reagent.

4.5. DIRECT ENANTIOMERIC SEPARATIONS ON CHIRAL STATIONARY PHASES

With the availability of numerous chiral stationary phases, direct enantiomeric separations by HPLC are now widely used for chiral assays of drugs and metabolites also in biological samples. More than 100 chiral stationary phases are commercially available, and several books on the application of chiral stationary phases have been published. One disadvantage of this method might be the high cost of these columns. Their stability is sometimes limited, the range of application of a column is often limited. Therefore, at least several different chiral stationary phases must be available for initial separation experiments. Quite often, these columns enable only enantiomeric separations, but not the separation of chemical species like the different metabolites. Then an achiral separation has to be combined with the chiral column either by collecting fractions from the achiral column and analyzing on the chiral one, or by column switching or column coupling.

Fortunately, it is sometimes possible to separate both the different metabolites as well as their enantiomers simultaneously on a chiral stationary phase.

4.6. CHIRAL ASSAY OF DRUGS AND METABOLITES: ETODOLAC

Etodolac, a nonsteroidal antiinflammatory drug, is rapidly metabolized to its acyl glucuronide and hydroxylated metabolites (Figure 4.5) that are also partially excreted as glucuronides. Etodolac was derivatized to diastereomers that are separable on silica gel. Also, direct separations on proteins like cross-linked bovine serum albumine or commercially available protein columns were obtained (6). In order to avoid derivatization steps, direct enantiomeric resolutions were used for the assays of etodolac and its metabolites in human urine after oral application of the drug. Etodolac was present in the urine only in the conjugated form as acyl glucuronide, the ester of etodolac with glucuronic acid. This ester was hydrolyzed rapidly after the addition of sodium hydroxide. After acidification, etodolac was extracted selectively with an unpolar organic solvent mixture, and the enantiomeric composition of etodolac could be determined directly on a chiral stationary phase. Small amounts of the more polar hydroxylated metabolites extracted together with etodolac

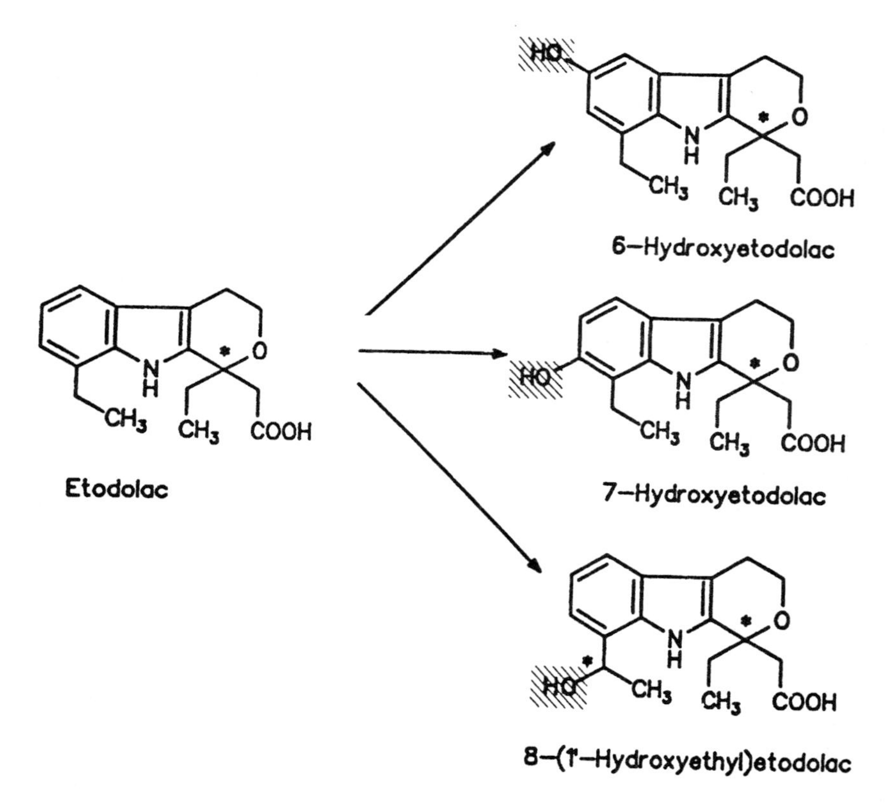

Figure 4.5. Structure of the analgesic etodolac and its three monohydroxylated metabolites.

were eluted with the solvent front and did not interfere with the chiral assay. In the first time intervals after oral application, the conjugated etodolac was excreted preferentially as *S*(+)-enantiomer, but in later intervals this stereoselectivity was reversed.

The more polar hydroxylated metabolites of etodolac (Figure 4.5) were present partially as free acids and partially as conjugates with glucuronic acid. The nonconjugated metabolites were extracted with a polar solvent and analyzed by HPLC on an ODS column (Figure 4.6, left). The very polar glu-

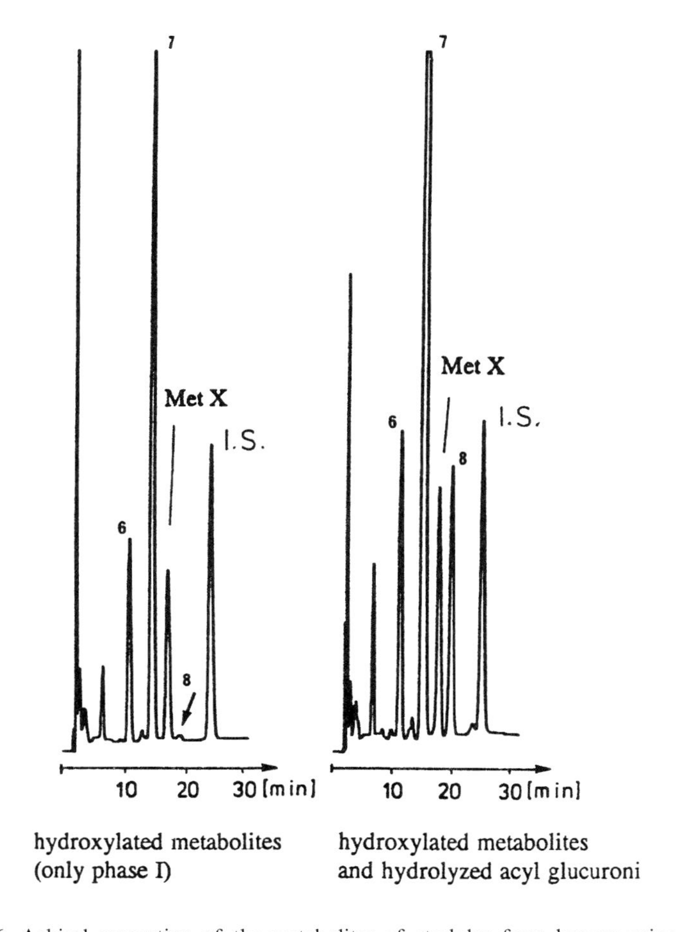

Figure 4.6. Achiral separation of the metabolites of etodolac from human urine, the time interval is 4–6 hours after oral application of 1 tablet of Lodine = racemic etodolac. Left: before deconjugation. Right: after deconjugation. HPLC column: LiChrospher 100 RP-18, guard column LiChrospher 60 CN, mobile phase 0.05 *M* phosphate buffer (pH 4.0)/acetonitrile 73:27. 6: 6-hydroxy etodolac. 7: 7-hydroxy etodolac. 8: 8-(1′-hydroxy etodolac). IS: Internal Standard suprofen. Met X: unknown metabolite.

curonides remained in the aqueous phase. In a parallel experiment, the aqueous phases were hydrolyzed, extracted, and analyzed again by HPLC (Figure 4.6, right). The peak areas after deconjugation now were much higher in comparison to the extraction step without hydrolysis. The differences of peak areas in both chromatograms correspond to the amount of free carboxylic acids liberated from the ester glucuronides by hydrolysis.

A direct chiral assay of this complex mixture of hydroxy isomers on chiral stationary phases was not possible due to peak overlapping of the enantiomers. Therefore, the metabolites were collected after achiral separation and each fraction was separately analyzed on a chiral stationary phase for enantiomeric composition. The 8-(1′-hydroxyethyl) metabolite with two asymmetric centers can exist in four optically active forms. Accordingly, four peaks with a synthetic sample of this metabolite are observed in the chromatogram on a chiral stationary phase. In human urine samples, however, mainly one of the components was detected after hydrolysis of the glucuronide, indicating high enantioselectivity and diastereoselectivity in the formation of this metabolite. Also, the formation of the other metabolites proved to be highly enantioselective, depending on the conjugation and time interval (6).

In this study, the amounts of metabolites and their enantiomeric composition proved to be very similar in all five human volunteers.

4.7. TRAMADOL: INTERINDIVIDUAL DIFFERENCES IN THE METABOLISM AND ASSAY OF GLUCURONIDES

Tramadol, an analgesic drug with two asymmetric centers, is used in therapy as the racemic Z-diastereomer. As the principal phase I metabolites, O- and N-demethyl tramadol were determined (Figure 4.7). The separation of these two metabolites and tramadol itself together with the Internal Standard, the ethoxy analogue of tramadol, was achieved on a reversed-phase column (7).

The cumulative excretion curves of tramadol and the O- and N-demethyl tramadol of five volunteers after the oral administration of 100 mg of racemic tramadol hydrochloride were determined. Four of the five volunteers exhibited very similar metabolic behavior, as shown for volunteer A in Figure 4.7. Only slight differences between all five volunteers were observed in the cumulative excretion of tramadol and N-demethyl tramadol. In contrast, the fifth volunteer *E* excreted only very small amounts of the phenolic metabolite, predominantly as a (−)-enantiomer. In contrast, the other four volunteers eliminated predominantly the (+)-enantiomer of O-demethyl tramadol.

The O-demethylated and N,O-didemethylated tramadol metabolites are partially present as ether glucuronides (Figure 4.8), where the phenolic OH group is covalently bound to optically active glucuronic acid. The diastere-

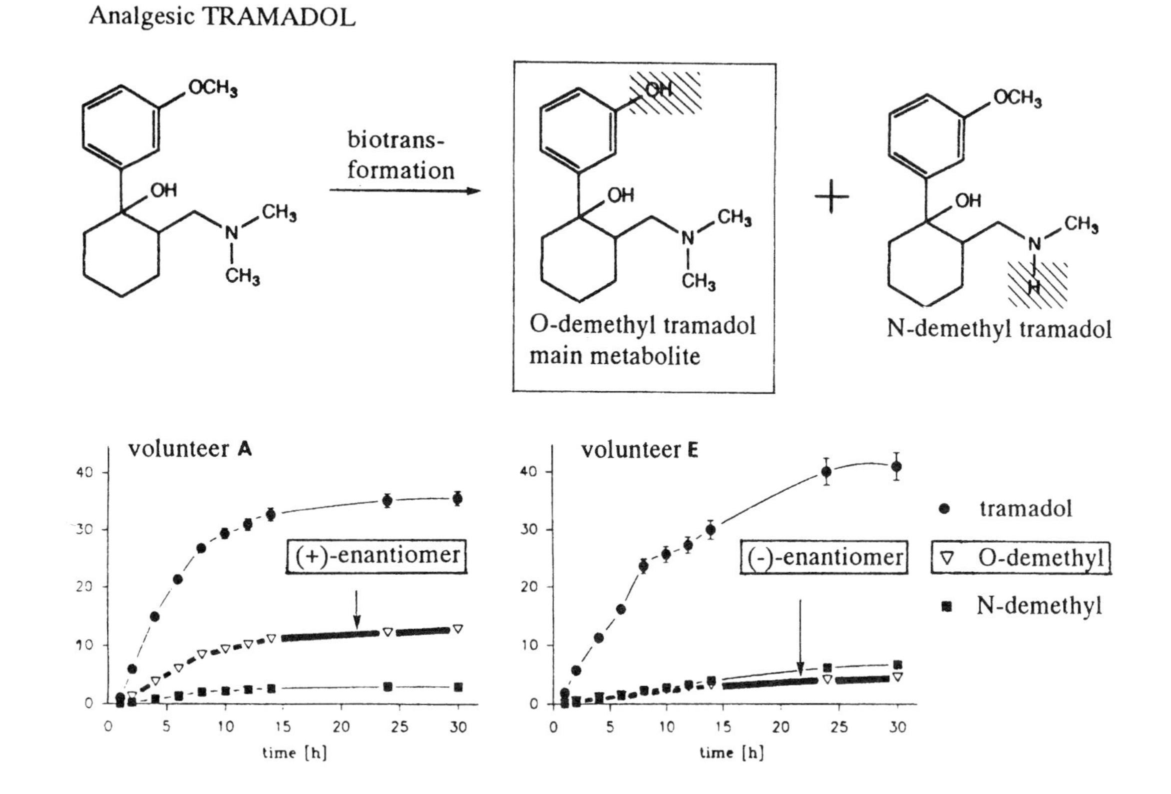

Figure 4.7. Structures of tramadol and its metabolites O-demethyl and N-demethyl tramadol. Cumulative excretion (mg) of tramadol, and O-demethyl and N-demethyl tramadol after the oral application of 100 mg of racemate, comparison of two volunteers.

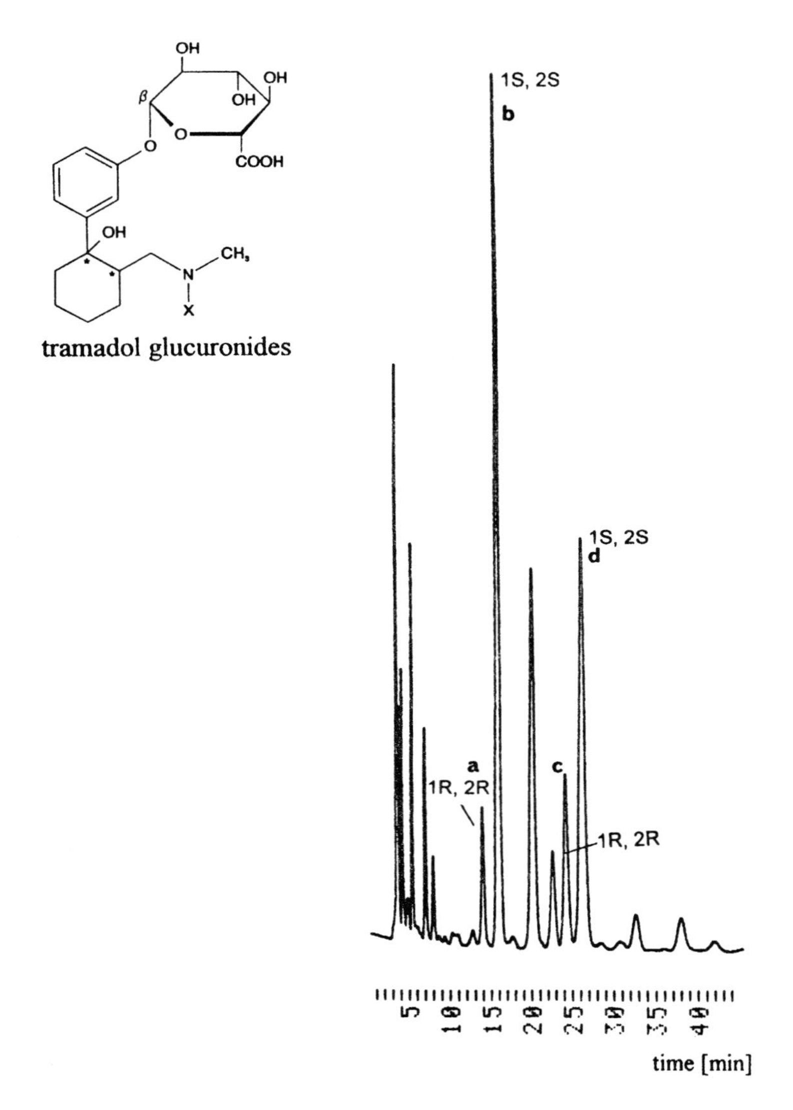

Figure 4.8. Assay of diastereomeric O-demethyl tramadol glucuronides (*a* and *b*: X= CH_3) and N,O-demethyl tramadol glucuronides (*c* and *d*: X= H) in human urine after the oral application of 100 mg of racemic tramadol. HPLC on silica gel, fluorimetric detection, λ_{em}= 300 nm.

omers thus formed can be analyzed in human urine after solid phase extraction directly by chromatography on the achiral stationary phase silica gel (Figure 4.8). 1*R*, 2*R*- and 1*S*, 2*S*-O-demethyl tramadol glucuronide are eluted first, followed by 1*R*, 2*R*- and 1*S*, 2*S*-N,O-didemethyl tramadol glucuronide. As a detection method, the very sensitive and specific fluorescence detection was used. Other peaks in the chromatogram are fluorescent constituents from urine and unidentified metabolites of tramadol (8).

4.8. CHIRAL ASSAYS BY CAPILLARY ELECTROPHORESIS

During the last few years, the new technique of capillary electrophoresis has become a powerful tool for the separation of drugs and its metabolites. Enantiomeric separations of chiral drugs can be simultaneously achieved easily using chiral additives like cyclodextrins. Some basic chiral drugs completely separated into the enantiomers by β-cyclodextrin and/or methyl, hydroxyethyl, and hydroxypropyl β-cyclodextrin (9), are listed in Table 4.1.

Important for the use of capillary electrophoresis in chiral assays of biotransformations is the very high separation efficiency, enabling the simultaneous separation also of metabolites of a similar structure and their enantiomers. In Figure 4.9*a*, such a separation of the antihistaminic drug dimethindene and two of its nonphenolic metabolites as racemates by capillary electrophoresis is demonstrated (10). The separation takes place in less than 12 minutes. No expensive chiral columns are necessary. This method was used to evaluate the enantioselective metabolism of dimethindene, *in vivo*, in rats and human volunteers [11, 12]. Figure 4.9*b* compares the results of a chiral assay of dimethindene by HPLC and capillary electrophoresis after oral application. A strongly time-dependent change of the enantiomeric ratios is observed with both volunteers. In the first time interval, (+)-dimethindene is

Table 4.1. Baseline Resolutions of Racemic Drugs by Capillary Electrophoresis Using β-Cyclodextrin Derivatives as Chiral Additives

Ambucetamide	Carvedilol	Clenbuterol
Ephedrine	Etilefrine	Imafen
Isoprenaline	Ketamine	Lofexidine
Mefloquine	Methylephedrine	Metomidate
Mianserin	Nefopam	Nomifensine
Norephedrine	Norfenefrine	Octopamine
Pholedrine	Salbutamol	Sotalol
Synephrine	Zopiclone	

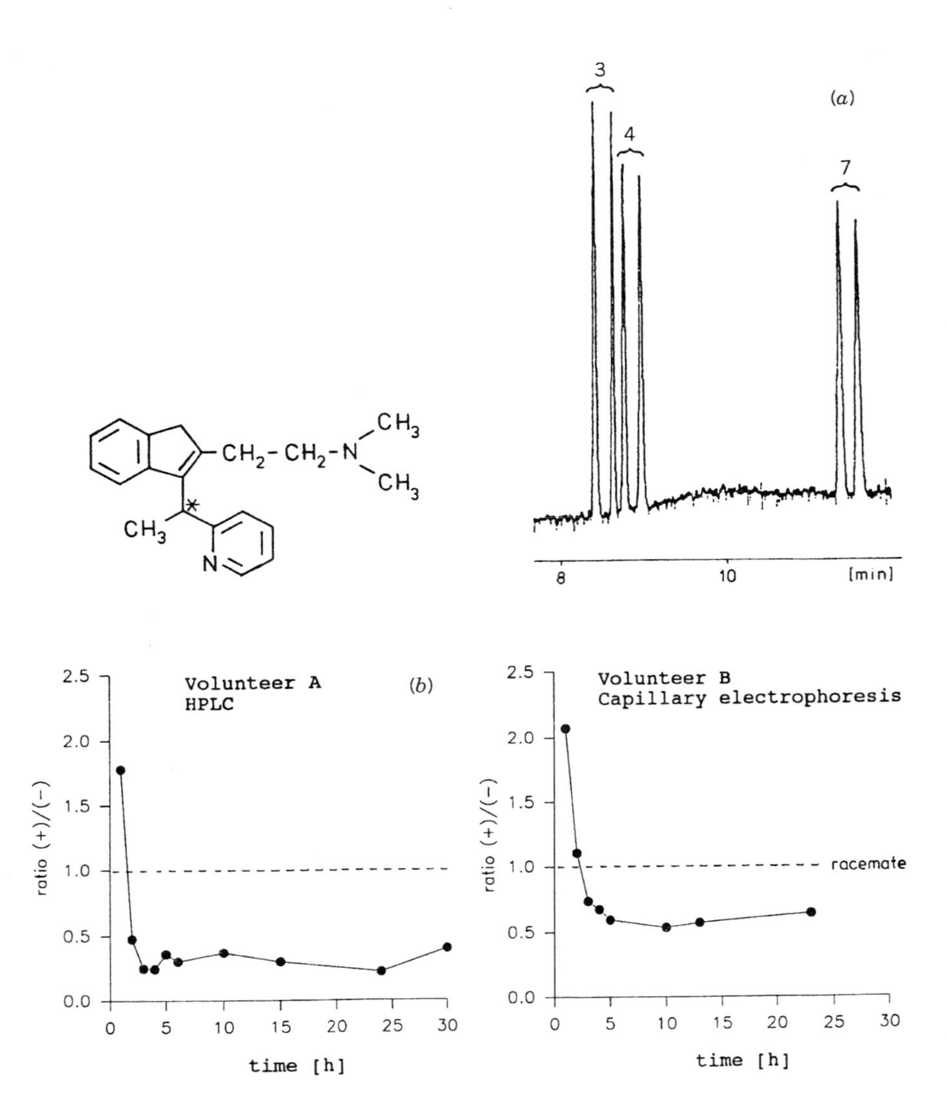

Figure 4.9. (*a*) Separation and additional chiral separation of a mixture of racemic dimethindene (3), N-demethyl dimethindene (4), and 6-methoxy dimethindene (7). Capillary electrophoresis, hydroxypropyl β-cyclodextrin as chiral additive. (*b*) Enantiomeric ratios of dimethindene in human urine after the oral application of 4 mg of racemic dimethindene. Volunteer A: determined by HPLC using a chiral stationary phase. Volunteer B: determined by capillary electrophoresis using hydroxypropyl β-cyclodextrin as the chiral additive.

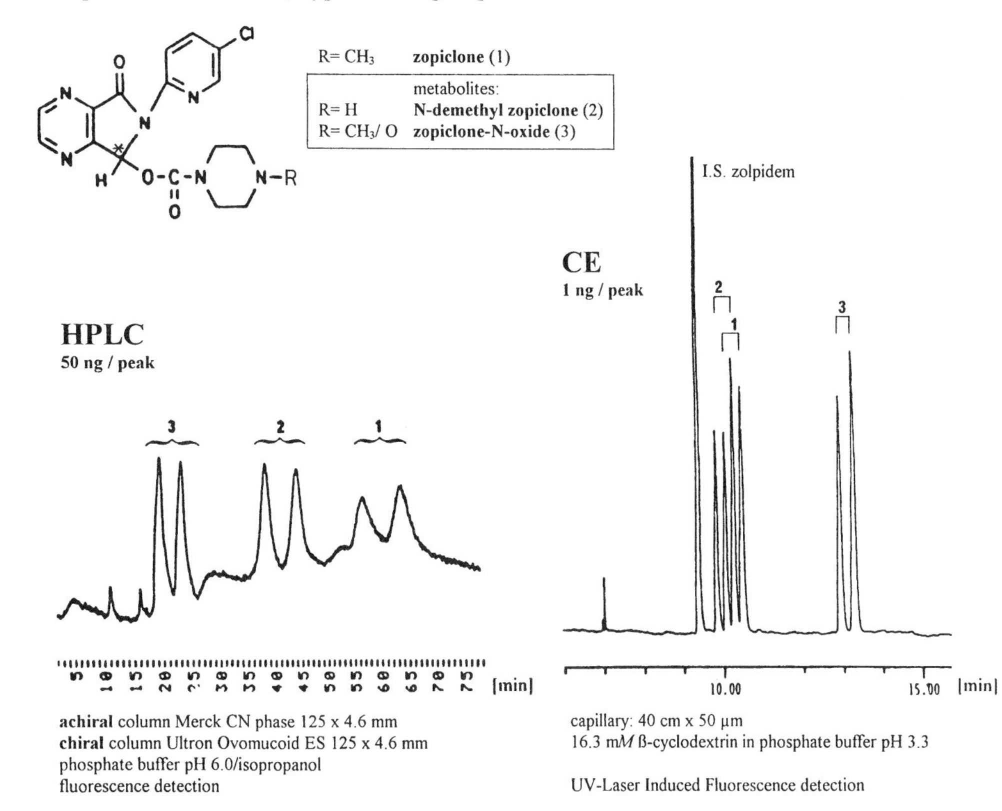

Figure 4.10. Chiral assay of a mixture of racemic zopiclone, N-demethyl zopiclone, and zopiclone N-oxide by HPLC (left) and capillary electrophoresis (right). Fluorescence detection.

enriched. About 2 hours after application of the drug, the enantiomeric ratio is reversed.

In our laboratory, capillary electrophoresis with chiral additives is replacing more and more often time-consuming chiral assays by HPLC. An example (13) is the separation of the chiral hypnotic zopiclone and its two main chiral metabolites, N-demethyl zopiclone and zopiclone N-oxide. Although a separation including chiral separations of these three components is possible by HPLC, using column coupling of an achiral and chiral stationary phase (a chromatogram on the left side), a much faster, more sensitive, more accurate, and simpler separation is possible by capillary electrophoresis using β-cyclodextrin as the chiral additive (Figure 4.10). Again, the mixture of the three components is separated into six different enantiomers. As the detection method of the fluorescent drug and its fluorescent metabolites, UV laser-induced fluorescence detection (UV-LIF) with a He-Cd-laser was used in capillary electrophoresis. With this method, the stereoselective metabolism of zopiclone was also evaluated (13).

Capillary electrophoresis proves to be a very elegant method also for the assay of very polar phase II metabolites like glucuronides that are separated as diastereomers. We are currently investigating the phase II biotransformations of chiral drugs like dimethindene using capillary electrophoresis [14].

4.9. CONCLUSIONS

Chiral assays mainly by HPLC have already enabled the evaluation of stereoselective biotransformations of numerous chiral drugs. In the future, capillary electrophoresis with chiral additives will become an additional valuable analytical method for these assays due to its high mass sensivity and versatility.

REFERENCES

1. Staudt, U., Schmahl, G., Blaschke G., and Mehlhorn, H. (1992). *Parasitol. Res.*, **103**, 392–397.
2. Blaschke, G. (1986). *J. Liq. Chromatogr.*, **9**, 341–368.
3. Kaup, A. (1989). Ph.D. Thesis, University of Münster.
4. Koch-Breuer, M. (1990). Ph.D. Thesis, University of Münster.
5. Pahlen, B. (1991). Ph.D. Thesis, University of Münster.
6. Becker-Scharfenkamp, U. and Blaschke, G. (1993) *J. Chromatogr.*, **621**, 199–207.
7. Elsing, B. and Blaschke, G. (1993). *J. Chromatogr*, **612**, 223–230.
8. Overbeck, P. (1995). Ph.D. Thesis, University of Münster.

9. Heuermann, M. and Blaschke, G. (1993). *J. Chromatogr.*, **648**, 267–274.
10. Heuermann, M. and Blaschke, G. (1994). *J. Pharmac. Biomed. Analysis*, **12**, 753–760.
11. Heuermann, M. (1993). Ph.D. Thesis, University of Münster.
12. Prien, D. and Blaschke, G. (1995). *J. Chromatogr. B*, in press.
13. Hempel, G. and Blaschke, G. (1990). *J. Chromatogr. B*, **675**, 139–146.
14. Rudolph, J. L. (1996). Ph.D. Thesis, University of Münster.

CHAPTER

5

KINETICS OF REACTIVE PHASE II METABOLITES: STEREOCHEMICAL ASPECTS OF FORMATION OF EPIMERIC ACYL GLUCURONIDES AND THEIR REACTIVITY*

HILDEGARD SPAHN-LANGGUTH

Department of Pharmacy,
Martin-Luther-University Halle-Wittenberg,
Wolfgang-Langenbeck-Strasse 4
D-06120 Halle/Saale, Germany

LESLIE Z. BENET, PARNIAN ZIA-AMIRHOSSEINI

Department of Pharmacy
University of California San Francisco
San Francisco, California 94143-0446

SEIGO IWAKAWA

Department of Hospital Pharmacy
School of Medicine
Kobe University
Kobe 650, Japan

and

PETER LANGGUTH

Astra Hässle AB, Kärragatan 5, S-43183 Mölndal, Sweden

For a considerable period, the evaluation of the properties of phase II conjugates of carboxylic acids with glucuronic acid (1-O-acyl-β-(D)-glucopyranosiduronates, 1-O-acyl glucuronides) and covalent binding via these reactive

* This paper is dedicated to Professor Richard Neidlein, Institute of Pharmaceutical Chemistry, Heidelberg, in commemoration of his 65th birthday.

The Impact of Stereochemistry on Drug Development and Use, Edited by Hassan Y. Aboul-Enein and Irving W. Wainer. Chemical Analysis Series, Vol. 142.
ISBN 0-471-59644-2

metabolites was merely descriptive. Then researchers interested in the kinetic properties of chiral carboxylic acid drugs realized that it might also be of interest to investigate stereochemical factors involved in the acyl migration of acyl glucuronides and covalent binding via these reactive phase II metabolites, since it was hypothesized that a toxicological potential may reside in their reactivity. This may be regarded as an important step with respect to an elucidation of correlations between stability and the reactivity of acyl glucuronides, with particular emphasis on covalent binding to proteins. Mainly as a result of studies of various chiral carboxylic acid drugs and their diastereomeric glucuronic acid conjugates, a hypothesis was established in which stability and covalent binding data were correlated (Benet et al., 1993). Such attempts to predict reacticity and, hence, potential toxicological properties resulted primarily from *in vitro* studies, where a difference in the extent of covalent binding was detected for those epimeric glucuronides, with differences between stereoisomers regarding stability, whereas compounds not exhibiting differences in stability did not show any discrimination in the extent of covalent binding to proteins *in vitro*. The purpose of this chapter is an elucidation of the properties of acyl glucuronides, with particular emphasis on epimeric acyl glucuronides, the products of conjugation of chiral carboxylic acids with glucuronic acid.

5.1. INTRODUCTION: GENERAL CHARACTERISTICS OF ACYL GLUCURONIDES FROM ACHIRAL AND CHIRAL SUBSTRATES

Phase II metabolites of drugs are usually anticipated to be rapidly excreted into urine and bile without exhibiting significant biological/pharmacological activity. However, today, it is largely accepted that they may be active and/or reactive (Kroemer and Klotz, 1992; Spahn-Langguth and Benet, 1992a). With respect to the phase II metabolism of carboxylic acid xenobiotics, conjugation with glucuronic acid plays an important role, yielding potentially reactive acyl glucuronides (ester glucuronides). These conjugates are of particular relevance, since they may undergo typical reactions such as hydrolysis, formation of β-glucuronidase-resistant isomers (acyl migration), and covalent binding to endogenous compounds (Faed 1984, Spahn-Langguth and Benet, 1992b).

5.1.1. Hydrolysis and Acyl Migration

Hydrolysis of an acyl glucuronide leads to regeneration of the aglycone. Rates of hydrolysis, which may occur *in vitro* and *in vivo*, are dependent on pH and temperature, with more rapid degradation of the enzymatically formed β-1-O-acyl glucuronide at higher pH—also at physiological pH—than at a more

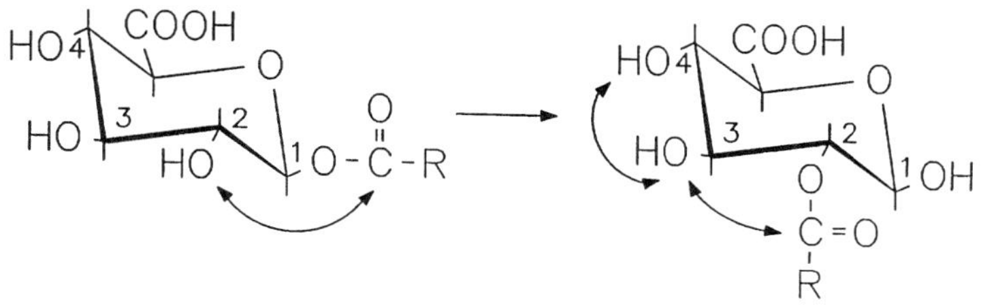

Figure 5.1. After formation of the C-1-O-acyl glucuronide from a carboxylic acid, the acyl residue can "migrate" within the glucuronic acid molecule, forming β-glucuronidase-resistant positional isomers at C-2, C-3, and C-4. The rearrangement is reversible with the exception of the enzymatically formed C-1 isomer, which cannot be regenerated via acyl migration.

acidic level. Acyl migration is the migration of the acyl residue from the C-1 position to the C-2, C-3, and C-4 positions with the formation of regioisomers. Backformation of the C-3 from the C-4 isomer as well as of the C-2 from the C-3 isomer is possible. However, the originally formed C-1 isomer cannot result from its regioisomers via acyl migration, that is, the high-energy 1-O-acyl bond is not regenerated under *in vitro* conditions (Figure 5.1). The studies of Bradow et al. (1989) indicated that there is no evidence for rearrangements beyond nearest-neighbor hydroxyl groups.

Like hydrolysis rates, the rates of acyl migration are pH-dependent. Hence, with respect to both *in vitro* enzyme kinetic studies and pharmacokinetic studies, sample treatment is of crucial importance. To avoid postsampling degradation, samples are usually immediately cooled and pH-stabilized (pH 3–4) (Hasegawa et al., 1982). Earlier studies not employing correct sample stabilization procedures yielded inaccurate measures of the pharmacokinetics of the carboxylic acid drugs as well as their glucuronides.

5.1.2. Glucuronide Formation and Isomerism Within the Glucuronic Acid Moiety

Stereochemical aspects with respect to acyl glucuronides include both the xenobiotic as well as the endogenous glucuronic acid, which enters the uridine diphospho glucuronosyltransferase-catalyzed reaction as uridine diphospho glucuronic acid (UDPGA). As for other metabolizing enzymes a variety of different isoenzymes, exist also for the membrane-bound UDPGTs (e.g., Burchell and Coughtrie, 1989). With respect to glucuronic acid and its activated form, stereochemical as well as nomenclature aspects have been the subject of a commentary by Dudley (1985). During the conjugation reaction, an anomeric inversion at the C-1 atom of glucuronic acid takes place, converting the α-D-glucuronic acid in UDPGA to β-D-glucuronic acid in the conjugate. While no interconversion between the two glucuronic acid anomers is possible for the enzymatically formed β-1-O acyl glucuronide, its regioisomers were found to occur as mixtures of the α- and β-anomers (Smith and Benet 1986). Furthermore, isomerization within the glucuronic acid moiety may result in furanose structures, open-chain and lactone forms as described for, for example, bilirubin glucuronides in the early work of Blanckaert et al. (1978).

Acyl migration is the predominating reaction *in vitro*, whereas *in vivo* hydrolysis is of higher relevance.

5.1.3. Kinetic Properties of the Glucuronides

The acidic glucuronic acid conjugates are more water-soluble than their precursors at physiological pH. Their renal clearance (CL_{ren}) frequently exceeds the glomerular filtration rate (GFR) significantly. For compounds with a CL_{ren} that is smaller than GFR, significant plasma protein binding was detected, that is, the unbound renal clearance calculated from these data is much higher. It may hence be assumed that renal excretion occurs preferentially via carrier-mediated tubular secretion. The carrier system responsible for this transport should be the organic anion transporter located in the proximal tubule of the kidney. Apparently, the active step is located in the basolateral membrane of the tubular cells. However, the exact mechanism is still controversial, since there are data supporting the theory of a sodium cotransport, whereas others have suggested organic anion exchange with endogenous organic anions as reviewed by Ott and Giacomini (1993).

Biliary secretion appears to be of minor relevance in humans, whereas significant amounts of glucuronides may be detected in the bile of rats, for example.

These elimination characteristics indicate that the kinetic behavior of acyl glucuronides is sensitive to competitive inhibition. Since the conjugation step may also be reversible *in vivo*, increased aglycone concentrations resulted, in addition to elevated acyl glucuronide concentrations, when probenecid was administered together with zomepirac (Smith et al., 1985). *In vivo* backformation of aglycone may either be base- or enzyme(esterase)-catalyzed. This phenomenon was first described in the work of Gugler et al. (1979) who found that—following clofibrate administration—clofibric acid concentrations were dependent on renal function, although its clearance is mainly metabolic, yet with an acyl glucuronide as the major metabolite. Gugler explained this finding by "futile cycling," that is, acyl glucuronide excretion is reduced in renal failure, leading to backformation of the respective aglycone and increased concentrations in the systemic circulation. With respect to the determination of kinetic parameters, we need to consider that clearance terms determined from the area under the concentration-time curve of the aglycone represent apparent values only. The determination of true clearances including the interconversion between aglycone and glucuronide requires the dosage (i.v.) of aglycone as well as glucuronide and quantification of both (Smith et al., 1990b).

5.1.4. Formation of Covalent Adducts: Mechanisms, Binding Sites, Extent of Binding to Albumin, Plasma Proteins, and Other Proteins (Homogenates, and Microsomal Proteins)

Wells et al. (1987) hypothesized that a correlation exists between reversible binding to plasma proteins and irreversible protein binding. Data about reversible plasma protein binding of glucuronides, however, are rare. With respect to acyl glucuronides, the lack of data mainly results from experimental problems, since the studies need to be carried out at physiological pH. By rapid ultrafiltration, the binding of carprofen as well as zomepirac and tolmetin to human serum albumin (HSA) was studied *in vitro* (Iwakawa et al., 1990; Ojingwa et al., 1994a). Interestingly, significant binding to HSA was found for the β-1-O-acyl glucuronides as well as for their regioisomers. Protein binding studies were also performed with ketoprofen glucuronides by Hayball et al. (1991a).

Covalent binding to HSA and plasma proteins has been demonstrated for numerous acyl glucuronides. Its extent was found to be clearly time- and pH-dependent in *in vitro* binding studies with HSA as the acceptor protein. The isomeric conjugates, but not the aglycone, were found to bind covalently as well (Smith et al., 1986).

Covalent binding may occur via two different mechanisms [as summarized by Benet and Spahn (1988)]: Direct nucleophilic displacement is possible and leads to the formation of an acylated protein and the release of glucuronic acid. The second mechanism includes an imine formation. For this reaction, acyl migration is a prerequisite. Subsequently, imine formation is possible between the free aldehyde of the open-chain glucuronic acid and a nucleophile, possibly lysine, in the albumin molecule with glucuronic acid being part of the adduct. The relevance of the second mechanism was confirmed by studies of different groups (Smith et al., 1990b; Ding et al., 1993, 1995; Grubb et al., 1993).

Studies performed by Dubois et al. (1993) as well as in our laboratory suggest that HSA is the major binding protein with respect to covalent binding. No explanation has been provided so far for the discrepancy between the higher extent of covalent binding to plasma proteins as compared to HSA, for example, because no covalent binding was detected with fibrinogen and gamma globulins, and only 0.14% of ketoprofen was bound to globulins after 3 h of incubation in the studies of Dubois et al. (1993).

In recent studies performed by Ding et al. (1993, 1995), Lys-199 (a lysine ε-amino group located in the hydrophilic pocket of subdomain II-A, which also plays a role in reversible binding to HSA) was identified as the most prominent covalent binding site for tolmetin glucuronide on the human serum albumin molecule, independent of whether binding occurred via the imine formation mechanism (Ding et al., 1993) or the nucleophilic displacement mechanism (Ding et al., 1995).

In vitro studies with blood constituents showed that plasma protein binding accounts for most of the observed covalent binding in blood. Apparently, acyl glucuronides do not permeate red blood cell membranes and, hence, do usually not have access to the intracellular hemoglobin, for which significant covalent binding was detected in *in vitro* studies (Ojingwa et al., 1994b).

In *in vitro* studies with cells, tissues, and tissue homogenates, as well as from *in vivo* studies, significant covalent binding to tissue proteins was detected (Kretz-Rommel and Boelsterli, 1994; Ojingwa et al., 1994; Dickinson and King, 1993; King and Dickinson, 1993; Ojingwa et al., 1994). Incubations of glucuronides with subcellular fractions revealed that covalent binding is negligible for the cytosolic fraction, but considerable for membrane-bound proteins (Ojingwa et al., 1994; Spahn-Langguth et al., 1994a; Hargus et al., 1995).

5.1.5. Adduct Kinetics

Covalent plasma protein adducts are detected when acyl glucuronides occur in plasma. The extent of formation is species-dependent, because of the differences in elimination pathways and activity of plasma esterases (Smith et al., 1990b; Iwakawa et al., 1989; 1991).

Few data are available about the *in vivo* adduct kinetics. However, for all investigated compounds, the persistence of covalent protein adducts in the systemic circulation was considerably higher than that of its precursors. The kinetics of an acyl glucuronide is usually rate-limited by its formation and the terminal half-lives of aglycone and glucuronide are similar, whereas the half-lives and residence times of adducts are much higher and kinetics are rate-limited by adduct elimination (Spahn-Langguth and Benet, 1992). The half-lives detected for the adducts of different glucuronides vary widely (with a range of 2–14 days), with the highest values approaching that for albumin turnover. An explanation for these differences between adducts has not been provided yet, but it was hypothesized that different adduct types may be involved. By using standard analytical procedures for adducts, the different adduct types are not discriminated, since only the total amount of aglycone that is released from protein via hydrolysis is measured.

Because of their long half-lives, adducts accumulate in the blood (and probably also in tissues) following chronic drug administration (Zia-Amirhosseini et al., 1994). Furthermore, when the extent of exposure to the glucuronide is increased, for example, with higher doses or reduced glucuronide clearance due to renal failure or drug–drug interactions, adduct concentrations are elevated as well (McKinnon and Dickinson, 1989). For clofibric acid, significant covalent binding was detected *in vivo*, in laboratory animals as well as patients. Covalently bound clofibric acid-protein adducts were detected in all 14 investigated patients (with a daily dose of 0.5–2.0 g), even in one subject in whom there was no measurable plasma clofibric acid (Sallustio et al., 1991). Adduct concentrations appeared to correlate with renal function in the patients. In rats, protein adducts were present in liver homogenates, and concentrations increased with the increasing duration of treatment.

Determinants of the potential toxicity of carboxylic acid drugs *in vivo* may, hence, be the ability of acyl glucuronides to lead to covalent binding with (for example, proteins), the plasma- or tissue concentration-time profile of the reactive glucuronide, and the residence time of the respective adduct in the organism (Benet and Spahn, 1988).

5.2. CHIRAL CARBOXYLIC ACIDS AND THEIR GLUCURONIDES

The group of nonsteroidal antiinflammatory 2-arylpropionic acids and related compounds has been extensively investigated with respect to kinetic differences between enantiomers and stereoinversion, that is, the biotransformation of the "distomer" (*R*-enantiomer), which was found to exhibit a central antinociceptive effect (Geisslinger et al., 1994), to the cyclooxygenase inhibiting eutomer (*S*-enantiomer) *in vivo* (e.g., Simmonds et al., 1980; Hutt et al., 1988). Structurally closely related 2-arylbutyric acids (indobufen) were evaluated as platelet aggregation inhibitors (Strolin-Benedetti et al., 1992; Grubb et al., 1993a).

However, chiral carboxylic acids (and their prodrugs) may be "located" in other groups of therapeutics as well. Examples are chiral lipid-regulating agents (e.g., beclobrate), chiral gyrase inhibitors (e.g., ofloxacin), and newly developed angiotensin-II receptor antagonists. Furthermore, acyl glucuronides may be formed from prochiral compounds via sequential metabolism (two or more steps). A well-known example is the formation of so-called "reduced fenofibric acid," which is formed from the fenofibrate metabolite fenofibric acid with species differences in both the extent of formation and product stereoselectivity (Weil et al., 1989).

Finally, acyl glucuronides may also result from such phase I metabolites, in which the carboxylic acid moiety is intact, such as in the fenoprofen metabolite 4-hydroxyfenoprofen or the naproxen metabolite desmethylnaproxen (Zia-Amirhosseini et al., 1988; Spahn-Langguth and Benet, 1992b).

5.3. ASSAY METHODS FOR EPIMERIC GLUCURONIDES AND ADDUCTS

5.3.1. Quantification of Acyl Glucuronides

In enzyme kinetic studies, direct quantification of product formation should be the method of choice and preferred to measuring the decrease of substrate concentration, which may result in considerable analytical errors in the determination of formation rates, especially when the substrate-to-product ratio is high. Conjugates with D-glucuronic acid are hydrophilic derivatives, which can easily be separated from the parent compound by chromatographic methods. Direct determination of the conjugate is advisable in both *in vitro* and *in vivo* kinetic studies, although conjugate levels may sometimes approximate the range of the detection limit.

Alternatively, the detection and quantification of conjugate levels are possible by indirect means, that is, via glucuronide cleavage with or without the preextraction of aglycone and subsequent quantification of the released aglycone. Hydrolysis of ester glucuronides may be accomplished by incubation with β-glucuronidase, which selectively cleaves the β-1-O-acyl glucuronides, or by alkaline hydrolysis with the additional cleavage of all positional isomers, if any are present in the sample. When enzymatic cleavage is performed, it is necessary to optimize the conditions in order to achieve a complete release of aglycone. Incomplete cleavage may lead to erraneous *R*/*S* ratios, since β-glucuronidase appears to release aglycones stereoselectively from glucuronides, as shown, for example, by el Mouelhi et al. (1988) for naproxen. Compared with enzymatic hydrolysis, cleavage is usually more rapid when performed under alkaline conditions (with a pH of 12–14). Hydrolysis of ester glucuronides also occurs via acid-catalysis (a pH of 2 or lower). Especially with the alkaline cleavage of acyl glucuronides, racemization may be a problem, as shown in preliminary studies with fenoprofen glucuronides performed by Volland and Benet (unpublished data) and the studies of Hayball et al. (1994).

Resolution of the epimeric glucuronides of 2-arylpropionic acids was usually achieved on octadecylsilane (ODS) stationary phases. We described the resolution of diastereomeric naproxen glucuronides as well as the glucuronides of various other 2-arylpropionic acids on Ultrasphere ODS using a mixture of acetonitrile (ACN) and 9 or 10 mM of tetrabutylammonium (TBA) buffer pH 2.5 (28:77, 23:77, 35:65, flow rates 1.5–2.0 ml/min; elution orders, *S*- before *R*-glucuronide) (Spahn and Benet, 1987; Spahn et al., 1988 andm 1989; Spahn, 1988; Iwakawa et al., 1989). Fournel-Gigleux et al. (1988) used a Lichrosorb Hibar RT column and ACN/trifluoroacetic acid/water (38:0.08:162) (see also Hamar-Hansen et al., 1986) to resolve the diastereomeric conjugates of 2-phenylpropionic acid (elution order *R*- before *S*-glucuronide). el Mouelhi et al. (1988) employed different ACN/buffer mixtures to resolve the epimeric conjugates of naproxen, ibuprofen, and benoxaprofen on Ultrasphere ODS (naproxen: 0.05M of ammonium acetate with 20% ACN, adjusted to pH 6 with glacial acetic acid; flow, 1.2 ml/min, retention times 14.2 and 15.2 min; ibuprofen: 0.1M of ammonium acetate with 35–80 % ACN as linear gradient within 10 min; flow rate 1.0 ml/min, retention times 5.4 and 6.0 min; benoxaprofen: 0.01 M of phosphate buffer pH 6.3 in ACN as gradient with 30–35 % within 3 min and then 35 % for 12 min, flow rate 1 ml/min, retention times 7.6 and 8.4 min; elution order *S*- before *R*-conjugate for all compounds). Additional analytical studies with naproxen glucuronides were published by Buszewski et al. (1990), studies with ketoprofen glucuronides by Chakir et al. (1994).

In summary, it is obvious that suitable chromatographic systems are available for the resolution of diastereomeric acyl glucuronides of various compounds, where mobile phases are usually acidic to assure stability of the metabolites during the chromatographic procedure.

However, because of various forms of isomerism, the analytical problems may—dependent on the nature of the glucuronide conjugate—become rather complex. Isomerization is not only possible via intramolecular rearrangement by acyl migration but also via isomerization of the sugar group, yielding furanose as opposed to pyranose structures. Except for the C-1 position (β-1-O-acyl glucuronides), α- and β-anomeric forms may occur in addition to open-chain forms and lactones (Blanckaert et al., 1978). As a consequence, numerous isomers of the original enzymatically formed β-1-O-acyl glucuronide may occur for each of the two drug enantiomers when a racemate was dosed. However, only very few approaches were made trying to resolve, for example, the epimeric positional isomers.

The isomeric conjugates of *S*-benoxaprofen were resolved via HPLC as described by Bradow et al. (1989). In stability studies with carprofen glucuronides, it was possible to separate migration products from the β-1-O-acyl glucuronide employing an ODS stationary phase (Ultrasphere ODS) and HPLC gradient elution with mixtures of acetonitrile (ACN) and tetrabutylammonium hydrogen sulfate (TBA) solution (0–15 min: 35% ACN, 10 mM TBA; 15–25 min: 35% ACN, 9 mM TBA; 25–35 min: 39% ACN, 8.5 mM TBA; 35–45 min: 42% ACN, 8 mM TBA; 45–55 min: 46% ACN, 7.5 mM TBA; 55–65 min: 50% ACN, 5 mM TBA; flow rate: 1 ml/min; 290/365 nm) (Iwakawa et al., submitted for publication). The retention times for *S*-

glucuronide, *R*-glucuronide, and carprofen were 38.5, 39.7, and 56.8 min, respectively. β-Glucuronidase-resistant migration products were observed at retention times of 37.5 and 39.6 min for the S-glucuronide and at 35.4 and 39.1 min for the *R*-glucuronide. The similarity in retention times did not permit a stereospecific assay of the migration products when *R*- and *S*-β-1-O-acyl glucuronides were both present (Iwakawa et al., submitted for publication).

However, there may also be a situation where, for example, subtracting free aglycone and the fraction released from the conjugate by β-glucuronidase treatment from concentrations found after alkaline hydrolysis does not result in the correct regioisomer concentrations. Other alkali-sensitive conjugates may occur in samples in addition to acyl glucuronides. Studies in mice revealed that pranoprofen acyl glucosides are formed *in vitro* and *in vivo*, and their formation is administration route- and dose-dependent as well as stereoselective (Arima 1990a; 1990b). While glucuronidation occurs mainly in the liver, the kidney was found to be the preferential site of glucosidation favoring formation of the glucoside of *S*(+)- as compared with *R*(–)-pranoprofen. The fraction of the dose converted to acyl glucoside decreased with increasing doses and was higher for p.o. than for i.p. administration (while the glucuronide fraction increased with increasing doses), indicating that this organ-selective elimination pathway is saturable (limited capacity of the enzyme or of the respective renal transport system).

5.3.2. Estimation of Adduct Concentrations

Generally, adducts are quantified following hydrolysis and release of the aglycone. In the first step, protein is precipitated by the addition of, for example, isopopropanol and acidified acetonitrile or acetonitrile/ethanol mixtures. The protein pellet obtained after centrifugation is washed several times with methanol/diethylether (3:1; vortexing and sonication, followed by centrifugation) to remove reversibly bound aglycone and conjugates from the protein binding sites. Adducts are incubated with 0.2–0.5 M of potassium hydroxide solution at 70–80°C for 30–120 min. Adduct concentrations are then defined as aglycone liberated from precipitated and extensively washed protein. As for (acid- and) base-catalyzed hydrolysis of glucuronides, an important prerequisite for a correct estimation of enantiomer concentrations is the configurational retention at the chiral center. Such stability studies were performed with beclobric acid, flunoxaprofen, and benoxaprofen, and showed negligible racemization during incubation under the above-mentioned conditions.

Approaches for direct measurement included tryptic digests followed by HPLC separation of the fragments as well as electrophoresis (SDS-PAGE) and blotting of adducts formed from, for example, the fluorescent benoxaprofen and flunoxaprofen acyl glucuronides (van Breemen and Fenselau, 1985; Spahn et al., 1990; Dahms and Spahn-Langguth, 1995). However, stereochemical factors have not yet been included into the development of a direct assay of adducts.

5.4. STABILITY OF THE β-1-O-ACYL GLUCURONIDES: STEREOCHEMICAL ASPECTS

Degradation of the diastereomeric β-1-O-acyl glucuronides—yielding positional isomers and aglycone—of various chiral carboxylic acids has been shown to be stereoselective. In studies with benoxaprofen and flunoxaprofen, carprofen, naproxen, and fenoprofen, the apparent first-order degradation half-lives were higher for the respective *R*-glucuronides, with half-lives always below 10 h Iwakawa et al., 1989; Spahn, 1989; Spahn et al., 1988a, 1989; Volland et al., 1991; Bischer et al., 1995). Under similar *in vitro* conditions, the differences between the apparent pseudo-first-order degradation half-lives of the diastereomeric glucuronides of beclobric acid were considerably higher, with 22.7 h for the levo- and 25.7 h for the dextrorotatory form (at a concentration of 5 μM in 150 mM phosphate buffer pH 7.4) (Mayer et al., 1993). The beclobric acid glucuronides exhibited a low tendency for acyl migration and hydrolysis, that is, a higher stability than has been observed for the acyl glucuronides of most other drugs.

Stereoselective degradation of carprofen glucuronides under different conditions and the influence of human serum albumin (HSA) were characterized in unpublished studies performed by our group (Iwakawa et al., submitted for publication). When *R*- and *S*-carprofen glucuronides (initial concentration: 5 μm) were incubated at pH 7.0, 7.4, and 8.0 at 37°C in phosphate buffer, degradation was highly stereoselective at pH 7.0. Stereoselectivity decreased while degradation velocity increased with higher pH, as summarized in Table 5.1. At all pH values, the *R*-glucuronide conjugate of carprofen degraded more rapidly than the *S*-glucuronide. Degradation of the β-1-O-acyl glucuronides occurred mainly via the formation of positional isomers. The hydrolysis to parent carprofen was slow for both glucuronides.

When HSA was added to the incubation medium, the stability of the *S*-glucuronide was decreased, whereas the apparent half-life of the *R*-glucuronide increased to some extent under these conditions. Interestingly, the effect of fatty-acid-free HSA was much greater than that of fraction V HSA. In the presence of HSA (2 mg/ml) carprofen glucuronides were readily hydrolyzed to parent drug, indicating an esterase-like activity of albumin, which was also observed with much lower HSA concentrations (0.2 mg/ml) (Iwakawa et al., 1989; Iwakawa et al., submitted for publication).

Table 5.1. Degradation Half-lives of Carprofen β-1-O-acyl Glucuronides: Influence of pH Conditions, Temperature, and Addition of Albumin on the velocities of Degradation and Their Enantioselectivities

	Half-life (h)		
	S-glucuronide	*R*-glucuronide	*S*/*R* Ratio
1. Various pH Values 37°C			
pH 7.0	6.42	2.60	2.43
pH 7.4	2.90	1.72	1.69
pH 8.0	0.85	0.60	1.41
2. Temperature Dependence at pH 7.4			
4°C	>100	>100	1.0
25°C	11.8	7.80	1.51
37°C	2.90	1.72	1.69
3. Effect of Human Serum Albumin (HSA), pH 7.4, 37°C			
Without HSA	2.90	1.72	1.69
With 30 μM HSA (Essentially fatty-acid-free)	1.55	2.80	0.55
With 30 μM fraction V HSA	1.82	1.78	1.02

Incubation period = 24 h.

The stereoselectivity of the degradation of fenoprofen β-1-O-acyl glucuronides was studied by Volland et al. (1991), who found enhanced stability (and decreased covalent binding) at lower pH levels, as was expected from studies with other compounds. The degradation rate of *R*-fenoprofen glucuronide was greater than that of *S*-fenoprofen glucuronide. The addition of HSA to the medium decreased the stability of both stereoisomers.

Stereoselective hydrolysis of flurbiprofen acyl glucuronides under various conditions was reported by Knadler and Hall (1991). They detected preferential hydrolysis of the conjugates of *S*-flurbiprofen in plasma and albumin solution and the significantly higher stability of both conjugates when no protein was present, which again suggests that albumin is the predominant source of hydrolytic activity in the case of flurbiprofen conjugates. The *in vivo* relevance of ester hydrolysis was demonstrated for clofibric acid and zomepirac glucuronides (Rowe and Meffin, 1984; Smith et al., 1990b), where the addition of esterase inhibitors reduced the apparent clearance considerably. With ketoprofen glucuronides, however, deconjugation in plasma was unchanged when esterase inhibitors were added to the incubation medium (Hayball et al.,

1992). Dubois et al. (1993) reported studies on the *in vitro* degradation of biosynthetic ketoprofen glucuronide after incubation with human plasma and HSA solutions at different concentrations (290 and 580 μM) and in protein-free buffer at pH 7.4 and 37°C (selected protein concentrations corresponding to that found in synovial fluid and plasma, respectively). Albumin catalyzed the hydrolysis of the glucuronide, but the extent of the reaction was not dependent on the protein concentration.

5.5. *IN VITRO* ENZYME-KINETIC STUDIES WITH LABILE GLUCURONIDES

5.5.1. Methodological Problems

When *in vitro* metabolic kinetics is reviewed, it is expected that the authors will characterize the saturable aspects of metabolic processes in terms of Michaelis–Menten parameters such as V_{max} and K_m. However, such analyses assume that metabolic processes are unidirectional and the metabolites formed are stable during analysis. In addition to acyl migration and hydrolysis of the epimeric glucuronides, studies of *in vitro* acyl glucuronidation—mostly performed using isolated membrane fractions (e.g., microsomes)—involve a further complication beyond that seen in *in vivo* studies. That is, disrupted cells and subcellular fractions and even washed microsomes contain hydrolytic enzymes that originate primarily from lysosomes. Thus, esterases and β-glucuronidase, which are most probably not present together with the glucuronidation enzyme in an intact cell, may be active during *in vitro* incubation. This problem was mentioned or discussed in few articles about *in vitro* glucuronidation (Nakamura and Yamaguchi, 1987; el Mouelhi et al., 1988; Spahn et al., 1989a; Magdalou et al., 1990) and may be an important source of error with respect to the estimation of rates of glucuronide formation, but also for the evaluation of enantioselectivities.

As discussed in a recent article (Spahn-Langguth and Benet, 1993), true metabolic formation rates may be approximated by:

Stabilization of the metabolic product via addition of inhibitors of hydrolytic enzymes, or by calculating the metabolite's degradation rate under similar conditions to that utilized in the formation experiments and by changing the pH of the medium from pH 7.4 to more acidic values to prevent acyl migration and hydrolysis, or

Estimation of the decomposition rate of the product in the medium and correction of the initial rates of formation.

The decomposition rate may be characterized by the respective rate constant k_d, which subsumes all processes that lead to a decrease in β-1-O-acyl glucuronide concentration as depicted in the scheme in Figure 5.2. The relationship between formation and decomposition may be of particular interest in characterizing the glucuronidation of enantiomers, since studies with benoxaprofen, flunoxaprofen, and naproxen (Spahn et al., 1988a; Spahn, 1989) indicated that the degradation rate constants as well as the difference between the two diastereomeric glucuronides were reduced when inhibitors of hydrolytic enzymes were added. With respect to these 2-arylpropionic acids, apparently, hydrolytic enzymes are selective for the respective *R*-glucuronides. For *S*/*R* ratios in product formation, it is obvious that if significant stereoselectivity in product decomposition occurs, then the concentrations in the medium and, of course, the *S*/*R* ratios are highly dependent on product loss, that is, the *S*/*R* ratios may not necessarily reflect the ratios of true formation. One might argue that prolonged incubation times should lead to complete conversion of substrate into product; however, since the plateau levels for the formation of *S*- and *R*-glucuronides are characterized by the ratio of the initial rate of formation and the pseudo-first-order decomposition rate constant (k_0/k_d) for each diastereoisomer, *S*/*R* ratios can differ from unity either because of different formation rates, decomposition rates, or both. And at the same time, it becomes obvious that differences in the formation rate of *S*- and

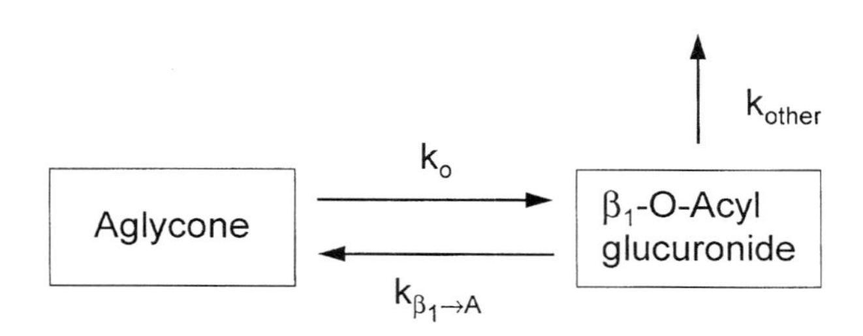

Figure 5.2. Simplified model describing reversible metabolism of acyl glucuronides during *in vitro* enzyme kinetic studies. Since for the determination, direct backformation of aglycone appears to be the most relevant step, the model that includes consecutive and reversible formation of the different positional isomers and subsequent aglycone formation as well as covalent binding was reduced. Then, for this situation, the change in the amount of product with time is described by $dG/dt = k_0 - k_d G$, where the apparent first-order product loss k_d is the sum of $k_{\beta_1\text{-A}}$ (the backformation of aglycone from the β-1-O-acyl glucuronide) and k_{other} (the other potential loss processes of isomerization and covalent binding). Solving the equation for G yields, $G = k_0\,(1 - e^{-k_d t})\,/\,k_d$, which can be rearranged to calculate the apparent zero-order initial formation rate, expressed in units of amount per time. The initial formation rate k_0 is dependent on aglycone and UDPGA concentrations, which are assumed to be constant during the initial experimental period.

R-glucuronide may exist in unstabilized systems and not be detected since the rates are potentially confounded by stereospecific product loss. The potential stereospecific difference in degradation rates significantly complicates the analysis of stereospecific glucuronides formation as compared to other stable, for example, ether, glucuronides.

Since the decomposition of acyl glucuronide product at acidic pH is mainly due to enzymatic hydrolysis, it appears advisable to block the enzymes rather than correct for the loss, since the hydrolytic enzymes also exhibit saturable kinetics, leading to decreased decomposition rates at high glucuronide concentrations. Studies from our laboratories indicated that UDPGT activities are not significantly changed when the pH of the incubation medium is shifted to 5.5 or 6.5 (Spahn et al., 1988a). Only for very unstable acyl glucuronides would it then be necessary to correct for product losses.

Employing such procedures, we have characterized the enzyme kinetic constants for UDPGTs using *S*(+)- and *R*(–)-benoxaprofen as aglycones in hepatic microsomes from sheep (Spahn et al., 1989, Spahn-Langguth and Benet, 1993) via determination of the true initial formation rates, before the latter were used for the calculation of bisubstrate-enzyme kinetic parameters (Table 5.2) as described by Zakim and Vessey (1973). (Fournel-Gigleux et al. (1988a) and others performed their glucuronidation studies at one fixed UDPGA concentration.)

In addition to the initial rates of formation, *S*/*R* ratios of glucuronide formation were also found to differ between substrates and to be species-dependent (e.g., Spahn, 1988; el Mouelhi et al., 1988, Fournel-Gigleux et al., 1988b).

Table 5.2. Kinetic Constants for UDPGT Using *S*(+)-Benoxaprofen and *R*(–)-Benoxaprofen as Aglycones in Hepatic Microsomes from Sheep

	S-(+)	R-(–)
V_{max} (nM min^{-1} mg protein^{-1})	4.39	5.51
K(UDPGA) (mM)	3.03	3.19
K′(UDPGA) (mM)	37.1	55.6
K(B) (mM)	0.53	0.37
K′(B) (mM)	2.88	4.82

Incubation conditions were as given by Spahn et al. (1989). Bisubstrate enzyme-kinetic constants were determined for an enzyme with a rapid-equilibrium random-order mechanism. K′(UDPGA) and K′(B) are the Michaelis–Menten constants for UDPGA and benoxaprofen; K(UDPGA) and K(B) are the true dissociation constants; and V_{max} is the maximal activity.

Source: Spahn-Langguth and Benet (1993).

5.5.2. Significant Species Differences in Glucuronide Formation and Substrate Enantioselectivities in Enzyme-Kinetic Studies

In vitro studies on the stereochemical aspects of acyl glucuronide formation in mammalians (different animal species and man) were performed with microsomes, solubilized microsomal protein, and immobilized protein. Stereoselective naproxen glucuronidation in rats as well as the possibility of interconversion between the two naproxen enantiomers (stereoinversion) in microsomal incubations was extensively evaluated by our group (Spahn and Benet, 1987a,b, 1988). While no stereoinversion was detected, formation differences between enantiomers were significant, with considerably higher glucuronide yields for *S*-naproxen.

With flunoxaprofen as a model substrate, glucuronide yields were compared between rat, guinea pig, rabbit, and sheep liver microsomes. Highest yields and negligible enantioselectivities were obtained for guinea pig and sheep liver microsomes; the lowest yields but highest enantioselectivities were obtained for rat liver microsomes. Glucuronidation was also detected with the kidney microsomes of rats and sheep, with stereoselectivities similar to those found with liver enzymes. The formation rates, however, were considerably lower than with liver microsomes. This is, in part, illustrated in Figure 5.3, in which the initial rates of glucuronide formation are depicted as obtained for the two enantiomers of flunoxaprofen in incubations with rat and sheep liver and kidney microsomes. Species differences were also reported

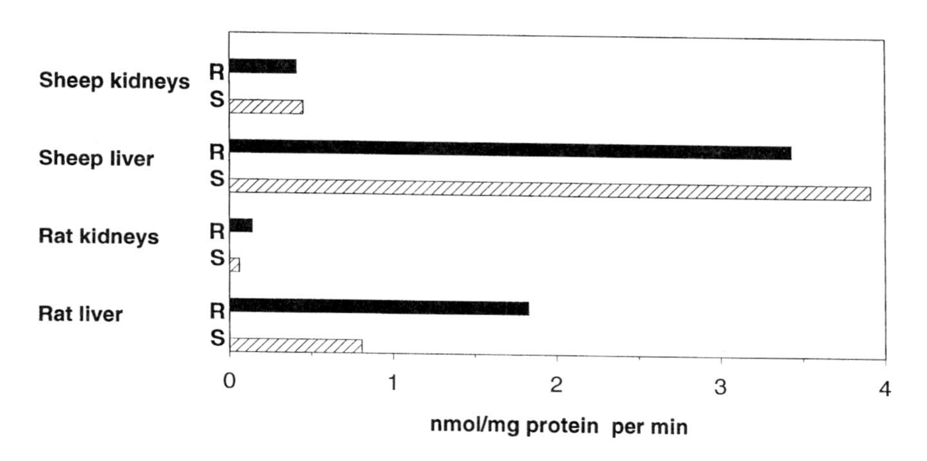

Figure 5.3. Initial rates of glucuronide formation when incubating *S*- and *R*-flunoxaprofen with UDPGTs originating from rat and sheep liver and kidney microsomes (solubilized; pH 6.6; 10 mM of UDPGA; 0.7 mM of flunoxaprofen enantiomer, $n = 6$).

for naproxen, ibuprofen, and benoxaprofen (el Mouelhi et al., 1988; Spahn et al., 1989a).

Preferential glucuronidation of the *R*-enantiomer with rat liver microsomes was observed with various 2-arylpropionic acids, including 2-phenylpropionic acid (Fournel-Gigleux et al., 1988a), flunoxaprofen, flurbiprofen, indoprofen, pirprofen, benoxaprofen, carprofen, and cicloprofen, whereas basically no difference was detected for ketoprofen enantiomers (Spahn, 1988). This was confirmed in the studies of Chakir et al. (1994), who performed microsomal incubations with ketoprofen at pH 5.5 under the addition of saccharic acid 1,4-lactone. They included dog, rabbit, and human liver microsomes in addition to those from rats and found relevant stereoselectivity (four-fold higher formation rate for *S*-ketoprofen) only for dog liver microsomes. With sheep liver microsome preparations, glucuronide yields were higher for *R*-flunoxaprofen (Spahn et al., 1988a,c) and *R*-fenoprofen (Volland and Benet, 1991) than for their respective *S*-enantiomer glucuronides.

The problem of stereoselective deconjugation had been discussed in the work of Nakamura and Yamaguchi (1987), who studied glucuronidation of 2-phenylpropionic acid by rat liver slices. The properties of unstable products formed in microsomal incubations (solubilized microsomes) were evaluated in our studies with the two fluorescent 2-arylpropionic acids possessing a benzoxazole structure (Spahn et al., 1988a, 1989b); to some extent included in the studies of el Mouelhi et al. (1988) (immobilized enzymes); continued in the work of Volland and Benet (1991), who showed formation and degradation differences between the two epimeric fenoprofen glucuronides; and later discussed with respect to the underlying mechanisms (Spahn-Langguth and Benet, 1993).

Glucuronidation studies with enzyme-induced liver microsomes performed, for example, in our laboratories with flunoxaprofen as the substrate and by Fournel-Gigleux et al. (1988a) with 2-phenylpropionic acid as the substrate clearly demonstrated that acyl glucuronide formation is significantly induced by phenobarbital, whereas other inducers (dexamethasone, 3-methylcholanthrene) were of minor relevance in our studies with flunoxaprofen (Spahn, unpublished data). *S*/*R* Ratios were not affected by any of the inducers.

5.5.3. Enzyme Induction and Inhibition: Do They Affect the Enantioselectivity of *In Vitro* Acyl Glucuronide Formation?

The studies with enzyme inducers suggest that both enantiomers are conjugated by the same or a closely related form of UDPGTs. Enantiomeric inhibition studies with flunoxaprofen as well as 2-phenylpropionic acid enantiomers indicated that the two enantiomers inhibit each other during glucuronidation and yield apparent inhibition rate constants in the range of 1–2 mM (Fournel-Gigleux et al., 1988a; Spahn, 1989). The mutual influence of the two substrate enantiomers was also detected for naproxen (el Mouelhi et al., 1988) as well as flunoxaprofen (Spahn, 1989) in liver microsomes from different mammalian species.

Competitive inhibition studies with probenecid showed a similar reduction in the glucuronidation of both enantiomers of naproxen (Spahn and Benet, 1987a). Furthermore, the formation of probenecid glucuronide was significantly reduced in the presence of naproxen, indicating mutual influence in glucuronidation kinetics in addition to alterations of other kinetic processes such as renal and biliary excretion.

5.6. BINDING OF ACYL GLUCURONIDES TO ALBUMIN

A reversible binding step of acyl glucuronides is hypothesized to play a prerequisite role in covalent binding (Wells et al., 1987). Hence, the extent of plasma protein binding was investigated for the two enantiomers of carprofen as well as for their respective epimeric conjugates using a rapid ultrafiltration technique, in order to avoid acyl migration and hydrolysis. In fact, carprofen glucuronides showed a considerable and stereoselective affinity to human serum albumin (HSA), although it was less than observed for the parent enantiomers. The *S*-glucuronide showed a higher binding affinity to HSA than the *R*-glucuronide. For unmetabolized carprofen, the *S*-enantiomer was bound to fatty-acid-free HSA to a much greater extent than the *R*-enantiomer. Warfarin reduced the binding of the glucuronides to a greater extent than did diazepam, whereas diazepam displaced the unconjugated enantiomers to a greater extent than did warfarin, suggesting differences in binding regions between carprofen enantiomers and their glucuronides on the albumin molecule (Iwakawa et al., 1990).

Similarly, naproxen glucuronides were investigated using a comparable experimental approach and were found to exhibit a considerable and stereoselective affinity to HSA, which was lower than for the respective aglycones. At physiological concentrations, the glucuronide of *R*-naproxen revealed a higher affinity to HSA than *S*-naproxen glucuronide, whereas at concentrations higher than 30 μM, the stereoselectivity was found to be reversed.

Reversible binding of acyl glucuronides was also investigated in the case of ketoprofen enantiomers (Hayball et al., 1992). These authors observed opposite stereoselectivity indices for the unbound fractions at physiological HSA concentrations compared to their degradation half-life ratio, which may be explained by separate sites for reversible binding and catalytic (esterase like) acitivity. The hypothesis of different binding areas on the albumin molecule for reversible binding and catalytic activity appears to be confirmed by the data reported for diflunisal by Williams and Dickinson (1994).

5.7. DETERMINATION OF ORGAN CLEARANCES VIA PERFUSION STUDIES: DETECTION OF GLUCURONIDES IN PERFUSATE AND BILE OR URINE

The highest capacity for glucuronide formation is located in the liver with hepatocytes as the smallest functional units. For xenobiotics mainly eliminated by the liver, concentration-time curves in plasma or blood as well as the amount of glucuronide excreted into urine are dependent on the various processes within the hepatocyte, which determine the kinetic pathways of a drug and its metabolites. Major determinants, as depicted in Figure 5.4, are the uptake of a compound into the hepatocyte, its elimination from the hepa-

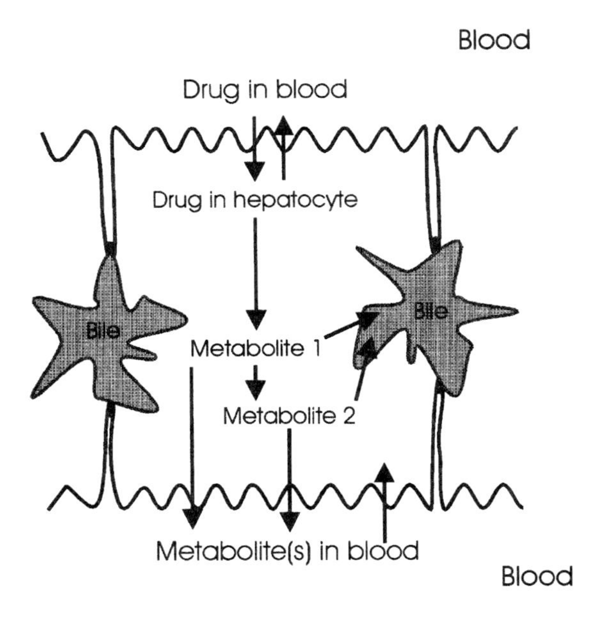

Figure 5.4. Simplified schematic representation for processes within the smallest functional unit of the liver, the hepatocyte. When a drug enters the hepatocyte, several clearance pathways are possible. It may diffuse back into the systemic circulation, be secreted into bile, or undergo phase I or phase II metabolism. A generated metabolite 1 may also reach the blood compartment or undergo biliary secretion, or it can then undergo sequential metabolism (phase I or, most probably, phase II) either immediately before exiting the liver or after metabolite 1 circulates in the plasma and then reenters the liver.

tocyte via diffusion or secretion into bile as well as biotransformation processes, their velocity, and the kinetic behavior of the products with the possibility of sequential metabolism (e.g., phase I metabolism as a first step and subsequent phase II conjugation of the metabolite). With respect to glucuronide kinetics, the molecular weight of the conjugate may be a determinant when studying species differences, since the molecular weight cut-off is different for mammalians, being lowest for rats (Klaasen and Watkins, 1984). Different glucuronide plasma concentrations should hence be observed in different mammalian species. An experimental model used to characterize various aspects of hepatic clearance processes is the perfused liver (Cox et al., 1985). Studies on the hepatic clearance of flunoxaprofen enantiomers in *in situ* perfused rat and rabbit livers—conducted with normal livers in the recirculating perfusion technique using synthetic plasma (3% BSA) with red blood cells as the perfusion medium—indicated that negligible glucuronide levels occur when perfusing rat livers, whereas higher glucuronide plasma concentrations result with rabbit livers. In rats, stereoinversion was detectable, but low. Hepatic clearance from the perfusate was calculated as 49.5 ml/h for *S*-flunoxaprofen and 64.2 ml/h for the *R*-enantiomer, which is in accordance with the higher glucuronidation rate for the *R*-enantiomer in rat liver microsomes. Interestingly, the addition of probenecid to the perfusate decreased the glucuronide concentrations in bile as well as bile flow. Hence, the total amount of glucuronide excreted into bile per unit of time was much lower. The absence of detectable glucuronide concentrations in the plasma of rats was confirmed in *in vivo* pharmacokinetic studies with i.v. carprofen, and flunoxaprofen as well as naproxen racemates (Iwakawa et al., 1991), in which a significant fraction of the dose was excreted into bile as the respective conjugate.

However, stereoselective clearance processes—including phase II metabolism—may also be observed in other organs, such as kidneys (Vree et al., 1992) or gut. The relevance of these clearance processes may be evaluated in studies with isolated cells or cell cultures or in perfusion studies (e.g., el Mouelhi and Schwenk, 1991; Taki et al., 1995).

5.8. KINETIC STUDIES *IN VIVO*

5.8.1. Stereoselectivities in Concentration Time Profiles of Acyl Glucuronides

In studies with drugs metabolized to acyl glucuronides, it may be a prerequsite for the interpretation of kinetic characteristics to simultaneously analyze aglycone and acyl glucuronide data, because of the *in vivo* reversibility of the biotransformation step. Stereoselectivities in concentration-time profiles of acyl glucuronides may result from a variety of potentially stereoselective processes—formation, disposition characteristics, and *in vivo* hydrolysis ("stability") as well as biliary and renal excretion of the glucuronides and, of course, the possibly stereoselective disposition of their precursors, the aglycone enantiomers. In many cases, only blood concentration-time profiles and urinary excretion data obtained from the p.o. dosage of a drug are available, which may not be sufficient to provide explanations for certain kinetic phenomena.

Glucuronide concentrations were measured together with aglycone concentrations in preclinical as well as clinical studies. In studies performed with different NSAIDs from the group of 2-arylpropionic acids (i.v. dosage of the individual enantiomers of naproxen, carprofen, flunoxaprofen) in bile-duct cannulated rats, glucuronide concentrations in plasma were below the detection limit, although the samples were pH-stabilized and immediately cooled. The renal excretion of aglycones and glucuronides of both enantiomers was negligible, whereas significant amounts of glucuronides were excreted via the bile. Biliary excretion of *R*-carprofen and its glucuronide was higher than that of the *S*-enantiomer and its glucuronide. In contrast, the biliary excretion of the *S*-enantiomers of flunoxaprofen and naproxen and their glucuronides was greater than that of their antipodes (Iwakawa et al., 1991). In this study, clearance values for *S*-flunoxaprofen were considerably smaller than in rats with an intact enterohepatic circulation ($t_{1/2}$: 8.8 vs. 32.4 and 69.3 h) (Segre et al., 1988; Pedrazzini et al., 1988). In the studies of Iwakawa et al. (1991), the enantiomeric inversion ratio (Hutt et al., 1988; Meffin et al., 1986), was 0.54 for *R*-flunoxaprofen and, hence, significantly higher than the ratios for *R*-carprofen (0.003) and *R*-naproxen (0.02).

The disposition of naproxen glucuronides and their regenerated aglycone enantiomers was investigated after intravenous dosing of *R*- or *S*-naproxen glucuronide in rats pretreated with or without the esterase inhibitor phenylmethylsulfonyl fluoride (Iwaki et al., 1995). The disposition characteristics of the diastereomeric glucuronides and their dependence on esterase inhibitors were evaluated in the course of these studies. Both naproxen glucuronides were rapidly hydrolyzed to naproxen after administration in rats. The formation of naproxen appeared to be greater from the *R*-glucuronide; however, no marked difference in the apparent clearances of *R*- and *S*-glucuronides was detected. Furthermore, the esterase inhibitor phenylmethylsulfonylfluoride did not significantly alter the disposition characteristics of the epimeric naproxen conjugates.

Data for *R*/*S*-benoxaprofen and *S*-flunoxaprofen (both of which have been subsequently withdrawn from the market) available from volunteers' and patients' samples collected in the last 14 years demonstrate that significant concentrations of glucuronides occur for both drugs. Presumably because of stereoinversion in humans, aglycone and glucuronide concentrations were higher for *S*- than for *R*-benoxaprofen. That is, *in vivo* aglycone/glucuronide ratios were smaller for *S*- than for *R*-benoxaprofen, although incubation with human liver microsomes had resulted in preferential glucuronidation of the R-enantiomer of benoxaprofen (el Mouelhi et al., 1988).

With a single dose of S-flunoxaprofen (100 mg), maximum plasma concentrations were detected approximately 2 h postdose and reached approximately 5% of the respective aglycone concentration. Data for benoxaprofen are in a comparable range; however, the dose range was higher. Plasma concentration-time curves of the glucuronides paralleled those of the respective aglycones in their terminal phase. For both compounds, a high fraction of the administered dose is excreted into urine as acyl glucuronides.

Carprofen represents another example for which significant concentrations of alkali-cleavable conjugates were detected in the plasma and urine of humans. Aglycone plasma concentrations following the dosage of carprofen racemate were only slightly higher for the respective *S*-enantiomer. Dosage of the single enantiomers of carprofen supported the hypothesis that stereoinversion is negligible in humans. Maximum glucuronide levels reached between 25 and 35% of the respective C_{max} values of the aglycones, however, with inverse stereoselectivity when compared with the aglycone, that is, with slightly higher concentrations of *R*-carprofen glucuronide in plasma (Spahn et al., 1988c; 1989b). Concentrations of the glucuronides were in a similar range as the aglycone levels and decayed in parallel. The total renal clearance of the glucuronide was higher for *S*- than for *R*-carprofen (35.6 vs. 26.4 ml/min). Since carprofen glucuronides are highly protein-bound, it may be extrapolated from the data that the renal clearance of the unbound glucuronide fraction significantly exceeds the glomerular filtration rate.

Maximum plasma concentrations of beclobric acid glucuronides obtained after a single p.o. dose of 100 mg of rac-beclobrate amounted to 0.36 μg equiv./ml for (–)- and 0.30 μg equiv./ml for (+)-beclobric acid (Mayer et al., 1993a) and were, hence, considerably lower than the respective aglycone levels, for which a more pronounced stereoselectivity was observed [2.3 μg/ml for (–)- and 3.7 μg/ml for (+)-beclobric acid]. The parallel decay of aglycone and glucuronide indicates that the kinetics of the glucuronide is rate-limited by its formation. Both aglycones and glucuronides accumulated upon repetitive administration. Urinary excretion rates were higher for (+)- than for (–)-beclobric acid glucuronide, and renal clearance appeared to be greater for the (+)-beclobric acid glucuronide.

5.8.2. Drug–drug Interactions and Renal Impairment May Influence *In Vivo* Acyl Glucuronide and Aglycone Levels

Both enzyme induction and inhibition are feasible mechanisms with respect to alterations in the kinetics of acyl glucuronides. For example, Hayball and

Meffin [1987] were able to show in rabbit studies that phenobarbital affects the extent of glucuronidation of fenoprofen enantiomers. Following phenobarbital pretreatment, the clearances of the individual enantiomers of fenoprofen to their respective glucuronides, which were 2.1-fold higher for the *R*-enantiomer in the control study, were increased by a mean of 1.6-fold for *R*- and 2.3-fold for S-fenoprofen.

As mentioned in the introduction, the coadministration of probenecid may reduce the velocity of aglycone and glucuronide elimination through different mechanisms. The competitive inhibition of the organic anion transporter appears to be most relevant with respect to the renal clearance of a compound. Renal excretion rates may also be reduced in renal impairment and elderly patients. Since a significant part of the clearance is represented by acyl glucuronide formation with its subsequent excretion and because aglycone and glucuronide elimination are interdependent, reduced glucuronide clearance will lead to elevated glucuronide levels and, as a consequence, elevated aglycone levels. For compounds subject to *in vivo* stereoinversion, the extent of formation of the *S*-enantiomer from the *R*-enantiomer may be increased, which will also lead to a shift in the *R*/*S* ratios of glucuronide AUCs (Figure 5.5). An increase in the fraction of the dose excreted into urine as *S*-glucuronide was detected in studies with rac-benoxaprofen and probenecid in humans (Spahn et al., 1987). The disposition of 2-phenylpropionic acid was studied in a renal-failure animal model. The clearances of both enantiomers were reduced. And again, the fraction of *R*-enantiomer being converted to the cyclooxygenase-inhibiting eutomer increased when compared to controls with normal renal function. For the interpretation of the data, changes in plasma protein binding may be of importance as well. A further complicating factor for data interpretation in renal dysfunction appears to be the involvement of the kidneys in the inversion process, with the possible consequence of reduced stereoinversion in this organ (Jones et al.,1986; Meffin et al., 1986; Yamaguchi and Nakamura, 1987).

Other clinical examples of the diminished clearance of carboxylic acids forming acyl glucuronides include the NSAIDs naproxen (Runkel et al., 1978; Anttila et al., 1980; Upton et al., 1984; Van den Ouwenland et al., 1988), ketoprofen (Stafanger et al., 1981; Upton et al., 1982; Advenier et al., 1983; Foster et al., 1989), and ximoprofen (Taylor et al., 1991), for which an accumulation of aglycone was observed when the renal clearance was reduced. Flurbiprofen was studied in uraemic patients (Knadler et al., 1992). Similar observations as for renal failure were made with probenecid coadministration, although the mechanism of this drug–drug interaction is not entirely renal. Results obtained for the stereoisomers of racemically administered carprofen when probenecid was dosed simultaneously (Spahn et al., 1989b) are summarized in Table 5.3. Probenecid coadministration led to a significant reduction of renal and nonrenal clearance processes. Although the highly variable and already

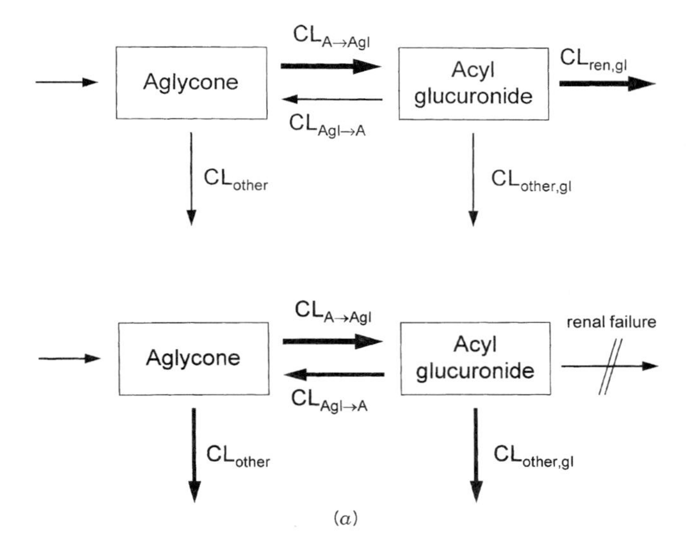

Figure 5.5(*a*). The meaning of "futile cycling" in clinical pharmacokinetics of chiral carboxylic acid drugs. *Top*: In the case of a reversible metabolic step, the kinetic characteristics for drug (or aglycone) and metabolite are interdependent. For many carboxylic acids, conjugation with glucuronic acid and subsequent excretion into urine represent the major elimination route in man, whereas other pathways, such as other metabolic routes as well as biliary excretion, are often of minor relevance. Changes in a particular clearance pathway may then affect the concentration-time curves of aglycone enantiomers (A) and glucuronide stereoisomers (Agl) to an extent that depends on the relevance of the respective pathway in the overall kinetic profile of each of the enantiomers. Except for compounds for which metabolic chiral inversion is observed, it is usually anticipated that the two enantiomers are independent of each other. However, *in vitro* glucuronidation data indicate that competitive inhibition occurs at higher concentrations.

Bottom: Reduction of renal excretion of the glucuronide (due to renal impairment or drug–drug interactions) was found to lead to an accumulation of glucuronide in the body and elevated aglycone levels due to the reversibility of this metabolic step (futile cycle). In summary, with respect to mass balance, increased aglycone backformation, increased alternative aglycone elimination, and increased nonrenal elimination of the glucuronide are observed without significant changes in the respective clearance values, except for the reduced renal clearance. When two aglycone enantiomers are present, they may behave differently in each of the steps.

Furthermore, an increase of acyl glucuronide levels will increase the extent of exposure of particular tissue structures to the reactive metabolites and increase the extent of covalent binding, which, again, may be different for the two stereoisomers.

[The changes of arrow sizes reflect the relative contribution of the respective clearance process in the overall mass transfer, not changes in clearance values, with normal and impaired renal function. $CL_{A \to Agl}$ = metabolic clearance of aglycone to the respective acyl glucuronide; $CL_{Agl \to A}$ = clearance value characterizing the regeneration of aglycone from the respective acyl glucuronide; CL_{other} = clearance of aglycone, which cannot be attributed to acyl glucuronide formation (renal and biliary excretion and other metabolic pathways); $CL_{ren,gl}$ = renal clearance of the acyl glucuronide; $CL_{other,gl}$ = clearance of the acyl glucuronide excluding renal excretion and aglycone backformation.]

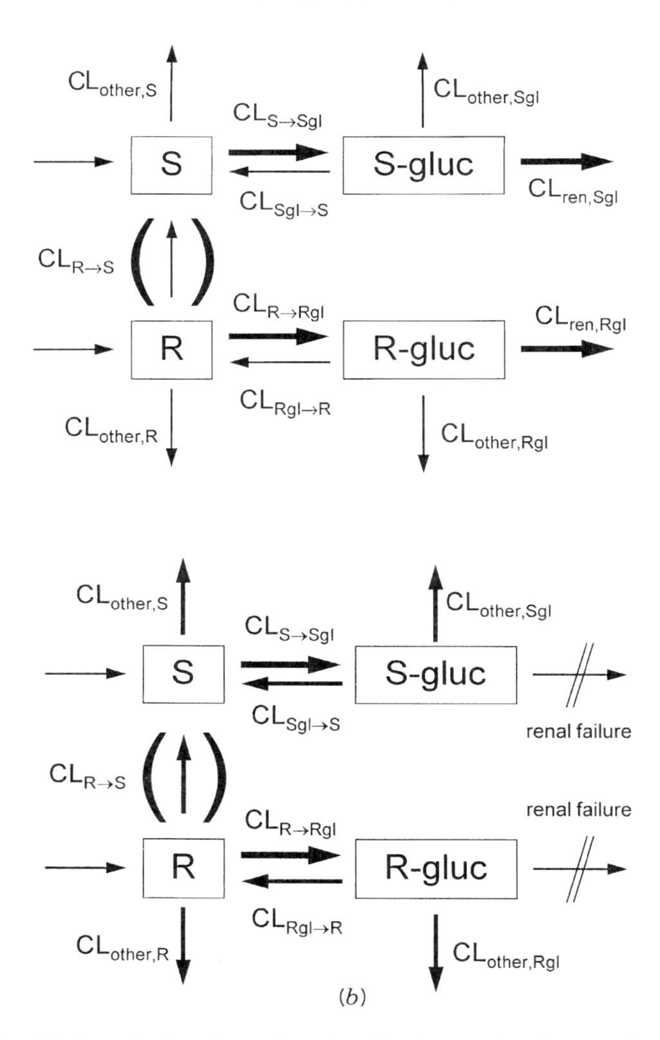

Figure 5.5(*b*): "Futile cycling" and stereoinversion. For drugs undergoing stereoinversion, the conversion of the *R* (−)-to the respective S(+)-enantiomer is possible in addition to the reversible glucuronidation step. As soon as stereoinversion occurs, which is the case for various 2-arylpropionic acid NSAIDs (species-dependent), both enantiomers have to be included in the kinetic model, to be able to understand the mechanisms of kinetic alterations under different conditions.

Top: As depicted in Figure 5.5(*a*), different elimination pathways are observed for the two enantiomers (*R*, *S*), the stereoselectivities of which may be different. Frequently, the major elimination pathway for these carboxylic acids in man is the formation of acyl glucuronides (*R*-gluc, *S*-gluc) with subsequent renal excretion of the respective glucuronides, whereas other elimination pathways for the aglycone enantiomers and excretion routes for the glucuronides are of minor relevance.

(*continued*)

Table 5.3. Kinetics of Carprofen and Its Glucuronide Conjugates With and Without Probenecid Coadministration: Clearances (True or Apparent Values) for *S*- and *R*-Carprofen and Their Glucuronides After Peroral Administration of 150 mg of Racemic Carprofen (arithmetic mean ± SD of three volunteers)

	S(+)-	*R*(–)-	*S*/*R* Ratio
A. Without Probenecid			
CL/F (ml/min)	45.0	35.1	0.77
CL_{ren} (ml/min)	0.44	0.39	1.02
$CL_{C\text{-}Cgl}$ (ml/min)	22.5	28.1	0.83
$CL_{C\text{-}other}$ (ml/min)	6.4	9.7	0.62
$CLC_{ren,gl}$ (ml/min)	35.6	26.4	1.37
B. With Probenecid			
CL/F (ml/min)	11.6	16.1	0.72
CL_{ren} (ml/min)	0.03	0.04	0.83
$CL_{C\text{-}Cgl}$ (ml/min)	1.99	3.24	0.72
$CL_{C\text{-}other}$ (ml/min)	9.6	12.9	0.75
$CLC_{ren,gl}$ (ml/min)	3.4	3.6	1.06

Concentrations of the diastereomeric glucuronides were calculated as *S*- or *R*-carprofen equivalents. Since only unconjugated carprofen was administered, the backformation of the aglyconce from the glucuronide was neglected, and the (apparent) kinetic parameters were calculated according to standard procedures. CL/F = apparent total clearance, where F is the bioavailability; CL_{ren} and $CL_{ren,gl}$ = renal clearances of parent drug and glucuronide; $CL_{C\text{-}Cgl}$ = metabolic clearance of carprofen to the glucuronide; $CL_{C\text{-}other}$ = clearance of carprofen, which cannot be attributed to renal excretion or glucuronidation.

Source: Spahn et al. (1989b).

Figure 5.5(*b*): (*contd.*)

Bottom: When the renal excretion of the two epimeric conjugates of the NSAID is significantly reduced due to renal failure, the glucuronides accumulate in the body and the fraction of the dose undergoing backformation of the aglycone (or possibly also other pathways) is increased. Furthermore, since the concentration of unconjugated *R*-enantiomer is higher than in subjects with normal renal function, the fraction undergoing stereoinversion is elevated to an extent that depends on the stereoinversion ratio.

[The changes of arrow sizes reflect the relative contribution of the respective clearance process in the overall mass transfer, not changes in clearance values, with normal and impaired renal function. $CL_{A \to Agl}$ = metabolic clearance of aglycone to the respective acyl glucuronide; $CL_{Agl \to A}$ = clearance value characterizing the regeneration of aglycone from the respective acyl glucuronide; CL_{other} = clearance of aglycone, which cannot be attributed to acyl glucuronide formation (renal and biliary excretion and other metabolic pathways); $CL_{ren,gl}$ = renal clearance of the acyl glucuronide; $CL_{other,gl}$ = clearance of the acyl glucuronide excluding renal excretion and aglycone backformation. Note that A and Agl have seen substituted by *R* and *S* and Rgl and Sgl, respectively; $CL_{R \to S}$ = stereoinversion clearance.]

low unchanged renal clearances of both enantiomers and their glucuronides were reduced to approximately one-tenth of their initial values, slightly higher clearance values were calculated for other clearance pathways. Interestingly, there was a tendency for *S*-carprofen glucuronide to be inhibited to a higher extent than *R*-glucuronide.

In addition to the increased exposure to the respective pharmacologically active aglycones, diminished renal clearance of the acyl glucuronides enhances the extent of exposure of the organism to these reactive metabolites.

5.9. *IN VITRO* AND *IN VIVO* STEREOSELECTIVITIES IN COVALENT BINDING TO PROTEINS VIA ACYL GLUCURONIDES: INCREASED ADDUCT LEVELS FOLLOWING INCREASED EXPOSURE

5.9.1. *In Vitro* Studies on the Extent of Covalent Binding to Proteins

Covalent binding of *R*- and *S*-carprofen glucuronides was examined employing the same incubation conditions as in other *in vitro* studies on covalent binding (Smith et al., 1986). The glucuronides were incubated with HSA (fatty-acid-free) at 37°C for 24 h. Carprofen was released from the adducts by treatment with 0.2 M potassium hydroxide at 80°C for 30 min. No covalent binding was observed when parent carprofen was incubated with HSA. After 1 h of incubation, the extent of covalent binding to HSA was higher for the *S*- than for the *R*-glucuronide, whereas after 24 h, covalent binding was significantly higher for *R*-carprofen glucuronide incubations.

Covalent binding was also found to be higher for *R*-fenoprofen (1.2%) than for *S*-fenoprofen (0.8%) when a 0.1-mM concentration of each conjugate enantiomer was incubated at pH 7.4 (37°C) (Volland et al., 1991). These data are in accordance with those obtained in studies with naproxen glucuronides (Bischer et al., 1995), where *in vitro* covalent binding was higher for the *R*- than for *S*-naproxen when a 50 μM concentration of each epimeric glucuronide was incubated under physiological conditions (pH 7.4, 37°C). This stereoselective difference was observed with an HSA-containing medium as well as in rat and human plasma, whereas incubation with unconjugated naproxen did not lead to covalent binding. Preincubation of HSA with acetylsalicylic acid and glucuronic acid as well as *S*-naproxen decreased the extent of covalent binding, again suggesting that lysine residues are important binding sites for covalent binding.

After incubation of beclobric acid β-1-O-acyl glucuronides with pooled plasma and HSA in pH 7.4 buffer, no significant difference between the two enantiomers was detected with respect to the magnitude of *in vitro* covalent binding. Similarly as with other acyl glucuronides (Dubois et al., 1993), the extent of covalent binding was clearly dependent on the concentration of the glucuronides, at least in a clinically relevant concentration range.

5.9.2. Extent of Covalent Binding to Proteins *In Vivo*

In three male volunteers, the extent of covalent binding of beclobric acid enantiomers was studied after single and multiple oral doses of racemic beclobrate (100 mg once daily) and covalent binding observed in all volunteers. The average maximum adduct densities for (–)- and (+)-beclobric acid were 0.147×10^{-4} and 0.177×10^{-4} mol/mol protein for a single beclobrate dose. Multiple dosing increased covalent binding three- to four-fold (Mayer et al., 1993b). From preliminary half-life calculations following single and multiple dosage, the terminal adduct half-life was found to be in the range of 2–3 days and was always shorter for the adduct of levorotatory beclobric acid.

In the case of fenoprofen, the analysis of plasma samples from a clinical study with the administration of 600 mg of rac-fenoprofen p.o. to healthy volunteers, revealed that—as opposed to *in vitro* conditions—the percentage of *S*-fenoprofen adduct was greater than that of the respective *R*-enantiomer. However, total adduct concentrations were fairly low (1.0 and 3.2 mol/mol protein $\times 10^{-4}$ for *R*- and *S*-fenoprofen) (Volland et al., 1991). An explanation for this discrepancy between *in vitro* and *in vivo* data is the stereoinversion that was detected in the case of fenoprofen with higher levels of *S*-fenoprofen and its glucuronide.

Accordingly, data on covalent binding obtained for benoxaprofen and flunoxaprofen in laboratory animals, volunteers, and patients showed that the extent of binding is higher for the S-enantiomer glucuronide, which is explained by the respective aglycone and glucuronide AUCs, which are elevated due to stereoinversion (Dahms and Spahn-Langguth, 1995 and 1996.)

Analysis of the plasma samples obtained from clinical studies following dosing of racemate as well as the single enantiomers indicated that the covalent binding of carprofen also occurs *in vivo* with maximum adduct densities of 0.83 ng/mg protein for a 25-mg dose of *R*-carprofen and 0.54 ng/mg protein for the same dose of *S*-carprofen. The discrepancy appears to be fully explained by a difference in the concentrations of the two epimeric glucuronides, where the AUC was higher for the *R*-carprofen glucuronide (13.2 vs. 9.0 $\mu g\ ml^{-1}\ h$). Fairly good correlations were found for the cumulative glucuronide AUCs vs. adduct AUCs, which appeared to be better than the correlations between glucuronide AUCs and maximum adduct levels (Iwakawa et al., submitted for publication. As observed with beclobric acid, adduct half-lives were different for the two carprofen stereoisomers and approximately 30% higher for the adduct derived from *R*-carprofen. They were, however, considerably longer than for beclobric acid, with values in the range of 14 days for the *R*- and 10 days for the *S*-carprofen adduct. Following the chronic dosage of carprofen (100 mg, 19 doses), maximum total adduct levels amounted to 4.3 ng/mg protein, whereas the total adduct after a single dose of enantiomer or racemate (25 or 50 mg, respectively) was in the range of 0.46–1.3 ng/mg protein (Iwakawa et al., submitted for publication).

As was expected, the extent of covalent binding of carprofen was also increased when probenecid was coadministered as summarized recently (Spahn-Langguth et al., 1994a), an observation that was also made with benoxaprofen. While probenecid did not significantly affect the *S*/*R* ratio of covalent binding in the case of carprofen, it was enhanced for benoxaprofen because of an increased stereoinversion (Spahn et al., 1987).

5.10. TREATMENT OF METABOLITE AND ADDUCT DATA

In general, when a potentially pharmacologically or toxicologically active compound is formed from a precursor, the input velocity and kinetic behavior of the precursor will affect the apparent residence time of the metabolite, also in the case of acyl glucuronides and the derived covalent adducts.

Hence, to analyze kinetic processes, in which acyl glucuronides are formed and covalent binding occurs, it may be beneficial to employ mean times analysis, since the various sequential metabolic steps may be regarded as a catenary chain. As for catenary chain compartments at the input site (von Hattingberg and Brockmeier, 1982), the mean residence times (MRTs) of the individual moieties are additive as summarized by Benet et al. (1992), Spahn-Langguth (1992) and Langguth (1994). When the drug and metabolite are both quantified, the systemic MRT of the metabolite—as if it was dosed as an i.v. bolus—is calculated from the ratio between the area under the first-moment curve and the area under the concentration-time curve, the AUMC/AUC ratio, of the metabolite obtained following dosage of the precursor and the AUMC/AUC ratio of the parent compound.

Therefore, when drug *A* is dosed and its metabolites *B* and *C* are measured in the catenary system depicted below, then the AUMC/AUC ratio (= the total mean residence time, MRT) for *C*, when *A* is dosed, will be equivalent to the sum of the MRTs of compounds *A*, *B*, and *C*. Hence, when drug or precursor *A* is dosed intravenously and *A*, *B*, and *C* are measured, then the MRTs of all compounds may be calculated as follows:

$$\mathrm{AUMC}_A^A/\mathrm{AUC}_A^A = \mathrm{MRT}_A^A$$

$$\mathrm{AUMC}_B^A/\mathrm{AUC}_B^A = \mathrm{MRT}_A^A + \mathrm{MRT}_B^B$$

$$\mathrm{AUMC}_C^A/\mathrm{AUC}_C^A = \mathrm{MRT}_A^A + \mathrm{MRT}_B^B + \mathrm{MRT}_C^C$$

where $AUMC_C^A$ and AUC_C^A represent the area under the first-moment curve and the area under the concentration-time curve for *C* when *A* is dosed. MRT_A^A, MRT_B^B, and MRT_C^C are the systemic mean residence times of compounds *A*, *B*, and C as obtained when they were or as if they would be dosed intravenously. With respect to the input of the metabolite, the AUMC/AUC ratio of the parent compound may be regarded as the mean input time (MIT) for the first metabolite (*B*).

For the oral dosage of *A* the AUMC/AUC ratio for *C* is equal to the sum of the AUMC/AUC ratio (= MIT + MRT) for *A* and the MRTs of *B* and *C*. Therefore, via mean time analysis it is possible to determine the systemic mean residence time of a metabolite—even with several metabolic steps and without the necessity to dose the metabolite *in vivo*—when the precursor and metabolite are both assayed. Important assumptions, however, are linear pharmacokinetics and the nonreversibility of the steps.

Sequential metabolism is frequently found with compounds undergoing phase-I and phase-II metabolism. Examples are the achiral lipid-regulating gemfibrozil or various β-adrenoceptor-blocking antagonists (e.g., *S*-penbutolol, *R*/*S*-carvedilol), for which there is evidence or it was clearly demonstrated that phase-I metabolites contribute to the overall pharmacokinetic profile (Brockmeier et al., 1988; Spahn-Langguth et al., 1994b; 1996, in preparation).

Another example for sequential metabolism with acyl glucuronide and adduct formation is the above-mentioned lipid-regulating agent beclobrate, for which the assumption of irreversibility of the steps within the catenary chain is regarded as correct, because the respective acyl glucuronides are rather stable, when compared with other acyl glucuronides and the extent of covalent binding is low. Furthermore, no stereoinversion or racemization was detected for beclobric acid *in vivo*. The respective catenary chain is depicted in Figure 5.6. The AUMC/AUCs for *A* after the dosage of *A*, and for *B* as well as *C* and *D* after the dosage of *A* were found to increase with each of the steps. The numbers obtained for the two enantiomers of beclobric acid or their metabolites following rac-beclobrate dosage are given in Table 5.4.

On the basis of the AUMC/AUC ratios of acyl glucuronides and the free carboxylic acid, the average MRTs of 1.1 h for the (–)- and 2.3 h for the (+)-beclobric acid glucuronide were found in the three investigated individuals. Very high MRTs were calculated for the covalent adducts, with that of (+)-beclobric acid being significantly higher (Spahn-Langguth, 1992; Mayer, 1993). In some cases, following the dosing of *A*, one might not know whether *B* procedes *C* or *C* procedes *B*. While the sequence is obvious for beclobrate metabolites, interpretation of the obtained mean times appears more difficult with fenoprofen and its metabolites as characterized by Benet and Volland [referenced in Benet et al. (1992)]. The AUMC/AUC values were not different

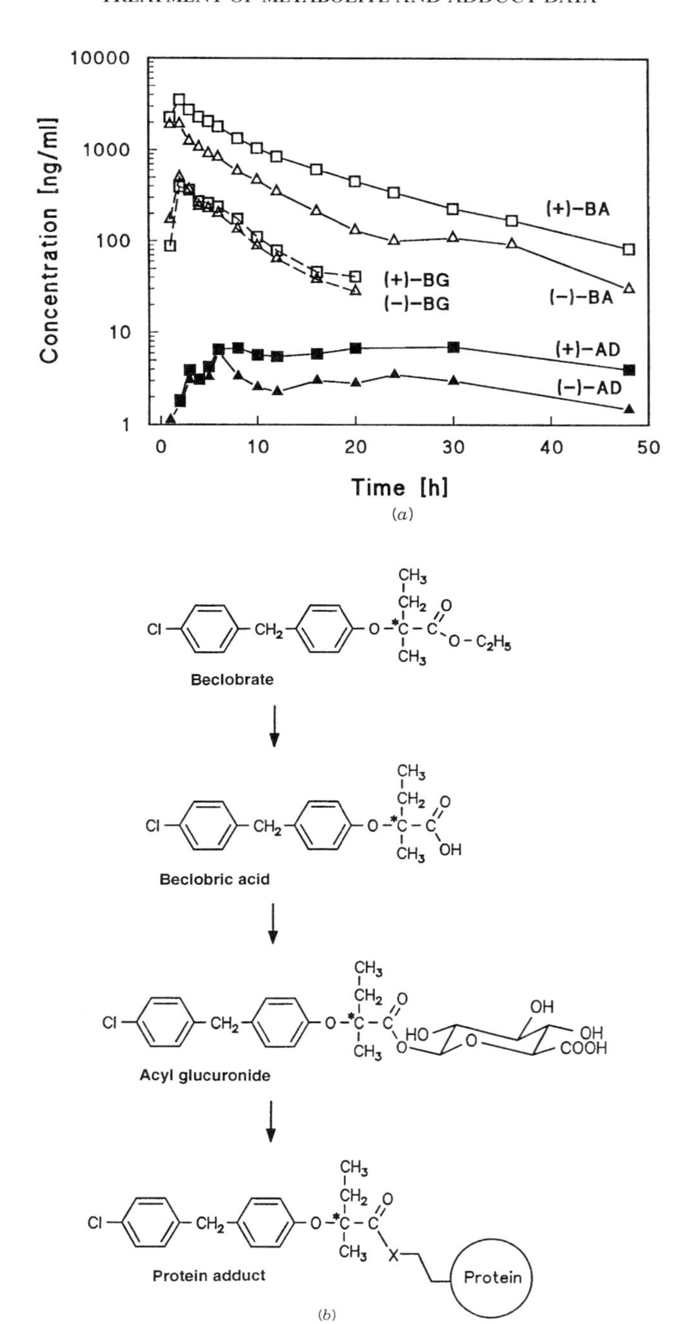

Concentration [ng/ml]
10000
1000
100
10
1
0
10
20
30
40
50
Time [h]
(+)-BA
(−)-BA
(+)-BG
(−)-BG
(+)-AD
(−)-AD
(a)
Beclobrate
Beclobric acid
Acyl glucuronide
Protein adduct
Protein
(b)

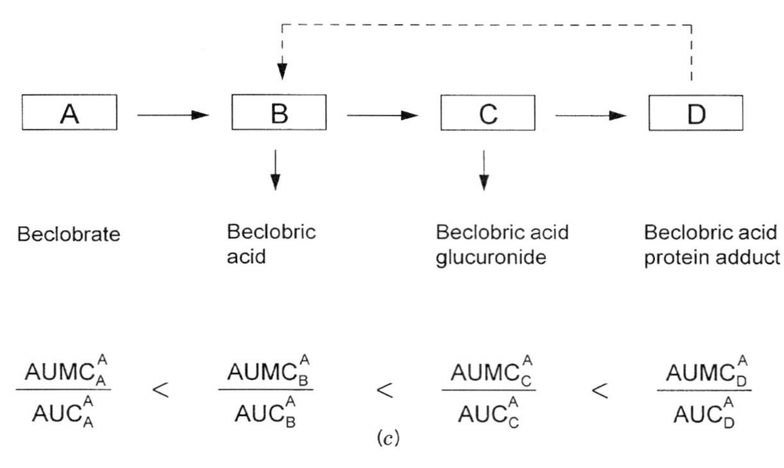

$$\frac{AUMC_A^A}{AUC_A^A} < \frac{AUMC_B^A}{AUC_B^A} < \frac{AUMC_C^A}{AUC_C^A} < \frac{AUMC_D^A}{AUC_D^A}$$

(*c*)

Figure 5.6. Treatment of glucuronide and adduct data—moment analysis: The metabolic steps observed following administration of rac-beclobrate may be treated as a catenary chain. Standard moment analysis procedures may be applied to calculate systemic residence times of aglycones and metabolites, since the backformation of aglycone from the glucuronide and the protein adduct were found to be negligible. Following p.o. dosage of rac-beclobrate, a lipid-regulating agent withdrawn from the Swiss market because of potential liver toxicity in 1990, beclobrate itself is hardly detected and only at very early sampling times. From the available data from preclinical and clinical studies, it has been concluded that a significant fraction of the beclobrate dose is hydrolyzed to beclobric acid before entering the systemic circulation. For beclobric acid, significant differences in the apparent clearances of the two enantiomers were detected, whereas the plasma concentration-time curves of their epimeric acyl glucuronides were very similar. The acyl glucuronides are reactive to some extent and form covalent adducts with plasma proteins, which appear to be eliminated in a stereoselective way. The plasma concentration-time curves of the consecutively formed beclobric acid enantiomers, the glucuronides, and covalent adducts obtained after a 100-mg single p.o. dose are depicted in (*a*) and the structural formulas of the respective compounds in (*b*). A scheme of the catenary chain is given in (*c*). The AUMC/AUC ratios, which are the total apparent mean residence times of the respective compounds following dosage of beclobrate, that is, the AUMC/AUCs for *A* after the dosage of *A*, for *B*, *C*, and *D* after the dosage of *A* were increasing and yielded the systemic mean residence times summarized in Table 5.4.

for fenoprofen, fenoprofen glucuronide, and 4-hydroxy fenoprofen glucuronide, whereas that for 4-hydroxy fenoprofen was significantly greater. This suggests that the measured concentrations for 4-hydroxy fenoprofen glucuronide are not derived from measured 4-hydroxy fenoprofen in plasma. Also, the plasma half-life ($t_{1/2}$) of 4-hydroxy fenoprofen was longer, indicating that the rate-limiting step for 4-hydroxy fenoprofen is not its formation. Apparently, hydroxy fenoprofen glucuronide formation occurs via sequential metabolic steps within the liver, whereas hydroxy fenoprofen that occurs in

Table 5.4. Average AUMC/AUC Ratios (total mean residence times following p.o. dosage of rac-beclobrate) and Calculated Systemic Mean Residence Times for Three Healthy Volunteers

		AUMC/AUC (h)	MRT (h)
Beclobric acid			
	(+)	12.7	
	(–)	11.7	
Beclobric acid glucuronide			
	(+)	15.1	2.3
	(–)	12.7	1.1
Beclobric acid protein adduct			
	(+)	119.6	104.5
	(–)	82.4	69.7

Source: Mayer (1992).

plasma following its formation in the liver is—most probably—not further glucuronidated. The kinetic analysis of the stereoisomer data is not yet available. Further complicating factors are the *in vivo* stereoinversion of fenoprofen, the lower stability of the acyl glucuronides (compared with beclobric acid), and the unknown stability of the glucuronide of hydroxy fenoprofen. [The role of first-pass metabolism for mean time calculations has been discussed by several authors, but only for a single metabolic step (Brockmeier and Ostrowski, 1985; Midha et al., 1983; Chan and Gibaldi, 1990).]

The objective of extended data analysis (on the basis of compartmental models) for different carboxylic acid drugs was the inclusion of a reversible step that is related to the hydrolysis of acyl glucuronide as well as the modeling of covalent binding. As opposed to gemfibrozil, in the case of beclobrate the compartmental analysis resulted in apparently negligible *in vivo* backformation of aglycone from the glucuronide. Employing a multi compartmental model (Figure 5.7), average transfer rates for both *in vivo* generation and elimination of the beclobric acid adducts differed slightly but significantly between the stereoisomers (Table 5.5) (Hermening et al., 1995). A similar approach is currently used to study the kinetics of carprofen and gemfibrozil, their acyl glucuronides and adducts.

The kinetic analysis may be further complicated by discontinuous drug input following p.o. dosage—probably due to gastric emptying phenomena (Langguth et al., 1994)—as described for example, for ibuprofen, flurbiprofen, furosemide, and etacrynic acid (Grahnen et al., 1984; Hammarlund et al., 1984; Parr et al., 1987; Dressman et al., 1992; Voith et al., 1993).

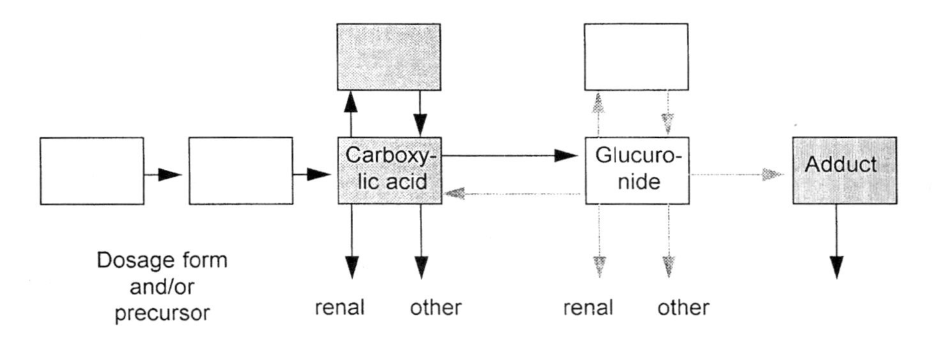

Figure 5.7. Compartmental approach including acyl glucuronide as well as adduct data: Multicompartmental model for carboxylic acids forming acyl glucuronides (modified from Hermening et al., 1995).

(For beclobric acid, where the two stereoisomers were treated independently because of the lack of stereoinversion, backformation of aglycone was negligible. For the aglycone enantiomers, a two-compartment disposition model yielded better-fitting criteria than a one- or more-compartment disposition model for the acyl glucuronides, whereas a one-compartment model was favorable for the adducts.)

Table 5.5. Rate Constants (h^{-1}) for the *In Vivo* Generation and Elimination of Beclobric Acid-Plasma Protein Adducts Determined Using the Compartmental Approach (Figure 5.7) for Single- and Multiple-Dose Data

	(+)	(–)	(+)/(–) Ratio
Generation	0.34 ± 0.08	0.56 ± 0.06	0.60
Elimination	0.010 ± 0.001	0.014 ± 0.004	0.71

Arithmetic means ± SD.
$n = 3$.
Source: Hermening et al. (1995).

5.11. IS THE EXTENT OF COVALENT BINDING PREDICTABLE?

As mentioned in the introductory part of this chapter, the extent of covalent binding for a particular compound corresponds to the extent of exposure to the reactive acyl glucuronide. However, when trying to compare different acyl glucuronides and their covalent binding to proteins *in vitro* or *in vivo*, the relationship becomes increasingly complex. When evaluating the *in vitro* covalent binding of the enantiomers of three chiral drugs via their glucuronides, carprofen, fenoprofen, and beclobric acid, it was found that long degradation half-lives of the β-1-O-acyl glucuronide appeared to correlate with low cova-

Table 5.6. Pseudo First-Order Degradation Rate Constants *k* for the β-1-O-Acyl Glucuronide of Different Chiral Carboxylic Acids (as determined in 150-mM pH 7.4 phosphate buffer at 37°C) and Moles of Aglycone Bound Covalently per Mole Protein (as determined following a 6-h incubation of 1 μM of the respective β-1-O-acyl glucuronide with 0.5-mM HSA at pH 7.4 and at 37°C)

	k (Buffer, h)	Adduct Density (moles / mole protein $\times 10^3$)
(+)-Beclobric acid	0.027	0.065
(–)-Beclobric acid	0.031	0.090
S(+)-Carprofen	0.22	1.00
R(–)-Carprofen	0.40	1.50
S(+)-Fenoprofen	0.36	1.51
R(–)-Fenoprofen	0.71	2.45
S(+)-Naproxen	0.40	1.03
R(–)-Naproxen	0.75	3.23

Source: Data from Mayer et al., 1993; Iwakawa et al., 1988, and Volland et al., 1991; as well as Iwaki et al. (submitted for publication).

lent binding (Table 5.6). By inclusion of the covalent binding data of acyl-glucuronides of achiral compounds into the data pool available in our laboratories, an excellent correlation was obtained between the moles of drug covalently bound per mole of protein of the adduct versus the degradation rate constant for the β-1-O-acyl glucuronide conjugates of each drug or enantiomer (Benet et al., 1993). That is, with respect to epimeric glucuronides of carboxylic acids, the least stable stereoisomer should usually exhibit the highest extent of covalent binding.

It may be expected that the relationship for *in vivo* reversible binding would be much more complex. However, the degree of covalent binding to plasma proteins should depend, at least, on the plasma concentrations of the acyl glucuronides and the degradation rate of each conjugate. The glucuronide concentrations are dependent on the rate of formation, its degradation and elimination as well as the administered dose. *In vivo* data available from different studies with achiral and chiral compounds showed a 25-fold variation in maximum adduct concentration, whereas AUCs for the glucuronide metabolites showed a 30-fold variation. Since for each drug there is a linear relationship between the amount covalently bound and the extent of glucuronide present (AUC), we normalize bound drug to AUC for comparison with *in vitro* degradation rates, yielding a significant linear correlation. Furthermore, there are indications that a structural relationship with the facility of covalent binding exists. The degree of substitution adjacent to the carboxyl group appears to determine the extent of covalent binding, with low binding, for fully substituted compounds (beclobric acid). However, this relationship needs to be further investigated, since some data available in the literature are not consistent with the obtained correlation (Benet et al., 1993).

5.12. THE POTENTIAL TOXICOLOGICAL SIGNIFICANCE OF THE REACTIVITY OF EPIMERIC ACYL GLUCURONIDES

Interestingly, of the 25 drugs withdrawn from the U.S. and British market during a period of 20 years (1964–1983) because of organ (liver and kidney) toxicity, six were carboxylic acid nonsteroidal antiinflammatory drugs, which are mainly metabolized via the formation of acyl glucuronides, that is, potentially reactive intermediates (Bakke et al., 1984). Examples of withdrawals that are chiral compounds include benoxaprofen, indoprofen,—very recently—suprofen, ketorolac, flunoxaprofen, and the lipid-regulating agent beclobrate, which is rapidly hydrolzed to the respective ester *in vivo*. Of course, a number of chiral carboxylic acid drugs are still on the market, although it is known in some cases that hypersensitivity reactions may occur.

As reviewed and hypothesized earlier, the mechanism of toxic side-effects may be of a different nature (Faed, 1984; Spahn-Langguth and Benet, 1992b). For nonsteroidal antiinflammatory drugs, one well-known hypothesis states that hypersensitivity reactions (anaphylactoid reactions as well as organotoxic reactions) are based on the pharmacologic profile of the compound. Inhibition of prostaglandin synthesis leads to an *increased production of leukotrienes.* Therefore, any compound that inhibits cyclooxygenase would then lead to such side-effects in predisposed patients, that is, the side-effects would be associated with the "eutomer."

The early studies of Stogniew and Fenselau (1982) on the reactivity of acyl glucuronides towards smaller endogenous nucleophiles (e.g., glutathione) were the basis for a second hypothesis, in which organ toxicities are explained by *glutathione depletion* and are hence independent of the drug's mode of action. Since gluthathione is chiral, it may be anticipated that the reactivity of two epimeric acyl glucuronides can be different. This stereochemical aspect is currently being investigated in our laboratories.

From studies reported in the literature, there is evidence that antibodies occur in the blood of patients, for example, at low levels following the dosage of acetylsalicylic acid and valproic acid (Amos et al., 1971; Williams et al., 1992). An immunological mechanism may be the basis for anaphylactic reactions as well as organ toxicity, since covalent binding to tissue proteins may be followed by subsequent antigen-antibody complex formation. Hence, the third hypothesis includes the theory that acyl glucuronides act as haptens, which, upon covalent binding to proteins, become *immunogens* and induce antibody production, as shown in the work of Zia-Amirhosseini (1994, Zia-Amirhosseini et al., 1995, as well as Williams et al., 1995). Antiadduct antibodies formed in mice following the adminstration of adducts with mice serum albumin appeared to be specific for the aglycone, but not for the glucuronic acid part of the molecule. Some cross-reactivity was observed for structurally related aglycones. The relevance of stereoisomerism in antibody-specificity with respect to compounds, which are adminstered as racemates, has not been elucidated yet.

In summary, it may be stated that for the first hypothesis for the toxicity of NSAIDs, only the concentrations of unconjugated eutomer are relevant. For the third hypothesis, the two determinants of potential toxicity are the reactivity of the acyl glucuronide and the extent of exposure of nucleophiles to the reactive glucuronide, reflected in its concentration-time profile in blood and tissues. Regarding their immunogenic potential, the structure of the aglycone as well as the persistence of the adduct in the organism are of importance. Although for these (kinetic) processes, stereochemical aspects have already been studied for various compounds, the question of the relevance of chirality has not yet been addressed in mechanistic studies of hypersensitivity.

ACKNOWLEDGMENTS

Studies performed at facilities of the School of Pharmacy, University of California in San Francisco, as well as of the Department of Pharmacology in Frankfurt/M. and the Department of Pharmaceutical Chemistry in Halle/Saale were in part supported by the National Institutes of Health grant GM 36633 (L. Z. B.) by grants from the Deutsche Forschungsgemeinschaft (H. S.-L.), Dr. Robert-Pfleger Stiftung (H. S.-L.), and the Fonds der Chemischen Industrie (H. S.-L.).

REFERENCES

Advenier, C., Roux, A., Gobert, C., Massias, P., Varoquaux, O., and Flouvat, B. (1983). Pharmacokinetics of ketoprofen in the elderly. *Br. J. Clin. Pharmac.*, **16**, 65–70.

Amos, H. E, Wilson, D. V, Taussig, M. J., and Carlton, S. J. (1971). Hypersensitivity reactions to acetylsalicylic acid. *Clin. Exp. Immunol.*, **8**, 563–572.

Anttila, M., Haataja, M., and Kasanan, A. (1980). Pharmacokinetics of naproxen in subjects with normal and impaired renal function. *Eur. J. Clin. Pharmacol.*, **18**, 263–268.

Arima, N. (1990). Acyl glucuronidation and glucosidation of pranoprofen, a 2-arylpropionic acid derivative, in mouse liver and kidney homogenates. *J. Pharmacobio.-Dyn.*, **13**, 724–732.

Arima, N. (1990). Stereoselective acyl glucuronidation and glucosidation of pranoprofen, a 2-arylpropionic acid derivative, in mice *in vivo*. *J. Pharmacobio.-Dyn.*, **13**, 733–738.

Bakke, O. M., Wardell, W. M., and Lasagna, L. (1984). Drug discontinuations in the United Kingdom and the United States: 1964–1983. *Clin. Pharmacol. Ther.*, **35**, 559–567.

Benet, L. Z. and Spahn, H. (1988). Acyl migration and covalent binding of drug glucuronides—potential toxicity mediators. In G. Siest, J. Magdalou, B. Burchell (eds.) *Molecular and Cellular Aspects of Glucuronidation*, Colloques INSERM/John Libbey Eurotext, Vol. 173, London: pp. 261–269.

Benet, L. Z., Bisher, A., Volland, C., Spahn-Langguth, H. (1992). The pharmaco- and toxicokinetics of reactive and active phase II metabolites. In D. J. A. Crommelin, K. K. Midha (eds.), *Topics in Pharmaceutical Sciences*, Med. Pharm. Scientific Publishers, Stuttgart, 533–544.

Benet, L. Z., Spahn-Langguth, H., Iwakawa, S., Volland, C., Mizuma, T., Mayer, S., Mutschler, E., and Lin, E. T. (1993). Prediction of covalent binding of acidic drugs in man. *Life Sci.*, **53**, PL141–146.

Bischer, A., Zia-Amirhosseini, P., Iwaki, M., McDonagh, A. F., and Benet, L. Z. (1995). Stereoselective binding properties of naproxen glucuronide diastereomers to proteins. *J. Pharmacokinet. Biopharm.* (in press).

Blanckaert, N., Compernolle, F., Leroy, P., Van Hautte, R., Fevery, J., and Heirwegh, K. P. M. (1978). The fate of bilirubin-IXα glucuronide in cholestasis and during storage *in vitro*. *Biochem. J.*, **171**, 203–214.

Bradow, G., Khan, L., and Fenselau, C. (1989). Studies of intramolecular rearrangements of acyl-linked glucuronides using salicylic acid, flufenamic acid, and (S)- and (R)-benoxaprofen and confirmation of isomerization in acyl-linked δ^9-11-carboxytetrahydrocannabinol glucuronide. *Chem. Res. Toxicol.*, **2**, 316–324.

Brockmeier, D. and Ostrowski, J. (1985). Mean-time and first-pass metabolism *Eur. J. Clin. Pharmacol.*, **29**, 45–48.

Brockmeier, D., Hajdu, P., Henke., W., Mutschler, E., Palm, D., Rupp, W., Spahn, H., Verho, M. T., and Wellstein, A. (1988). Pharmacokinetics of penbutolol, its effect on exercise heart rate and *in vitro* inhibition of radioligand binding using plasma samples *Eur. J. Clin. Pharmacol.*, **29**, 45–48.

Burchell, B. and Coughtrie, M. W. H. (1989). UDP-glucuronosyltransferases. *Pharmacol. Ther.*, **43**, 261–289.

Buszewski, B., el Mouelhi, M., Albert, K., and Bayer, E. (1990). Influence of the structure of chemically bonded C18 phases on HPLC separation of naproxen glucuronide diastereomers. *J. Liq. Chromatogr.*, **13**, 505–524.

Castillo, M., and Smith, P. C. (1995). Disposition and reactivity of ibuprofen and ibufenac acyl glucuronides *in vivo* in the Rhesus monkey and *in vitro*, with human, serum albumin. *Drug Metab. Disp.*, **23**, 566–572.

Chakir, S., Maurice, M. H., Magdalou, J., Leroy, P., Dubois, N., Lapicque, F., Abdelhamid, Z., and Nicolas, A. (1994). High-performance liquid chromatographic enantioselective assay for the measurement of ketoprofen glucuronidation by liver microsomes, *J. Chromatogr.*, **654**, 61–68.

Chan, K. K. H. and Gibaldi, M. (1990). Effects of first-pass metabolism on metabolite mean residence time determination after oral administration of parent drug. *Pharm. Res.*, **7**, 59–63.

Cox, J. W., Cox, S. R., VanGiessen, G., and Ruwart, M. J. (1985). Ibuprofen stereoisomer hepatic clearance and distribution in normal and fatty *in situ* perfused rat liver. *J. Pharmacol. Exp. Ther.*, **232**, 636–643.

Dahms, M. and Spahn-Langguth, H. (1995). Acyl glucuronides as reactive intermediates: Detection of flunoxaprofen protein adducts in biological material. *Naunyn Schmiedeberg's Arch. Pharmacol.*, **351** (suppl.), R5.

Dahms, M. and Spahn-Langguth, H., (1996). Covalent binding of acidic drugs via reactive intermediates: Detection of benoxaprofen and flunoxaprofen adducts in biological material. *Die Pharmazie* (in press).

Dickinson, R. G. and King, A.R. (1993). Studies on the reactivity of acyl glucuronides–V. Glucuronide-derived covalent binding of diflunisal to bladder tissue of rats and its modulation by urinary pH and β-glucuronidase. *Biochem. Pharmacol.*, **46**, 1175–1182.

Ding, A., Ojingwa, J.C., McDonagh, A.F., Burlingame, A.L., and Benet, L. Z. (1993). Evidence for covalent binding of acyl glucuronides to serum albumin via an imine mechanism as revealed by tandem mass spectrometry. *Proc. Nat. Acad. Sci. USA*, **90**, 3797–3801.

Ding, A., Zia-Amirhosseini, P., McDonagh, A. F., Burlingame, A. L., and Benet, L. Z. (1995). Reactivity of tolmetin glucuronide with human serum albumin: Identification of binding sites and mechanisms of reaction by tandem mass spectrometry. *Drug Metab. Disp.*, **23**, 369–376.

Dressman, J. B., Beradi, R. R., Elta, G. H., Gary, T. M., Montgomery, P. A., Lau, H. S., Pelekoudas, K. L., Szpunar, H. S., and Wagner, J. G. (1992). Absorption of flurbiprofen in the fed and fasted state. *Pharm. Res.*, **9**, 901–907.

Dubois, N., Lapicque, F., Maurice, M. H., Pritchard, M., Fournel-Gigleux, S., Magdalou, J., Abiteboul, M., Siest, and G., Netter, P., (1993). *In vitro* irreversible binding of ketoprofen glucuronide to plasma proteins. *Drug Metab. Disp.*, **21**, 617–623.

Dudley. K. H. (1985). Commentary: Stereochemical formulas of β-D-glucuronides. *Drug Metab. Disp.*, **13**, 524–528.

el Mouelhi, M., Ruelius, H. W., Fenselau, C., and Dulik, D. M. (1988). Species-dependent enantioselective glucuronidation of three 2-arylpropionic acids: Naproxen, ibuprofen and benoxaprofen. *Drug Metab. Disp.*, **16**, 627–634.

el Mouelhi, M. and Schwenk, M. (1991). Stereoselective glucuronidation of naproxen in isolated cells from liver, stomach, intestine, and colon of the guinea pig. *Drug Metab. Disp.*, **19**, 844–845.

Faed, E. M. (1984). Properties of acyl glucuronides: Implications for studies of the pharmacokinetics and metabolism of acidic drugs. *Drug Metab. Rev.*, **15**, 1213–1249.

Foster, R. T., Jamali, F., and Russell, A. S. (1989). Pharmacokinetics of ketoprofen enantiomers in cholecystectomy patients: Influence of probenecid. *Eur. J. Clin. Pharmacol.*, **37**, 589–594.

Fournel-Gigleux, S., Hamar-Hansen, C., Motassim, N., Antoine, B., Mothe, O., Decolin D., Caldwell, J., and Siest, G. (1988). Substrate-specificity and enantioselectivity of arylcarboxylic acid glucuronidation. *Drug Metab. Disp.*, **16**, 627–634.

Fournel-Gigleux, S., Magdalou, J., Lafaurie, C., Siest, G., Grislain, L., Garnier, M. H., Dabé, J. F., Luijten, W., Bromet, N., and Devissaget, M. (1988). Glucuronidation of perindopril by hepatic microsomes: Interspecies comparison, In G. Siest, J. Magdalou, and B. Burchell (eds.), *Molecular and Cellular Aspects of Glucuronidation*, Colloques INSERM/John Libbey Eurotext, Vol. 173, London pp. 305–309.

Geisslinger, G., Ferreira, S. H., Menzel, S., Schlott, D., and Brune, K. (1994). Antinociceptive actions of R (–)-flurbiprofen—a non-cyclooxygenase inhibiting 2-arylpropionic acid—in *rats*. *Life Sci.*, **54**, PL173–177.

Grahnen, A., Hammarlund, M., and Lundquist, T. (1984). Implications of intraindividual variability in bioavailability studies of furosemide. *Eur. J. Clin. Pharmacol.*, **27**, 595–602.

Grubb, N., Caldwell, J., and Strolin-Benedetti, M. (1993). Excretion balance and urinary metabolism of indobufen in rats and mice. *Biochem. Pharmacol.*, **46**, 759–761.

Grubb, N., Weil, A., and Caldwell, J. (1993). Studies on the *in vitro* reactivity of clofibryl and fenofibryl glucuronides. Evidence for protein binding via a Schiff's base mechanism. *Pharmacology.*, **46**, 357–364.

Gugler, R., Kurten, J. W., Jensen, C. J., Klehr, U., Hartlapp, J. (1979). Clofibrate disposition in renal failure and acute and chronic liver disease. *Eur. J. Clin. Pharmacol.*, **230,** 237–241.

Hamar-Hansen, C., Fournel-Gigleux, S., Magdalou, J., Butin, J. A., and Siest, G. (1986). Liquid-chromatographic assay for the measurement of glucuronidation of arylcarboxylic acids using uridine diphospho [V-^{14}C] glucuronic acid. *J. Chromatogr.*, **383**, 51–60.

Hammarlund, M. M., Paalzow, L. K., and Odlind, B. (1984). Pharmacokinetics of furosemide in man after intravenous and oral administration. Application of moment analysis. *Eur. J. Clin. Pharmacol.*, **26**, 197–207.

Hargus, S. J., Martin, B. M., and Pohl, L. R. (1995). Covalent modification of rat liver dipeptidyl peptidase IV by diclofenac metabolites. *The International Symposium on Biological Reactive Intermediates*, Munich, Jan. 1995, abstract vol., p. 74.

Hasegawa, J., Smith, P. C., and Benet, L. Z. (1982). Apparent intramolecular acyl migration of zomepirac glucuronide. *Drug Metab. Disp.*, **10**, 469–473.

von Hattingberg, H. M., and Brockmeier, D. (1982). A concept for the assessment of bioavailability in complex systems in terms of amounts and rates. In G. Bozler, and J. M. van Rossum (eds.), *Pharmacokinetics During Drug Development: Data Analysis and Evaluation Techniques*, Stuttgart: G. Fischer Verlag, 315–323.

Hayball, P. J., and Meffin, P. J. (1987). Enantioselective disposition of 2-arylpropionic acid nonsteroidal anti-inflammatory drugs. III. Fenoprofen disposition. J. *Pharmacol. Exp. Ther.*, **240**, 631–636.

Hayball. P. J., Nation, R. L., Bochner, F., Newton, J. L., and Massy-Westropp, R. A. (1991). Plasma protein binding of ketoprofen enantiomers in man: Method development and its application. *Chirality*, **3**, 460–466.

Hayball, P. J., Nation, R. L., and Bochner, F. (1992). Stereoselective interactions of ketoprofen glucuronides with human plasma protein and serum albumin. *Biochem. Pharmacol.*, **44**, 291–299.

Hayball, P. J., Nation, R. L., Bochner, F., Sansom, L. N., Ahern, M. J., and Smith, M. D. (1991). The influence of renal function on the enantioselective pharmacokinetics and pharmacodynamics of ketoprofen in patients with rheumatoid arthritis. *Br. J. Clin. Pharm.*, **31**, 546–550.

Hayball, P. J., Wrobel, J., Tamblyn, J. G., and Nation, R. L. (1994). The pharmacokinetics of ketorolac enantiomers following intramuscular administration of the racemate. *Br. J. Clin. Pharm.*, **37**, 75–78.

Hermening, A., Gräfe, A. K., Mayer, S., Mutschler, E., and Spahn-Langguth, H. (1995). Pharmacokinetics of fibrates and their acyl glucuronides: Inclusion of a reversible metabolic step and covalent binding into kinetic modelling. *Naunyn Schmiedeberg's Arch. Pharmacol.*, **351** (suppl.), R5.

Hutt, A. J., Caldwell, J., and Fournel-Gigleux, S. (1988). The metabolic chiral inversion and dispositional enantioselectivity of the 2-arylpropionic acids and their biological consequences. *Biochem. Pharmacol.*, **37**, 105–114.

Iwakawa, S., Spahn, H., Benet, L. Z., and Lin, E. T. (1988). Carprofen glucuronides: Stereoselective degradation and interaction with human serum albumin. *Pharm. Res.*, **5** (suppl.), S214.

Iwakawa, S., Suganuma, T., Lee, S. F., Spahn, H., Benet, L. Z., and Lin, E. T. (1989). Direct determination of diastereomeric carprofen glucuronides in human plasma and urine and preliminary measurements of stereoselective metabolic and renal elimination after oral administration of carprofen in man. *Drug Metab. Disp.*, **17**, 474–480.

Iwakawa, S., Spahn, H., Benet, L. Z., and Lin, E. T. (1990). Stereoselective binding of the glucuronide conjugates of carprofen enantiomers to human serum proteins. *Biochem. Pharmacol.*, **39**, 949–953.

Iwakawa, S., Spahn, H., Benet, L. Z., and Lin, E. T. (1991). Stereoselective disposition of carprofen, flunoxaprofen, and naproxen in rats. *Drug Metab. Disp.*, **19**, 853–857.

Iwakawa, S., Spahn-Langguth, H., Benet, L. Z., and Lin, E. T. (1997). Carprofen acyl glucuronides: Stereoselective degradation and interaction with human serum proteins (submitted for publication.)

Iwaki, M., Bischer, A. C., Nguyen, A. C., McDonagh, A. F., and Benet, L. Z. (1995). Stereoselective disposition of naproxen glucuronide in the rat. *Drug Metab. Disp.* (in press).

Jones, M. E., Sallustio, B. C., Purdie, Y. J., and Meffin. P. J. (1986). Enantioselective disposition of 2-arylpropionic acid nonsteroidal anti-inflammatory drugs. II. 2-Phenylpropionic acid protein binding. *J. Pharmacol. Exp. Ther.*, **238**, 288–294.

King, A. R., and Dickinson, R. G. (1993). Studies on the reactivity of acyl glucuronides. IV. Covalent binding of diflunisal to tissues of the rat. *Biochem. Pharmacol.*, **45**, 1043–1047.

Klaasen, C. D., and Watkins, J. B. (1984). Mechanisms of bile formation, hepatic uptake and biliary excretion. *Pharmacol. Rev.*, **36**, 1–67.

Knadler, M. P., and Hall, S. D. (1991). Stereoselective hydrolysis of flurbiprofen conjugates. *Drug Metab. Disp.*, **19**, 280–282.

Knadler, M. P., Brater, D. C., and Hall, S. D. (1992). Stereoselective disposition of flurbiprofen in uraemic patients. *Br. J. Clin. Pharm.*, **33**, 377–383

Kretz-Rommel, A. and Boelsterli, U. A. (1994). Selective protein adducts to membrane proteins in cultured rat hepatocytes exposed to diclofenac: Radiochemical and immunochemical analysis. *Mol. Pharmacol.*, **45**, 237–244.

Kroemer, H. K., and Klotz, U. (1992). Glucuronidation of drugs. A re-evaluation of the pharmacological significance of the conjugates and modulating factors. *Clin Pharmacokinet.*, **23**, 292–310.

Langguth, P. (1994). Compartmental approaches and integral input parameters. Ist International Workshop on Strategies for Oral Drug Delivery, Uppsala, Sept. 1994.

Langguth, P., Lee, K. M., Spahn-Langguth, H., and Amidon, G. L. (1994). Variable gastric emptying and discontinuities in drug absorption profiles: Dependence of rates and extent of cimetidine absorption on motility phase and pH. *Biopharm. Drug Disp.*, **15**, 719–746.

Magdalou, J., Chajes, V., Lafaurie, C., and Siest, G. (1990). Glucuronidation of 2-arylpropionic acids pirprofen, flurbiprofen, and ibuprofen by liver microsomes. *Drug Metab. Disp.*, **18**, 692–697.

Mayer, S. (1992). Analytik und Pharmakokinetik des chiralen Lipidsenkers Beclobrat. Ph.D. Thesis, Department of Pharmacology, Johann Wolfgang Goethe-University Frankfurt/M.

Mayer, S., Spahn-Langguth, H., Gikalov, I., and Mutschler, E. (1993a). Pharmacokinetics of beclobric acid enantiomers and their glucuronides after single and multiple p.o. dosage of rac-beclobrate. *Arzneim-Forsch./Drug Res.*, **42–2**, 1354–1358.

Mayer, S., Mutschler, E., Benet, L. Z., and Spahn-Langguth, H. (1993b). *In vitro* and *in vivo* irreversible plasma protein binding of beclobric acid enantiomers. *Chirality*, **5**, 120–125.

McKinnon, G. E., and Dickinson, R. G. (1989). Covalent binding of diflunisal and probenecid to plasma protein in humans: Persistence of the adducts in the circulation. *Res. Commun. Chem. Pathol. Pharmacol.*, **66**, 339–354.

Meffin, P. J., Sallustio, B. C., Purdie, Y. J., and Jones, M. E. (1986). Enantioselective disposition of 2-arylpropionic acid nonsteroidal antiinflammatory drugs: I. 2. Phenylpropionic acid disposition. *J. Pharmacol. Exp. Ther.*, **238**, 280–287.

Midha, K. K., Roscoe, R. M. H., Wilson, T. W., and Cooper, J. K. (1983). Pharmacokinetics of glucuronidation of propranolol following oral administration in humans. *Biopharm. Drug Disp.*, **4**, 331–338.

Nakamura, Y., and Yamaguchi, T. (1987). Stereoselective metabolism of 2-phenylpropionic acid in rats: I. *In vitro* studies on the stereselective isomerization and glucuronidation of 2-phenylpropionic acid. *Drug Metab. Disp.*, **15**, 529–534.

Ojingwa, J., Spahn-Langguth, H., and Benet, L. Z. (1994). Reversible binding of tolmetin, zomepirac and their glucuronide conjugates to human serum albumin. *J Pharmacokin. Biopharm.*, **22**, 19–40.

Ojingwa, J., Spahn-Langguth, H., and Benet, L. Z. (1994). Irreversible binding of tolmetin to macromolecules via its glucuronide: Binding to blood constituents, tissue homogenates and subcellular fractions *in vitro*. *Xenobiotica*, **24**, 495–506.

Ott, R. J., Giacomini, K. M. (1993). Stereoselective transport of drugs across epithelia. In: I. W. Wainer, (ed), *Drug Stereochemistry: Analytical Methods and Pharmacology*, New York: Marcel Dekker, 281–314.

Parr, A. F., Beihn, R. M., Franz, R. M., Sypunar, G. J., and Jay, M. (1987). Correlation of ibuprofen bioavailability with gastrointestinal transit by scintigraphic monitoring of ^{171}Er-labeled sustained release tablets. *Pharm. Res.*, **4**, 486–491.

Pedrazzini, S., De Angelis, M., Zanoboni Muciaccia, Z., Sacchi, C., and Forgione, A. (1988). Stereochemical pharmacokinetics of the 2-arylpropionic acid non-steroidal antiinflammatory drug flunoxaprofen in rats and man. *Arzneim-Forsch./Drug Res.*, **38**, 1170–1175.

Rowe, B. J., and Meffin, P. J. (1984). Diisopropylfluorophosphate increases clofibric acid clearance: Supporting evidence for a futile cycle. *J. Pharmacol. Exp. Ther.*, **230**, 237–241.

Runkel, R., Mroszczak, E., Chaplin, M., Sevelius, H., and Segre, E. (1978). Naproxen–probenecid interaction. *Clin. Pharmacol. Ther.*, **24**, 706–713.

Sallustio, B. C., Knights, K, M., Roberts, B. J., and Zacest, R. (1991). *In vivo* covalent binding of clofibric acid to human plasma proteins and rat liver proteins. *Biochem. Pharmacol.*, **42**, 1421–1425.

Segre, G., Bianchi, E., and Zanolo, G. (1988). Pharmacokinetics of flunoxaprofen in rats, dogs, and monkeys. *J. Pharm. Sci.*, **77**, 670–673.

Simmonds, R. G., Woodage, T. J., Duff, S. M., and Green, J. N. (1980). Stereospecific inversion of R-(–)-benoxaprofen in rat and man. *Eur. J. Drug Metab. Pharmacokinet.*, **5**, 169–172.

Smith, P. C., and Benet, L. Z. (1986). Characterization of the isomeric esters of zomepirac glucuronide by proton NMR. *Drug Metab. Disp.*, **14**, 503–505.

Smith, P. C., Langendiijk, P. N. J., Hasegawa, J., and Benet, L. Z. (1985). Effect of probenecid on the formation and elimination of acyl glucuronides: Studies with zomepirac. *Clin. Pharmacol. Ther.*, **38**, 121–127.

Smith, P, C., McDonagh, A. F., and Benet, L. Z. (1986). Irreversible binding of zomepirac to plasma protein *in vitro* and *in vivo*. *J. Clin. Invest.*, **77**, 934–939.

Smith, P. C., Benet, L. Z., and McDonagh, A. F. (19890a). Covalent binding of zomepirac glucuronide to proteins: Evidence for a Schiff base mechanism. *Drug Metab. Disp.*, **18**, 639–644.

Smith, P. C., McDonagh, A. F., and Benet, L. Z. (1990b). Effect of an esterase inhibitor on the disposition of zomepirac and its covalent binding to plasma proteins in the guinea pig. *J. Pharmacol. Exp. Ther.*, **230**, 218–224.

Smith, P. C., Song, W. Q., and Rodriguez, R. J. (1992). Covalent binding of etodolac acyl glucuronide to albumin *in vitro*. *Drug. Metab. Disp.*, **20**, 962–965.

Spahn, H. (1988). Assay method for the product formation in *in vitro* enzyme kinetic studies of uridine diphosphoglucuronosyltransferases: 2-Arylpropionic acids. *J. Chromatogr.*, **430**, 368–375.

Spahn, H. (1989). Characterization of stereoselective processes in drug metabolism and pharmacokinetics. Habilitation Thesis, Department of Pharmacology, Johann Wolfgang Goethe-University, Frankfurt/M.

Spahn, H, and Benet, L. Z. (1987a). Influence of probenecid on the glucuronidation of naproxen enantiomers by hepatic UDP-glucuronosyltransferases in rat liver microsomes. *Pharm. Res.*, **4** (suppl.), S111.

Spahn, H., and Benet, L. Z. (1987b). Enantioselectivity of hepatic UDP-glucuronyltranserases in rat liver microsomes towards 2-arylpropionic acids: Glucuronidation of naproxen enantiomers. In J. M. Aiache, J. Hirtz (eds.), *3rd European Congress of Biopharmaceutics and Pharmacokinetics Proceedings*, Vol II, *Experimental Pharmacokinetics*, Freiburg 1987, 261–268.

Spahn, H., Hale, V., Iwakawa, S., and Benet, L. Z. (1989c). Effects of probenecid on the hepatic elimination of flunoxaprofen. *Naunyn Schmiedeberg's Arch. Pharmacol.* (suppl.), 14.

Spahn, H., Iwakawa, S., Lin, E. T., and Benet, L. Z. (1989a): Procedures to properly characterize *in vivo* and the *in vitro* enantioselective glucuronidation: studies with benoxaprofen glucuronides. *Pharm. Res.*, **6**, 125–132.

Spahn, H., Iwakawa, S., Lin, E. T., and Benet, L. Z. (1987). Influence of probenecid on the urinary excretion rates of the diastereomeric benoxaprofen glucuronides. *Eur. J. Drug Metab. Pharmacokinet.*, **12**, 233–239.

Spahn, H., Iwakawa, S., and Benet, L. Z. (1988). Stereoselective formation and degradation of flunoxaprofen glucuronides in microsomal incubations. *Pharm. Res.*, **5** (suppl.), S200.

Spahn, H., Iwakawa, S., Ojingwa, J., and Benet, L. Z. (1988b). Glucuronidation of flunoxaprofen enantiomers by UDPGTs from different sources. International Conference on Pharmaceutical Sciences and Clinical Pharmacology, Jerusalem, May/June 1988.

Spahn, H., Spahn, I., Pflugmann, G., Mutschler, E., and Benet, L. Z. (1988c). Measurement of carprofen enantiomers in human plasma and urine using L-leucinamide as chiral coupling component. *J. Chromatogr.*, **433**, 331–338.

Spahn, H., Spahn, I., and Benet, L. Z. (1989b). Probenecid-induced changes in the clearance of carprofen enantiomers—a preliminary study. *Clin. Pharmacol. Ther.*, **45**, 500–505.

Spahn, H., Zia-Amirhosseini, P., Näthke, I., Mohri, K., and Benet, L. Z. (1990). Characterization of drug adducts formed with proteins via acyl glucuronides—by electrophoresis and blotting., *Pharm. Res.*, **7** (suppl.), S257.

Spahn-Langguth, H. (1992). The use of mean residence time concepts to define precursor-product relationships in drug delivery and metabolism. 2nd Jerusalem Conference on Pharmaceutical Sciences and Clinical Pharmacology, Jerusalem, May 1992.

Spahn-Langguth, H., and Benet, L. Z. (1992a): Active and reactive phase-II metabolities: The glucuronides pathway. In Crommelin, D. J. A., Midha, K. K. (eds.) Topics in Pharmaceutical Sciences 1991, Proc. Intl. Congress of Pharmaceutical Sciences, Washington DC. Medpharm Scientific Publishers, Stuttgart, pp. 505–516.

Spahn-Langguth, H., and Benet, L. Z. (1992). Acyl glucuronides revisited: Is the glucuronidation process a toxification as well as a detoxification mechanism? *Drug Metab. Rev.*, **24**, 5–48.

Spahn-Langguth, H., and Benet, L. Z. (1993). Microsomal acyl glucuronidation: Enzyme kinetic studies with labile glucuronides. *Pharmacology*, **46**, 268–273.

Spahn-Langguth, H., Zia-Amirhosseini, P., Iwakawa, S., Ojingwa, J., Büschges R., and Benet, L. Z. (1994a). Aktive und reaktive Glucuronide: Reversible und irreversible Interaktionen mit Proteinen. In, H. J. Dengler, and E. Mutschler (eds.) Fremdstoffmetabolismus und Klinische Pharmakologie, Stuttgart: Gustav Fischer Verlag, pp. 29–50.

Spahn-Langguth, H., Gräfe, A. K. , Mayer, S., and Büschges, R. (1994b). Fibrates: The occurrence of active and reactive metabolites. In, Reid, E., Hill, H. M., and Wilson, I. D. (eds.), *Biofluid and Tissue Analysis for Drugs Including Hyperlipidaemics*, Methodological Surveys in Bioanalysis of Drugs, Vol. 23, Cambridge. The Royal Society of Chemistry, 87–102.

Spahn-Langguth, H., Benet, L. Z., Möhrke, W., and Langguth, P. (1996). First-pass phenomena: Sources of stereoselectivities and variabilities of concentration-time profiles after oral dosage.

Stafanger, G., Larsen, H. W., Hansen, H., and Sorensen, K. (1981). *Scand. J. Rheumatol.*, **10**, 189–192.

Stogniew, M., and Fenselau, C. (1982). Electrophilic reactions of acyl-linked glucuronides: Formation of clofibrate mercapturate in humans. *Drug Metab. Disp.*, **10**, 609–613.

Strolin-Benedetti, M., Frigerio, E., Tamassia, V., Noseda G., and Caldwell J. (1992). The dispositional enantioselectivity of indobufen in man. *Biochem. Pharmacol.*, **43**, 2032–2034.

Taylor, I. W., Chasseaud, L. F., Taylor, T., James, I., Dorf, G., and Darragh, A. (1991). Pharmacokinetics of the anti-inflammatory drug ximoprofen in healthy young and elderly subjects: Comparison with elderly rheumatic patients. *Br. J. Clin. Pharmac.*, **32**, 242–245.

Taki, Y., Sakane, T., Nadai, T., Sezaki, H., Langguth, P., and Yamashita, S. (1995). Gastrointestinal absorption of peptide drugs: Quantitative evaluation of the degradation and permeation of metkephamid in rat small intestine. *J. Pharmacol. Exp. Ther. J. Pharmacol. Exp. Ther.*, **274**, 373–377.

Upton, R. A., Williams, R. L., Buskin, J. N., and Jones, R. M. (1982). Effects of probenecid on ketoprofen kinetics. *Clin. Pharmacol. Ther.*, **31**, 705–712.

Upton, R. A., Williams, R. L., Kelly, J., and Jones, R. M. (1984). *Br. J. Clin. Pharmac.*, **18**, 207–214.

Weil, A., Caldwell, J., Guichard, J. P., and Picot, G. (1989). Species differences in the chirality of the carbonyl reduction of [^{14}C]-fenofibrate in laboratory animals and humans. *Chirality*, **1**, 197–201.

Wells, D. S., Janssen, F. W., and Ruelius, H. W. (1987). Interactions between oxaprozin glucuronide and human serum albumin. *Xenobiotica*, **17**, 1437–1449.

Williams, A.M., and Dickinson, R. G. (1994). Studies on the reactivity of acyl glucuronides-VI. Modulation of reversible and covalent interaction of diflunisal acyl glucuronide and its isomers with human plasma protein *in vitro*. *Biochem. Pharmacol.*, **47**, 457–467.

Williams, A. M., Worrall, S., DeJersey, J., and Dickinson, R. G. (1992). Studies on the reactivity of acyl glucuronides: Glucuronide-derived adducts of valproic acid and plasma proteins and anti-adduct antibodies in humans. *Biochem. Pharmacol.*, **43**, 745–755.

Williams, A. M. Worralls, S., De Jersey, J., and Dickinson R. G. (1995). Studies on the reactivity of acyl glucuronides-VIII. Generation of an antiserum for the detection of diflunisal-modified proteins in diflunisal dosed rats. *Biochem. Pharmacol.*, **49**, 209–217.

van Breemen, R. B., and Fenselau, C. (1985). "Acylation of albumin by 1-O-acyl glucuronides. *Drug Metab. Disp.*, **13**, 318–320.

Van der Ouweland, F. A., Jansen, P. A. F., Tan, Y., Van de Putte, L. B. A., Van Ginneken, C. A. M., and Gribnau, F. W. J. (1988). Pharmacokinetics of high dosage naproxen in elderly patients. *Int. J. Clin. Pharmacol. Ther. Toxicol.*, **26**, 143–147.

Voith, B., Spahn-Langguth, H., Paliege, R., Knauf, H., and Mutschler, E. (1993). Etacrynic acid: Evaluation of the kinetics/effect relationship. *Arch. Pharm. (Weinheim)*, **326**, 629 (abstract).

Volland, C., and Benet, L. Z. (1991). *In vitro* enantioselective glucuronidation of fenoprofen. *Pharmacology*, **43**, 53–60.

Volland, C., Sun, H., Dammeyer, J., and Benet, L. Z. (1991). Stereoselective degradation of the fenoprofen acyl glucuronide enantiomers and irreversible binding to plasma proteins. *Drug. Metab. Disp.*, **19**, 1080–1086.

Vree, T. B., Hekster, Y. A., and Anderson, P. G. (1992). Contribution of the human kidney to the metabolic clearance of drugs. *Ann. Pharmacother.*, **26**, 1421–1428.

Yamaguchi, T, and Nakamura, Y. (1987). Stereoselective metabolism of 2-phenylpropionic acid in rat. II. Studies on the organs responsible for the optical isomerization of 2-phenylpropionic acid in rat *in vivo*. *Drug Metab. Disp.*, **15**, 535–539.

Zakim, D., and Vessey, D. A. (1973): Techniques for the characterization of UDP-glucusonyl transferase, glucose-6-phosphatase, and other tightly bound microsomal enzymes. *Methods Biochem. Anal.*, **21**, 1–37.

Zia-Amirhosseini, P. (1994). Hypersensitivity to nonsteroidal anti-inflammatory drugs: Exploration of a theory. Ph.D. Thesis, School of Pharmacy, University of California, San Francisco.

Zia-Amirhosseini, P., Spahn, H., and Benet, L. Z. (1988). Measurement of naproxen, desmethylnaproxen and their conjugates in plasma and urine. *Pharm. Res.*, **5**, (suppl.), S-200.

Zia-Amirhosseini, P., Ojingwa, J., Spahn-Langguth, H., McDonagh, A. F., and Benet, L. Z. (1994). Enhanced covalent binding of tolmetin to proteins in humans after multiple dosing. *Clin. Pharmacol. Ther.*, **55**, 21–27.

Zia-Amirhosseini, P., Harris, R. Z., Brodsky, F. M., Benet, L. Z. (1995). Hypersensitivity to nonsteroidal antiinflammatory drugs. *Nature Med.*, **1**, 2–4.

CHAPTER

6

STUDIES ON THE STEREOSELECTIVITY OF THE BIOLOGICALLY IMPORTANT METABOLISM OF ALKENE–ALKENE OXIDE (OXIRANE)-ALKANEDIOL

DOROTHEE WISTUBA

Institut für Organische Chemie der Universität
72076 Tübingen, Germany

6.1. INTRODUCTION

The metabolism of aliphatic xenobiotics in mammals with olefinic double bonds occurs preferentially in the endoplasmic reticculum of liver cells. The metabolic conversion of these substances leading to their elimination from the organism is caused by the reduction of their lipid solubility. Often, reactive intermediates are formed that have a mutagenic or carcinogenic potential due to their covalent binding to cellular macromolecules. Thus, highly reactive oxiranes are formed as initial metabolites in the epoxidation reaction of alkenes catalyzed by cytochrome P-450-dependent monooxygenases (1) . The alkane epoxides, possessing alkylating properties (2), are converted to vicinal alkane diols by epoxide hydrolase-catalyzed hydration (3,4) and/or conjugated with glutathione, catalyzed by glutathione *S*-transferase (5) (see Figure 6.1).

Like most enzymes, cytochrome P-450 and epoxide hydrolase may function as inherently chiral catalysts. Whereas, according to Figure 6.1, the epoxidation of the alkene by cytochrome P-450 represents a *prochiral* recognition process (*product enantioselectivity*), the subsequent oxirane transformation is a *chiral* recognition process (kinetic resolution, *substrate enantioselectivity*). The striking differences in the biological activities between oxirane enantiomers [e.g., phenyloxirane (6,7)] underline the importance of studies devoted to the determination of enantioselectivities in the formation and transformation of epoxides catalyzed by enzymes.

Depending on the substrate enantioselectivity of epoxide hydrolase and glutathione *S*-transferase and on the product enantioselectivity of

The Impact of Stereochemistry on Drug Development and Use, Edited by Hassan Y. Aboul-Enein and Irving W. Wainer. Chemical Analysis Series, Vol. 142.
ISBN 0-471-59644-2

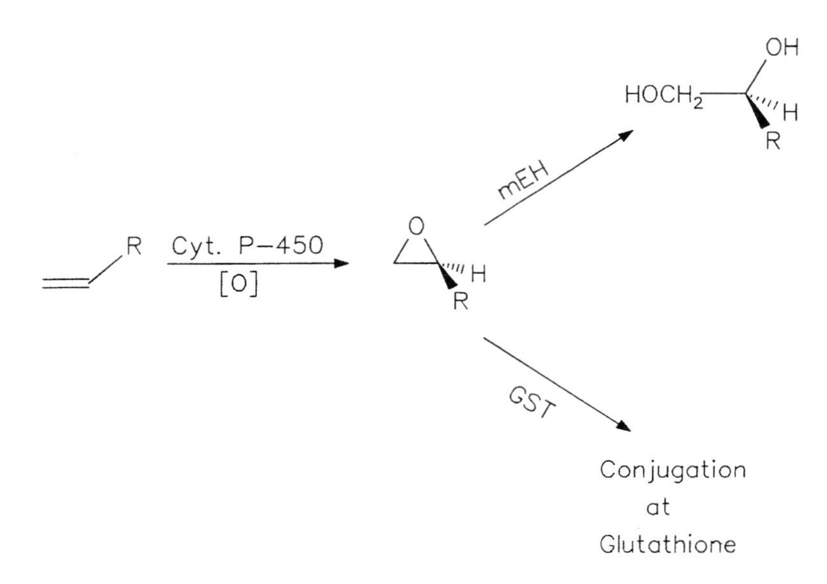

Figure 6.1. Mechanistic pathway of alkene–alkene oxide (oxirane)–alkane diol metabolism (only one enantiomer is shown) (mEM-microsomal epoxide hydrolase; GST-glutathione *S*-transferase).

cytochrome P-450, it is possible to obtain the accumulation of one enantiomer of the enzymatically formed oxiranes. This can have important implications since the biological activity of oxiranes corresponded on their absolute configuration.

Cytochrome P-450 consists of multiple isozymes with different, but generally overlapping substrate specificities (8). The pattern of cytochrome P-450 isozymes appears to be genetically controlled and is altered by exposure to xenobiotics (9). The striking effect of enzyme induction on enantioselectivity has been observed (10–15).

Epoxide hydrolase consists of two main forms: the microsomal and the cytosolic epoxide hydrolase. Both catalyze the antiperiplanar addition of water to the oxirane ring and have broad but different substrate selectivities. The reaction proceeds with the inversion of the configuration at the oxirane carbon atom, to which the addition of water takes place. It has recently been shown that the catalytic mechanism of microsomal epoxide hydrolase involves a covalent ester intermediate (16) and not the commonly accepted mechanism of a direct, general-base-catalyzed attack of water (17). At present, nothing is known about the catalytic mechanism of cytosolic epoxide hydrolase.

6.2. ALIPHATIC NONFUNCTIONALIZED OLEFINES

6.2.1. Cytochrome P-450-Catalyzed Epoxidation

6.2.1.1. Prochiral Olefines

Prochiral alkenes have *enantiotopic* faces, whereas chiral alkenes have *diastereotopic* faces. Hence, the product stereoselectivity of the microsomal epoxidation of alkenes is determined by the preferential orientation of the oxygen attack at the prostereogenic carbon–carbon double bond. In microsomal epoxidations, cytochrome P-450 can differentiate between the enantiotopic faces of the double bond of small, aliphatic, nonfunctionalized, prochiral olefins such as propene, 1-butene, 1-octene, 2-methyl-1-butene, *trans*- 2-butene, *cis*/*trans*-2-pentene, 2-methyl-2-butene (18–20) (see Table 6.1).

The amount and, in a few cases, the sign of enantioselectivity are remarkably species-dependent for the three mammalian species, rat, mouse, and human, with the rat showing the highest enantiomeric bias (19). In control rat liver microsomes, the extent of product enantioselectivity depends critically on the structure of olefines. The highest enantioselectivity of the cytochrome P-450-dependent monooxygenases was observed in the epoxidation of olefins of terminal double bonds. Neither extension of the length of the carbon chain nor alternation of its degree of hybridization has a decisive effect on the enantiomeric composition. However, the enantioselectivity of the enzyme system is markedly influenced by the degree of substitution of the olefinic double bond. The increase of the steric congestion of the alkene by di- or trialkyl substitution at the carbon–carbon double bond reduces prochiral recognition considerably, as evidenced by the low enantiomeric excess of the corresponding oxiranes. The epoxidation of the alkene containing a trialkyl-substituted carbon–carbon double bond, that is, 2-methyl-2-butene (trimethylethene), yields essentially a racemic oxirane metabolite. It is important to note that all oxirane enantiomers formed in excess possess the *S* configuration at the chiral carbon atom(s), suggesting a common prochiral recognition mode at the active site of cytochrome P-450 isozymes of rat liver microsomes (18, 19). The remarkable species dependence of the microsomal enantioselective epoxidation of prochiral alkenes by cytochrome P-450-dependent monooxygenases is observed when both the rat and mouse are compared. Thus, all oxirane metabolites produced by control mouse liver microsomes are essentially racemic (19). Only slight or no product enantioselectivity was found in the epoxidation of 1,1-dialkylethenes with rat, mouse, or rabbit microsomes. No significant enantioselectivity ($<2\%$) was found in the formation of 2-*t*-butyl-2-methyloxirane and 2-(1′-dimethylpropyl)-2-methyloxirane, and a very low e.e. (enantiomeric excess) (8%) of 2-butyl-2-methyloxirane with rabbit enzymes (23).

Table 6.1. Enantiomeric Ratio of Oxiranes Formed in the Epoxidation of Prochiral Alkenes with Untreated (Control) or Induced [phenobarbital (PB) or benzo[a]pyrene (BP)] Rat or Mouse Liver Microsomes (18–21)

Substrate	Metabolite	Abs.	Rat		Mouse		
		config. (%)	Control	PB Ind.	Control	PB Ind.	BP Ind.
1-Propene	Methyloxirane	*R*	31	30	45	55	42
		S	69	70	55	45	58
1-Butene	Ethyloxirane	*R*	25	30	49	62	
		S	75	70	51	38	
1-Octene	Octyloxirane	*R*		40[47](20)			
		S		60[53](20)			
1,3-Butadiene	Vinyloxirane	*R*	29	30	46	61	48
		S	71	70	54	39	52
Isoprene	Isopropenyl	*R*	32		46		
	oxirane	*S*	68		54		
	2-Methyl-2-	*R*	44		46		
	vinyloxirane	*S*	56		54		
2-Methyl-1-	2-Ethyl-2-	*R*	45	39	47	49	
butene	methyloxirane	*S*	55	61	53	51	
trans -2-	*trans*-2,3-Di-	2*R*, 3*R*	36	43	45	50	
Butene	methyloxirane	2*S*, 3*S*	64	57	55	50	
trans -2-	*trans* -2-Ethyl-3-	2*R*, 3*R*	43	48	54	57	54
Pentene	methyloxirane	2*S*, 3*S*	57	52	46	43	46
cis-2-	*cis*-2-Ethyl-3-	2*R*, 3*S*	39	48	49	50	52
Pentene	methyloxirane	2*S*, 3*R*	61	52	51	50	48
2-Methyl-2-	Trimethyl-	*R*	51	50	51	50	50
butene	oxirane	*S*	49	50	49	50	50

Very low enantiomeric excess had been reported for the epoxidation product of 2-methyl-1-butene with both rat (10%) and mouse (2%) microsomes (19). In contrast to rat and in analogy to mouse, human liver microsomes epoxidize terminal aliphatic alkenes, 2-methyl-2-butene (trimethylethene), as well as *trans*-2-butene to almost racemic oxiranes (see Table 6.2).

cis-2-Pentene is enantioselectively epoxidized by human microsomes to (2*R*, 3*S*)-2-ethyl-3-methyloxirane, which has the opposite configuration of that preferentially formed by microsomes of the rat. No significant differences in the *in vitro* enantioselective oxirane formation by microsomes of human individuals have been observed (19).

Pretreatment of rats and mice with hepatic enzyme inducers such as phenobarbital has a decisive effect on the enantioselective formation of aliphatic

Table 6.2. Enantiomeric Composition of Oxiranes Formed in the Epoxidation of Prochiral Alkenes with Human Liver Microsomes (19)

Substrate	Metabolite	Abs. Config. (%)	Individuals			
			1	2	3	4
1-Propene	Methyloxirane	*R*	48	44	43	50
		S	52	56	57	50
1,3-Butadiene	Vinyloxirane	*R*	52	54	56	53
		S	48	46	44	47
trans -2-Pentene	*trans* -2-Ethyl-3-methyloxirane	2*R*, 3*R*	48	51	53	51
		2*S*, 3*S*	52	49	47	49
cis -2-Pentene	*cis* -2-Ethyl-3-methyloxirane	2*R*, 3*S*	63	59	61	61
		2*S*, 3*R*	37	41	39	39
2-Methyl-2-butene	Trimethyloxirane	*R*	50	49	50	52
		S	50	51	50	48

oxiranes by hepatic microsomes. In rats, the effect of phenobarbital on the e.e. of the oxirane metabolite is negligible for linear terminal alkenes. Small changes by phenobarbital induction cause reduction in the case of *trans*- and *cis*-2 alkene(s) and lead to an increase of e.e. in the case of alkenes with the geminal 2,2-dialkyl-substituted double bond (see Table 6.1). In mice, phenobarbital induction only shows distinct effects on the product enantioselectivity in the epoxidation reaction of terminal alkenes. Most significantly, in contrast to the results obtained with rat liver microsomes, the sign of enantioselectivity is reversed, that is, the opposite enantiotopic face of the carbon–carbon double bond is preferentially epoxidized, leading to antipodal R-oxirane enantiomers formed in excess (see Table 6.1). Benzo[α]pyrene induction exhibits no significant influence on the e.e. (19).

6.2.1.2. *Chiral Olefines*

The enzymatic epoxidation of a racemic chiral alkene such as 3-methyl-1-pentene, and *cis*- or *trans*-4-methyl-2-hexene represents a competitive process between both enantiomers. Three stereoselective processes may be considered (see Figure 6.2) (24, 25):

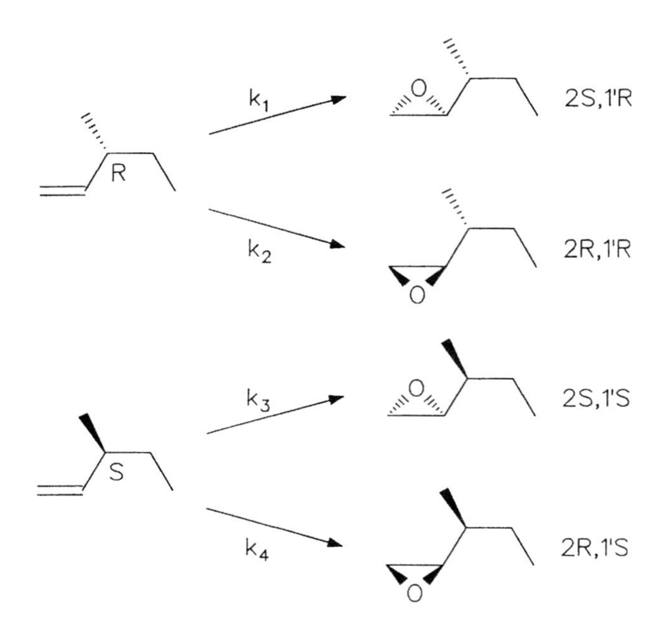

Figure 6.2. Substrate enantioselectivity = $(k_1 + k_2)$ vs. $(k_3 + k_4)$; product diastereoselectivity = k_1 vs. k_2 and k_3 vs. k_4; product enantioselectivity = k_1 vs. k_4 and k_2 vs. k_3.

1. *Substrate enantioselectivity* ("chiral recognition" by kinetic enantiomer differentiation)
2. *Product diastereoselectivity* (diastereoface differentiation of an individual alkene enantiomer)
3. *Product enantioselectivity* ("prochiral recognition" by *external* enantioface differentiation of the respective alkene antipodes)

Only a slight *substrate enantioselectivity* (1.) in favor of the *R*-configured alkenes was found on the incubation of 3-methyl-1-pentene and *cis*- and *trans*-4-methyl-2-hexene with rat liver microsomes (control and phenobarbital-induced), resulting in a small excess of 1′R-configured oxirane metabolite (see Table 6.3). There is, however, a distinct and different *product diastereoselectivity* (2.) in the rat (control and phenobarbital-induced) microsomal epoxidation of the individual enantiomers of 3-methyl-1-pentene and of *trans*- and *cis*-4-methyl-2-hexene that is strikingly different from that of the nonenzymatic epoxidation of the alkenes with achiral *m*-chloroperbenzoic acid (the Prilezaev reaction) (see Table 6.3). The results contained in Table 6.3 may also be discussed in terms of *product enantioselectivity* (3.), whereby the differenti-

Table 6.3. Composition of Oxiranes Formed in the Epoxidation of Racemic Alkenes with *m*-Chloroperbenzoic Acid (mCPBA) and Untreated (control) or Induced [phenobarbital (PB)] Rat or Mouse Liver Microsomes (19)

Substrate	Metabolite	Abs. Config. (%)	m-CPBA	Rat Control	Mouse PB Ind.	Control	PB Ind.
3*R*-3-Methyl-1-	2-(1′-Methyl	2*S*, 1′*R*	30	46	43	21	14
pentene	propyl)oxirane	2*R*, 1′*R*	20	8	10	28	32
3*S*-3-Methyl-1-	2-(1′-Methyl-	2*S*, 1′*S*	20	35	30	20	16
pentene	propyl)oxirane	2*R*, 1′*S*	30	11	17	31	38
trans-4R-4-	*trans* - 2-Methyl-3-(1′-	2*S*, 3*S*, 1′*R*	31	36	32		
Methyl-2-hexene	methylpropyl)oxirane	2*R*, 3*R*, 1′*R*	19	18	22		
trans-4*S*-4-	*trans* -2-Methyl-3-(1′-	2*S*, 3*S*, 1′*S*	19	24	24		
Methyl-2-hexene	methylpropyl)oxirane	2*R*, 3*R*, 1′*S*	31	22	22		
cis-4R-4-	*cis*-2-Methyl-3-(1′-	2*R*, 3*S*, 1′*R*	32	46	47		
Methyl-2-hexene	methylpropyl)oxirane	2*S*, 3*R*, 1′*R*	18	6	5		
cis-4S-4-	*cis*-2-Methyl-3-(1′-	2*R*, 3*S*, 1′*S*	18	32	17		
Methyl-2-hexene	methylpropyl)oxirane	2*S*, 3*R*, 1′*S*	32	16	31		

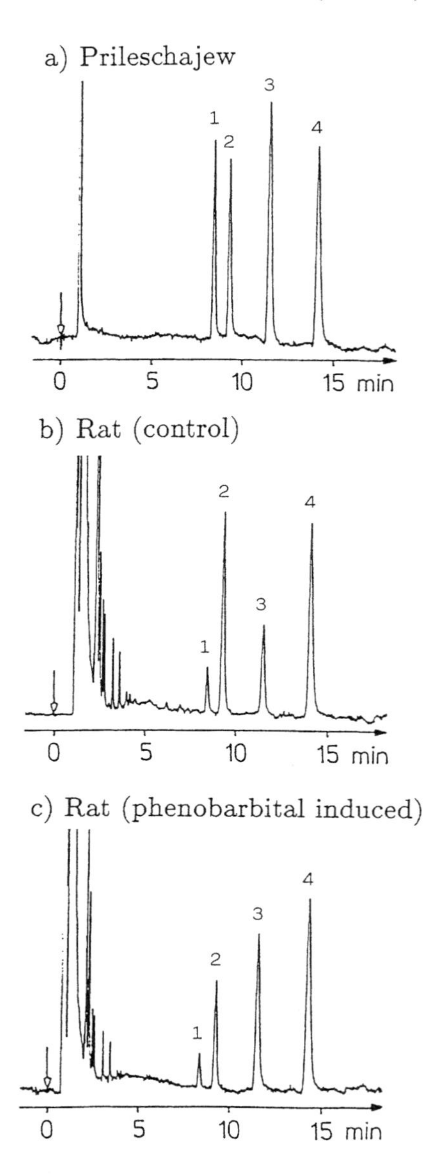

Figure 6.3. Analysis of the enantiomers (1: 2*S*, 3*R*, 1′*R*; 2: 2*R*, 3*S*, 1′*S*; 3: 2*S*, 3*R*, 1′*S*; 4: 2*R*, 3*S*, 1′*R*) of *cis*-2-methyl-3-(1′-methylpropyl)oxirane formed in the epoxidation of *cis*-4-methyl-2-hexene with *m*-chloroperbenzoic acid (*a*), by control (*b*), and by phenobarbital-induced rat liver microsomes (*c*) by complexation gaschromatography. Column: 25 m × 0.3 mm glass capillary coated with 0.125-m nickel (II) bis[(3-heptafluorobutanoyl)-(1*R*)-camphorate] in SE 30; 90°C, 1.5 bar N_2 (19).

ation of the isozymes between the *externally enantiotopic* faces of the carbon–carbon double bond of the respective alkene enantiomers is considered. As is evident from Figure 6.3, the chiral nature of the catalytic center of cytochrome P-450 of rats leads to a high degree enantioselectivity, whereas such a differentiation is impossible in the nonenzymatic epoxidation via the achiral Prilezaev reaction (see Figure 6.3, top). In agreement with the results for prochiral alkenes (see Table 6.1), oxirane enantiomers with an *S* configuration at the carbon atom carrying the bulkiest alkyl substituent are formed in excess on epoxidation with microsomes of the rat. This effect is pronounced (and similar) for the enantiomeric pairs (2*S*, 1′*R*)/(2*R*, 1′*S*)- and (2*S*, 1′*S*)/(2*R*, 1′*R*)-2-(1′-methylpropyl)oxirane (4:1), highest for *cis*-(2*R*, 3*S*, 1′*S*/(2*S*, 3*R*, 1′*R*)-2-methyl-3-(1′-methylpropyl)oxirane (5:1), and lowest for *trans*-(2*S*, 3*S*, 1′*R*)/(2*R*, 3*R*,1′*S*)- and (2*S*, 3*S*, 1′*S*)/(2*R*, 3*R*, 1′*R*)-2-methyl-3-(1′-methylpropyl)oxirane (<3:2) (19). Species dependence was found in the cytochrome P-450-catalyzed epoxidation of 3-methyl-1-pentene. In contrast to rat liver microsomes, 2-(1′-methylpropyl)oxirane with an *R* configuration at the ring carbon atom 2 is preferentially formed with the liver microsomes of untreated and phenobarbital-induced mice (see Table 6.3) (19).

6.2.1.3. *Prochiral Diolefines*

Studies on the hepatic microsomal metabolism of 1,3-butadiene and isoprene (2-methyl-1,3-butadiene), important products of the petrochemical industry, have shown that the primary metabolites are highly reactive and mutagenic monooxiranes. 1,3-Butadiene and isoprene possess two prochiral carbon–carbon double bonds that, in the case of isoprene, are not identical. Thus, the epoxidation of isoprene catalyzed by cytochrome P-450-dependent monooxygenases (see Figure 6.4) can occur with regioselectivity and/or product enantioselectivity, this being determined by the preferential orientation of the oxygen attack at the enantiotopic double bonds (see Figure 6.5).

1,3-Butadiene was epoxidized *in vitro* by the cytochrome P-450 of rats and mice to vinyloxirane (26). Isopropyloxirane and 2-methyl-2-vinyloxirane were formed from isoprene (27, 28). Both reactions occur with slight but different product enantioselectivity (18, 19, 22) (see Table 6.1). 2-Methyl-2-vinyloxirane possesses a very high reactivity toward water and, thus, a half-life of only 75 min at pH 7.4 (27). The fact that this spontaneous hydrolysis is a racemic process (both enantiomers react at the same rate) and competes with the enzymatic formation of 2-methyl-2-vinyloxirane leads to problems in the quantitative determination of regioselectivity and product enantioselectivity for this substrate. Only with mouse liver microsomes was distinct regioselectivity observed (at steady state, the ratio of isopropyloxirane and 2-methyl-2-vinyloxirane was 1:2.6) (22).

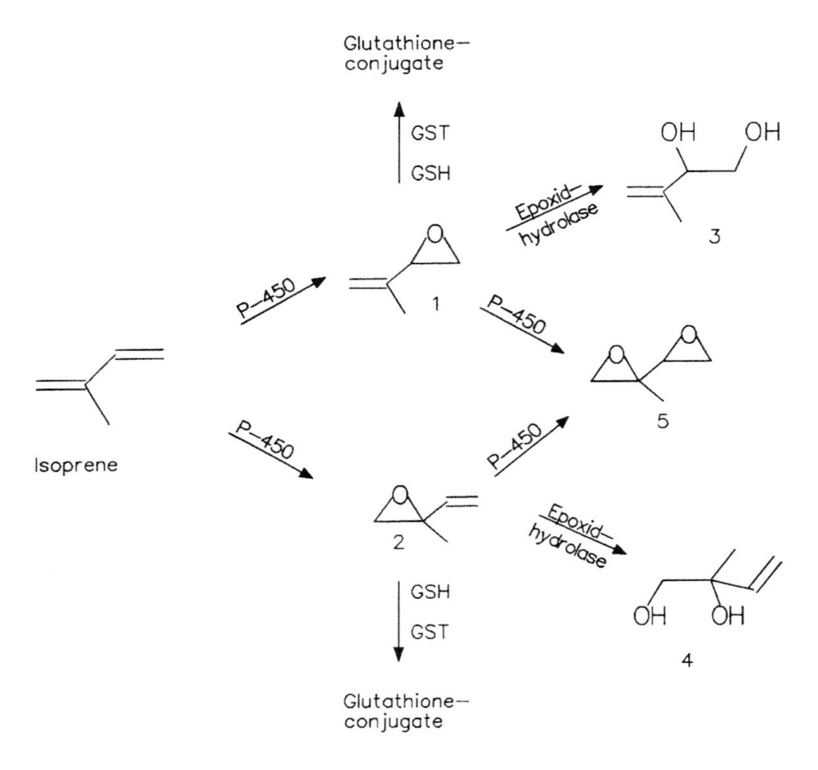

Figure 6.4. Metabolism of isoprene (GST: glutathione *S*-transferase; GSH: glutathione; P-450: cytochrome P-450; 1-isopropenyloxirane; 2:2-methyl-2-vinyloxirane; 3:3-methylbutane-1,2-diol; 4:2-methylbutane-1,2-diol; 5:2-methyl-2,2′-bioxirane).

Both monooxiranes of isoprene were further epoxidized *in vitro* to 2-methyl-2,2′-bioxirane (22) (see Figure 6.4).

Whereas the two carbon–carbon double bonds of prochiral isoprene have enantiotopic faces, the remaining double bond of the chiral monooxiranes, isopropenyloxirane and 2-methyl-2-vinyloxirane, possesses diastereotopic faces. In both monooxiranes, two elements of (pro)stereogenicity are inherently combined: (1) prochirality (re, si) and (2) chirality (*R*, *S*). In the analogy of chiral alkenes, substrate enantioselectivity, product diastereoselectivity, and/or product enantioselectivity can appear (see Table 6.4). Only slight substrate enantioselectivity was found in the case of isopropenyloxirane in favor of the (2*R*)-configured oxirane. However, there is distinct and different product diastereoselectivity in the rat and mouse microsomal epoxidation of the individual enantiomers of isopropenyloxirane and 2-methyl-2-vinyloxirane

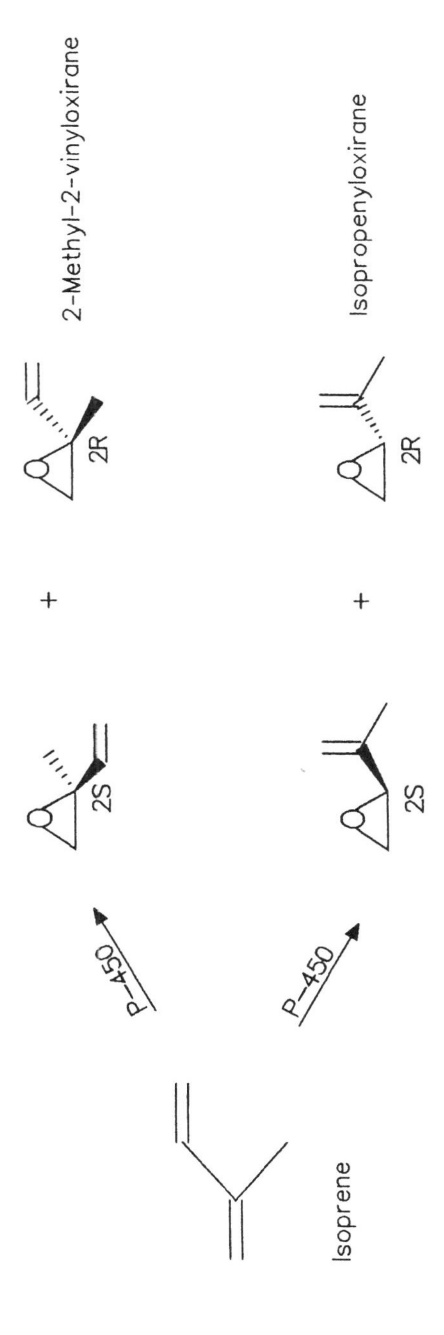

Figure 6.5. Epoxidation of isoprene catalyzed by cytochrome P-450 to the monooxiranes isopropenyloxirane and 2-methyl-2-vinyloxirane.

Table 6.4. The Enantiomeric and Diastereomeric Composition of 2-Methyl-2,2′-bioxirane 5 Formed by Epoxidation of Racemic 2-Isopropenyloxirane 1 and of Racemic 2-Methyl-2-Vinyloxirane 2 with Rat and Mouse Liver Microsomes [Substrate Enantioselectivity ($k_1 + k_2$) vs. ($k_3 + k_4$); Product Diastereoselectivity k_1 vs. k_2 and k_3 vs. k_4; Product Enantioselectivity k_1 vs. k_3 and k_2 vs. k_4] (22)

Substrate		2-Methyl-2,2′-bioxirane	Rat (%)	Mouse (%)
2*R* (re)	k_{1a}	2*R*, 2′*S*	46.6	37.0
(si)	k_{2a}	2*S*, 2′*S*	11.8	17.1
2*S* (si)	k_{3a}	2*S*, 2′*R*	26.4	29.0
(re)	k_{4a}	2*R*, 2′*R*	15.9	16.9
2*R* (re)	k_{1b}	2*S*, 2′*R*	44.0	32.6
(si)	k_{2b}	2*S*, 2′*S*	11.3	14.8
2*S* (si)	k_{3b}	2*R*, 2′*S*	21.9	39.4
(re)	k_{4b}	2*R*, 2′*R*	22.8	13.2

(exception: the rat microsomal epoxidation of 2*S*-2-methyl-2-vinyloxirane). Whereas the epoxidation of the *R*-configured monooxiranes occurs preferentially at the re face, the *S*-enantiomers will be attacked by oxygen preferentially from the si face. When one considers the differentiation of the enzyme between the externally enantiotopic faces of the carbon–carbon double bond of the monooxiranes, a distinct product enantioselectivity is found. The enantiomeric pair (2*R*, 2′*S*)/(2*S*, 2′*R*)[1] is formed by the epoxidation of the re face of

[1]The carbon atom C-2 of the substrate isopropenyloxirane becomes C-2′ of the product 2-methyl-2,2′-bioxirane because of the formal change of the carbon atom numbering caused by the IUPAC rule. Because of the formal change in the descriptor caused by the priority rule of Cahn, Ingold, and Prelog, the 2*R* (or 2*S*) stereochemistry of the substrate isopropenyloxirane becomes 2′*S* (or 2′*R*) of the product 2-methyl-2,2′-bioxirane, and the 2*S* (or 2*R*) stereochemistry of the substrate 2-methyl-2-vinyloxirane becomes 2*R* (or 2*S*) of the product 2-methyl-2,2′-bioxirane.

the *R*-isopropenyloxirane and the si face of the *S*-enantiomer with an enantiomeric excess (e.e.) of 27.7% with rat liver microsomes and 12.1% with mouse liver microsomes. (2*S*, 2′*S*)- and (2*R*, 2′*R*)-2-methyl-2,2′-bioxirane is formed as either racemic (mouse) or with an enantiomeric excess of e.e. = 14.8% (rat). The microsomal epoxidation of 2-methyl-2-vinyloxirane leads for both enantiomeric pairs of the diastereomer 2-methyl-2,2′-bioxiranes to an enantiomeric excess [(2*R*, 2*S*)/(2*S*, 2′*R*): e.e. = 33.5% (rat) and 9.4 % (mouse); (2*S*, 2′*S*)/(2*R*, 2′*R*): e.e. = 8.6% (rat) and 5.7% (mouse)]. In summary, it has been found that the oxygen attack occurs preferentially at the re face of the *R*-enantiomer during the rat and mouse microsomal epoxidation of isopropenyloxirane and during the rat microsomal epoxidation of 2-methyl-2-vinyloxirane. Only in the case of the mouse microsomal epoxidation of 2-methyl-2-vinyloxirane was the si face of the *S*-enantiomer epoxidized preferentially (see Table 6.4) (22).

6.2.2. Epoxide-Hydrolase-Catalyzed Hydrolysis of Aliphatic Oxiranes

The oxirane hydrolysis catalyzed by the epoxide hydrolase occurs by nucleophilic attack of water on the oxirane carbon atom to form vicinal diols (3, 29). Studies with ^{18}O-labeled oxiranes or $H_2{}^{18}O$ have shown that enzyme-catalyzed hydrolysis proceeds with a high degree of regioselectivity favoring the attack by water on the less sterically hindered oxirane carbon atom (30, 31). An alternative method for determining regioselectivity is the comparison of the enantiomeric excess of one oxirane enantiomer and that of the corresponding diol (32). Regioselectivities from 96–100% were found.

Epoxide hydrolase possesses both substrate enantioselectivity in the hydrolysis of racemic alkyl-substituted aliphatic oxiranes and product enantioselectivity in the hydrolysis of *meso*-oxiranes.

The substrate enantioselectivity of chiral mono-substituted oxiranes with unbranched alkyl groups like methyloxirane, vinyloxirane, *n*-butyloxirane, *n*-octyloxirane, and epichloro- and epibromohydrin is generally low and has the same sign. It was found that *S*-methyloxirane, *S*-vinyloxirane, and *R*-epichloro- and *R*-epibromohydrin, possessing the same relative stereochemistry,[2] were preferentially hydrolyzed by rat liver microsomes (32, 33). Contrary to these results, Bellucci et al. (34), using rabbit liver microsomes, determined the preferential hydrolysis of *R*-*n*-butyloxirane. In the hydrolysis of mono-substituted oxiranes with branched alkyl groups, especially *t*-butyloxirane or isopropenyloxirane, inhibitory effects lead to a complex pattern of enantioselectivity. The *R*-oxiranes were consumed preferentially in the first stage of the reaction; in the second stage the hydrolysis of the *S*-enantiomers

[2] Formal change of the descriptor caused by the rule of Cahn, Ingold, and Prelog.

occurred at a faster rate, than as that of the *R*-enantiomers in the first part of the reaction (22, 32, 34). Similar results were obtained in the hydrolysis of phenyloxirane (35) and *p*-nitrophenyloxirane (36) due to the inhibitory effects of the *R*-enantiomer that had a higher affinity for the microsomal epoxide hydrolase active site toward the *S*-enantiomer.

The introduction of a second alkyl substituent leads to 2,2-dialkyl-substituted oxiranes and suppresses the enantioselectivity of the rabbit microsomal epoxide hydrolase reaction of 2-*t*-butyl-2-methyloxirane and 2-(2′,2′-dimethylpropyl)-2-methyl-oxirane. The hydrolysis of these substrates is practically nonenantioselective. Only 2-butyl-2-methyloxirane was hydrolyzed with distinct enantioselectivity favoring the hydrolysis of the *S*-enantiomer (37). In the case of reactive 2-methyl-2-vinyloxirane, a superposition of the spontaneous (racemic process) and enzymatic (enantioselective process) hydrolysis was found. For two species, rat and mouse, 2*S*-2-methyl-2-vinyloxirane was hydrolyzed preferentially and the reaction occurred with slight substrate enantioselectivity (22).

cis-2,3-Dimethyloxirane represents the smallest *meso*-oxirane containing enantiotopic ring carbon atoms. The microsomal epoxide hydrolase discriminates between the opposite configurations of the two ring carbons. Thus, the water attack occurs with high regioselectivity (93%) at the *S*-configured oxirane carbon atom, leading to the major metabolite *threo*-(2*R*, 3*R*)-butane-2,3-diol (e.e. = 86%) (38). This result shows that high prochiral recognition of the enantiotopic ring carbon atoms, which had been demonstrated for sterically more crowded *meso*-substrates such as *cis*-diphenyloxirane (39) and epoxycyclohexane (30), already occurs in the case of the smallest aliphatic *meso*-oxirane.

The enzymatic hydrolysis of chiral *cis*-oxiranes, containing at least one methyl substituent, shows complete or nearly complete substrate enantioselectivity and regioselectivity, with nucleophilic attack by water occurring with the inversion of the configuration at the methyl-substituted ring carbon atom of the *S* configuration. Racemic *cis*-2-ethyl-3-methyloxirane, the smallest aliphatic *cis*-oxirane, was hydrolyzed with complete regioselectivity and enantioselectivity (e.e. > 99%) to pentane-2,3-diol. Only the 2*R*, 3*S*-configured oxirane was metabolized, and *threo*-2*R*, 3*R*-pentane-diol was formed exclusively by regioselective water attack at the methyl-substituted oxirane carbon atom (33, 38). Since practically no product enantioselectivity is observed in the cytochrome P-450-catalyzed epoxidation of *cis*-2-pentene (19), the enantiomer (2*S*, 3*R*)-2-ethyl-3-methyloxirane, inactive toward microsomal epoxide hydrolase, must be detoxified via another pathway in the *in vivo* metabolization of *cis*-2-pentene. The highly interesting result of complete kinetic resolution was found for all three mammalian species, human, mouse, and rat (38) (see Figure 6. 6). The quantitative substrate enantioselectivity

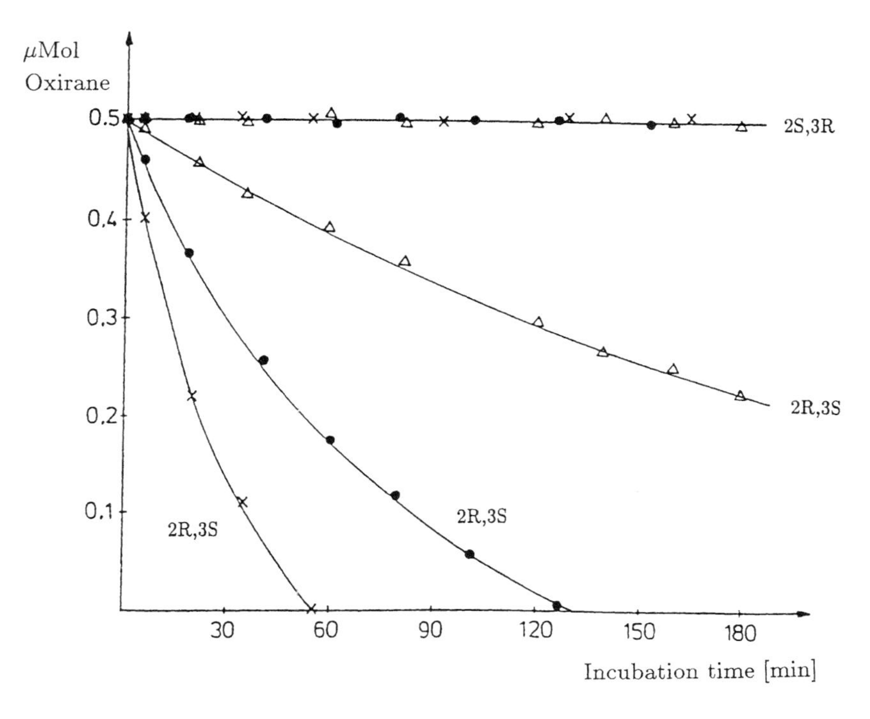

Figure 6.6. Substrate enantioselective hydrolysis of racemic *cis*-2-ethyl-3-methyloxirane catalyzed by microsomal epoxide hydrolase of human (x), mouse (Δ), and rat (•) (38).

implies that only one enzyme is active, or all isozymes exert the same chiral recognition.

The same high regioselectivity and enantioselectivity were found in the enzymatic hydrolysis of *cis*-2-methyl-3-(1′-methylpropyl)oxirane to the metabolite *threo*-4-methylhexane-2,3-diol. Absolute configuration at the chiral atom in the side chain is not important for the hydrolytic process. A small amount of (2*S*, 3*S*, 4*R*)- and (2*S*, 3*S*, 4*S*)-4-methylhexane-2,3-diol due to the water attack at the ring carbon atom bearing the 1′-methylpropyl group was formed only after an extremely long incubation (8 h) (38). Drastic reduction of substrate enantioselectivity and/or regioselectivity has been observed for *cis*-2-ethyl-3-propyloxirane in comparison with the methyl-substituted *cis*-oxiranes, that is, *cis*-dimethyloxirane, *cis*-2-ethyl-3-methyloxirane, and *cis*-2-methyl-3-(1′-methylpropyl)oxirane. Both enantiomers of 2-ethyl-3-propyloxiranes are extremely poor substrates for the microsomal epoxide hydrolase (38).

In the *in vitro* hydrolysis of the racemic *trans*-2,3-dialkyl-substituted oxiranes, *trans*-2,3-dimethyloxirane, *trans*-2-ethyl-3-methyloxirane, and *trans*-2-methyl-3(1′-methylpropyl)oxirane, all configuration isomers are metabolized with a higher rate for the (2*S*, 3*S*)-isomers (33, 38).

Only one example of the microsomal epoxide hydrolase-catalyzed hydrolysis of a trialkyl-substituted oxirane was described. The conversion of trimethyloxirane leading to (2*R*)-2-methylbutane-2,3-diol (96% e.e. after 2-h incubation time) occurs with high substrate enantioselectivity in favor of the *S*-enantiomer and with complete regioselectivity at the monomethyl-substituted ring carbon atom (38). A structural hypothesis for the active site of the rat microsomal epoxide hydrolase was described (38) that explains the substrate enantioselectivity and regioselectivity of aliphatic oxiranes. Orienting the oxirane ring in the plane of the paper with the oxygen atom on the top, the water attack always occurs at the ring carbon atom, with the hydrogen atom situated right behind the plane and the hydrogen atom or methyl group right in front of the plane. If no possibility for arranging the oxirane ring in such a manner exists, hydrolysis is extremely slow or does not take place. However, the hypothesis cannot rationalize inhibitory effects and is not valid for the rabbit enzyme system (37).

Both forms of epoxide hydrolase, the microsomal and cytosolic epoxide hydrolase, are commonly considered to be endowed with distinct and complementary substrate specificity (40). G. Bellucci et al. (41) found that the cytosolic enzyme exhibits regioselectivity and substrate enantioselectivity toward racemic mono-substituted alkyloxiranes (octyloxirane, butyloxirane, neopentyloxirane, *t*-butyloxirane), similar to that observed for microsomal epoxide hydrolase.

6.3. PHENYL-SUBSTITUTED OLEFINES

6.3.1. Metabolism of Styrene and Ring-Substituted Derivatives

6.3.1.1. Cytochrome P-450-Catalyzed Epoxidation of Styrene

Styrene and its metabolite styrene oxide (phenyloxirane), a potential carcinogen, have received considerable attention as a potential health hazard due to their widespread industrial use in the production of polymer materials. Stereoselective metabolism could modulate the genotoxicity of styrene because the *R*-styrene oxide is more mutagenic to *Salmonella typhimurium* strain TA100 than the *S*-styrene oxide (42). Enantioselectivity was observed in both *in vitro* and *in vivo* metabolism.

Table 6.5. Enantioselectivity of Cytochrome P-450-Catalyzed Epoxidation of Styrene to Styrene Oxide and the Influence of Treatment with Phenobarbital or β-Naphtoflavone

Treatment	Enantiomeric Ratio of Styrol Oxide		Microsomes	References
	R (%)	*S* (%)		
Untreated	43	57	Rat liver	(35)
	40	60	Rat liver	(43)
	61	39	Rabbit lung	(44)
Phenobarbital (PB)	48	52	Rat liver	(43)
	63	37	Rabbit lung	(44)
β-Naphtoflavone (βNF)	56	44	Rat liver	(43)
	63	37	Rabbit lung	(44)

Cytochrome P-450-dependent monooxygenases catalyze the epoxidation of styrene to styrene oxide with distinct product enantioselectivity (see Table 6.5). Whereas rat liver microsomes epoxidized styrene preferentially to the *S*-styrol oxide (35, 43, 45), epoxidation with rabbit pulmonary microsomes favored the formation of *R*-styrol oxide (44). Pulmonary metabolism of styrene is of high importance because inhalation is a primary route of styrene exposure in mammals (46). Liver microsomes from rat pretreated with phenobarbital (PB) or β-naphtoflavone (βNF) caused changes in enantioselectivity (43). While PB induction leads to a slight decrease of e.e., the sign of enantioselectivity is changed on βNF induction (see Table 6.5). These results differ from earlier studies that found no appreciable alteration in the *R* and *S* ratio with liver microsomes of rats pretreated with PB, 3-methylcholantrene, or polychlorinated biphenyls (35, 45). The discrepancy is most likely due to the sensitivity of the methods used. No significant difference from the control enantiomeric ratio was observed in pulmonary microsomes from rabbits treated with PB or βNF (44) (see Table 6.5).

There are different possibilities for the participation of the cytochrome P-450 isozymes in the enantioselective process of alkene epoxidation:

1. Only one isozyme, catalyzing oxirane formation with defined enantioselectivity, is involved.
2. More than one isozyme is involved, whereby
   (a) All involved isozymes catalyze oxirane formation with complete enantioselectivity (e.e. = 100%) and (at least one) with opposite chirality.

Table 6.6. Enantioselective Epoxidation of Styrene to Styrene Oxide by Rat Liver Cytochrome P-450 Isozymes (43)

Treatment	P-450 Isozyme	pmol of P-450 per Incubate	Enantiomeric Ratio of Styrol Oxide	
			R(%)	S(%)
Untreated	UT-A	10	37	63
	UT-F	5	51	49
	UT-H	10	49	51
	UT-I	5	52	48
Phenobarbital	PB-B	5	41	59
	PB-C	50	53	47
	PB-D	50	44	56
	PCN-E	50	50	50
β-Naphtoflavone	βNF-B	50	56	44
	ISF-G	5	50	50

(b) All involved isozymes catalyze oxirane formation with the same enantioselectivity (e.e. 100%).

(c) The involved isozymes catalyze oxirane formation with different enantioselectivities (not excluding e.e. = 0%).

G. L. Foureman et al. (43) used purified rat liver P-450 enzymes and observed that the nature and degree of enantioselectivity during the conversion of styrene to styrene oxide varied considerably among the isozymes (see Table 6.6). Cytochrome P-450$_{UT-A}$ had the highest degree of enantioselectivity from the untreated group, whereas P-450$_{UT-F}$, P-450$_{UT-H}$, and P-450$_{UT-I}$ were not enantioselective for this reaction. Only two isozymes (P-450$_{PB-B}$, and P-550$_{PB-D}$) isolated from the liver of PB-induced rats showed remarkable enantioselectivity. P-450$_{\beta NF-B}$, a βNF-induced enzyme, favored *R*-styrol oxide formation, whereas P-450$_{ISF-G}$ had no enantiopreference.

The enantioselectivities for styrene epoxidation in rabbit pulmonary microsomes are also due to the combined contributions of the individual P-450 isozymes (44). Enantiomeric ratios for the major forms of P-450 are identical, whether measured in microsomes plus monospecific inhibitory antibodies or in a reconstituted monooxygenase system (see Tables 6.7 and 6.8).

Likewise, enantioselectivity occurs during the *in vivo* metabolism and/or excretion of styrene because isomeric thioether metabolites of unequal ratios are present in the urine of rats (47, 48) and humans (49). In analogy to the *in vitro* conditions, the *R*-styrol oxide was preferentially formed (48).

Table 6.7. Effect to Antibodies to Cytochrome P-450, Forms 2 or 5, on the Rate and Enantioselectivity of Styrene Metabolism by Rabbit Lung Microsomes (44)

Incubation Conditions	Rate of Metabolism (nmol styrene oxide/min/nmol P-450)	Abs. Config.	
		R (%)	S (%)
Control	9.3	63	37
Anti-P-450, form 2	4.2	50	50
Anti-P-450, form 5	6.4	66	33
Anti-P-450, forms 2 and 5	0.34	54	56

Table 6.8. Specific Activity and Enantioselectivity for the Epoxidation of Styrene to Styrene Oxide by Purified Rabbit Cytochrome P-450 Isozymes in Reconstituted Monooxygenase Systems (44)

Cytochrome P-450 Isozymes	Specific Activity (nmol styrene oxide/min/nmol P-450)	Abs. Config.	
		R (%)	S (%)
Form 2	10.0	66	33
Form 5	4.7	50	50
Form 6	4.5	47	53

6.3.1.2. *Epoxide Hydrolase Catalyzed Hydrolysis of Monophenyl-Substituted Oxiranes*

Substrate enantioselectivity was observed during the microsomal epoxide hydrolase catalyzed hydrolysis of racemic styrene oxide (phenyloxirane) (35, 37, 45, 50) and several ring substituted derivatives (51) and of *cis*- and *trans*-β-alkyl substituted styrene oxides (2-alkyl-3-phenyloxiranes) (37, 50, 52) (see Table 6.9). In the most cases water attack occurs with high regioselectivity at the non-benzylic oxirane carbon atom. Hydrolysis of styrene oxirane proceeded by the highly selective (>98%) introduction of a hydroxy group at the non-benzylic carbon atom (45, 35, 50) leading to the retention of configuration. Whereas the ring-opening of *trans*-(2R, 3R)- and *cis*-(2S, 3R)-2-methyl-3-phenyloxirane occurred by a >98% regiospecific attack at the methyl substituted ring carbon atom, the opening of *trans*-(2S, 3S)- and *cis*-(2R, 3S)-2-methyl-3-phenyloxirane involved an only 88–90% attack at the same C-2. A 10–20% opening occurred by an unusual attack at the more hindered phenyl substituted oxirane carbon atom (50). *Cis*-2-alkyl-3-phenyloxiranes with alkyl groups larger than methyl were hydrolyzed by microsomal epoxide hydrolase in an enantioconvergent way. The oxirane ring opening occurs essentially at the S carbon of both enantiomers (52). The enzyme can discriminate between the steric effect due to a phenyl and a small methyl group (50), but this ability is lost with larger alkyl groups, so that only the preference for nucleophilic attack at the S oxirane carbon determines the regio- and stereochemistry of the ring opening of both enantiomers of oxiranes (52). Thus, in the case of 2-methyl-3-phenyloxirane the corresponding diol obtained at complete conversion was racemic. However, nearly enantiomerically pure (1*R*, 2*R*)-1-phenylalkane-1,2-diols were formed from oxiranes with longer alkyl chains (>90– >98% e.e.) (52).

S-styrene oxide (substrate: single enantiomer) was hydrolyzed by rat liver microsomes to *S*-phenylethanediol four times as fast as *R*- or racemic styrene oxide (35). Racemic styrol oxide was hydrolyzed with a bi-phase reaction profile. In the earlier stage of the reaction, the *R*-enantiomer is preferentially hydrolyzed until about 50% of the substrate is consumed; then the reaction rate of the *S*-enantiomer increases rapidly. S-phenylethanediol was formed at higher rates than *R*-phenylethanediol in the first stage of the reaction caused by the inhibitory effect of the *R*-oxirane enantiomer on the hydrolysis of the *S*-oxirane enantiomer. This inhibitory effect is a result of the higher affinity of the *R*-enantiomer for the epoxide hydrolase active site (35, 45). Similar results have been observed with *p*-nitrophenyloxirane (36).

Substrate enantioselectivity was affected by substituents on the phenyl ring of styrol oxide. The extent to which it was affected depended on the orientation and substrate concentration. Methyl, chloro, and nitro groups in the *meta*-position caused a considerable decrease in enantioselection at 5 mM but not at 0.5-mM substrate concentration. The presence of *para*-substituents produced either a complete loss of enantioselection, or a modest low enantiomer discrimination. One or two *ortho*-chloro substituents produced a remarkable loss of enantioselection only at 5-mM substrate concentration. Lowering the substrate concentration led to a better kinetic resolution (51).

The hydrolysis of 2-methyl-3-phenyloxirane occurs with low enantioselectivity for the *trans*-isomer and with very high enantioselectivity for the *cis*-isomer. A preferential consumption of the 2*S*, 3*S*-enantiomer of *trans*-2-methyl-3-phenyloxirane and of the 2*S*, 3*R* enantiomer of *cis*-2-methyl-3-phenyloxirane was found in rabbit liver microsomes (50) (see Table 6.9). In analogy to the structural related oxirane *cis*-2-ethyl-2-methyloxirane (38), a complete kinetic resolution was achieved with *cis*-2-methyl-3-phenyloxirane (50). Only one of the enantiomers of *cis*-2-ethyl-3-methyloxirane has been reported to be hydrolyzed, and this enantiomer has the same absolute configuration as the first consumed enantiomer of *cis*-2-methyl-3-phenyloxirane (50). Also, a complete substrate enantioselection can be observed in the hydrolysis of *cis*-2-ethyl-2-phenyloxirane (52). Longer alkyl chains caused a drastic decrease or a practically lost of the substrate enantioselectivity (52).

Table 6.9. Enantioselectivity in the Microsomal Epoxide Hydrolase-Catalyzed Hydrolysis of Racemic Oxiranes

Oxirane (Substrate)	Preferential consumption	Species	Reference
Styrene oxide	R	Rabbit	(51, 50)
	R	Rat	(35)
m-Methylstyrene oxide	R[a]	Rabbit	(51)
m-Chlorostyrene oxide	R[a]	Rabbit	(51)
m-Nitrostyrene oxide	R[a]	Rabbit	(51)
o-Chlorostyrene oxide	R[a]	Rabbit	(51)
o,o′-Dichlorostyrene oxide	R[a]	Rabbit	(51)
p-Methylstyrene oxide	R[a]	Rabbit	(51)
p-Chlorostyrene oxide	R[a]	Rabbit	(51)
p-Nitrostyrene oxide	R[a]	Rabbit	(51)
trans-2-Methyl-3-phenyloxirane	2S,3S	Rabbit	(50)
cis-2-Methyl-3-phenyloxirane	2S,3R	Rabbit	(50)
cis-2-Ethyl-3-phenyloxirane	2S,3R	Rabbit	(52)
cis-2-Phenyl-3-propyloxirane	2R,3S	Rabbit	(52)
Benzyloxirane	[b]	Rabbit	(51)
Phenoxymethyloxirane	[c]	Rabbit	(51)

[a] Abs. config. not determined, but presumed in analogue to other mono-substituted oxiranes.
[b] Very slight enantioselection observed.
[c] No enantioselection observed.

Racemic α-methyl substituted styrene oxide (2-methyl-2-phenyloxirane) and several ring substituted derivatives were hydrolyzed slowly by pig liver microsomes to provide R-diols with an enantiomeric excess between 13 and 34% at 63–70% conversion (53). But nothing was reported about the enantiomeric excess of the diols at 50% or 100% conversion and the enantiomeric composition of the remaining oxiranes.

Both main forms of epoxide hydrolase, the microsomal epoxide hydrolase and the cytosolic epoxide hydrolase, catalyze the hydrolysis of styrene oxide and its derivatives. For these substrates, the cytosolic and microsomal epoxide hydrolase show a different regioselectivity and enantioselectivity (37). Whereas cytosolic epoxide hydrolase results in a nonregioselective and non-enantioselective water attack on the styrene oxide and a regiospecific and non-enantioselective water attack on the benzylic carbon atom of *trans*-2-methyl-3-phenyloxirane, microsomal epoxide hydrolase always leads to a regiospecific and enantioselective ring-opening at the non benzylic oxirane carbon atom (37). It is noteworthy that cytosolic epoxide hydrolase catalyzed the ringopening of *trans*-2-methyl-3-phenyloxirane exclusively at the oxirane ring carbon atom bearing the larger substituent (37). E. C. Dietze et al. reported on the inhibition of cytosolic epoxide hydrolase with chiral oxiranes, especially 3-phenylglycidol and 3-(4-nitrophenyl)glycidol and their derivatives (54). With one exception, they found that cytosolic epoxide hydrolase binds *S*, *S*-enantiomers more tightly than *R*, *R*-enantiomers.

6.3.2. Metabolism of Stilbene

cis/*trans*-Stilbene was epoxidized by cytochrome P-450 monooxygenases to give *cis*/*trans*-stilbene oxide. While the epoxidation of *cis*-stilbene leads to a *meso*-compound, the prochiral olefin *trans*-stilbene could be epoxidized with product enantioselectivity. To our best knowledge, nothing has been reported about the enantioselectivity of this reaction. The microsomal epoxide hydrolase-catalyzed hydrolysis of racemic *trans*-stilbene oxide to *meso*-1,2-diphenyl-1,2-ethanediol occurs with substrate enantioselectivity (39), whereby *S*,*S*-*trans*-stilbene oxide was hydrolyzed at a faster rate than the *R*, *R*-enantiomer. The enzymatic ring-opening of *cis*-stilbene oxide, a *meso*-epoxide, occurs preferentially at the *S*-configured oxirane carbon atom to yield nearly optically pure *threo*-1*R*, 2*R*-diphenyl-1,2-ethanediol (39, 55–58) (see Table 6.10). Symmetrically *p*,*p*′-di-substituted *cis*-stilbene oxides, such as *cis*-4,4′-dimethylstilbene oxide, cis-4,4′-diethylstilbene oxide, *cis*-4,4′-diisopropylstilbene oxide, and *cis*-4,4′-dichlorostilbene oxide, were hydrolyzed by microsomal epoxide hydrolase with high product enantioselectivity to give *R*, *R*-diols with e.e. ≥90% (58). The high product enantioselectivity is independent of the presence and nature of the substituents in the *para*-positions of

Table 6.10. Product Enantioselectivity of the Epoxide-Hydrolase-Catalyzed Hydrolysis of *cis*-Stilbene Oxide

Substrate	Diol		Enzyme	Reference
	Enantiomeric Ratio	Abs. Config.		
cis-Stilbene oxide	>99:<1	*R*, *R*	mEH	(39)
	94:6	*R*, *R*	mEH	(57, 58)
	85:15	*R*, *R*	cEH	(57)
cis-4,4′-Dimethylstilbene oxide	>98:<2	*R*, *R*	mEH	(58)
cis-4,4′-Diethylstilbene oxide	>98:<2	*R*, *R*	mEH	(58)
cis-4,4′-Diisopropylstilbene oxide	>98:<2	*R*, *R*	mEH	(58)
cis-4,4′-Dichlorostilbene oxide	95:5	*R*, *R*	mEH	(58)

mEH = microsomal epoxide hydrolase.
cEH = cytosolic epoxide hydrolase.

the phenyl ring. All *p,p′*-di-substituted compounds served as much poorer substrates for the microsomal epoxide hydrolase than stilbene oxide (58).

Para-ring-substituted *cis*-stilbene oxides are hydrated by *trans*-opening to highly optically active *threo*-1*R*, 2*R*-diphenylethane diols (59). Whether the substituent was chloro, methyl, or nitro, the epoxide hydrolase-catalyzed hydrolysis of either racemic or optically pure *para*-ring-substituted *cis*-stilbene oxides resulted in diols of the same high optical purity. Selective water attack at the *S*-ring carbon atom was observed with preferential consumption of the 1*R*, 2*S*-oxirane (59). Similar high substrate enantioselectivities were found in the case of stucturally related substrates, for example, *cis*-2,3-dialkyl-substituted oxiranes [2-ethyl-3-methyloxirane, 2-methyl-3-(1′-methylpropyl)oxirane] (38) and *cis*-2-methyl-3-phenyloxirane) (50). Microsomal and cytosolic forms of epoxide hydrolase exhibit a qualitatively similar product enantioselectivity. With both enzymes, *threo*-1*R*, 2*R*-diphenyl-1,2-ethanediol was formed, but the cytosolic epoxide hydrolase-promoted reaction led to lower e.e. (34) (see Table 6.10).

CONCLUSION

The stereoselective formation of metabolites and/or enantioselective transformation of the primary metabolites can have important consequences for biological activity such as mutagenicity or cancerogenicity. If there is no complementary behavior in the formation of highly reactive oxiranes, catalyzed by cytochrome P-450, and in the detoxification reactions, catalyzed by epoxide hydrolase and glutathione *S*-transferase, an accumulation of one oxirane enantiomer can occur.

ACKNOWLEDGMENT

Sincere thanks are expressed to Prof. V. Schurig, University of Tübingen, Germany for helpful discussions and inspiring support of this work.

REFERENCES

1. Maynert, E. W., Foreman, R. L., and Watabe, T. (1970). Epoxides as obligatory intermediates in metabolism of olefins to glycols. *J. Biol. Chem.*, **245**, 5234–5238.
2. Ehrenberg, L. and Hussain, S. (1981). Genetic toxicity of some important epoxides. *Mut. Res.*, **86**, 1–113.
3. Oesch, F. (1973). Mammalian epoxide hydrases: Inducible enzymes catalyzing the

inactivation of carcinogenic and cytotoxic metabolites derived from aromatic and olefinic compounds . *Xenobiotica*, **3**, 305–340.

4. Seidegård, J., and DePierre, J. W. (1983). Microsomal epoxide hydrolase. Properties, regulation and function. *Biochim. Biophys. Acta*, **695**, 251–270.
5. Fjellstedt, T. A., Allen, R. H., Duncan, B. K., and Jacoby, W. B. (1973). Enzymatic conjugation of epoxides with glutathione. *J. Biol. Chem.*, **248**, 3702–3707.
6. Seiler, J. P. (1990). Chirality-dependent DNA reactivity as the possible cause of the differential mutagenicity of the two components in an enantiomeric pair of epoxides. *Mut. Res.*, **245**, 165–169.
7. Gadberry, M. G., DeNicola, D. B., and Cardson, G. P. (1996). Pneumotoxicity and Hepatoxicity of styrene and styrene oxide. *J. Toxicol. Environ. Health*, **48**, 273–294.
8. Black, S. D. and Coon, M. J. (1986). Comparative structures of P-450 cytochromes. In *Cytochrome P-450, Structure, Mechanism and Biochemistry*. P. R. de Montellano (ed.), New York: Plenum, p. 161.
9. Nebert, D. W., and Jenson, N. M., (1979). The ah locus: Genetic regulation of the metabolism of carcinogens, drugs, and other environmental chemicals by cytochrome P450-mediated monooxygenases. *CRC Crit. Rev. Biochem.*, **6**, 401–437.
10. vanBladeren, P. J., Armstrong, R. N., Cobb, D., Thakker, D. R., Ryan, D. E., Thomas, P. E., Sharma, N. D., Boyd, D. R., Levin, W., and Jerina, D. M. (1982). Stereoselective formation of benz[a]anthracene (+)-(5*S*, 6*R*)-oxide and (+)-(8*R*, 9*S*)-oxide by a highly purified and reconstituted system containing cytochrome P-450c. *Biochem. Biophys. Res. Commun.*, **106**, 602–609.
11. Weems, H. B., Fu, P. P., and Yang, S. K. (1986). Stereoselective metabolism of chrysene by rat liver microsomes. Direct separation of diol enantiomers by chiral stationary phase H.P.L.C. *Carcinogenesis*, **7**, 1221–1230.
12. Yang, S. K. and Bao, Z.-P. (1987). Steroselective formations of K-region and non-K-region epoxides in the metabolism of chrysene by rat liver microsomal cytochrome P-450 isozymes. *Mol. Pharmacol.*, **32**, 73–80.
13. Yang, S. K., Mushtag, M., Chiu, P. L., and Weems, H. B. (1986). Stereoselectivity of rat liver cytochrome P-450 isozymes: Direct determination of enantiomeric composition of K-region epoxides formed in the metabolism of benz[a]anthracene and 7,12-dimethylbenz[a]anthracene. *Adv. Exp. Med. Biol.*, **197**, 809–817.
14. Yang, S. K., Mushtag, M., and Weems, H. B. (1987). Stereoselective formation and hydration of benzo[c]phenanthrene 3,4- and 5,6-epoxide enantiomers by rat liver microsomal enzymes. *Arch. Biochem. Biophys.*, **255**, 48–63.
15. Yang, S. K., Mushtag, M., Weems, H. B., and Miller, D. W., (1987). Stereoselective formation and hydration of 12-methylbenz[a]anthracene 5,6-epoxide enantiomers by rat liver microsomal enzymes. *Biochem. J.*, **245**, 191–204.
16. Lacourciere, G. M., and Armstrong, R. N., (1993). The catalytic mechanism of microsomal epoxide hydrolase involves an ester intermediate. *J. Am. Chem.*, **115**, 10466–10467.

17. Armstrong, R. N. (1987). Enzyme-catalyzed detoxication reactions: Mechanisms and stereochemistry. *CRC Crit. Rev. Biochem.*, **22**, 39–88.
18. Schurig, V. and Wistuba, D. (1984). Asymmetric microsomal epoxidation of simple prochiral olefins. *Angew. Chem. Int. Ed. Engl.*, **23**, 796–797.
19. Wistuba, D., Nowotny, H.-P., Träger, O., and Schurig, V. (1989). Cytochrome P-450-catalyzed asymmetric epoxidation of simple prochiral and chiral aliphatic alkenes: Species dependence and effect of enzyme induction on enantioselective oxirane formation. *Chirality*, **1**, 127–136.
20. Ortiz de Montellano, P. R., Mangold, B. L. K., Wheeler, C., Kunze, K. L., and Reich, N. O. (1983). Stereochemistry of cytochrome P-450-catalyzed epoxidation and prosthetic heme alkylation. *J. Biol. Chem.*, **258**, 4208–4213.
21. Wistuba, D. (1986). Enantioselektive Metabolisierung kleiner aliphatischer Olefine und Oxirane durch Cytochrome-P-450-abhängige Monooxygenasen, Epoxid-Hydrolasen und Glutathion S-Transferasen. Ph.D. thesis, University of Tübingen, Department of Organic Chemistry, Tübingen.
22. Wistuba, D., Weigand, K., and Peter, H. (1994). Stereoselectivity of *in vitro* isoprene metabolism. *Chem. Res. Toxicol.*, **7**, 336–343.
23. Bellucci, G., Chiappe, C., Cordoni, A., and Marioni, F. (1994). The rabbit liver microsomal biotransformation of 1,1-dialkylethylenes. *Chirality.*, **6**, 207–212.
24. Prelog, V. (1962). The stereospecificity of the enzymic reduction of carbonyl groups. *Ind. Chim. Belg.*, **11**, 1309–1318.
25. Jenner, P. and Testa, B. (1973). The influence of stereochemical factors on drug disposition. *Drug Metab. Rev.*, **2**, 117–184.
26. Malvoisin, E., Hoest, G. L., Poncelet, F., Roberfroid, M., and Mercier, M. (1979). Identification and quantitation of 1,2-epoxybutene-3 as the primary metabolite of 1,3-butadiene. *J. Chromatogr.*, **178**, 419–425.
27. del Monte, M., Citti, L., and Gervasi, P. G. (1985). Isoprene metabolism by liver microsomal monooxygenases. *Xenobiotica*, **15**, 591–597.
28. Dahl, A. R., Birnbaum, L. S., Bond, J. A., Gervasi, P. G., and Henderson, R. F. (1987). The fate of isoprene inhaled by rats: Comparison to butadiene. *Toxicol. Appl. Pharmacol.*, **89**, 237–248.
29. Oesch, F., Kaubisch, N., Jerina, D. M., and Daly, J. W. (1971). Hepatic epoxide hydrase. Structure-activity relationships for substrates. *Biochemistry*, **10**, 4858–4866.
30. Jerina, D. M., Ziffer, H., and Daly, J. W. (1970). The role of the arene-oxide-oxepin system in the metabolism of aromatic substrates. IV. Stereochemical considerations of dihydrodiol formation and dehydrogenation. *J. Am. Chem. Soc.*, **92**, 1056–1061.
31. Hanzlik, R. P., Edelman, M., Michaely, W. J., and Scott, G. (1976). Enzymatic hydration of [^{18}O]-epoxides. Role of nucleophilic mechanisms. *J. Am. Chem. Soc.*, **98**, 1952–1955.
32. Wistuba, D. and Schurig, V. (1992). Enantio- and regioselectivity in the epoxide-hydrolase-catalyzed ring opening of simple aliphatic oxiranes. Part I: Monoalkylsubstituted oxiranes. *Chirality*, **4**, 178–184.

33. Wistuba, D. and Schurig, V. (1986). Complementary epoxide hydrolase–vs. glutathione S-transferase-catalyzed kinetic resolution of simple aliphatic oxiranes—complete regio- and enantioselective hydrolysis of *cis*-2-ethyl-3-methyl-oxirane. *Angew. Chem. Int. Ed. Engl.*, **25**, 1032–1034.
34. Bellucci, G., Chiappe, C., Conti, L., Marioni, F., and Pierini, G. (1989). Substrate enantioselection in the microsomal Epoxide hydrolase catalyzed hydrolysis of monosubstituted oxiranes. Effect of branching of alkyl chains. *J. Org. Chem.*, **54**, 5978–5983.
35. Watabe, T., Ozawa, N., and Hiratsuka, A. (1983). Studies on metabolism and toxicity of styrene-VI. Regioselectivity in glutathione S-conjugation and hydrolysis of racemic, R- and S-phenyloxiranes in rat liver. *Biochem. Pharmacol.*, **32**, 777–785.
36. Westkaemper, R. B. and Hanzlik, R. P. (1981). Mechanistic studies of epoxide hydrolase utilizing a continous spectrophotometric assay. *Arch. Biochem. Biophys.*, **208**, 195–204.
37. Bellucci, G., Chiappe, C., Cordoni, A., and Marioni, F. (1994). Different enantioselectivity and regioselectivity of the cytosolic and microsomal epoxide hydrolase catalyzed hydrolysis of simple phenyl substituted epoxides. *Tetrahedron Lett.*, **35**, 4219–4222.
38. Wistuba, D., Träger, O., and Schurig, V. (1992). Enantio- and regioselectivity in the epoxide-hydrolase-catalyzed ring opening of aliphatic oxiranes. Part II. Dialkyl- and trialkylsubstituted oxiranes. *Chirality*, **4**, 185–192.
39. Watabe, T. and Akamatsu, K. (1972). Stereoselective hydrolysis of acyclic olefin oxides to glycols by hepatic microsomal epoxide hydrolase. *Biochim. Biophys. Acta*, **279**, 297–305.
40. Wang, P., Meijer, J., and Guengerich, F. P. (1982). Purification of human liver cytosolic epoxide hydrolase and comparison to the microsomal enzyme. *Biochemistry*, **21**, 5769–5776.
41. Bellucci, G., Chiappe, C., Marioni, F., and Benetti, M. (1991). Regio- and enantioselectivity of the cytosolic epoxide hydrolase-catalyzed hydrolysis of racemic monosubstituted alkyloxiranes. *J. Chem. Soc. Perkin Trans.*, **1**, 361–363.
42. Pagano, D. A., Yagen, B., Hernandez, O., Bend, J. R., and Zeiger, E. (1982). Mutagenicity of (R)- and (S)-styrene 7,8-oxide and the intermediary mercapturic acid derivatives formed from styrene 7,8-oxide. *Environ. Mutagen.*, **4**, 575–584.
43. Foureman, G. L., Harris, C., Guengerich, F. P., and Bend, J. R. (1989). Stereoselectivity of styrene oxidation in microsomes and in purified cytochrome P-450 enzymes from rat liver. *J. Pharmacol. Exp. Ther.*, **248**, 492–497.
44. Harris, C., Philpot, R. M., Hernandez, O., and Bend, J. R. (1986). Rabbit pulmonary cytochrome P-450 monooxygenase system: isozyme differences in the rate and stereoselectivity of styrene oxidation. *J. Pharmacol. Exp. Ther.*, **236**, 144–149.
45. Watabe, T., Ozawa, N., and Yoshikawa, K. (1981). Stereochemistry in the oxidative metabolism of styrene by hepatic microsomes. *Biochem. Pharmacol.*, **30**, 1695–1698.

46. Ramsey J. C. and Young, J. D. (1978). Pharmacokinetics of inhaled styrene in rats and humans. *Scand. J. Work Environ. Health*, **4** (suppl. 2), 84–91.
47. Delbressine, L. P. C., Van Bladeren, P. J., Smeets, F. L. M., and Seutter-Berlage, F. (1981). Stereoselective oxidation of styrene to styrene oxide in rats as measured by mercapturic acid excretion. *Xenobiotica.*, **11**, 589–594.
48. Watabe, T., Ozawa, N., and Yoshikawa, K. (1982). Studies on metabolism and toxicity of styrene. V. The metabolism of styrene, racemic, R(+)-, and (S)-(–)-phenyloxiranes in the rat. *J. Pharm. Dyn.*, **5**, 129–133.
49. Korn, M., Wordarz, R., Drysch, R., and Schmahl, F. W. (1987). Stereometabolism of styrene in man. *Arch. Toxicol.*, **6**, 86–88.
50. Bellucci, G., Chiappe, C., Cordoni, A., and Marioni, F. (1993). Substrate enantioselectivity in the rabbit liver microsomal epoxide hydrolase catalyzed hydrolysis of *trans* and *cis* 1-phenylpropene oxides. A. Comparison with styrene oxide. *Tetrahedron. Asym.*, **4**, 1153–1160.
51. Bellucci, G., Chiappe, C., and Marioni, F. (1992). Enantioselective hydrolysis of Substituted phenyloxiranes by rabbit liver microsomal epoxide hydrolase. *Ind. J. Chem.*, **31B**, 828–831.
52. Bellucci, G., Chiappe, C., and Cordoni, A. (1996). Enantioconvergent transformation of racemic *cis-β*-alkyl substituted styrene oxides to (R,R) threo diols by microsomal epoxide hydrolase catalysed hydrolysis. *Tetrahedron. Asymmetry*, **7**, 197–202.
53. Basavaiah, D., Raju, S. B. (1995). Enantioselective hydrolysis of 2,2-disubstituted oxiranes mediated by microsomal epoxide hydrolase. *Synth. Commun.*, **25**, 3293–3306.
54. Dietze, E. C., Kuwano, E., and Hammock, B. D. (1993). The interaction of cytosolic epoxide hydrolase with chiral epoxides. *Int. J. Biochem.*, **25**, 43–52.
55. Watabe, T., Akamatsu, K. and Kiyonara, K. (1971). Steroselective hydrolysis of *cis*- and *trans*-stilbene oxides by hepatic microsomal epoxide hydrolase. *Biochem. Biophys. Res. Commun.*, **44**, 199–204.
56. Bellucci, G., Berti, G., Chiappe, C., Fabri, F., and Marioni, F. (1989). Product enantioselectivity in the microsomal epoxide hydrolase catalyzed hydrolysis of 10,11-dihydro-10,11-epoxy-5H-dibenzo[a,d]-cycloheptene. *J. Org. Chem.*, **54**, 968–970.
57. Bellucci, G., Capitani, I., Chiappe, C., and Marioni, F. (1989). Product enantioselectivity of the microsomal and cytosolic epoxide hydrolase catalyzed hydrolysis of meso epoxides. *J. Chem. Soc., Chem. Commun.*, 1170–1171.
58. Bellucci, G., Chiappe, C., and Ingrosso, G. (1994). Kinetics and stereochemistry of the microsomal epoxide hydrolase-catalyzed hydrolysis of *cis*-stilbene oxide. *Chirality*, **6**, 577–582.
59. Dansette, P. M., Makedonska, V. B., and Jerina, D. (1978). Mechanism of catalysis for the hydration of substituted styrene oxides by hepatic epoxide hydrase. *Arch. Biochem. Biophys.*, **187**, 290–298.

CHAPTER

7

CHIRAL BARBITURATES: SYNTHESIS, CHROMATOGRAPHIC RESOLUTIONS, AND BIOLOGICAL ACTIVITY

JACEK BOJARSKI

Department of Organic Chemistry, College of Medicine, Jagiellonian University, 30-688 Krakow, Poland

7.1. INTRODUCTION

Derivatives of barbituric acid [2,4,6-(1H,3H,5H)-pyrimidinetrione] and 2-thiobarbituric acid are well-known drugs with sedative, hypnotic, and anticonvulsant activity. The chirality of these compounds results from the chiral centers being situated in the C-5 substituent and/or at the C-5 carbon atom of the pyrimidine ring. Metabolic transformation in the body may also create a chiral center in the molecule of the parent drug (Figure 7.1).

More than 60 years have elapsed since the first report on the synthesis of enantiomers of optically active barbiturate (5-ethyl-5-*sec*-octylbarbituric acid) and a comparison of their biological activity in terms of minimum effective and fatal doses (1). In the second half of this century, both the chemistry and pharmacology of chiral barbiturates have been vividly developed. Early conclusions about the low importance of the stereoisomerism of barbiturates in their pharmacological applications (2) are now considerably changed, and research on chiral barbiturates may serve as a good example of the stereochemical concepts in medicinal chemistry and pharmacology.

This chapter provides actual information on (and should substantially enrich our understanding of) the optical isomers of barbiturates, which constituted only a small paragraph in a recent monograph on progress in barbituric acid chemistry (3).

The Impact of Stereochemistry on Drug Development and Use, Edited by Hassan Y. Aboul-Enein and Irving W. Wainer. Chemical Analysis Series, Vol. 142.
ISBN 0-471-59644-2

2-oxo barbiturates:

hexobarbital

mephobarbital

butabarbital

pentobarbital

secobarbital

methohexital

talbutal

2-thioxo barbiturates:

thiopental

thiamylal

thiohexital

metabolites:

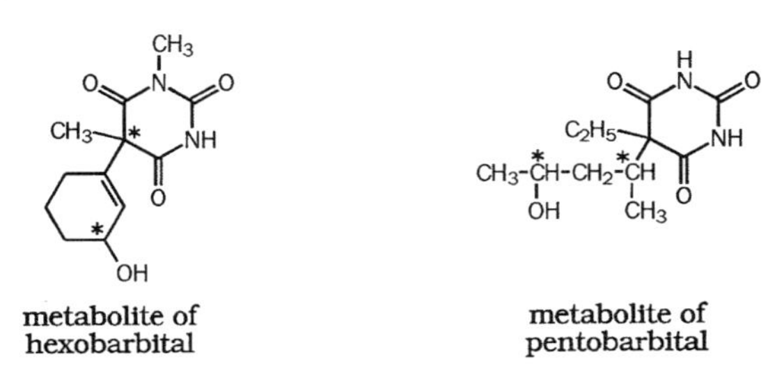

Figure 7.1. Chemical structures of selected chiral barbiturate and thiobarbiturate drugs and their metabolites. Chiral centers are marked with asterisks.

7.2. SYNTHESIS OF CHIRAL BARBITURATES

Optically active forms of barbiturates have been synthesized by general methods of synthesis for this class of compounds, that is, by the condensation of urea, thiourea, or guanidine and their derivatives with enantiomers of esters of substituted malonic or cyanoacetic acids. Synthetic routes for (*S*)(–)-pentobarbital (4) (Scheme 1) and both enantiomers of hexobarbital (5) (Scheme 2) illustrate this approach. In this last case, the same optically active substrate served for the synthesis of both enantiomers and dicyandiamide was used for the preparation of one of them.

Several enantiomers of barbiturates and 2-thiobarbiturates with chiral centers in the C-5 substituents have been prepared similarly (4, 6–9), while Knabe and co-workers prepared a lot of optically active N-substituted barbiturates (10, 11). The synthesis of a 2-thio analog of hexobarbital was accomplished by the condensation of N-methylthiourea with appropriate cyano- acetate, although with a lower yield (5), but the same procedure failed for the synthesis of enantiomers of 1-methyl-5-phenyl-5-propyl-2-thiobarbituric acid. They were obtained instead by the reaction of free phenylpropylcyanoacetic acid with N-methylthiourea in the presence of dicyclohexylcarbodiimide (12).

The stereospecific synthesis of (*R*)(–)-hexobarbital with a trideuteromethyl group at the nitrogen atom was carried out similarly using N-trideutero-methylurea and (*S*)(+)-methyl-2-cyano-2-(1-cyclohexenyl)propionate (13). Both antipodes of this last compound, condensed with carbonyl ^{14}C-labeled N-methylurea, yielded both enantiomers of hexobarbital with the ^{14}C-labeled 2-CO group (14).

Enantiospecific condensation of optically active substituted cyanoacetates with dicyandiamide, followed by methylation and hydrolysis, yielded several chiral barbiturates with substituted N-aminoethyl substituents on the nitrogen atom (15).

Recently, new synthesis of N-substituted chiral barbiturates was reported by Murata et al. (16, 17). The substrate—5,5-di-substituted N,N′-bisacyloxymethylbarbiturates were asymmetrically hydrolyzed in the presence of a lipase catalyst in diisopropyl ether saturated with water. Interestingly, hydrolyses catalyzed by lipase from *Candida rugosa* yielded optically active N-acyloxymethylbarbiturates, whereas with lipase from *Humicola lanuginosa* their antipodes were obtained. These products were converted into corresponding chiral N-methyl barbiturate drugs (mephobarbital and hexobarbital) by the methylation and hydrolysis of N-acyloxy moiety. In a similar manner, all four optical isomers of N-{2-[(aminocarbonyl)oxy]-3-butoxypropyl}-5-ethyl-5-phenylbarbituric acid (febarbamate) were synthesized (17).

The metabolic transformation of pentobarbital and its 5-allyl analog (secobarbital), as well as their 2-thio counterparts (thiopental and thiamylal,

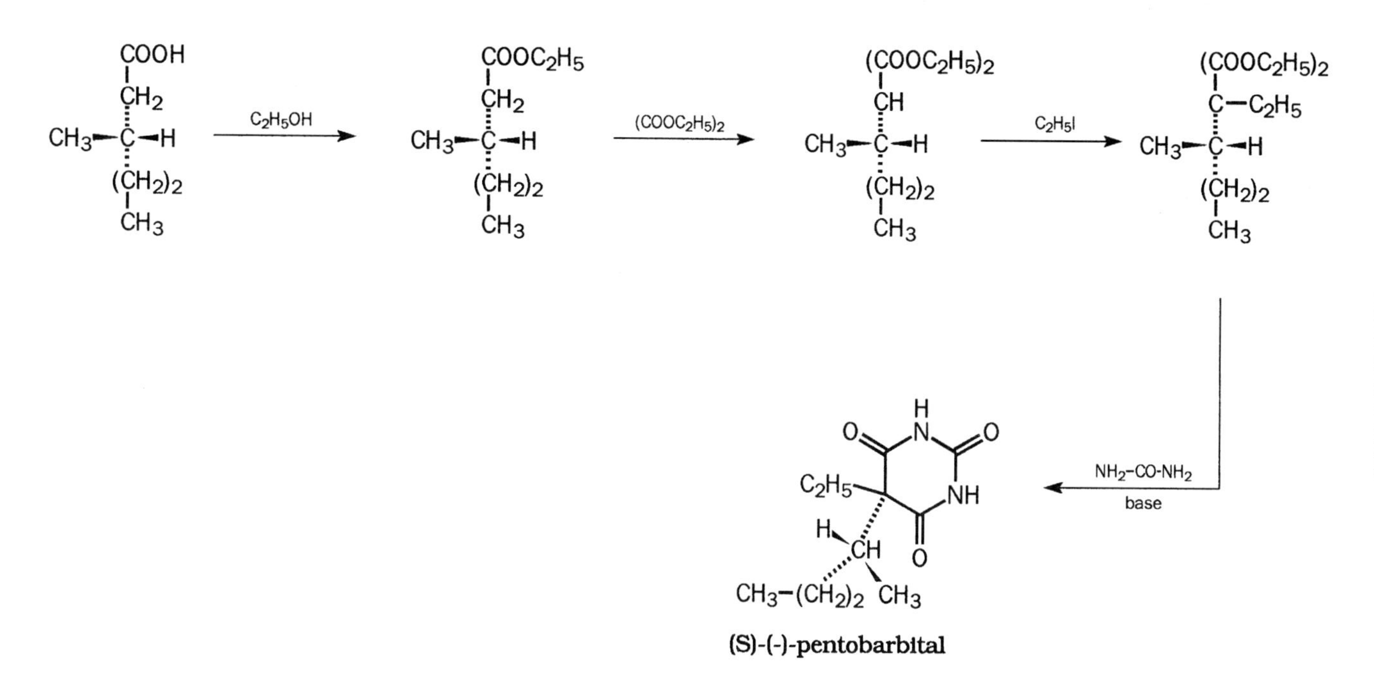

Scheme 1. Synthetic route to (*S*)(−)-pentobarbital (14).

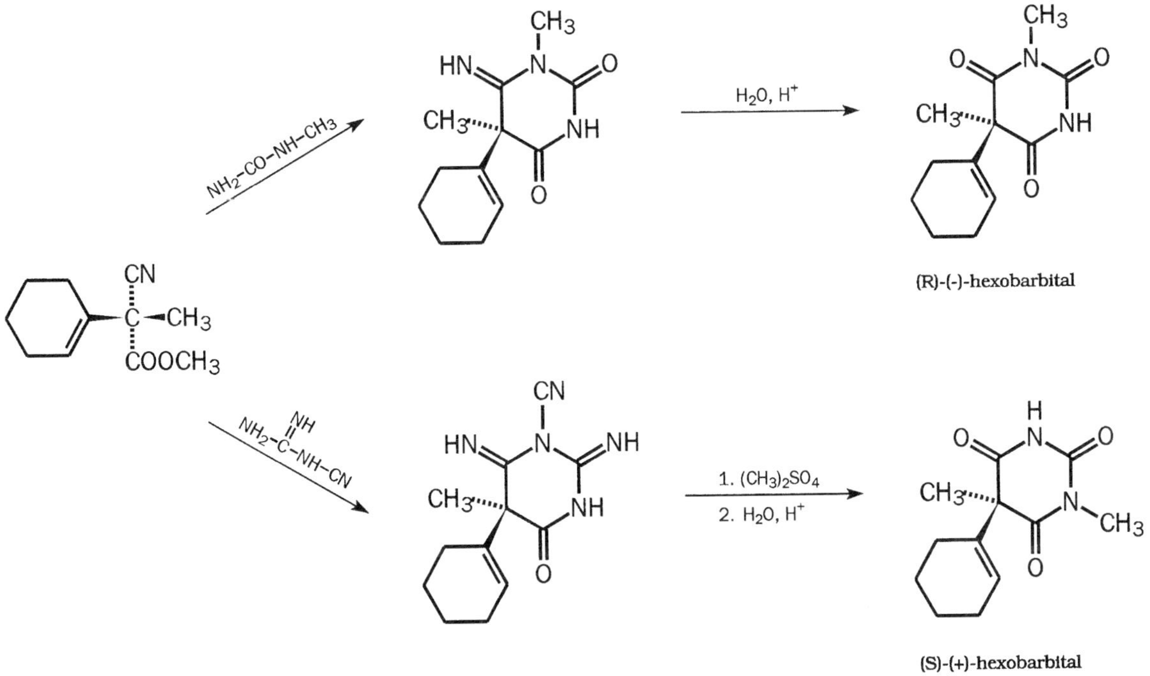

Scheme 2. Synthetic routes to (*R*)(−)- and (*S*)(+)-hexobarbital (5).

respectively) yields, among others, the products with the hydroxyl group situated at the C-3 atom of the 1-methylbutyl substituent. All possible enantiomeric and diastereomeric pairs of these compounds were obtained synthetically by Carroll and Mitchell (18), by the condensation of urea or thiourea with appropriate stereoisomers of 2-ethoxycarbonyl-2-ethyl-(or allyl-) 3,5-dimethylvalerolactone.

7.3. RESOLUTIONS OF RACEMIC MIXTURES

7.3.1. Classical Methods

The resolution of a racemic mixture of optical antipodes is an alternative synthetic approach, competitive with asymmetric synthesis, to obtain pure enantiomers. In spite of its limitations, this method, based on the formation of diastereomeric salts, their separation, and subsequent cleavage to parent compounds, proved successful for some barbituric acid derivatives. The first chiral barbiturate drug resolved using N-methylquininium hydroxide was hexobarbital (19).

The same chiral resolving agent was also used for the resolution of enantiomers of mephobarbital (20, 21), vinylbital [5-ethenyl-5-(1-methylbutyl)barbituric acid], pentobarbital, and 5-(2-bromo-2-propenyl)-5-(1-methylpropyl), barbituric acid (21). Further, it has been shown that the levorotatory optical isomer of pentobarbital with an S configuration, obtained in that way, was only 28% optically pure, and 12 recrystallizations of diasteromeric salt were needed to obtain the pure form of (*S*)(−)-pentobarbital with a 3% yield (4). In spite of this fact, the method was used for the resolution of other barbiturates, such as 5-(1-cyclohexenyl)-5-ethyl-l-methyl (22) and various 5-methyl- and 5-dimethylbutyl-5-ethylbarbituric acids (23). There are also several reports on the application of this method for the resolutions of enantiomers of hexobarbital (24), mephobarbital (25), pentobarbital (26), and 5-(1,3-dimethylbutyl)-5-ethylbarbituric acid (26–28) for further applications in pharmacological investigations.

(1*S*)(+)-10-Camphorsulfonic acid and brucine were used as resolving agents by Doran (29) during the synthesis of different optical isomers of methohexital. This first reagent was also used for the successful resolution of 17 out of 27 synthesized racemic barbiturates with basic C-5 substituents (30), but failed for the resolution of racemic 5-allyl-5-(2-hydroxypropyl)barbituric acid (31). This last barbiturate was resolved after esterification with phthalic acid and the formation of diastereomeric salt with (−)-quinine. The salt was resolved by fractional crystallization from acetone, yielding pure levorotatory monophthalate. Since the (+)-ester was contaminated with the antipode, its purification was done via the diastereomeric salt with (+)-quinidine, less soluble in acetone. The final cleavage of esters was carried out by boiling with 6% HCl and yielded pure enantiomers of the barbiturate.

Racemic N-substituted barbituric acids obtained by the N-aminoethylation of phenobarbital and norhexobarbital were resolved mainly with (*R*)(–)- and (*S*)(+)-1,1′-binaphthyl-2,2′-diyl hydrogen phosphate, but complete resolution was achieved only for derivatives of this first barbiturate, whereas for the second only enantiomeric enrichment was observed for these resolving agents, as well as for (1*S*)(+)-10-camphorsulfonic acid (32).

7.3.2. Chiral Chromatography of Barbiturates

Since the first reports on the successful chromatographic resolution of several racemic mixtures of N-substituted barbiturates by column chromatography on microcrystalline cellulose triacetate (33, 34) and optically active polyacrylic and polymethacrylic amides (35) [for a review, see (36)], many techniques of chiral chromatography have been applied for similar purposes and pharmacological applications of their results reported.

7.3.2.1. Liquid Chromatography

Liquid chromatography, especially in its high-performance version (HPLC), is the most frequently used technique for the resolution of the racemates of barbiturates in a direct mode on various chiral stationary phases (CSPs), or with an addition of a resolving reagent to the mobile phase. Among other drugs, hexobarbital and mephobarbital (5-ethyl-l-methyl-5-phenylbarbituric acid) were often used as probes for testing the resolving power of new CSPs and the effect of various chromatographic parameters (temperature, mobile phase composition, etc.) on the resolution of enantiomers, expressed in terms of a separation factor (α). Armstrong and co-workers (37–40) and others (41–47) have reported the separation of these compounds on various CSPs prepared from β-cyclodextrin, a cyclic oligomer built of seven α-D-glucose units, or its derivatives. Studies on the effect of temperature on the enantioseparation of mephobarbital on such a column revealed that this was an entropy-controlled separation and the increase in temperature improved the chiral resolution (48). The cyclodextrin chiral selectors as the additives to the mobile phase, forming the dynamically generated CSPs, were also used for this purpose (49–57). Recently, carboxymethyl- and carboxyethyl-β-cyclodextrin have proved to be effective chiral mobile phase additives for the separation of enantiomers of hexobarbital. Strong effects of the type of cyclodextrin derivative, pH, and the type of base (NaOH and triethylamine) used to adjust the pH were observed in these experiments (58). The resolution of enantiomers of other N-substituted chiral barbiturates was accomplished using this method (59), which was also employed for the quantification of enantiomers of hexobarbital in rat blood (60). It was found that the simultaneous addition of β-cyclodextrin and its permethylated derivative to the mobile phase improved the resolution of enantiomers of mephobarbital (61). A β-cyclodextrin chiral HPLC method for this compound was adopted for studies on stereoselective metabolism, pharmacokinetics (62, 63), and protein binding (64). The 2-thio analog of mephobarbital and relative chiral barbiturates were resolved with β-cyclodextrin added to the mobile phase (65). Thuaud and Sebille (66) studied the resolution of barbiturate and thiobarbiturate drug enantiomers with the chiral center in the side chain—the type of compounds hardly undergoing enantioseparation, using different cyclodextrin columns. Interestingly, they also found a higher chiral recognition for the 2-thiocarbonyl group, in comparison with the 2-carbonyl moiety observed earlier for mephobarbital and its 2-thio analog (65). They also stated that the substitution of β-cyclodextrin with hydroxypropyl substituents favors the enantiomeric discrimination of barbiturates and thiobarbiturates investigated. The examples of resolutions with cyclodextrins are shown in Figure 7.2.

Polysaccharide derivatives were often used for chiral HPLC of barbiturates. Mephobarbital was successfully resolved on cellulose tricinnamate CSP (67). Striking differences were observed for the chiral resolution of mephobarbital and hexobarbital on benzoylcellulose and *o*-, *m*-, and *p*-methylbenzoylcelluloses (68). Under the same chromatographic conditions, the first compound was resolved only on the first and last CSP, whereas its counterpart was resolved only on the second and third CSP. These results lead to the conclusion that even for CSPs modified by homologous molecules or positional isomers, their enantioselectivity toward N-methyl barbiturates considerably depends on rather small structural differences of C-5 substituents. Rizzi (69–72) investigated thoroughly the enantioselectivity of swollen microcrystalline cellulose triacetate and its dependence on different parameters of the chromatographic process and analyte structure using hexobarbital as one of the test compounds. Optically active barbiturates obtained by lipase-catalyzed enantioselective synthesis (16, 17), mephobarbital and another antiepileptic drug—benzonal (1-benzoyl-5-ethyl-5-phenylbarbituric acid) (73), were well resolved on a cellulose tris-(4-methylbenzoate) (Chiralcel OJ) commercially available column, whereas for hexobarbital and mephobarbital, the baseline resolution on cellulose tris-3,5-dimethylphenylcarbamate (Chiralcel OD) was not achieved (46). Further experiments with a Chiralcel OJ column for the enantioseparations of different 5,5-di- and 1,5,5-tri-substituted barbiturates showed that this column gives much better results for these last compounds. Besides, it was found that ethanol was a better mobile phase modifier than methanol. Better resolutions were observed for barbiturates with one

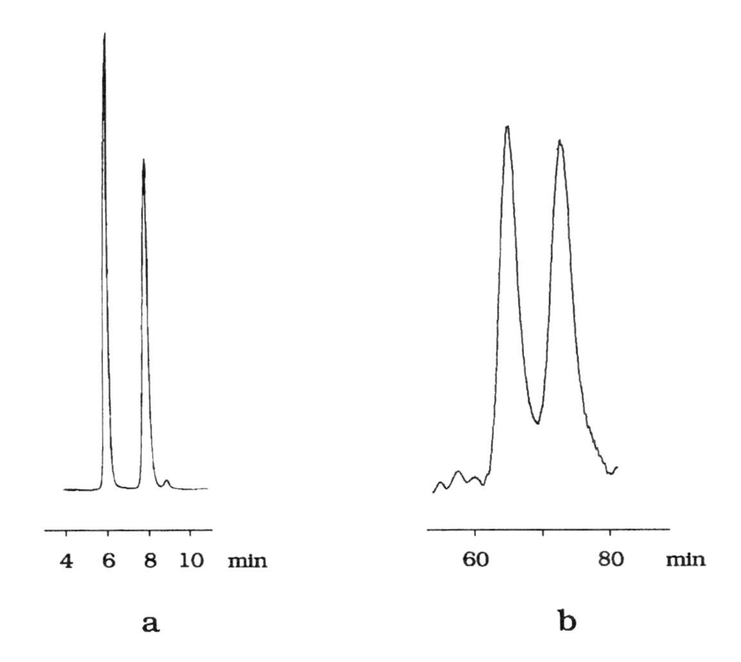

Figure 7.2. Examples of chiral liquid chromatographic separations at room temperature. (a) Hexobarbital enantiomers CSP, permethylated β-cyclodextrin; column, NUCLEODEX β-PM (Macherey-Nagel, 200 mm); mobile phase, methanol:0.1% triethylammonium acetate, pH = 4 (55:45 v/v), flow rate 0.7 ml/min. [Courtesy of Dr. H. Riering (Macherey-Nagel)]; (b) 2-Thiomephobarbital enantiomers: column, ChromSil C18, 10 μm, 250 mm; mobile phase aqueous saturated solution of β-cyclodextrin:96% ethanol:0.1 M NaH_2PO_4: 1 M Na_2HPO_4 (80:10:7.5:2.5 v/v), flow rate 1.5 ml/min.

cyclic and one aliphatic substituent at the C-5 carbon atom, whereas no separation was observed for those with two aliphatic groups. Both steric and electronic effects seem to contribute to the stereodifferentiation of barbiturates on that column (74).

Protein CSPs were often used for the resolution of enantiomers of different chiral barbiturates. The first report dealt with hexobarbital resolved on a chiral α_1-acid glycoprotein column (Enantio Pac) (75). It was found that the separation factors increased with an increase in the pH of the mobile phase (phosphate buffer with 1% of isopropanol). Benzonal and related compounds were resolved on the bovine serum albumin (BSA) bound to silica (Resolvosil column), but pH dependence was not so explicit (76). The effect of various organic modifiers added to the mobile phase (77), and cross-linking reagents for BSA (78), on the enantioselectivity for these compounds was described further.

Mephobarbital was resolved on a new generation α_1-acid glycoprotein column (Chiral-AGP) with a phosphate buffer (pH 7.0) and 2% isopropanol as the mobile phase, but for the buffer without the organic modifier, no resolution was observed (79). The effect of different modifiers on the enantioselectivity of the same column for various barbiturates was also investigated (80). This column was also used for the assay of hexobarbital enantiomers in rat plasma (81).

Hexobarbital and mephobarbital were resolved on the ovomucoid column but not on the cellulase CSP(46), and Haginaka et al. used modified ovomucoid columns for the successful resolution of hexobarbital enantiomers (82, 83). For the mobile phase consisting of 10% ethanol and 20 mM of phosphate buffer (90:10 v/v), the best separation factors were found at a pH of 6.9.

Yang et al. (84) investigated the separation of several barbiturates on ionically and covalently bonded Pirkle CSPs with (*R*)-N-(3,5-dinitrobenzoyl) phenylglycine and (*S*)-N-(3,5-dinitrobenzoyl)leucine. Hexobarbital was resolved on all of them, but mephobarbital and secobarbital only on two, while butabarbital was not resolved at all. In general, the separation factors were rather low ($\alpha \leq 1.12$). This first CSP covalently bonded (marketed as ChiraSep) proved useful in the enantioseparations of hexobarbital and secobarbital (85).

A special CSP with a derivative of optically active N,N′-2,6-pyridinediyl-bis[(*S*)-2-phenylbutanamide] was designed to separate racemic mixtures of barbiturates (and other structurally related compounds), taking into account the chiral recognition principle based on the differential hydrogen bond association of enantiomers (86). Nine N-substituted compounds (including hexobarbital, mephobarbital, and methohexital) were resolved, with α values between 1.06 and 1.16. The same principle was considered responsible for the resolution of hexobarbital and mephobarbital on the CSP derived from optically active N,N′-dialkyltartramide (87).

Different CSPs operating on the ligand-exchange principle were used for the separation of enantiomers of barbituric acid derivatives. Sinibaldi et al. (88) reported the successful resolution of enantiomers of some N-methyl-substituted drugs of this family (hexo- and mephobarbital, methohexital), while N-unsubstituted seco- and pentobarbital remained unresolved, on the CSP obtained from L-valinamide and ethylene glycol diglycidyl ether and loaded with copper(II) ions. In an alternative approach, polymerized (–)-*trans*-1,2-diaminocyclohexane complexed with the same ions was used as the chiral additive to the eluent, resulting in the resolution of hexobarbital and 5-butyl-5-cyclohexenyl-l-methylbarbituric acid enantiomers, whereas methohexital and mephobarbital racemates were not resolved (89). Recently, Gübitz et al. applied CSP obtained by binding a chiral amino alcohol (*S*-prolinol) (90) or L-proline (91) to silica gel and loading it with Cu(II) ions for the optical

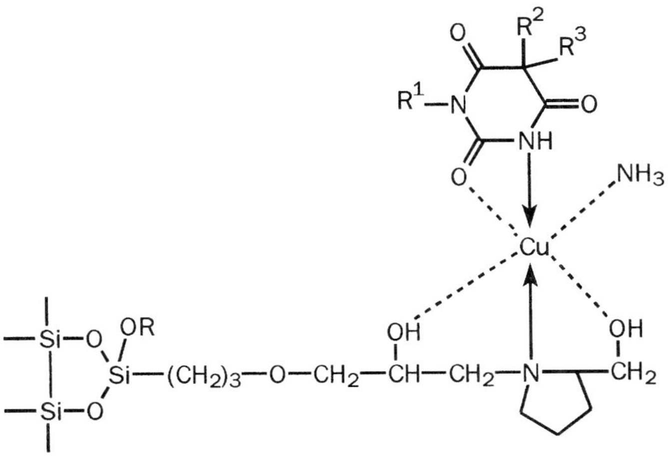

Figure 7.3. Model for the structure of the mixed complex between S-prolinol as the selector ligand and the barbiturate. From (90) with permission.

resolution of hexobarbital, mephobarbital methohexital, and 1,5-dimethyl-5-phenylbarbituric acid (Figure 7.3). Secobarbital was not resolved under similar conditions, demonstrating once again that the good optical resolution of barbiturates with the chiral center in the C-5 substituent is much harder to achieve than that for the N-substituted barbiturates. Hexobarbital was used as a positive probe of the enantioselectivity of a synthetic chiral polymer, poly[N-acryloyl-(*S*)-phenylalanine ethyl ester] (92). The same drug was baseline resolved on the CSP obtained by the copolymerization of urea with formaldehyde, in the presence of L-leucinamide, but for mephobarbital and thiopental the separation of enantiomers was less satisfactory (93). Both hexo- and mephobarbital enantiomers were successfully separated on CSPs obtained by the covalent binding of macrocyclic antibiotic vancomycin and its 3,5-dimethylphenyl derivative (94).

Although a nonchiral compound, 5,5-diethylbarbituric acid (barbital) proved indispensable to the enantiomeric separation of underivatized aliphatic β-amino alcohols by chiral ligand-exchange chromatography with N-*n*-dodecyl-L-proline as the chiral selector (95).

7.3.2.2. *Gas Chromatography*

The separation of enantiomers of hexobarbital and mephobarbital was also achieved by capillary gas chromatography on such CSPs as XE-60-L-valine-

Table 7.1. Selected Stationary and Mobile Phases, Capacity Factors of the First Eluted Enantiomer (k'_1), and Separation Factors α for the Chiral Liquid Chromatographic Resolution of Hexobarbital Enantiomers at Room Temperature

Chiral Stationary Phase	Mobile Phase Components (Volume Ratios)	k'_1	α	Reference
α_1-acid glycoprotein (chiral-AGP)	3% 2-propanol in 0.01M phosphate buffer, pH = 7.0	1.26	1.59	
	0.01M phosphate buffer, pH = 4.5	9.39	1.44	(79)
	0.01M phosphate buffer, pH = 7.5	11.6	2.10	
	0.02 M phosphate buffer, pH = 6.5.2-propanol (95:5)	0.72	1.31	(46)
Ovomucoid	0.02 M phosphate buffer, pH = 6.9:ethanol (90:10)	0.63	1.36	(83)
	0.02 M phosphate buffer, pH = 6,9:ethanol (95:5)	2.59	1.78	(82)
Tribenzoylcellulose	hexane:2-propanol (90:10)	19.87[a]	1.46	(68)
Triacetylcellusose	ethanol:methanol:water (67.2:20:12.8)	0.67	1.57	(72)
β-cyclodextrin	methanol:1% aqueous triethylammonium acetate (15:85)	9.39	1.14	(39)
LiChrosorb RP-18 dynamically coated with heptakis (2,3,6-tri-O-methyl)-β-cyclodextrin (TM-β-CD)	40% methanol + 0.2% H_3PO_4+ 0.64 nM TM-β-CD	4.87	1.21	(52)
LiChrosorb Si60 dynamically coated with TM-β-CD	hexane:*n*-propanol (75:1)+ 1 mg/ml of TB-β-CD	6.54	1.18	(52)
β-cyclodextrin/diol	0.05M phosphate buffer, pH = 6.9:acetonitrile (80:10)	4.20	1.13	(45)

[a] Capacity factor for the second eluted enantiomer.

(*R*)-α-phenylethylamide (96) [the method used for stereospecific analysis of racemic hexobarbital in rat blood (97)], perpentylated α-cyclodextrin (effective also for other N-methylated barbiturates) (98), heptakis (2,6-di-O-pentyl)-β-cyclodextrin (99), and permethylated β-cyclodextrin (100). Racemic hexobarbital was also resolved using this method on Chirasil-Val, a CSP with the L-valine *t*-butyl amide selector covalently bonded to the polysiloxane polymer (101). Hexakis (3-O-acetyl-2,6-di-O-pentyl)-α-cyclodextrin was claimed effective for the gas chromatographic enantioseparation of barbiturates, but without presentation of experimental results (102). Recently, a CSP with heptakis-(2,6-di-O-methyl-3-O-pentyl)-β-cyclodextrin was used for the preparative separation of enantiomers of hexobarbital (103).

Gas chromatography combined with mass spectrometry (GC-MS) was applied for an assay of pseudoracemic hexobarbital, consisting of equimolar amounts of (*S*)(+)-enantiomer and its deuterated antipode (13). The same technique, but using negative ion chemical ionization mass spectrometry (NICI-MS), was employed for the enantiospecific determination of hexobarbital and its metabolites in biological fluids (104).

An interesting example of the enantioseparation of hexobarbital on the same fused-silica capillary column by capillary gas chromatography, supercritical fluid chromatography, and capillary electrochromatography was recently demonstrated by Schurig et al. (105, 106).

Selected representative experimental details of the chiral chromatography of hexobarbital enantiomers are summarized in Table 7.1.

7.3.2.3. *Capillary electrophoresis*

In the beginning of this decade, Japanese authors first reported successful enantioseparations of some barbiturates by electrophoretic techniques. Thus, resolutions of secobarbital, pentobarbital, and hexobarbital enantiomers were obtained by capillary zone electrophoresis (CZE) based on complexation with various cyclodextrins (107). Better results, that is complete or near baseline separations, were observed for the methylated derivatives than for the unsubstituted α- and β-cyclodextrins.

Nishi et al. (108) used cyclodextrin-modified micellar electrokinetic chromatography (MECC) with sodium dodecyl sulfate (SDS) for the resolution of 95 enantiomers of different drugs, among them pentobarbital and thiopental. These barbiturates were rather poorly resolved and only with γ-cyclodextrin. The addition of 10% of methanol as an organic modifier made the separation worse; on the other hand, the addition of chiral additives, such as (+)-camphor-10-sulfonic acid or (−)-menthoxyacetic acid, improved the results.

The enantiomers of hexobarbital and pentobarbital were separated also by capillary electrophoresis using cappillaries packed with the α_1-acid glycoprotein chiral stationary phase (109). The effects of organic modifiers (lower alcohols, acetonitrile), pH, and concentrations of modifer (2-propanol) and electrolyte (sodium phosphate) were investigated and compared with the results of HPLC separations. Hexobarbital was also used as one of the test compounds for studies on the chiral separations by capillary electrophoresis in fused-silica capillaries packed with a β-cyclodextrin CPS (110).

Very interesting enantioseparations of four chiral barbiturates (butabarbital, secobarbital, hexobarbital, and mephobarbital) under MECC conditions with various cyclodextrins as chiral selectors were reported by Francotte et al. (111). It was found that the enantiomers of N-methyl barbiturates with the chiral center at the C-5 ring atom were resolved with β-cyclodextrin and only those of hexobarbital were separated with heptakis-(2,6-di-O-methyl-β-cyclodextrin). Only secobarbital enantiomers were resolved with γ-cyclodextrin, whereas butabarbital was not resolved at all. This clearly suggests the important effect of the barbiturate and cyclodextrin structure on the enantioselectivity. The effects of cyclodextrin and SDS concentrations, the addition of an organic modifier (15% methanol), temperature, and pH were also investigated. The strong influence of pH (the investigated range being 7–9) on selectivity was found only for N-methyl barbiturates and the effect of methanol was contrary to that reported earlier (108). Preliminary results suggest the possible application of this method for the determination of hexobarbital in clinical analysis.

Ward reports resolutions of mephobarbital and hexobarbital enantiomers using native β-cyclodextrin as the chiral mobile phase additive. These barbiturates were poorly resolved in the absence of an organic modifier, but exhibited excellent enantioselectivity when methanol was added to the β-cyclodextrin buffer (112).

7.4. OTHER ANALYTICAL ASPECTS OF BARBITURATE CHIRALITY

One of the important problems, from the chemical and pharmacological point of view, associated with enantiomers of chiral barbiturate drugs is their optical purity. The criterion of unchanged optical rotation after repeated crystallization for optically pure preparation sometimes may be misleading, and independent methods are needed for its verification. Methods of chiral chromatography may be used to estimate this parameter. It was claimed that enantiomeric contributions of less than 1% could be quantified, and in favorable instances the detection limit may be as low as 0.1% (98). An alternative analytical determination of optical purity is based on an NMR method using chiral lanthanide shift reagents. Thus, tris[3-(-heptafluoropropylhydroxymethylene)-(+)-camphorato]praseodymium(III) was used for an assay of synthetic thiohexital enantiomers, which were found to be at least 98% optically pure (8). Knabe et al. used tris[3-(heptafluoropropylhydroxymethylene-(+)-camphorato]europium(III) and tris[3-(trifluoromethylhydroxymethylene)-(+)-camphorato]europium(III) for the optical purity determination of hexobarbital enantiomers (113) and a series of chiral N-substituted barbiturates with basic substituents at the C-5 atom (114), whereas Rothchild et al. used these reagents for an assay of enantiomeric shift differences for thiamylal (115), mephobarbital (116), methohexital (117), and other chiral barbiturate and thiobarbiturate drugs (thiopental, pentobarbital, butabarbital, secobarbital, and talbutal) (118). The results of these studies demonstrated significant differences in the complexation site between thio- and oxobarbiturates and showed that the enantiomeric shift differences are larger for these first compounds.

An alternative estimation of the optical purity of hexobarbital and l-methyl-5-phenyl-5-propylbarbituric acid enantiomers was done by the isotope dilution method using 2-^{14}C-labeled specimens (20).

The optical purity of the compound may be determined when the optical purity of the substrate for its synthesis is known, and racemization is not involved in the reaction course. In that way, Rice (9) found better than 99% optical purity for enantiomers of 5-(1,3-dimethylbutyl)-5-ethylbarbituric acid synthesized from enantiomers of 3,5-dimethylhexanoic acid of that purity. This result, in turn, served for the estimation of the optical purity of earlier obtained less pure preparations of enantiomers of this barbiturate (23, 28), from the reported values of their optical rotations.

It is also of interest to correlate the signs of optical rotations of individual enantiomers with their absolute configurations. Here again, the knowledge of the absolute configuration of synthetic substrates and the stereochemical course of the process (the retention or inversion of configuration) may be of assistance, as was proved for N-unsubstituted (4, 7, 9, 18) and N-substituted (121, 122) chiral barbiturates. On the other hand, the conversion of an enantiomer to the product with the defined absolute configuration may indicate the absolute configuration of this former compound. Knabe et al. established the absolute *S* configuration of dextrorotatory 1,5-dimethyl-5-phenylbarbituric acid by its alkaline hydrolysis to (+)-α-methyltropic acid and the stereospecific transformation of this last compound into (*R*)(−)-2-amino-2-phenylpropanol (123, 124). With this result, it was possible to correlate absolute configurations of other enantiomers of barbiturates by the analysis of their optical rotatory dispersion and circular dichroism spectra. Thus, the same *S* configuration was ascribed to dextrororatory enantiomers of mephobarbital and hexobarbital (123, 124).

The analysis of such spectra proved to be useful for optically active barbiturates with basic C-5 substituents (114) and 5-(1-cyclohexenyl) tri- and tetra-substituted barbiturates (125), but failed for N-substituted 5,5-dialkylbarbiturates. It was then proposed to determine the absolute configuration of enantiomers of 5-ethyl-5-hexyl-l-methylbarbituric acid by the correlation with that of 5-(1-cyclohexenyl) analogs by the sequence of reactions shown in

Scheme 3 (126). The absolute configuration of the enantiomers of 1-methyl-5-phenyl-5-propyl-2-thiobarbituric acid, being (*R*) and (*S*) for levorotatory and dextrorotatory antipodes, respectively, was deduced from the known configuration of the appropriate optically active synthetic substrates and confirmed by correlation with that of the 2-oxo analog, obtained by oxidation with hydrogen peroxide in sulfuric acid solution (127). Similarly, it was found that the levorotatory enantiomer of vinylbital, after hydrogenation with plantinum on activated carbon, yields (–)(*S*)-pentobarbital, whereas its antipode was obtained after the desulfuration of (+)-thiopental with Na_2O_2 in an aqueous medium (128).

The X-ray analysis was also occasionally used for the determination of absolute configuration of dextrorotatory 5-allyl-5-isopropyl-l-methylbarbituric acid (10) and one of the diastereomeric N-glucosides of amobarbital (129).

7.5. STEREOSELECTIVE CHEMISTRY OF BARBITURATES

A few stereospecific reactions were mentioned above, discussing the correlation of the absolute configuration of optically active barbiturates with chemical conversions, but it is generally true that, contrary to pharmacological investigations, the chemical reactions of individual enantiomers of chiral barbiturates were seldom explored. Knabe and co-workers (130–132) investigated the reduction of racemic and enantiomerically pure barbiturates and found that N-substituted compounds (among others, hexobarbital and mephobarbital) were reduced with $LiAlH_4$ to hexahydropyrimidine derivatives, while in the presence of $AlCl_3$ 2-oxotetrahydropyrimidines were formed (Scheme 3). These last compounds were also obtained when N-unsubstituted barbiturates were reduced with $LiAlH_4$, while thiopental gave a 2-thioxotetrahydropyrimidine derivative under these conditions. It was reported that no racemization occurs during these processes for pure enantiomers.

The stereospecific reaction course was also found for N-alkylation of the hexobarbital enantiomers with ethyl iodide (133), but all these reactions did not involve bond breaking at the chiral center of the molecule.

After it was found that racemic N-methyl-substituted barbiturates irradiated by 254-nm UV light in an alkaline medium yield hydantoin derivatives (134, 135), this reaction was run with the (*S*)(+)-hexobarbital enantiomer. The main product, (*R*)(+)-(1-cyclohexenyl)-3,5-dimethyl-2,4-imidazolidinedione, was formed by the photochemical breakdown of the C-4, C-5 bond, that is, the bond at the C-5 chiral center, an extrusion of carbon oxide and a closure of the ring. It had the same retained configuration of substituents at the C-5 atom as the substrate, and the inversion of the configuration of the substrate (and the product) did not exceed 5% (136).

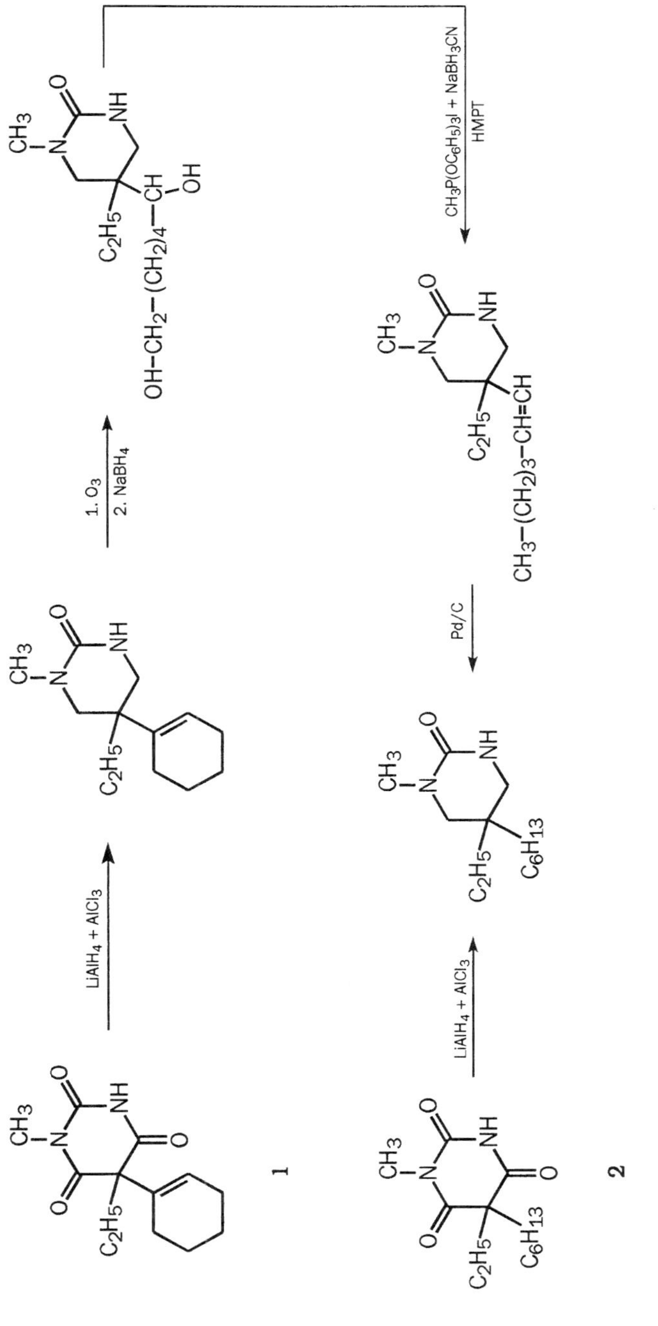

Scheme 3. Chemical transformations for assessment of absolute configuration of 5-ethyl-5-hexyl-1-methylbarbituric acid (2) by correlation with the known configuration of 5-(1-cyclohexenyl)-5-ethyl-1-methylbarbituric acid (1) (126).

7.6. DIFFERENTIAL BIOLOGICAL ACTIVITY OF BARBITURATE STEREOISOMERS

This section is restricted solely to the discussion of differences in the activities of optical isomers (and occasionally diastereoisomers) of physiologically active barbiturates. For more general information on biological and therapeutic effects of barbiturates, the reader is directed to selected review articles published recently (137–144).

Early pharmacological tests of the anesthetic activity of enantiomers of several barbituric acids, both N-substituted and unsubstituted, reported none (1) or moderate (6) differences, but these results were further questioned due to rather low optical purity or large doses required to produce an effect.

When four possible stereoisomers and two racemic mixtures of methohexital, a barbiturate with two chiral centres, were investigated significant qualitative and quantitative differences were observed (145), depending on the specimens and also the experimental animals used.

Further studies revealed differences in the potencies of enantiomers of many chiral barbiturates with chiral centres in the C-5 substituent and at the C-5 atom of the ring, and those of pentobarbital and hexobarbital were the most extensively investigated.

7.6.1. Pentobarbital Enantiomers

The first comparisons of the activity of pentobarbital enantiomers were made in rats and revealed that (*S*)(–)-antipode is four times more potent, taking into account the rapidity of onset and sleeping time (146). The levorotatory enantiomer was also more quickly metabolized. These findings were confirmed in studies in mice (147) and some qualitative differences were noted. The (*R*)(+)-enantiomer caused anesthesia preceded by hyperirritability and spasticity, whereas its counterpart produced a smoother and more rapid anesthetic effect.

These studies were repeated and expanded on optical isomers of other closely related barbiturate drugs with a 1-methylbutyl substituent (secobarbital, thiopental, and thiamylal) (148). The levorotatory enantiomers with an *S* configuration were always more toxic and potent as anesthetics in comparison with their antipodes and racemates, but therapeutic indices were found to be similar for all the compounds investigated.

The possible relation between differences in the anesthetic potency of pentobarbital and secobarbital enantiomers and the concentrations of these stereoisomers in different brain regions was investigated by Freudenthal and Martin (149), and their results demonstrated a lack of stereoselectivity in the transport of these enantiomers across the blood-brain barrier, since no significant differences were found in the concentrations of optical isomers of the same barbiturate in brain stem, hypothalamus, cerebellum, and cortex.

The metabolism of pure pentobarbital enantiomers studied in dogs (150, 151) revealed the formation of two pairs of metabolites hydroxylated at the 3 position of the l-methylbutyl substituent. Their absolute configurations corresponded with those of (3*R*)-OH and (3*S*)-OH, with the retained configuration at the C-l atom of the substituent. Their excreted amounts were the same in the first pair of diastereomers after the administration of (*R*)(+)-pentobarbital, but differed markedly in favor of 5-ethyl-5-[(3*S*)-hydroxy-(1*S*)-methylbutyl]barbituric acid, when its antipode was used.

Similar studies conducted with hepatic microsomes from male rats (152) confirmed these findings and additionally demonstrated differences in the Michaelis constants for the formation of these metabolites, thus suggesting that there are at least two enzymes involved in the catalysis of hydroxylation of pentobarbital enantiomers.

Further confirmation of qualitative differences in the central nervous system (CNS) activity of pentobarbital enantiomers came from studies at the cellular level when their effects on the electrophysiological properties of the membrane (potential and conductance) of the mouse spinal neurons were compared (153). It was found that (*R*)(+)-pentobarbital was predominantly excitatory and its counterpart produced inhibitory responses. These results evidenced the existence of several stereospecifically distinct sites of barbiturate action explored in further research. It was demonstrated that the (*S*)(–)-isomer opens chloride ion channels in neurons, a property similar to that demonstrated by γ-aminobutyric acid (GABA), a well-known inhibitory neurotransmitter [154]. It was also demonstrated that this enantiomer potentiates GABA function depresses the voltage-activated conductance of calcium ions and directly activates inhibitory Cl^- conductance (155).

The optical isomers of pentobarbital and other chiral barbiturates were extensively used in research aimed at the identification of the barbiturate receptor sites in the CNS, and the exploration of the mechanism for their anesthetic, anticonvulsive, and sometimes observed convulsive activities. Several papers reported their differential interactions with binding sites of other substances, which mediate the activity of the GABA receptor-ionofore complex, such as diazepam (156–158) dihydropicrotoxinin (159), GABA (160, 161) , and *t*-butylbicyclophosphorothionate (162)

Roth et al. (163, 164) demonstrated again quantitative differences in the neuronal activity of pentobarbital enantiomers and provided further evidence for the concept of specific cellular and membrane recognition sites for barbiturate activities.

The IC_{50} values for the inhibition of potassium ions-stimulated release of tritium-labeled acetylcholine determined *in vitro* in brain slices from mice for the stereoisomers of pentobarbital were not significantly different, whereas those for secobarbital indicated the higher potency of the (*R*)(+)-antipode. However, this stereospecificity stood in contrast to the hypnotic action of these enantiomers determined in the *in vivo* experiments (165).

The enantiomers of pentobarbital also showed stereoselectivity in the competition for acetylcholine binding sites in membranes from *Torpedo californica* with ^{14}C-labeled racemate, and the levorotatory isomer was found to be about three times less potent in this activity than its counterpart (166). A similar experimental model was further used in the investigation of the actions of pentobarbital enantiomers on nicotinic cholinergic receptors (167). The IC_{50} values for the displacement of ^{14}C-labeled (+)-pentobarbital from the acetylcholine sites were four-fold higher for (*S*)(−)-pentobarbital, and the modulation of [^{3}H] acetylcholine binding by pentobarbital was stereoselective at low, but not high, concentrations. These results were discussed in terms of the stereoselectivity of the binding site and its comparison with stereoselectivity in other systems.

Pharmacokinetics of pentobarbital enantiomers in humans and rabbits was investigated by Cook et al. (168) by the enantioselective radioimmunoassay method. The estimated values of pharmacokinetic parameters (elimination constants, volumes of distribution, clearance) showed some quantitative differences, depending on the species investigated. For example, in man, the (*S*)-enantiomer had lower median clearance by 25% and was more strongly (by approximately 10%) bound to plasma proteins, but in their final conclusions, Cook et al. stated that these differences do not appear sufficient to account for the observed differences in potency and duration of action.

Several papers reported the differentiation of behavioral effects of pentobarbital enantiomers in various animal species, such as rats (169, 170), mice (170), and pigeons (171, 172). Occasionally, isomers of secobarbital were also included in these investigations (171,172). The results revealed none or moderate differences of potency in favor of the more potent (*S*)(−)-enantiomer. On the other hand, the effects of the social isolation of mice significantly lowered the brain levels of both enantiomers of pentobarbital in comparison with those obtained for aggregated mice (173).

Recent contributions reported the similar potency of pentobarbital enantiomers at depressing voltage-dependent sodium channels from human brain cortex (174), but they differed in the effects on calcium channel current and GABA receptor responses, where the (*S*)(−)-enantiomer was significantly more potent (175).

7.6.2. Hexobarbital Enantiomers

Differences between the anesthetic potency of hexobarbital enantiomers were reported by Wahlström in 1966 (176). The (+)-antipode was about three times more potent in experiments with rats. The differences in the concentrations in different tissues (brain, liver, serum) were observed for these antipodes and also for the enantiomers of the demethylated metabolite (177). In other experiments, however, the same concentrations of both enantiomers in the brain were found to be responsible for the specific anesthetic response (178). Enzymes from rat liver microsomes more quickly metabolized (+)-hexobarbital in rats both untreated and pretreated with phenobarbital (179–182), and sex (180) and kinetic differences (181) in this respect were observed. Further detailed studies on the stereoselective metabolism of hexobarbital enantiomers, especially the hydroxylation process (24, 183) and the formation of glucuronides in hydroxylated metabolites (184), confirmed these finding and offered some new differentiating data.

When the structure-anesthetic potency relationships for a series of enantiomers of N-methylated barbiturates were investigated, it was found that the nature of the cyclic hydrocarbon substituent at the C-5 atom does not affect the anesthetic potency ratio of the respective enantiomers. However, this potency was dependent on the length of the second, aliphatic substituent and changed from higher for the (+)-enantiomers for 5-methyl substitution (among others, for hexobarbital), to higher for the (–)-enantiomer for the 5-ethyl group (185, 186). Moreover, the (+)-enantiomer of 5-phenyl-5-propyl-l-methylbarbituric acid caused convulsions, whereas its optical counterpart showed anesthetic properties (186). This change of potency with C-5 substitution was confirmed in another study (187) and served as the basis for the theoretical design of a model for a barbiturate receptor (188–190).

The pharmacokinetics of hexobarbital enantiomers studied in man (191) and rat (192) showed differences in some parameters. The elimination rate was always higher for (–)-hexobarbital, but the volumes of distribution did not differ significantly. It was concluded that the results do not suggest the stereoselective distribution of hexobarbital antipodes in the rat, but that the stereospecific mechanism at the active sites in the CNS may be responsible for the anesthetic properties of hexobarbital. Further differences in the pharmacokinetic parameters of hexobarbital enantiomers were reported in the papers of Van der Graaff et al. (193, 194).

The differences in activities of hexobarbital enenatiomers were also found for their interaction with atropine in rats (195), enhancement of [^{3}H] diazepam binding in rat brain (156, 157, 196), inhibition of binding of [^{3}H]picrotoxinin (159), effect on binding of [^{35}S]*t*-butylbicyclophosphorothionate (162, 197), [^{3}H]oxotremorine or [^{3}H]quinuclidinylbenzilate (198).

The effect of aging on the stereoselective drug disposition in man was studied using the enantiomers of hexobarbital, and the mean oral clearance for the (*R*)(–)-isomer was about twice as great in young subjects as old ones, whereas the other antipode did not differ significantly between these groups (199). The metabolic clearance of (*S*)-hexobarbital was about six times higher than for the (*R*)-enantiomer without induction with phenobarbital (5-ethyl-5-phenylbarbituric acid), when the effect of aging on the metabolism of simultaneously administered antipyrine and hexobarbital enantiomers administered to male rats was studied (200). The phenobarbital pretreatment significantly influenced clearance.

7.6.3. Enantiomers of Other Barbiturates

Several other enantiomers of barbituric acid derivatives showed differential biological activity. Thus, among enantiomers of 5-(2-bromoallyl)-5-isopropyl-1-methylbarbituric acid, the levorotatory isomer was more potent as the anesthetic and had a higher tendency to produce convulsions (201). Similarly, the (–)-enantiomer of mephobarbital had hypnotic properties (202), increased the survival time of mice exposed to oxygen (203), and enhanced diazepam binding in rat brain (157, 196), whereas its antipode was not active (202) although both proved to be good metabolic inducers (25). They also differentially enhanced the GABA binding in cerebral cortex (204), inhibited the binding of dihydropicrotoxinin (159), and potentiated responses to muscimol, while only (–)-enantiomer reduced the effect of picrotoxin (205).

Preliminary data indicating the stereoselective metabolic hydroxylation of mephobarbital enantiomers were also published (206) and further confirmed and extended with some pharmacokinetic parameters (62). The stereoselective binding of mephobarbital to proteins in human plasma (64), and the effects of aging and gender on its stereoselective metabolism and changes in pharmacokinetic parameters (63), were also demonstrated.

Stereoselective differences were revealed in the binding of mephobarbital and its other N-alkyl derivatives to human serum albumin (207).

Several studies were devoted to enantiomers of 5-(1,3-dimethylbutyl)-5-ethylbarbituric acid, of which the (+)-isomer was found to be convulsive, while its antipode had anesthetic properties (28, 208, 209).

A similar qualitative difference in activity for enantiomers of l-methyl-5-phenyl-5-propylbarbituric acid, mentioned above, stimulated various electrophysiological (210), distribution (211), and binding studies (156, 157, 159, 162, 196, 204, 212), to account for this effecct. The result suggested differential actions of enantiomers on different sites of the GABA receptor complex.

Such qualitative differences were also observed for enantiomeric pairs of 5-butyl-5-phenyl-1-methylbarbituric acid (213), its 5-pentyl analog (214), and in a series of N-alkyl-substituted barbiturates, with the 1-pyrrolidinyl and 1-piperidinyl substituents at the C-5 atom (215, 216). Quantitative differences in the CNS activities of enantiomers of barbiturates were reported for secobarbital (217, 218), thiopental (218, 219), thiohexital (8), 5-allyl-5-(2-hydroxypropyl) barbituric acid (31), and several N-aminoethylbarbiturates (216,

Table 7.2. Differential Values for Biological Activities of Enantiomers of Some Barbiturate Drugs

Biological Activity	Barbiturate Drug						Reference
	Hexobarbital		Pentobarbital		Mephobarbital		
	(*S*)(+) =	(*R*)(−) =	(S)(−) =	(*R*)(+) =	(*S*)(+)	(*R*)(−) =	
Inhibition of [^{3}H]dihydropicrotoxinin binding (IC_{50} in μM)[a]	2 ± 1	6 ± 2	22 ± 7	82 ± 10	2 ± 1	4 ± 2	(59)
Enhancement of [^{3}H]diazepam binding (in %)[a]	46	12	65	38	0	62	(196)
Enhancement of [^{3}H]GABA binding (in %)[a]			28 ± 4	6 ± 3	2 ± 2	30 ± 6	(204)
Inhibition of [^{35}S]TBPT[b] binding (IC_{50} in μM)[a]	145	380	70	125			(162)
Median anesthetic dose (mg/kg)[c]			30.0 ± 4.8	56 ± 3.4			(148)
Intrinsic clearance (ml $min^{-1}kg^{-1}$)[d]	2947 ± 358	411 ± 65					(193)
Mean survival time (min) after 5% O_2 exposition[e]					3.84 ± 0.45	12.64 ± 2.01	(203)

[a] In rat cortex membranes.
[b] *t*-Butylbicyclophosphorothionate.
[c] In mice.
[d] In rats.
[e] In mice, after a dose of 100 mg/kg.

220). Differences in the distribution in the serum and brain of rats of the enantiomers of 1-diethylaminoethyl-5-ethyl-5-phenylbarbituric acid were also described (221). It was noted that the higher concentration in the brain of the (*S*)(+)-enantiomer was in contrast with the higher CNS potency of its (*R*)(–)-counterpart.

N-glucosylation, one of the pathways of barbiturate metabolism, produces diastereomeric derivatives with a different configuration at the C-5 atom of barbiturate moiety, and fixed stereochemical configurations of the chiral centers of the glucose unit bound to the nitrogen atom. Several studies of this process for amobarbital (129, 222, 223) and phenobarbital (224) proved that the formation and/or urinary excretion of their N-glucosides is stereoselective.

Table 7.2 lists several selected numerical values illustrating the differential biological activity of barbiturate enantiomers.

7.7. CONCLUSION

Chiral barbiturates continue to be used as probes of the resolving capacity of new chromatographic and electrophoretic systems. Hexobarbital enantiomers were resolved by HPLC on peralkylated β-cyclodextrins (225) and *s*-triazine derivative of tripeptide CSP (226). Mephobarbital enantiomers were resolved by HPLC on permethylated 6-mono-*O*-alkenyl-β-cyclodextrin CSP (227) and among some other antiepileptic drugs, by a column-switching technique, using β-cyclodextrin as a mobile phase additive (228). Hexobarbital and mephobarbital enantiomers were also separated with native and cationic β-cyclodextrins as chiral mobile phase additives (229) and on polysaccharide CSPs in narrow-bore columns (benzonal enantiomers were also resolved) (230), as well as on the covalently bonded permethylated cyclodextrins (231). Schurig et al. (232) reported separation of hexobarbital, mephobarbital, and pentobarbital enantiomers on fused-silica column coated with Chirasil-Dex by GC at 130°C.

Thiopental enantiomers in sheep (233) and human (234) plasma were resolved by HPLC on α_1-acid glycoprotein column, while those of thiamylal in human serum were separated with β-cyclodextrin in chiral mobile phase (235). These methods were claimed suitable for pharmacokinetic studies for enantiomers of these thiobarbiturates.

The enantiomers of hexobarbital were resolved by electrochromatography using the previously mentioned chiral phase Chirasil-Dex (236), by micellar electrokinetic capillary chromatography with glucopyranoside-based surfactants (237), and by capillary electrophoresis with rifamycin SV as a chiral selector (238), while the order of their electrophoretic migration was reversed by addition of cetyltrimethylammonium bromide to the buffer containing hydroxypropyl-β-cyclodextrin (239).

Molecular mechanics calculations were applied for prediction of chirally discriminating chromatographic behavior of β-cyclodextrin complexes with mephobarbital (240, 241) and hexobarbital (241) enantiomers.

Soine et al. further studied diastereomeric N-glucosides and N-glucuronides of some 5,5-disubstituted barbiturates and, interestingly, product enantioselectivity was observed for the barbiturates N-glucosides, while substrate enantioselectivity was found for N-glucosylation of pentobarbital enantiomers (242–244).

The effect of conditions (albumin species and concentration, pH, and temperature) on the stereoselective binding of enantiomers of 1-methyl-5-phenyl-5-propylbarbiturate to serum albumin was studied by Büch et al. (245). It was found that thermodynamics and kinetics of *t*-butylbicyclophosphorothionate binding differentiate convulsant and depressant barbiturate stereoisomers acting via $GABA_A$ ionophores (246).

Although optical isomers of barbiturates differ sometimes markedly in their activities, none of the barbiturate drugs used in therapy is marketed as an enantiomer. Instead, racemic mixtures or mixtures of diastereomers are manufactured and prescribed. One reason for this lies probably in not so simple and economically feasible stereospecific synthesis of these compounds. Besides, there are other drugs of similar or even better therapeutic properties and the popularity of barbiturate drugs today is on the decline.

On the other hand, the enantiomers of barbiturates were indispensable in research on the stereochemical features of CNS active agents and their sites of action, and their future use may bring about other interesting applications.

REFERENCES

1. Huesh, C. M. and Marvel, C. S. (1928). *J. Am. Chem. Soc.*, **50**, 855.
2. Doran, W. J. (1959). *Med. Chem.*, **4**, 1.
3. Bojarski, J. T., Mokrosz, J. L., Barton, H. J., and Paluchowska, M. H. (1985). *Adv. Heterocycl. Chem.*, **38**, 229.
4. Carroll, F. I. and Meck, R. (1969). *J. Org. Chem.*, **34**, 2676.
5. Knabe, J. and Strauss, D. (1968). *Angew. Chem. Int. Ed. Engl.*, **7**, 463.
6. Kleiderer, E. C. and Shonle, H. A. (1934). *J. Am. Chem. Soc.*, **56**, 1772.
7. Cook, C. E. and Tallent, C. R. (1969). *J. Heterocycl. Chem.*, **6**, 203.
8. Carroll, F. I., Philip, A., Naylor, D. M., Christensen, H. D., and Goad, W. C. (1981). *J. Med. Chem.*, 24, 1241.
9. Rice, K. C. (1982). *J. Org. Chem.*, **47**, 3617.

10. Knabe, J., Rummel, W., Buch, H. P., and Franz, N. (1978). *Arzneim. Forsch.*, **28**, 1048 (and references therein).
11. Knabe, J. and Wunn, W. (1980). *Arch. Pharm.*, **313**, 93.
12. Knabe, J. and Schamber, L. (1982). *Arch. Pharm.*, **315**, 878.
13. Van der Graaff, M., Hofman, P. H., Breimer, D. D., Vermeulen, N. P .E., Knabe, J., and Schamber, L. (1985). *Biomed. Mass Spectrom.*, **12**, 464.
14. Knabe, J., Junginger, H., Strauss, D., and Schmidt, H.-L. (1972). *Arch. Pharm.*, **305**, 277.
15. Knabe, J. and Lampen, P. (1987). *Arch. Pharm.*, **320**, 756.
16. Murata, M. and Achiwa, K. (1991). *Tetrahedron Lett.*, **32**, 6763.
17. Murata, M., Uchida, H. L., and Achiwa, K. (1992). *Chem. Pharm. Bull.*, **40**, 2605.
18. Carroll, M. and Mitchell, G. N. (1975). *J. Med. Chem.*, **18**, 37.
19. Knabe, J. and Kräuter, R., (1965). *Arch. Pharm.*, **298**, 1.
20. Knabe, J., Kräuter, R., and Philipson, K. (1965). *Tetrahedron Lett.*, **1965**, 571.
21. Knabe, J. and Philipson, K. (1966). *Arch. Pharm.*, **299**, 231.
22. Knabe, J., Strauss, D., and Urbahn, C. (1968). *Pharmazie*, **23**, 522.
23. Sitsen, J. M. A. and Fresen, J. A. (1973). *Pharm. Week.*, **108**, 1053.
24. Miyano, K., Fujii, Y., and Toki, S. (1980). *Drug Metab. Disp.*, **8**, 104.
25. Gordis, E. (1971). *Biochem. Pharmacol.*, **20**, 246.
26. Andrews, P. R., Jones, G. P., and Poulton, D. B. (1982). *Eur. J. Pharmacol.*, **79**, 61.
27. Perry, R., Downes, H., and Karler, R. (1969). *Fed. Proc.*, **28**, 666.
28. Downes, H., Perry, R. S., Oslund, R. E., and Karler, R. (1970). *J. Pharmacol. Exp. Ther.*, **175**, 692.
29. Doran, W. J. (1960). *J. Org. Chem.*, **25**, 1737.
30. Knabe, J. and Reinhardt, J. (1982). *Arch. Pharm.*, **315**, 706.
31. Bobranski, B., Wilimowski, M., Barczynska, J., Seniuta, R., Sedzimirska, B., and Witkowska, M. (1973). *Arch. Immunol. Ther. Exp.*, **21**, 299.
32. Knabe, J. and Lampen, P. (1987). *Arch. Pharm.*, **320**, 719.
33. Blaschke, G. (1980). *Angew. Chem. Int. Ed. Engl.*, **19**, 13.
34. Blaschke, G. and Markgraf, H. (1984). *Arch. Pharm.*, **317**, 465.
35. Blaschke, G., Kraft, H. P., and Markgraf, H. (1983). *Chem. Ber.*, **116**, 3611.
36. Blaschke, G. (1986). *J. Liq. Chromatogr.*, **9**, 341.
37. Armstrong, D. W. and DeMond, W. (1984). *J. Chromatogr. Sci.*, **22**, 411.
38. Hinze, W. L., Riehl, T. E., Armstrong, D. W., DeMond, W., Alak, A., and Ward, T. (1985). *Anal Chem.*, **57**, 237.
39. Armstrong, D. W., Ward, T. J., Amstrong, R. D. and Beesley, T. E. (1986). *Science*, **232**, 1132.
40. Berthod, A., Jin, H. L., Beesley, T. E., Duncan, J. D., and Armstrong, D. W. (1990). *J. Pharm. Biomed. Anal.*, **8**, 123.

41. Vigh, G., Quintero, G., and Farkas, G. (1990). *J. Chromatogr.*, **506**, 481.
42. Haginaka, J. and Wakai, J. (1990). *Anal. Chem.*, **62**, 997.
43. Rizzi, A. M. and Plank, C. (1991). *J. Chromatogr.*, **557**, 199.
44. Eto, S., Noda, H., and Noda, A. (1992). *J. Chromatogr.*, **579**, 253.
45. Haginaka, J. and Wakai, J. (1992). *Anal. Sci.*, **8**, 137.
46. Vandenbosch, C., Massart, D. L., and Lindner, W. (1992). *J. Pharm. Biomed. Anal.*, **10**, 895.
47. Thuaud, N., Sébille, B., Deratani, A., Pöpping, B., and Pellet, C. (1993). *Chromatographia*, **36**, 373.
48. Cabrera, K. and Lubda, D. (1994). *J. Chromatogr.*, A **666**, 433.
49. Sybilska, D., Zukowski, J., and Bojarski, J. (1986). *J. Liq. Chromatogr.*, **9**. 591.
50. Zukowski, J., Sybilska, D., and Bojarski, J. (1986). *J. Chromatogr.*, **364**, 225.
51. Zukowski, J., Sybilska, D., Bojarski, J., and Szejtli, J. (1988). *J. Chromatogr.*, **436**, 381.
52. Pawlowska, M. and Zukowski, J. (1990). *Mikrochim. Acta*, **11**, 55.
53. Pawlowska, M. (1991). *J. Liq. Chromatogr.*, **14**, 2273.
54. Pawlowska, M. (1991). *Chirality*, **3**, 136.
55. Pawlowska, M. and Zukowski, J. (1991). *J. High Res. Chromatogr.*, **14**, 139.
56. Pawlowska, M. and Lipkowski, J. (1991). *J. Chromatogr.*, **457**, 59.
57. Pawlowska, M. and Zukowski, J. (1991). *Chem. Anal.* (Warsaw), **36**, 447.
58. Szeman, J. and Ganzler, K. (1994). *J. Chromatogr.*, **668**, 509.
59. Zukowski, J. and Nowakowski, R. (1989). *J. Liq. Chromatogr.*, **12**, 1545.
60. Chandler, M. H. H., Guttendorf, R. J., Blouin, R. A., and Wedlund, P. J. (1987). *J. Chromatogr.*, **419**, 426.
61. Sybilska, D., Bielejewska, A., Nowakowski, R., Duszczyk, K., and Jurczak, J. (1992). *J. Chromatogr.*, **625**, 349.
62. Lim, W. H. and Hooper, W. D. (1989). *Drug. Metab. Dispos.*, **17**, 212.
63. Hooper, W. D. and Qing, M. S. (1990). *Clin. Pharmacol. Ther.*, **48**, 633.
64. O'Shea, N. J. and Hooper, W. D. (1990). *Chirality*, **2**, 257.
65. Bojarski, J., Kubaszek, M., Barton, H., and Chmiel, E. (1994). *J. Chromatogr.*, A **668**, 481.
66. Thuaud, N. and Sebille, B. (1994). *J. Chromatogr.*, A, **685**, 15.
67. Ichida, A., Shibata, T., Okamoto, I., Yuki, Y., Namikoshi, H., and Toga, Y. (1984). *Chromatographia*, **19**, 280.
68. Francotte, E. and Wolf, R. M. (1992). *J. Chromatogr.*, **595**, 63.
69. Rizzi, A. M. (1989). *J. Chromatogr.*, **478**, 71.
70. Rizzi, A. M. (1989). *J. Chromatogr.*, **478**, 87.
71. Rizzi, A. M. (1989). *J. Chromatogr.*, **478**, 101.
72. Rizzi, A. M. (1990). *J. Chromatogr.*, **513**, 195.
73. Aboul-Enein, H. Y., Serignese, V., and Bojarski, J. (1993). *J. Liq. Chromatogr.*, **16**, 2741.

74. Aboul-Enein, H. Y., Serignese, V., Aboul-Basha, L.I., and Bojarski, J. (1996). *Pharmazie*, **51**, 159.
75. Hermansson, J. and Eriksson, M. (1986). *J. Liq. Chromatogr.*, **9**. 621.
76. Allenmark, S., Andersson, S., and Bojarski, J. (1988). *J. Chromatogr.*, **436**, 479.
77. Andersson, S. and Allenmark, S. (1989). *J. Liq. Chromatogr.*, **12**, 345.
78. Andersson, S., Thompson, R. A., and Allenmark, S. G. (1992). *J. Chromatogr.*, **591**, 65.
79. Hermansson, J. (1989). *Trends Anal. Chem.*, **8**, 251.
80. Enquist, M. and Hermansson, J. (1990). *J. Chromatogr.*, **519**, 271.
81. Vermeulen, A. M., Rosseel, M. T., and Belpaire, F. M. (1991). *J. Chromatogr.*, **567**, 472.
82. Haginaka, J., Seyama, C., Yasuda, H., Fujima, H., and Wada, H. (1992). *J. Chromatogr.*, **592**, 301.
83. Haginaka, J., Murashima, T., Seyama, C., Fujima, H., and Wada, H. (1993). *J. Chromatogr.*, **631**, 183.
84. Yang, Z-Y., Barken, S., Brunner, C., Weber, J. D., Doyle, T. D., and Wainer, I. W., (1985). *J. Chromatogr.*, **324**, 444.
85. *Merck Spectrum*, **11**(1), 13 (1994).
86. Feibush, B., Figueroa, A., Charles, R., Onan, K. D., Feibush, P., and Karger, B. L. (1986). *J. Am. Chem. Soc.*, **108**, 3310.
87. Dobashi, Y. and Hara, S. (1987). *J. Org. Chem.*, **52**, 2490.
88. Sinibaldi, M., Carunchio, V., and Messina, A. (1988). *Analusis*, **16** (suppl. no. 9–10), 92.
89. Masia, P., Nicoletti, I., Sinibaldi, M., Attanasio, D., and Messina, A. (1988). *Anal. Chim. Acta*, **204**, 145.
90. Gübitz, G., Wintersteiger, R., Mihellyes, S., and Kobinger, G. (1993). *Quim. Anal.*, **12**, 45.
91. Gubitz, G., Mihellyes, S., Kobinger, G., and Wutte, A. (1994). *J. Chromatogr. A*, **666**, 91.
92. Hargitai, T., Reinholdsson, P., Törnell, B., and Isaksson, R. (1991). *J. Chromatogr.*, **540**, 145.
93. Castellani, L., Federici, F., Sinibaldi, M., and Messina, A. (1992). *J. Chromatogr.*, **602**, 21.
94. Armstrong, D. W., Tang, Y., Chen, S., Zhou, Y., Bagwill, C., and Chen, J.-R. (1994). *Anal. Chem.*, **66**, 1473.
95. Yamazaki, S., Nagaya, S., Saito, K., and Tanimura, T. (1994). *J. Chromatogr. A*, **662**, 219.
96. König, W. A. and Ernst, K. (1983). *J. Chromatogr.*, **280**, 135.
97. Van der Graaff, M., Vermeulen, N. P. E., Hofman, P. H., and Breimer, D.D., (1986). *J. Chromatogr.*, **375**, 411.
98. König, W. A., Lutz, S., Evers, P., and Knabe, J. (1990). *J. Chromatogr.*, **503**, 256.

99. Armstrong, D. W., Li, W., Stalcup, A. M., Secor, H. V., Izac, R. R., and Seeman, J. I., (1990). *Anal. Chim. Acta*, **234**, 365.
100. *Biotext* (Supelco) (1992). **5**. 2, 6.
101. Vermeulen, N. P. E., and Breimer, D. D., (1983). In *Stereochemistry and Biological Activity of Drugs*, Ariëns E. J., Soudijn, W., and Timmermans P. B. M. W. M. (eds.), Oxford: Blackwell Scientific Publications, p. 33.
102. König, W. A., Lutz, S., Colberg, C., Schmidt, N., Wenz, G., von der Bey, E., Mosandl, A., Günther, C., and Kusterman, A. (1988). *J. High Res. Chromatogr.*, **11**, 621.
103. Hardt, I. and König,W. A. (1994). *J. Chromatogr. A*, **666**, 611.
104. Prakash, C., Adedoyin, A., Wilkinson, G. R., and Blair, I. A. (1991). *Biol. Mass Spectrom.*, **20**, 559.
105. Schurig, V. (1994). *J. Chromatogr. A.*, **666**, 111.
106. Jung, M., Mayer S., and Schurig V. (1994). *LC.GC Int.*, **7**, 340.
107. Tanaka, M., Asano, S., Yoshinaga, M., Kawaguchi, Y., Tetsumi, T., and Shono, T. (1991). *Fresenius J. Anal. Chem.*, **339**, 63.
108. Nishi, H., Fukuyama, T., and Terabe, S. (1991). *J. Chromatogr.*, **553**, 503.
109. Li, S. and Lloyd, D. K. (1993). *Anal Chem.*, **65**, 3684.
110. Li, S. and Lloyd, D. K. (1994). *J. Chromatogr. A*, **666**, 321.
111. Francotte, E., Cherkaoui, S., and Faupel, M. (1993). *Chirality*, **5**, 516.
112. Ward, T. J. (1994). *Anal Chem.*, **66**, 633A.
113. Knabe, J. and Gradmann, V. (1977). *Arch. Pharm.*, **310**, 468.
114. Knabe, J. and Reinhardt, J. (1982). *Arch. Pharm.*, **315**, 772.
115. Eberhart, S. and Rothchild, R. (1984). *Appl. Spectrosc.*, **38**, 74.
116. Rothchild, R. and Simons, P. (1984). *Spectochim. Acta*, **40A**, 881.
117. Avolio, J. and Rothchild, R. (1984). *J. Magnet. Res.*, **58**, 328.
118. Eberhart, S. T., Hatzis, A., Jimenez, J., Rothchild, R., and Simons, P. (1987). *J. Pharm. Biomed. Anal.*, **5**, 233.
119. Knabe, J. and Gradmann, V. (1973). *Arch. Pharm.*, **306**, 306.
120. Knabe, J. and Franz, N. (1976). *Arch. Pharm.*, **309**, 173.
121. Knabe, J., Büch, H. P., Gradmann, V., and Wolff, I. (1977). *Arch. Pharm.*, **310**, 421.
122. Knabe, J. and Gradmann, V. (1977). *Arch. Pharm.*, **310**, 515.
123. Knabe, J., Junginger, H., Geismar, W., and Wolf, H. (1970). *Liebigs Ann. Chem.*, **739**, 15.
124. Knabe, J., Hunginger, H., and Geismar, W. (1971). *Arch. Pharm.*, **304**, 1.
125. Knabe, J. and Wunn, W. (1982). *Arch. Pharm.*, **315**, 997.
126. Knabe, J. and Wunn, W. (1976). *Arch. Pharm.*, **309**, 601.
127. Knabe, J. and Schamber, L. (1982). *Arch. Pharm.*, **315**, 878.
128. Knabe, J. and Geismar, W. (1968). *Arch. Pharm.*, **301**, 682.

129. Soine, W. H., Soine, P. J., Wireko, F. C., and Abraham, D. J. (1990). *Pharm. Res.*, **7**, 794.

130. Knabe, J., Geismar, W., and Urbahn, C. (1969). *Arch. Pharm.*, **302**, 468.

131. Knabe, J. and Wunn, W. (1979). *Arch. Pharm.*, **312**, 973.

132. Knabe, J., Büch, H. P., and Biwersi, J. (1993). *Arch. Pharm.*, **326**, 79.

133. Knabe, J., Büch, H. P., Franz, N., and Wolff, I. (1976). *Arch. Pharm.*, **309**, 681.

134. Barton, H., Bojarski, J., and Mokrosz, J. (1982). *Tetrahedron Lett.*, **23**, 2133.

135. Barton, H., Zurowska, A., Bojarski, J., and Welna, W. (1983). *Pharmazie*, **38**, 268.

136. Barton, H. J., Bojarski, J. T., Zurowska, A., and Ekiert, L. (1990). *J. Photochem. Photobiol.*, **A54**, 187.

137. Ho, I. K. and Harris, A. R. (1981). *Ann. Rev. Pharmacol. Toxicol.*, **21**, 83.

138. Richter, J. A. and Holtman, J. R. (1982). *Prog. Neurobiol.*, **18**, 275.

139. Andrews, P. R. and Mark, L. C. (1982). *Anesthesiology*, **57**, 314.

140. Willow, M. and Johnston, G. A. R. (1983). *Int. Rev. Neurobiol.*, **24**, 15.

141. Macdonald, R. L. (1983). Barbiturate and hydantoin anticonvulsant mechanisms of action In *Basic Mechanisms of Neuronal Hyperexcitability*, New York: Alan, R. Liss, p. 361.

142. Olsen, R. W., Fischer, J. B., and Dunwiddie, T. V. (1986). Barbiturate enhancement of γ-aminobutyric acid receptor binding and function as a mechanism of anesthesia. In Roth, S. H. and Miller, K. W. (eds.), *Molecular and Cellular Mechanisms of Anesthetics*, New York: Plenum Press, p. 165.

143. Van der Graaff, M, Vermeulen, N. P. E., and Breimer, D. D. (1988). *Drug Metab. Rev.*, **19**, 109.

144. Macdonald, R. L. and Kelly, K. M. (1993). *Epilepsia*, **34** (suppl. 5), Sl.

145. Gibson, W. R., Doran, W. J., Wood, W. C., and Swanson, E. E. (1959). *J. Pharmacol. Exp. Ther.*, **125**, 23.

146. Buch, H., Grund, W., Buzello, W., and Rummel, W. (1969). *Biochem. Pharmacol.*, **18**, 1005.

147. Waddell, W. J. and Baggett, B. (1973). *Arch. Int. Pharmacodyn.*, **205**, 40.

148. Christensen, H. D. and Lee, I. S. (1973). *Toxicol. Appl. Pharmacol.*, **26**, 495.

149. Freudenthal, R. I. and Martin, J. (1975). *J. Pharmacol. Exp. Ther.*, **193**, 664.

150. Palmer, K. H., Fowler, M. S., Wall, M. E., Rhodes, L. S., Waddell, W. J., and Baggett, B. (1969). *J. Pharmacol. Exp. Ther.*, **170**, 355.

151. Palmer, K. H., Fowler, M. S., and Wall, M. E. (1970). *J. Pharmacol. Exp. Ther.*, **175**, 38.

152. Holtzmann, J. L. and Thompson J. A. (1975). *Drug Metab. Disp.*, **3**, 113.

153. Huang, L-Y. M. and Barker, J. L. (1980). *Science*, **207**, 195.

154. Mathers, D. A. and Barker, J. L. (1980). *Science*, **209**, 507.

155. Owen, D. G., Barker, J. L., Segal, M., and Study, R. E. (1986). Postsynaptic actions of pentobarbital in cultured mouse spinal neurons and rat hippocampal

neurons. In Roth, S. H., and Miller, K. W. (eds.), *Molecular and Cellular Mechanisms of Anesthetics*, New York: Plenum Press, p. 27.

156. Leeb-Lundberg, F., Snowman, A., and Olsen, R. (1980). *Proc. Natl. Acad. Sci. USA*, **77**, 7469.
157. Ticku, M. K. (1981). *Biochem. Pharmacol.*, **30**, 1573.
158. Skolnick, P., Rice, K. C., Barker, J. L., and Paul, S. M. (1982). *Brain Res.*, **233**, 143.
159. Ticku, M. K. (1981). *Brain Res.*, **211**, 127.
160. Leeb-Lundberg, L. M. F. and Olsen, R. W. (1983). *Mol. Pharmacol*, **23**, 315.
161. Stephenson, F. A. and Olsen, R. W. (1983). Biochemical pharmacology of the GABA receptor-ionofore protein complex. In Mandel, P. and De Feudis, F. V. (eds.), *CNS Receptors – From Molecular Pharmacology to Behavior*, New York: Raven Press, p. 71.
162. Ticku, M. K. and Rastogi, S. K. (1986). Barbiturate-sensitive sites in the benzodiazepine-GABA receptor-ionofore complex. In Roth S. H., and Miller, K. W. (eds.), *Molecular and Cellular Mechanisms of Anesthetics*, New York: Plenum Press, p. 179.
163. Roth, S. H., Tan, K.-S., and MacIver, M. B., (1986). Selective and differential effects of barbiturates on neuronal activity. In Roth S. H. and Miller K. W. (eds.), *Molecular and Cellular Mechanisms of Anesthetics*, New York: Plenum Press, p. 43.
164. MacIver, M. B. and Roth, S. H. (1987). *Can. J. Physiol. Pharmacol.*, **65**, 385.
165. Holtman, J. R. and Richter, J. A. (1981). *Biochem. Pharmacol.*, **30**, 2619.
166. Miller, K. W., Sauter, J. F., and Braswell, L. M. (1982). *Biochem. Biophys. Res. Comm.*, **105**, 659.
167. Roth, S. H., Forman, S. A., Braswell, L. M., and Miller, K. W. (1989). *Mol. Pharmacol.*, **36**, 874.
168. Cook, C. E., Seltzman, T. B., Tallent, C. R., Lorenzo, B., and Drayer, D. D. (1987). *J. Pharmacol. Exp. Ther.*, **241**, 779.
169. Rastogi, S. K., Wenger, G. R., and McMillan, D. E. (1985). *Arch. Int. Pharmacodyn.*, **276**, 247.
170. Wenger, G. R. (1986). *Pharmacol. Biochem. Behav.*, **25**, 375.
171. Wenger, G. R., Donald, J. M., and Cunny, H. C. (1986). *J. Pharmacol. Exp. Ther.*, **237**, 445.
172. Wessinger, W. D. and Wenger, G. R. (1987). *Psychopharmacology*, **92**, 334.
173. Watanabe, H., Ohdo, S., Ishikawa, M., and Ogawa, N. (1992). *J. Pharmacol. Exp. Ther.*, **263**, 1036.
174. Frenkel, C. E., Duch, D. S., and Urban, B. W. (1990). *Anesthesiology*, **72**, 640.
175. ffrench-Mullen, J. M. H., Barker, J. L., and Rogawski, M. A. (1993). *J. Neurosci.*, **13**, 3211.
176. Wahlström, G. (1966). *Life Sci.*, **5**, 1781.
177. Rummel, W., Brandenburger, U., and Büch, H., (1967). *Med. Pharmacol. Exp.*, **16**, 496.

178. Wahlström, G., Büch, H., and Buzello, W. (1970). *Acta Pharmacol. Toxicol.*, **28**, 493.

179. Degkwitz, E., Ullrich, V., Staudinger, H., and Rummel, W. (1969). *Hoppe-Seylers, Z. Physiol. Chem.*, **350**, 547.

180. Furner, R. L., McCarthy, J. S., Sitzel, R. E., and Anders, M. W. (1969). *J. Pharmacol. Exp. Ther.*, **169**, 153.

181. McCarthy, J. S. and Stitzel, R. E. (1971). *J. Pharmacol. Exp. Ther.*, **176**, 772.

182. Feller, D. R. and Lubawy, W. C. (1973). *Pharmacology*, **9**, 129.

183. Miyano, K. and Toki, S. (1980). *Drug Metab. Disp.*, **8**, 111.

184. Mijano, K., Ota, T., and Toki, S. (1981). *Drug Metab. Disp.*, **9**, 60.

185. Büch, H., Knabe, J., Buzello, W., and Rummel, W. (1970). *J. Pharmacol. Exp. Ther.*, **175**, 709.

186. Büch, H. P., Schneider-Affeld, F., and Rummel, W. (1973). *Naunyn-Schmiedeberg's Arch. Pharmacol.*, **277**, 191.

187. Wahlström, G. and Norberg, L. (1984). *Brain Res.*, **310**, 261.

188. Höltje H. D. (1977). *Arch. Pharm.*, **310**, 650.

189. Andrews, P. R., Mark, L. C., Winkler, D. A., and Jones, G. P. (1983). *J. Med. Chem.*, **26**, 1123.

190. Höltje, H. D. (1986). *Arch. Pharm.*, 319, 570.

191. Breimer, D. D. and Van Rossum, J. M. (1973). *J. Pharm. Pharmacol.*, **25**, 762.

192. Breimer, D. D. and Van Rossum, J. M. (1974). *Eur. J. Pharmacol.*, **26**, 321.

193. Van der Graaff, M., Vermeulen, N. P. E., Joeres, R. P., and Breimer, D. D., (1983). *Drug Metab. Disp.*, **11**, 489.

194. Van der Graaff, M., Vermeulen, N. P. E., and Breimer, D. D. (1986). *Pharm. Week. Sci. Ed.*, **8**, 139.

195. Wahlstrom, G. (1979). *Eur. J. Pharmacol.*, **59**, 219.

196. Leeb-Lundberg, F. and Olsen R. (1982). *Mol. Pharmacol.*, **21**, 320.

197. King, R. G., Nielsen, M., Stauber, G. B., and Olsen, R. W. (1987). *Eur. J. Biochem.*, **169**, 555.

198. Nordberg, A., and Wahlström, G., (1984). *Brain Res.*, **310**, 198.

199. Chandler, M. H. H., Scott, S. R., and Blouin, R. A. (1988). *Clin. Pharmacol. Ther.*, **43**. 436.

200. Groen, K., Breimer, D. D., Jansen, E. J., and Van Bezooijen, C. F. A. (1994). *J. Pharmacol. Exp. Ther.*, **268**, 531.

201. Wahlström, G. (1968). *Acta Pharmacol. Toxicol.*, **26**, 81.

202. Büch, H., Buzello, W., Neurohr, O., and Rummel, W. (1968). *Biochem. Pharmacol.*, **17**, 2391.

203. Steen, P. A. and Michenfelder, J. D. (1978). *Stroke*, **9**, 140.

204. Ticku, M. K. (1983). *Neuropharmacology*, **22**, 1459.

205. Harrison, N. L. and Simmonds, M. A. (1983). *Br. J. Pharmacol.*, **80**, 387.

206. Kupfer, A. and Branch, R. A. (1985). *Clin. Pharmacol. Ther.*, **38**, 414.
207. Krug, R., Altmayer, R., and Büch, H. P. (1994). *Arzneim. Forsch.*, **44**, 109.
208. Sisten, J. M. A. and Fresen, J. A. (1974). *Pharm. Week.*, **109**, 1.
209. Nicholson, G. M., Spence, I., and Johnston, G. A. R. (1988). *Neuropharmacology*, **27**, 459.
210. Grossman, W., Jurna, I., and Theres, C. (1974). *Naunyn-Schmiedeberg's Arch. Pharmacol.*, **282**, 367.
211. Schombert, I., Schneider-Affeld, F., and Büch, H. P. (1985). *Arzneim. Forsch.*, **29**, 38.
212. Ticku, M. K., Rastogi, S. K., and Thygarajan, R. (1985). *Eur. J. Pharmacol.*, **112**, 1.
213. Knabe, J., Büch, H. P., Gradmann, V., and Wolff, I. (1977). *Arch. Pharm.*, **310**, 421.
214. Knabe, J., Büch, H. P., and Kirsch, G. A. (1987). *Arch. Pharm.*, **320**, 323.
215. Knabe, J., Büch, H. P., and Reinhardt, J. (1982). *Arch. Pharm.*, **315**, 832.
216. Knabe, J. (1989). *Arzneim. Forsch.*, **39**, 1379.
217. Haley, T. J. and Gidley, J. T. (1970). *Eur. J. Pharmacol.*, **9**, 358.
218. Mark, L. C., Brand, L., Perel, J. M., and Carroll, F. I., (1978). *Excerpta Med. Int. Congr. Ser.*, **399**, 144.
219. Haley, T. J. and Gidley, J. T. (1976). *Eur. J. Pharmacol.*, **36**, 211.
220. Knabe, J., Büch, H. P., and Lampen, P. (1987). *Arch. Pharm.*, **320**, 807.
221. Knabe, J., Büch, H. P., and Lampen, P. (1987). *Arch. Pharm.*, **320**, 1103.
222. Soine, W. H., Soine, P. J., Overton, B. W., and Garrettson, L. K. (1986). *Drug Metab. Disp.*, **14**, 619.
223. Mongrain, S. E. and Soine, W. H. (1991). *Drug Metab. Disp.*, **19**, 1012.
224. Soine, W. H., Soine, P. J., England, T. M., Welty, D. F., and Wood, J. H. (1990). *J. Pharm. Biomed. Anal.*, **8**, 365.
225. Ciucianu, I. and König, W. A. (1994). *J. Chromatogr. A*, **685**, 166.
226. Ôi, N., Kitahara, H., Matsushita, Y., and Kisu, N., (1996). *J. Chromatogr. A*, **722**, 229.
227. Ciucanu, I. (1996). *J. Chromatogr. A*, **727**, 195.
228. Eto, S., Noda, H., and Noda, A. (1994). *J. Chromatogr. B*, **658**, 386.
229. Roussel, C. and Favrou, A. (1995). *J. Chromatogr. A*, **704**, 67.
230. Chankvetadze, B., Chankvetadze, L., Sidamonidadze, Sh., Yashima, E., and Okamoto, Y. (1995). *J. Pharm. Biomed. Anal.*, **13**, 695.
231. Riering, H. and Sieber, M. (1996). *J. Chromatogr. A*, **728**, 171.
232. Schurig, V., Jung, M., Mayer, S., Fluck, M., Negura, S., and Jakubetz, H. (1995). *J. Chromatogr. A*, **694**, 119.
233. Huang, J. L., Mather, L. E., and Duke, C. C. (1995). *J. Chromatogr. B*, **673**, 245.
234. Jones, D. J., Nguyen, K. T., McLeish, M. J., Crankshaw, D. P., and Morgan, D. J. (1996). *J. Chromatogr. B*, **675**, 174.

235. Sueyasu, M., Ikeda, T., Otsubo, K., Taniyama, T., Aoyama, T., and Oishi, R. (1995). *J. Chromatogr. B*, **665**, 133.

236. Schurig, V. and Mayer, S. (1994). *Electrophoresis*, **15**, 835.

237. Tickle, D. C., Okafo, G. N., Camilleri, P., Jones, R. F. D., and Kirby, A. J. (1994). *Anal. Chem.*, **66**, 4121.

238. Ward, T. J., Dann III, C., and Blaylock, A. (1995). *J. Chromatogr. A*, **715**, 337.

239. Schmitt, T. and Engelhardt, H. (1995). *J. Chromatogr. A*, **697**, 561.

240. Durham., D. G. and Liang, H. (1994). *Chirality*, **6**, 239.

241. Durham, D. G. (1996). *Chirality*, **7**, 58.

242. Soine, W. H., Soine, P. J., England, T. M., Graham, R. M., and Capps, G. (1994). *Pharm. Res.*, **11**, 1535.

243. Soine, W. H., Yu, C.-F., Thomas, D., Nayak, V., Cao, S. L., and Westkaemper, R. B. (1995). *Med. Chem. Res.*, **5**, 462.

244. Neighbors, S. M. and Soine, W. H. (1995). *Drug Metab. Disp.*, **23**, 548.

245. Büch, H. P., Knabe, J., and Krug, R. (1995). *Arzneim-Forsch.*, **45(II)**, 1049.

246. Macksay, G., Molnar, P., and Simonyi, M. (1996). *Naunyn-Schmiedeberg's Arch. Pharmacol.*, **353**, 306.

CHAPTER

8

ETHAMBUTOL AND TUBERCULOSIS, A NEGLECTED AND CONFUSED CHIRAL PUZZLE

BERNARD BLESSINGTON

Bradford University
Pharmaceutical Chemistry Department
Bradford BD7 1DP
United Kingdom

8.1. INTRODUCTION

Ethambutol had the misfortune to be discovered around 1961 at a time when earlier, major advances in the battle against tuberculosis (TB) were starting to show dramatic results. It was consequently routinely incorporated into the arsenal of miraculous new TB treatments, discovered in the period 1945–1960, which apparently reduced this most dreadful of diseases to near insignificance, at least in developed countries (1). From 1945–1960, a 100-fold reduction in deaths from TB was recorded in England and Wales. To all intents, TB, the "Captain of all these men of Death" (2) and probably the world's greatest slaughterer, had been defeated by a combination of public health education, contact tracing, improved living standards, cattle and milk-pasteurization controls, vaccination, and most importantly a string of spectacular drug developments, starting with streptomycin (1944), PAS(para-amino salicylic acid, 1946), thiacetazone (1946), isoniazid (1952), ethionamide (1956), ethambutol (1961), and ending with rifampicin (1966). Reading the scientific literature, one can almost hear the sighs of relief and imagine the associated drying-up of research funding, perhaps making way for the swinging 1960s and permissive 1970s.

In stark contrast was, and still is, the situation in developing countries, where vast reservoirs of endemic TB hardly flinched, in fact often thrived because of miniscule health budgets, ignorance, poverty, drought, and endless armed conflicts. Health programs, aimed at TB eradication, were inevitably

The Impact of Stereochemistry on Drug Development and Use, Edited by Hassan Y. Aboul-Enein and Irving W. Wainer. Chemical Analysis Series, Vol. 142.
ISBN 0-471-59644-2

half-successful attempts to tackle so fearsome a threat and contrived only to make matters worse. One whole issue (3) of the World Health Organizations (WHO) newsletter has been devoted to illustrating the rampant progress of TB in Africa, China, and India and the efforts to undo past mistakes. WHO estimates that worldwide 8 million new cases of TB occur annually and 2 million deaths per year are due to TB. The majority occur in developing parts of the world and even now, most would have been preventable with existing treatments.

Today, however, developed nations have awoken to the threat TB poses to them and stand in dread of a resurgence of "the white plague" (4) and its modern ally, HIV-AIDS, with virtually all their "pharmaceutical armor" blunted. Urgent action is required and, as part of this, an updated reappraisal of ethambutol is both timely and pertinent to this volume.

8.2. ETHAMBUTOL

Ethambutol(I) was discovered as a result of an extensive research program, specifically for new antitubercular drugs, at Lederle Laboratories Division, American Cyanamid, Pearl River, New York. A whole series (5–15) of clear, explicit, and elegant papers by R. G. Shepherd outlined the scope, methods, and results of this undertaking. A very readable account of this work has also been given by Ira Ringler (16).

Strictly speaking, ethambutol is the approved name for the dextrorotatory (+) optical isomer of 2,2′-(ethylenediimino)-di-1-butanol. Its Chemical Abstract (C.A.) index name is 1-butanol, 2,2′-(1,2-ethanediyldiimino)bis-, [S-(R*, R*)]-(9CI) and its C.A. registry numbers are 74-55-5 for the free base and 1070-11-7 for its dihydrochloride salt. Corresponding numbers for the free base form of the (–)-isomer (II) and the optically inactive *meso*-isomer (III) and racemate are 10054-05-4, 10054-06-5, and 36697-71-9, respectively. Ethambutol must be one of the earliest examples of a man-made, racemic drug that displays such specific chiral pharmacological activity that it must be resolved and administered as a single (+)-enantiomer. This work was done in the 1960s, long before chiral chromatography and even the Cahn–Ingold–Prelog (CIP, *R*/*S*) system of nomenclature were widely used. During the intervening years, little was added to the chiral aspects of (+)-ethambutol except confusion. Eventually, the (*R*/*S*) system was applied to ethambutol, but incorrectly because *R*, *R*; *S*, *S*; *S*-(*R**, *R**), and *R*-(*R**, *R**) configurations have all been quoted (17–20, 137, 138). Not until 1990 was the ensuing confusion clarified by the use of chiral-chromatography and the correct *S*, *S* configuration established (93). It is small wonder that little light has been shed on the molecular mystery producing the chiral action of this important and fascinating drug.

Two reviews of the research on ethambutol up until 1977 have been published: The first (21) emphasizes analytical methods (34 refs.) and the second (22) discusses the biological aspects of ethambutol (127 refs.). Chemical Abstracts Service (C.A.S.) online searches, using ethambutol as the sole key word, revealed 409 additional references between 1977 and February 1994. Of these papers, over half discuss the clinical effects of ethambutol treatment, usually in conjunction with other antitubercular drugs. This aspect will not be properly addressed in this chapter. What is notable is that papers prior to 1985 were heavily weighted toward a discussion of ethambutol's main side-side effect, ocular neuropathy (23–30). The emphasis changed sharply thereafter and recent papers (31–36) concentrate on its efficacy against atypical mycobacteria, particularly mycobacterium avium complex (MAC) associated with HIV-AIDS.

Few papers have added new methodology for the analysis and molecular study of ethambutol or contributed to an understanding of its mode of action. The mechanism of mycobacterial resistance to this drug or of its synergistic interaction with other drugs used with it in combination therapy has barely been probed (38–40).

8.3. TUBERCULOSIS AND ITS TREATMENT

Tuberculosis, commonly called TB or known by its earlier names of consumption or phthisis, has been a major human bacterial disease since antiquity (37). It is caused by two pathogens: Mycobacterium (M.) *tuberculosis* transmitted from human to human host and M. *bovis* transmitted to humans from cattle and their milk. A related pathogen, M. *leprae*, is the organism causing leprosy. In contrast, other common M. species (e.g., M. *avium*, M. *kansasii*, M. *xenopi*, etc.) were loosely classified as commensals since they are frequently found on or in humans but do not normally produce disease. However, later discussion of opportunistic infections should be noted.

Mycobacteria are unusual among bacteria because of the complexity (38) (Figure 8.2) of their thick, waxy coats, rich in mycolic acids, which make them unusually hardy and resistant to drying out, compared to other bacteria. Because of this, they can better survive on dust particles, once spread by coughing, sneezing, spitting, etc., and this facilitates their transmission via dust inhalation into the small passages of the lungs. They can also be spread by eating food contaminated in this way.

Mycobacteria's lipid coating made them initially difficult to detect by standard microscopy staining methods until Robert Koch's brilliant demonstration of their role in TB. Subsequent work by Paul Ehrlich took advantage of the unusual properties of these "acid and alcohol fast" (AAF) bacilli, and his selective staining technique is essentially the same one used today (the Ziehl–Neelsen stain) to identify active cases of pulmonary TB.

The same complex mycobacterial envelopes cause very serious problems for the human immune system. The immune response to initial infection is complex, but a major part of it is to activate phagocytic cells (mainly macrophages) to engulf and destroy the intruding mycobacteria. However, this is not easy because of the complex mycobacterial envelope, and a stalemate situation can arise in which the mycobacteria actually exist as intracellular parasites within the macrophages (39). When the immune status of the host is compromised, as would occur as a result of malnutrition, HIV-AIDS, or during transplant or oncochemotherapy, then the mycobacteria can start to proliferate within their macrophage hosts, eventually bursting free, multiplying extracellularly, and releasing the aggressive contents of the macrophages to damage surrounding tissues and so facilitate further spreading of the mycobacterial pathogen. It is these destructive processes that characterize clinical TB and lead to its miliary phase, when skin, bones, and glands can be infected.

Another characteristic of mycobacteria is their very slow growth rate, compared to other bacteria, and their facile ability to undergo mutations. This is particularly important in the chemotherapy of TB, where treatment is prolonged because actively dividing bacilli are mainly affected. Typical treatments extend from 6 months to 2 years. Should the treatment be incomplete, then it will ensure that the surviving organisms are those which had, or have developed by mutation during treatment, the most resistance to the antitubercular drug regime being used (40).

This became apparent soon after the discovery of streptomycin (41), the first of the spectacular antitubercular drugs, when patients showed remarkable recovery in the short term but over a longer period relapsed because strains resistant to streptomycin started to proliferate. Once other effective drugs (e.g., PAS) were discovered, then combination therapy was introduced (42). In modern practice (43), this means that a combination of four or five of the most effective drugs (first-line drugs) are given simultaneously, at near maximum dose for each, for about 3 months. This "initial phase" has the objective of reducing as quickly as possible the number of proliferating bacilli and minimizing the possibility of resistance developing. Then follows the "continuing phase" when two first-line drugs are given at lower doses for up to 2 years. These regimes have varied over time and from country to country, but WHO and U.K. recommendations now favor shorter treatment periods.

During treatment, microbiological checks should be conducted to see that resistant species do not emerge. The objective is to join forces with the host's natural immune defense mechanism and totally overwhelm the invading organisms, thus preventing the emergence of any resistant strains. When resistance is found, then referral to specialist facilities should be made to enable treatment with a wider range of drugs, including second-line drugs, those which are generally less effective or subject to adverse side-effects and so are not in widespread use.

In addition and of crucial importance, a range of contact-tracing and educational measures are also instituted during the initial phase. Failure to implement these procedures will inevitably lead to the emergence of resistant strains, even multi-drug-resistant strains of (MDR) mycobacteria. In reality, poor prescribing, irregular patient compliance, and inadequate supervision are at the root of present problems, particularly in developing countries (44, 45).

8.4. ETHAMBUTOL DISCOVERY AND USE

A series of papers (5–15) describe the research undertaken at the Pearl River Laboratories of Lederle directed at discovering new antitubercular drugs. The initial lead compounds were N,N′-dialkylethylenediamines (Figure 8.1, IV)

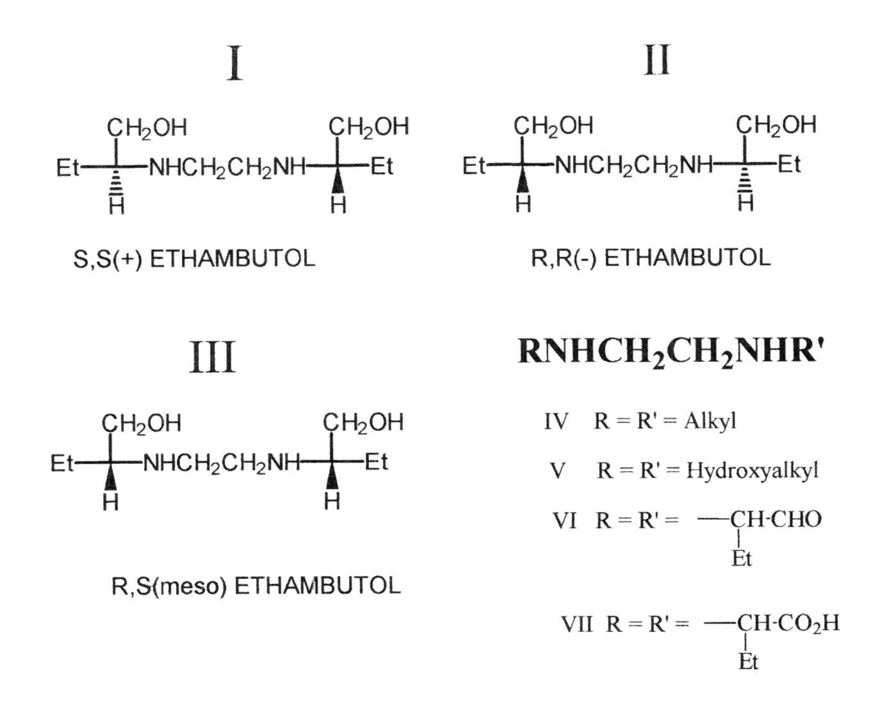

Figure 8.1. Structure of (+)ethambutol, its stereomers and related compounds.

originally developed as rubber additives by a different division of American Cyanamid, but fortunately then entered into the TB screening program. Once antimycobacterial activity was discovered, over 600 related compounds were produced and fully tested both *in vitro* and *in vivo*. The latter studies commenced with guinea pigs, but quite soon the superior mouse test was developed and adopted. Not only did this provide a considerable saving in costs, but the mouse is a much better model since its immune response is much closer to that of humans, in contrast to guinea pigs that are particularly vulnerable to TB infection.

The molecular structures of this series of test compounds were fascinating because they were so simple and unlike any other anti-TB drug known. Another striking feature was their specificity for mycobacteria, since they were found to have no activity against a wide range of other bacterial species, fungi, and viruses. This was in marked contrast to previous experience with mycobacterial species that are notorious for their invulnerability to such a wide range of potent, broad-spectrum antibacterials (chloramphenicol, cephalosporins, penicillins, sulphonamides, etc.), presumably because their unique and complex lipid membranes (Figure 8.2) present insurmountable permeability barriers.

This program was developed for over two years and had advanced to

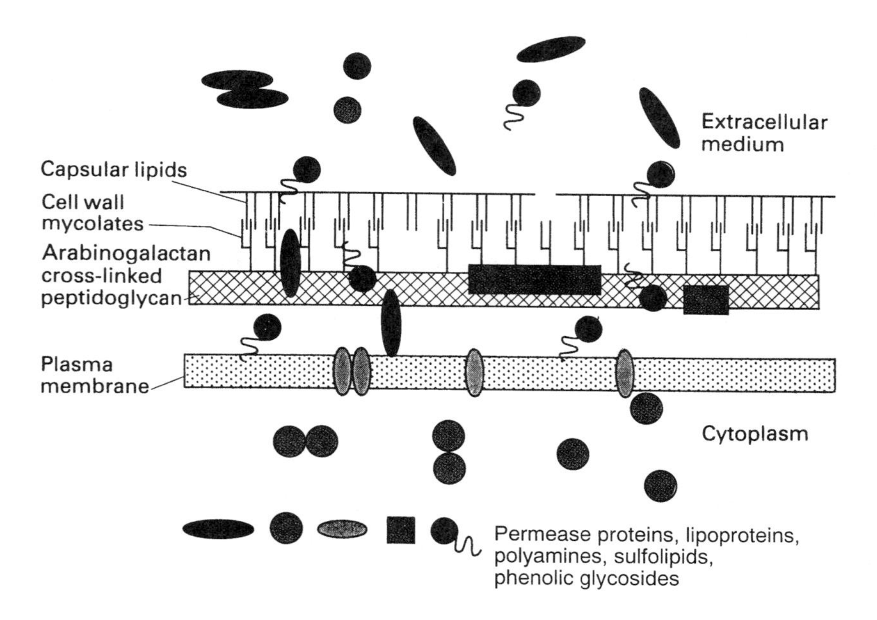

Figure 8.2. Mycobacterial cell wall.

clinical trials when an even more interesting series of related compounds, the N,N′-hydroxyalkylethylendiamines (Figure 8.1, V) were discovered. The decision was made to develop these compounds and abandon the earlier N,N′-dialkyl series. From this research, (+)-ethambutol emerged (5).

In both series of compounds, the possible existence of chiral isomers, each capable of different pharmacological activities, had been recognized at an early stage. In fact, it was both the increased antimicrobial activity together with significant reductions in toxicity that dictated the hydroxyalkyl series be prioritized. For a short period, the drug was produced and used clinically as a racemate, but this very quickly changed and the (+)-isomer alone, which strictly speaking is ethambutol, was marketed by Lederle Laboratories as Myambutol. The (–)-isomer had about 1/500th of the antibacterial activity of the (+)-isomer, whereas the optically inactive *meso*-isomer had 1/12th of (+)-ethambutol's antitubercular activity.

In contrast, the major clinical, human adverse side-effect of (+)-ethambutol, quite serious ocular toxicity (23–30), was associated equally with all three stereomers. This optical neuropathy was both dose- and duration-dependent. Early manifestations, the occurrence of blue-green color blindness, could be detected relatively easily using color sensitivity charts and were found when dosage exceeded 25 mg of ethambutol/kg/day (22). Fortunately, this side-effect was reversible in most cases, once treatment was discontinued. However, prolonged exposure to high blood levels does eventually lead to permanent eye damage and even blindness. Modern practice (43) is to discourage the treatment of very young children whose eyesight cannot be easily monitored and to treat patients with impaired renal function with care, using regular eye checks. Subject to these provisos, ethambutol has proven to be one of the safest and most acceptable of TB treatments. Toxicity in the mouse screening program was only found with high dosage, producing death or pronounced weight loss, at comparable levels for all three stereomers.

Ethambutol is usually presented as its dihydrochloride salt, which is freely soluble in water and readily absorbed following oral dosage. Myambutol tablets of only 100 mg (yellow) and 400 mg (grey) are officially approved (43) in the United Kingdom, but both generic and parallel imported white tablets can be found in use. Recommended dosage, if unsupervised, is 25 mg/kg body weight daily during the initial phase, dropping to 15 mg/kg/day during the continuing phase, to give target plasma levels of 3–5 mcg/ml. Because ethambutol is often used to treat resistant cases and these often arise due to poor patient compliance, an alternative dosing regime of 30 mg/kg three times per week can be used when given under supervision. Combination dosage forms [e.g., Mynah300 is Lederle's proprietary product containing ethambutol dihydrochloride (300 mg) and isoniazid (100 mg)] are also available (43) in the United Kingdom to try to facilitate patient compliance.

Technical aspects of the formulation of ethambutol tablets will not be discussed in detail here. Suffice it to say that several allotropic forms of the hydrochloride salt exist. These and other standard physical properties of the bulk drug such as IR, H-NMR spectra, powder X-ray, routine thermogravimetric and microbiological analysis, titration, colorimetric and gravimetric methods of analysis have been referenced in a standard monograph (21). This same article reviewed the analytical methods published up to July 1977 and paid special attention to the application of methods to pharmacokinetic studies. Much of this work relied on ^{3}H and ^{14}C radiochemical work. The absorption, distribution, and excretion kinetics for ethambol are presented and discussed. The human metabolism of ethambutol, following an intravenous dose of ^{14}C ethambutol, first by oxidation to a dialdehyde (VI, 5%) and then to the diacidic compound (VII, 3%) is described along with their ion-exchange chromatographic separation and ninhydrin detection. These and other studies established that none of the metabolites had antibacterial action and ethambutol did not accumulate in the body and was not appreciably protein bound. It was 90% cleared within 24 h., 67–74% via the kidneys,12–20% via feces. Metabolites accounted for 8–15% with unchanged and unbound drug accounting for more than 85% of the dose. It was also reported that this drug can penetrate human eryrthrocytes and attain intracellular levels in excess of plasma levels. However, it does not appear to penetrate intact meninges in normal subjects, but apparently can do so and appear at a therapeutic level of 1–2 mcg/ml in the cerebrospinal fluid of patients with tubercular meningitis (47). A review of this work has been presented (46). Further studies on both animal and human response to ethambutol have also been published (48–55).

8.5. ETHAMBUTOL SYNTHESIS

The major industrial synthesis (Figure 8.3) of (+)-ethambutol is by direct coupling of (+)-2-aminobutan-1-ol and 1,2-dichloroethane (56). An eight-fold excess of the amino alcohol is used and the economics of the process hinge on the efficient recovery and recycling of the excess during latter stages. This method is virtually the same as that used in the laboratory (8) for the original discovery of the drug. Statistically, the major chiral impurity in the reactant, the (–)-amino alcohol, will lead to the production of the *meso* impurity in the final product, which can easily be separated from the desired (+)-product because of its markedly different solubility properties. The possibility of producing significant levels of the (–)-isomer of ethambutol by this route is consequently small. Final recrystallization will produce a good-quality product, provided the starting amino alcohol is of reasonable chemical and optical purity.

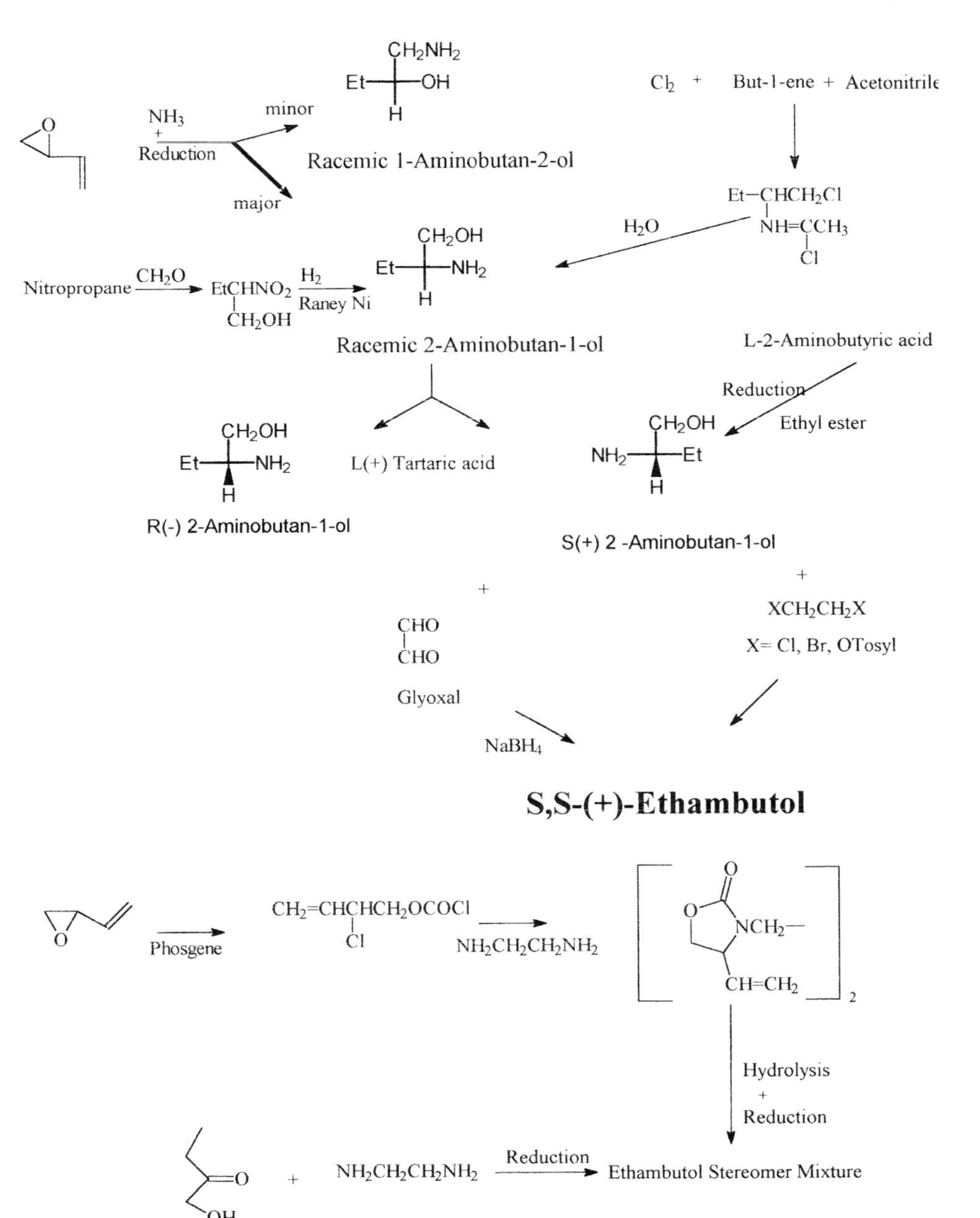

Figure 8.3. Synthetic routes to (+)ethambutol.

Essentially, the same mechanism applies when 1,2-dibromoethane or ditosylglycol is employed to couple two molecules of amino alcohol. Modification to the method (8, 57, 58) to permit microsynthesis has been used to produce ^{14}C and ^{3}H radiolabeled drug and to incorporate ^{2}H stable isotopes into ethambutol for internal standard purposes or biological studies.

Two additional different synthetic approaches (Figure 8.3) have been described, the first of which has been used in the laboratory and adapted in the patent literature. It involves reduction of the Schiff base, produced from ethylenediamine and 3-hydroxybutan-2-one, using either hydride agents (Figure 8.3) or platinum oxide/hydrogen gas. Both these approaches could be modified for chiral synthesis, but would produce both the (–)-isomer and *meso*-isomer of ethambutol as significant impurities. Schiff bases produced from glyoxal and *S*(+)-2-aminobutan-1-ol have similarly been reduced (8, 59) (Figure 8.3) to (+)-ethambutol.

Another method involves the ring-opening of an epoxide (8, 60) (Figure 8.3) followed by reaction with ethylendiamine and again could produce both the *meso*- and (–)-isomer in addition to the required (+)-ethambutol. Two possible epoxide opening mechanisms can operate, so both structural and chiral isomeric products need to be considered.

Opening 3,4-epoxybut-1-ene with ammonia (Figure 8.3) has been used to prepare the racemic amino alcohol (61, 62) prior to its resolution and conversion to (+)-ethambutol with dichloroethane. The ethambutol isomers produced were N-TFA-L-prolyl diasterioderivatized (Figure 8.4). Subsequent chromatographic analysis (63) showed that 1-aminobutan-2-ol related products were a significant impurity in addition to the expected chiral isomers. Other laboratory variants on this theme have been reported and others appear in the patent literature (64, 65) but without comment as to how the optical purity of their products could be analyzed.

At present, the key to ethambutol production is good-quality (+)-2-aminobutan-1-ol. This is produced by classical optical resolution of the racemic material using L (+)-tartaric acid (66). Two routes to the racemic starting material are shown (Figure 8.3). The resolution patent reveals that if anhydrous methanol is used as solvent, then (+)-2-aminobutanolic tartrate preferentially crystallizes. By varying the water content, the other isomer can be selectively precipitated. As of this writing (56), there is no published method for the racemization, recycling, or conversion of the unwanted (–)-2-aminobutan-1-ol into (+)-ethambutol (67).

Other patented methods of producing the (+)-amino alcohol include reducing a chiral amino acid (68) (Figure 8.3.) and classical resolution using (+)-N-benzoyl-*trans*-2-aminocyclohexanecarboxylic acid (69).

A variety of adverse effects have been reported during ethambutol chemotherapy (22). Hazards associated with the industrial production of ethambutol are mainly dermatitis and skin neurological (tingling) sensations experienced by some operatives. These can be quite serious in some cases, producing sensitization and the inability to continue working with ethambutol (56). Analytical studies, using gas chromatography (GC), have been used to monitor airborne contamination during industrial production (70–73).

8.6. ETHAMBUTOL ANALYSIS

Microbiological methods were originally used for the assay of ethambutol, but have been replaced by chromatographic and spectroscopic techniques. Improvements in microbiological methods have been directed mainly at detection, species characterization, and obtaining sensitivity data (80–82).

The survey (21) to 1977 included many methods designed for bulk drug assay. To these can be added more recent, nonspecific colorimetric assays (74–78). Improved isotopically labeled bioassays have since been reported (57, 58, 79). Stable isoptopes incorporated into internal standards are discussed later under mass spectrometry.

Significant improvements (83–91) in gas chromatography (GC) with FID and EC detection; combined gas chromatography and mass spectrometry (GC-MS); high-performance liquid chromatography (HPLC) alone (93) and coupled to circular dichroism (139) (HPLC-CD) and capillary zone electrophoresis (56) (CZE) have been reported.

Different silylation methods have been used: some to give only O-silyl products, others to fully silylate ethambutol prior to GC. Both electron impact (EI) and chemical ionization (CI). GS-MS has been used to characterize these derivatives and then to provide quantitative data in biological assays using selective ion monitoring (SIM); usually for pharmacokinetic studies (57,58). Trifluoroacetyl (TFA) derivatives have also been reported and compared to the silylation procedures. One report (63) used chiral diasterioderivatization with N-trifluoroacetyl-L-prolyl chloride and subsequent achiral GC separation of the products on packed OV-1 columns. This method gave baseline resolution of the N-TFA-L-prolyl amide derivative for each enantiomer of racemic 2-aminobutan-1-ol, but would not work directly with racemic ethambutol. However, when it was subsequently silylated with trimethylsilylimidazole (Figure 8.4) then the di-N-TFA-L-prolyl-(di-O-TMS)-ethambutol derivatives were produced and good GC separations on packed OV-1 columns were obtained for all three stereomers. The structures of these derivatives were established by GC-MS.

HPLC has not been applied directly to ethambutol because the drug contains no chromophores for easy UV detection. Ligand exchange HPLC has been attempted but showed only poor resolution, although the visible color

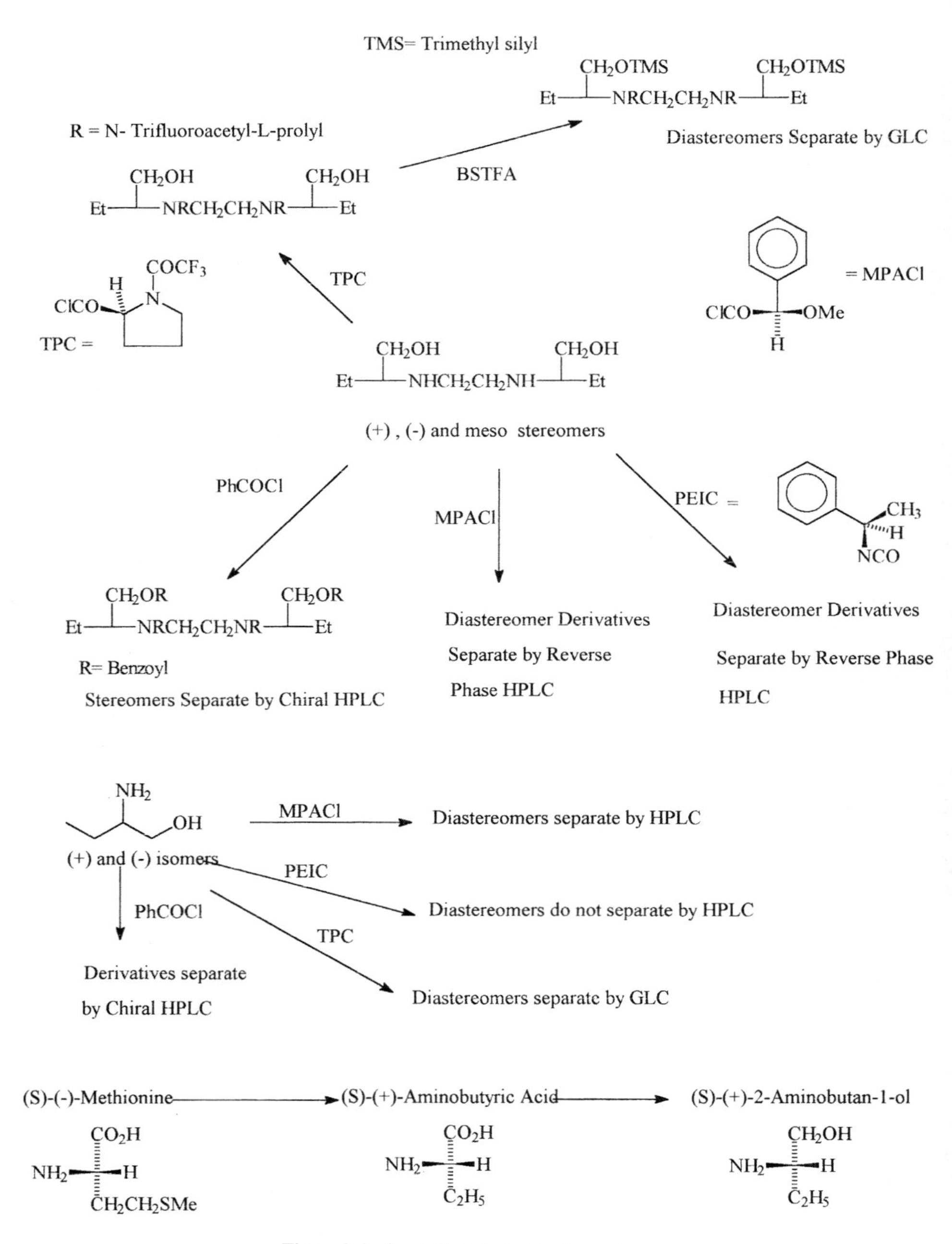

Figure 8.4. Analytical derivatization schemes.

generated by several metal chelates (96) does allow sensitive detection. Attempts to develop chiral LEC methods using chiral mobile phase additives have also been disappointing (94, 95).

The successful HPLC separation of chromophoric derivatives of ethambutol has been reported. Extension to chiral separations has broken new ground. Italian workers first reported (97) the preparation and HPLC separation of stable diastereomeric derivatives using two chiral derivatizing agents (Figure 8.4, PEIC and MPACl). R(−)-α-methoxyphenylacetamide diasteriomers of all stereomers of ethambutol and 2-aminobutan-1-ol were separated on an achiral LiChrosorb Si60 column with a chloroform: ethylacetate (90:10 v/v) mobile phase. The R(+)-1-phenylethyl isocyanate derivatives of all ethambutol stereomers were separated on an achiral, octadecyl reverse phase column with a methanol:water(65:35 v/v) mobile phase. However, no separation of the derivatives of racemic 2-aminobutan-1-ol was obtained. The method was not adapted for quantitative analysis and the structures of the derivatives were not established.

Another advance (Figure 8.4) was perbenzoyl derivatization followed by chiral separation using a Pirkle 3,5-dinitrobenzoyl-D-phenylglycine covalent chiral column (93). The benzoyl derivatives were cheap, simple, and easy to prepare and had both strong chromophores for UV detection and amide dipolar groups that were capable of dipole interactions, thus assisting separation by the chiral column. All three stereomers of ethambutol were separated from each other and also from the two, separated enantiomers of the precursor 2-aminobutan-1-ol, which can be simultaneously derivatized. The procedure was improved and adapated to yield a single-step, quantitative derivatization (140) and direct chiral separation, capable of being applied at the microgram level. Improvements to this method and biological applications should follow.

These studies clarified the confusion existing over the absolute stereochemistry of (+)-ethambutol, as discussed later. The method was also extended (139, 103) to coupled chiral HPLC-CD and this produced unexpected results, as discussed later, which threw considerable doubt on claims that exciton coupled circular dichroism is a reliable method for the determination of absolute stereochemistry.

8.7. ETHAMBUTOL STEREOCHEMISTRY

The profound influence of stereochemistry on the biological response of the three stereomers of ethambutol has already been described. In early papers (5), this presented no problem since each optical isomer was described by its physical property (dextro, levo, or meso). Early entries in the United States

Pharmacopoeia (USP) and the British Pharmacopoeia (BP) both use this description. Later papers (9) did say that the stereochemistry of (+)-ethambutol was the same as that of the naturally occurring amino acids (L in Fischer terminology). However, around 1980 the introduction of Cahn–Ingold–Prelog's (CIP) designation of absolute stereochemistry became the norm and at this point in time matters went wrong.

Thus, the BP (1980) and European Pharmacopoeia (EP) (1987) both explicitly drew the *R*, *R* structure and also named (+)-ethambutol with *R*, *R* stereochemistry, whereas the USP (1980) drew it incorrectly as the *R*, *R* while naming it as *R*-(*R**, *R**). In contrast, Klyne's atlas (137) named and showed it as the (*S*,*S*)-isomer, although Clark's reference work (138) showed and named it as *R*, *R*. How this came about is not clear; probably it owes much to the confusing (*R***R**) text system introduced by Chemical Abstracts to designate absolute and relative stereochemistry. This confusion should not arise in the future because Chemical Abstracts are moving toward a major review of stereochemistry recognition (98) using explicit computer graphics for display/retrieval, so text-based systems will be subservient.

This confusion was first resolved using the chiral HPLC methods described above (93). The absolute stereochemistry of *S*-(+)-2-aminobutan-1-ol is sound, since the (+)-enantiomer of this compound can be derived directly from the L-natural amino acids, by reduction of both L-methionine and L-α-aminobutyric acid as shown in Klyne's atlas (88). Unambiguous synthesis, not involving the chiral centers, was carried out for both the (+)- and (–)-isomers of ethambutol, as described by Wilkinson, using authentic samples of *S*(+)-2-aminobutan-1-ol and *R*(–)-2-aminobutan-1-ol from a number of different commercial sources. Each product was characterized by chiral HPLC, as were their respective starting materials. Medicinal tablets containing the biologically active drug were then extracted, and the active ingredient was shown by chiral HPLC to correspond to the *S*, S-isomer of (+)-ethambutol. This chiral chromatography was used to simultaneously assay the drug and detect and quantify the levels of impurities corresponding to both the enantiomer (*R*, R) and the *meso*-isomer (*S*,*R*) of ethambutol and both enantiomers of 2-aminobutan-1-ol.

Soon after this work was published, a definitive X-ray study of (+)-ethambutol as its dihydrobromide salt was published (99) that confirmed that (+)-ethambutol did indeed have the *S*, *S* configuration. Of interest were two earlier X-ray studies of (+)-ethambutol (100) and the *meso*-isomer (101) as dihydrochloride salts that gave all the angles and bond lengths but did not comment on absolute stereochemistry. Perhaps later improvements in X-ray sensitivity and the higher scattering power of the bromide ion over the chloride meant that absolute stereochemistry could not be established in the earlier work.

There are virtually no methods of establishing absolute stereochemistry that are capable of confirming or challenging X-ray conclusions. One very specific and rather unusual method (102) that has been proposed is exciton coupled circular dichroism (ECCD). This method relies on a chiral compound having two chromophoric groups, of comparable energy, flanking its chiral center. The phase change of the resulting CD curve, it has been claimed, is governed solely by the chirality of the chiral center.

Both dibenzoyl 2-aminobutan-1-ol and tetrabenzoyl-ethambutol enantiomers satisfy these criteria and, in fact, all display exciton coupled spectra. Unfortunately, the spectra for the tetrabenzoylethambutol enantiomers are the opposite of those shown by the precursor dibenzoyl aminobutanols. Conformational factors were proposed (103) to explain this finding and also the apparent reversal of exciton coupled CD spectra subsequently found with dibenzoyl-N-methyl-aminobutanol enantiomers when compared to dibenzoyl-aminobutanol enantiomers. If conformational effects do influence exciton coupled CD spectra, that would complicate analysis to such an extent that ECCD could not be regarded as a reliable method for the assignment of absolute stereochemistry.

8.8. ETHAMBUTOL OPTICAL PURITY

The standard procedure used to monitor optical purity is polarimetry, although it suffers from the serious limitation of poor precision, due to the low specific rotations, +13.7° for the free base and +7.7° for the hydrochloride salt when measured in water using a sodium light source. To improve this method, the European Pharmacopoeia (1987) has modified the polarimetry by generating a copper chelate of ethambutol and working at 436-nm with a mercury lamp, rather than the 589-nm D line from a sodium lamp.

A more knotty problem arises from the interpretaion of these measurements as the percentage of enantiomeric excesses (% e.e.), since the major impurity in the ethambutol used arises from the optically inactive *meso*-isomer, not the (–)-isomer, as a direct consequence of the synthetic method so widely employed in production. The commercial procedure (56) is to employ a supplementary GC or CZE method to estimate this *meso* form, but these would not enable the (–)-isomer to be quantified. If a significant amount of amino butanol precursor is present, then the analysis is even more complex. Should alternative chiral synthesis be developed, then methods of assaying all three stereomers simultaneously would be required, for significant amounts of the (–)-enantiomer could then be involved. Chiral HPLC methods would be capable of dealing with these problems while offering improved sensitivity.

8.9. ETHAMBUTOL'S MODE OF ACTION

The mode of action of antitubercular drugs, in general, is poorly understood (39, 104) and ethambutol's mode of action has an additional chiral dimension. A number of mechanisms have been proposed. The first invokes a metal chelation role and was proposed for the earlier N,N′-dialkylethylenediamine series (7) and has been the subject of several papers (105–109). It was this concept and its rational extension that actually led to the development of the hydoxyalkylethylenediamine series that include ethambutol.

Ethylenediamines are known to be capable of forming two distinct types of metal complex (Figure 8.5). One involves solvation giving a 1:1 metal-to-ligand ratio and the second showing a 1:2 metal-to-ligand ratio. It was argued that stronger complexation would ensue if additional intramolecular forces could be involved by, for example, hydroxy side-chain groups. Hundreds of such compounds were prepared, tested, and reported. Increased antitubercular activity was found, as well as reduced toxicity in the mouse test. (Toxicity means lethal toxicity or serious weight loss in this instance, not optical disorders.) From this series, (+)-ethambutol was selected and developed and the earlier N,N′-alkyl compounds were abandoned. The marked influence of chirality on antimycobacterial activity and the lack of chiral influence on mouse toxicity were quickly discovered in the laboratory. Soon after clinical trials with racemic drug commenced, the optical toxicity in humans became apparent and from then only the optically pure (+)-isomer was marketed.

In some very early papers, Wilkinson discussed the significance of these findings and he pointed out that metal binding capability was not alone sufficient to explain the chiral significance of this work. The metal or its chelate must play some crucial chiral role, perhaps as part of an enzyme, co-factor, or cell wall permease.

The second startling characteristic of ethambutol was its specific antimycobacterial action and ineffectiveness against a wide range of other bacteria, fungi, and viruses. An obvious explanation for this would be if it were involved in some specific aspect of mycobacterial cell wall function, stability, or biosynthesis (38). This was particularly attractive because this mode of action had been proposed for some antitubercular drugs, whereas others penetrated the wall and damaged ribosomal protein brosynthesis (104). Specific inhibition of mycolic acid biosynthesis by isoniazid and the role of β lactams in normal cell wall disruption are both known. Arabinoglycan polymers were also considered targets since they were known to provide the structural framework stabilizing the lipid components. Many of these theories are discussed in Begg's review (22) covering the literature to July 1977.

A crucial but brief report (110) in 1962 from Lederle Laboratories pointed to involvement with natural polyamines such as spermine, spermidine,

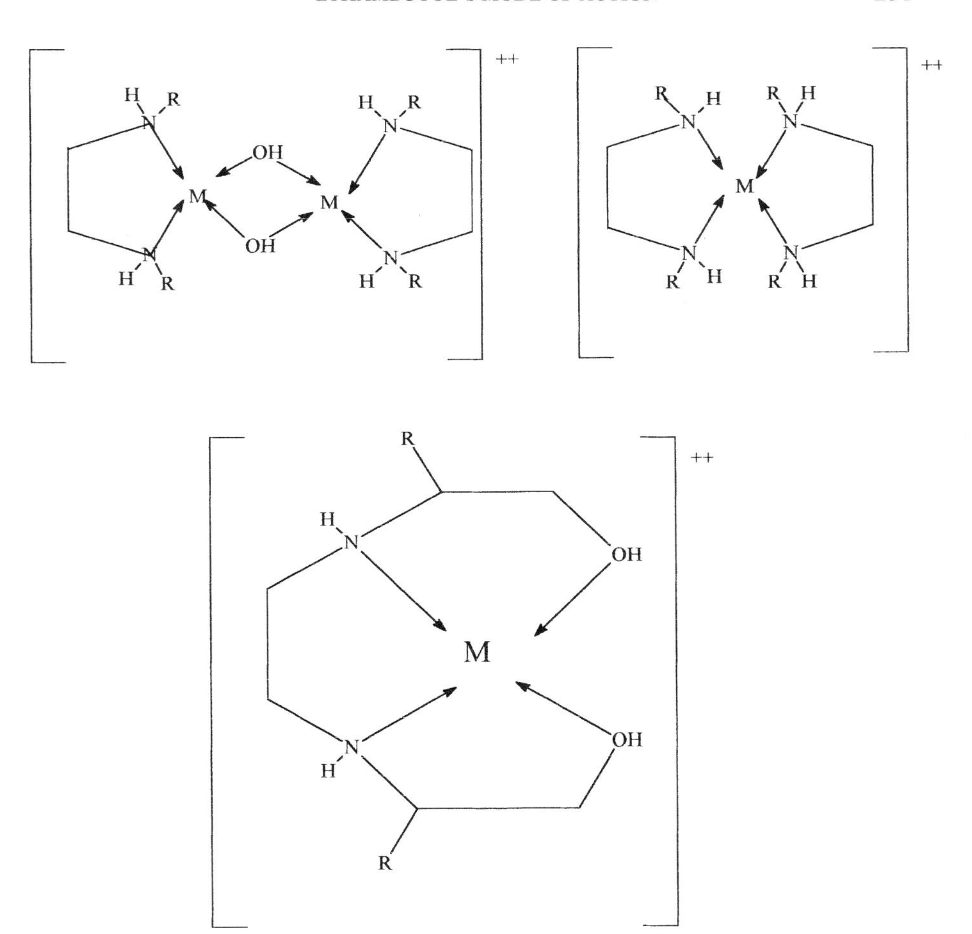

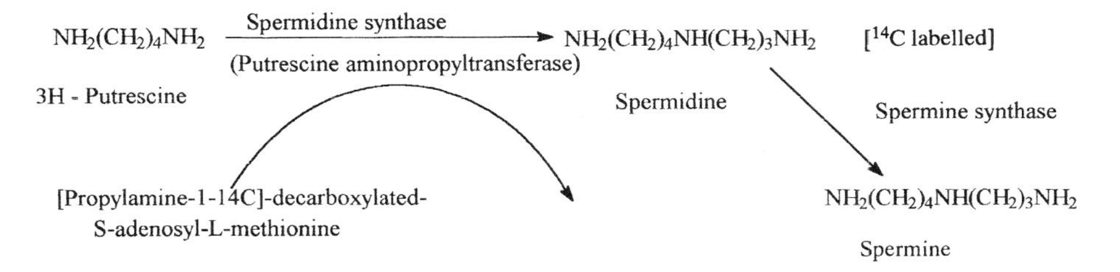

Figure 8.5. Structures relating to the mode of action of (+)ethambutol.

putrescine, and cadaverine. By studying the effect of ethambutol on the inhibition of growth of M. *smegmatis*, it was shown that some 4 h elapsed, almost twice the doubling time, before growth ceased. Furthermore, this growth inhibition could be reversed by the addition of polyamines, particularly spermine, and divalent metal ions. A more extensive paper in 1965 (111) extended this and showed that magnesium ions, but not sodium, calcium, copper, zinc, or iron, had this effect. The polyamines involved were shown to influence protein production, and both RNA and DNA turnover, although to different extents and after different time lapses. However, the differential action of the enantiomers of ethambutol was not examined. Not until 1983 was the specific influence of (+)-ethambutol on mycobacterial polyamine biochemistry probed. A brief note, based on radiochemical techniques (112), indicated that selective blocking of the conversion of putrescine to spermidine was found for spermidine synthase from mycobacterium *bovis*, in marked contrast to (+)-ethambutol's inability to block the conversion of putrescine to spermidine *in vivo* and in cell-free enzyme preparations from Pseudomonas *aeruginosa* and Eschericia *coli*. This paper did not include the crucial experiment to see if the (–)-isomer of ethambutol was devoid of this inhibitory activity. A later paper (113) did show that the effect is stereospecific; no inhibition of spermidine production was shown by (–)-ethambutol. No further references extending this work to M. *tuberculosis* or ethambutol-resistant strains of mycobacteria, or any enzyme characterization studies have appeared.

The significance of polyamines in stabilizing both ribosomes and mycobacterial cell walls was discussed and further general references were given. Direct experimental evidence, for example, chiral chromatography studies, for cell wall binding, the penetration and intracellular localization of ethambutol, particularly during the very early phase of contact would be invaluable. This might improve our understanding of synergistic interactions. Several claims (114–118) have been made from *in vitro* and clinical data that early damage to the mycobacterial wall or permease (porrin) channels through the wall, caused by brief contact with ethambutol, facilitates the penetration of other antibacterials. The lack of sensitive, stereospecific, direct methods of ethambutol analysis has been a serious handicap to the investigation of these hypotheses.

Evidence for ethambutol interference with phospholipids, fatty acids (particularly mycolic acids), glucose, amino acids, and arabinogalactan biosynthesis have been presented (119–131), but none appear to have established directly, by using both enantiomers of ethambutol, that stereospecific factors are involved. Most of these results are based on radiochemical procedures.

Since this manuscript was prepared papers (141, 142, and refs. therein) proposing a specific cell wall inhibition mechanism for ethambutol have appeared. These demonstrate, in Mycobacterium *smegmatis*, inhibition by

(+)ethambutol of polymerization steps in the biosynthesis of the arabinan component of cell wall arabinogalactan (AG) and subsequently the inhibition of the synthesis of the arabinan of lipoarabinomannan (LAM). These works did not include a control study, using (–)ethambutol to show that these inhibitory effects are stereopspecific, so matching the known clinical of action of (+)ethambutol. Collaborative work is now in hand to test and clarify this essential point (143)

Studies (132, 133) directed toward understanding the visual and tingling neuropathy side-effects are difficult to evaluate since no characteristic chiral influence is displayed.

It is, in fact, the very specific features required by this molecule for antibacterial activity that make it so fascinating and a worthy subject for further study using modern analytical and biotechnology methods. Not only are chiral factors crucial, even apparently minor changes in the overall molecular structure by increasing the N–N separation, altering the branching on either the ethylenediamine or N-alkyl substituents, or modifying the N basicity lead to a substantial loss of activity. Many hundreds of structural types were prepared and tested both *in vitro* and *in vivo* to illustrate the point. One interesting observation was that N-methylation but not N-ethylation or larger N-alkylation gave analogs active *in vivo* but not *in vitro*, presumably due to biological N-demethylation but not general N-dealkylation. The extent of these studies probably accounts for the paucity of subsequent analog studies. Cyclic amino alcohols provided products with activity similar to ethambutol but a different spectrum of action against a range of mycobacterium species (134, 135). The synthesis of N-stearyl and N-palmityl prodrugs did not progress to biological testing (136). Earlier studies had shown that N-acyl analog were inactive both *in vitro* and *in vivo* (9).

REFERENCES

1. Ryan, F. (1992). Tuberculosis: The Greatest Story Never Told. Bromsgrove, England: Swift Publishers.
2. Bunyan, J. (1680). *The Life and Death of Mr. Badman.*
3. *TB. a Global Emergency* (1993). *World Health*, No. 4 (July–Aug).
4. Dubos, R. and Dubos, J. (1952). *The White Plague*. New Brunswitch, N.J.: Rutgers Univ. Press.
5. Wilkinson, R. G., Shepherd, R. G., Thomas, J. P., and Baughn, C. (1961). Stereospecificity in a new type of synthetic antituberculous agent. *J. Am. Chem. Soc.*, **83**, 2212.
6. Thomas, J. P., Baughn, R. G., Wilkinson, R. G., and Shepherd R. G. (1961). A

new synthetic compound with antituberculous activity in mice: Ethambutol. *Am. Rev. Resp. Dis.*, **83**, 891.

7. Shepherd, R. G. and Wilkinson, R. G. (1962). Antituberculous agents, II. N,N′-diisopropylethylenediamine and analogs. *J. Med. Pharm. Chem.*, **5**, 823.
8. Wilkinson, R. G., Cantrall, M. B., and Shepherd, R. G. (1962). Antituberculous agents III. (+)-2,2′-(Ethylenediimino)-di-1-butanol and some analogs. *J. Med. Pharm. Chem.*, **5**, 835.
9. Shepherd, R. G., Baughn, C., Cantrall, M. L., Goodstein, B., Thomas, J. P., and Wilkinson, R. G. (1966). Antituberculous agents IV. *Ann. New York Acad. Sci.*, **135**, 686–710.

*10. Cantrall, M. B., Wilkinson, R. G., and Shepherd, R. G. (1966). Antituberculous agents V. Hydroxy and ether analogs of ethambutol. Selective O-alkylation of aminoalcohols. *J. Med. Chem.*, **9**.

*11. Cantrall, M. B., Wilkinson, R.G., and Shepherd, R. G. (1966). Antituberculous agents VI. Varied branching of the alkyl group in the antimycobacterial N,N′-dialkylethylenediamines. *J. Med. Chem.*, **9**.

*12. Goodstein, B. B. and Shepherd, R. G. (1966). Antituberculous agents VII. Modifications of the alkylene chain and of the basic centres of the antimycobacterial ethylenediamines. *J. Med. Chem.*, **9**.

*13. Cantrall, M. B., Wilkinson, R.G., and Shepherd, R. G. (1966). Antituberculous agents VIII. Alkyl-substituted N,N′-di-sec-butyl ethylenediamines. *J. Med. Chem.*, **9**.

*14. Cantrall, M. B., Wilkinson, R.G., Shepherd, R. G., Goodstein, B. B., and Sassiver, M. L. (1966). Antituberculous agents IX. Neighboring group effects in synthesis of ethylenediamines. *J. Med. Chem.*, **9**.

*15. Cantrall, M. B., Wilkinson, R.G., and Shepherd, R. G. (1966). Antituberculous Agents X. Stereospecificity In N,N′-di-sec-butyl ethylenediamines. Absolute configuration of ethambutol. *J. Med. Chem.*, 9.

16. Ringler, I. (1977). Splitting the antipodes. In F. H. Clarke (ed.), *How Modern Medicines are Discovered*, New York: Futura Pub. Co.
17. *British Pharmacopoeia*, Vol. I (1980). London: HMSO.
18. U.S. Pharmacopoeia, revision XXI (1995). Rockville, MD: USP Convention.
19. *European Pharmacopoeia*, 2nd ed. (1987). 553, Maisonneuve SA. St. Ruffine.
20. *British Pharmacopoeia*, Vol. I (1993). London: HMSO.
21. Lee, C. S., and Benet, L. Z. (1978). Ethambutol. In K. Florey (ed.), *Analytical Profiles of Drug Substances*, Vol. 7. New York: Academic Press, pp. 231–249.
22. Beggs, W. H. (1979). Ethambutol. In F. F. Hahn (ed.), *Antibiotics*, V, New York: Springer-Verlag, pp 43–46.
23. Smith, J. L. (1987). Should ethambutol be barred. *J. Clini. Neuro-Ophthalmol.*, **7**, 84–86.

*cited reference in references 7–9, above.

24. Alvarez, K. L. and Krop, L. C. (1993). Ethambutol-induced ocular toxicity revisited. *Annals Pharmacother.*, **27**, 102–103.
25. Schild, H. S. and Fox, B. C. (1991). Rapid-onset reversible ocular toxicity from ethambutol therapy. *Am. J. Med.*, **90**, 404–406.
26. Kumar, A., Sandramoul, S., Verma, L., Tewari, H. K., and Khosla, P. K. (1993). Ocular ethambutol toxicity—is it reversible. *J. Clin. Neuro-Ophthalmol.*, **13**, 15–17.
27. Polak, B. C. P., Leys, M., and Yanlith, G. H. M. (1985). Blue-yellow color-vision changes as early symptoms of ethambutol oculotoxicity. *Ophthalmologica*, **191**, 223–226.
28. Fledelius, H. C., Petrera, J. E., Skjodt, K., and Trojaborg, W. (1987). Ocular ethambutol toxicity—a case-report with electrophysiological considerations and a review of Danish cases 1972–81. *Acta Ophthalmologica*, **65**, 251–255.
29. Harcombe, A., Kinnear, W., Britton, J., and Macfarlane, J. (1991). Ocular toxicity of ethambutol. *Resp. Med.*, **85**, 151–153.
30. Kahana, L. M. (1987). Toxic ocular effects of ethambutol. *Can. Med. Assoc. J.*, **137**, 213–216.
31. Zimmer, B. L., Deyoung, D. R., and Roberts, G. D. (1982). Invitro synergistic activity of ethambutol, isoniazid, kanamycin, rifampin, and streptomycin against mycobacterium-avium-intracellulare complex. Antimicrob. *Agents Chemother.*, **22**, 148–150.
32. Heifets, L. B., (1982). Synergistic effect of rifampin, streptomycin, ethionamide, and ethambutol on mycobacterium-intracellulare. *Am. Rev. Resp. Dis.*, **125**, 43–48.
33. Hoffner, S. E., Kratz, M., Olssonliljequist, B., Svenson, S. B., and Kallenius, G. (1989). *In vitro* synergistic activity between ethambutol and fluorinated quinolones against mycobacterium-avium complex. *J. Antimicrob. Chemother.*, **24**, 317–324.
34. Horsburgh, C. R. (1991). Current concepts–mycobacterium-avium complex infection in the acquired-immunodeficiency-syndrome. *New Engl. J. Med.*, **324**, 1332–1338.
35. Yajko, D. M., Nassos, P. S., Sanders, C. A., and Hadley, W. K. (1991). Effects of antimicrobial agents on survival of mycobacterium-avium complex inside alveolar macrophages obtained from patients with human-immunodeficiency-virus infection. *Antimicrob. Agents Chemother.*, **35**, 1621–1625.
36. Kent, R. J., Bakhtiar, M., and Shanson, D. C. (1992). The *in vitro* bactericidal activities of combinations of antimicrobial agents against clinical isolates of mycobacterium-avium-intracellulare. *J. Antimicrob. Chemother.*, **30**, 643–650.
37. Manchester, K. (1984). Tuberculosis and leprosy in antiquity. *Med. History*, **28**, 162–73.
38. Rastogi, N. (ed.) (1991). 7th Forum: Structure and function of the cell envelope. *Res. Microbiol.*, **142**, 419–481.
39. Rastogi, N. (ed.) (1990). 5th Forum: Killing intracellular mycobacteria. *Res. Microbiol.*, **141**, 191–270.

40. Rastogi, N. (ed.) (1993). 9th Forum: Emergence of multiple-drug resistant tuberculosis. *Res. Micrbiol.*, **144**, 103–159.
41. Streptomycin treatment of pulmonary tuberculosis: A medical research council investigation (1948). *Br. Med. J.*, **II**, 769–782.
42. Combined treatment of tuberculosis with streptomycin and PAS (1949). *Br. Med. J.*, **II**,1521.
43. *British National Formulary* (1993). London: Br. Med. Assoc. and Royal Pharm. Soc. GB.
44. Iseman, M. D. (1985). Tailoring a time bomb. *Am. Rev. Resp. Dis.*, **132**, 735–736.
45. Iseman, M. D. (1993). Treatment of multi-drug-resistant tuberculosis. *New Engl. J. Med.*, 784.
46. Place, V. A., Peets, E. A., Buyske, D. A., and Little, L. R. (1966). Metabolic and special studies of ethambutol in normal volunteers and tuberculous patients. *Annal N.Y. Acad. Sci.*, **135**, 775–795.
47. Place, V. A., Pyle, M. M., and De La Huerga, J. (1969). Ethambutol in tuberculous meningitis. *Am. Rev. Resp. Dis.*, **99**, 783–785.
48. Kelly, R. G., Kaleita, E., and Eisner, H. J. (1981). Tissue distribution of ethambutol-C-14 in mice. *Am. Rev. Resp. Dis.*, **123**, 689–690.
49. Varughese, A., Brater, D. C., Benet, L. Z., and Lee, C. S. C., (1986). Ethambutol kinetics in patients with impaired renal-function. *Am. Rev. Resp. Dis.*, **134**, 34–38.
50. Liss, R. H., Letourneau, R. J., and Schepis, J. P. (1981). Distribution of ethambutol in primate tissues and cells. *Am. Rev. Resp. Dis.*, **123**, 529–532.
51. Thomas, J. P. and Durr, F. E. (1983). Ethambutol dose-plasma level correlation studies in guinea-pigs. *Am. Rev. Resp. Dis.*, **127**, 352–353.
52. Lee, C. S. E. and Varughese, A. (1984). A disposition kinetics of ethambutol in nephrectomized dogs. *J. Pharmaceut. Sci.*, **73**, 787–789.
53. Chen, M. M., Lee, C. S., and Perrin, J. H. (1984). Absorption and disposition of ethambutol in rabbits. *J. Pharmaceut. Sci.*, **73**, 1053–1055.
54. Varughese, A., Brater, D. C., Benet, L. Z., and Lee, C. (1984). Pharmacokinetics of ethambutol in renal-failure. *J. Clin. Pharmacol.*, **24**, 410.
55. Sokolova, G. B., Ziya, A. V., Abramovich, A. G., Bessarabova, T. N., and Ivleva, A. Y. (1986). Pharmacokinetics of ethambutol in patients with pulmonary tuberculosis. *Farmakologiya I Toksikologiya*, **49**, 40–42.
56. McDougall, J. I. (1991). Lederle Laboratories, Cyanamid GB, Gosport (personal communication).
57. Ohya, K., Shintani, S., and Sano, M. (1980). Determination of ethambutol in plasma using selected ion monitoring. *J. Chromatog.*, **221**, 293–299.
58. Holdiness, M. R., Israili, Z. H., and Justice, J. B. (1981). Gas chromatographic-mass spectrometric determination of ethambutol in human-plasma. *J. Chromatog.*, **224**, 415–422.
59. Billian, J. H., and Diesing, A. C. (1957). *J. Org. Chem.*, **22**, 1068.

60. Bernadi, L., Foglio, M., and Temperilli, A. (1974). *J. Med. Chem.*, **17**, 555.
61. US 3 953 513 (1976). Gruppo Lepetit, Ert 27.4.
62. DE 2 263 715 (1972). Soc. Farma. Italia, Anm. 28. 12.
63. Ye-Sook, K., Jeong-Rok, Y., Man-Ki, P., and Nam-Ho, P. (1981). Determination of the optical isomers of ethambutol (myambutol) and 2-amino-1-butanol by GLC. *Arch. Pharm. Res.*, **4**, 1–8.
64. BE-P 862 627 (1978). American Cyanamid, Anm. 4.1.
65. DE 2 547 654 (1975). Basf, Anm. 24.10.
66. US 3 553 257 (1971). American Cyanamid, Ert 5.1.
67. McDougall, J. I. (1994). Lederle Laboratories, Cyanamid GB, Gosport (personal communication).
68. DE 2 446 320 (1974). Denki Kagaku Kogyo, Anm. 27.9.
69. GB 1 471 838 (1975). Nippon Soda, Anm. 26.3.
70. Holdiness, M. R. (1986). Contact-dermatitis to ethambutol. *Contact Derm.*, **15**, 96–97.
71. Kerremans, A., Majoor, C. L. H., and Gribnau, F. W. J. (1981). Hypersensitivity to ethambutol. *Tubercle*, **62**, 215–217.
72. Prasad, R. and Mukerji, P. K. (1989). Ethambutol induced thrombocytopaenia. *Tubercle*, **70**, 211–212.
73. Tabak, D., (1988). Determination of ethambutol hydrochloride in air by using fluoropore filter sampling and a derivatisation GC procedure. *Am. Indus. Hyg. Assoc. J.*, 49, 620–623.
74. Shingbal, D. M. and Naik, S. D. (1982). Colorimetric determination of ethambutol hydrochloride. *J. Assoc. Off. Analy. Chemists*, **65**, 899–900.
75. Ng, T. L. (1982). An investigation into the various factors influencing the pharmacopoeial assay of ethambutol hydrochloride tablets. *Analyst*, **107**, 695–700.
76. Hassan, S. S. M. and Shalaby, A. (1992). Determination of ethambutol in pharmaceutical preparations by atomic-absorption spectrometry. *Spectrophotom. Potentiom. Mikrochimica Acta*, **109**, 193–199.
77. Mbay, M. T. and Bosly, J. (1981). Periodate-oxidation of ethambutol. *Journal de Pharmacie de Belgique*, **36**, 253–254.
78. Lacroix, C., Cerutti, F., Nouveau, J., Menager, S., and Lafont, O. (1987). Determination of ethambutol in plasma by liquid-chromatography and ultraviolet spectrophotometric detection. *J. Chromatog.-Biomed. Appl.*, **415**, 85–94
79. Causse, J. E., Pasqualini, R., Cypriani, B., Weil, R., Vandervalk, R., Bally, P., Dupuy, A., Couret, I., Benbarek, M., and Descomps, B., (1990). Labelling of ethambutol with TC-99M using a new reduction procedure—pharmacokinetic study in the mouse and rat. Applied radiation and isotopes. *Internat. J. Rad. Appl. Intrument., Part A*, **41**, 493–496.
80. Nilsson, L. E., Ansehn, S., and Hoffner, S. E. (1988). Rapid susceptibility testing of mycobacterium tuberculosis by bioluminescence assay of mycobacterial ATP. *Antimicrob. Agents Chemother.*, **32**, 1208–1212.

81. Hoel, T. and Eng, J. (1991). Radiometric and conventional drug susceptibility testing of mycobacterium-tuberculosis. *Apmis*, 99, 977–980.
82. Brisson-Noel, A. et al. (1991). Diagnosis of tuberculosis by DNA amplification in clinical practise. *Lancet*, **338**, 364–366.
83. Ng, T. L. (1982). A gas-chromatographic mass-spectrometric study of the trimethylsilylation of ethambutol and a tablet assay-method based on the trimethylsilyl derivative. *J. Chromatog. Sci.*, **20**, 479–482.
84. Ye-Sook, K., Jeong-Rok, Y., Man-Ki, P., Nam-Ho, P. (1981). Determination of the optical isomers of ethambutol (myambutol) and 2-amino-1-butanol by GLC. *Arch. Pharm. Res.*, **4**, 1–8.
85. Lee, C. S. and Benet, L. Z. (1976). GLC determination of ethambutol in plasma and urine of man and monkey. *J. Chromatog*, **128**, 188–192.
86. Bennet, R. M., Manno, J. E., and Manno, B. R. (1974). GC determination of ethambutol. *J. Chromatog.*, **89**, 80–83.
87. Lee, C. S. and Bennet, L. Z. (1987). Micro and macro GLC determination of ethambutol in biological fluids. *J. Pharm. Sci.*, **67**, 470–473.
88. Bessarabova, T. N., Linberg, L. F., Popov, S. A., and Mirimskii, A. S. (1987). The method of determination of ethambutol in blood serum. *Otkrytiya, Izobretaniya*, **32**, 175.
89. Ohya, K., Shintani, S., and Sano, M. (1980). Determination of ethambutol in plasma using selected ion monitoring. *J. Chromatog.*, **221**, 293–299.
90. Lee, C. S. and Wang, L. H. (1980). Improved GLC determination of ethambutol. *J. Pharm. Sci.*, **69**, 362–363.
91. Holdiness, M. R., Israili, Z. H., and Justice, J. B. (1981). Gas chromatographic-mass spectrometric determination of ethambutol in human-plasma. *J. Chromatog.*, **224**, 415–422.
92. Sen, A. K., Bandyopadhyay, A., Podder, G., and Chowdhury, B. (1990). Reversed-phase high-performance liquid-chromatographic determination of ibuprofen and ethambutol in pharmaceutical dosage form. *J. Ind. Chem. Soc.*, **67**, 443–444.
93. Blessington, B. and Beiraghi, A. (1990). Study of the stereochemistry of ethambutol using chiral liquid-chromatography and synthesis. *J. Chromatog.*, **522**, 195–203.
94. Belov, Y. (1990). Inst. Physiologically Active Subst., Chernogolovka, Moscow, Russia (personal communication).
95. Gubitz, G. (1993). Inst. fur Pharmazeutische Chemie der Karl-Franzens Univ., Graz, Austria (personal communication).
96. Kitigawa, T., Oie, S., and Taniyama, H. (1976). Analysis of d-ethambutol by circular dichroism of its copper chelates. *Chem. Pharm. Bull.*, **24**, 3019–3024.
97. Gambertini, G. and Ferioli, V. (1988). Determination of optical purity by HPLC of compounds of pharmaceutical interest. *Farmaco Ed. Prat.*, **43**:11, 357–363.
98. Blackwood, J. E, Blower, P. E., Layten, S. W., Lillie, D. H., Lipkus, A. H., Peer, J. P., Qian, C., Staggenborg, L. M., and Watson, C. E. (1991). Chemical abstracts

service chemical registry system 13. Enhanced handling of stereochemistry. *J. Chem. Inf. Comput. Sci.*, **31**, 205–212.

99. Godfrey, R., Hargreaves, R., and Hitchcock, P. B. (1992). Absolute configuration of (+) ethambutol hydrobromide. *Acta. Crystallog., Section C*, **48**, 79–81.
100. Sorensen, A. M. and Simonsen, O. (1989). (R,S)-2,2′-(1,2-ethanediyldi-imino)-bis(1-butanol)dihydrochloride 45. *Acta Cryst., C* (cr. str. comm.), **45**, 506.
101. Hamalainen, M., Lehtinen, M., and Ahlgren, M. (1985). (S,S)-ethambutol dihydrochloride. *Arch. Pharm.*, **318**, 26.
102. Harada, N. and Nakanishi, K. (1983). *Circular Dichroic Spectroscopy*. London: Oxford Univ. Press.
103. Blessington, B. and Lo, T. W. (1993). A failure of exciton coupled circular dichroism to predict absolute stereochemistry. A preliminary computer modelling explanation. 4th Int. Symposium on Chiral Discrimination, Montreal.
104. Coulson, C. J. (1994). *Molecular Mechanisms of Drug Action*. London: Taylor and Francis.
105. King, A. and Schwartz, R. (1985). Effects of a chelating drug, ethambutol, on zinc-absorption and tissue distribution in zinc-adequate and zinc-marginal rats. *Fed. Proc.*, **44**, 542.
106. Bogden, J. D., Zadzielski, E., Alrabiai, S., and Aviv, A. (1983). Effect of chelating drugs on metabolism of essential metals. 1. Ethambutol. *Fed. Proc.*, **42**, 827.
107. King, A. B. and Schwartz, R. (1987). Effects of the antituberculous drug ethambutol on zinc-absorption, turnover and distribution in rats fed diets marginal and adequate in zinc. *J. Nutr.*, **117**, 704–708.
108. Bhattacharyya, R. G., Paul, U. K., Chatterjee, A. B., and Bag, S. P. (1990). Interaction of ethambutol with transition-metal ions in solution-formation-constants and stereochemical configurations of the Cu(II), Ni(II), Co(II) and Zn(II) complexes and underlying biological implications. *Ind. J. Chem., Sec. A: Inorg. Bio.-Inorg. Phys.*, **29**, 986–995.
109. King, A. and Schwartz, R. (1986). Possible potentiation of ethambutol effects on serum and urine Ca, Mg, and Zn by induction of drug-metabolism. *Fed. Proc.*, **45**, 1097.
110. Forbes, M., Kuck, N. A., and Peets, E. A. (1962). Mode of action of ethambutol. *J. Bacteriol.*, **84**, 1099–1103.
111. Forbes, M., Kuck, N. A., and Peets, E. A. (1965). Effect of ethambutol on nucleic acid metabolism in mycobacterium smegmatis and its reversal by polyamines and divalent cations. *J. Bacteriol.*, **89**, 1299–1305.
112. Poso, H., Paulin, L., and Brander, E. (1983). Specific-inhibition of spermidine synthase from mycobacteria by ethambutol. *Lancet*, **2**, 1418.
113. Paulin, L. G., Brander, E. E., and Poso, H. J. (1985). Specific-inhibition of spermidine synthesis in mycobacteria spp by the dextro isomer of ethambutol. Antimicrob. Agents Chemother., **28**, 157–159.
114. Hoffner, S. E., Kratz, M., Olssonliljequist, B., Svenson, S. B., Kallenius, G. (1989). *In vitro* synergistic activity between ethambutol and fluorinated

quinolones against mycobacterium-avium complex. *J. Antimicrob. Chemother.*, **24**, 317–324.

115. Rastogi, N., Labrousse, V., and Desousa, J. P. C. (1993). Ethambutol potentiates extracellular and intracellular activities of clarithromycin, sparfloxacin, amikacin, and rifampin against mycobacterium-avium. *Current Microbiol.*, **26**, 191–197.
116. Yates, M. D. and Collins, C. H. (1981). Sensitivity of opportunist mycobacteria to rifampicin and ethambutol. *Tubercle*, **62**, 117–121.
117. Sareen, M. and Khuller, G. K. (1990). Cell-wall and membrane-changes associated with ethambutol resistance in mycobacterium-tuberculosis H37RA. *Antimicrob. Agents Chemother.*, **34**, 1773–1776.
118. Cheema, S., Asotra, S., and Khuller, G. K. (1985). Ethambutol induced leakage of phospholipids in mycobacterium-smegmatis. *IRCS Med. Sci.-Biochem.*, **13**, 843–844.
119. Takayama, K. and Datta, A. K. (1991). Structure-to-function relationship of mycobacterial cell-envelope components. *Res. Microbiol.*, **142**, 443–448.
120. Kilburn, J. O. and Takayama, K. (1981). Effects of ethambutol on accumulation and secretion of trehalose mycolates and free mycolic acid in mycobacterium-smegmatis. *Antimicrob. Agents Chemother.*, **20**, 401–404.
121. Kilburn, J. O., Takayama, K., Armstrong, E. L., and Greenberg, J. (1981). Effects of ethambutol on phospholipid-metabolism in mycobacterium-smegmatis. *Antimicrob. Agents Chemother.*, **19**, 346–348.
122. Cheema, S., Asotra, S., and Khuller, G. K. (1986). Correlation of amino-acid uptake and susceptibility to ethambutol (EMB) with phospholipid-composition of EMB-sensitive and EMB-resistant strains of mycobacterium-smegmatis. *Ind. J. Exp. Biol.*, **24**, 705–709.
123. Sareen, M. and Khuller, G. K. (1990). Effect of ethambutol on the phospholipids of ethambutol susceptible and resistant strains of mycobacterium-smegmatis ATCC-607. *Ind. J. Biochem. Biophys.*, **27**, 39–42.
124. Cheema. S. and Khuller, G. K. (1985). Phospholipid-composition and ethambutol sensitivity of mycobacterium-smegmatis ATCC-607. *Ind. J. Exp. Bio.*, **23**, 511–513.
125. Cheema. S. and Khuller, G. K. (1985). Metabolism of phospholipids in mycobacterium-smegmatis ATCC 607 in the presence of ethambutol. *Ind. J. Med. Res.*, **82**, 207–213.
126. Sareen, M. and Khuller, G. K. (1988). Phospholipids of ethambutol-susceptible and resistant strains of mycobacterium-smegmatis. *J. Biosci.*, **13**, 243–248.
127. Cheema, S., Asotra, S., and Khuller, G. K. (1987). Effect of exogenous fatty-acids on ethambutol susceptibility of sensitive and resistant mycobacterium-smegmatis ATCC-607. *Ind. J. Exp. Biol.*, **25**, 230–232.
128. Silve, G., Valeroguillen, P., Quemard, A., Dupont, M. A., Daffe, M., and Laneelle, G., (1993). Ethambutol inhibition of glucose-metabolism in mycobacteria—a possible target of the drug. *Antimicrob. Agents Chemother.*, **37**, 1536–1538.

129. Takayama, K. and Kilburn, J. O. (1989). Inhibition of synthesis of arabinogalactan by ethambutol in mycobacterium-smegmatis. *Antimicrob. Agents Chemother.*, **33**, 1493–1499.

130. Sareen, M. and Khuller, G. K. (1990). Cell-wall composition of ethambutol susceptible and resistant strains of mycobacterium-smegmatis ATCC-607. *Lett. Appl. Microbiol.*, **11**, 7–10.

131. Hoffner, S. E. and Svenson, S. B. (1991). Studies on the role of the mycobacterial cell-envelope in the multiple-drug resistance of atypical mycobacteria. *Res. Microbiol.*, **142**, 448–451.

132. Gigon, S. and Dehaller, R. (1983). Retinal damage caused by ethambutol and serum levels of zinc and vitamin A. *Klinische Monatsblatter fur Augenheilkunde*, **182**, 469–473.

133. Yiannikas, C. and Walsh, J. C. (1982). The use of visual evoked-responses in the detection of subclinical optic neuritis secondary to ethambutol. *Neurology*, **32**, 4.

134. Cremieux, A., Baghdadi, N., Berthelot, P., and Debaert, M. (1983). Antimycobacterial effects of some analogs of ethambutol. *Annales de Microbiologie*, **2**, 177–182.

135. Berthelot, P., Debaert, M., Cremieux, A., and Baghadi, N. (1983). Cyclic analogs of ethambutol active against mycobacteria. *Farmaco-Edizione Scientifica*, **38**, 73–80.

136. Mazumdar, U. K. and Dey, D. C. (1985). Preparation and evaluation of ethambutol derivatives. *Ind. J. Pharm. Sci.*, **47**, 179–180.

137. Klyne, W. and Buckingham, J. (1974). *Atlas of Stereochemistry*. London: Chapman and Hall.

138. Moffat, A. C. (e.d.) (1986). *Clarke's Isolation and Identification of Drugs*, 2nd ed. London: Pharmaceutical Press.

139. Blessington, B, Beiraghi, A., Lo, T. W., Drake, A., and Jonas, G. (1992). Chiral HPLC-CD studies of the antituberculosis drug (+) ethambutol. *Chirality*, **4**, 227–229.

140. Blessington, B. and Beiraghi, A. (1991). A Method for the quantitative enantioselective HPLC analysis of ethambutol and its stereoisomers. *Chirality*, **3**, 139–144.

141. Deng., L. Y., Mikusova, K., Robuck, K. G., Scherman, M., Brennan, P. J., and McNeil, M. R. (1995). Recognition of multiple effects of ethambutol on metabolism of Mycobacterial cell envelope. *Antimicrobial Agents and Chemotherapy.*, **39**(3), 694–701.

142. Mikusova, K., Slayden, R. A., Besra, G. S., and Brennan, P. J. (1995). Biogenesis of the Mycobacterial cell wall and the site of action of ethambutol. *Antimicrobial agents and chemotherapy*. **39**(11), 2484–2489.

143. Brennan, P. J., Colorado State University, College of Veterinary Medicine and biomedical Sciences, Department of Microbiology, Fort Collins, Colorado, Personal communication, Jan 1996.

CHAPTER

9

STEREOGENIC ELEMENTS OF PHARMACEUTICAL COMPOUNDS: SOME ASPECTS ON ISOMERISM, RESOLUTION, AND STEREOCHEMICAL INTEGRITY

STIG G. ALLENMARK

Department of Organic Chemistry
University of Göteborg
S-41296 Göteborg, Sweden

9.1. INTRODUCTION

The development of drug therapy has accelerated during the last few decades. Whereas previously drugs were synthesized on a rather vague rational basis, today they are often designed from detailed information on their target molecules that may be receptor proteins, enzymes, etc. A rational drug design relies heavily on modern tools like computerized molecular modeling techniques, multidimensional NMR methods for the structural analysis of large molecules, X-ray crystallography, and so on. The fundamental role of stereochemistry in molecular interactions involving biological macromolecules has become quite obvious. However, stereochemistry does not simply deal with static atom configurations in space, but covers also the dynamic behavior of molecules, that is, the interconversion of various atom configurations as they take place in flexible molecules as a result of a variety of elementary processes.

The fate of a drug in a biological system is often dependent on its participation in a large number of chemical equilibria and reaction processes. Since the compound is constantly present in a chiral environment, it is not surprising that if it is composed of two enantiomeric forms, these may exhibit different pharmacokinetic behavior and result in unequal pharmacological effects. In fact, it has been shown that enantioselectivity in pharmacokinetics may arise from differences in one or more of the processes of absorption, distribution, metabolism, and excretion (1). Since the issue of drug chirality is still associated with some confusion and controversy, though, it is the purpose of

The Impact of Stereochemistry on Drug Development and Use, Edited by Hassan Y. Aboul-Enein and Irving W. Wainer. Chemical Analysis Series, Vol. 142.
ISBN 0-471-59644-2

this chapter to review some fundamental aspects of chirality and stereochemical behavior and, it is hoped, provide some better insight into the particular problems linked to chiral drugs.

9.2. CHIRALITY AND OPTICAL ISOMERISM: A BRIEF OVERVIEW

By definition, a chiral object is one that has a nonidentical mirror image. This is equivalent to saying that it is lacking inverse symmetry elements. It is customary to distinguish between asymmetric molecules that lack any symmetry element (except for the trivial C_1 axis) and dissymmetric molecules that do not possess any S_n symmetry elements. This means dissymmetric molecules may possess a C_n axis ($n > 1$, e.g., often a C_2 axis).

It follows, of course, from these general definitions that the by far most common source of molecular chirality, i.e., that of an asymmetric carbon atom as a stereogenic center, represents only one special case. Any molecular arrangement of atoms in space that lacks inverse symmetry elements is chiral and gives rise to two enantiomeric forms. The possible separation and isolation of these forms are therefore, in principle, only related to their stability under the conditions present. If their barrier to interconversion is too low, they will not be possible to obtain. Thus, if a tertiary amine, bearing three different substituents, is considered, it is a chiral molecule existing as a pair of enantiomers due to its pyramidal configuration (**I**, Figure 9.1). It is not resolvable at room temperature, however, due to its rapid inversion around

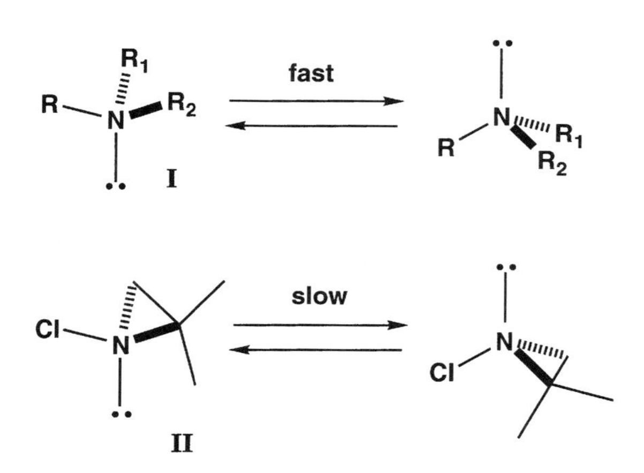

Figure 9.1. Illustration of the generation of enantiomer stability in an aziridine due to the increased inversion barrier created by an unfavorable geometry in the transition state.

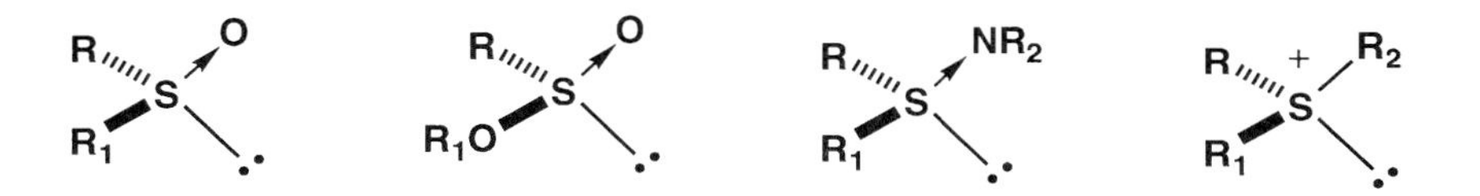

Figure 9.2. Examples of tricoordinated sulfur compounds with stable pyramidal configuration.

the nitrogen atom. If, on the other hand, the nitrogen atom is incorporated into a ring system (**II**, Figure 9.1), which prevents the formation of normal 120° angles in the equatorial plane of the transition state of the inversion process, the enantiomers will be sufficiently stable to possibly isolate (2).

It is interesting that on going from nitrogen to its corresponding second-row element, phosphorus, the tertiary phosphines are much more stable to inversion and therefore resolvable. Similarly, quite a number of tricoordinated sulfur compounds (such as sulfoxides, sulfinates, sulfimides, and sulfonium salts) are configurationally stable (Figure 9.2) and a large number of them have been prepared in optically active form.

In many cases, the chirality of a molecule cannot be ascribed to the presence of an asymmetric atom, however, but rather to a molecular dissymmetry created by the particular molecular geometry present. Classical examples of chiral structures of this type are given in Figure 9.3. A chiral molecule will

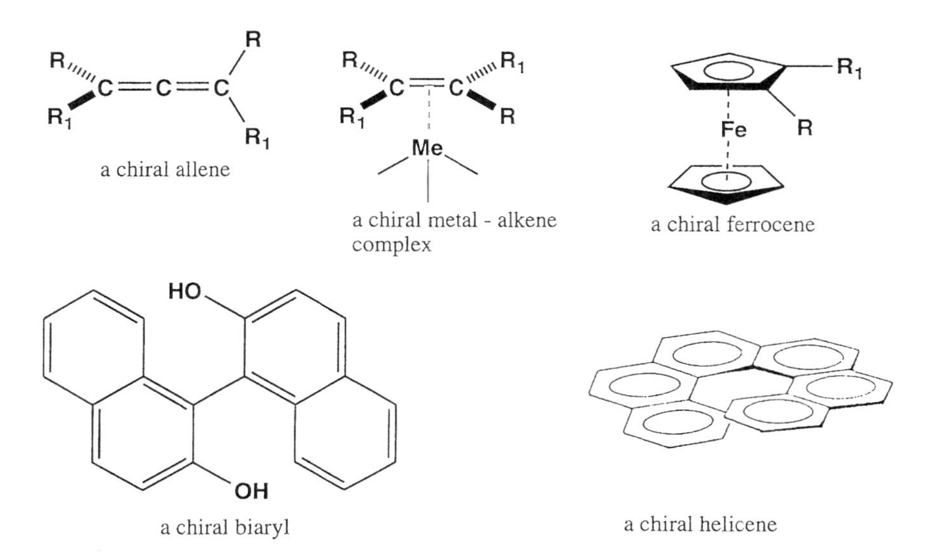

Figure 9.3. Various types of molecular dissymmetry causing chirality.

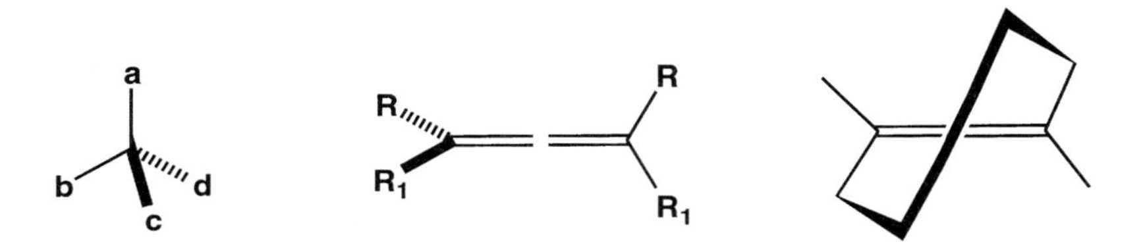

Figure 9.4. The meaning of central, axial, and planar chirality.

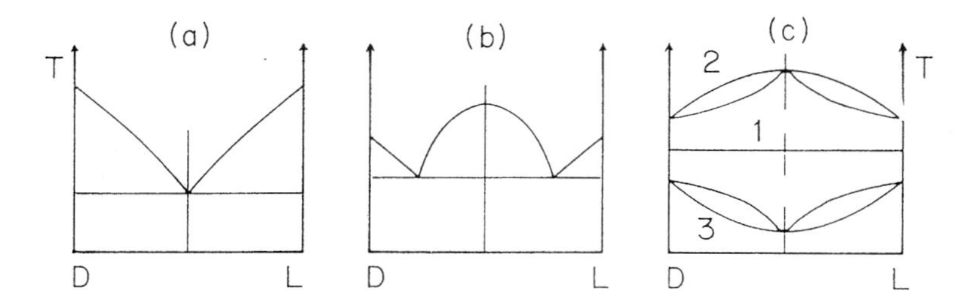

Figure 9.5. Binary phase diagrams of three different types of crystalline racemates. (Reprinted from Jacques, J., Collet, A., and Wilen, S., *Enantiomers, Racemates and Resolutions*, New York: Wiley-Interscience.)

always belong to one of three different categories in which the three-dimensional space is occupied asymmetrically about a chiral (stereogenic) center, a chiral axis, or a chiral plane, respectively (Figure 9.4).

The two enantiomers of a racemate can behave in different ways during crystallization of a racemate, a phenomenon best described by means of binary phase (melting point) diagrams. The three fundamental types of phase diagrams that can be found in mixtures of enantiomers are given in Figure 9.5. The most common situation is that of Figure 9.5*b*, where the racemate crystallizes as a racemic compound. This means that the unit cell of a crystal is made up from the two antipodes in equal amounts, forming the crystal lattice. In Figure 9.5*a*, however, the enantiomers form a eutectic point at 50% of each antipode and the most energetically favored crystal lattice is formed from one and the same antipode. In this case, the racemate crystallizes as a conglomerate, that is, a mixture of crystals, each composed of molecules of identical configuration. This is therefore a case of spontaneous resolution. In Figure 9.5*c*, the enantiomers form a solid solution, yielding a so-called pseudoracemate (3). Three different types have been identified: the ideal solid solution (1) and solid solutions with a maximum (2) or minimum (3). It is important to be aware of the fact that organic compounds often exhibit polymorphism, that is, they are able to crystallize in several forms, and consequently interconversion of the different types of racemates described above may occur.

9.3. OPTICAL ISOMERISM IN COMPOUNDS USED AS PHARMACEUTICALS

9.3.1. Carbon as Stereogenic Center

Numerous drugs of this type have been synthesized as racemates and also used as such. Some of the more important structures used are given in Table 9.1.

Table 9.1. Some Fundamental Structure Types of Pharmaceuticals Incorporating a Stereogenic Carbon Center

Class of Compounds	Basic structure
2-Aminoalkanols	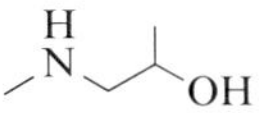
Benzodiazepinones	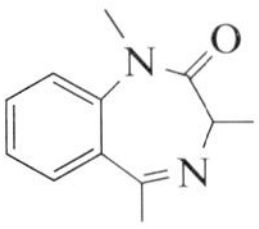
Benzothiadiazines and related structures	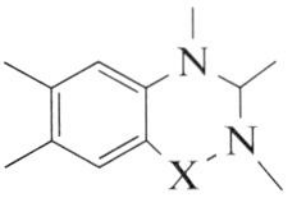 $X = SO_2$ benzothiadiazines X = CO tetrahydroquinazolinones
1,4-Dihydropyridines	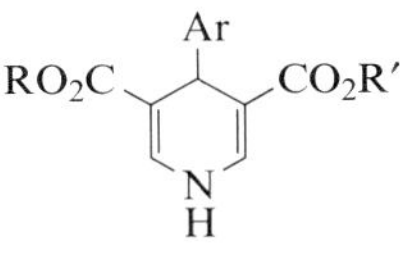

9.3.2. Heteroatom Stereogenic Centers

Only a few groups of drugs belong to this category. A series of compounds with anticancer potency containing an oxazaphosphorine skeleton (**III**) (Figure 9.6) and having a phosphorus atom as the stereogenic center are of interest, since **IIIa** is the cytostatic drug "ifosfamide" ("holoxan"), actually a kind of nitrogen mustard.

The analgesic sulindac (Z)-5-fluoro-2-methyl-1-[p-(methylsulfinyl)-benzylidene]-indene-3-acetic acid (**IV**) is a sulfoxide and the asymmetric sulfur atom gives rise to a pair of enantiomers. Similarly, the gastric acid secretion

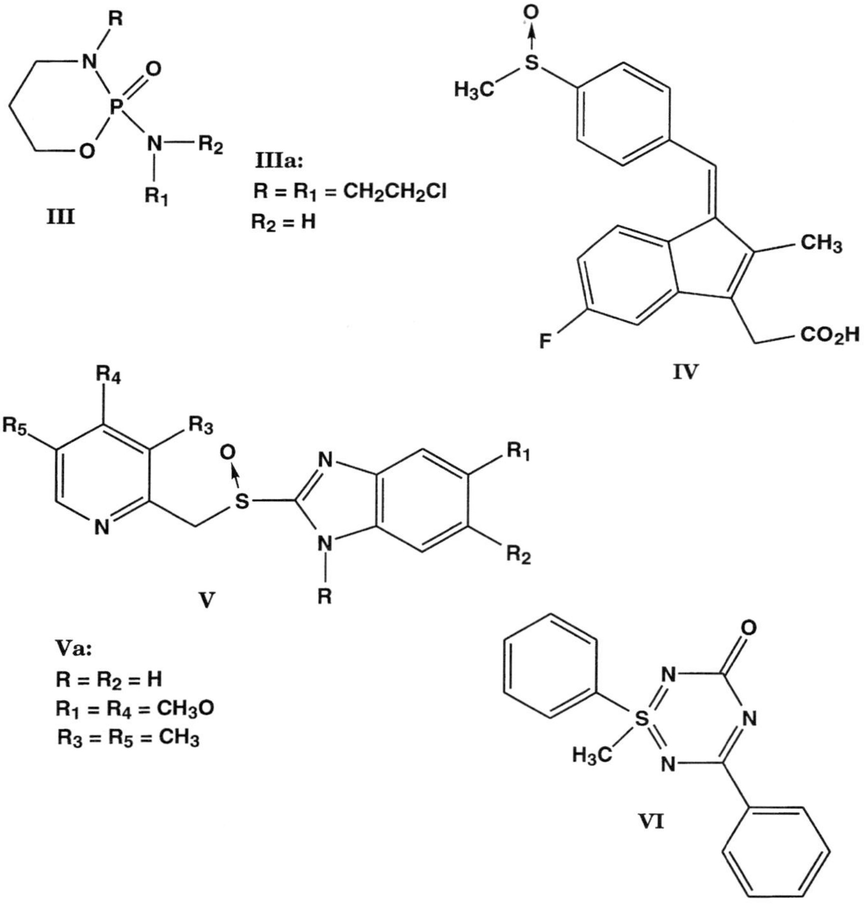

III

IIIa:
$R = R_1 = CH_2CH_2Cl$
$R_2 = H$

IV

V

Va:
$R = R_2 = H$
$R_1 = R_4 = CH_3O$
$R_3 = R_5 = CH_3$

VI

Figure 9.6. Some chiral drugs with heteroatom stereogenic centers.

inhibitors of the structure **V**, notably omeprazole (Losec) (**Va**), are chiral due to the presence of a sulfoxide group. Another drug with sulfur as a stereogenic center is **VI**, a cyclic sulfodiimide. Compounds of this type possess spasmolytic activity (4).

9.3.3. Molecular Dissymmetry

The few examples of drugs of this type are all chiral due to steric barriers of interconversion, that is, they show atropisomerism. Stoll and coworkers were the first to show (5) that dibenzo[a,d]cycloheptene derivatives of type **VII** (Figure 9.7) were resolvable into enantiomers due to atropisomerism. The seven-membered ring in **VII** is nonplanar, but ring-inversion is restricted by

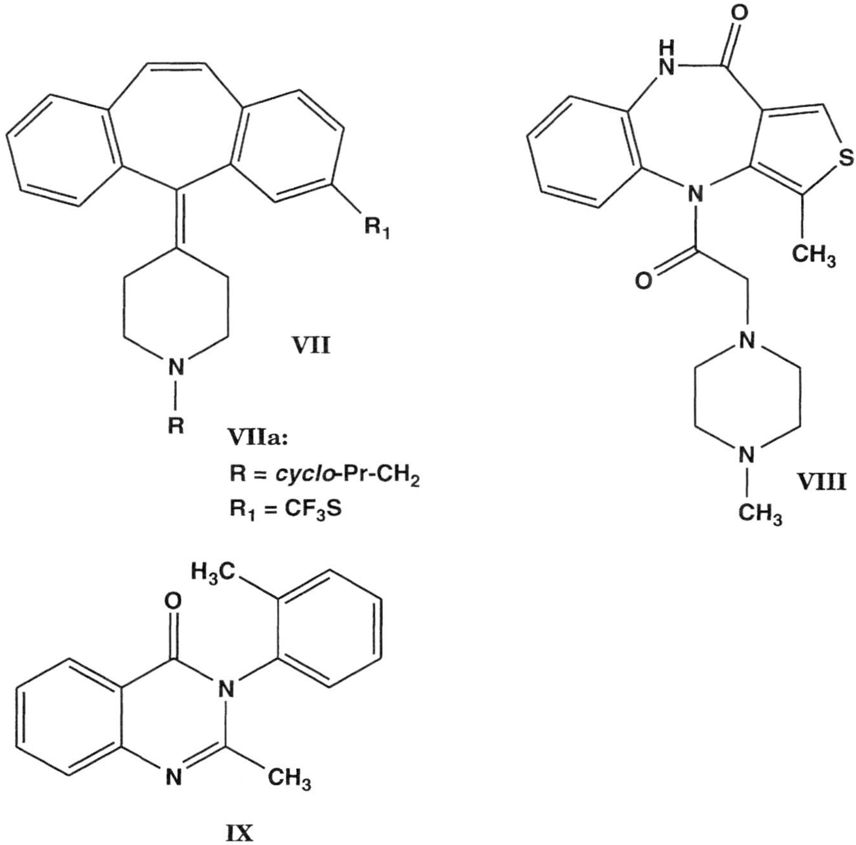

VII

VIIa:
R = *cyclo*-Pr-CH_2
R_1 = CF_3S

VIII

IX

Figure 9.7. Chiral drug structures resulting from molecular dissymmetry.

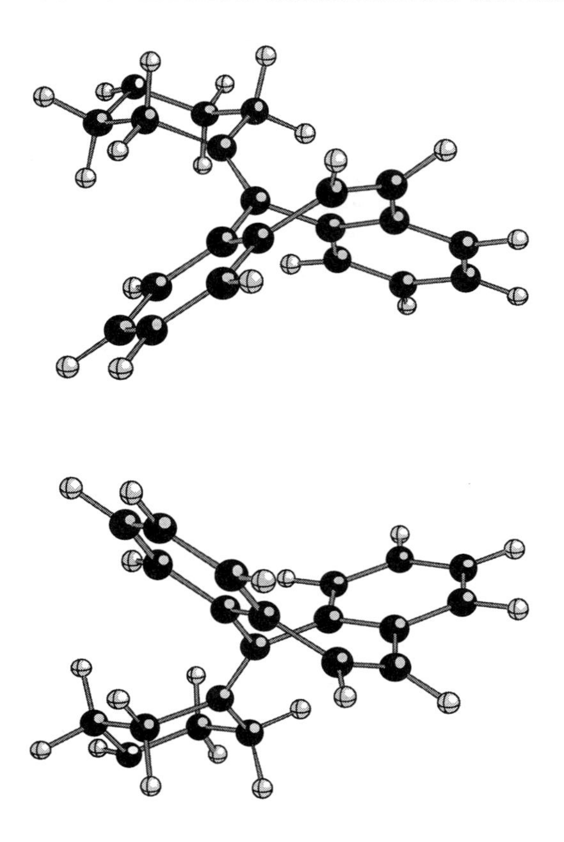

Figure 9.8. The two noninterconvertible conformations of **VII** (here, $R = R_1 = H$). For $R_1 \neq H$, these correspond to stable enantiomeric forms.

interaction of the protons in the aromatic ring positions 4 and 6 with the closest piperidine ring protons. Compounds of this category (like **VIIa**) were later found to be active as neurolepics (6), with specific antidopaminergic and anticholinergic activities in the respective enantiomers (7).

A calculation of the conformational states of the ring system **VII** using a PM3 semiempirical method, gives the result shown in Figure 9.8. The calculations also yield a high potential energy barrier for interconversion.

A similar situation is present in telenzepine (**VIII**) for which compound a potential energy barrier of 35 $kcal mol^{-1}$ has been calculated (8).

Another example of an atropisomeric structure of a drug is found in methaqualone (2-methyl-3-(o-tolyl)-4[3H]quinazoline) (**IX**). This represents a case of classical atropisomerism, where rotation of the tolyl group about the C—N bond is sterically hindered. Since the barrier to internal rotation is high

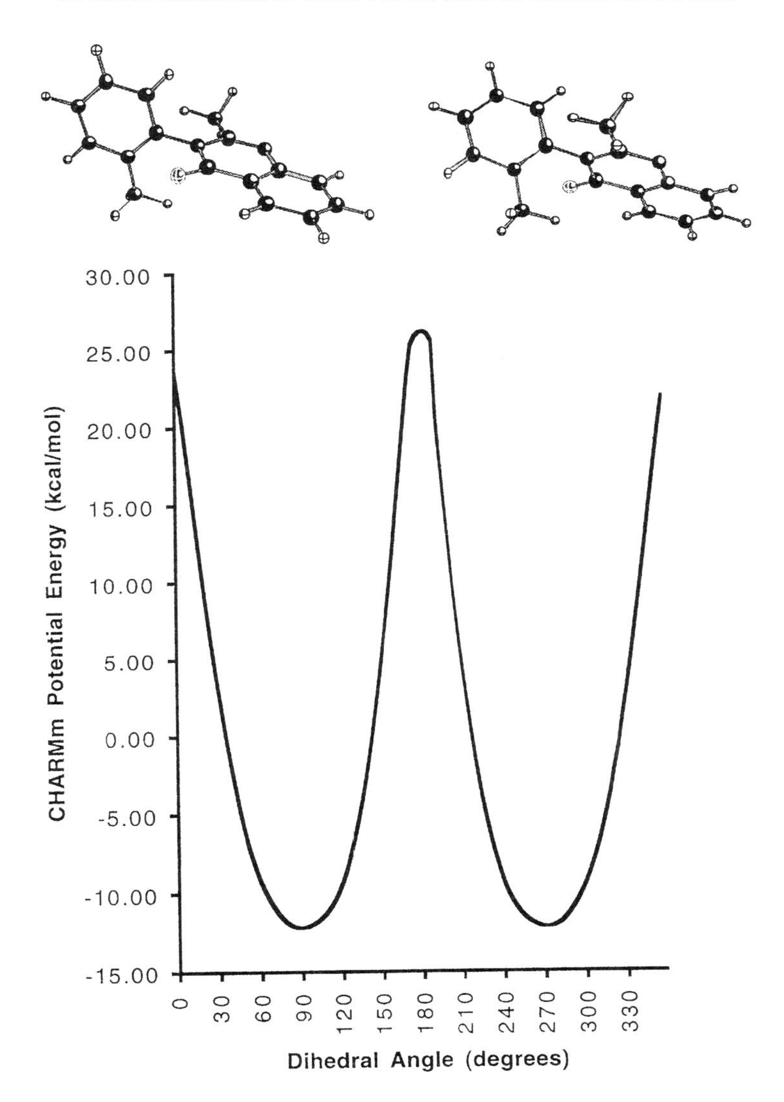

Figure 9.9. Potential energy minima corresponding to the two atropisomeric forms of methaqualone (**IX**).

even at room temperature, the compound exists in two stable enantiomeric forms. Figure 9.9 shows the potential energy as a function of the dihedral angle upon rotation about the C—N bond in methaqualone, as calculated by a CHARMm molecular mechanics program. The two enantiomers correspond to the minima at 90° and 270°, respectively. The lowest barrier to interconversion was estimated to be around 38 $kcalmol^{-1}$, which is more than sufficient for complete stability.

9.4. ENANTIOMERICALLY PURE PHARMACEUTICALS: BY OPTICAL RESOLUTION OR ASYMMETRIC SYNTHESIS ?

9.4.1. Some Aspects on Available Methods to Achieve Optical Resolution

A number of different methods are available today for optical resolution purposes (3). In addition to the classical procedures of making diastereomeric salts of acids or bases, a number of other potentially useful techniques exist, not the least of those relying on chromatographic separation (9). Furthermore, methods based on resolution by preferential crystallization (also called resolution by entrainment and applicable to racemates crystallizing as conglomerates), recrystallization of cleavable diastereomeric derivatives, and kinetic resolution by enzymes or microorganisms may also be worth considering. Table 9.2 gives an overview of the various possibilities.

Table 9.2. Potentially Useful Methods for the Optical Resolution of Drug Racemates

Method	Possible Applicability
Classical resolution by recrystallization of diastereomeric salts	Racemic acids and bases. The technique has also been used after derivatization of the target compound, for example resolution of alcohols as monophthalates.
Separation of covalent derivatives obtained from an optically active reagent	Reported cases include resolutions of aldehydes and ketones as menthylhydrazones, of alcohols as various carbamates and carbonates, of alkenes, sulfoxides, phosphines, and other metal coordinating compounds as zero-valent metal complexes, etc. (3).
Preferential crystallization (resolution by entrainment)	Racemates crystallizing as conglomerates.
Kinetic resolution	Mainly hydrolyzable substrates for use with esterases, lipases, amidases, etc.
Chromatographic resolution	In the indirect mode to achieve the separation of diastereomeric derivatives or in the direct mode employing a chiral phase system.

9.4.1.1. Classical Resolution Methods

In the search for useful resolution methods, one should be aware of the fact that very simple chemical modifications of the compound of interest may produce large effects with respect to the crystallization properties and melting point diagrams. Thus, it has been shown fairly recently, that whereas (±)-

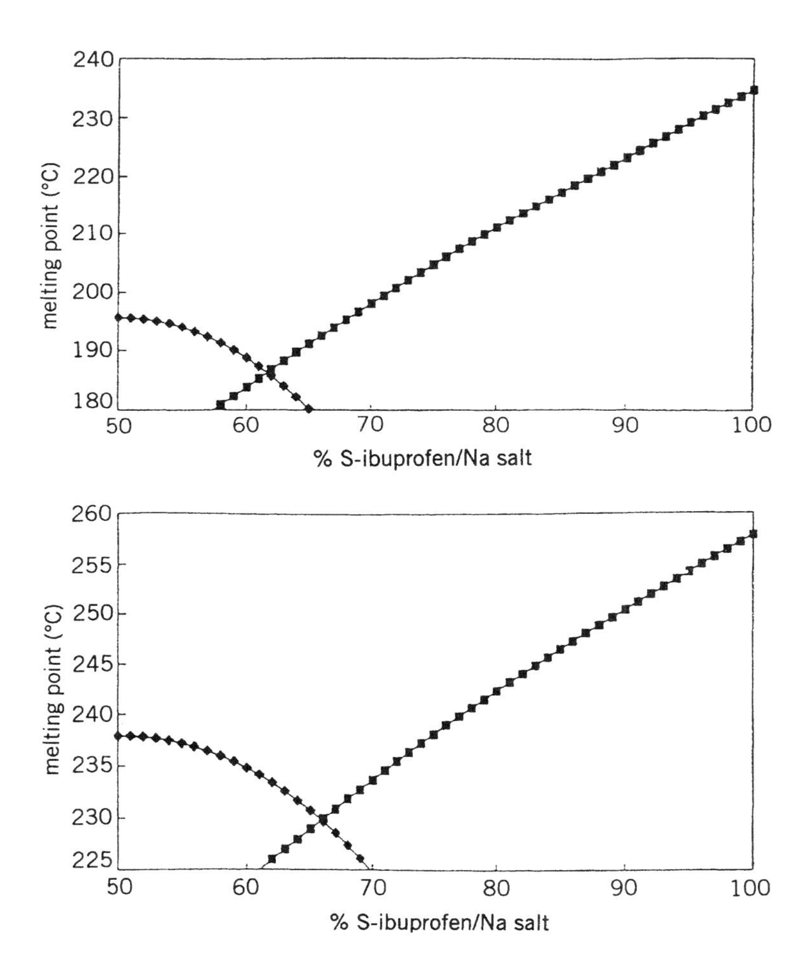

Figure 9.10. Calculated melting point diagrams for the sodium salts of ibuprofen and naproxen, respectively. (Reprinted from Manimaran, T. and Stahly, G. P. (1993). *Tetrahedron: Asymmetry*, **4**, 1949 with permission, Pergamon Press Ltd.)

naproxen[1] [2-(6-methoxy-2-naphthyl)propionic acid] crystallizes as a racemic compound (i.e., shows a melting point diagram corresponding to Figure 9.5b), its methyl and ethyl esters are typical conglomerates (10) and therefore separated into enantiomers by techniques of preferential crystallization. Of perhaps even greater interest, from the point of view of finding a simple resolution method, is the fact that some metal salts of ibuprofen [(±)-2-(4-isobutylphenyl)propionic acid] and (±)-naproxen give melting point diagrams showing eutectic points rather close to the racemic composition (11) (Figure 9.10). This means that optically enriched material can be easily resolved by

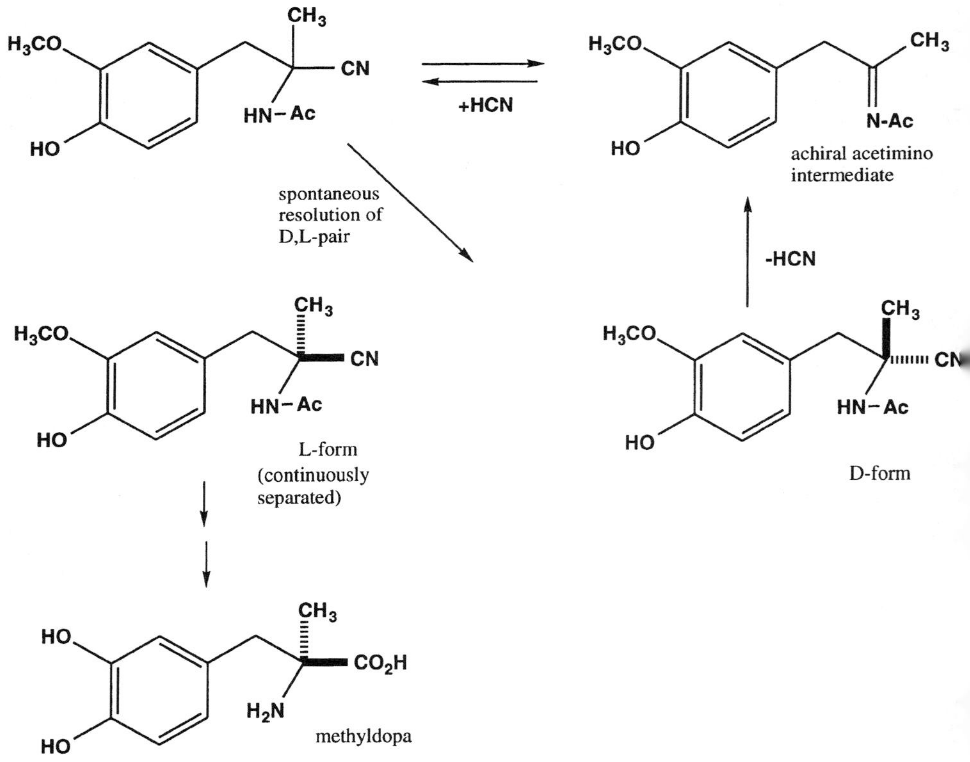

+HCN

achiral acetimino intermediate

spontaneous resolution of D,L-pair

-HCN

L-form (continuously separated)

D-form

methyldopa

Figure 9.11. Utilization of a spontaneous resolution and a racemization step to completely convert a racemic aminonitrile intermediate into its L-antipode, a precursor of L-methyldopa.

[1]Naproxen is the trade name for the pure (+)-enantiomer. For the sake of simplicity, however, the name naproxen is also used for the racemate in this text.

simple recrystallization. Figure 9.10*a* tells us that recrystallization of a sodium salt of ibuprofen containing at least 62% of one enantiomer (24% e.e.) will yield increased optical purity. Similarly (Figure 9.10*b*) a sodium salt of naproxen of 32% e.e. can be used for recrystallization to obtain an enantiomerically pure compound. Experiments have shown (11) that (*S*)-ibuprofen of virtually 100% e.e. can be obtained in 95% yield from a sample of 76% e.e. by a simple precipitation of the sodium salt from acetone (performed by dissolving sodium hydroxide pellets in an acetone solution of the acid) followed by acidification.

Another illustrative example of the use of spontaneous resolution in the process of making enantiopure drugs is found in a commercial procedure for the hypertensive drug L-metyldopa. The key steps are carried out via an aminonitrile intermediate, which after a few subsequent steps yields methyldopa (Figure 9.11).

Techniques of the type described above are of great importance, since a majority of kinetic resolutions and asymmetric syntheses are still far from being totally stereoselective.

Today, there is also great interest in the use of enzymes for kinetic resolutions (12), particularly in esterification reactions in organic solvent media (13). Hydrolytic enzymes have been successfully used to obtain the partial resolution of a number of racemic pharmaceuticals and the profens belong to one of the most investigated classes of compounds. The rates and enantioselectivities were found to be highly dependent on the aryl group in the profen, however, and the best results have been obtained with naproxen (14), suprofen, and ibuprofen (15).

9.4.1.2. Chromatographic Methods

Although both indirect and direct[2] methods have been successfully applied for analytical purposes (8, 16, 17), much remains to be done in the field of preparative applications. One common problem in using chiral phase systems is their limited sample capacity. If the sample capacity is defined as the maximum sample/sorbent ratio possible without a reduction of the k'- and α-values, the limit, which is set by the density of available chiral binding sites, is roughly of the order of 1 mg/g of sorbent or less. In preparative chromatography, the sample capacity has to be exceeded by column overloading (18). The degree of overloading is then a function of the separation factor achieved and will always be a compromise between desired throughput rate (g/h) and enantiomeric purity (e.e.). Thus far, the most promising results have been obtained

[2] The terms *indirect* and *direct* here denote chromatographic separations *without* and *with* a chiral phase system, respectively.

with the use of chiral sorbents based on microcrystalline triacetylcellulose (MCTA), polyacrylamide and -methacrylamide, and Pirkle's aromatic π-π-interacting selectors.

Since many of the most important pharmaceuticals are protolytes, often special care has to be taken in order to avoid nonselective interactions, which not only reduce α-values but also may cause severe peak tailing. Especially when chiral straight-phase systems are used to separate the enantiomers of amine bases, very dramatic changes in the chromatographic appearance can

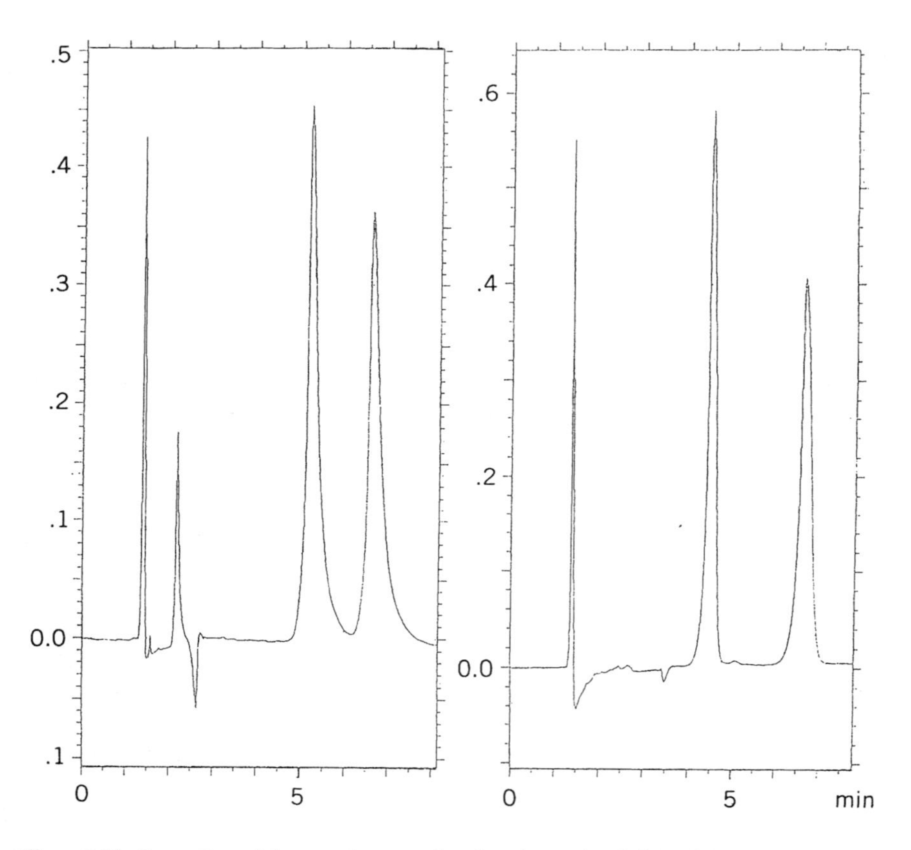

Figure 9.12. Separation of the enantiomers of mefloquine under slightly different mobile phase conditions. A 4.6 × 250 mm column packed with a 7-μ Kromasil-based chiral sorbent (available as Kromasil CHI-DMB from EKA Nobel AB, S-44580 Bohus, Sweden) was used with hexane +5% of 2-propanol as the mobile phase at 2.0 ml/min. Sample amount injected-20 μg. Additions to the mobile phase: (*a*) 0.1% TFA and (*b*) 0 .1% TFA + 0.05% TEA.

be obtained from small additions of ion-pairing agents to the mobile phase, as has been demonstrated recently by Schleimer and Pirkle (19). Investigations of a new type of chiral stationary phase (20), operating via selective hydrogen bond formation in nonpolar mobile phase systems, have displayed similar phenomena as illustrated by Figure 9.12 for the antimalarial drug mefloquine [(*R*, *S*)-*erythro*-α-(2-piperidyl)-2,8-*bis*(3-fluoromethyl)-4-quinolinemethanol].

9.4.2. Some Important Strategies in Asymmetric Synthesis

The other approach to enantiomerically pure pharmaceuticals is based on asymmetric synthesis from a suitable precursor, either via an asymmetric reaction with a prochiral compound or via reactions involving asymmetric induction in a chiral starting material.

A variety of asymmetric reactions have been carried out on prochiral substrates with the use of optically active reagents or catalysts. Among the most important reactions used to prepare synthetically useful chiral building blocks are the dihydroxylation of alkenes according to K. B. Sharpless and the asymmetric hydroboration reactions developed by H. C. Brown. These highly successful reaction types, which often give very high e.e.'s, both make use of optically active reagents derived from the chiral pool. In the first case, (+)(*R*)-tartaric acid is transformed into a derivative (such as the diethyl ester) that is then allowed to form a titanium complex. An epoxidation reaction by the use of a peroxide reagent is then actually catalyzed by the titanium complex, that is, the complex is used as a stereocontrolling chiral auxiliary during many reaction cycles. The interest in the use of homochiral metal complexes as catalysts for asymmetric syntheses has grown enormously during the last years and many preparatively useful routes to enantiopure strategic intermediates for the pharmaceutical industry can be expected in the future.

Complementary to the chemical methods in use to prepare enantiopure intermediates of pharmaceutical interest, a large number of enzymatic reactions have come into use. Carbonyl reductions can be performed with almost complete stereospecificity using yeasts, and hydrolytic enzymes, present in lipases and esterases, are now widely used for kinetic resolutions of ester racemates. An example of how the latter type of reaction has been utilized to produce an early stage chiral intermediate in a drug synthesis is given in Figure 9.13. The lipase here gives a > 99.9% e.e. of the strategic (1*R*, 2*S*)-2-phenylcyclohexanol, used in subsequent asymmetric reaction steps, to provide the target molecule, naltiazem, with the correct configuration at its two stereogenic centers.

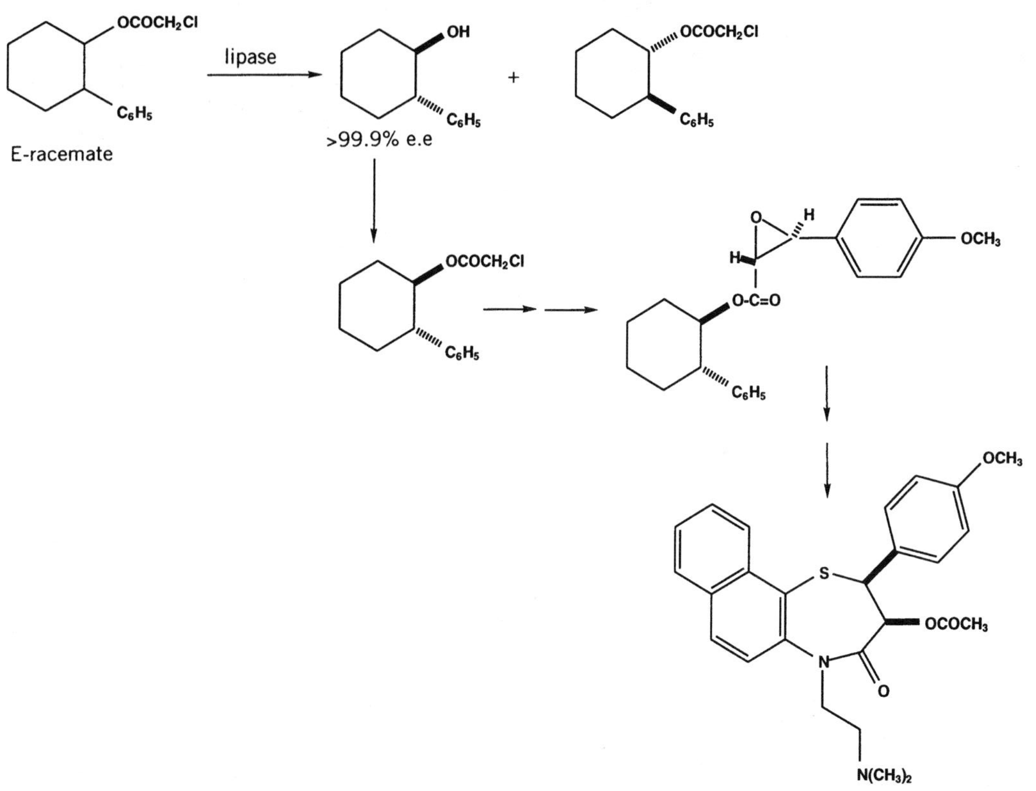

Figure 9.13. Outline of the route to naltiazem via the (1*R*, 2*S*)-2-phenylcyclohexanol chiral building block obtained by enzymatic kinetic resolution.

9.5. BARRIERS TO ENANTIOMER INTERCONVERSION

As quite recently pointed out by Testa et al. (21), the concept of "interconversion of stereoisomers" is often used in an undefined sense due to a certain "semantic confusion" or even poor understanding. In the following section, this important topic is treated very briefly from the point of view of mechanistic classification.

9.5.1. Thermal Enantiomerization or Racemization

One important aspect of chiral pharmaceuticals is their stereochemical integrity, that is, the stability of an enantiomer towards racemization. Traditionally, the ease by which two enantiomers interconvert under a given set of conditions has been studied as the rate of racemization of one of the optical antipodes. With the advent of chiral chromatographic techniques, however, an optically active form of the compound to be studied is no longer necessary in many cases. A few criteria have to be fulfilled though: (1) the racemate should be resolvable by liquid chromatography on a chiral stationary phase, (2) the interconversion rate should be fast enough on the chromatographic timescale, and (3) the interconversion rate should not be influenced by the contact between the compound and the stationary phase.

Under conditions leading to racemization, the equilibrium state is a racemate with interconverting enantiomers according to eq. (9.1).

$$(R)\text{-A} \underset{k_{-1}}{\overset{k_1}{\rightleftharpoons}} (S)\text{-A} \tag{9.1}$$

The rate constants for the forward and reverse reactions, k_1 and k_{-1}, are, of course, equal. A theoretical treatment of the relation between these rate constants and the chromatographic appearance under given conditions has been presented (2, 22, 23) and will not be repeated here. It is sufficient to point out that, as a rule of thumb, peak coalescence in LC is observed if the half-lifetime of the enantiomers is of the order of magnitude of their retention times (24). An illustration of the influence of flow rate and temperature on the chromatographic appearance of two interconvertible enantiomers is given in Figure 9.14.

Interconversions of this kind are not uncommon among chiral pharmaceuticals. Typical examples (Figure 9.15) are found in some of the benzodiazepinones, like oxazepam (**X**) and its analogs (25). In these compounds the enantiomerization is acid-base-catalyzed, which is often seen during chiral reversed phase chromatography of the racemates in nonneutral buffer media.

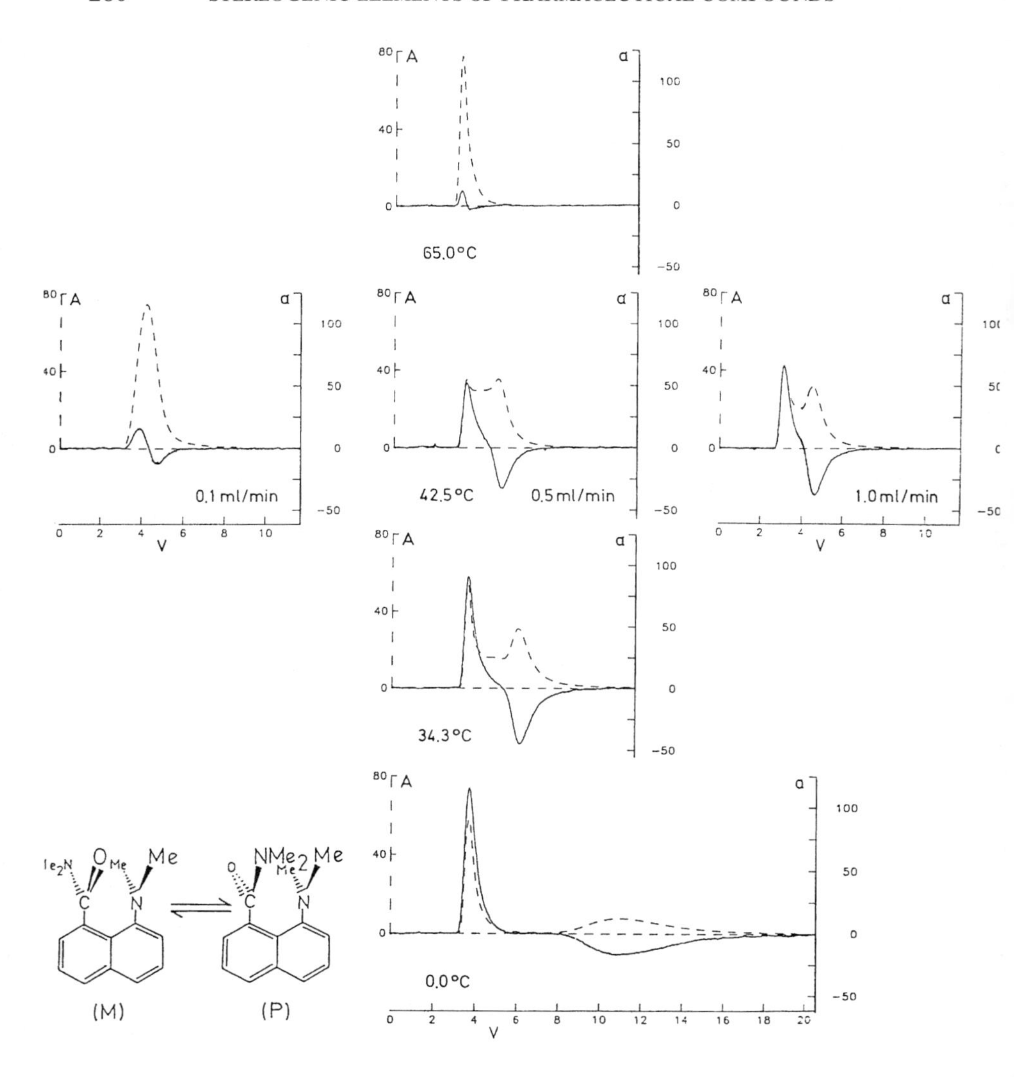

Figure 9.14. Dependence of chromatographic peak shape upon the time scale and temperature for a pair of interconverting enantiomers. (Reprinted from: Mannschreck, A., Zinner, H., and Pustet, N., (1989). *Chimia*, **43**, 165–166. With permission from Schweizerischer Chemiker-Verband.)

Other known examples include thalidomide (**XI**) (26) and chlorthalidone (**XII**) (27); the latter is found to be racemized via a carbonium ion mechanism (Figure 9.16) . The half-life of optically active **XII** has been estimated to 4 min in water at 37°C (27).

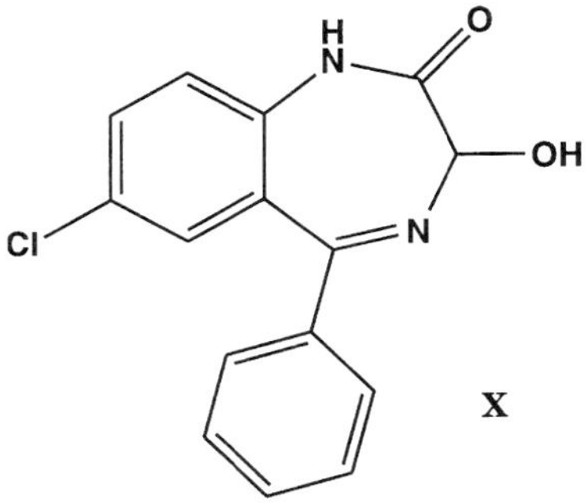

X

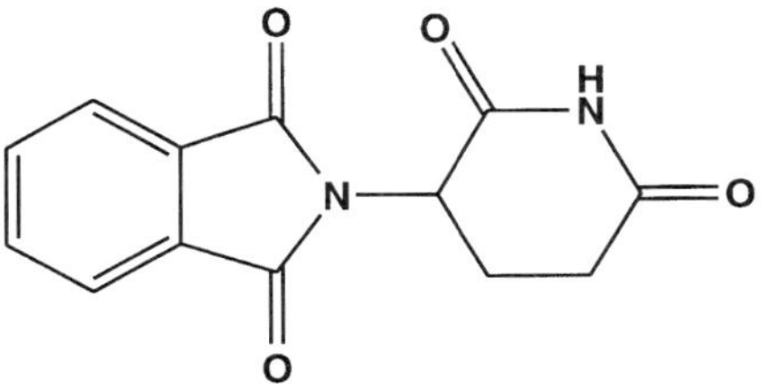

XI

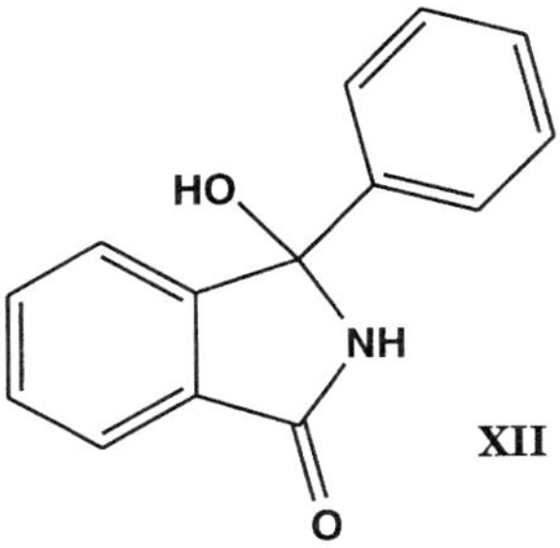

XII

Figure 9.15. Some stereochemically labile structures.

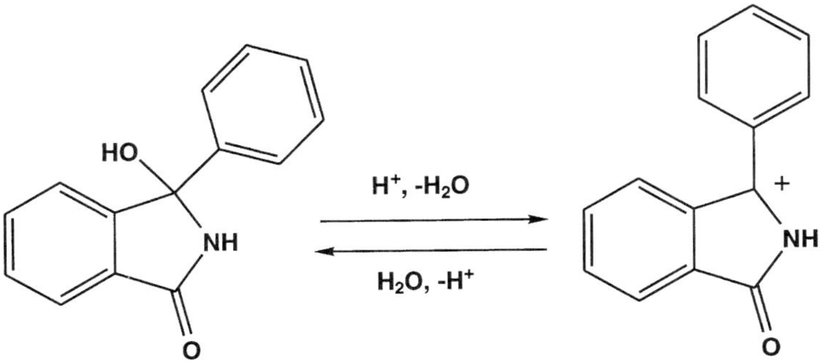

H^+, $-H_2O$

H_2O, $-H^+$

Figure 9.16. Enantiomerization of chlorthalidone via a planar, achiral carbonium ion.

9.5.2. Enzymatically Mediated Interconversion

Enzymes may also catalyze enantiomer interconversions, one well-known example being the amino acid racemates present in certain microorganisms.

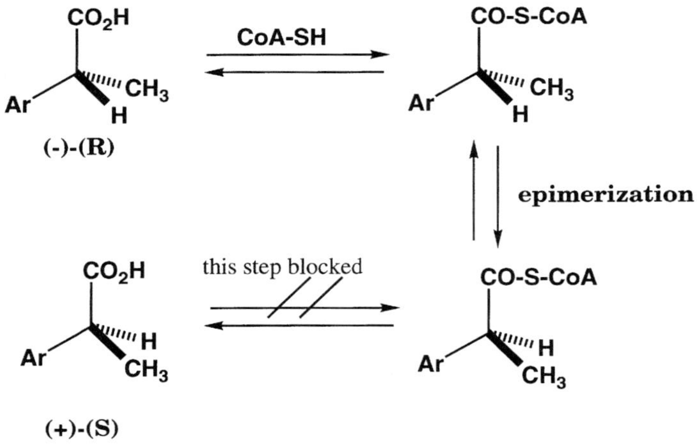

Figure 9.17. Suggested mechanism of the unidirectional inversion at the stereogenic center of certain profens *in vivo*.

Enzymes can further be involved in the transformation of one enantiomer of a drug into its antipode *in vivo*, that is, after administration. Such a metabolic chiral inversion was observed as early as 1973 (28) when it was found that in man, the urinary metabolites of both the (–)(*R*)- and (+)(*S*)-forms of ibuprofen were dextrorotatory. It soon become clear that a *unidirectional* inversion process from (–)(*R*)- to the active (+)(*S*)-form of the profen drug took place (29). It has been suggested that the key step in this process is the ability of the (*R*)-profen but not its (*S*)-antipode to become enzymatically transformed into a CoA thioester. Since the latter readily undergoes an epimerization reaction [an inversion at the stereogenic carbon atom in the acyl (profen) part, probably enzyme-catalyzed] and each thioester is hydrolyzed back to the free acid, the net result is a conversion of the (*R*)-profen to its (*S*)-antipode. The overall reaction scheme is outlined in Figure 9.17. It should be remembered that a scheme of this kind is generally valid, meaning that if equilibrium is not attained, due to the blockage of an enzymatic step from one of the enantiomers (the enantiomer being a nonsubstrate), a unidirectional transformation will result.

9.6. SOME FINAL REMARKS

The question concerning drug chirality and its relevance to pharmaceutical effects is complicated in so far as it is to a large extent a question of the stereochemical integrity or lifetime under the conditions used. Enantiopure drugs may undergo substantial racemization prior to the onset of the biological effect. Whether a compound should be regarded as a "single" species, exhibiting two (or more) chiral states, or as a pair of discrete enantiomeric forms is a matter of rather arbitrary definition. Traditionally, enantiomers have been characterized by chiroptical data ([α], sign of the Cotton effect, etc.), which requires their isolation, usually carried out at room temperature. Enantiomeric states are present in a variety of rapidly equilibrating molecular structures, however, and can be studied by NMR and chromatographic techniques at low temperature. One should also remember that a racemic drug may actually be a prodrug, yielding an achiral active compound by means of a nonenzymatic and hence nonstereoselective process.

ACKNOWLEDGMENT

The author is indebted to Dr. Ian McEwen for valuable help with the molecular modeling work. Grants K-KU 2508-313 and -315 from the Swedish Natural Science Research Council are also gratefully acknowledged. Thanks are additionally owed to Astra Hässle AB and EKA Nobel AB for generous financial support.

REFERENCES

1. Tucker, G. T. (1990). Enantiomer-specific pharmacokinetics. *Acta Pharm. Nord.*, **2**:3, 193–196.
2. Bürkle, W., Karfunkel, H., and Schurig, V. (1984), Dynamic phenomena during enantiomer resolution by complexation gas chromatography. A kinetic study of enantiomerization. *J. Chromatogr.*, **288**, 1–14.
3. Jacques, J., Collet, A., and Wilen, S. (1981). *Enantiomers, Racemates and Resolutions*; New York: Wiley-Interscience.
4. Haake, M., Georg, G., Fode, H., Eichenauer, B., Ahrens, K. H., and Szelenyi, I., (1983). Spasmolytically active N-(aminoalkyl)sulfur diimides. *Pharm. Ztg.*, **128**, 1529–1533.
5. Ebnother, A., Juckner, E., and Stoll, A. (1945). Atropisomerie in der Dibenzo[a,d]-cyclohepten-Reihe. *Helv. Chim. Acta*, **48**, 1237–1239.
6. Remy, D. C., Little, K. E., Hunt, C. A., Anderson, P. S., Arison, B. H., Englehardt, E. L., Hirschmann, R., Clineschmidt, B. V., Lotti, V . J., Bunting, P. R., Ballentine, R. J., Papp, N. L., Flataker, L., Witoslawski, J. J., and Stone, C. A. (1977). Synthesis and stereospecific antipsychotic activity of (–)-1-cyclopropyl-methyl-4-(3-trifluoromethyl-thio-5H-dibenzo[a,d]cyclohepten-5-ylidene)piperidine. *J. Med. Chem.*, **20**, 1013–1019.
7. Clineschmidt, B. V., McKendry, M. A., Papp, N. L., Plueger, A. B., Stone, C. A.,

Totaro, J. A., and Williams, M. (1979). Stereospecific antidopaminergic and anticholinergic actions of the enantiomers of (±)-1-cyclopropylmethyl-4-(3-trifluoromethyl-thio-5H-dibenzo[a,d]cyclohepten-5-ylidene)piperidine (CTC), a derivative of cyproheptadine. *J. Pharmacol. Exp. Ther.*, **208**, 460–467.

8. Eveleigh, P., Hulme, E. C., Schudt, C., and Birdsall, N. J. M. (1989). The existence of stable enantiomers of telenzepine and their stereoselective interaction with muscarinic receptor subtypes. *Mol. Pharmacol.*, **35**, 477–483.
9. Allenmark, S. (1991). *Chromatographic Enantioseparation*, 2nd ed. Chichester/New York: Horwood/Wiley.
10. Kazutaka, A., Ohara, Y., and Takakuwa, Y. (1983). U.S. Patent 4 417 070.
11. Manimaran, T. and Stahly, G. P. (1993). Optical purification of profen drugs. *Tetrahedron: Asymmetry*, **4**, 1949–1954.
12. Faber, K. (1992). *Biotransformations in Organic Chemistry*. New York: Springer-Verlag.
13. Sih, C. J. and Wu, S. H. (1989). Resolution of enantiomers via biocatalysis. *Topics Stereochem.*, **19**, 63–125.
14. Gu, Q.-M., Chen, C.-S. and Sih, C. J. (1986). A facile enzymatic resolution process for the preparation of (+)-S-2-(6-methoxy-2-naphthyl)propionic acid (Naproxen). *Tetrahedron Lett.*, **27**, 1763–1766.
15. Sih, C. J., Gu, Q.-M., Fülling, G., Wu, S.-H., and Reddy, D. R. (1988). The use of microbial enzymes for the synthesis of optically active pharmaceuticals. *Develop. Indust. Microbiol.*, **29**, 221–229.
16. Lough, W. J. (ed.) (1989). *Chiral Liquid Chromatography*. Glasgow and London: Blackie and Sons Ltd.
17. Krstulovic, A. M. (ed.) (1989). *Chiral Separations by HPLC: Applications to Pharmaceutical Compounds*: Chichester: Ellis Horwood.
18. McDonald, P. D. and Bidlingmeyer, B. A. (1987). Strategies for successful preparative liquid chromatography. In B. A. Bidlingmeyer (ed.), *Preparative Liquid Chromatography*, Amsterdam: Elsevier.
19. Schleimer, M. and Pirkle, W. H. (1992). Enantiomer separation by HPLC and capillary-SFC on polysiloxane-based chiral stationary phases. 3rd International Symposium on Chiral Discrimination, Oct. 5–8, Tübingen, Germany, abstract no. 73.
20. Allenmark, S., Andersson, S., Möller, P., and Sanchez, D. (1995). A new class of network-polymeric chiral stationary phases. *Chirality*, **7**, 248–256.
21. Testa, B., Carrupt, P.-A. and Gal, J. (1993). The so-called "interconversion" of stereoisomeric drugs: An attempt at clarification. *Chirality*, **5**, 105–111.
22. Melander, W. R., Lin, H.-J., Jacobson, J., and Horváth, C. (1984). Dynamic effect of secondary equilibria in reversed-phase chromatography. *J. Phys. Chem.*, **88**, 4527–4536.
23. Hanai, R. and Wada, A. (1987). Analysis of the *cis-trans* isomerization kinetics of L-alanyl-L proline by the elution-band relaxation method. *J. Chromatogr.*, **394**, 273–278.

24. Mannschreck, A., Zinner, H., and Pustet, N. (1989). The significance of the HPLC time scale: An example of interconvertible enantiomers. *Chimia*, **43**, 165–166.
25. Yang, S. K. and Lu, X.-L. (1992). Resolution and stability of oxazepam enantiomers. *Chirality*, **4**, 443–446.
26. Shealy, Y. F., Opliger, C. E., and Montgomery, J. A. (1968). Synthesis of D- and L-thalidomide and related studies. *J. Pharm. Sci.*, **57**, 757–764.
27. Severin, G. (1992). Spontaneous racemization of chlorthalidone: Kinetics and activation parameters. *Chirality*, **4**, 222–226.
28. Mills, R. F. N., Adams, S. S., Cliffe, E. E., Dickinson, A., and Nicholson, J. S. (1973). The metabolism of Ibuprofen. *Xenobiotica*, **3**, 589–598.
29. Mayer, J. M. (1990). Stereoselective metabolism of anti-inflammatory 2-arylpropionates. *Acta Pharm. Nord.*, **2**:3, 197–216.

CHAPTER

10

THE IMPORTANCE OF CHIRAL SEPARATIONS IN PHARMACEUTICALS

SUT AHUJA

Ahuja Consulting
Monsey, New York 10952

Stereoisomers are isomeric molecules with an identical constitution but a different spatial arrangement of atoms. The symmetry factor classifies stereoisomers as either diastereoisomers or enantiomers. Enantiomers are molecules that relate to each other as an object, with a mirror image that is not superimposable. These molecules are commonly called chiral (from the Greek word *cheir* meaning hand, i.e., they are like a pair of hands). A pair of enantiomers is possible for all molecules containing a single chiral carbon atom (one with four different groups attached). Diastereoisomers or diastereomers are basically stereoisomers that are not enantiomers of each other. Although a molecule may have only one enantiomer, it may have several diastereomers. However, two stereoisomers cannot be both enantiomers and diastereomers of each other simultaneously.

Stereoisomerism can result from a variety of sources, including the single chiral carbon (or chiral center), for example, a chiral atom that is a tetrahedral atom with four different substituents. It is not necessary for a molecule to have a chiral carbon in order to exist in enantiomeric forms, but it is necessary that the molecule as a whole be chiral. A detailed discussion of these topics may be found in several books and review articles (1–8).

In articles or textbooks that discuss stereochemistry, there is widespread use of synonymous and sometimes incorrect terminology (stereogenic center, chiral center, and asymmetric center; nonracemic mixture and nonequimolar mixture; optical purity and stereoisomeric purity; enantioselective, enantiospecific, stereoselective, and stereospecific, etc.). For chemical names, International Union of Pure and Applied Chemistry (IUPAC) nomenclature may be preferable. For generic names, different conventions are currently used

The Impact of Stereochemistry on Drug Development and Use, Edited by Hassan Y. Aboul-Enein and Irving W. Wainer. Chemical Analysis Series, Vol. 142.
ISBN 0-471-59644-2

[International Nonproprietary Name (INN), United States Adopted Names (USAN)]; a single convention should be established. Scientists engaged in the development of chiral drugs should endeavor to use standardized terminology and definitions, and a uniform terminology convention should be established in the near future.

The importance of determining the stereoisomeric composition of chemical compounds, especially those of pharmaceutical importance, is now well recognized (9). Dextromethorphan provides a dramatic example in that it is an over-the-counter antitussive, whereas levomethorphan, its stereoisomer, is a controlled narcotic. Similarly, it has been reported that the teratogenic activity of thalidomide may reside exclusively in the *S*-enantiomer (10). This conclusion remains controversial. However, it emphasizes the need of careful investigations, because 12 of the 20 most prescribed drugs in the United States and 114 of the top 200 possess one or more asymmetric centers in the drug molecule (11). About half of the 2050 drugs listed in the U.S. Pharmacopeial Dictionary of Drug Names contain at least one asymmetric center, and 400 of them have been used in racemic or diastereomeric forms (12). The market size for the chiral compounds has been estimated to be $5 billion by the next century. The differences in the physiologic properties between enantiomers of these racemic drugs have not yet been examined in many cases, probably because of the difficulties of obtaining both enantiomers in optically pure forms. Some enantiomers may exhibit potentially different pharmacologic activities, and the patient may be taking a useless, or even undesirable, enantiomer when ingesting a racemic mixture. To ensure the safety and effect of currently used and newly developing drugs, it is important to isolate and examine both enantiomers separately. Furthermore, it is necessary to measure and control the stereochemical composition of drugs in at least three situations. Each situation may present a specific technical problem during (a) manufacture, where problems of preparative scale separations may be involved; (b) quality control (or regulatory analysis), where analytical questions of purity and stability predominate; and (c) pharmacologic studies of plasma disposition and drug efficacy, where ultratrace methods may be required (2).

Accurate assessment of the isomeric purity of substances is critical since isomeric impurities may have unwanted toxicologic, pharmacologic, or other effects. Such impurities may be carried through a synthesis and preferentially react at one or more steps and yield an undesirable level of another impurity. Frequently, one isomer of a series may produce a desired effect, whereas another may be inactive or even produce some undesired effect. Large differences in activity between stereoisomers point out the need to accurately assess the isomeric purity of pharmaceuticals. Often, these differences exist between enantiomers, the stereoisomers being most difficult to separate. Some

Table 10.1. Activities of Some Stereoisomers (8)

Compound	Stereoisomers and Activities
Amphetamine	*d*-Isomer is a potent CNS stimulant, whereas *l*-isomer has little, if any, effect.
Diethylstilbestrol (DES)	*trans*-Isomer is much more estrogenic than *cis*-.
Ascorbic acid	(+)-Isomer is a good antiascorbutic, whereas the (–)-isomer has no such properties.
Propoxyphene	α-*l* is an active antitussive, α-*d* is a potent analgesic, but β-*d* and β-*l* are substantially inactive.
Epinephrine	(–)-Isomer is more than 10 times more active a vasoconstrictor than the (+)-isomer.
Synephrine	(–)-Isomer has 60 times the pressor activity of the (+)-isomer.
Quinine/quinidine	Quinidine is the (+)-enantiomer, with cardiac suppressant effects; quinine is the (–)-enantiomer with other medicinal uses.
Propranolol	Racemic propranolol is administered, but only the *S*(–)-isomer has the desired β-adrenergic blocking activity.
Warfarin	Racemic warfarin is administered, but the *S*(–)-isomer is 5 times more potent as a blood anticoagulant than the *R*(+)-isomer.

examples of activity differences between stereoisomers are noted in Table 10.1.

10.1. REGULATORY PERSPECTIVE

Some guidance on regulatory practices has been provided in Canada, the EEC, Japan, and the United States in documents at various stages of development. All the requirements are based on the same scientific basis. These requirements evolved after discussions between various authorities with industry and with international colleagues. As a result, the texts are comparable to each other in content. The current regulatory position of the Food and Drug Administration is summarized below with regard to the approval of racemates and pure stereoisomers (13). Circumstances in which stereochemically sensitive analytical methods are necessary to ensure the safety and efficacy of a drug have been discussed (14, 15). Regulatory guidelines for new drug applications (NDAs) are interpreted for the approval of a pure enantiomer in which the racemate is marketed, for the approval of either a racemate or pure enantiomer in which neither is marketed, and for clinical investigations to compare the safety and efficacy of a racemate and its enantiomers. The basis for such regulation has been drawn from historical situations (thalidomide and benoxaprofen) as well as currently marketed drugs (arylpropionic acids, disopyramide, and indacrinone).

The primary regulatory focus of the Food and Drug Administration is on considerations of both the clinical efficacy and consumer safety of a potential drug. Because the chiral environment found *in vivo* affects the biological activity of a drug, the approval of stereoisomeric drugs for marketing can present special challenges. The case of thalidomide is an example of a problem that may have been complicated by ignorance of stereochemical effects. The use of racemates can lead to erroneous models of pharmacokinetic behavior and to the potential for opportunities to manipulate pharmacologic activity. It is technically feasible to design experiments that will unambiguously determine whether or not a stereochemically pure drug is more effective and/or less toxic than the racemate.

The FD&C Act requires a full description of the methods used in the manufacture of the drug, which includes testing to demonstrate its identity, strength, quality, and purity. The question of stereochemistry was approached directly in the guidelines issued by the FDA in 1987 on the submission of NDA for the manufacture of drug substances (15). Therefore, the submissions should show the applicant's knowledge of the molecular structure of the drug substance. For chiral compounds, this includes the identification of all chiral centers. The enantiomer ratio, although 50:50 by definition for a racemate, should be defined for any other admixture of stereoisomers. The proof of structure should consider stereochemistry and provide appropriate descriptions of the molecular structure. The guidelines do not discuss conditions under which a determination of absolute configuration is desirable or essential. Obviously, it would be appropriate data for supporting the manufacture of optically pure drugs.

U.S. regulatory requirements demand that the bioavailability of the drug be demonstrated (16). When pharmacokinetic models differ between enantiomers, it seems obvious that establishing the bioavailability of the drug from a racemate is a much more complex task, which cannot be accomplished without separation of the enantiomers and investigation of their pharmacokinetics as individual molecular entities.

It is expected that the toxicity of impurities, degradation products, and residues from manufacturing processes will be investigated as the development of a drug is pursued. The same standards should, therefore, be applied to the enantiomeric molecules in a racemate (15). Whenever a drug can be obtained in a variety of chemically equivalent forms (such as enantiomers), it makes sense to explore the potential *in vivo* differences between these forms.

Although it is now technologically feasible to prepare purified enantiomers, the development of racemates may continue to be appropriate. However, the following should be considered in product development:

1. Appropriate manufacturing and control procedures should be used to assure the stereoisomeric composition of a product with respect to identity, strength, quality and purity. Manufacturers should notify compendia of these specifications and tests.
2. Pharmacokinetic evaluations that do not use a chiral assay will be misleading if the disposition of the enantiomers is different. Therefore, techniques to quantify individual stereoisomers in pharmacokinetic samples should be available early. If the pharmacokinetics of the enantiomers are demonstrated to be the same or to exist as a fixed ratio in the target population, an achiral assay or an assay that monitors one of the enantiomers may be used subsequently.

10.2. INDUSTRIAL PERSPECTIVE

The present position of the Pharmaceutical Manufacturers Association is that the development of either a racemate or single enantiomer should be made on a case-by-case basis, depending on pharmacologic and toxicologic considerations and technical feasibility, together with the view that clinical and preclinical data obtained on racemates may be used to support the development and marketing of a single isomer (17).

A sponsor's decision to develop, register, and market a racemate must be scientifically justified. The risk/benefit to the patient will weigh significantly in this decision process. Although not all-inclusive, the following specific examples illustrate typical situations where the decision to develop a racemate might be made:

- The enantiomers have been shown to have pharmacologic and toxicologic profiles similar to the racemate.
- The enantiomers are rapidly interconverted *in vitro* and/or *in vivo* on a timescale such that administering a single enantiomer offers no advantage.
- One enantiomer of the racemate is shown to be pharmacologically inactive and the racemate is demonstrated to be safe and effective.
- Synthesis or isolation of the preferred enantiomer is not practical. This assumes that meaningful effort has been directed to synthesis and isolation of the preferred enantiomer without success, or that even though

isolation and synthesis may have been realized on a small scale, large-scale application of these methods may not be feasible.

- Individual enantiomers exhibit different pharmacologic profiles, and the racemate produces a superior therapeutic effect relative to either enantiomer alone.

A drug product that contains a chiral drug substance, regardless of its stereochemical form (racemate or enantiomer), which has not previously undergone registration in the same or in another stereochemical form, should be regarded as a new chemical entity (18). The registration of any NCE requires comprehensive preclinical and clinical study; the details of the study, are generally established on a case-by-case basis.

The choice of the stereochemical form of a chiral drug should be based on scientific data relating to quality, safety, efficacy, and risk/benefit. The responsibility for the decision as to which stereochemical form of a drug to be developed should clearly reside with the applicant. Regulatory requirements concerning the development of chiral drugs should be consistent with these principles.

After a drug has received approval to be marketed in a particular stereochemical form, a subsequent comparative clinical study of individual enantiomers should not be required automatically unless a new safety issue that may be dependent on stereochemical configuration arises or new claims are envisaged that are to be based on the pharmacologic or toxicologic activities of one or both of the enantiomers. An exception to this might be an agreement by the applicant to conduct such studies as a condition to obtaining marketing authorization.

When a stereochemical form of a chiral drug is chosen for development, a validated enantiospecific assay for use in dosage form development and the assay of enantiomers in biological fluid should be developed at an early stage and used unless the use of a nonenantioselective assay(s) provides results equivalent to those obtained with the enantiospecific assay. The pharmacokinetic assays used in the study of chiral drugs should be capable of accurately measuring levels of individual enantiomers in biological fluids, unless it has been shown that stereoselectivity effects are not significant in the pharmacokinetics of the drug.

For a single enantiomer, susceptibility to stereochemical instability via either *in vivo* or *in vitro* inversion, racemization, or epimerization should be established. For a racemate, the potential for *in vivo* stereochemical inversion of one of the enantiomers must be assessed. The *in vitro* stereochemical stability of the drug substance should be evaluated throughout the claimed shelf life of the product.

The assays and purity tests that are applied to the bulk drug substance and to the dosage form, and used in release and stability testing, should be capable of accurately measuring both the content of the drug substance and the content of all relevant stereochemically related and chemically related substances unless the absence of stereoselective degradation has been demonstrated.

Purity evaluation of a drug containing a single enantiomer, the optical antipode, should be treated as a potential related substance, impurity, or degradation product and addressed as such in accordance with normal practices. The purity level required for a chiral reference standard should depend on its intended use. An absolute measure of enantiopurity should be stated when a reference standard will be used as an assay standard.

It is important to demonstrate the batch-to-batch consistency by evaluating the impurity profile of a chiral drug substance, especially with respect to the content of chemically and stereochemically related substances in batches used for toxicology studies and clinical trials and the relationship to that of the marketed drug product.

Enantiospecific identity tests preferably should be used for the identification of a chiral drug, regardless of its stereochemical form. A validated optical rotation measurement is desirable unless other methods have been found to be more suitable. When describing the production process for a chiral drug substance, the synthetic steps in which the stereogenic center is generated, maintained, affected, or manipulated should be provided in relevant detail in the marketing application.

During development of a formulation form of chiral drug, regardless of its stereochemical form, it is important to consider the effect of its stereochemical form on the physical-chemical properties of the drug substance (polymorphism, rate of dissolution, crystallinity). For a racemate, it may be useful to determine if the drug substance is a racemic compound (i.e., a true racemate) or a racemic mixture (a conglomerate). Generic drug products that contain a chiral drug substance should be held to the same manufacturing and control standards that were originally applied in the approval of the innovator's product. Discussed below is the development of a racemate and racemate enantiomer switch (18).

10.2.1. Development of Racemates

Let us consider the case of a racemate that, in spite of being composed of two optical antipodes, need not be differentiated from other NCEs containing a single chemical species. Although production of a single enantiomer may be achieved on a small scale, scale-up may not always be possible; if scale-up problems contribute to the decision to develop a racemate, a discussion of the work undertaken to attempt scale-up should be provided in the marketing application. Otherwise, the justification for the development of a racemate should generally not be required.

The pharmacodynamic profile of each enantiomer should be established for primary pharmacologic effect with the establishment of secondary effects as required to assure safety. If a significant and/or unexpected toxicity is observed in preclinical studies with a racemate, its relationship to one or both of the enantiomers has to be investigated.

Animal safety studies should be supported by toxicokinetics that employ an enantiospecific assay in order to assess enantiomer levels to assure that the animals' exposure to each enantiomer supports the anticipated exposures in human studies. If the pharmacokinetic profile of the enantiomers is the same or a fixed ratio in various animal species, the use of a nonenantiospecific assay may be appropriate for subsequent studies.

The clinical pharmacokinetics of both enantiomers should be defined at an early stage, after administration of the racemate, and this data should be evaluated in the context of results or preclinical toxicokinetic studies.

The use of nonenantiospecific assays in bioequivalence studies may be justified only when studying single stable enantiomers or, in the case of a racemate, when it has been shown that the rate of release of the enantiomers from the formulation does not influence the relative AUC values.

10.2.2. Racemate–Enantiomer Switches

A drug product that contains a chiral drug substance, regardless of its stereochemical form, that has been approved in another stereochemical form may represent a special case requiring less comprehensive preclinical and clinical development than would ordinarily be required for an NCE. The details of such an abbreviated development program must be established on a case-by-case basis and justified by the applicant. For example, for a racemate-to-enantiomer switch, it may not be necessary to repeat certain studies that would normally be required with the selected enantiomer when relevant studies have previously been conducted with the racemate and the data from these studies are available to the applicant of the enantiomer. Additional studies of relevant stereochemical species (bridging studies), which are determined on a case-by-case basis, may be required to support the elimination of certain normally required studies. The feasibility of conducting such studies will depend on the stereochemical integrity of the enantiomers to be studied. Examples of these bridging studies are:

- Comparisons of the profiles of significant pharmacologic and toxicologic activities of the selected enantiomer and the racemate (it may also be necessary to study the optical antipode in order to investigate the possibility of interactions).

- Comparisons of the pharmacokinetic profiles of the selected enantiomer and the racemate (it may also be necessary to study the pharmacokinetics of the optical antipode in order to assure that its removal does not alter the kinetics of the single enantiomer).

An acceptable preclinical toxicology bridging strategy might consist of a 3-month repeat dose comparison of the selected enantiomer and the racemate in the most appropriate species and a segment II reproductive toxicity comparison in the most appropriate species. When a single enantiomer shows evidence of increased toxicity compared to that of the racemate, this should be further investigated and the implications of this toxicity for therapeutic use must be considered.

Since the effects of the addition of the optical antipode on the overall characteristics of a single enantiomer-based drug product would be unknown, the switch of an approved single enantiomer-based drug to a racemate-based drug should be approached as if it were the development of an unrelated NCE.

It is necessary that regulatory agencies provide clear regulatory guidance to facilitate the efficient development of new drugs. However, science should lead the development of regulations and not vice versa. As the result of recent activity concerning drug stereochemistry on the part of regulatory agencies around the world and the global approach to drug development that many transnational companies now pursue, there exists a need and an opportunity to achieve a sensible consensus in the establishment of regulatory requirements for chiral drugs.

10.3. MARKETING STATUS OF SINGLE ISOMERS

The results of one study (19) indicate that the use of single-isomer chiral drugs has increased in the period from 1982–1991. The main reason for this increase appears to be a greater emphasis on the development of synthetic single isomer chiral drugs. This is exemplified by the introduction of a number of angiotensin-converting enzyme inhibitors during this period. Two results of this survey appeared to be particularly striking:

- The overall reduction in the number of compounds of natural and semi-synthetic origin in 1991 compared to 1982
- The large increase in the number of synthetic single isomer chiral drugs

This led to consideration of the classification of the origin of the drugs studied in this survey. The majority of previous surveys of this type gave no clear indication of the classification employed other than natural, semisynthetic, or synthetic origin. In this survey designations were based on the origin of the category considered to be most appropriate in terms of the present-day source of the drug used in the production of formulated products, for example, ascorbic acid and caffeine were both defined as being of synthetic origin. It is possible that in previous surveys, these two compounds (and others included in this survey) were classified as being of natural origin. This may have resulted in some minor differences in the reported results in comparison to previous studies; however, it can be concluded that the presented results are a true reflection of the changes that have occurred in the use of single-isomer chiral drugs over the last 10 years.

10.4. ABSOLUTE CONFIGURATION

Discussed here is an example of the determination of absolute configuration of a series of compounds in the pursuit of a potent and selective bronchodilator (20). The biological properties of a series of β-adrenoreceptor stimulants with the general structure 3-acylamino-4-hydroxy-α-[(N-substituted amino)-methyl] benzyl alcohol are well known. Like many other members of the series, 3-formamido-4-hydroxy-α-[N-(p-methoxy-α-methylphenethyl)amino]-methylbenzyl alcohol(I) has two asymmetric carbons in its molecule and is a mixture of two pairs of enantiomers IA and IB. These were separated by selective crystallization, and one of them (IA) was found to be highly promising as a potent and selective bronchodilator.

Compounds IA and IB have been now separated further into their respective optical isomers to determine the absolute configurations of these four isomers: (–)-IA, (+)-IA, (–)-IB, and (+)-IB. IA, IB, and N-(*p*-methoxy-α-methylphenethyl)amine (II) were resolved using the (+)- and (–)-forms of tartaric acid as resolving agents in their optical isomers as shown in Table 10.2.

Two isomers [(–)-IA and (–)-IB] were also obtained from (–)-II and 4′-benzyloxy-3′-nitro-2-bromoaceto-phenone by several steps. The configurations of the enantiomers of II were determined by chemical correlation with those of N-p-hydroxy-α-methylphenethyl-amine, whose configuration is known. The configurations of the asymmetric carbon (α) carrying the OH group and the asymmetric carbon (β) located in the amine moiety of (+)-IA were respectively determined by correlation with (+)-4-methoxy-3-nitrobenzoic acid and (+)-II, whose configurations are known, through (+)-3-amino-4-methoxy-α-[N-(*p*-methoxy-α-methylphenethyl)amino]methylbenzyl alcohol. On the basis of these experiments, the configurations of the four isomers can

Table 10.2. Optical Rotations and Melting Points of Isomers of I (20)

Compound	mp (°C)	$[\alpha]^{2}_{D}$
(+)-IA	[a]	+29.3
(+)-IA(+)-tartrate[b]	184	+40.4
(–)-IA	[a]	+30.1
(–)-IA(–)-tartrate[b]	185	–42.6
(+)-IB	150	+8.9
(+)-IB(+)-tartrate[b]	172	+12.2
(–)-IB	150	–9.0
(–)-IB(–)-tartrate[b]	172	–12.3

[a] An amorphous solid.
[b] The rotations of the tartrates were measured as aqueous solutions and those of the bases as methanolic solution.

be depicted as follows: (–)-IA = αR, βR, (+)-IA = αS, βS, (–)-IB = αS, βR, and (+)-IB = αR, βS. The bronchodilator activity of these compounds was found to decrease in the order of (–)-IA > (+)-IB > (+)-IA > (–)-IB.

The bronchodilator activity of the isomers was compared using isolated tracheal preparations of guinea pigs, and the data are given in Table 10.3. Compounds (–)-IA and (+)-IB, which have the *R* configuration of the α-carbon, were more potent than the corresponding isomers (–)-IB and (+)-IA, respectively, which have the *S* configuration. These data are in general agreement with those reported for sympathomimetic amines such as norephinephrine, epinephrine, isoproterenol, and salbutamol, the activity of which is known to

Table 10.3. Relative Potencies of the Isomers of I on Tracheobronchial Muscle (20)

Isoproterenol Compound	Configurations		Relative Bronchodilator Potencies[a] dose ratios (Isoproterenol = 1)
	α	β	
Racemic IA[b]			0.1
(–)-IA[b]	*S*	*S*	0.08
(+)-IA[b]	*S*	*S*	0.31
Racemic IB[b]			0.91
(–)-IB[b]	*S*	*R*	1.1
(+)-IB[b]	*R*	*S*	0.2

[a] Based on the effective dose required to give 50% relaxation of histamine-induced constriction of isolated guinea pig tracheal preparations.
[b] The compound was used as its fumarate; the molar ratio of the compound to fumaric acid was 2:1.

reside almost exclusively in one of their isomers having the *R* configuration. However, the difference in potency between the isomers of I was far smaller than might have been expected from that of the sympathomimetic amines cited above. (–)-1A with an αR, βR configuration was only 14 times more potent than (–)-IB that has an αS, βR configuration, and the difference in potency between (+)-IB and (+)-IA, which have the configurations αR, βS and αS, βB, respectively, was only marginal.

Another interesting feature that may be noted from Table 10.3 is the effect of configuration around the β-carbon on the potency of the isomers. The isomer (–)-IA with an αR, βR configuration was about 3 times as potent as (+)-IB, whose configuration is αR, βS. On the other hand, the potency of (+)-IA with an αS, βS configuration was about three times that of (–)-IB, whose configuration is αS, βR. These data show that the configuration at the β-carbon also influences the bronchodilator potency of compound I, although it is impossible to correlate a particular configuration with the increase in potency. In this connection, it is pertinent to quote the β-blocking activity reported for the four isomers of 2-(α-methyl-2-phenethyl-amino)-1-(2-naphthyl)ethanol, which, like compound I, has two asymmetric carbons α and β, the former neighboring the naphthyl group and the latter carrying the amine moiety. The β-blocking activity of the isomers, as reported, decreases in the order αR, βR > αR, βS >> αS, βR > αS, βB. This shows that with this compound, the *R* configuration at the β-carbon, as well as at the α-carbon seems to be the preferred configuration for the potency. The relationship between the stereochemistry and the β-stimulant activity of the series of compounds including I is quite complicated and much remains to be elucidated before a general rule can be worked out.

10.5. BIOLOGICAL SIGNIFICANCE

The enantiomers of promethazine are almost equivalent in terms of their antihistaminic properties and toxicity, yet the β-blocking activity of (–)*S*-propanolol is considerably greater than that of (+)*R*-propanolol (19). With the intravenous anesthetic ketamine, which is routinely administered as the racemate, reports indicate that the (+)*S*-isomer is superior to the (–)*R*-isomer in terms of the provision of adequate anesthesia. Additionally, the (–)*R*-isomer of ketamine has been shown to be the major cause of the postoperative side-effects (hallucinations and other transient psychotic sequelae) observed with the use of racemic ketamine. Two examples of compounds that had been developed as single-isomer drugs, namely, D-penicillamine and L-dopa, have resulted in decreased toxicity compared to their racemates (21). As both these compounds are relatively old, the decision to use single isomers must surely relate to the technology available at the time of their development, both being based on amino acid chemistry, and hence adds to the argument that with current technology single-isomer drugs should become the norm rather than the exception. This suggests that stereochemical differences in pharmacology and toxicology should be studied early in drug development so that a rational decision can be made at that stage on the material to be finally developed and marketed.

A potential commercial problem can arise if a single isomer of a previously marketed racemate could be patented by a company other than the compound's originator. In this situation, obviously the competitive position of the originator in the marketplace would be difficult. The significance of chirality in the assessment of human and veterinary marketing authorizations needs to be addressed. It may be emphasized that in all cases, stereoisomers are different compounds rather than different forms of the same compound.

The knowledge that enantiomers of chiral compounds may differ widely in biological activity, qualitatively as well as quantitatively, is well known (22). Nevertheless, most of the pharmacologic data available to date on chiral drugs is obtained from experiments with racemates, which assume that the biological activity generally resides in one of the enantiomers. With the advancements made in the stereospecific synthesis and stereoselective analysis of drugs, pharmacologists are now offered new possibilities to explore the steric aspects of drug action. Unfortunately, the degree of resolution is seldom specified in published work on the stereoselectivity of drugs. Discussed below are examples of derivatives of phenylethylamine that act with adrenergic mechanisms.

10.5.1. Compounds with One Chiral Center

Terbutaline is a β_2-selective adrenoceptor agonist with one chiral center. The (–)-enantiomer of this compound is a potent relaxant of tracheal smooth muscle, whereas the (+)-enantiomer is ≈3000 times less active in this respect (Table 10.4). To achieve this result, the distomer must not be contaminated with more than 0.03% of the eutomer. The effect of both enantiomers was blocked by *rac*-propranolol, but the distomer markedly less than the eutomer. This may indicate an unspecific relaxing power of the distomer at the very high concentrations in question. Alternatively, an atypical β-adrenoceptor is involved. It seems that the validity of the eudismic ratio obtained experimentally is limited by both the enantiomeric purity of the drug and problems with specificity at very high drug concentrations. No interaction was found between the two enantiomers of terbutaline on tracheal smooth muscle, even while the distereomer is in a hundred-fold excess (Table 10.4).

Table 10.4. Steric Aspects of Agonism and Antagonism at β-Adrenoceptors in Guinea Pig Trachea (22)

Compound	Potency	Eudismic Ratio
Agonism, pD_2		
(–)-Terbutaline	7.28	3300
(+)-Terbutaline	3.76	
(–)-Terbutaline with 5 mmol/L of (+)-terbutaline	7.78	
(–)-Clenbuterol	8.54	>10,000
(+)-Clenbuterol	<5.5	

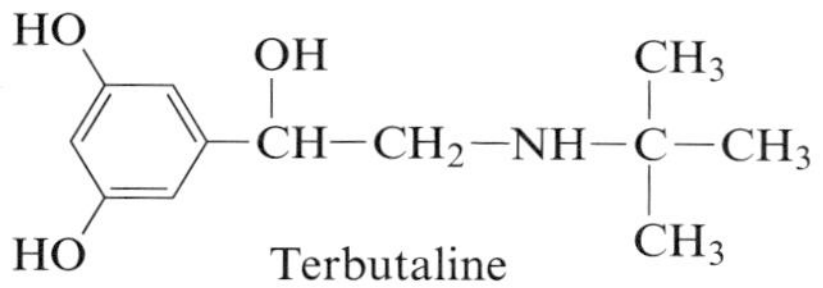

HO
OH CH₃
CH–CH₂–NH–C–CH₃
HO CH₃

Terbutaline

H₂N–SO₂
OH
H₃C– –CH–CH₂–NH–CH₂–CH₂–O–
OCH₃

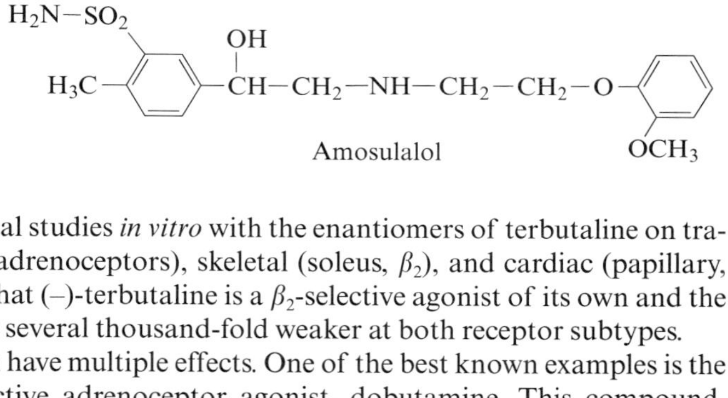

Amosulalol

Detailed functional studies *in vitro* with the enantiomers of terbutaline on tracheal (mainly β_2-adrenoceptors), skeletal (soleus, β_2), and cardiac (papillary, β_1) muscle show that (–)-terbutaline is a β_2-selective agonist of its own and the (+)-enantiomer is several thousand-fold weaker at both receptor subtypes.

Racemates can have multiple effects. One of the best known examples is the reputedly β_1-selective adrenoceptor agonist, dobutamine. This compound, widely used as a tool for characterizing receptors, consists of one β-agonist, the (+)-enantiomer, and a partial α-agonist, the (–)-enantiomer. Another example of the dual properties of a chiral adrenoceptor ligand is the antagonist amosulalol. The (–)-enantiomer inhibits β-adrenoceptors unselectively and, in the same concentration range, α_1-adrenoceptors, whereas a hundred-fold higher concentration is required to inhibit α_2-adrenoceptors (Table 10.5). The (+)-enantiomer, however, is a potent and highly α_1-selective antagonist with much lower affinity for β-adrenoceptors. Interestingly, the eudismic ratio is much higher for interaction with β-receptors than for α-receptors. It is important to note that for β-effects the (–)-isomer is the eutomer while the condition is reversed for α-effects.

Table 10.5. pA_2 Values for Amosulalol and Its Desoxy Derivative (22)

Receptor	(−)	(+)	Desoxy	Eudismic Ratio
$_1$, Rat right atrium	7.71	6.03	6.19	48
β_2, Guinea pig trachea	7.38	5.71	5.76	47
α_1, Rabbit aorta	7.17	8.31	8.14	14
α_2, Rat vas deferens	4.92	5.36	6.05	3

Removal of the β-hydroxyl group of amosulalol results in a marked drop in its affinity for β-receptors, while the affinity for a-receptors is unchanged (α_1) or even increased (α_2). This finding does not fully comply with the Easson–Stedman hypothesis, which postulates the three-point interaction of the drug with the receptor. When adopted for phenylethanolamines, this hypothesis says that if the β-hydroxyl group is in the "wrong" position (*S* configuration), the potency of the compound equals that of the desoxy derivative. Since both adrenaline and noradrenaline appear to comply with the Easson–Stedman hypothesis for both receptor types, presumably it is the phenoxyethylamine rather than the phenylethanolamine part of amosulalol that interacts with the α-adrenoceptors. Thus, α- and β-adrenoceptors may interact with different parts of the same molecule. This suggestion is supported also by the "anti-Pfeiffer" behavior of amosulalol. Pfeiffer's rule says, in principle, that the eudismic ratio for enantiomer pairs increases with the increased affinity of the eutomer. Amosulalol has a high affinity and a low eudismic ratio when it interacts with β-adrenoceptors. This anomaly can be explained by proposing that the phenoxy-ethylamine part of the molecule interacts with the α-adrenoceptor, in which case the β-hydroxyl group will have a more remote position. Similar results and a similar interpretation have been given for the enantiomers of carvedilol, another α- and β-adrenoceptor antagonist.

10.5.2. Compounds with Two Chiral Centers

With two chiral centers, four different stereoisomers (two diastereomer pairs) are possible. Although many drugs have two chiral centers, there are few examples of a complete pharmacologic examination of the different enantiomers. Among the N-substituted phenylethanolamines, there are three structurally related adrenoceptor ligands, two agonists, and one antagonist, whose individual stereoisomers have been investigated. These compounds are the β_2-selective agonist formoterol, a *p*-trifluoromethyl anilide derivative (PTFMA) with β-agonistic properties, and labetalol, which blocks α- and β-adrenoceptors.

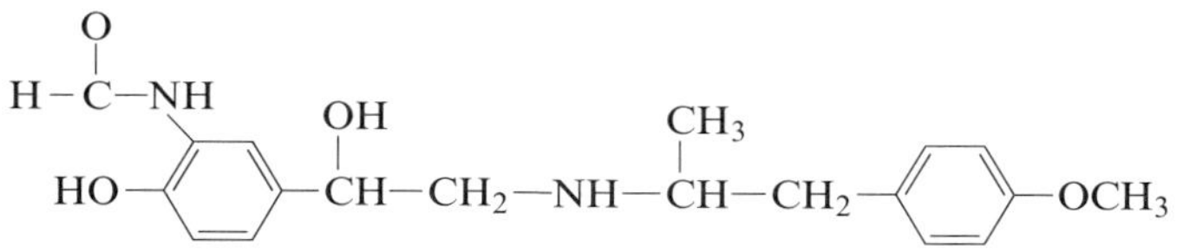

Formoterol

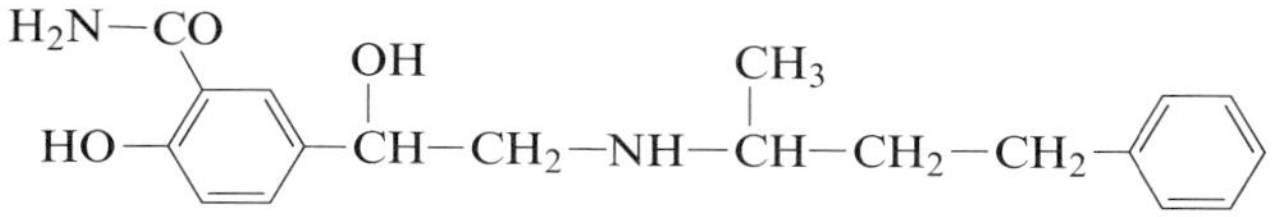

Labetalol

For the enantiomers of formoterol and their diastereomers, the order of potency is (*R*; *R*) >> (*R*; *S*) = (*S*; *R*) > (*S*; *S*) with respect to relaxation of the airway smooth muscle. Comparable results were obtained with PTFMA for the activation of adenylate cyclase in turkey erythrocytes and for the stereoisomers of labetalol with respect to the inhibition of β_1-adrenoceptors in guinea pig heart.

The (*R*; *R*)-enantiomer is the most potent for both compounds, followed by the (*R*; *S*)-diastereomer. A common feature is also the relatively high eudismic ratio (*R*; *R*)/*S*; *S*) and the low ratio of (*R*; *S*)/*S*; *R*). This suggests that the configuration of the nitrogen substituent is critical for the interaction of the phenylethanolamine moiety of the molecule with the β-adrenoceptor.

There appears to be quite different steric requirements for the interaction of labetalol with the α-adrenoceptor. In this respect, the (*S*; *R*)-isomer is the most potent, and the eudismic ratio (*S*; *R*)/(*R*; *S*) is about 50. There is no clear difference in potency between the (*R*; *R*)- and (*S*; *S*)-enantiomers, which is in sharp contrast to the condition for interaction with β-adrenoceptors. It appears that α- and β-adrenoceptors interact with different parts of the molecule.

From these examples of structurally related compounds, it appears that the *R*-configuration at the carbon atom carrying the hydroxyl group is essential for interaction with β-adrenoceptors. The influence of the configuration at the carbon atom attached to the nitrogen is variable, but in no case is the (*R*; *S*)-isomer more potent than the (*R*; *R*)-isomer. It should be noted, however, that between the (*R*; *R*)- and (*R*; *S*)-diastereomers, there may be not only potency differences but also differences in selectivity for β-adrenoceptors in different tissues.

10.5.3. Biological Consequences

The reported eudismic ratio may differ from one laboratory to another. Thus, the eudismic ratio for the relaxant effect of salbutamol on guinea pig tracheal smooth muscle has been reported to be 70:300. Part of this variation may be due to traces of the more active isomer in the less active one. This factor becomes more important with the higher true eudismic ratio, as was pointed out by Barlow et al. (23) over 20 years ago. From this equation, it follows that the higher the observed eudismic ratio, the more uncertain the true ratio, since minute traces of the eutomer in the distomer are difficult to detect by analytical methods.

The degree of resolution is rarely specified in the pharmacologic literature, and when it is, there is often just a note on the optical rotation. This measure of enantiomeric purity is far from reliable when the enantiomeric contamination is below a few percent. Since eudismic ratios observed in biological experiments often fall in the range of 100–1000, more sensitive analytical methods have to be utilized. When the degree of enantiomeric purity is uncertain, conclusions regarding chiral aspects of structure-activity relationships, including an evaluation of Pfeiffer's rule, should be made with care. Particularly, subtle differences between diastereomers, for example, must be interpreted with caution when the enantiomeric purity is not known. This is illustrated by the following example.

In the first report on the stereoisomers of formoterol, the highest potency ratio obtained for the relaxation of tracheal smooth muscle was about 14, far smaller than what might be expected, and probably a result of incomplete resolution (22). When the contamination with the (*R*; *R*)-isomer was reduced from 1–0.1%, the eudismic ratio (*R*; *R*)/(*S*; *S*) increased from 50–850, and the order of potency between the (*S*; *R*)- and (*S*; *S*)-isomers was reversed. At the same time, the potency difference between the (*R*; *S*)- and (*S*; *R*)-enantiomers disappeared. A further increase in purity might reduce the potency of the (*S*; *S*)-enantiomer even more, but the potency of the (*S*; *R*)-enantiomer appears to have approached its true value.

In the pharmacologic characterization of chiral compounds, it is essential to find out whether the distomer may have adverse effects. In experiments addressing this question, very high doses are usually employed. If the distomer in this case is contaminated with the eutomer, the effects observed may be due to the traces of the eutomer or may be the result of an interaction between the enantiomers at a fixed ratio different from the racemate rather than the effect of the distomer, per se. This possibility must be considered in a safety assessment.

A controversy exists concerning the therapeutic use of nonsteroidal antiinflammatory drugs (NSAIDs) as to whether these drugs should be produced, marketed, and used clinically as the racemate or a single enantiomer (24). Traditionally, the therapeutic and major toxic effects of NSAIDs have been attributed to the ability of these drugs to inhibit the synthesis of stable prostaglandins, through the direct inhibition of prostaglandin H synthetase, which serves both as a cyclooxygenase and peroxidase. These properties have been largely determined from the results of *in vitro* studies and, in turn, have shown a marked stereoselectivity for chiral NSAIDs, in favor of the *S*-enantiomers (30–100 times), as opposed to their *R*-antipodes. Therefore, it has been suggested that the *R*-enantiomers of NSAIDs are unnecessary impurities or isomeric ballasts and a stereochemically pure enantiomer is superior to its respective racemate.

Malmberg and Yaksh (25) have suggested that the powerful antiinflammatory effects of NSAIDs have most likely diverted attention away from many other properties of therapeutic relevance. For example, indomethacin is twice as potent in the inhibition of cyclic AMP-dependent protein kinase than in the inhibition of prostaglandin H synthetase. Subsequently, many other non-prostaglandin-dependent mechanisms, including the uncoupling of oxidative phosphorylation, inhibition of renal anion transport, changes in neutrophil and leukocyte function, and interruption of signal transduction through G proteins, have been identified as therapeutically relevant.

The most significant adverse effects of NSAID therapy occur in the gastrointestinal (G.I.) system. The G.I. effects of NSAIDs have predominately been presented as ulceration of the stomach and duodenum. There is a growing body of evidence, however, that more distal (small intestinal) damage may be more widespread, persistent, and serious than previously thought. In addition, it has been demonstrated that NSAIDs cause increased small intestinal permeability at the level of the mucosal tight junction and that this may lead to intestinal inflammation, which in turn has been implicated in the genesis of the more serious intestinal sequelae. The permeability of tight junctions may be partly regulated by prostaglandins. Therefore, one can expect NSAIDs to cause increased intestinal permeability. Further, it may be reasonable to expect that the effect on intestinal permeability is attributed almost exclusively to the *S*-enantiomers.

Although etodolac is used clinically as a racemate, its prostaglandin synthetase inhibition has been attributed almost exclusively to the *S*-enantiomer. Due to metabolic inversion to the active *S*-enantiomer, the intrinsic effect of the *R*-enantiomer is difficult to examine for many chiral NSAIDs. However, because of the fixed nature of its asymmetric center, the *R*-enantiomer of etodolac does not undergo chiral inversion. The unexpected increase in the urinary excretion of Cr-EDTA in rats, following (*R*)-etodolac administration, is direct evidence for its intrinsic activity in increasing small intestinal permeability. This suggests that the inhibition of prostaglandin synthetase may not be the only mechanism through which etodolac affects the G.I. tract. Indeed, (*R*)-etodolac is only one order of magnitude less potent in increasing intestinal permeability than the *S*-enantiomer, which contrasts with the up to 150-fold difference in potency observed *in vitro* against prostaglandin synthetase.

Further, the increase in urinary excretion of Cr-EDTA by the *S*-enantiomer (6 mg/kg) appears to be equal to, if not greater than, that of the racemate [12 mg/kg containing 6 mg/kg (*S*)- and 6 mg/kg (*R*)-etodolac]. This may suggest that despite the apparent activity of both enantiomers, their presence in the racemate may exert an ameliorating effect on the extent of intestinal permeability observed. The discrepancy between the effects of individual enantiomers and their racemate may result, in part, from differences in their physicochemical properties, which in turn may alter their biodistribution and even drug-receptor interaction. It has been shown that the enantiomers of ibuprofen have substantially different solubility parameters than racemic ibuprofen. In addition, the octanol/water partition coefficient of etodolac enantiomers has been shown to be affected by the presence of the antipode.

The data suggest that mechanisms other than peripheral prostaglandin synthetase inhibition may be responsible for the effects of etodolac on intestinal permeability. Further, the data demonstrating the activity of the *R*-enantiomer in increasing intestinal permeability are congruent with the observations of other investigators who have shown that the *R*-enantiomers of ibuprofen and flurbiprofen produce both analgesia and changes in polymorphonuclear lymphocyte function. From the therapeutic perspective the decision to employ either the racemate or stereochemically pure enantiomer must be based on examination of the relevant and possibly diverse mechanisms of action rather than the single most obvious effect.

10.6. PHARMACOKINETICS

Pirprofen is a member of the 2-arylpropionic acid (2-APA) class of nonsteroidal antiinflammatory drugs (NSAIDs). The drug has a chiral center and was marketed as the racemate prior to its removal from the pharmaceutical market due to several reported cases of hepatotoxicity (26). Similar to other chiral NSAIDs, the *S*-enantiomer of pirprofen possesses much more pharmacologic activity than its *R*-antipode.

Stereoselective pharmacokinetics have been observed for several chiral NSAIDs. Furthermore, the degree of stereoselectivity is both drug- and species-specific. As examples, ketoprofen and flurbiprofen each display a considerable degree of stereoselectivity in their pharmacokinetics in the rat, yet show little enantioselective disposition in humans. On the other hand, both humans and rats share similar patterns and degrees of stereoselectivity in the pharmacokinetics of etodolac, ketorolac, and fenoprofen.

For an important class of chiral NSAIDs, the 2-APA derivatives, part of the interspecies variation in the disposition of the enantiomers, can be explained on the basis of the abilities of the species to bioinvert the *S*-enantiomer to its antipode. This pathway of metabolism may occur both systemically and presystemically. Therefore, the route of administration and the properties of the dosage form may influence the pharmacokinetics of 2-APA NSAIDs that undergo inversion.

The enantioselective disposition of pirprofen has been studied in 11 healthy human volunteers. Although pharmacokinetic indices were not reported, it appeared that after oral doses the plasma concentrations of *S*-pirprofen were higher than those of the respective *R*-enantiomer. Likewise, in a preliminary study involving three female rats dosed orally, the area under the plasma concentrations time curve (AUC) of *S*-pirprofen were 22% higher than those of the *R*-enantiomer.

The pharmacokinetics of the enantiomers of the NSAID drug pirprofen were studied in male Sprague-Dawley rats after oral and intravenous (i.v.) doses of the racemate. No significant differences were detected between the enantiomers after oral or i.v. dosing in t, Vd, or ΣXu. However, the *R*:*S* area under the plasma concentration (AUC) ratio after oral doses (0.92 ± 0.13) was slightly but significantly lower than matching i.v. doses (1.05 ± 0.036). The absolute bioavailability of the active *S*-enantiomer (78.5%) after oral doses was higher than that of the inactive *R*-enantiomer (69.3%). The plasma protein binding of both enantiomers was saturable over a five-fold range of plasma concentrations. At higher plasma concentrations, the *S*-enantiomer was less bound than the *R*-enantiomer. In an *in vitro* experiment using everted rat jejunum, no chiral inversion was discernible. The dependency of the AUC ratio of the enantiomers on the route of administration may be due to stereoselective first-pass metabolism.

Deuterium labeling techniques and stereoselective GC/MS methodology have been employed to investigate the mechanism by which R-ibuprofen undergoes metabolic chiral inversion in the rat *in vivo* (27). Following oral administration of a mixture of *R*-ibuprofen (7.5 mg kg^{-1}) and *R*-[ring-$^2H_4$; 2-^2H]ibuprofen (*R*-[$^2H_5$]ibuprofen) (7.5 mg kg^{-1}) in male Sprague-Dawley rats, the enantiomeric composition and deuterium excess of the drug were determined in serial plasma samples and pooled urine collected over 10 h. The results demonstrate that:

1. *R*-ibuprofen undergoes extensive inversion of configuration to its *S*-antipode in the rat.
2. Chiral inversion of *R*-[$^2H_5$] ibuprofen yields *S*-[$^2H_4$]ibuprofen in a process that involves the quantitative loss of the deuterium atom present originally at C-2.
3. Labeling of *R*-ibuprofen with deuterium at C-2 does not introduce a measurable kinetic deuterium isotope effect on the chiral inversion reaction.

4. Metabolism of *R*-[$^2H_5$]ibuprofen leads to the appearance in plasma and urine of molecules of *R*-ibuprofen labeled with four atoms of deuterium.

On the basis of these findings, a mechanism is proposed for the chiral inversion reaction that invokes the stereoselective formation of the coenzyme A thioester of *R*-ibuprofen as a key metabolite; conversion of this species to the corresponding enolate tautomer affords a symmetrical intermediate through which the racemization of ibuprofen occurs *in vivo*.

The influence of aging on the pharmacokinetics and tissue distribution of (*R*)- and (*S*)-propranolol was studied in 3-, 12-, and 24-month-old rats. After both i.v. and oral administration of *rac*-propranolol, the plasma concentrations were higher for the (*R*)- than (*S*)-enantiomer (28). For the tissue concentrations, the reverse was true. The free fraction of (*S*)-propranolol in plasma was about four times larger than that of (*R*)-propranolol, and this is the main factor responsible for the differences in kinetics between the two enantiomers. There was a suggestion for a difference in tissue binding between the two enantiomers. With aging, the plasma and tissue concentrations of both enantiomers increase, probably due to a decrease in blood clearance. Tissue binding did not change much with aging. Notwithstanding the marked differences between the kinetics of the propranolol enantiomers, the changes that occur with aging affect both enantiomers to the same degree.

The plasma disposition of the enantiomers of ibuprofen has been investigated following the oral administration of the racemic drug (400 mg) to 24 healthy male volunteers (29). The plasma elimination of (*R*)-ibuprofen was found to be more rapid than that of the *S*-enantiomer [plasma half-life: (*R*) 2.03 h; (*S*) 3.05 h; $2P < 0.001$], resulting in a progressive enrichment in the plasma content of this isomer, some 64% of the total area under the plasma concentration time curves (AUC) being due to the pharmacologically active enantiomer. The influence of dose on the pharmacokinetic characteristics of the enantiomers of ibuprofen, over the range of 200–800 mg, was investigated in three subjects. Examination of dose-normalized AUC values and oral clearance indicate the dose dependence of (*R*)-ibuprofen disposition.

The enantioselective protein binding of mephobarbital (MPB) was investigated in human plasma and human serum albumin solutions by equilibrium dialysis (30). A small but statistically significant difference was observed in the *in vitro* plasma protein binding of the enantiomers; (*S*)-MPB was ~59% bound and (*R*)-MPB 67% bound. The binding to albumin [(*S*)-MPB: 29% bound and (*R*)-MPB ~41% bound] was less than to plasma proteins but showed somewhat greater enantioselectivity, suggesting that albumin binding is a major source of the enantioselectivity in plasma. The effects of MPB concentration, varying enantiomeric concentration ratio, and phenobarbital on the enantioselective binding of MPB were studied. The effect of age was also investigated by measuring the binding in plasma from eight young (18–25 years) and eight elderly (>60 years) male subjects who took single doses of MPB. The results were in close agreement with the *in vitro* binding data, and the binding of both enantiomers was marginally but significantly lower in the young compared with the elderly subjects. These differences in binding were consistent with previously observed pharmacokinetic differences between the two subject groups.

Mefloquine (MQ) is a chiral antimalarial agent effective against chloroquine-resistant *Plasmodium falciparum* (31). It is commercially available as a racemic mixture of the (+)- and (–)-enantiomers for oral administration. The pharmacokinetics of the (+)- and (–)-enantiomers of MQ were studied in eight healthy volunteers after administration of a first oral dose of 250 mg of racemic MQ and at steady state after 13 repeated doses of 250 mg given at 1-week intervals. Plasma samples were collected, and concentrations of each enantiomer were determined using a previously described achiral–chiral double column-switching liquid chromatographic method. At each time point, higher plasma concentrations values were found for the (–)-enantiomer ($p < 0.001$). At steady state, C_{max} values of (–)-MQ were higher than those of (+)-MQ (1.42 ± 0.19 vs. 0.26 ± 0.05 mg/l; $p < 0.001$). Similarly, the plasma concentrations 7 days after the final dose were higher for (–)-MQ (1.01 ± 0.26 vs. 0.11 ± 0.04 mg/L; $p < 0.001$). AUC values at steady state were also higher for (–)-MQ (197.3 ± 36.7 vs. 30.1 ± 8.9 mg/l·h; $p < 0.001$). The terminal half-life values ($T_{1/2\beta}$) were longer for (–)-MQ (430.4 ± 225.2 vs. 172.8 ± 56.5 h; $p < 0.001$). This study shows that the pharmacokinetics of MQ are highly stereoselective.

5-Dimethylsulfamoyl-6,7-dichloro-2,3-dihydrobenzofuran-2-carboxylic acid (DBCA), a promising uricosuric, diuretic, and antihypertensive agent, was administered intravenously to rats (32). The levels of DBCA in plasma, and the areas under the curve of concentration vs. time (AUC values) of the *S*(–)-enantiomer were higher than those of the *R*(+)-enantiomer. Total body clearance was significantly greater for the *R*(+)-enantiomer. This stereoselective elimination was due to a difference in the nonrenal clearance, which seemed to reflect hepatic metabolism or biliary excretion. Hepatic metabolism seemed more likely because AUC and the amount of urinary excretion of the N-monodemethylated metabolite of DBCA were greater for the *R*(+)-enantiomer. The plasma had higher free fractions of the *S*(–)-enantiomer, a result suggesting that this enantiomer is distributed more readily to the tissues, including the liver. This result indicates that protein binding was not responsible for the stereoselective metabolism of (*R*)(+)-DBCA. Although there was no difference in the renal clearances of the enantiomers, the renal clearance of free (*R*)(+)-DBCA exceeded that of the *S*(–)-enantiomer, a result indicating the preferential excretion of the *R*(+)-enantiomer into the urine.

Comparison of the pharmacokinetics of individual enantiomers after intravenous administration of each enantiomer or its racemate showed that the enantiomers interact with one another; dosing with the racemate delayed the elimination of each enantiomer because of the mutual inhibition of hepatic metabolism and renal excretion for (*R*)(+)-DBCA and of renal excretion for (*S*)(–)-DBCA.

10.7. DETOXIFICATION

The detoxification of the enantiomers of glycidyl 4-nitrophenyl ether (GNPE), (–)(*R*)- and (+)(*S*)-GNPE, and glycidyl 1-naphthyl ether (GNE), (–)(*R*)- and (+)(*S*)-GNE, by rat liver glutathione transferase and epoxide hydrolase has been studied (33). Enantioselectivity is observed with both enzymes favoring the (*R*)-isomers as determined by the formation of conjugate, diol, and remaining substrate measured by HPLC. Enantiomers of GNE are detoxified by cytosolic epoxide hydrolase but those of GNPE were not. Substantial nonenzymatically formed conjugates of enantiomers of GNPE are detected, showing (*S*)-GNPE to be the more reactive of the pair.

The influence of a single oral dose of 30 mg of nicardipine on the pharmacokinetics of (*R*)- and (*S*)-propranolol, given orally as 80 mg of *rac*-propranolol, was studied in 12 healthy volunteers (34). The plasma concentrations were higher for the (*S*)-enantiomer than for the (*R*)-enantiomer. The Cl_o and Cl'_{intr} of (*S*)-propranolol were significantly lower than the Cl_o and Cl'_{intr} of (*R*)-propranolol. The unbound fraction of (*R*)-propranolol was significantly higher than that of (*S*)-propranolol. These changes were more important for (*R*)- than for (*S*)-propranolol. The protein binding was not altered by nicardipine. The enantioselective effect of nicardipine on the metabolic clearance of propranolol appears to be due to an interaction at the level of the metabolizing enzymes. The effect on blood pressure of *rac*-propranolol was little affected when nicardipine was coadministered with *rac*-propranolol, and its bradycardic effect was reduced.

A surprisingly large number of marketed drugs are racemic mixtures. The pharmacokinetic literature on racemic drugs contains a vast amount of information on drug–drug interactions derived from the measurement of total drug concentrations in plasma and urine. The appreciation of the role of stereochemistry in drug interactions with racemic warfarin resulted in long-overdue scientific rigor being applied to the study of drug interactions. It is compelling that much of the literature was uninterpretable. A better understanding of oxidative metabolism, particularly the complexity of the cytochrome P-450 family of enzymes, has also strengthened the scientific basis of drug interactions. The investigator and clinician alike must consider both stereoselectivity and isozyme selectivity in the study of drug interactions to understand the nature of the interaction so as to more effectively use new and potent drugs (35).

A solvent mixture containing dioxane, acetonitrile, and hexane was found to be suitable as a mobile phase to resolve oxazepam enantiomers by chiral stationary phase high-performance liquid chromatography using covalent Pirkle columns (36). The resolved oxazepam OX enantiomers in this solvent mixture had a racemization half-life greater than 3 days at 23°C. When desiccated at 0°C as dried residue, OX enantiomers were stable for at least 50 days with less than 2% racemization. The conditions that stabilized OX enantiomers significantly facilitated the determination of racemization half-lives of OX enantiomers in a variety of aqueous and nonaqueous solvents and at different temperatures.

Chlorthalidone (CTD), a diuretic and antihypertensive agent, is a 3-hydroxy-3-phenylphthalimidine (HPP) carrying substituents on the phenyl group (37). Chromatographic investigation of the degradation pathways of CTD showed the formation of a typical set of products. With one exception, the resulting chromatographic peaks were traced to previously identified compounds; the unknown component was found to be in equilibrium with CTD in aqueous media and has been identified as a dehydrated form, Δ^2-CTD.

The presence of the minor equilibrium product (approximately 0.7% at room temperature) spurred its investigation as a potential intermediate in the racemization of CTD. On the basis of structure alone, the facile racemization of the enantiomers of CTD was postulated and corroborated by a recent report of CTD racemization catalyzed by acid or base. The study focused on the kinetics and activation parameters of a tripartite equilibrium found to exist between the enantiomers of CTD and achiral Δ^2-CTD through a planar carbonium ion intermediate; all reactions proceed under conditions approximating pure water at room temperature.

The enantiomers of chlorthalidone (CTD) and a minor achiral dehydration product, Δ^2-CTD, have been shown to exist in dynamic equilibrium in aqueous media through a carbonium ion intermediate. The barrier to inversion at carbon is low for an uncatalyzed system: $\Delta G = 21.6$ kcal/mol. This behavior extends to other 3-hydroxy-3-phenylphthalimidines (HPPs), with formation of the analogous D^2-HPP blocked by alkyl substitution at the 2-nitrogen.

The presence of one or several elements of chirality (centers, axes or planes of chirality, and generally helicity) in drug molecules generates specific properties that may be advantageous in some cases, but inevitably require special attention. Examples of advantages include the possibility of increased selectivity and the fact that chirality, per se, is an invaluable probe in molecular pharmacology and biochemistry. In contrast, problems generated by stereoisomerism include the need for stereospecific analytical methods, the influence of the degree of resolution on activity, and the increased complexity of metabolic, pharmacologic, and clinical studies.

One major problem arising from the presence of elements of chirality is their possible lack of configurational stability, that is, the danger of interconversion of stereoisomeric drugs. At a recent symposium ("Chirality at the Crossroads," April Paris, 26–29, 1992), the problem of the interconversion of stereoisomers was often mentioned and discussed (38). During the symposium, it rapidly became obvious that the use of this term suffers from semantic confusion and even poor understanding, resulting in unproductive discussions and suggesting the possibility of inappropriate regulations. We draw from our experience as chemists and pharmacologists and attempt to clarify for nonspecialists the issue of configurational instability. By classifying and discussing recognized examples, a few predictive rules can be offered, which should be an incentive for much-needed, systematic investigations.

All cases considered involve the configurational instability of a single element of chirality either in compounds that contain only one such element (that is, interconversion between enantiomers, chiral inversion), or in compounds that contain two or more such elements (that is, interconversion between epimers, epimerization). This is the first criterion of classification used here. The second criterion separates nonenzymatic from enzymatic reactions. In addition, a distinction is made based on the nature of the unstable element of chirality.

A number of stereogenic elements in molecules may not be configurationally stable. Thus, although as a rule, tetracoordinate chiral centers are stable, tricoordinate chiral centers of first-row atoms (C, O, and N) are not. Chiral axes and chiral planes cover a continuum of possibilities between high lability and high stability. Attention needs to be focused on (1) tetracoordinate chiral carbon atoms displaying decreased stability, and (2) helicity resulting from a nonsymmetrical, nonplanar ring system. Two recent studies may be consulted regarding the stereodynamics of axes of chirality.

A variety of reactions can be categorized under the global concept of the interconversion of stereoisomers. Thus, racemization or epimerization can result from the inversion of labile chiral centers. From the examples available, some predictive rules are suggested for a chiral center of the type R″R′RC—H undergoing base-catalyzed inversion and a provisional table of affecting groups is presented. Unimolecular inversion of nonsymmetrical, nonplanar ring systems can also result in racemization or epimerization, but no generalization can yet be offered. Besides these cases of nonenzymatic reactions, a limited variety of enzymatic reactions can operate to interconvert stereoisomers, the outcome rarely being a racemic mixture. An important aspect of stereoisomer interconversion is the timescale in which the phenomenon is observed. Thus, several reactions to nonenzymatic racemization or epimerization are fast compared to the duration of action of the drug and therefore have pharmacologic significance, whereas others are slower and are of pharmaceutical relevance only.

It must be remembered that configurational stability and instability are relative phenomena. Given proper conditions (temperature and pH), no stereoisomer is configurationally stable. However, only two timescales and related sets of conditions are relevant as far as drugs are concerned. The pharmaceutical timescale and conditions imply that drugs remain (configurationally) stable during the whole manufacturing process and for the shelf life, whereas the pharmacologic timescale is concerned with stability under physiological conditions (37°C, pH 7.4) and for the time of residence of the drug in the body. Emphasis is generally on the pharmacologic timescale, but most of the information available is to be found in reports of pharmaceutical investigations.

Stereoisomeric enrichment occurring *in vivo*, in particular, enrichment of a stereoisomeric impurity, should not be confused with the interconversion of stereoisomers. Indeed, differences in rates of metabolism or excretion may profoundly affect the stereoisomeric ratio of the drug in blood or urine. Assessment of mass balance, or the use of compounds of very high stereoisomeric purity, should allow such confusions to be avoided.

To bring the problem of racemization into an even broader context, it must be recalled that the stereoisomeric purity of some amino acid residues in a number of proteins is known to be decreased in aged humans. Thus, both collagen and proteoglycan show an age-dependent accumulation of D-aspartic acid. This problem of "protein racemization" as it is (not quite properly) labeled is now beginning to attract attention as a possible mechanism or marker of tissue aging.

10.8. STABILITY

An example of a drug displaying extreme chiral instability is oxazepam, a drug whose kinetics of racemization are studied in organic solvents and aqueous solutions (38). These studies are interesting and relevant from a pharmacologic viewpoint, since they indicate that oxazepam racemizes at ambient temperature and in the neutral pH range with a pseudo-first-order rate constant of 0.1 ± 0.05 min^{-1} (a half-life of racemization of approximately 10 ± 5 min), suggesting a half-life of racemization of 1–4 min at 37°C. This rate of racemization is extremely fast compared to the duration of action of the drug, indicating that oxazepam is correctly viewed as a single compound existing in two very rapidly interconverting chiral states.

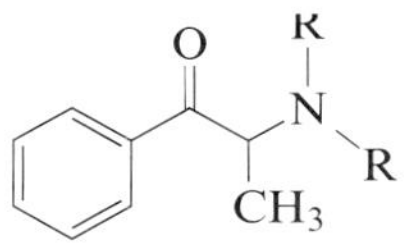

Amfepramone congeners

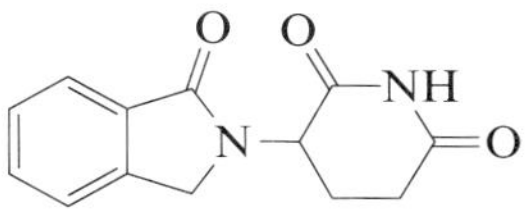

Thalidomide

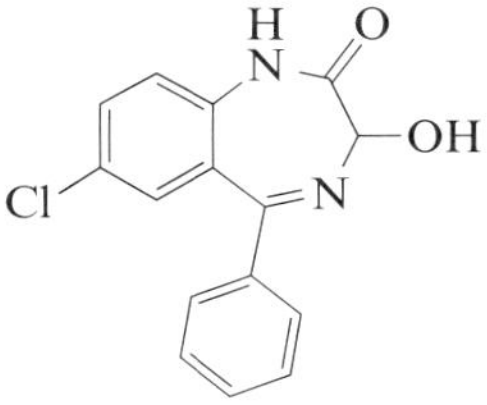

Oxazepam

Kinetic information is also available for the anoretic agent amfepramone (II, R = R′ = ethyl, also known as diethylpropion) and its two N-deethylated metabolites, N-ethylamino-propiophenone (II, R = ethyl, R′ = H) and amino-propiophenone (II, R = R′ = H, also known as *rac*-cathinone). With a common intermediate for racemization and α-carbon deuteration assumed, the rate of the latter reaction was measured in D_2O (35°C and pD 7.3–7.4), revealing $t_{1/2}$ values of 15 and 21 h for the tertiary and secondary amine, respectively. The half-life of racemization of the primary amine was estimated to be about 30 h. These values suggest that configurational inversion should have only a marginal influence on the pharmacologic activity of these compounds, their biological half-lives being considerably shorter.

In thalidomide (2-phthalimidoglutarimide), teratogenic activity on mice and rats was reported to reside in the (–)(*S*)-enantiomer, but racemization was seemingly overlooked. Indeed, the compound was found to racemize with relative ease (residual enantiomeric excess of 2/3 after 5 days in DMF or 80% DMF at room temperature). Studies under biomimetic conditions yielded a half-life of racemization of 2.5 h in phosphate buffer at pH 7.4 and 37 °C, with serum albumin markedly accelerating the reaction. In fact, thalidomide isolated from the plasma of rabbits 2 h after i.v. injection was completely racemized. Facile racemization has also been demonstrated by two analogs of thalidomide, 2-phthalimidinoglutarimide (EM 12) and 2-phthalimidoadipinimide. The former compound, for example, racemized in marmosets with a $t_{1/2}$ of about 3 h. Such results have generated a lively debate, the consensus being that it is practically impossible to demonstrate stereoselectivity in any *in vivo* biological effect of thalidomide. Only *in vitro* tests of short duration can be expected to yield reliable indications.

The extent of racemization of (+)-chlorthalidone as a function of pH has been examined. The minimum of the log *K*/pH curve is pH 3 (39). The reaction mechanism of inversion is postulated to involve a carbenium cation over the entire pH range and a ring-opening reaction in the alkaline range.

Because of the constantly increasing demand for optically pure drugs, it is of great importance to elucidate factors affecting stereochemistry in order to provide a stable formulation with high chiral quality of the desired isomer (39). Isomerization may be influenced by factors such as pH, buffer salts, ionic strength, solvents, temperature, and light. One of the aims of stability studies of optical isomers in drug formulations is therefore to determine the extent of isomerization. Suitable and specific methods have to be employed, and limitation of the undesired isomers should be specified.

Due to stereoselective interactions with receptors, it is known that enantiomers can exhibit different biological actions and the distomer can even give rise to adverse effects, as the example of penicillamine demonstrates.

Guidelines and drafts have now been issued by the EEC and FDA, which include rules concerning, for example, batch-to-batch consistency of the enantiomeric ratio in the various batches (enantiomer purity) and a full description of the stability of the drug substance (enantiomeric stability). These requirements are also applicable to pure stereoisomeric drug substances in pharmaceutical dosage forms, since optical isomers are not invariably stable compounds. However, to date only a few systematic studies have been performed to elucidate the stability of the configuration of chiral drugs in pharmaceutical dosage forms.

REFERENCES

1. Wainer, I. W. (1993). *Drug Stereochemistry*. New York: Marcel Dekker.
2. Ahuja, S. (1991). *Chiral Separations by Liquid Chromatography*, ACS Symposium Series 471. Washington, DC, American Chemical Society.
3. Allenmark, S. (1991). *Chromatographic Enantioseparation*. Chichester: Ellis Horwood.
4. Ahuja, S. (1989). Selectivity and Detectability Optimizations in HPLC. New York: Wiley.
5. Lough, W. J. (1989). *Chiral Liquid Chromatography*. Glasgow: Blackie and Son Ltd.
6. Gal, J. (1987). *LC-GC*, **5**, 106.
7. Hara, S., and Cazes, J. (1986). *J. Liq. Chromatogr.*, **9**:2 & 3.
8. Souter, R. W. (1985). *Chromatographic Separations of Stereoisomers*, Boca Raton, FL: CRC Press.
9. Ahuja, S. (1988). 1st International Symposium on Separation of Chiral Molecules, Paris, France, May 31–June 2.

10. Blaschke, G., Kraft, H. P., Fickentscher, K., and Koehler, F. (1979). *Arzneim.-Forsch*, **29**, 1640.
11. "Top 200 Drugs in 1982" (April 1982). *Pharmacy Times*, p. 25.
12. Okamoto, Y. (March, 1987). *CHEMTECH*, p. 176.
13. Rauws, A. G. and Gruen, K. (1994). *Chirality*, **6**, 72.
14. DeCamp, W. H. (1989). *Chirality*, **1**, 2.
15. Guidelines for Submitting Supporting Documentation in Drug Applications for the Manufacture of Drug Substances (1987). Office of Drug Evaluation and Research (HFD-100) Food and Drug Aministration, Rockville, MD.
16. Code of Federal Regulation (1988). Title 21, Government Printing Office, Sect. 314.5(d)3, Washington, DC.
17. PMA Ad Hoc Committee on Racemic Mixtures (May 1990). *Pharm. Technol.*, p. 46.
18. Gross, M. et al. (1993). *Drug Inform. J.*, **27**, 453.
19. Millership, J. S., and Fitzpatrik, A. (1993). *Chirality*, **5**, 573.
20. Murase, K., Masc, T., Ida, H., Takahashi, K., and Murakami, M. (1978). *Chem. Pharm. Bull.*, **26**, 1123.
21. Hutt, A. J. (1991). *Chirality*, **3**, 161.
22. Waldeck, B. (1993). *Chirality*, **5**, 355.
23. Barlow, B. B., Franks, F. M., Pearson, J. D. M. (1972). *J. Pharm. Pharmacol.*, **24**, 753.
24. Wright, M. R., Davies, N. M., and Jamali, F. (1994). *J. Pharm. Sci.*, **83**, 911.
25. Malmberg, A. B. and Yaksh, T. L. (1992). *Science*, **257**, 1276.
26. Brocks, D. R., Liang, W. T. C., and Jamali, F. (1993). *Chirality*, **5**, 61.
27. Sannins, S. M., Adams, W. J., Kaiser, D. G., Halstead, G. W., Hosley, J., Barnes, H., and Baillie, T. A. (1991). *Drug Metab. Disp.*, **19**, 405.
28. Vermulen, A. M., Belpaire, F. M., Moerman, E., DeSmet, F., and Bogaert, M. G. (1992). *Chirality*, **4**, 73.
29. Avgerinos, A., and Hutt, A. J. (1990). *Chirality*, **2**, 249.
30. O'Shea, N. J. and Hooper, W. D. (1990). *Chirality*, **2**, 257.
31. Gimenez, F., Pennie, R. A., Koren, G., Crevoisier, C., Wainer, I. G., and Farinotti, R. (1994). *J. Pharm. Sci.*, **83**, 824.
32. Higaki, K., Kadono, K., and Nakano, M. (1992). *J. Pharm. Sci.*, **81**, 935.
33. Chen, R., Nguyen, P., You, Z., and Sinsheimer, J. E. (1993). *Chirality*, **5**, 501.
34. Vercruysse, L., Belpaire, F., Wynant, P., Massart, D. L., and Dupont, A. G. (1994). *Chirality*, **6**, 5.
35. Gibaldi, M. (1993). *Chirality*, **5**, 407.
36. Yang, S. K. and Lu, X. L. (1992). *Chirality*, **4**, 446.
37. Severin, G. (1992). *Chirality*, **4**, 226.
38. Testa, B., Carrupt, P. A., and Gal, J. (1993). *Chirality*, **5**, 105.
39. Lamparter, E., Blaschke, G., and Schluter, J. (1993). *Chirality*, **5**, 370.

CHAPTER

11

SEPARATION OF OPTICALLY ACTIVE PHARMACEUTICALS USING CAPILLARY ELECTROPHORESIS

TIMOTHY J. WARD and KAREN D. WARD

Department of Chemistry
Millsaps College, Jackson, Mississippi 39210

11.1. INTRODUCTION

Recently, capillary electrophoresis (CE) has experienced explosive growth, creating many new applications in the field of analytical separations. The most frequently used modes of CE have been capillary zone electrophoresis (CZE), capillary electrokinetic chromatography (EKC), capillary gel electrophoresis (CGE), capillary isotachophoresis (ITP), and capillary isoelectric focusing. Of the various modes of CE available, only EKC is capable of separating nonionic or neutral solutes. Since the principle of separation for CE and other methods such as HPLC differ, CE provides an excellent complementary technique.

There are a number of reasons for the increased interest in CE for the separation of chiral compounds. The small-diameter capillaries used in CE proficiently dissipate heat, allowing the use of high voltages, which results in rapid and efficient separations. Zone broadening from convection is minimized since the sample moves as a discrete plug in the narrow bore tube. This allows one to generate many theoretical plates, which is especially attractive for chiral separations where selectivity factors (α) are often quite small. Since CE uses an extremely small volume, the use of exotic and expensive chiral selectors becomes feasible due to the small amounts consumed. In addition, the small injection volume permits the analysis of individual enantiomers that are often difficult to obtain and scarce in quantity. Another advantage is the ease with which separation media can be changed in the capillary column. In terms of method development, one can quickly and efficiently alter the run buffer to screen various separation media at a minimum cost.

The Impact of Stereochemistry on Drug Development and Use, Edited by Hassan Y. Aboul-Enein and Irving W. Wainer. Chemical Analysis Series, Vol. 142.
ISBN 0-471-59644-2

There are two basic approaches to the enantiomeric separation of chiral compounds: indirect that involves the derivatization of the enantiomers with a pure chiral reagent to form a pair of diastereomers and direct where a chiral environment is used to resolve the individual enantiomers. Of the many different media used for direct chiral separations in CE, most of the methods utilized have their origins in HPLC. This chapter will examine several of the direct methods in detail and give a short overview of several miscellaneous techniques. We will begin by examining the cyclodextrin-based techniques which are the most widely used, followed by several new methods that use macrocyclic antibiotics. It is not possible to extensively discuss every technique currently used due to space considerations; thus, we will conclude with a short overview that will summarize these methods. Methods that use chiral micelles will be discussed in detail in Chapter 13 and therefore will not be covered in this chapter.

11.2. CYCLODEXTRINS

Cyclodextrins (CD) are nonionic, cyclic chiral oligosaccharides composed of glucopyranose units bound through α-(1,4) linkages in the C-1 (D) chair conformation. They contain secondary hydroxyl groups on C-2 and -3 of the glucose unit and primary hydroxyl groups attached to C-6 of the glucose unit on the opposite side of the CD. The secondary hydroxyl groups are located on the wider side of the cavity and the primary hydroxyl groups on the narrower end. Thus, the CD structure is shaped like a truncated cone with a relatively hydrophobic interior and a hydrophilic exterior as depicted in Figure 11.1. Greek letters are used to denote the number of glucose units in the cyclodextrin molecule: α for six, β for seven, and γ for eight. Table 11.1 lists some of the physico-chemical properties of α, β and γ cyclodextrin. These are the three most common cyclodextrins used in chemical separations and are commercially available. The hydroxyl groups on the CD cavity can be derivatized to change their physico-chemical properties as well as their chiral recognition properties, as will be demonstrated later.

Cyclodextrins are to date the most extensively used chiral selector in CE. Several factors have contributed to the extensive use of CDs in liquid chromatography and CE. Cyclodextrins are extremely stable, do not appreciably absorb ultraviolet (UV) or visible light, and their physical and chemical properties have been thoroughly studied and reported. CDs are excellent chiral selectors due to their ability to form inclusion complexes with a wide variety of compounds. Chiral recognition is achieved by the inclusion of an aromatic or alkyl group into the CD cavity with interactions between the hydroxyl groups at the rim of the CD cavity and a substituent group on or near the chiral center of the analyte.

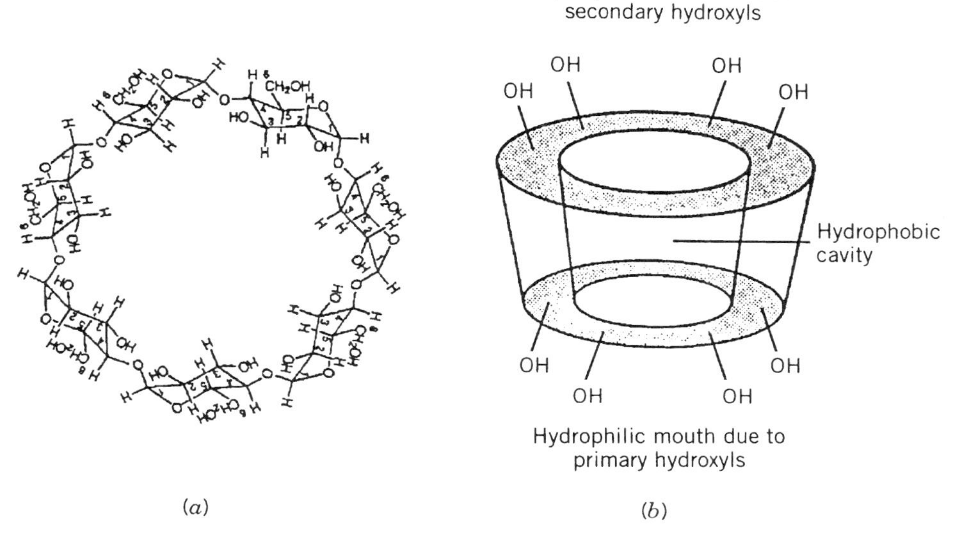

Figure 11.1. Structural diagram of β-cyclodextrin (*a*) that illustrates details of the α-(1,4) glycosidic linkage, C-1 (D) chair conformation, and the numbering system employed to describe the ring system. (*b*) Illustration of the hydrophilic faces and hydrophobic cavity present in cyclodextrin.

Cyclodextrins have been used in GC, LC, TLC, CE and SFC to resolve a wide variety of carboxylic acids, amines, drugs, and other compounds. CDs have been used as a bonded stationary phase in both liquid and gas chromatography as well as a mobile phase modifier in LC and TLC to separate optical isomers. In LC and TLC bonded phase packings, cyclodextrin is attached via a spacer group that may, in addition to inclusion complex formation, be involved in the distribution process. It was shown previously that

Table 11.1. Physical Properties of Cyclodextrins (CD)

CD	Number of Glucose Units	Molecular Weight	Inner Cavity Diameter, (Å)	UV Transparent	Solubility (% w/v) at 25°C)
α-CD	6	973	5.7	Yes	14.5
β-CD	7	1135	7.8	Yes	1.8
γ-CD	8	1297	9.5	Yes	23.2

Source: Adapted from (69)

longer retention of a compound does not necessarily produce better enantiomeric resolution in LC(1). Although inclusion complexation may be necessary for chiral recognition, it is not the only factor involved in retention. The use of cyclodextrins in CE has the advantage that the cyclodextrin is free in solution and inclusion complex formation should be operating as the primary distribution process, excluding wall effects. Moreover, the possibility exists for multiple complex formation between the solute and cyclodextrin (2).

β-cyclodextrin, which is composed of seven glucopyranose units, has been the most widely used cyclodextrin for chiral separations. This is due to the fact that β-CD is of the optimal size to form inclusion complexes with a significant number of chiral compounds. Although β-CD's low solubility hinders its use at higher concentrations, the secondary hydroxyls at the rim of the cyclodextrin can be derivatized with various functional groups to improve solubility and enhance its selectivity for different compounds.

Cyclodextrins added to the buffer either migrate at the velocity of the electroosmotic flow (EOF) if they are neutral, faster than the electroosmotic flow if they have been derivatized, resulting in a positively charged species; or slower than the electroosmotic flow if they have been derivatized, resulting in a negatively charged species. The migration time of the solutes is controlled by their degree of binding with the cyclodextrin. In the case of charged solutes, they migrate according to their electrophoretic mobility when uncomplexed and at the electrophoretic velocity of the inclusion complex when bound to the cyclodextrin. The mobility of the solute-cyclodextrin complex is less than the uncomplexed solute since the mass-to-charge ratio of the complex is greater than the uncomplexed solute. Even though the electrophoretic mobilities of enantiomers are identical, when their binding constants to the cyclodextrin are different, they can be resolved.

11.2.1. Parameters That Affect Separation

11.2.1.1. CD Size and Type: Neutral CDs

To achieve the optimal resolution in chiral CE separations using CDs, the three most important parameters to be considered are the cyclodextrin type and concentration, pH, and organic modifier. Secondary consideration should be given to EOF, background electrolyte (BGE), temperature, current, and applied field strength. Since native CDs vary in the size of their cavity, different analytes often show specific affinities for the various native CDs. The CD size and shape can also be changed by chemical modification of the hydroxyl groups. For example, α-, β-, γ-, and some methylated CDs exhibit different enantiomeric selectivity for the dansylated amino acids (DNS-AA) (3). Native β- and γ-CD have a high enantiomeric selectivity for the DNS-AA, but α-CD does not, probably due to the small cavity size that precludes the inclusion of analytes. Methylating β-CD reduced its enantioselectivity, but trimethylated-α-CD exhibited considerably increased chiral resolution for the DNS-AA. The migration order of the enantiomers also can be reversed upon chemical modifications of CDs. It is believed that the change from a hydrophilic rim to a more hydrophobic rim upon methylation of the CD results in a change in the affinity of the individual enantiomers to form an inclusion complex. There have been a number of derivatized CDs used for chiral separations, with the methylated and hydroxypropyl CDs being the most successful. Table 11.2 lists selected separations of neutral and neutral derivatized CDs, demonstrating the different selectivity and applications for each CD.

Table 11.2. Neutral Cyclodextrins Used in Chiral CE Separations

Cyclodextrin	Class of Analyte	References
Native CDs		
α-CD	Amino acids	(38, 51)
	Amino acids (N-blocked)	(3, 70)
	Barbiturates	(70)
	CNS-active agents	(33)
β-CD	Adrenergic drugs	(31)
	Amino acids (N-blocked)	(3, 70)
	Anesthetics	(71)
	Antiasthmatics	(28)
	Anticholinergics	(24)
	Antihistaminics	(31, 71)
	Antiinflammatory drugs	(25, 72)
	Barbiturates	(9)
	β-Blockers	(4, 21, 31, 70, 71, 73)
	Bronchodilators	(31, 32)
	CNS-active agents	(33)
	Dopamine agonist	(51)
γ-CD	Amino acids (N-blocked)	(3, 70)
	Antiarrhythmics	(30)
	Antiasthmatics	(28)
γ-CD	Antimalarials	(74)
	Antipsychotics	(28, 74)
Neutral Derivatized CDs		
DM-α-CD	Amino acids (N-blocked)	(3, 70)
TM-α-CD	Amino acids (N-blocked)	(3, 70)
	Barbiturates	(70)

(*continued*)

Table 11.2. (*contd.*)

Cyclodextrin Native CDs	Class of Analyte	References
DM-β-CD	Adrenergic drugs	(17, 33, 38, 39, 71, 74)
	Amino acids (N-blocked)	(3, 70)
	Anesthetics	(29, 75)
	Antibacterials	(28)
	Anticancer agents	(76)
	Barbiturates	(70)
	β-Blockers	(4, 14, 17, 21, 29, 31)
	Bronchodilators	(21, 37, 71, 74)
TM-β-CD	Amino acids (N-blocked)	(3, 70)
	Antidepressants	(29)
	Barbiturates	(70)
	β-Blockers	(31)
	Ca-channel blockers	(29)
HP-β-CD	Adrenergic drugs	(71)
	Antiarrhthmics	(71)
	Anticholinergics	(71)
	Antidepressants	(71)
	Antifungals	(15, 16)
	Antihistamines	(71)
	Antihypertensives	(71)
	Antiinflammatory drugs	(25, 72)
	Antimalarials	(71)
	Barbiturates	(9)
	β-Blockers	(17, 31)
	Bronchodilators	(40, 71)
	Hypnotics	(71)

CD = Cyclodextrin.
DM-α-CD = Heptakis(2,6-di-O-methyl)-α-CD.
TM-α-CD = Heptakis(2,3,6-tri-O-methyl)-α-CD.
DM-β-CD = Heptakis(2,6-di-O-methyl)-β-CD.
TM-β-CD = Heptakis(2,3,6-tri-O-methyl)-β-CD.
HP-β-CD = Hydroxypropyl-β-CD.

11.2.1.2. CD Structure: Charged CDs

Neutral CDs have a limited time window to achieve separation, dependent on the difference in electrophoretic mobility of the free analyte and the neutral CD. An anionic CD derivative should be advantageous due to its electrophoretic mobility that will be opposite to the electroosmotic flow (4–6).

Any analyte complexed with the anionic CD derivative will experience slowed electrophoretic mobility due to its increased effective negative charge. Figure 11.2 depicts the situation in which the analyte's migration time is altered by its binding to CD; this time in which the analyte elutes is called the separation or migration window. The neutral CDs travel at the velocity of the electroosmotic flow and therefore the separation is limited as shown in Figure 11.2a. The anionic CD moves against the EOF, thus expanding the migration window, and the analyte-CD complex migrates at a velocity somewhere between that of the EOF and uncomplexed CD, depending on the degree of binding as shown in Figure 11.2b. Therefore, the separation window would be much larger as depicted in Figure 11.2b by using an anionic derivatized CD. This increased residence time can help separate analytes with weak interactions with CD.

The first use of charged cyclodextrins in CE involved the sodium salt of 2-O-carboxymethyl-β-cyclodextrin (β-CMCD) to separate a series of nonionic structural isomers (6). A novel polyanionic β-CD derivative, sulfobutyl ether β-CD, was used by Tait et al. (7) for the baseline separation of racemic ephedrine, pseudoephedrine, and several related compounds. The separations were superior to those obtained using the neutral compounds β-CD or heptakis (2,6-dimethyl)-β-CD, and much lower concentrations of the sulfobutyl β-CD were necessary for the separation, probably due to the large counter-current mobility of the sulfobutyl β-CD that resulted in a wide separation window.

Lurie et al. (8) separated illicit cationic drugs by chiral CE, with some separations achieved using mixtures of heptakis (2,6-dimethyl)-β-CD and the polyanionic CD sulfobutyl ether β-CD (β-CD SBE IV) which again acts as a counter-migrating complexing reagent. The resolution and migration time were adjusted by using various ratios of the two added CDs. Also, a model for the effect of the CD mixture on migration time and chiral resolution was presented in the study. Although symmetrical peaks were obtained when CD was absent or only the neutral CD derivative was present, the use of anionic CDs alone resulted in significant tailing. The authors postulated that this tailing was due to electrodispersion resulting from a mobility mismatch between the β-CD-SBE (IV)-drug complex and the background electrolyte (BGE). To reduce the tailing, as little of the anionic β-CD was used as possible to achieve resolution. Another probable cause of band broadening is the heterogeneity of the anionic β-CD itself.

Other anionic CDs used in CE are those with carboxy groups (9,10), such as carboxymethylated β-CD, carboxyethylated β-CD, and succinylated β-CD. The carboxy groups on the CD are protonated at $pH < 4$, and the CD acts like a neutral CD. At $pH > 5$, the carboxy groups are deprotonated, giving the CD a negative charge. These CDs have been used to separate adrenergic drugs, antihistaminics, barbiturates, and β-blockers.

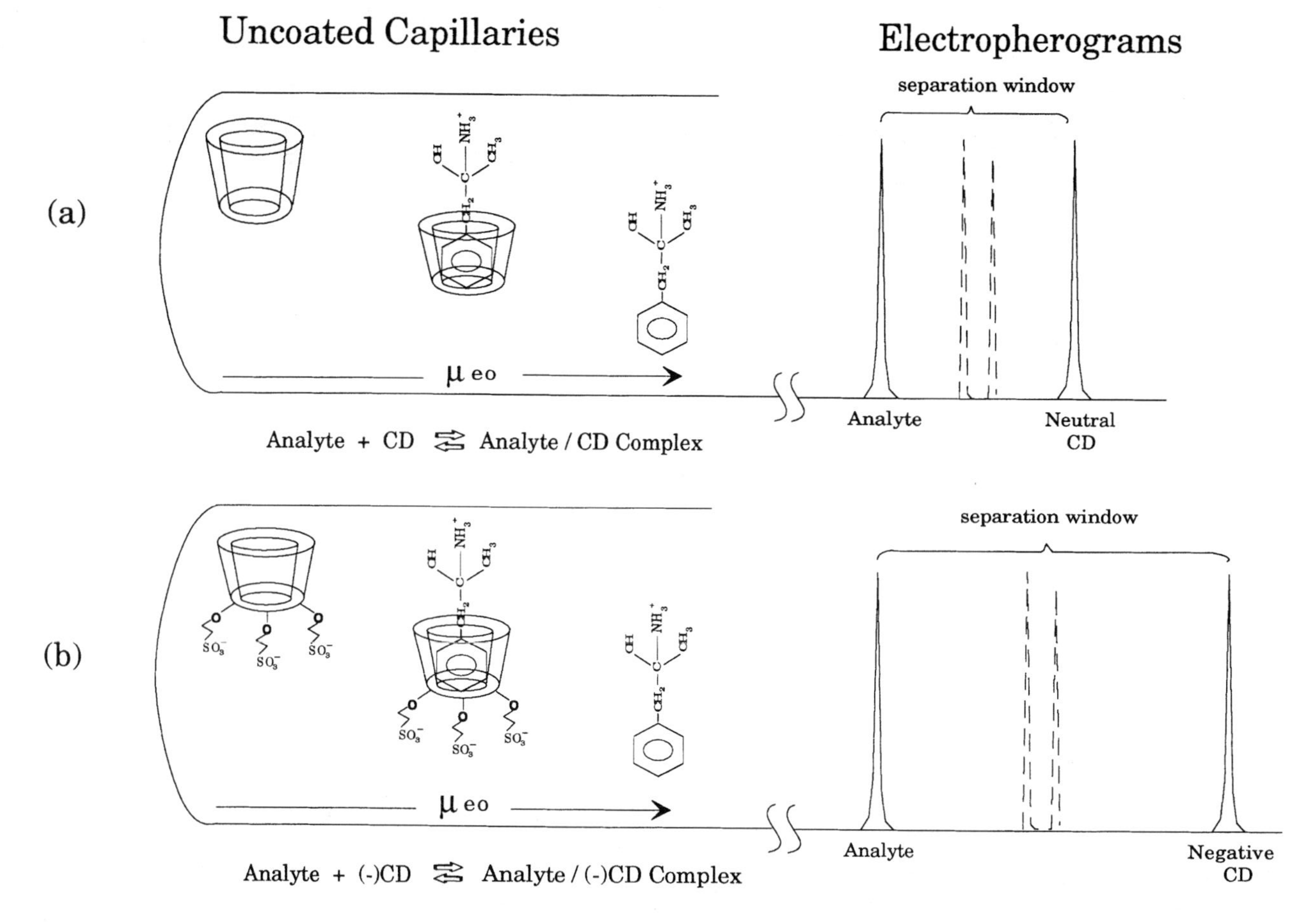

Figure 11.2. Schematic depicting the partitioning of solutes with CD and the effect of the inclusion complex formed on the relative migration windows with (a) β-CD and (b) sulfobutyl ether-β-CD.

Table 11.3. Charged Cyclodextrins Used in Chiral CE Separations

Cyclodextrin (CD)	Class of Analyte	References
Anionic Deriviatized CDs		
Carboxymethylated β-CD	Adrenergic drugs	(9)
	Antihistaminics	(9, 10)
Carboxyethylated β-CD	Adrenergic drugs	(10)
Carboxysuccinated β-CD	Barbiturates	(10)
	β-Blockers	(10)
Sulfobutyl ether β-CD	Adrenergic drugs	(7, 8)
Cationic Derivatized CDs		
Mono-(6-β-aminoethylamino-6-deoxy)-β-CD	Amino acids (N-blocked)	(12)
6^A-Methylamino-β-CD	Mandelic acid	(11)
6^A,6^D-Dimethylamino-β-CD	Lactic acid derivatives	(11)
	Mandelic acid derivatives	(11)

Only a few cationic CDs have been reported in the literature. For example, 6^A,6^D-dimethylamino-β-CD and 6^A-methylamino-β-CD were used to resolve some 2-hydroxy acids such as derivatives of mandelic and lactic acid (11). Dansylated amino acids were separated using mono-(6-β-aminoethylamino-6-deoxy)-β-CD (12). Table 11.3 summarizes selected applications using charged cyclodextrins.

11.2.1.3. CD Concentration

The selection of the appropriate CD concentration is an important step in optimization. In the absence of CD, analytes migrate through the run buffer with a relatively high mobility. When small amounts of CD are added and form inclusion complexes with the analyte, the analyte's mobility is considerably decreased. Increasing the CD concentration causes an increase in viscosity, which reduces the EOF and further increases migration times. It has been shown that the CD concentration can strongly affect separations, with enantiomeric resolution possible only in a narrow range of concentrations. For example, Schmitt and Engelhardt (13) showed that the separation of dansylated racemic phenylalanine occurred at a hydroxypropyl β-CD concentration of 0.3% (w/v), but no resolution was observed at lower or slightly higher CD concentrations. At very high CD concentrations [15% (w/v)], resolution was again observed, but the migration order of the enantiomers had switched. The authors postulated that different separation mechanisms were occurring at low CD concentration and high CD concentration. At low CD concentration, the separation resulted from the different analyte binding constants to the CD and thus different residence times in the CD complex. At high CD concentration, stable diastereomeric complexes were formed, and separation resulted from the different mobilities of these stable diastereomeric complexes.

A theoretical model was developed by Wren and Rowe (4, 14) in which enantioresolution is shown to be dependent on the concentration of chiral selector. They found that an optimum concentration of CD that results in the greatest enantioselectivity exists and is dependent on the affinity of the analyte to the CD. The model successfully applied β-blockers as analytes with native and methylated β-CD as chiral selectors. Penn et al. (15, 16) studied the binding constant of the antifungal agent tioconazole to hydroxypropyl-β-CD to find that maximum selectivity occurred at the cyclodextrin concentration equal to the reciprocal of the average binding constant. A systematic approach was outlined for the optimization of enantiomeric resolution using CE with added CDs, and thermodynamic parameters were determined for the complexation of tioconazole with a series of CD derivatives. Cyclodextrin concentration and its effect on resolution have been studied in various other reports, including those by Peterson (17), Schutzner and Fanali (18), and Shibukawa et al. (19), who used the dependence of enantiomeric separation on CD concentration to quantitatively determine binding constants.

11.2.1.4. pH and Organic Modifiers

Since it affects the ionization of functional groups of charged analytes and derivatized CDs as well as influencing EOF, pH is an extremely critical parameter in CE. The advantage of working at low pH is that the electroosmotic flow can be suppressed in uncoated capillaries and thus increase the migration window of the solutes being separated. At high pH, the high electroosmotic flow can be used to an advantage to decrease the migration times of negatively charged solutes that migrate away from the detector, or sweep anionic chiral selectors through the column. Using native β-CD at high pH, Ward et al. observed excellent enantioselectivity with a number of compounds (20). Dansylated D, L-aspartic and D, L-glutamic acid were completely separated using native β-CD and upon the addition of 20% methanol to the mobile phase, the enantiomers of nine other dansylated amino acids were baseline-resolved.

Several groups have attempted to improve resolution in CD-based systems by the addition of methanol to the run buffer (4, 21, 22). Ward et al. (20) examined various organic modifiers to enhance enantiomeric selectivity and their effect on the resolution, efficiency, and retention time of dansylated amino acids. All amino acids that contained a substituent capable of hydrogen bonding or a branched aliphatic side chain showed an increase in resolution with increasing amounts of methanol. The addition of methanol increases viscosity, decreases the electroosmotic flow, and increases migration times. However, longer migration times did not always result in better resolutions, indicating that factors other than residence time in the column affect enantioselectivity. Other organic modifiers studied were acetonitrile and dimethylformamide. Generally, lower concentrations of acetonitrile and dimethylformamide were required to achieve the same resolution (R_s) for a given amount of methanol.

A multiple-equilibria-based model has been developed by Ranjee et al. (23–26) that accounts for the separation of charged analytes as a function of pH and β-CD concentration. By matching mobilities of the analyte and the co-ion of the BGE, peak efficiencies can be improved due to reduction of electromigration dispersion. Fenoprofen was used as a test analyte, and predictions of the model were determined experimentally, with good separations and symmetric peaks obtained.

11.2.1.5. BGE and Buffer Additives

BGE can have a strong influence on the chiral CE separation, since the mobility of the BGE ions influences analyte peak shape and thus resolution. Bandbroadening effects due to electrodispersion may be seen when the conductivities of the analyte and BGE differ. Kuhn et al. (27) showed that an increase in the ionic strength of the buffer may enhance enantiomeric resolution due to an increased hydrophobic interaction between the analyte and the CD cavity.

Urea is often added to the run buffer to increase the solubility of β-CD, but its effect on enantiomeric resolution cannot always be predicted, which is also true for the other buffer additives including complexing agents, detergents, organic solvents, etc. Often, additives are included in the run buffer to suppress EOF. Snopek et al. (28) resolved racemic mixtures of basic drugs using CDs with cellulose derivatives added to the acidic run buffer to suppress the electroosmotic flow and to prevent adsorption of the analytes to the capillary wall. A significant decrease in peak width and tailing was observed. Heptakis(2,3,6-tri-O-methyl)-β-CD has been used for the separation of basic analytes such as verapamil, fluoxetine, and bupivacaine with cationic surfactants and methylhydroxyethylcellulose to improve reproducibility and peak shape (29). In another study (30), γ-CD was used in conjunction with polyvinyl alcohols and cellulose derivatives to resolve four tocainide compounds. The addition of these additives resulted in an increase in migration time due to the reduction of the electroosmotic flow, and peak symmetry and peak width were improved due to the decrease of analyte–wall interactions.

Tetraalkylammonium cations were used with β-CD (31) for the separation of basic compounds including metaphrine, isoproterenol, doxylamine, and ephedrine alkaloids. Again, the added cations suppressed analyte adsorption to the capillary wall, resulting in the reduction of the electroosmotic flow. The short-chain tetraalkylammonium compounds could be used at higher concentrations without micelle formation; however, analysis times were long and the resolution of enantiomers was less than 1.5. Polymerized β-CD has been used by Nishi et al. (32) for the resolution of enantiomers of trimetoquinol and some related compounds. The high molecular mass of the polymerized β-CD results in a much reduced mobility.

Another parameter to be considered in optimization is the applied field strength. Nielen discussed the chiral CE separation of basic drugs using α, β, and γ-CD, as well as a neutral CD derivative, heptakis (2,6-di-O-methyl)-β-CD (33). Most of the drugs separated were cardiovascular or CNS-active compounds. The effect of applied field strength on the electrophoretic mobilities, the coefficient of electroosmotic flow, and thus resolution was examined. Resolution was found to be strongly dependent on field strength, with the optimum being around 215 V/cm. This was believed to be due to the influence on resolution of the increase in the coefficient of EOF and in the electrophoretic mobilities at higher field strengths. This paper also showed the feasibility of increasing sample throughput by the injection of next samples while the previous separation is still in progress.

11.2.2. Use of Micelles with Cyclodextrin

Another way to increase solute solubility and separation selectivity is to use CDs in a mixed system with micelles in what is termed cyclodextrin-modified micellar electrokinetic chromatography (CD-MEKC). In this system, cyclodextrins create a third phase along with the aqueous buffer and micelle. Sodium dodecyl sulfate (SDS) is the most common and extensively used surfactant in CD-MEKC. The hydrophilic surface of cyclodextrin precludes it from appreciably partitioning into the micelle core. This has the effect of decreasing the time that a solute may spend in the micelle due to the increased competition with the CD. Hydrophobic solutes are either incorporated into the micelle or complexed with cyclodextrin but cannot solubilize into the aqueous phase. Solutes that are less hydrophobic can partition among the three phases: micelle, cyclodextrin, and aqueous phase. Thus, the addition of cyclodextrin to a micellar system reduces the apparent distribution coefficient of the solute to the micelle, enabling the separation of highly hydrophobic species. In this case, the cyclodextrin behaves as another phase in competition with the micelle for the solute. In the absence of cyclodextrin, hydrophobic molecules are almost totally incorporated into the micelle and exhibit little to no selectivity. Since the micelle migrates at a different velocity than the cyclodextrin or the aqueous phase, and an analyte molecule complexed with cyclodextrin migrates at the velocity of the bulk solutions, the separation of enantiomers is possible in each case. The addition of cyclodextrin to the micellar mobile phase provides the necessary chiral selector and increases the selectivity of the system.

Pharmaceutical compounds such as the barbiturates thiopental and pentobarbital can be successfully separated using γ-CD and SDS (34). The addition of chiral *d*-camphor-10-sulfonate or *l*-menthoxyacetic acid to the micellar solutions enhanced enantioselectivity. Ueda and co-workers separated a series of 2,3-dicarboxaldehyde amino acids using CD-MEKC and reported that γ-CD generally exhibited better enantioselectivity than β-CD with SDS micellar systems (35). This increased selectivity associated with γ-CD is believed to be due to the fact that the larger cavity of the γ-CD can include the surfactant monomers that exist in solution along with the analyte. Systems employing other surfactants such as the chiral bile salt taurodeoxycholate with β-CD have also been used to separate a number of racemic compounds (36).

11.2.3. Quantitative Considerations

Altria (37) reported the need to correct peak areas for accurate quantification when peak area normalization is used because the later migrating enantiomer spends a longer time in the detector cell which results in an overestimated peak area for the later eluting enantiomer. Previously, another phenomenon had been reported by Fanali and Bocek (38) in which the absorption spectra of (+)- and (–)-ephedrine in the absence of CD were different than when the enantiomers formed an inclusion complex with heptakis (2,6-di-O-methyl)-β-CD in the run buffer, necessitating a correction factor to be used for quantitative analysis.

An internal standard can be used to improve reproducibility in the quantification of enantiomers by CE. With an internal standard, *d*-epinephrine in pharmaceutical formulations containing $\cong$ 1% *l*-epinephrine was quantitatively measured by CE using heptakis(2,6-di-O-methyl)-β-CD in the run buffer (39). It is preferable that the undesired enantiomer (which is usually the minor constituent) migrate earlier than the other enantiomer for ease of detection. Although the above techniques used CDs, it is important to point out that the same considerations for quantification also apply to other chiral selectors.

Besides being able to quantitatively measure the concentrations of individual enantiomers, chiral techniques must be reproducible from laboratory to laboratory. An intercompany cross-validation study by Altria et al. (40) demonstrated that CE chiral separations of racemic clenbuteral using hydroxypropyl-β-CD had good repeatability among seven pharmaceutical companies, indicating the reproducibility of the technique between different laboratories.

11.3. MACROCYCLIC ANTIBIOTICS

Although cyclodextrins have proved enormously successful as a chiral selector in CE, the need for more diverse and selective resolving agents to be developed continues. To enhance the scope of CE for enantiomeric separations, biopolymers with multiple binding sites are a relatively unexplored group of chiral selectors.

Macrocyclic antibiotics are one of the newest and perhaps most varied class of chiral selectors. These compounds generally have multiple stereogenic centers and a variety of functional groups that are known to provide the multiple interactions (e.g., hydrogen-bonding groups, hydrophobic pockets, aromatic groups, amide linkages, etc.) necessary for enantioselectivity. As a result of these functional groups, most macrocyclic antibiotics are ionizable and sufficiently soluble in aqueous buffers and solvents commonly used in CE. These chiral selectors have been evaluated in CE, HPLC, and TLC to separate a wide variety of enantiomers (41–47). They are often complementary in the types of

Table 11.4. Macrocyclic Antibiotics Used in Chiral CE Separations

Macrocyclic Antibiotic	Class of Analyte	References
Ristocetin A	Amino acids (N-blocked)	(45)
	Nonsteroidal antiinflammatory drugs	(45)
Vancomycin	Amino acids (N-blocked)	(44)
	Nonsteroidal antiinflammatory drugs	(44)
	Antineoplastic drugs	(44)
Rifamycin B	Adrenergic drugs	(42, 43)
	β-Blockers	(43)
	Bronchodilators	(43)
	Vasoconstrictors	(43)
	Vasodilators	(43)
Rifamycin SV	Amino acids (N-blocked)	(42)
	Barbiturates	(42)

compounds they can separate. For example, rifamycin B, an ansamycin, is enantioselective for many positively charged analytes, whereas vancomycin, a glycopeptide, can resolve a variety of chiral compounds containing free carboxylic acid functional groups. Table 11.4 summarizes the various macrocyclic antibiotics and classes of compounds that have been separated using them as a chiral selector in the run buffer. There are many different macrocyclic antibiotics, each with distinct properties and chiral selectivities. We will examine the ansamycins rifamycin B and rifamycin SV, and the glycopeptides vancomycin and risocetin A.

11.3.1. Ansamycins

Rifamycins have a characteristic ansa structure, a ring structure or chromophore spanned by an aliphatic chain. Rifamycins differ from one another in the type and location of the substituents on their naphthohydroquinone ring. Figure 11.3 shows how rifamycin B differs from rifamycin SV by the *R* group attached to the naphthohydroquinone ring at the 9 position. In rifamycin B, this group is an oxy-acetic acid ($—OCH_2COOH$), whereas in rifamycin SV, the *R* group is a hydroxyl ($—OH$). Since the carboxylic and hydroxyl groups on rifamycin B are ionizable, it can exist as a dibasic acid. Rifamycin SV is essentially neutral at pHs commonly employed. In addition to the functional groups mentioned above, each rifamycin has nine stereogenic centers, four hydroxy groups, one carboxymethyl group, and one amide bond.

Both rifamycin B and SV absorb strongly in the UV and visible spectral regions. They each exhibit maxima at approximately 220, 304 and 425 nm and minima at 275 and 350 nm. Because of their absorption characteristics, most compounds are monitored by indirect detection when using these chiral selectors. This also places a limit on how high a concentration of chiral selector can be used in the run buffer. When concentrations of 30 mM rifamycin B or SV are used, the background absorbance is very high, which results in unacceptable signal-to-noise ratios.

11.3.1.1. Rifamycin B

Rifamycin B is negatively charged and enantiomerically resolves cationic compounds such as protonated amines. Armstrong and co-workers successfully resolved a series of pharmacologically active amino alcohols using rifamycin B as the chiral selector (43). These included adrenergics, vasoconstrictors, bronchiodilators,vasodilators, and β-adrenergic blockers. The separation of terbutaline by rifamycin B using indirect detection is shown in Figure 11.4. In general, Armstrong et al. found that compounds in which a hydroxy group was α to the aromatic ring resolved better than those where

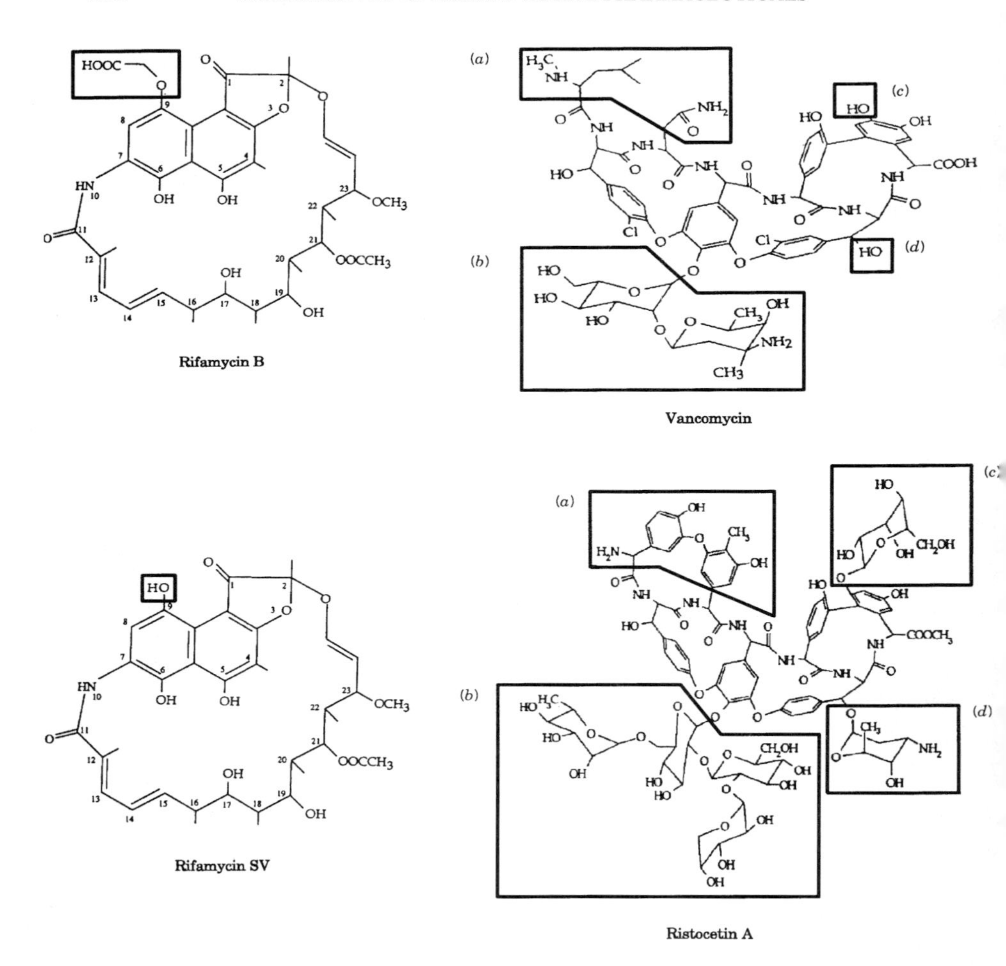

Figure 11.3. Structures of selected macrocyclic antibiotics showing the differences and similarities between the ansamycins rifamycin B and rifamycin SV, and the glycopeptides vancomycin and ristocetin A.

it was further removed from the ring. Secondary amines exhibited better enantioselectivity than primary amines and chiral compounds containing two or more aromatic rings showed little enantioresolution, indicating a possible size selectivity with this chiral selector. Ward later demonstrated that compounds containing multiple rings such as propranolol could be resolved using optimized conditions (42).

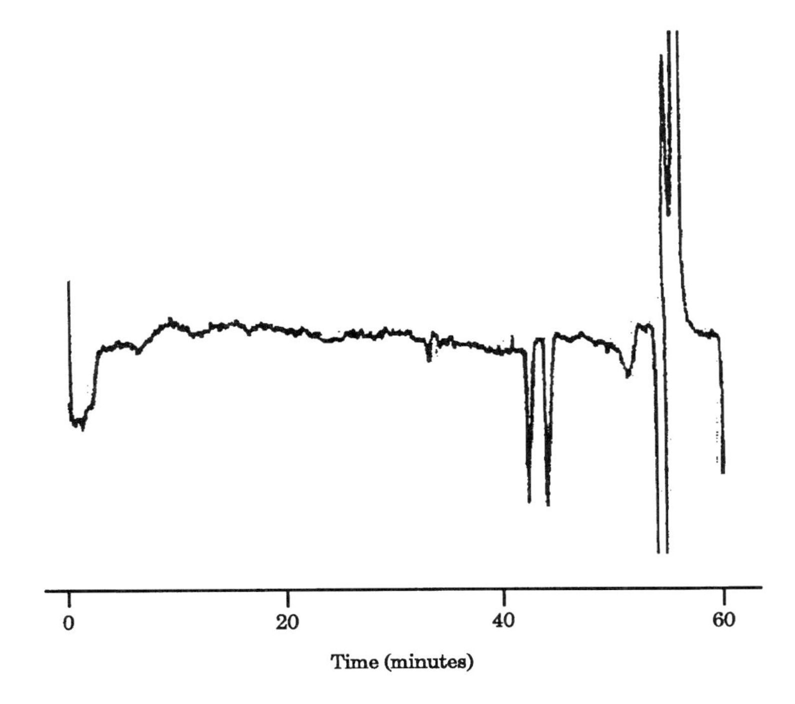

Figure 11.4. CE separation of the enantiomers of terbutaline using indirect detection at 254 nm. A solution of 30% 2-propanol; 70% 0.1 M, pH 7, phosphate buffer (by volume) containing 20 mM of rifamycin B. Applied potential was +8 kV.

11.3.1.1.1. Parameters that Affect Separation. There are several factors to consider in optimizing separation with these chiral selectors. As was previously discussed, the three most common parameters to consider are chiral selector concentration, pH, and organic modifier. Other parameters that bear consideration with these selectors are buffer concentration, buffer type, amount of solute injected, and detection wavelength.

Increasing chiral selector concentration increases enantioresolution, but tends to slightly increase migration times due to the small decrease in electrophoretic mobility. Increasing the chiral selector concentrations much above 25 mM produces an extremely small signal-to-noise ratio due to the strong UV absorption of the chiral selector. Interestingly, no enantioselectivity is observed in the absence of organic modifier in the run buffer for this chiral selector. Of the organic modifiers investigated, 2-propanol provides the greatest enhancement to enantioresolution. Increasing the 2-propanol concentration in the run buffer increases enantioresolution, decreases electrophoretic mobilities, and increases migration times.

Enantioresolution increases with pH, reaches a maximum at pH 7, and decreases with increasing pH. This can be explained by examining the charge on the chiral selector and analyte being separated. Rifamycin B, a dibasic acid, loses some fraction of the negative charge present on the molecule as pH is lowered, which precludes a strong charge–charge interaction with the positively charged amine-containing analyte. As pH is increased to pH 7, rifamycin B exists primarily as a di-anion while the analyte is positively charged, providing a strong electrostatic interaction. At higher pH's, the amine group on the analyte is deprotonated, again precluding a strong charge–charge interaction. Although other parameters such as electroosomotic flow are also affected, clearly charge–charge interactions play a prominent role in enantioresolution.

Ward et al. examined the effect of wavelength detection on sensitivity, specifically looking at regions where an absorption minima occurred (42). When epinephrine was injected under identical conditions at 275 and 350 nm, they found the sensitivity was much greater at 350 nm than 275 nm. This can be attributed to the fact that at absorption minima, the baseline noise is substantially reduced, resulting in improved sensitivity. Since resolution and peak-to-peak separation are affected by the amount of analyte loaded on the capillary column, working at 350 nm (greatest UV minima) and loading less analyte (approximately 80% less) on the column substantially improved separation resolutions for five of the seven solutes studied.

11.3.1.2. Rifamycin SV

Several negatively charged solutes were resolved using rifamycin SV as the chiral selector in the run buffer (42). Interestingly, rifamycin SV seems particularly suited for separating systems containing at least two rings. Hexobarbital and glutethimide each contain two rings that are not conjugated whereas dansyl aspartic acid has a conjugated ring system. Dansyl aspartic acid absorbs at longer wavelengths and was monitored by direct UV detection at 350 nm, whereas hexobarbital and glutethimide were both monitored by indirect UV detection. Rifamycin SV separates negatively charged solutes and is a complementary chiral selector to rifamysin B, which resolves positively charged solutes.

11.3.2. Glycopeptides

The glycopeptides appear to be more enantioselective and widely applicable for the resolution of anionic solutes than rifamycin SV. Vancomycin and ristocetin A are structurally related macrocyclic glycopeptide antibiotics that are active against gram-positive bacteria. As depicted in Figure 11.3, each has a aglycone portion consisting of fused macrocyclic rings that forms a characteristic "basket" shape, various pendant sugar moities, along with numerous functional groups. The areas indicated by a–d in Figure 11.3 highlight the different functional groups attached to the glycopeptide backbone for each macrocyclic peptide. As might be expected from the similarity they share in structure, both are adept at separating similar solute types. Although there are many properties that they have in common, they also differ in several ways that will be discussed as we examine each chiral selector.

11.3.2.1. Vancomycin

Vancomycin has a aglycone portion consisting of three fused macrocyclic rings with two side chains: a carbohydrate dimer and N-methyl leucine. It has a total of 18 stereogenic centers and among its functional groups are nine hydroxyl groups, two amines, seven amido groups, and five aromatic rings (two have Cl substituents). It is soluble in water and aqueous buffers, polar aprotic solvents such as dimethyl formamide, slightly soluble in methanol, and relatively insoluble in less polar organic solvents. As can be expected from the aromatic rings present in its structure, vancomycin absorbs strongly in the UV spectral region with a minima at approximately 260 nm at acidic pH. This minima is lost and there is a general shift to slightly longer wavelengths at higher pH values.

Vancomycin possesses a number of ionizable functional groups and has six reported pK_a values ≅ 2.9, 7.2, 8.6, 9.6, 10.5, and 11.7, with the last three values believed to be from the phenolic groups. It exhibits an effective electrophoretic mobility at pH 7.2 in 0.1-M phosphate buffer. Although it contains many ionizable groups, pH ranges between 5 and 7 are normally employed since vancomycin is unstable at pH values outside of this range and at elevated temperatures. It is believed that the macrocylic rings are opened via the hydrolysis of the amide bonds as well as possible cleaving of the sugar moities from the ring structures. Thus, by avoiding elevated temperatures (> 23°C) and pH extremes, solutions are stable for approximately 1 week.

Armstrong et al. resolved over 100 racemates including N-blocked amino acids, nonsteroidal antiinflammatory drugs, and antineoplastic compounds using vancomycin in the run buffer (44). The N-derivatized amino acids included highly fluorescent groups such as 6-aminoquinolyl-N-hydroxysuccinimidyl (AQC), dansyl, phenylthiohydantion (PHTH); strong UV-absorbing like N-3,5-dinitrobenzoyl (DNB), N-3,5-dinitropyridyl (DNP_{yr}); and simple N-blocking groups like benzoyl, acetyl, or formyl. The enantioselectivity for many of these compounds was quite large, with R_s values greater than 10 not uncommon and most separations having resolutions greater than 2.

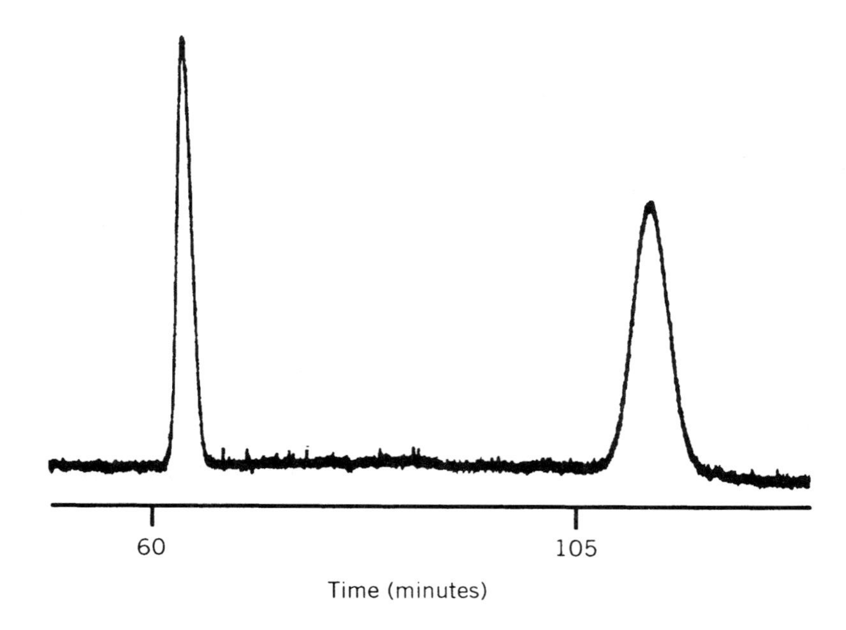

Figure 11.5. Capillary electropherogram showing the resolution of racemic dansyl aspartic acid. The conditions were as follows: 2-mM vancomycin in 0.1-M phosphate buffer, pH 5. The applied potential was +10 kV and direct detection at 340 nm.

Figure 11.5 shows the separation of dansyl aspartic acid by our laboratory, demonstrating the enormous selectivity that vancomycin exhibits for many compounds that contain a carboxylic acid moiety.

11.3.2.1.1. Parameters that Affect Separation. Similar to the ansamycins, the concentration of a chiral selector has a profound impact on resolution and migration times. One very important consideration when compared with either rifamycin is that relatively dilute solutions of vancomycin are used (1–5 mM), coupled with the fact that a minima exists at ≅ 260 nm allows direct detection of most solutes. Both enantiomeric resolutions and migration times increased when the chiral selector concentration was increased from 1–5 mM. Although vancomycin is positively charged at all pHs used in this study, the analytes all contain carboxlic acid groups and have mobilities opposite the electroosmotic flow. It would be expected that the migration times of solutes should decrease as the chiral selector concentration is increased and complexation between analyte and selector increased. This was not the case with all solutes studied. There are several factors that contribute to increasing migration times as chiral selector concentration is increased.

Increasing vancomycin concentration in the run buffer substantially decreases the electroosmotic flow velocity. There are two factors responsible for this phenomenon. The predominant factor is that vancomycin adsorbs onto the wall of the capillary and suppresses to some degree the electroosmotic flow, as well as acting as a pseudostationary phase in retarding the solutes migration. Second, increasing vancomycin can also increase solution viscosity although at the solution concentrations used (1–5 mM), this effect is expected to be minimal.

Resolution and migration times of the test analytes are substantially affected by changing pH. In general, increasing pH from 5–8 decreases enantioselectivity for most solutes, with the most significant decrease occurring between pH 6 to pH 8. Note that at pH 8, vancomycin (zero mobility at pH 7.2) now acquires a negative charge and migrates in the same direction as the test analytes. Since pH affects the charge and mobility of the chiral selector as well as the analyte, resolutions can be altered by affecting charge interactions or changing the relative mobility difference between the analyte-chiral selector complex.

Organic modifiers have less of an effect on vancomycin-based separations as they did with the rifamycins. Of the organic modifiers investigated, 2-propanol appeared to exhibit the greatest influence on separations. Although organic modifiers did enhance resolutions for some analytes, the effects were small in comparison to enhancements obtained by increasing the chiral selector concentration or optimizing the run buffer pH.

11.3.2.2. Ristocetin A

Ristocetin A consists of an aglycone portion of four fused macrocyclic rings to which several sugars are attached. It has more stereogenic centers (38 total) than either of the ansamycins or vancomycin, as well as seven aromatic rings, 21 hydroxyl groups, two primary amines, one methyl ester, and six amido groups. It is soluble in acidic aqueous and polar aprotic solvents like dimethylformamide and dimethyl sulfoxide. It is relatively insoluble in water and nonpolar organic solvents.

Although ristocetin A and vancomycin have a number of properties in common, they also differ significantly in several ways. Ristocetin A has four fused macrocyclic rings versus three for vancomycin, a greater molecular mass of 2066 versus 1449 for vancomycin, and its phenyl rings are not chloro-substituted. Ristocetin A exhibits a zero electrophoretic mobility at pH $\cong$ 7.5 in 0.1-M phosphate buffer which is slightly higher than that found for vancomycin. Therefore, at pHs less than 7.5, it has a positive charge and, like vancomycin, interacts electrostatically with anionic compounds. The absorbance profile of ristocetin A is very similar to vancomycin, and at the concentrations used in the run buffer (1–5 mM), the direct detection of most solutes is possible.

Over 120 racemic compounds were separated using dilute solutions of this chiral selector, including various N-blocked amino acids, nonsteroidal anti-inflammatories, and a variety of other carboxylic acid-containing compounds (45). Given the structural similarities between vancomycin and ristocetin, it is not unusual that their enantioselectivities are similar as well. However, ristocetin A does differ from vancomycin in several aspects and exhibits enantioselectivity with a number of compounds not resolved by vancomycin. These include compounds such as mandelic acid and several of its derivatives, β-phenyllactic acid, tropic acid, 2-bromo-3-methyl butyric acid, and 1-benzocyclobutenecarboxylic acid. In terms of separation characteristics, ristocetin A and vancomycin have several other features in which they differ. At chiral selector concentrations between 2–5 mM, ristocetin A-based separations displayed significantly smaller migration times than analogous separations using vancomycin. This is believed to be due to the fact that vancomycin adsorbs to the capillary wall much more strongly than ristocetin A. Ristocetin A with its bulkier pendant sugar moieties tends to adsorb less strongly to the capillary wall.

11.3.2.2.1. Parameters that Affect Separation. In general, increasing the chiral selector concentration resulted in increased enantioselectivity and longer migration times. This increase is also attributable to wall adsorption of the chiral selector but as noted above does not increase migration times as nearly as much as vancomycin-based separation. Similar to vancomycin, lower pHs generally produce higher enantioresolutions than higher pHs,

Table 11.5. Parameters Affecting CE Separations with Macrocyclic Antibiotics

	Increasing Antibiotic Concentration	Increasing pH	Adding Organic Modifier
Ristocetin			
R_s	Increases	Decreases	Increases
EOF	Decreases	Increases	Decreases
Vancomycin			
R_s	Increases	Decreases	Small increase
EOF	Decreases	Increases	Decreases
Rifamycin B			
R_s	Increases	Varies	Increases
EOF	Small decrease	Increases	Decreases

although like vancomycin at pHs below 4–5, significant degradation of the chiral selector occurs. Unlike vancomycin, organic modifiers can significantly enhance enantioselectivity when using ristocetin A. For example, the resolution of ketoprofen in 2-mM ritocetin A, 0.1-M phosphate buffer (pH 7) increases from 2.6–5.7 upon the addition of 20% 2-propanol to the run buffer. This also caused the migration time to more than double (12.6–26.3 min) due to the increase in solution viscosity and decreased electroosmotic flow. As organic modifier concentration is increased, a leveling effect with respect to resolution is observed and further increases in organic modifier (> 40%) begin to decrease enantioresolution. Table 11.5 gives a short summary of the various parameters that affect separation for each macrocyclic antibiotic.

11.4. MISCELLANEOUS CHIRAL SELECTORS

11.4.1. Crown Ethers

Crown ethers are macrocyclic polyether ring systems capable of forming stable inclusion complexes with potassium, ammonium, and performylated primary amines. Chiral crown ethers previously have been used in HPLC to resolve enantiomers by a number of workers (48, 49). The first use of crown ethers as a chiral selector in CE was reported by Kuhn et al. (50). One condition necessary for enantioselectivity is that the protonated primary amine form an inclusion complex with the crown ether. Thermodynamic studies of the complexes confirm this condition. Although the inclusion of the amine is essential for enantioselectivity, it is not the only condition necessary for chiral resolution. Additional chiral interaction between substituents of the crown ether and the analyte are necessary for chiral discrimination. Separation is based on the formation of diastereomeric complexes of the individual enantiomers with the crown ether based on their binding constants and changes in their electrophoretic mobilities upon forming these diastereomeric complexes.

The macrocyclic crown ether, 18-crown-6-tetrocarboxylic acid, has been used to resolve various chiral amines (50, 51) and amino alcohols (52). The enantiomers of analogues of DOPA, tyrosine, and γ-amino-butyric acid (GABA) were also resolved using crown ethers by Walbroehl and Wagner (53). Although the best results are usually obtained when the chiral center is adjacent to the amine group, resolution is possible when the chiral center is in the δ-position. Walbroehl and Wagner found that analogs of DOPA and tyrosine with a C=C on the carbon β to the amine group were separated better than when the carbon α to the amine group contained a substituent. They attributed this difference in selectivity to a lack of complexation of the amine to the crown ether due to steric hinderance. As with other chiral selecters discussed, chiral selector concentration, pH, organic modifiers, and buffer composition affect separation.

11.4.2. Proteins

Proteins are well established as successful chiral stationary phases in HPLC for the separation of enantiomers. They have the advantage that they can be very substrate-specific, resulting in excellent enantioselectivity for many compounds. The first work using proteins in CE as a chiral selector was reported by Birnbaum and Nillson (54). They called this technique "capillary gel affinity electrophoresis" although most manuscripts now refer to protein-based separations as affinity EKC. The use of proteins as chiral selectors has been limited somewhat due to their absorption on UV light, their adsorption onto the capillary wall, and often the relatively low purity of the proteins used. These effects often lead to bandbroadening and irreproducible separations.

The enantiomers of leucovarin were separated using a 0.1% solution of bovine serum albumin (BSA) (55). Human serum albumin (HSA) was used as the chiral selector in CE to resolve racemic tryptophan and warfarin (56). As is commonly found with these selectors, HSA was distributed in two locations in the system: adsorbed to the capillary wall and free in the buffer solution. The wall-adsorbed HSA was found to play the most important role in the separation of warfarin, whereas the free solution HSA significantly affected the separation of tryptophan. Various other proteins have been employed for enantiomeric separation in affinity EKC, including AGP, OVM, and fungal cellulase (57).

11.4.3. Heparin

Heparin is naturally occurring polydisperse, polyanionic glycosaminoglycan used in the prevention of thrombosis following surgery. The basis subunit of heparin is either a di- or tetrasaccharide composed of uronic acid and 2-amino-2-deoxy-D-glucose, structured with α-1,4 linkages. Heparin has a strong anionic charge due to its variable but always large number of sulfate groups. The α-1,4 linkages result in a helical configuration in aqueous solution, and its highly anionic character enhances its aqueous solubility. Due to its high degree of sulfonation, heparin has an electrophoretic migration opposite to the electroosmotic flow.

Various antimalarial drugs, antihistimines, and other compounds of pharmaceutical interest were resolved using heparin as a chiral mobile phase additive (58). All resolved compounds contained an aromatic heterocyclic nitrogen-containing ring. The separations were highly pH-dependent, although this seemed to be related to the pK_as of the separated compounds. Although the mechanism of chiral separation has not been completely elucidated, it appears to be based on a combination of ionic, hydrogen-bonding, and hydrophobic interactions, as well as possible inclusion within the heparin superstructure.

11.4.4. Maltooligosaccharides

A variety of commercial maltooligosaccharide mixtures, corn syrups, and some individual oligomers were screened for their chiral recognition properties in CE (59). A more recent work (60) assessed different maltooligosaccharides as CE chiral selectors in the separation of the drugs ibuprofen, warfarin, ketoprofen, and simendan, among others. The effect of zwitterionic buffers and organic modifiers were investigated and NMR spectroscopic complexation studies were conducted to determine the chiral recognition mechanisms.

11.5. IMMOBILIZED CHIRAL PHASES

A number of groups have investigated immobilizing an appropriate chiral support in a capillary column to achieve enantioselectivity. Three approaches for the immobilization of the chiral phase are as follows: gels containing chiral agents immobilized in a capillary column (5, 6), silica gel particles with bound chiral agents packed into a capillary (62, 63), and chiral agents chemically bonded to the wall of the capillary column (64–66).

11.6. CONCLUSION

Capillary electrophoresis has proven to be a very efficient and versatile technique for the rapid separation of enantiomers. As our understanding of the mechanism of separation for current chiral selectors improves and additional chiral selectors are developed, the scope and applicability of enantiomeric separations will continue to expand. The advantages of CE-based chiral separations are numerous: high efficiency, ease of method development, small quantities of reagents consumed, and ease of changing the migration order of enantiomers. The general disadvantage is that CE is limited to analytical applications.

Although this chapter was limited in the number of chiral selectors that could be discussed in detail, there are several excellent reviews available that provide numerous references and applications of chiral selectors in CE (41, 67, 68).

REFERENCES

1. Armstrong, D. W., Ward, T. J., Czech, A., Czech, B. P., and Bartsch, R. A. (1985). *J. Org. Chem.*, **50**, 5556–5559.
2. Armstrong, D. W., Nome, F., Spino, L. A., and Golden, T. D. (1986). *J. Am. Chem. Soc.*, **108**, 1418–1421.
3. Tanaka, M., Yoshinaga, M., Asano, S., Yamashoji, Y., and Kawaguchi, Y. (1992). *Fresenius J. Anal. Chem.*, **343**, 896–900.
4. Wren, S. A. C. and Rowe, R. C. (1992). *J. Chromatogr.*, **603**, 235–241.
5. Guttman, A., Paulus, A., Cohen, A. S., Grinberg, N., and Karger, B. L. (1988). *J. Chromatogr.*, **448**, 41–53.
6. Terabe, S., Ozaki, H., Otsuka, K., and Ando, T. (1985). *J. Chromatogr.*, **332**, 211–217.
7. Tait, R. J., Thompson, D. O., Stella, V. J., and Stobaugh, J. F. (1994). *Anal. Chem.*, **66**, 4013–4018.
8. Lurie, I. S., Klein, R. F. X., Cason, T. A. D., LeBelle, M. J., Brenneisen, R., and Weinberger, R. E. (1994). *Anal. Chem.*, **66**, 4019–4026.
9. Schmitt, T., and Engelhardt, H. (1993). *J. High Res. Chromatogr.*, **16**, 525–529.
10. Schmitt, T. and Engelhardt, H. (1993). *Chromatographia*, **37**, 475–481.
11. Nardi, A., Eliseev, A., Bocek, P., and Fanali, S. (1993). *J. Chromatogr.*, **638**, 247–253.
12. Terabe, S. (1989). *Trends Anal. Chem.*, **8**, 129–134.
13. Schmitt, T., and Engelhardt, H. (1993). *J. High. Res. Chromatogr.*, **16**, 525–529.
14. Wren, S. A. C., and Rowe, R. C. (1993). *J. Chromatogr.*, **635**, 113–118.
15. Penn, S. G., Goodall, D. J., and Loran, J. S. (1993). *J. Chromatogr.*, **636**, 149–152.
16. Penn, S. G., Bergstrom, E. T., Goodall, D. J., and Loran J. S. (1994). *Anal. Chem.*, **66**, 2866–2873.
17. Peterson, T. E. (1993). *J. Chromatogr.*, **630**, 353–361.
18. Schutzner, W., and Fanali, S. (1992). *Electrophoresis*, **13**, 687–690.
19. Shibukawa, A., Lloyd, D. K., and Wainer, I. W. (1993). *Chromatographia*, **35**, 419–429.
20. Ward, T. J., Nichols, M., Sturdivant, L., and King, C. C. (1995). *Amino Acids.*, **8**, 337–344.
21. Fanali, S. (1991). *J. Chromatogr.*, **545**, 437–444.
22. Nardi, A., Ossicini, L., and Fanali, S., (1992) . *Chirality*, **4**, 56–61.
23. Rawjee, Y. Y., Williams, R. L., and Vigh, G. (1993). *J. Chromatogr.*, **635**, 291–306.
24. Rawjee, Y. Y., Williams, R. L., and Vigh, G. (1993). *J. Chromatogr. A*, **652**, 233–245.
25. Rawjee, Y. Y. and Vigh, G. (1994). *Anal. Chem.*, **66**, 619–627.
26. Rawjee, Y. Y., Williams R. L., and Vigh, G. (1994). *Anal. Chem.*, **66**, 3777–3781.
27. Kuhn, R., Stoecklin, F., and Erni, F. (1992). *Chromatographia*, **33**, 32–36.

28. Snopek, J., Soini, H., Novotny, M., Smolkova-Keulemansova, E., and Jelinek, I. (1991). *J. Chromatogr.* **559**, 215–222.
29. Soini, H., Riekkola, M. L., and Novotny, M. (1992). *J. Chromatogr.*, **608**, 265–274.
30. Belder, D., and Schomburg, G. (1992). *J. High. Res. Chromatogr.*, **15**, 686–693.
31. Quang, C., and Khaledi, G. (1993). *Anal. Chem.*, **65**, 3354–3358.
32. Nishi, H., Nakamura, K., Nakai, H., and Sato, T. (1994). *J. Chromatogr. A*, **678**, 333–342.
33. Nielen, M. W. (1993). *Anal. Chem.*, **65**, 885–893.
34. Nishi, H., Fukuyama, T., and Terabe, S. (1991). *J. Chromatogr.*, **553**, 503–516.
35. Ueda, T., Mitchell, R., Kitamura, F., Metcalf, T., Kuwana, T., and Nakamoto, A. (1992). *J. Chromatogr.*, **593**, 265–274.
36. Okafo, G. N. and Camilleri, P. (1993). *J. Microcol. Sep.*, **5**, 149–153.
37. Altria, K. D. (1993). *Chromatographia*, **35**, 177–182.
38. Fanali, S. and Bocek, P. (1990). *Electrophoresis*, **11**, 757–760.
39. Peterson, T. E. and Trowbridge, D. (1992) *J. Chromatogr.*, **603**, 298–301.
40. Altria, K. D., Harden, R. C., Hart, M., Hevizi, J., Hailey, P. A., Makwana, J. V., and Portsmouth, M. J. (1993). *J. Chromatogr.*, **641**, 147–153.
41. Ward, T. J. (1994). *Anal. Chem.*, **66**, 633A–640A.
42. Ward. T. J., Dann, C., and Blaylock, A. (1995). *J. Chromatogr. A*, **715**, 337–344.
43. Armstrong, D.W., Rundlett, K., and Reid, G. L. (1994). *Anal. Chem.*, **66**, 1690–1695.
44. Armstrong, D. W., Rundlett, K. L., and Chen, J. (1994). *Chirality*, **6**, 496–509.
45. Armstrong, D. W., Gasper, M. P., and Rundlett, K. L. (1995). *J. Chromatogr. A*, **689**, 285–304.
46. Armstrong, D. W., Tang, Y., Chen, S., Zhou, Y., Bagwill, C., and Chen, J. (1994). *Anal. Chem.*, **66**, 1473–1484.
47. Armstrong, D. W. and Zhou, Y. (1994). *J. Liq. Chromatogr.*, **17**:8, 1695–1707.
48. Dotsevi, G., Sogah, E., and Cram, D. J. (1975). *J. Am. Chem. Soc.*, **97**, 1259.
49. Sousa, L. R., Sogah, G. D. Y., Hoffamn, D. H., and Cram, D. J. (1978). *J. Am. Chem. Soc.*, **100**, 4569.
50. Kuhn, R., Erni, F., Bereuter, T., and Hausler, J. (1992). *Anal. Chem.*, **64**, 2815–2820.
51. Kuhn, R., Stoecklin, F., and Erni, F. (1992), *Chromatographia*, **33**, 32–36.
52. Hohne, E., Krauss, G. J., and Gubitz, G., (1992). *J. High. Res. Chromatogr.*, **15**, 698.
53. Walbroehl, Y. and Wagner, J. (1994). *J. Chromatogr. A*, **685**, 321–329.
54. Birnbaum, S. and Nillson, S. (1992). *Anal. Chem.*, **64**, 2872–2874.
55. Barker, G. E., Russo, P., and Hartwick, R. A. (1992). *Anal. Chem.*, **64**, 3024–3028.
56. Yang, J. and Hage, D. S. (1994). *Anal. Chem.*, **66**, 2719–2725.
57. Busch, S., Kraak, J. C., and Poppe, H. (1993). *J. Chromatogr.*, **635**, 119–126.

58. Stalcup, A. M. and Agyei, N. M. (1994). *Anal. Chem.*, **66**, 3054–3059.
59. D'Hulst, A. and Verbeke, N. (1992), *J. Chromatogr.*, **608**, 275.
60. Soini, H., Stefansson, M., Riekkola, M. L., and Novotny, M. V. (1994). *Anal. Chem.*, **66**, 3477–3484.
61. Cruzado, I. and Vigh, G. (1992). *J. Chromatogr.*, **608**, 421–425.
62. Li, S. and Lloyd, D. K. (1993). *Anal. Chem.*, **65**, 3684–3690.
63. Li, S. and Lloyd, D. K. (1994). *J. Chromatogr.*, **666**, 321–335.
64. Mayer, S. and Schurig, V. (1992). *J. High Res. Chromatogr.*, **15**, 129–131.
65. Armstrong, D. W., Tang, Y., Ward, T. J., and Nichols, M. (1993). *Anal. Chem.*, **65**, 1114–1117.
66. Mayer, S., and Schurig, V. (1993). *J. Liq. Chromatogr.*, **16**, 915–917.
67. Nishi, H., and Terabe, S. (1995). *J. Chromatogr.*, **694**, 245–276.
68. Kuhn, R., and Hoffstetter-Kuhn, S. (1992). *Chromatographia*, **34**, 505–512.
69. Bender, M. L., and Komiyama, M. (1978). *Cyclodextrin Chemistry*. Berlin: Springer-Verlag.
70. Tanaka, M., Asano, S., Yoshinago, M., Kawaguchi, Y., Tetsumi, T., and Shono, T. (1991). *Fresenius' J. Anal. Chem.*, **339**, 63–64.
71. Heuermann, M., and Blaschke, G. (1993). *J. Chromatogr.*, **648**, 267–274.
72. Rawjee, Y. Y., Staerk, D. U., and Vigh, G. (1993). *J. Chromatogr.*, **635**, 291–306.
73. Quang, C. and Khaledi, G. (1994). *J. High. Res. Chromatogr.*, **17**, 99–101.
74. Nishi, N., Kokusenya, Y., Miyamota, T., and Sato, T. (1994). *J. Chromatogr. A*, **659**, 449–457.
75. Schutzner, W. and Fanali, S. (1992). *Electrophoresis*, **13**, 687–690.
76. Cellai, L., Desiderio, C., Filippetti, R., and Fanali, S. (1993). *Electrophoresis*, **14**, 823–825.

CHAPTER

12

CHIRAL RECOGNITION MECHANISM OF POLYSACCHARIDES CHIRAL STATIONARY PHASES

EIJI YASHIMA and YOSHIO OKAMOTO

Department of Applied Chemistry
School of Engineering
Nagoya University
Nagoya, Japan

12.1 INTRODUCTION

Optically active compounds have recently been attracting great attention in many fields dealing with drugs, natural products, agrochemicals, ferroelectric liquid crystals, and so on. Therefore, their preparation and analysis are of increasing importance. Since enantiomers of chiral drugs often show quite different physiological behaviors in chiral living systems (1,2), a detailed investigation of the pharmacokinetic, physiological, toxicological, and metabolic activities of both enantiomers should be necessary before use. As a result of these trends, an increasing number of drugs are marketed as single-isomer forms in the pharmaceutical industry. Annual sales of enantiomerically pure drugs have reached up to $35 billion and will increase with expectations (3).

Chromatographic enantioseparations, particularly resolution by high-performance liquid chromatography (HPLC), have advanced considerably in the past decade and become a practically useful method for not only determining their optical purity but also obtaining optical isomers. Especially in the pharmaceutical industry, chiral HPLC is well recognized to be useful for the research and development of chiral drugs (4). The preparation of a chiral stationary phase (CSP) capable of effective chiral recognition is the key to this separation technique. Therefore, many CSPs for HPLC have been prepared (5–9) and about 100 of them have been commercialized. Large -scale, preparative HPLC systems have already been put on the market as a process for the isolation and purification of chiral drugs and natural products.

The Impact of Stereochemistry on Drug Development and Use, Edited by Hassan Y. Aboul-Enein and Irving W. Wainer. Chemical Analysis Series, Vol. 142.
ISBN 0-471-59644-2

The CSPs are classified into two types: One consists of chiral small molecules that are usually immobilized on a support such as silica gel, and the other is based on chiral polymers. The polymers can be used as porous gel or supported on silica gel. Details of the former CSPs are reviewed (6–9). The latter CSPs include polyacrylamides (10), one-handed helical polymethacrylates (11, 12), polyamides (13), proteins (14, 15), and polysaccharide derivatives, most of which are thoroughly reviewed elsewhere. This chapter focuses on the resolution of enantiomers on the CSPs consisting of polysaccharide derivatives, which appear to be among the most useful CSPs (16–18). In particular, cellulose triacetate, tribenzoates, and phenylcarbamate derivatives of cellulose and amylose show high chiral recognition.

The mechanism of chiral discrimination on polysaccharide phases has not yet been satisfactorily elucidated, but some interesting approaches to understanding the mechanism have been carried out by chromatographic elucidation, computational methods, and spectroscopic methods. These results should be useful for researchers using the polysaccharide phases not only to select a suitable column, but also to develop a novel polysaccharide-based CSP. Some significant mechanistic aspects are also reviewed briefly in this chapter.

12.2. CELLULOSE AND CELLULOSE ESTERS

12.2.1. Cellulose

Polysaccharides, such as cellulose **1** and amylose **2**, are the most readily available polymers with optical activity and are known to exhibit resolution as a CSP, although their chiral resolving ability is not so high. The first utilization of the chirality of cellulose for resolution was demonstrated by Kotake et al. (19). They separated racemic amino acid derivatives into two spots of enantiomers by paper chromatography. Later, Dalgliesh extended their work and proposed the three-point rule (20). The rule that three simultaneous interactions are necessary for chiral discrimination to occur has often been quoted for an explanation of the chiral discrimination mechanism.

Although thoroughly purified native cellulose with a high degree of crystallinity (crystal form I; microcrystalline cellulose) is capable of the complete resolution of amino acids by liquid chromatography (21), these materials themselves do not give practically useful CSPs because of low resolving abilities and difficulty in handling. Cellulose has been easily derivatized by the reaction on active hydroxy groups with suitable reagents to afford, for instance, triacetate, tribenzoate, and trisphenylcarbamate derivatives. The resulting derivatized polysaccharide phases offer great improvements in the chromatographic and enantioselective properties compared with native

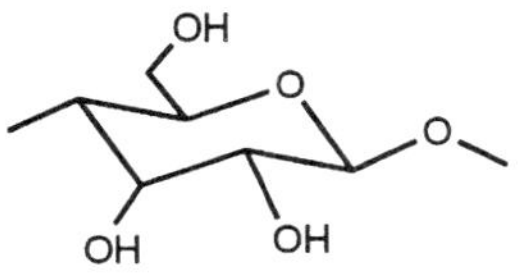

1: Cellulose 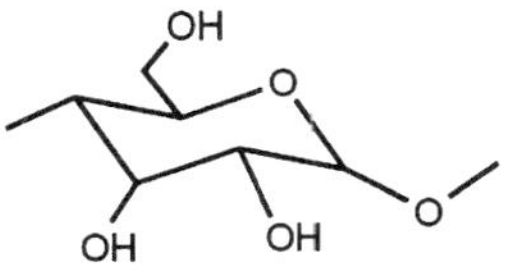 **2: Amylose** 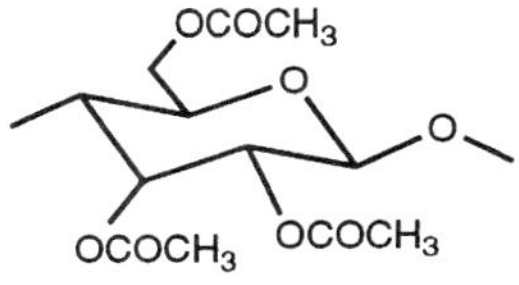 **3: Cellulose triacetate**

cellulose. Presently, many CSPs consisting of polysaccharide derivatives have been on the market and extensively used for both the analytical and preparative separation of enantiomers (16–18, 22).

12.2.2. Cellulose Esters

The first successful CSP derived from cellulose was microcrystalline cellulose triacetate (**3**, CTA-I), which is obtained by the acetylation of native microcrystalline cellulose (form I) under heterogeneous conditions and preserves the original structure of native cellulose (23). The CSP shows interesting chiral resolving properties in liquid chromatography, although the chiral resolution power of a partially acetylated cellulose is poor (24). Hesse and Hagel pointed out that the microcrystallinity of CTA-I was essential for chiral recognition, since the resolving ability was substantially reduced and the reversal of elution order occurred in some cases after the dissolution and reprecipitation of CTA-I (23). Various racemic compounds, including aromatic and aliphatic compounds, have been resolved on CTA-I in both an analytical and preparative scale, usually using an ethanol–water mixture as the eluent. The high loading capacity of CTA-I makes it one of the most commonly used stationary phases for the large-scale separation of enantiomers, particularly pharmaceutically important chiral drugs (10, 25). Some stereochemically interesting racemic compounds and chiral drugs resolved on CTA-I are shown in Figure 12.1 (10, 25–36).

The mechanism for chiral recognition on CTA-I is not yet satisfactorily elucidated probably because of the difficulty in determining the complex structure of CTA-I, which seems to be characterized by the presence of many adsorbing sites for interactions, although the structure of CTA-I was analyzed by means of X-ray (37–39), solid-state ^{13}C NMR (38), and electron microscopy (37). Hesse and Hagel (23) and later Francotte et al. (37) proposed an inclusion mechanism by which enantiomers may be adsorbed in the chiral cavities of the CTA-I matrix. Other theoretical (40) and X-ray studies of the model compound, fully acetylated D-glucopyranose-(*R*)-phenylethyl amine inclusion complex (41) also support the inclusion mechanism. Francotte et al. (37) extensively studied the relation between the crystallinity of CTA and

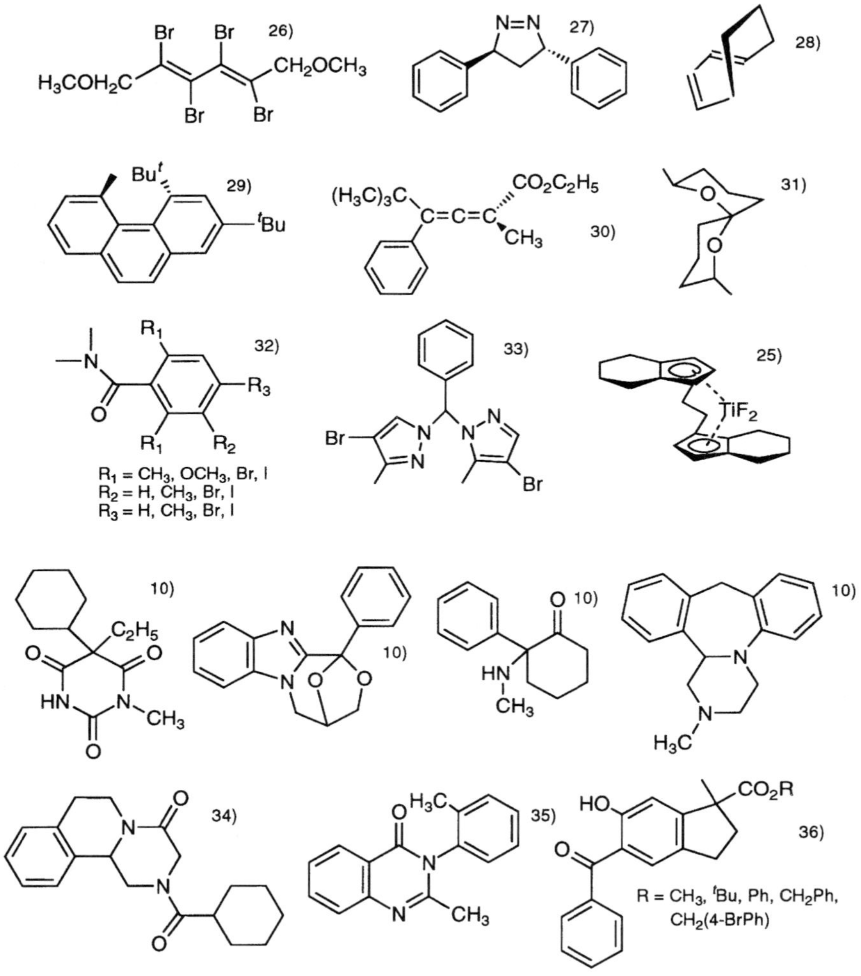

Figure 12.1. Compounds resolved on CTA-I (**3**). The numbers in parentheses next to the structures represent references cited in this chaper.

resolution power; CTA-I obtained by the heterogeneous acetylation of native cellulose showed the best resolution power, whereas CTA-II prepared by the reprecipitation of CTA-I from solution exhibited poor resolution and showed a reversal of elution order for Tröger base. Moreover, an enhancement of crystallinity of CTA-I by annealing resulted in lowering the resolution power of the CTA-I, and the chromatographic behavior was rather similar to that of CTA-II (37). Heterogeneous acetylation may provide a certain supramolecular structure for CTA-I having multiple interaction sites with specific surface and cavities inside the matrix that may be responsible for high chiral recognition for a wide range of enantiomers (42). The existence of such a supramolecular structure for CTA was also proposed by using NMR (43).

When CTA-I is coated on macroporous silica gel from a solution, it affords a useful CSP (Chiralcel OA) (44, 45). As expected, its chiral recognition ability is completely different from that of CTA-I. The new CSP is of greater advantage than CTA-I in terms of column efficiency, durability, and a choice of eluents as the mobile phase. X-ray crystallographic analysis of the material suggested that it has an almost amorphous structure rather than a crystal one (38), indicating that microcrystallinity is not essential for chiral recognition. Enantiomers of Tröger base and *trans*-stilbene oxide were eluted in reversed elution order on the two triacetate column (39, 44). This may be ascribed to the different higher-order structures or different supramolecular structure of the two triacetates. Therefore, the chiral recognition of CSP greatly depends on conditions of the preparation of CSP, for instance, the coating solvent and molecular weight of cellulose (38).

These results have aroused wide interest in the use of derivatized polysaccharides for enantioseparation. Scheme 1 shows the structures of cellulose tribenzoates prepared by Okamoto's group, some of which are also useful CSPs when coated on silica gel (44, 46). The effect of substituents on the phenyl groups of cellulose tribenzoate (CTB, **10**, Chiralcel OB) has been systematically studied (46). Alkyl, halogen, trifluoromethyl, and methoxy groups were selected as the substituents. Resolution ability was greatly affected by the inductive effect of such substituents. The benzoate derivatives having electron-donating substituents, such as the methyl group, showed better chiral recognition ability than those having electron-withdrawing substituents, such as the halogen groups. However, the most electron-donating methoxy group was not suitable because of the high polarity of the substituent itself. Among

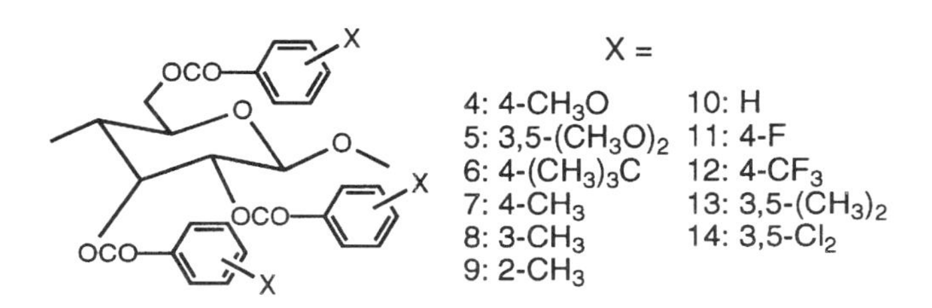

Scheme 1. Structures of cellulose tribenzoates (44, 46).

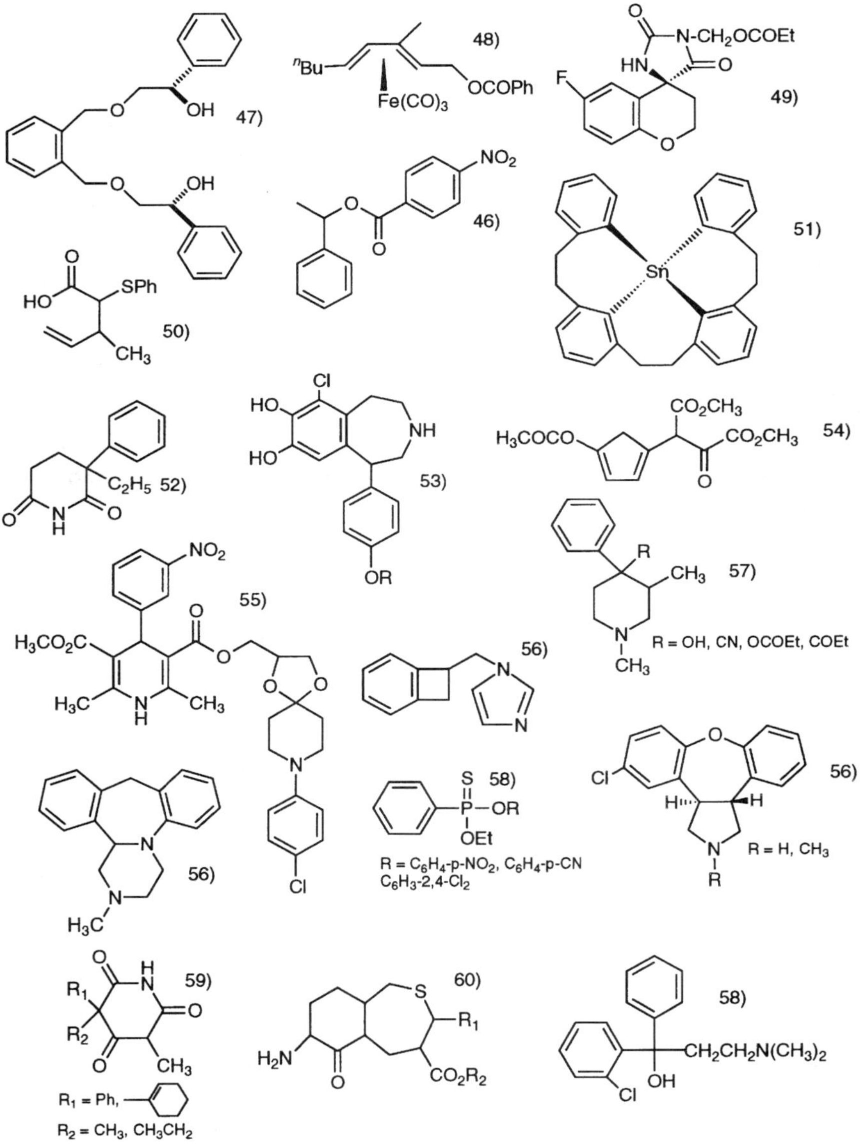

Figure 12.2. Compounds resolved on cellulose tris(4-methylbenzoate) (7, Chiralcel OJ). The numbers in parentheses next to the structures represent references cited in this chapter.

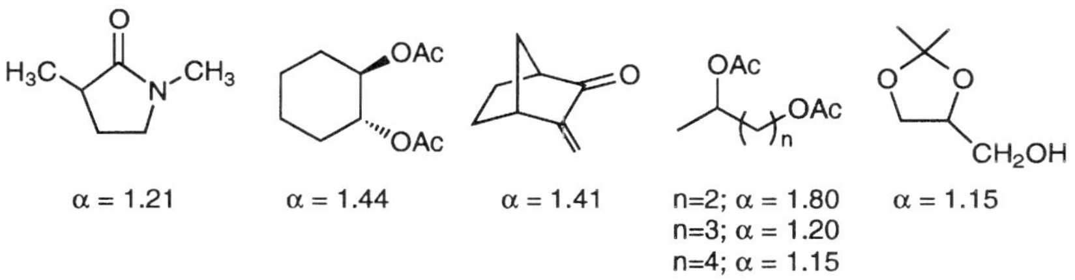

$\alpha = 1.21$ $\alpha = 1.44$ $\alpha = 1.41$ n=2; $\alpha = 1.80$ $\alpha = 1.15$
n=3; $\alpha = 1.20$
n=4; $\alpha = 1.15$

Figure 12.3. Nonaromatic compounds resolved on cellulose tribenzoate (**10**, Chiralcel OB) (39).

the benzoates, cellulose tris(4-methylbenzoate) (**7**, Chiralcel OJ) exhibited very high chiral recognition for various racemic compounds including drugs and appears to be a practically useful CSP. Some stereochemically interesting compounds and chiral drugs resolved on **7** are shown in Figure 12.2 (46–60). Several nonaromatic compounds (Figure 12.3) are also resolved on CTB (39).

Mannschreck's group (61) and later Francotte's group (62–64) reported that spherical beads of cellulose tribenzoate and its derivatives with a methyl substituent on the phenyl moiety are useful CSPs, particularly for preparative purposes, because of their high loading capacity. Francotte and co-workers resolved many drugs on the CSPs in large quantity (Figure 12.4) (25, 62, 64). They claimed that the benzoyl cellulose beads showed complementary chiral recognition to the CTA phase (62). The CTB-derivative beads show almost the same enantioseparation ability compared with the corresponding coated-type CTB derivatives on silica gel.

The main chiral adsorbing sites are considered to be the polar carbonyl groups of esters, which can interact with racemic compounds through hydrogen bonding and dipole–dipole interaction for chiral discrimination (46). Wainer and co-workers proposed a similar mechanism (65) based on the separation properties of a series of enantiomeric aromatic amides (66) and alcohols (67) on the CTB phase; the mechanism of retention is an attractive binding-steric fit formulation involving hydrogen bonding and dipole–dipole interaction rather than inclusion. However, chiral recognition ability was greatly dependent on the conditions of the preparation of CSP (46), particularly on the solvent used to dissolve CTB derivatives in the coating process, as observed in the case of cellulose triacetate. Therefore, other factors such as the morphology of CTB derivatives may be related to enantioseparation (46, 62). Steinmerier and Zugenmaier proposed a left-handed 3/2 helical structure for CTB independent of the preparation conditions (68). Similar results were also reported by Francotte et al. by means of X-ray analysis (62). Recently, Oguni et al. investigated the chiral discrimination mechanism of cellulose tris (4-methylbenzoate) by means of ^{13}C NMR spectroscopy (69). Several carbon

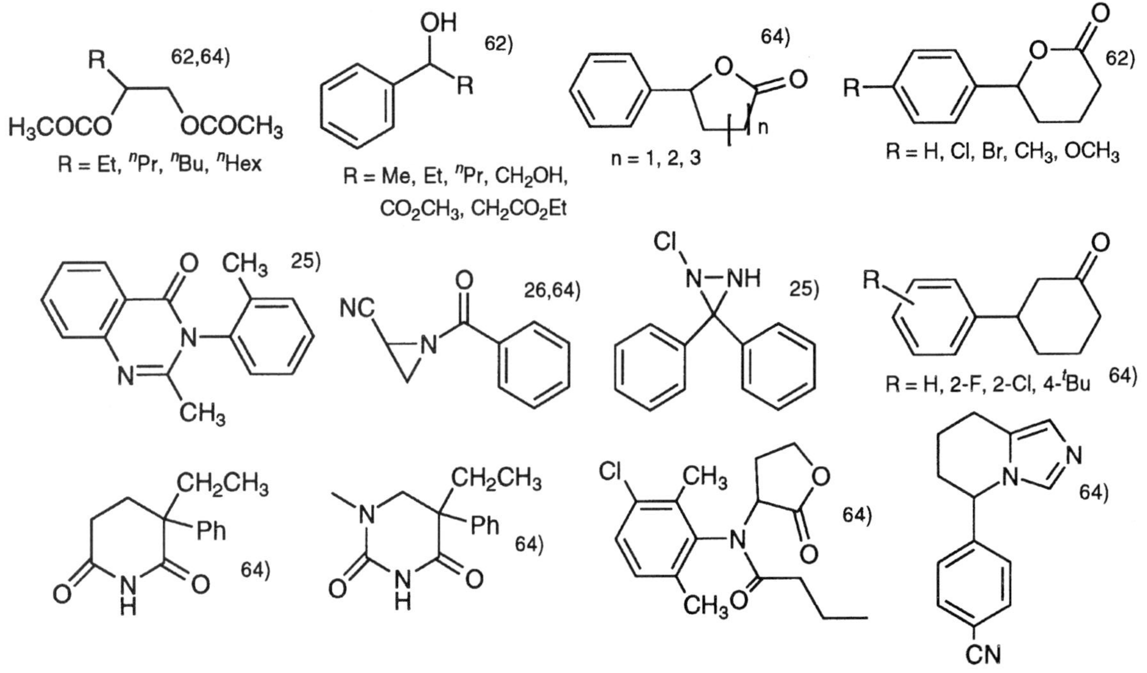

Figure 12.4. Compounds resolved on cellulose tribenzoates beads. The numbers in parentheses next to the structures represent references cited in this chapter.

resonances of 1-phenylethanol were split into enantiomers in the presence of the polymer. The approach may serve to elucidate the chiral discrimination mechanism at a molecular level.

Among these cellulose triester, triacetate, tribenzoate, tris(4-methylbenzoate), and tricinnamate have been commercialized as CSPs.

Tribenzoates of amylose showed low chiral recognition (46).

12.3. PHENYLCARBAMATES OF CELLULOSE AND AMYLOSE

Cellulose trisphenylcarbamate derivatives are the most deeply investigated polysaccharide phases with respect to enantioseparation and the mechanism of chiral discrimination (16–18). Cellulose trisphenylcarbamate derivatives are readily prepared by the reaction of microcrystalline cellulose with substituted phenyl isocyanates, and show very high resolving power to a wide range of racemates having various functional groups when coated on silica gel. The chiral recognition abilities of a series of cellulose phenylcarbamate derivatives have been evaluated extensively (Scheme 2) (18,70) and the chiral recognition mechanism has been proposed based on chromatographic, computational, and spectroscopic methods. The chiral resolving power of CSPs was greatly influenced by the introduction of substituents on the phenyl groups (70). The introduction of an electron-donating methyl group or an electron-withdrawing halogen at the 3- and/or 4-position improved the resolution ability for many racemates, but 2-substituted derivatives showed low chiral recognition. The derivatives with heteroatom substituents, such as methoxy and nitro groups, show poor chiral recognition (70) since racemic compounds can interact with the polar substituents far from a chiral glucose residue. Therefore, bulky alkoxy substituents such as isopropoxy and isobutoxy improve resolving power (71). Recently, phenylcarbamate derivatives having both an electron-donating methyl group and an electron-withdrawing chloro or fluoro group on the phenyl moieties were found to exhibit high enantioseparation for many racemates, for example, 3,4- or 3,5-chloro-methylphenylcarbamates of cellulose showed particularly high chiral recognition ability (72,73).

Similarly, the chiral resolving power of amylose derivatives was improved by introducing methyl or chloro groups on the phenyl moieties, (74, 75). However, in contrast to the cellulose derivatives, tris(4-methoxyphenylcarbamate) (71) and tris(5-chloro-2-methylphenylcarbamate) (76) of amylose showed high chiral recognition. Possible structures are 3/2 helical chain conformation for cellulose tris(phenylcarbamate) (**26**) (68, 77) and 4/1 helical chain conformation for amylose tris(phenylcarbamate) (78). These different higher-order structures seem to be responsible for the different influence of the substituents on the resolving power of cellulose and amylose.

X =

15: 4-CH_3O	25: 2-CH_3	35: 3,5-$(CH_3)_2$	45: 2-Cl-6-CH_3
16: 4-C_2H_5O	26: H	36: 2,6-$(CH_3)_2$	46: 3-Cl-2-CH_3
17: 4-$(CH_3)_2CHO$	27: 4-F	37: 3,4,5-$(CH_3)_3$	47: 3-Cl-4-CH_3
18: 4-$(CH_3)_2CHCH_2O$	28: 4-Cl	38: 3,5-Cl_2	48: 4-Cl-2-CH_3
19: 4-$(CH_3)_3Si$	29: 2-Cl	39: 3,4-Cl_2	49: 4-Cl-3-CH_3
20: 4-CH_3	30: 3-Cl	40: 2,6-Cl_2	50: 3-F-4-CH_3
21: 4-CH_3CH_2	31: 4-Br	41: 3,5-F_2	51: 5-F-2-CH_3
22: 4-$(CH_3)_2CH$	32: 4-CF_3	42: 3,5-$(CF_3)_2$	
23: 4-$(CH_3)_3C$	33: 4-NO_2	43: 2-Cl-4-CH_3	
24: 3-CH_3	34: 3,4-$(CH_3)_2$	44: 2-Cl-5-CH_3	

Scheme 2.

Among many cellulose and amylose trisphenylcarbamate derivatives prepared so far, 3,5-di-substituted derivatives such as 3,5-dimethyl- (**35**) (CDMPC, Chiralcel OD) and 3,5-dichlorophenylcarbamates (**38**) of cellulose and amylose tris(3,5-dimethylphenylcarbamate) (**52**) (ADMPC, Chiralpak AD) show particularly interesting and efficient optical resolving abilities for a variety of racemic compounds (70, 74). Table 12.1 summarizes the results of the optical resolution of some racemic compounds (**53**~**62**) on these CSPs. The CDMPC resolves a variety of racemic compounds including aromatic hydrocarbons, amines, carboxylic acid (79), alcohols, amino acid derivatives (80), and many drugs (81) including β-adrenergic blocking agents (β-blockers) (82). The chromatographic resolution results of β-blockers on the 3,5-dimethylphenylcarbamate are shown in Table 12.2 and Figure 12.5. In all separations, (*R*)-(+)-isomers eluted first followed by (*S*)-(–)-isomers showing complete separation, except for acebutanol. Sotalol was not resolved on CDMPC, but instead on ADMPC (81). Some stereochemically interesting compounds, drugs, and pharmaceuticals resolved on the CSP are shown in Figures 12.6 (85–95) and 12.7 (81, 96–109), respectively.

The CDMPC often exhibits high resolving ability for many compounds, but several drugs were separated better on other phenylcarbamate derivatives than CDMPC (81). For instance, a calcium antagonist, nicardipine, and nitredipine were completely resolved on cellulose tris(4-*tert*-butylphenylcarbamate) (**23**) (81), where CDMPC and ADMPC could not resolve them completely. Figure 12.8 shows the resolution of several racemic drugs on the phenylcarbamates of polysaccharides.

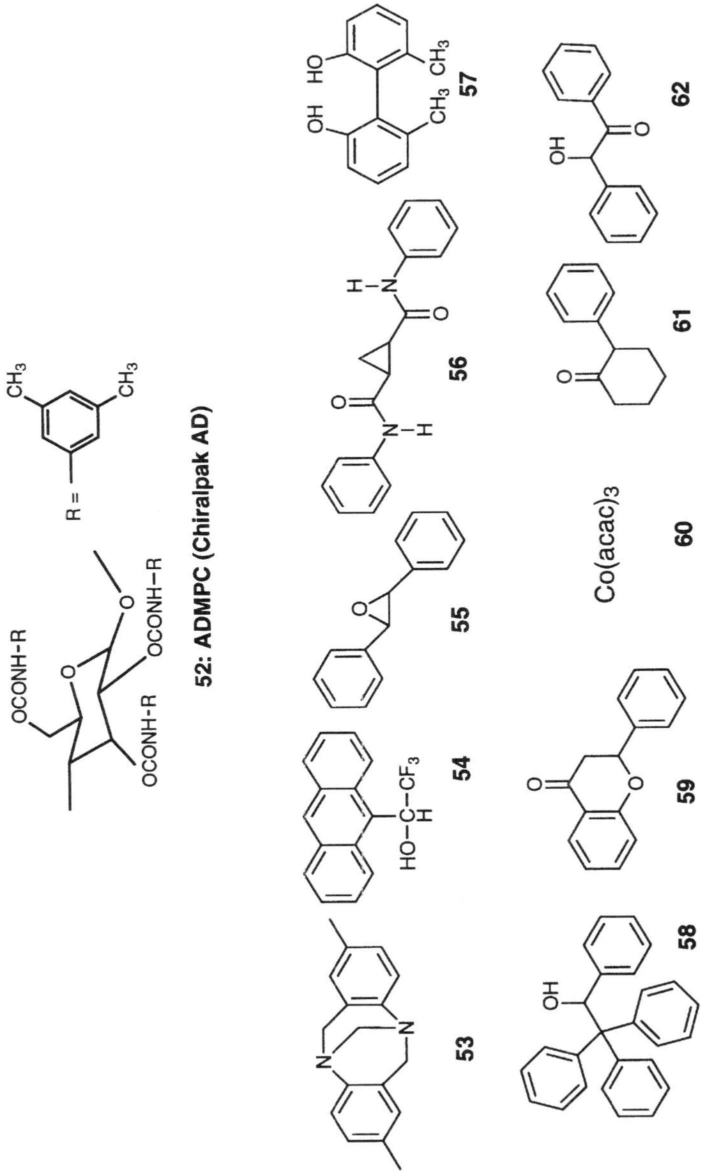

52: ADMPC (Chiralpak AD)

OCONH-R, OCONH-R, OCONH-R, R = 3,5-dimethylphenyl (CH_3, CH_3)

53 **54** **55** **56** **57**

58 **59** $Co(acac)_3$ **60** **61** **62**

Table 12.1. Resolution on 3,5-Disubstituted Phenylcarbamates of Cellulose and Amylose[a]

Compound	**35**		**38**		**52**	
	k'_1	α	k'_1	α	k'_1	α
53	0.97(+)	1.32	0.87(+)	1.65	0.53(+)	1.58
54	2.13(–)	2.59	0.28(–)	1.38	1.30(+)	1.15
55	0.74(–)	1.68	0.56(+)	1.84	0.42(+)	3.04
56	0.83(+)	2.17	0.59(+)	1.41	3.25(+)	2.01
57	2.36(–)	1.83	1.62(+)	1.11	2.46(–)	2.11
58	1.37(+)	1.34	0.40(+)	1.29	2.65(+)	1.98
59	1.47(–)	1.41	1.55(–)	1.20	0.93(+)	1.12
60	0.42(+)	Approx.1	0.76(+)	1.82	0.25(–)	Approx.1
61	1.17(–)	1.15	2.65(–)	1.26	0.61(–)	Approx.1
62	2.43(+)	1.58	3.08(–)	1.21	3.14(–)	2.01

[a] Eluent, hexane-2-propanol (90:10); flow rate, 0.5 mL/min; temperature, 25°C.

Table 12.2. Resolution of β-Blockers on Cellulose Tris(3,5-Dimethylphenylcarbamate) (CDMPC, Chiralcel OD)[a]

Ar–O–CH₂–CH(OH)–CH₂–NH–CH(CH₃)₂ (Ar)		Separation Factor (α)	Resolution Factor (Rs)
2-($CH_2CH{=}CH_2$)phenyl	Alprenolol[b]	3.87(+)	6.88
2-($OCH_2CH{=}CH_2$)phenyl	Oxprenolol	6.03(+)	8.69
4-(CH_2CONH_2)phenyl	Atenolol	1.58(+)	1.97
2-($COCH_3$)-4-($NHCOCH_2CH_2CH_3$)phenyl	Acebutolol	1.00	—
4-($CH_2CH_2OCH_3$)phenyl	Metoprolol (83)	2.95(+)	—

(*continued*)

Table 12.2. (*contd.*)

Ar–O–CH₂–CH(OH)–CH₂–NH–CH(CH₃)₂		Separation Factor (α)	Resolution Factor (Rs)
$CH_2CH_2OCH_2$–	Betaxolol (84)[c]	>3	3.58
$OCH_2CH_2OCH_2$–	Cicloprolol (84)[d]	~1.5	2.38
	Propranolol	2.29(+)	5.56
N H	Pindolol	5.07(+)	>3
OH OH	Nadolol (84)[d]	~1.2	1.68

[a] Eluent: hexane-2-propanol-$HNEt_2$ (80:20:0.1), 0.5 mL/min (82).
[b] Eluent: hexane-2-propanol (90:10), 0.5 mL/min (82).
[c] Eluent: hexane-2-propanol-$HNEt_2$ (92:8:0.05), 1.5 mL/min (84).
[d] Eluent: hexane-2-propanol-ethanol-$HNEt_2$ (80:5:15:0.05), 1.0 mL/min (84).

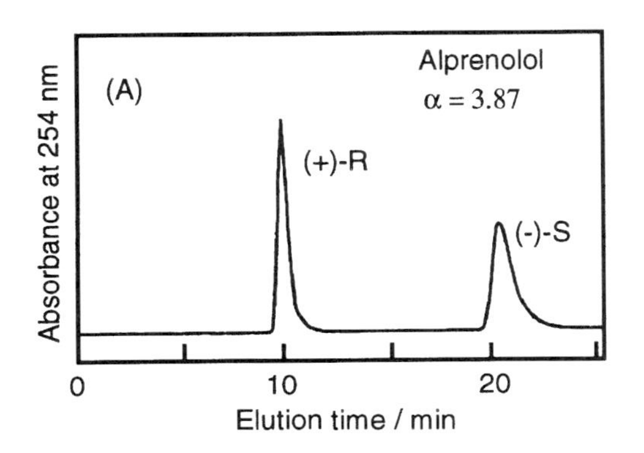

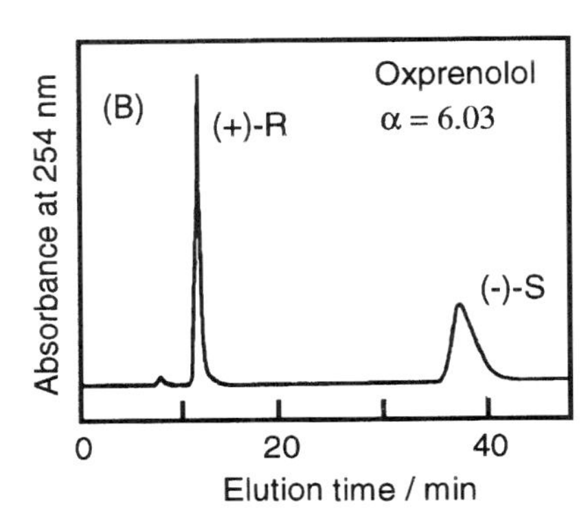

Figure 12.5. Resolution of alprenolol and oxprenolol on cellulose tris(3,5-dimethylphenylcarbamate) (**CDMPC**, Chiralcel OD) (82). Eluent: (A) hexane-2-propanol (90:10) and (B) hexane-2-propanol-$HNEt_2$ (80:20:0.1).

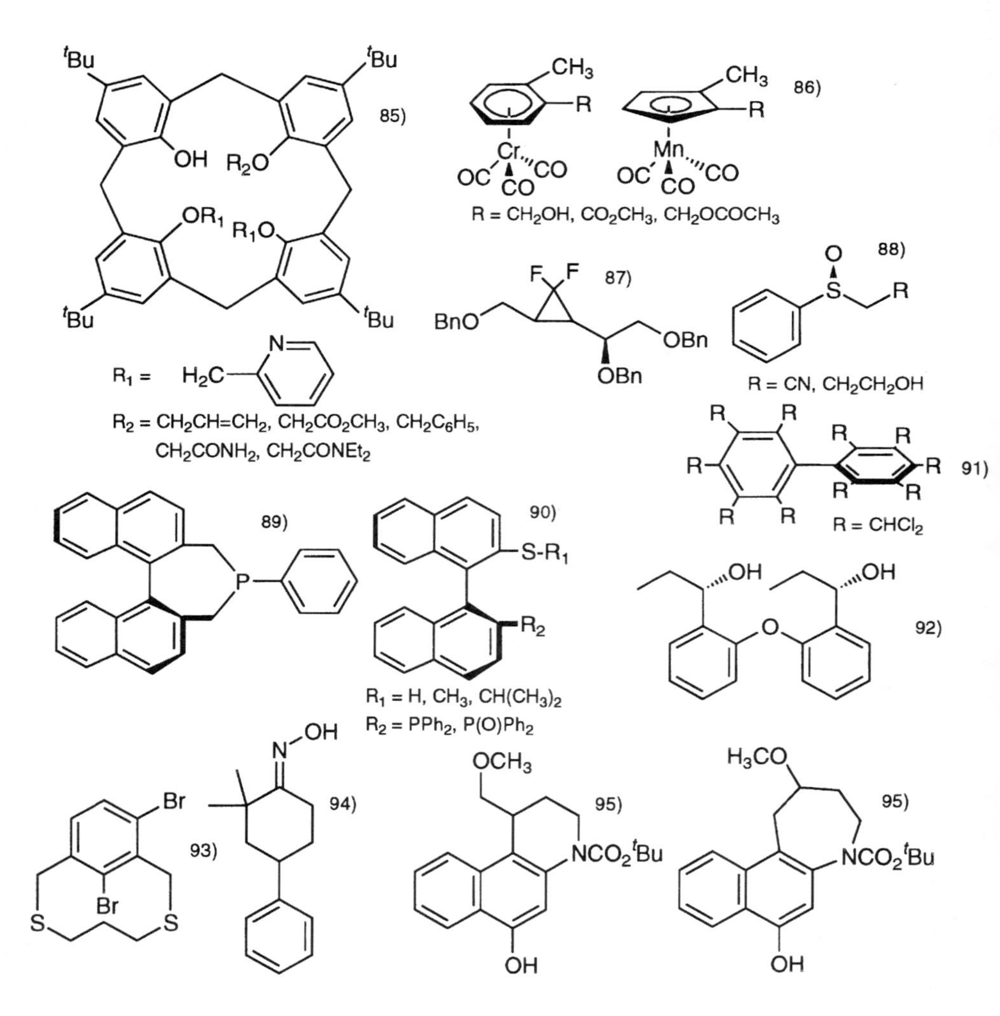

Figure 12.6. Compounds resolved on cellulose tris(3,5-dimethylphenylcarbamate) (CDMPC, Chiralcel OD). The numbers in parentheses next to the structures represent references cited in this chapter.

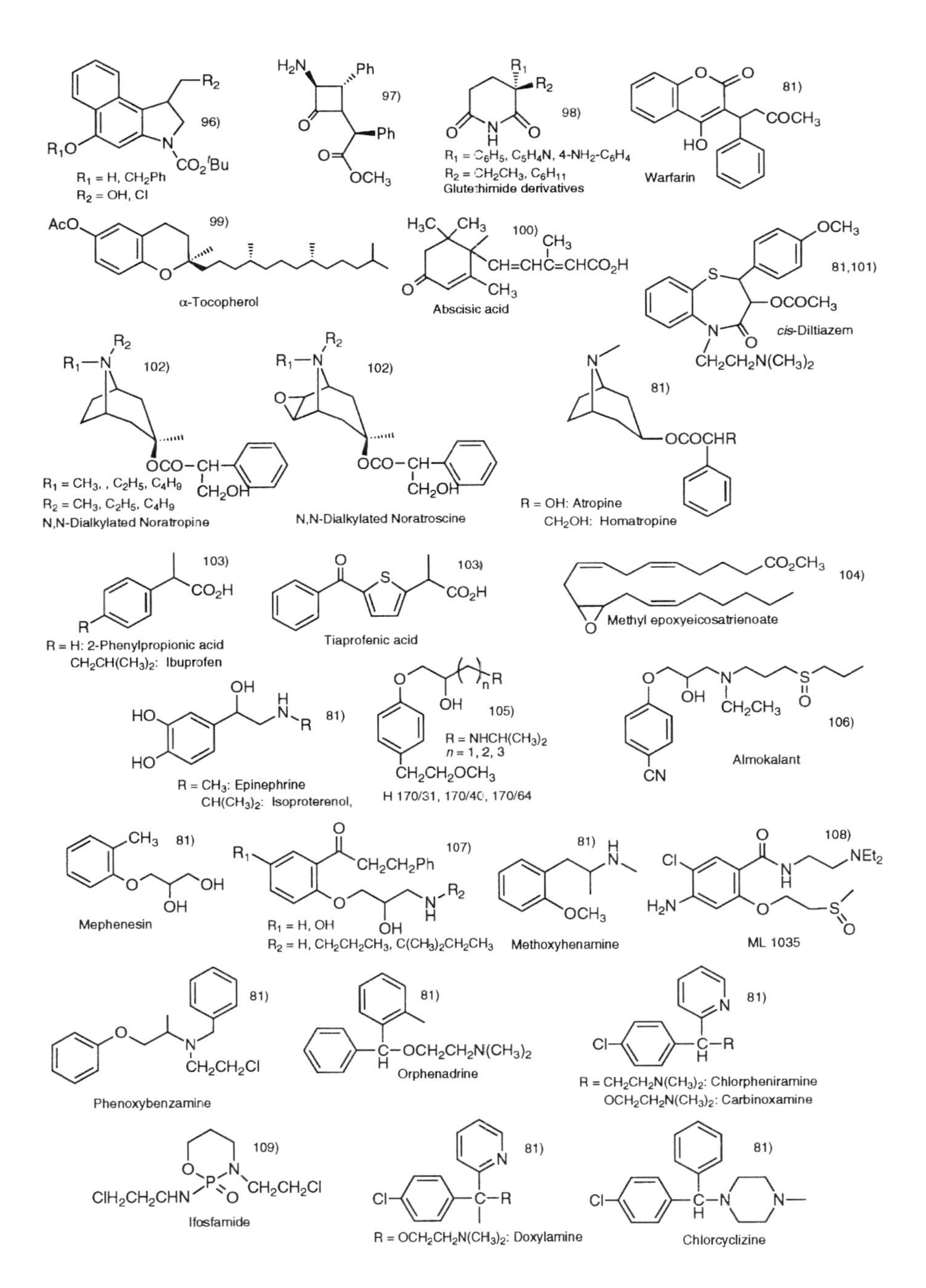

Figure 12.7. Drugs and pharmaceuticals resolved on cellulose tris(3,5-dimethylphenylcarbamate) (**CDMPC**, Chiralcel OD). The numbers in parentheses next to the structures represent references cited in this chapter.

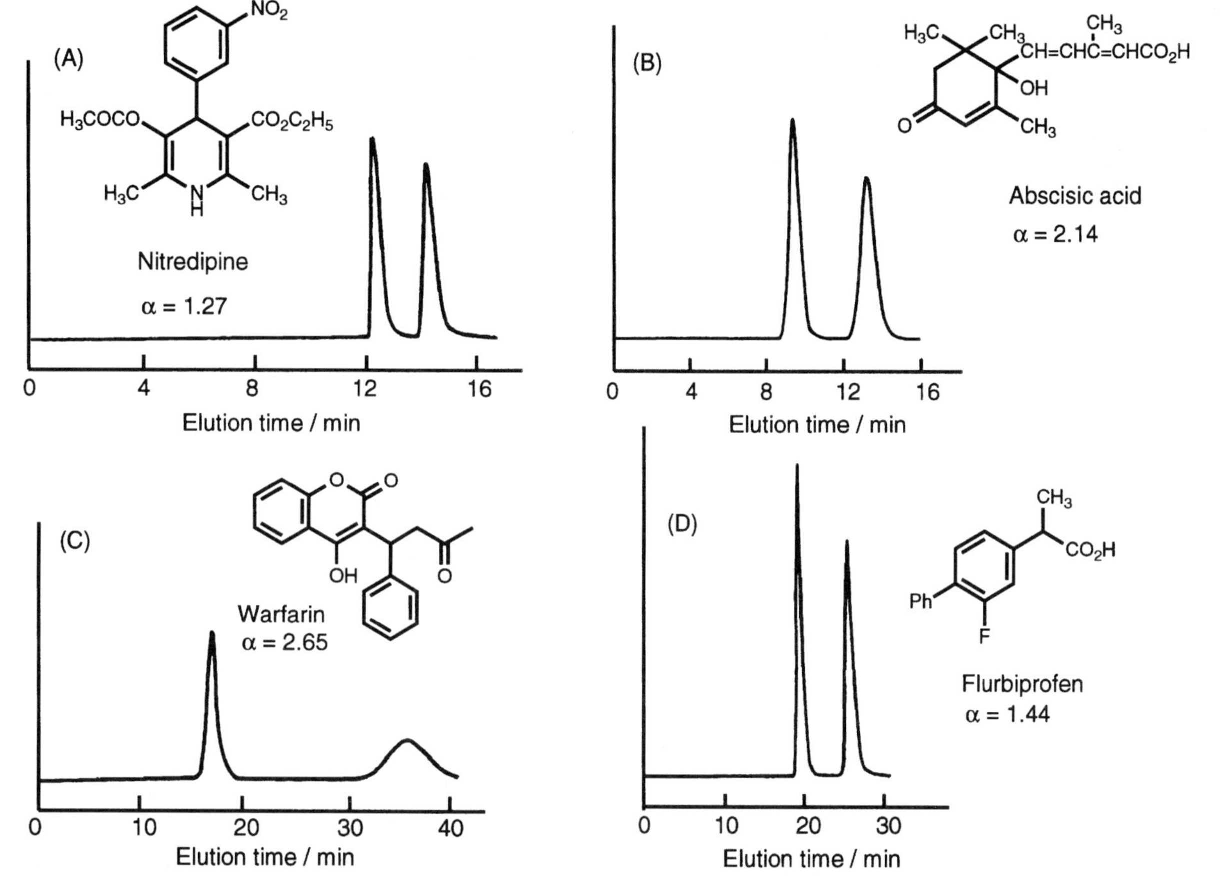

Figure 12.8. Resolution of nitredipine (A), abscisic acid (B), warfarin (C), and flurbiprofen (D). Column: (A) cellulose tris(4-*t*-butylphenylcarbamate); (B) and (C), CDMPC; (D) ADMPC (81, 100, 103, 115). Eluent: (A) hexane-2-propanol-chloroform (85:10:5); (B) and (D), hexane-2-propanol-trifluoroacetic acid (80:20:1); (C) hexane-2-propanol-HCO_2H (80:20:1). Flow rate: 0.5 mL/min.

For comparison, we examined 510 racemic compounds for enantioseparation on CDMPC; 229 of them were completely resolved and 86 were partially resolved, showing two peaks (18). This means that about 62% of 510 racemic compounds were resolved on the CSP. The cellulose tris(3,5-dimethylphenylcarbamate) may be one of the most powerful CSPs. We also examined the enantioseparation of 384 racemic compounds on an ADMPC column; 107 of them were completely resolved and 102 partially resolved with two overlapped peaks. Consequently, when two tris(3,5-dimethylphenylcarbamate)s of cellulose and amylose were used for 510 racemic compounds, 186 racemates were separated only on the cellulose derivative, 85 only on the amylose derivative, and 129 on both columns. This means that 400 racemic compounds, about 78% of 510 racemic compounds, were resolved at least on either two columns. The chiral recognition of the amylose derivatives differs from that of the corresponding cellulose derivatives. Some enantiomers elute in reversed order on the two CSPs. The two CSPs show complementary chiral recognition for many racemates, indicating that some enantiomers not resolved on CDMPC may be resolved on the ADMPC, and vice versa (18). Comparison between the two phases for enantioseparation of a series of amidotetralins (110) and chiral sulfoxide (111) was performed and the complementary chromatographic behavior observed. Some drugs and natural products resolved on the amylose tris(3,5-dimethylphenylcarbamate) are shown in Figure 12.9 (81, 96, 103, 106, 107, 112–116).

As an eluent, a hexane-2-propanol mixture is often used for the efficient separation of enantiomers. The structure of alcohols as an additive influences enantioseparation (22). When an analyte has a basic amino group, the addition of a small amount of an amine such as diethylamine or isopropylamine gives better separation without tailing of peaks (82). For acidic compounds, the addition of a small amount of a strong acid such as CF_3COOH is recommended (79). Aqueous eluents are also usable to resolve drugs and this mobile phase system is widely applicable to the investigation of the pharmacokinetic, physiological, toxicological, and metabolic activities of both enantiomers in living systems (117, 118). However, other solvents such as chloroform and tetrahydrofuran (THF), in which polysaccharides are dissolved or swollen, cannot be used as main mobile phases. To improve this defect, the phenylcarbamate derivatives of cellulose were chemically bonded to silica gel using diisocyanate as a spacer (119). However, the chiral recognition ability was reduced compared to that of coated-type CSPs when 10% of hydroxy groups of cellulose were chemically bonded nonregioselectively to amino-functionalized silica gel with diisocyanate.

Recently, CDMPC and ADMPC regioselectively bonded to silica gel were prepared with 4,4′-diphenylmethane diisocyanate as a spacer (120). ADMPC regioselectively bonded at the 6-position to silica gel by using a small amount

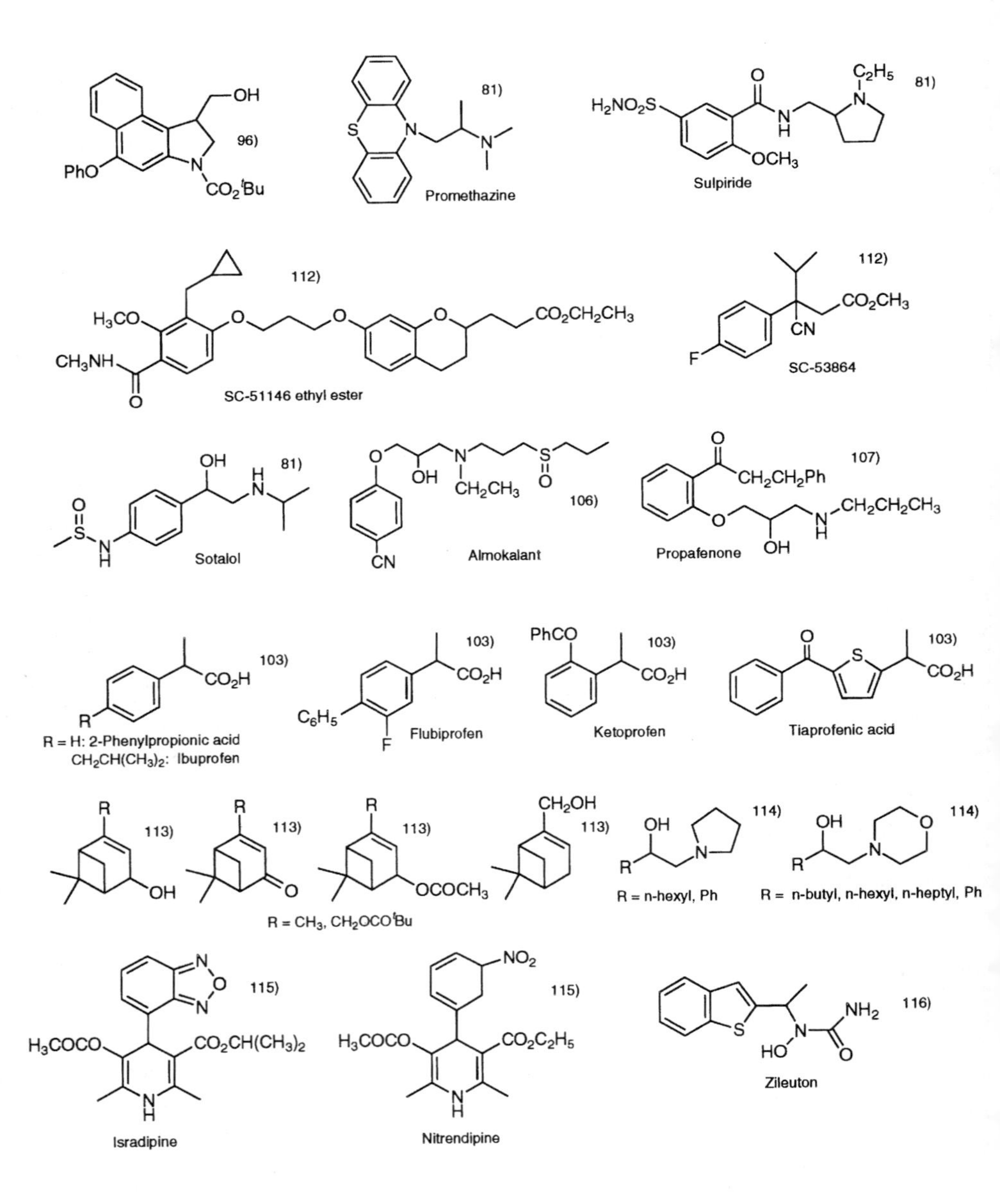

Figure 12.9. Compounds resolved on amylose tris(3,5-dimethylphenylcarbamate) (**ADMPC**, Chiralpak AD). The numbers in parentheses next to the structures represent references cited in this chapter.

of diisocyanate showed a higher resolving power than that bonded at the 2- or 3-position. For CDMPC, the position of glucose in immobilization on silica gel hardly affected chiral recognition. Some racemic compounds were more efficiently resolved on the chemically bonded-type CSP using chloroform as a component of the mobile phase. The above mentioned CSPs may be chemically bonded to silica gel through plural hydroxy groups of polysaccharides, which will cause an alternation in the higher-order structure of the polymers, giving rise to a decrease in chiral recognition ability. Very recently, 3,5-dimethylphenylcarbamate of amylose was successfully chemically bonded to silica gel only at the reducing terminal residue of amylose (121). Amylose with a desired chain length was prepared by the polymerization of α-D-glucose 1-phosphate dipotassium salt with functionalized maltooligosaccharides using a phosphorylase from potato. The amylose was successfully bonded to silica gel and allowed to react with 3,5-dimethylphenyl isocyanate to afford CSPs with excellent resolving ability and high durability against solvents such as THF and chloroform (Scheme 3) (122).

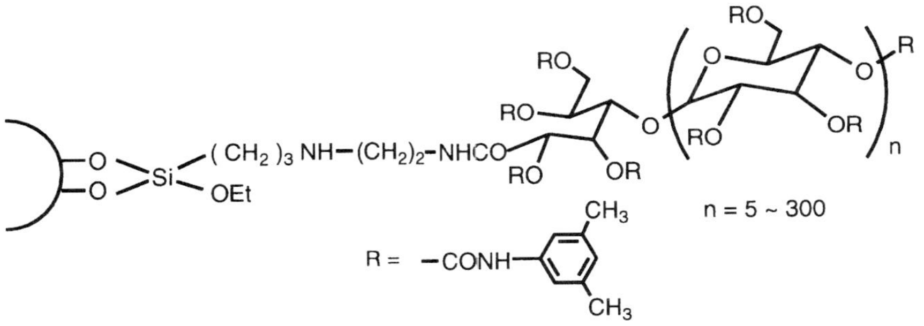

Scheme 3. CSPs show excellent resolving ability and high durability against solvents such as THF and chloroform.

12.4. MECHANISM OF CHIRAL DISCRIMINATION ON PHENYLCARBAMATES OF POLYSACCHARIDES

Most of the cellulose trisphenylcarbamate derivatives form a lyotropic liquid crystalline phase in a highly concentrated solution (70, 123) and show high crystallinity under a polarizing microscope when they are cast from a solution. This indicates that the carbamates coated on silica gel from a solution also have an ordered structure in which phenylcarbamate groups are regularly arranged. Such an ordered structure seems to be very important for efficient chiral recognition on CSPs derived from polymers. A few cellulose phenylcar-

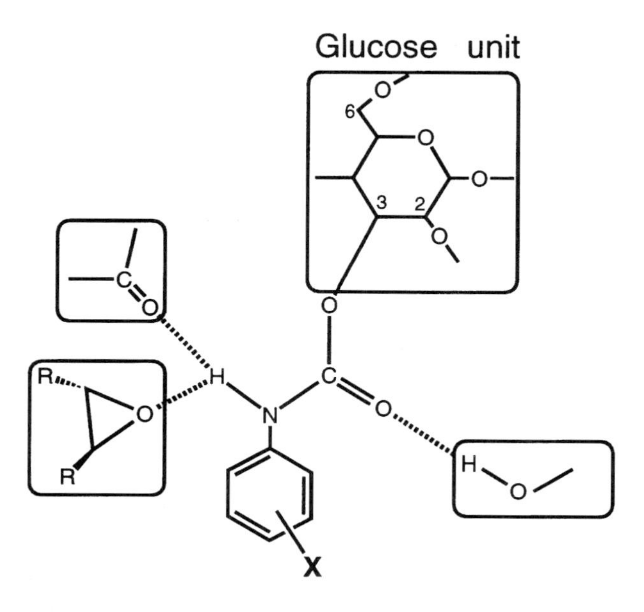

Figure 12.10. Sites for hydrogen bonding on substituted phenylcarbamates.

bamate derivatives and alkylcarbamates do not form such a liquid crystalline phase showing low chiral resolving power.

The most important adsorbing sites for chiral discrimination on phenylcarbamate derivatives are probably the polar carbamate groups as shown in Figure 12.10. These groups are capable of interacting with a racemic compound via hydrogen bonding with NH and C=O groups and the dipole-dipole interaction on C=O (18,70). Therefore, the nature of the substituents on the phenyl groups affects the polarity of carbamate residues, which must lead different chiral resolving power.

Although phenylcarbamate derivatives of polysaccharides become widely used CSPs, the chiral discrimination mechanism at a molecular level is not clear. The exact structures of the phenylcarbamate derivatives should be determined in order to reveal the mechanism. Figure 12.11 shows the one stable structure obtained by molecular mechanics calculation based on the proposed structure for cellulose tris(phenylcarbamate) (CTPC) by X-ray analysis (68,77). CTPC possesses a left-handed three-fold (3/2) helix and glucose residues are regularly arranged along the helical axis. A chiral helical groove or ditch with polar carbamate residues exists along the main chain. The polar carbamate groups are favorably located inside and hydrophobic aromatic groups are placed outside the polymer chain so that polar enantiomers may be

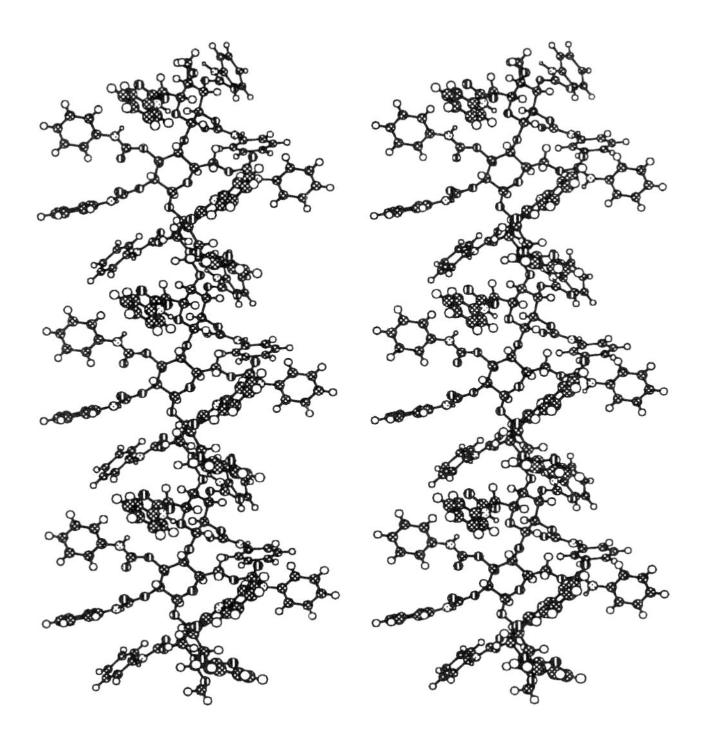

Figure 12.11. Optimized structure of cellulose trisphenylcarbamate (CTPC) (stereoview).

inserted into the groove and interact with the carbamate residues via hydrogen bonding formation. This interaction appears to lead the efficient chiral discrimination.

Besides these polar interactions, the $\pi-\pi$ interaction between the phenyl group of a CSP and aromatic group of a solute may play some role for chiral recognition because several nonpolar aromatic compounds have also been resolved (124, 125).

CTPC appears to maintain its helical structure even in solution (126). However, most of the phenylcarbamate derivatives with high chiral resolving power as CSPs are soluble only in polar solvents such as pyridine and THF. In such polar solvents, the chiral discrimination of enantiomers by NMR is hardly detected due to the strong interaction of the solvents with the polar carbamate residues; therefore, it had been difficult to elaborate on the chiral discrimination mechanism by NMR. Recently, we have found that several new phenylcarbamate derivatives, for instance, tris(4-trimethylsilylphenylcarbamate) (**19**) (127) and tris(5-fluoro-2-methylphenylcarbamate) (**51**) (128) of cellulose, are soluble in chloroform and show chiral discrimination in ^{1}H and ^{13}C NMR spectroscopies as well as HPLC. The phenylcarbamate derivatives exhibited high chiral recognition and resolved many racemic compounds by HPLC.

OCONH-R

OCONH-R OCONH−R

R = —C_6H_4—$Si(CH_3)_3$ **19**

F, H_3C **51**

Figure 12.12 shows the 500-MHz ^{1}H NMR spectra of (±)-*trans*-stilbene oxide (**55**) in the presence and absence of **19**. The methine proton of **55** was enantiomerically separated into two singlet resonances in the presence of **19** (127). This indicates that **19** can discriminate the enantiomers even in solution. From the measurement of enantiomerically pure (+)- and (–)- **55**, it was found that only the methine proton of the (–)-isomer shifted downfield, whereas that of the (+)-isomer scarcely shifted. In the chromatographic enantioseparation of (±)-**55** on the CSP **19**, the (+)-isomer eluted first, followed by the (–)-isomer, and complete baseline separation was attained. This indicates that the (–)-isomer adsorbs more strongly on **19**. This stronger interaction is related with the downfield shift of the (–)-isomer observed in NMR.

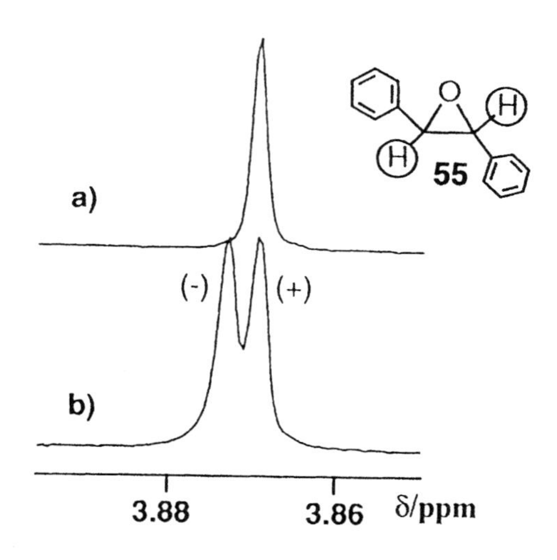

Figure 12.12. 500 MHz ^{1}H NMR spectra of *trans*-stilbene oxide (5 mg, **55**) in the absence (*a*) and presence (*b*) of cellulose tris(4-trimethylsilylphenylcarbamate) (20 mg, **19**) in $CDCl_3$ at 20 °C.

As described above, the most important adsorbing site for chiral discrimination on phenylcarbamate derivatives is the carbamate residue, which can interact with enantiomers via hydrogen bonding formation (Figure 12.10). In the case of **55**, the cyclic ether oxygen may interact with the NH proton of the carbamate residue through hydrogen bonding. Therefore, the addition of acetone capable of hydrogen bonding with the NH proton resulted in no splitting of the methine proton (127). Analogous change in the ^{1}H NMR of the methine proton of **55** was also induced by the addition of 2-propanol in place of acetone. However, interestingly, the methyl groups of 2-propanol split into a pair of doublets, indicating that the two methyl groups were magnetically nonequivalent in the presence of **19** (127). The ^{1}H NMR signals of other enantiomers of the Tröger base (**53**), benzoin (**62**), and several *sec*-alcohols were also separated into two sets of peaks in the presence of **19** in $CDCl_3$. The present results imply that the phenylcarbamate derivatives of the polysaccharides can be used as a chiral shift reagent.

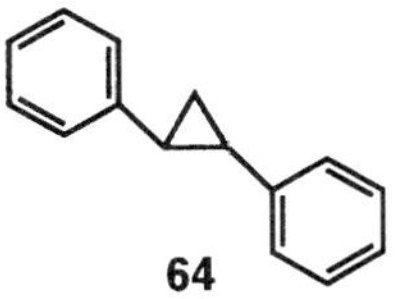

64

The force-field calculation of interaction energies between cellulose trisphenylcarbamate (CTPC) (**26**) and **55** or *trans*-1,2-diphenyl cyclopropane (**64**) was carried out using QUANTA/CHARMm and MOLECULAR INTERACTION programs to gain insight into the chiral recognition mechanism of phenylcarbamate derivatives (129). In chromatographic enantioseparation, **55** was completely resolved ($\alpha = 1.46$) on CTPC and the (+)-(*R*,*R*)-isomer eluted first, followed by the (–)-isomer, but **64** was not separated ($\alpha = \sim 1$). The result of the calculation of interaction energies between CTPC and the enantiomers suggested that (–)-(*S*,*S*)-**55** may more closely interact with CTPC than (+)-(*R*, *R*)-**55**, and significantly different interaction energy was not observed for the enantiomers of compound **64**. These calculations agreed well with the observed chromatographic resolution on CTPC (129). Computational studies will be useful and appreciable to the fields in chiral discrimination (130).

12.5. OTHER PHENYLCARBAMATES OF POLYSACCHARIDES

Phenylcarbamates of other polysaccharides such as chitosan (**65**), xylan (**66**), curdlan (**67**), dextran (**68**), and inulin (**69**) showed unique and interesting chiral properties (131).

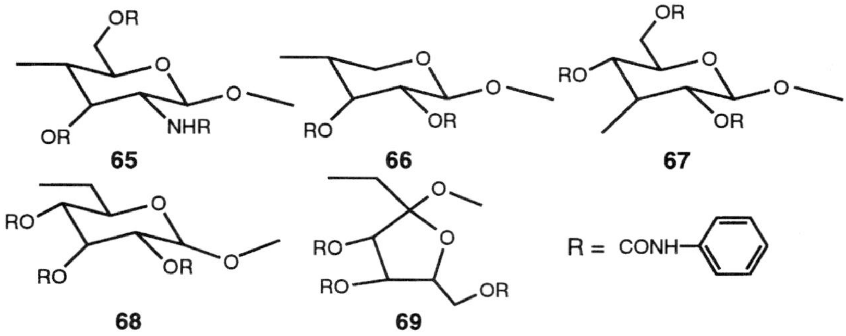

65 **66** **67**

68 **69** R = CONH–Ph

Although tris(alkylcarbamate) such as methylcarbamate and cyclohexylcarbamate of cellulose showed very low chiral recognition abilities, some of the tris(aralkylcarbamate) of cellulose and amylose showed characteristic chiral recognition for many racemic compounds (132). Among several aralkylcarbamates of cellulose and amylose, cellulose tris[(*R*)- or (*RS*)-1-phenylethylcarbamate] and amylose tris[(*S*)- or (*RS*)-1-phenylethylcarbamate] exhibit high resolving power and some of the racemic compounds shown in Figure 12.13 are better resolved on amylose tris[(*S*)-1-phenylethylcarbamate] (**70**) (Chiralpak AS) than other polysaccharides carbamates including phenylcarbamate derivatives. Figure 12.13 (58, 132–137) shows some compounds including drugs resolved on **70**.

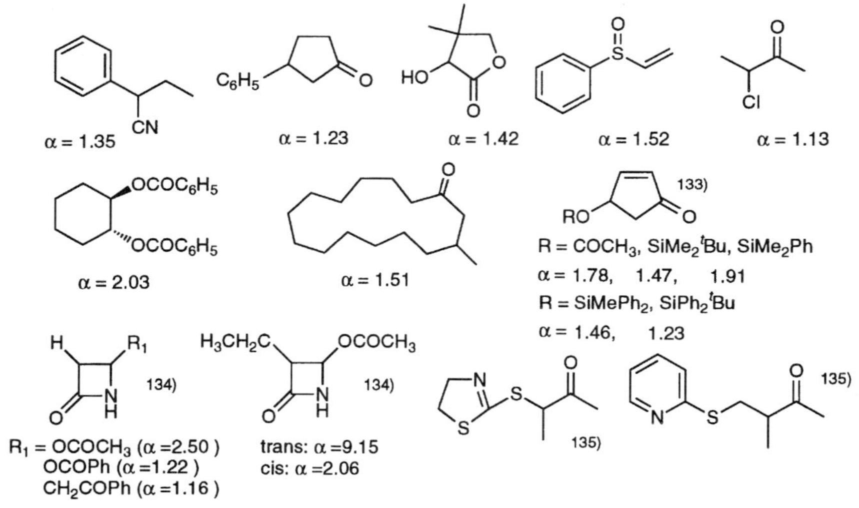

$\alpha = 1.35$ $\alpha = 1.23$ $\alpha = 1.42$ $\alpha = 1.52$ $\alpha = 1.13$

$\alpha = 2.03$ $\alpha = 1.51$

133)
R = $COCH_3$, $SiMe_2{}^tBu$, $SiMe_2Ph$
$\alpha = 1.78$, 1.47, 1.91
R = $SiMePh_2$, $SiPh_2{}^tBu$
$\alpha = 1.46$, 1.23

134)
$R_1 = OCOCH_3$ ($\alpha = 2.50$)
OCOPh ($\alpha = 1.22$)
CH_2COPh ($\alpha = 1.16$)

134)
trans: $\alpha = 9.15$
cis: $\alpha = 2.06$

135) 135)

Figure 12.13. (*continued*)

Figure 12.13. *(contd.)*

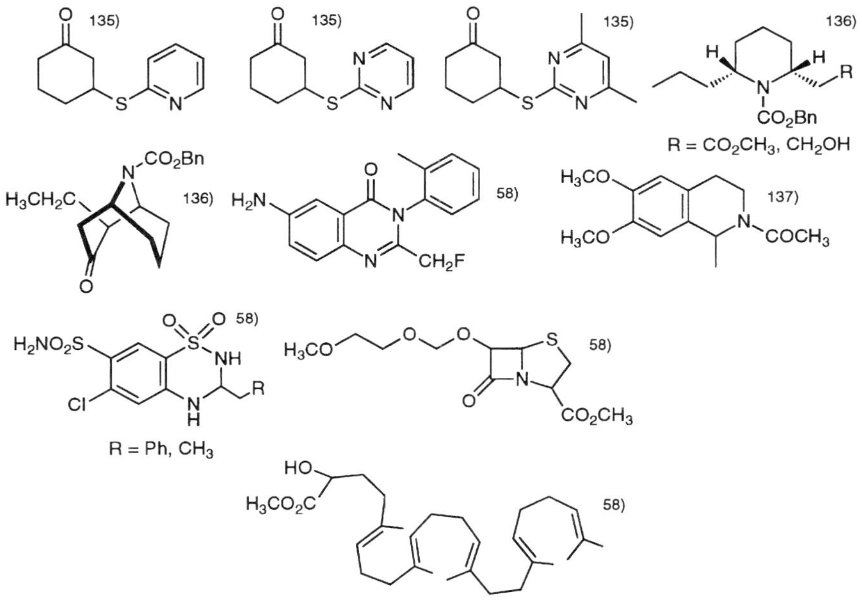

135) 135) 135) 136)

CO_2Bn

R = CO_2CH_3, CH_2OH

CO_2Bn H_3CH_2C 136) H_2N CH_2F 58) H_3CO H_3CO 137) $COCH_3$

H_2NO_2S Cl R 58) R = Ph, CH_3

H_3CO 58) CO_2CH_3

HO H_3CO_2C 58)

Figure 12.13. Compounds resolved on amylose tris [(*S*)-1-phenylethylcarbamate] (**70**, Chiralpak AS) (132). The numbers in parentheses next to the structures represent references cited in this chapter.

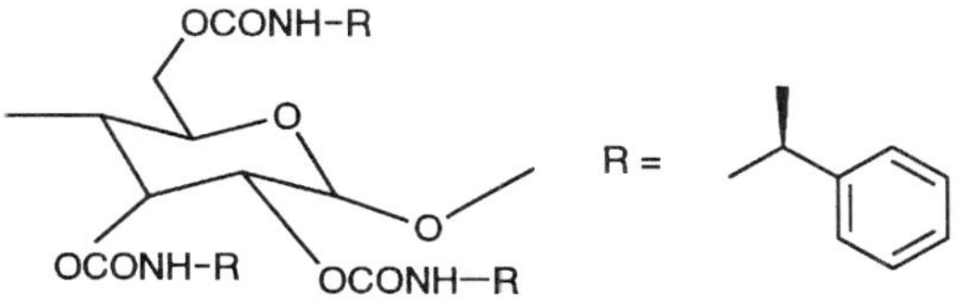

OCONH–R OCONH–R OCONH–R R =

70: Amylose tris((S)-1-phenylethylcarbamate) (Chiralpak AS)

12.6. CONCLUSION

We have prepared many polysaccharides phases and evaluated them as CSPs for enantioseparation. Some of them are commercially available, collected in Scheme 4. A wide range of racemic compounds including aliphatic and aromatic compounds with or without functional groups and many drugs are

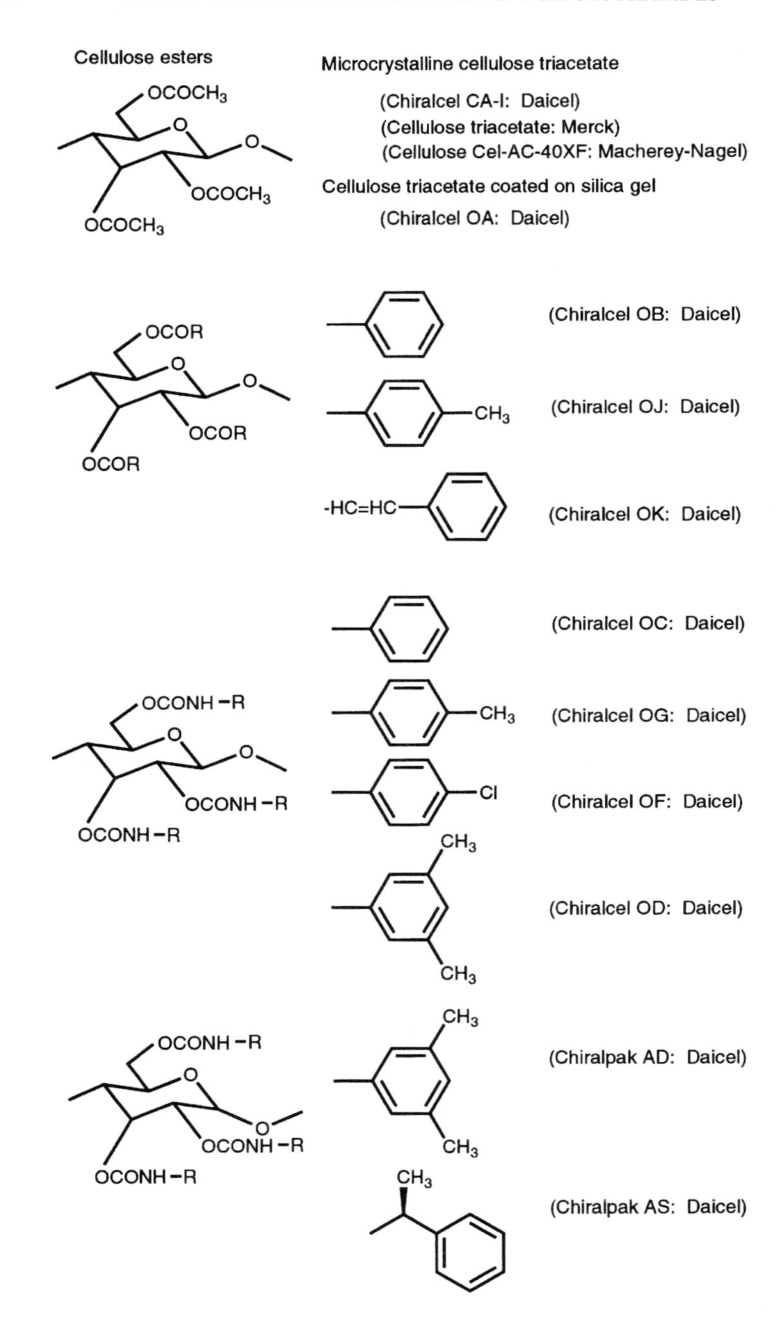

Scheme 4. Commercially available polysaccharide-based CSPs.

resolved on CSPs. Among the CSPs, 3,5-dimethylphenylcarbamates of cellulose (Chiralcel OD) and amylose (Chiralpak AD) and cellulose tris(4-methylbenzoate) (Chiralcel OJ) show high chiral recognition, which allows the resolution of more than 80% of the racemic compounds tested. These CSPs can be used for both the analytical and preparative separation of enantiomers. Enantioseparation using a membrane consisting of cellulose tris(3,5-dimethylphenylcarbamate) has been examined (138). The membrane can enrich oxprenolol in two enantiomers by enantioselective permeation. This technique seems to be important for the preparative scale separation of enantiomers. Elucidation of the exact chiral discrimination mechanism on the CSPs must serve to predict the elution order of enantiomers and develop a more excellent polysaccharide-based CSP.

REFERENCES

1. Waldeck, B. (1993). *Chirality*, **5**, 350.
2. Millership, J. S. and Fitzpatrick, A. (1993). *Chirality*, **5**, 573.
3. Stinson, S. C. (1994). *Chem. Eng. News*, **72**, 38.
4. Mutton, I. M. (1994). '*Chiral Separations by Liquid Chromatography*.' In G. Subramanian (ed.), Weinheim: VCH Publisher, Chap. 11.
5. Okamoto, Y. (1987). *Chemtech*, 176.
6. Pirkle, W. H. and Pochapsky, T. C. (1989). *Chem. Rev.*, **89**, 347.
7. Taylor, D. R. and Maher, K. (1992). *J. Chromatogr. Sci.*, **30**, 67.
8. Allenmark, S. G. (1988). '*Chromatographic Enantioseparation: Methods and Application*.' New York: John Wiley & Sons.
9. Subramanian, G. (1994). '*Chiral Separations by Liquid Chromatography*.' Weinheim: VCH Publisher.
10. Blaschke, G. (1986). *J. Liq. Chromatogr.*, **9**, 341.
11. Okamoto, Y., Okamoto, I., Yuki, H., Murata, S., Noyori, R., and Takaya, H. (1981). *J. Am. Chem. Soc.*, **103**, 6971.
12. Okamoto, Y. and Hatada, K. (1986). *J. Liq. Chromatogr.*, **9**, 369.
13. Saigo, K. (1992). *Prog. Polym. Sci.*, **17**, 35.
14. Allenmark, S. G. (1994). *J. Chromatogr. A*, **666**, 167.
15. Hermansson, J. and Hermansson, I. (1994). *J. Chromatogr. A*, **666**, 181.
16. Okamoto, Y. and Aburatani, R. (1990). *J. High Res. Chromatogr.*, 708.
17. Okamoto, Y., Kaida, Y., Aburatani, R., and Hatada, K. (1991). In. Ahuja, S. (ed.), *Chiral Separations by Liquid Chromatography*. ACS Symposium Series 471, Washington DC: ACS, Chap. 5.
18. Okamoto, Y. and Kaida, Y. (1994). *J. Chromatogr. A*, **666**, 403 (1994).
19. Kotake, M., Sakan, T., Nakamura, N., and Senoh, S. (1951). *J. Am. Chem. Soc.*, **73**, 2975.

20. Dalgliesh, C. (1952). *J. Chem. Soc.*, 137
21. Fukuhara, T., Isoyama, M., Shimada, A., Itoh, M., and Yuasa, S. (1987). *J. Chromatogr.*, **387**, 562.
22. Dingene, J. (1994). '*Chiral Separations by Liquid Chromatography*'. In G. Subramanian (ed.), Weinheim: VCH Publisher, Chap. 6.
23. Hesse, G. and Hagel, R. (1973). *Chromatographia*, **6**, 227; Hesse G. and Hagel, R. (1976). *Liebigs Ann. Chem.*, 966.
24. Luttringhaus, A., Hess, V., and Rosenbaum, H. J. (1967). *Z. Naturforsch.*, **B22**, 1296.
25. Francotte, E. (1994). *J. Chromatogr. A*, **666**, 565.
26. Becher, G. and Mannschreck, A. (1981). *Chem. Ber.*, **114**, 2365.
27. Schneider, M. P. and Bippi, H. (1980). *J. Am. Chem. Soc.*, **102**, 7363.
28. Isaksson, R., Rochester, J., Sandström, J., and Wistrand, L.-G. (1985). *J. Am. Chem. Soc.*, **107**, 4074.
29. Scheübl, H., Fritzche, U., and Mannschreck, A. (1984). *Chem. Ber.*, **117**, 336.
30. Krause, N. and Hnadke, G. (1991). *Tetrahedron Lett.*, **32**, 7225.
31. Isaksson, R., Liljefors, T., and Reinholdsson, P. (1984). *J. Chem. Soc., Chem. Comm.*, 137.
32. Cuyegkeng, M. A. and Mannschreck, A. (1987). *Chem. Ber.*, **120**, 803.
33. Ballestros, P., Clarmunt, R. M., Elguero, J., Roussel, C., and Chemlal, A. (1988). *Heterocycle*, **27**, 351.
34. Ching, C. B., Lim, B. G., Lee, E. J. D., and Ng, S. C. (1993). *J. Chromatogr.*, **634**, 215.
35. Mannschreck, A., Koller, H., Stühler, B., Davies, M. A., and Traber, J. (1984). *Eur. J. Med. Chem.-Chim. Ther.*, **19**, 381.
36. Francotte, E., Stierlin, H., and Faigle, J. W. (1985). *J. Chromatogr.*, **346**, 321.
37. Francotte, E., Wolf, R. M., Lohmann, D., and Mueller, R. (1985). *J. Chromatogr.*, **347**, 25.
38. Shibata, T., Sei, T., Nishimura, H., and Deguchi, K. (1987). *Chromatographia*, **24**, 552.
39. Shibata, T., Okamoto, I., and Ishii, K. (1986). *J. Liq. Chromatogr.*, **9**, 313.
40. Wolf, R. M., Francotte, E., and Lohmann, D. (1988). *J. Chem. Soc. Perkin Trans., II*, 893.
41. Francotte, E. and Rihs, G. (1989). *Chirality*, **1**, 80.
42. Francotte, E. and Wolf, R. M. (1990). *Chirality*, **2**, 16.
43. Buchanan, C. M., Hyatt, J. A., and Lowman, D. W. (1989). *J. Am. Chem. Soc.*, **111**, 7312.
44. Okamoto, Y., Kawashima, M., Yamamoto, K., and Hatada, K. (1984). *Chem. Lett.*, 739.
45. Ichida, A., Shibata, T., Okamoto, I., Yuki, Y., Namikoshi, H., and Toda, Y. (1984). *Chromatographia*, **19**, 280.

46. Okamoto, Y., Aburatani, R., and Hatada, K. (1987). *J. Chromatogr.*, **389**, 95.
47. Takeuchi, S., Ohira, A., Miyoshi, N., Mashio, H., and Ohga, Y. (1994). *Tetrahedron Asym.*, **5**, 1763.
48. Uemura, M., Nishimura, H., Yamada, S., Hayashi, Y., Nakamura, K., Ishihara, K., and Ohno, A. (1994). *Tetrahedron Asym.*, **5**, 1673.
49. Mizuguchi, E., Achiwa, K., Wakamatsu, H., and Terao, Y. (1994). *Tetrahedron Asym.*, **5**, 1407.
50. Corey, E. J. and Lee, D.-H. (1991). *J. Am Chem. Soc.*, **113**, 4026.
51. Okada, K., Mikami, M., and Oda, M. (1993). *Chem. Lett.*, 1999.
52. Aboul-Enein, Y. and Islam, M. R. (1990). *J. Chromatogr. Sci.*, **28**, 307.
53. Camiller, P., Dyke, C. A., Paknoham, S. J., and Senior, L. A. (1990). *J. Chromatogr.*, **498**, 414.
54. Mori, M., Nukui, S., and Shibasaki, M. (1991). *Chem. Lett.*, 1797.
55. Siret, L., Macaudiere, P., Bargmann-Leyder, N., Tambute, A., Caude, M., and Gougeon, E. (1994). *Chirality*, **6**, 440.
56. Maris, F. A., Verboort, P. J. M., and Hindriks, H. (1991). *J. Chromatogr.*, **547**, 45.
57. Yin, D., Khanolkar, A. D., Makriyannis, A., and Froimowitz, M. (1994). *J. Chromatogr. A*, **678**, 176.
58. Daicel Chemical Industry, (1993). Tokyo.
59. Murata, M. and Achiwa, K. (1991). *Tetrahedron Lett.*, **46**, 6763.
60. Nagai, U. and Pavone, V. (1989). *Heterocycles*, **28**, 589.
61. Rimbock, K.-H., Kastner, F., and Mannschreck, A. (1986). *J. Chromatogr.*, **351**, 346.
62. Francotte, E. and Wolf, R. M. (1991). *Chirality*, **3**, 43.
63. Juvancz, Z., Grolimund, K., and Francotte, E. (1992). *Chirality*, **4**, 459.
64. Francotte, E. and Wolf, R. M. (1992). *J. Chromatogr.*, **595**, 63.
65. Wainer, I. W. and Alembik, M. C. (1986). *J. Chromatogr.*, **358**, 85.
66. Wainer, I. W., Alembik, M. C., and Smith, E. (1987). *J. Chromatogr.*, **388**, 65.
67. Wainer, I. W., Stiffin, R. M., and Shibata, T. (1987). *J. Chromatogr.*, **411**, 139.
68. Steinmeier, H. and Zugenmaier, P. (1987). *Carbo. Res.*, **164**, 97.
69. Oguni, K., Matsumoto, A., and Isokawa, A. (1994). *Polym. J.*, **11**, 1257.
70. Okamoto, Y., Kawashima, M., and Hatada, K. (1986). *J. Chromatogr.*, **363**, 173.
71. Okamoto, Y., Ohashi, T., Kaida, Y., and Yashima, E. (1993). *Chirality*, **5**, 616.
72. Chankvetadze, B., Yashima, E., and Okamoto, Y. (1992).*Chem. Lett.*, 617.
73. Chankvetadze, B., Yashima, E., and Okamoto, Y. (1994). *J. Chromatogr. A*, **670**, 39.
74. Okamoto, Y., Aburatani, R., Fukumoto, T., and Hatada, K. (1987). *Chem. Lett.*, 1857.
75. Okamoto, Y., Aburatani, R., and Hatada, K. (1990). *Bull. Chem. Soc. Japan*, **63**, 955.

76. Chankvetadze, B., Yashima, E., and Okamoto, Y. (1995). *J. Chromatogr. A*, **694**, 101.
77. Vogt, U. and Zugenmaier, P. (1985). *Ber. Bunsenges, Phys. Chem.*, **89**, 1217.
78. Vogt, U. and Zugenmaier, P. (1983). European Science Foundation Workshop on Specific Interaction in Polysaccharide Systems, Uppsala, Sweden.
79. Okamoto, Y., Aburatani, R., Kaida, Y., and Hatada, K. (1988). *Chem. Lett.*, 1125.
80. Okamoto, Y., Kaida, Y., Aburatani, R., and Hatada, K. (1989). *J. Chromatogr.*, **477**, 367
81. Okamoto, Y., Aburatani, R., Hatano, K., and Hatada, K. (1988). *J. Liq. Chromatogr.*, **11**, 2147.
82. Okamoto, Y., Kawashima, M., Aburatani, R., Hatada, K., Nishiyama, T., and Masuda, M. (1986). *Chem. Lett.*, 1237.
83. Ching, C. B., Lim, B. G., Lee, E. J. D., and Ng, S. C. (1992). *Chirality*, **4**, 174.
84. Krstulovic, A. M., Fouchet, M. H., Burke, J. T., Gillet, G., and Durand, A. (1988). *J. Chromatogr.*, **452**, 477.
85. Caccameste, S. and Pappalardo, S. (1993). *Chirality*, **5**, 159.
86. Yamazaki, Y., Morohashi, N., and Hosono, K. (1991). *J. Chromatogr.*, **542**, 129.
87. Taguchi, T., Shibuya, A., Sasaki, H., Endo, J., Morikawa, T., and Shiro, M. (1994). *Tetrahedron Asym.*, **5**, 1423.
88. Secundo, F., Carrea, G., Dallavalle, S., and Franzosi, G. (1993). *Tetrahedron Asym.*, **4**, 1981.
89. Gladiali, S., Dore, A., Fabbri, D., de Lucchi, O., and Manassero, M. (1994). *Tetrahedron Asym.*, **5**, 511.
90. Gladiali, S., Dore, A., and Fabbri, D. (1994). *Tetrahedron Asym.*, **5**, 1143.
91. Biali, S. E., Kahr, B., Okamoto, Y., Aburatani, R., and Mislow, K. (1988). *J. Am. Chem. Soc.*, **110**, 1917.
92. Soai, K., Hayase, T., Shimada, C., and Isobe, K. (1994). *Tetrahedron Asym.*, **5**, 789.
93. Grimme, S., Pischel, I., Vögtle, F., and Nieger, M. (1995). *J. Am. Chem. Soc.*, **117**, 157.
94. Murakata, M., Imai, M., Tamura, M., and Hoshino, O. (1994). *Tetrahedron Asym.*, **5**, 2019.
95. Boger, D. L. and Mesini, P. (1994). *J. Am. Chem. Soc.*, **116**, 11335.
96. Boger, D. L. and Yun, W. (1994). *J. Am. Chem. Soc.*, **116**, 7996.
97. van Maanen, H. L., Jastrzbski, J. T. B. H., Verweij, J., Kieboom, A. P. G., Spek, A. L., and van Koten, G. (1993). *Tetrahedron Asym.*, **4**, 1441.
98. Aboul-Enein, H. Y. and Serignese, V. (1994). *Chirality*, **6**, 378.
99. Mizuguchi, E., Takemoto, M., and Achiwa, K. (1993). *Tetahedron Asym.*, **4**, 1961.
100. Okamoto, Y., Aburatani, R., and Hatada, K. (1988). *J. Chromatogr.*, **448**, 454.
101. Ishii, K., Minato, K., Nishimura, N., Miyamoto, T., and Sato, T. (1994). *J. Chromatogr. A*, **686**, 93.

102. Hampe, T. R. E., Schlüter, J., Brandt, K. H., Nagel, J., Lamparter, E., and Blaschke, G. (1993). *J. Chromatogr. A*, **634**, 205.

103. Okamoto, Y., Aburatani, R., Kaida, Y., Hatada, K., Inotsume, N., and Nakano, M. (1989). *Chirality*, **1**, 239.

104. Zhang, J. Y. and Blair, I. A. (1994). *J. Chromatogr. B*, **657**, 23.

105. Balmer, K., Lagerström, P.-O., and Persson, B.-A. (1992). *J. Chromatogr.*, **592**, 331.

106. Balmer, K., Lagerström, P.-O., Larsson, S., and Persson, B.-A. (1993). *J. Chromatogr.*, **631**, 191.

107. Hollenhorst, T. and Blaschke, G. (1991). *J. Chromatogr.*, **585**, 329.

108. Butler, B. T., Silvey, G., Houston, D. M., Borcherding, D. R., Vaughn, V. L., McPhail, A. T., Radzik, D. M., Wynberg, H., Hoeve, W. T., van Echten, E., Ahmed, N. K., and Linnik, M. T. (1992). *Chirality*, **4**, 155.

109. Masurel, D. and Wainer, I. W. (1989). *J. Chromatogr.*, **490**, 133.

110. Witte, D. T., Bruggeman, F. J., Franke, J. P., Copinga, S., Jansen, J. M., and de Zeeuw, R. A. (1993). *Chirality*, **5**, 545.

111. Matlin, S. A., Tiritan, M. E., Crawford, A. J., Cass, Q. B., and Boyd, D. R. (1994). *Chirality*, **6**, 135.

112. Miller, L., Honda, D., Fronek, R., and Howe, K. (1994). *J. Chromatogr. A*, **658**, 429.

113. Abu-Lafi, S., Sterin, M., Levin, S., and Mechoulam, R. (1994). *J. Chromatogr. A*, **664**, 159.

114. Nicholson, L. W., Pfeiffer, C. D., Goralski, C. T., Singaram, B., and Fisher, G. B. (1994). *J. Chromatogr. A*, **687**, 241.

115. Okamoto, Y., Aburatani, R., Hatada, K., Honda, M., Inotsume, N., and Nakano, M. (1990). *J. Chromatogr.*, **513**, 375.

116. Thmas, S. B. and Surber, B. W. (1992). *J. Chromatogr.*, **623**, 390.

117. Ikeda, K., Hamasaki, T., Kohno, H., Ogawa, T., Matsumoto, T., and Sakai, J. (1989). *Chem. Lett.*, 1089.

118. Ishikawa, A. and Shibata, T. (1993). *J. Liq. Chromatogr.*, **16**, 859.

119. Okamoto, Y., Aburatani, R., Miura, S., and Hatada, K. (1987). *J. Liq. Chromatogr.*, **10**, 1613.

120. Yashima, E., Fukaya, H., and Okamoto, Y. (1994). *J. Chromatogr. A*, **677**, 11.

121. Enomoto, N., Furukawa, S., Ogasawara, Y., Yashima, E., and Okamoto, Y. (1994). *In 66th Annual Meeting of the Chemical Society of Japan*, 2B233, P. 140.

122. Enomoto, N., Furukawa, S. Ogasawara, Y., Akano, H., Kawamura, Y., Yashima, E., and Okamoto, Y. (1996). *Anal. Chem.*, **68**, 2798.

123. Vogt, U. and Zugenmaier, P. (1983). *Makromol. Chem. Rapid Commun.*, **4**, 759.

124. Okamoto, Y., Hatano, K., Aburatani, R., and Hatada, K. (1989). *Chem. Lett.*, 715.

125. Hopf, H., Grahn, W., Barrett, D. G., Gerdes, A., Hilmer, J., Hucker, J., Okamoto, Y., and Kaida, Y. (1990). *Chem. Ber.*, **123**, 841.

126. Danhelka, J., Netopilik, M., and Bohdanecky, M. (1987). *J. Polym. Sci. Part B: Polym. Phys.*, **25**, 1801.
127. Yashima, E., Yamada, M., and Okamoto, Y. (1994). *Chem. Lett.*, 579.
128. Yashima, E., Yamamoto, C., and Okamoto, Y. (1995). *Polym. J.*, **27**, 856; Yashima, E., Yamamoto, C., and Okamoto, Y. (1996). *J. Am. Chem. Soc.*, **118**, 4036.
129. Yashima, E., Yamada, M., Kaida, Y., and Okamoto, Y. (1995). *J. Chromatogr. A*, **694**, 347.
130. Lipkowitz, K. B. (1994). In G. Subramanian (ed.), '*Chiral Separations by Liquid Chromatography*'. Weinheim: VCH Publisher, Chap. 2.
131. Okamoto, Y., Kawashima, M., and Hatada, K. (1984). *J. Am. Chem. Soc.*, **106**, 5357.
132. Kaida, Y. and Okamoto, Y. (1993). *J. Chromatogr.*, **641**, 267.
133. Kaida, Y. and Okamoto, Y. (1992). *Chem. Lett.*, **1992**, 85.
134. Kaida, Y. and Okamoto, Y. (1992). *Chirality*, **4**, 122.
135. Toda, F., Tanaka, K., and Sato, J. (1993). *Tetrahedron Asym.*, **4**, 1771.
136. Muraoka, O., Okumura, K., Maeda, T., Tanabe, G., and Momose, T. (1994). *Tetrahedron Asym.*, **5**, 317.
137. Morimoto, T., Nakajima, N., and Achiwa, A. (1995). *Tetrahedron Asym.*, **6**, 75.
138. Yashima, E., Noguchi, J., and Okamoto, Y. (1994). *J. Appl. Polym. Sci.*, **54**, 1087.

CHAPTER

13

MICELLE-MEDIATED CAPILLARY ELECTROPHORETIC SEPARATION OF ENANTIOMERIC COMPOUNDS

MICHAEL E. SWARTZ

Waters Corporation
Milford, Massachasetts 01757

PHYLLIS R. BROWN

Department of Chemistry
University of Rhode Island
Kingston, Rhode Island 02881

13.1. INTRODUCTION

Chirality arises from an element of asymmetry in a molecule that may be a center, an axis, or a plane. Molecules that are not superimposable on their mirror images are chiral, a class of stereoisomers called enantiomers. Enantiomers have identical physical properties, except for the rotation of optically active or polarized light. Chemical properties are identical as well, except in a chiral environment, for example, in biological systems. Enantiomers are frequently distinguished by biological systems and may possess significantly different pharmacokinetic, pharmacologic, and toxicologic properties. It is for this reason that the development of drugs as pure enantiomers rather than racemic mixtures is an emerging trend in many pharmaceutical companies (1). Technological advances in asymmetric synthesis and the process scale separation of chiral compounds now permit the commercial production of many enantiomers. As a result, the U.S. Food and Drug Administration (FDA) has issued a policy statement for the development of new stereoisomeric drugs (2). In this policy statement, it is the position of the FDA that

The Impact of Stereochemistry on Drug Development and Use, Edited by Hassan Y. Aboul-Enein and Irving W. Wainer. Chemical Analysis Series, Vol. 142.
ISBN 0-471-59644-2

"the stereoisomeric composition of a drug with a chiral center should be known and the quantitative isomeric composition of the material used in pharmacologic, toxicologic, and clinical studies known. Specification for the final product should assure identity, strength, quality and purity from a stereochemical viewpoint."

As a result of this policy statement, there is an immense need for the development of new and/or improved methods for the determination of enantiomeric purity. Given that most single-enantiomer drugs are developed with enantiomeric purity in excess of 95%, enantioselective chromatographic methods are routinely employed.

The predominate chromatographic method used for the separation of enantiomeric mixtures is high-performance liquid chromatography (HPLC) utilizing chiral stationary phases (3). At present, there are currently in excess of 50 different commercial chiral stationary phases available. The proliferation of chiral HPLC stationary phases is a result of the fact that the separations are characterized by compound specific methods and low efficiency. Often, changes in the mobile phase result in unpredictable changes in enantioselectivity, making method development a tedious and time-consuming process. The limitation in efficiency dictates the goal of developing HPLC phases with higher alpha values. However, in view of the large number of chiral phases available, it is difficult to find a single chiral phase or selectand that shows a sufficient alpha for a broad range of compounds. What is needed is a technique that offers increased efficiency to take advantage of the low alphas often encountered. High efficiencies decrease the number of selectands needed and a broader applicability is realized. It is for this reason (high efficiencies) that capillary electrophoresis (CE) has significantly increased in popularity.

From its beginning in the early 1980s (4–6), the use of CE for the separation of pharmaceutical compounds has increased dramatically. The use of high currents and a narrow bore capillary for effective heat dissipation has led to fast, highly efficient separations. Therefore, CE has been investigated almost since its inception for the analysis of enantiomeric mixtures. The format and scale allow for the use of reagents that would prove either too expensive, or problematic (e.g., a high detector background that limits HPLC sensitivity), for use in other techniques. Many different chiral selectands have been successfully employed for enantioselective separations in the free solution mode of CE, as summarized by Ward elsewhere in this volume. However, much of the success realized by CE for the analysis of pharmaceutical compounds would not have been possible without a significant breakthrough that occurred in the mid-1980s. This is when Terabe et al. first reported the use of a mode of CE that has come to be known as micellar electrokinetic capillary chromatography (MECC) (7). By the addition of a surfactant (sodiumdodecylsulfate, SDS) to the buffer at solution concentrations above its critical micelle concentration (CMC), micelles are formed that allow the separation of both ionic and neutral compounds simultaneously. By logical extension, the MECC mode has been successfully applied to the separation of a number of enantiomeric mixtures. Both indirect and direct methods have been used. Indirect methods employ derivatization of the solute by an enantiomerically pure reagent followed by the separation of the resulting diastereomers utilizing typical MECC conditions. Direct methods utilize the solute in its native form and provide a chiral environment to induce the separation of enantiomers. In the MECC mode, direct separations can be accomplished in one of two ways, either by the addition of a separate and distinct chiral selectand along with SDS, or by the use of a chiral surfactants in the place of or in combination with SDS in the buffer. Chiral surfactants can be further subdivided into two classes: natural products and synthetic.

If properly designed and implemented, micelle-mediated CE separations of enantiomeric mixtures have several distinct advantages when compared to free solution enantiomeric separations, or in comparison with other analytical techniques in general (e.g., HPLC). First and foremost is the ability to separate complex mixtures of both chiral and achiral components. This separation is possible due to the solute's general hydrophobic interaction with the micelle in addition to the enantiomeric recognition made possible by the chiral selectand. This advantage greatly extends the applicability of the technique to a broad range of compound classes. Second, in some instances, it is possible to reverse exactly the migration order of the enantiomers. This reversal is important for the quantitation of trace levels of enantiomeric impurities, where it is desirable to have the trace enantiomer peak elute first before the larger, often tailed peak of the major enantiomer. All the inherent advantages of CE in general can also be maintained, such as high efficiencies, low volumes of predominately aqueous solvents, low sample volume requirements, and a fast, simple, and straightforward method development process.

Following an introduction to the theory of MECC, the use of micelle-mediated CE separation of enantiomeric compounds by both indirect and direct methods will be discussed. In each section, the relative advantages and disadvantages of the technique will be highlighted, and representative examples from the literature provided.

13.2. THEORY OF MECC

To systematically develop methods and optimize separations in the MECC mode of CE, it is necessary to have at least a basic understanding of the underlying principles. Since several good reviews of the principles of MECC and their application are available (8–11), only a brief presentation as it pertains to enantiomeric separations is necessary here.

MECC is essentially a chromatographic technique, and with minor modifications, all the values associated with chromatographic calculations such as capacity factor (k'), selectivity (α), and resolution (R_s) can be used to describe interactions with the micelles. In MECC, k' is defined as (5, 12)

$$k' = \frac{(t_r - t_{\text{aq}})}{[t_{\text{aq}}(1 - t_r/t_{\text{mc}})]} \tag{13.1}$$

where t_r is the observed migration time of the solute, t_{mc} the migration time of a solute completely partitioned into the micelle (a k' of infinity, usually measured with a highly hydrophobic compound like Sudan 3 or sulconazole), and t_{aq} the migration time of the solute if it does not interact with the micelle. For neutral compounds, t_{aq} is equal to the electroosmotic flow time (t_0), determined by injecting methanol or another organic solvent. For ionic compounds, the calculation is more complex, given the negative charge of the micelles most commonly employed, but the basic premise is still that the apparent mobility of the solute is equal to the sum of the electroosmotic mobility and electrophoretic mobility:

$$\mu_{\text{apparent}} = \mu_{\text{osmosis}} + \mu_{\text{electrophoretic}} \tag{13.2}$$

For charged solutes, t_{aq} is calculated as follows:

$$t_{\text{aq}} = \frac{1}{(1/t_{\text{CZE}} + 1/t_{\text{os,MECC}} - 1/t_{\text{os,CZE}})} \tag{13.3}$$

where t_{CZE} is the migration time of the compound in the free zone mode without surfactant (all other conditions identical), $t_{\text{os,MECC}}$ the electroosmotic flow migration time in the MECC experiment, and $t_{\text{os,CZE}}$ the electroosmotic flow time in the free zone experiment. The enantioselectivity α can then be calculated:

$$\alpha = k'_2/k'_1 \tag{13.4}$$

In addition, the MECC equivalent to chromatographic resolution can then be obtained by the following (5):

$$R_s = \frac{N^{1/2}}{4(\alpha - 1/\alpha)(k'_2/1 + k'_2)[(1 - t_0/t_{\text{mc}})/(1+(t_0/t_{\text{mc}})k'_1)]} \tag{13.5}$$

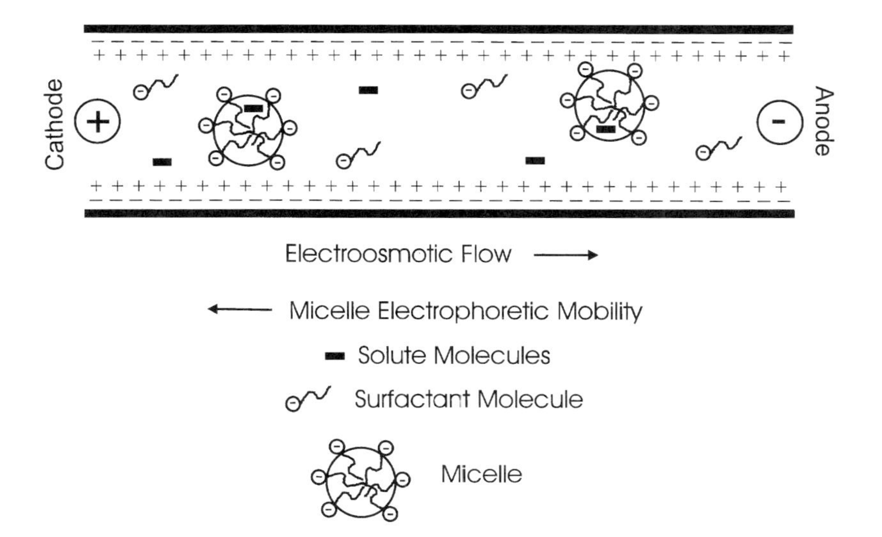

Figure 13.1. Schematic representation of a capillary longitudinal cross section illustrating the separation principle of MECC. In this instance, the anionic micelles consist of aggregations of SDS molecules. The positive and negative charges at the wall of the capillary represent electrolyte ions and ionized silanol groups, respectively.

where N is the theoretical plate number. As seen in Eq. (13.5), MECC differs from standard chromatographic resolution calculations in the addition of the final term (t_0/t_{mc}). This is referred to as the "elution window," and it has a significant effect on resolution, particularly if t_0/t_{mc} is large, or negative.

These underlying principles of MECC are illustrated schematically in Figure 13.1. In MECC, the key variables used to optimize selectivity include the type of surfactant, its concentration, pH, and organic solvent or an other modifier concentration. To obtain and optimize an enantiomeric MECC separation, it is necessary to optimize solute partitioning into the micelle using both its hydrophobicity (surfactant and organic modifier concentration) and/or its charge (pH). Partitioning is measured by k' (Eq. 13.1) and reaches an optimum value according to the following equation (13):

$$k'_{opt} = \left(\frac{t_{mc}}{t_{aq}}\right)^{1/2} \tag{13.6}$$

From a practical standpoint, for any given pair of enantiomers, the free solution mobility is first determined experimentally. Next, the MECC experiment is performed, and the above relationships are calculated to determine

k' for the enantiomers and k'_{opt} for the MECC conditions. If the enantiomers $k' < k'_{opt}$, steps must be taken to increase partitioning by either increasing the surfactant concentration, or adjusting the pH to mediate charge attraction or repulsion. Likewise, if the enantiomers $k' > k'_{opt}$, steps must be taken to decrease partitioning by either decreasing the surfactant concentration, adding organic modifiers to compete with the surfactants, or, again, adjusting the pH. If, following this optimization scheme, resolution of the enantiomers is not obtained, an alpha problem is indicated and a different chiral selectand should be investigated.

13.3. APPLICATIONS

13.3.1. Indirect Separations of Enantiomers by MECC

Indirect separations are common in HPLC, where conventional reverse phase packing materials can be used to separate diastereomers (14). The underlying principles of the indirect separation technique are shown in Figure 13.2.

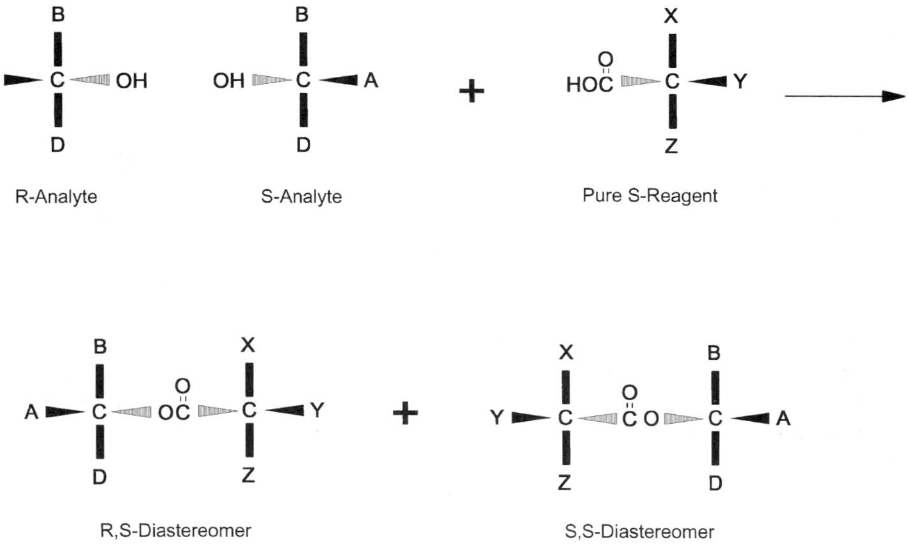

Figure 13.2. The underlying principles of the indirect separation technique. Reaction of pure enantiomers with an enantiomerically pure derivatizing reagent results in the corresponding diastereomer.

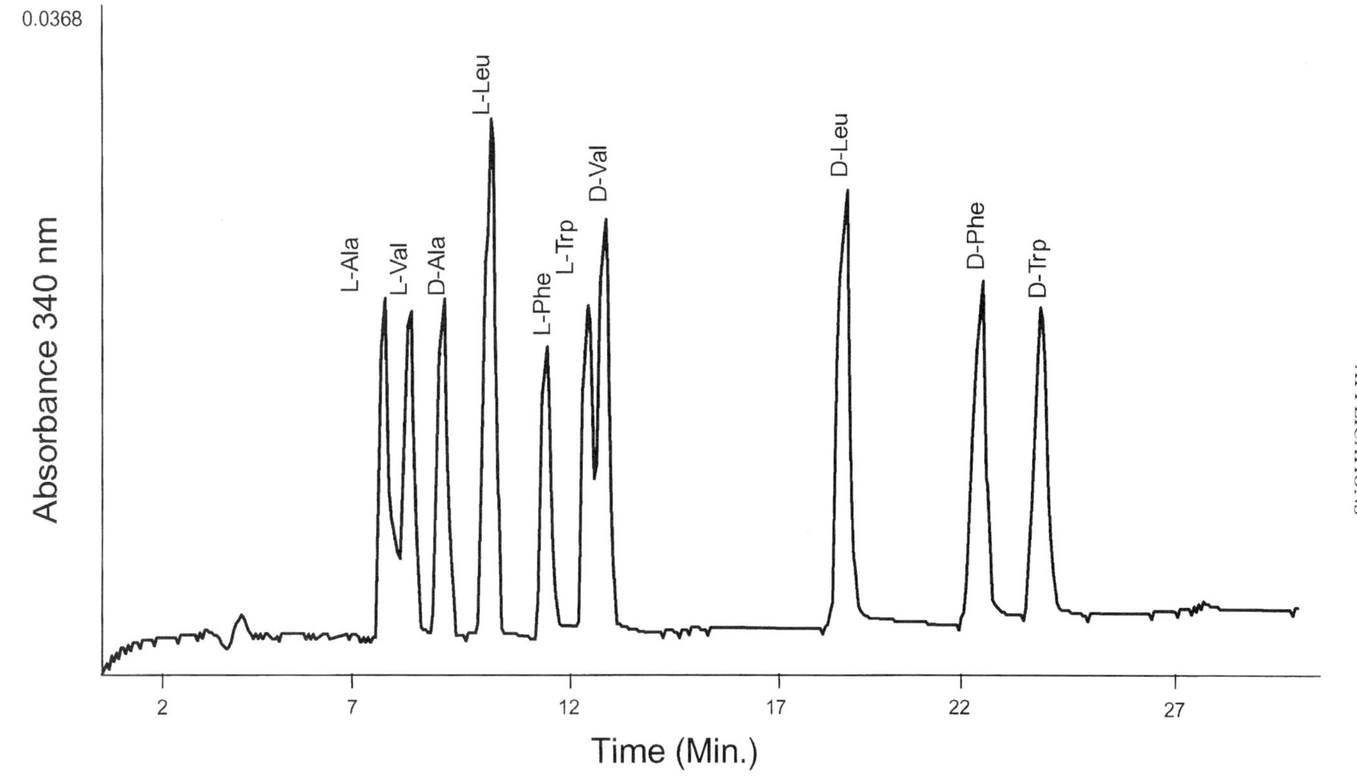

Figure 13.3. Electropherogram of racemic mixtures of Ala, Asp, Glu, and Leu derivatized with L-Marfey's reagent. The electropherogram was obtained by free solution capillary electrophoresis using a 75-μm I.D. capillary: 50 mM of ammonium phosphate, pH 3.3, 20 KV, 120 mA, 25°C, detection at 214 nm. [Reprinted from (17) with permission].

Essentially, enantiomers are converted to diastereomers by a chemical reaction with an enantiomerically pure reagent. Since MECC is essentially a chromatographic technique and arguably at least a distant cousin of reverse phase HPLC, it is no surprise that these HPLC separations also work in the MECC mode. In order to be successful, however, there are certain requirements on the part of the derivatization reagent, derivatization reaction, and enantiomers of interest. First and foremost, the enantiomer must have an appropriate functional group that lends itself to the derivatization reaction. The reagent must be of high enantiomeric purity and have suitable MECC properties. The reaction should be selective and proceed with rapid kinetics; however, without racemization of either reactants or products. In addition, both enantiomers should react at the same rate, and the reaction should proceed to completion, or at least be reproducible. Reagents that satisfy these requirements include 2,3,4,6-tetra-*o*-acetyl-β-D-glucopyranosyl isothiocyanate (often referred to as GITC), and 1-fluoro-2,4-dinitrophenyl-5-L- or D-alanine amide (L- or D-Marfey's reagent).

The indirect approach has been applied to the separation of amino acid enantiomers (15–17), and amphetamine, methamphetamine, and their hydroxyphenethylamine precursors (18). An example utilizing L-Marfey's reagent is shown in Figure 13.3. In practice, indirect approaches are usually used only as a last resort, if all other approaches fail. The added complexity of the sample preparation, derivatization, and the very real likelihood that racemization can occur (affecting quantitative results) make it less attractive than other techniques. In the case of amino acid analysis, however, sample preparation and derivatization may be much less of a concern since derivatization is often performed for detectability. However, the potential for racemization still must be addressed. In the rare instances where this is the technique of choice, simplified chromatographic conditions can be used, and often increased sensitivity is obtained. In addition, although it has not been reported, migration-order reversal of the enantiomeric pairs as an aid in quantitation should be possible due to the availability of both isomers of some derivatizing reagents.

13.3.2. Direct Separations of Enantiomers by MECC

13.3.2.1. Cyclodextrin-Modified MECC

Cyclodextrins (CDs) are toroidal cyclic oligomers consisting of *d*-glucose units. The most common of these are α-, β-, and γ-cyclodextrin consisting of six, seven, and eight *d*-glucose units, respectively. Cyclodextrins can also be derivatized at the 2-, 3-, or 6-position rim hydroxyl groups, with methyl or hydroxypropyl groups commonly employed. A more thorough review of the chemistry and free solution applicability of these molecules to enantioselective CE separations appears elsewhere in this volume (Chapter 11) and will not be repeated here. However, cyclodextrins have also been employed in combination with SDS and bile salt micelles in the MECC mode of CE to provide a third phase imparting chirality to the buffer. When enantiomers form inclusion complexes with CDs, the stability of the complex is different for each enantiomer, resulting in differential solute migration and chiral recognition.

The first reference to the use of CD- modified MECC was reported by Nishi et al. in 1990 using γ-CD in combination with SDS (19). Since then, several different CDs have been employed in the same basic approach for the enantioselective separation of a diversity of compounds (20–32). The effects

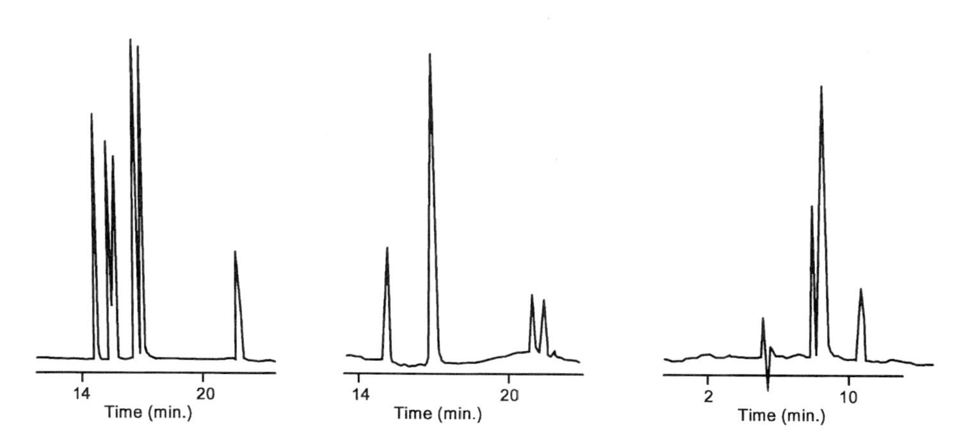

Figure 13.4. MECC electropherogram of the enantiomeric resolution of the racemic barbiturates butabarbital, secobarbital, hexobarbital, and mephobarbital (peaks 1–4 respectively) in the prescence of (*a*) 30 mM of β-CD, (*b*) 30 mM of γ-CD, and (*c*) 30 mM of β-CD and 30 mM of γ-CD. [Reprinted from (21) with permission].

Compound	R_1	R_2	R_3
Butabarbital	H	CH_2CH_3	$CH(CH_3)CH_2CH_3$
Secobarbital	H	$CH_2CH{=}CH_2$	$CH(CH_3)CH_2CH_2CH_3$
Hexobarbital	CH_3	CH_3	1-Cyclohexene
Mephobarbital	CH_3	CH_2CH_3	Phenyl

of the pH, CD type and concentration, as well as the effects of both chiral and achiral buffer additives have been studied for barbiturate enantiomers (20), and barbiturates and nonsteroidal aromatase inhibitor enantiomers as illustrated in Figure 13.4 (21). The resolution of *R*- and *S*-chlorpheniramine obtained both with and without SDS using β-CD in combination with urea has been reported (22). The use of β-CD in combination with SDS and 1-propanol for the separation of racemethorphan and racemorphan enantiomers, in addition to several other related compounds, from a urine matrix has been utilized for the screening of banned substances in athletic competitions (23). Additional applications of the use of β-CD in combination with SDS for the separation of β-blocker (24) and anticonvulsant (25) drugs also exist in the literature. The use of γ-CD in association with SDS for the separation of the enantiomers of diniconazole, uniconazole, and structurally related compounds (26, 27), and amino acid enantiomers (28, 29), as well as the resolution of the enantiomers of a new cholesterol-lowering drug utilizing hydroxypropyl β-CD, has also been reported (30). These applications are further highlighted in Figures 13.5 and 13.6. Bile salts (see below) have also been utilized in combination with various CDs (19, 29, 31,32). This approach has been most commonly employed for the separation of *d*- and *l*-amino acids (29, 19, 31), however, other applications for compounds such as mephenytoin and fenoldopam also exist (32).

The use of CDs in combination with either SDS or bile salts is very advantageous for the separation of complex mixtures of both achiral and chiral

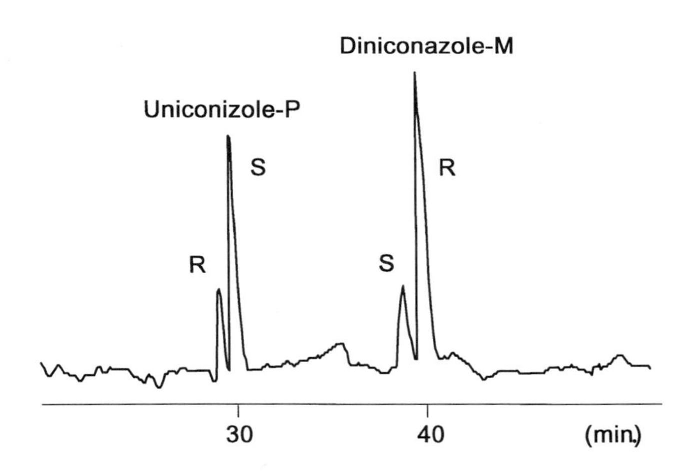

Figure 13.5. Enantiomeric separation of diniconazole-M and uniconazole-P. Separation solution: 100 mM of SDS and 2 M of urea in 100-mM borate buffer (pH 9.0) containing 50 mM of γ-CD and 2-methyl-2-propanol (95:5, v/v). [Reprinted from (27) with permission].

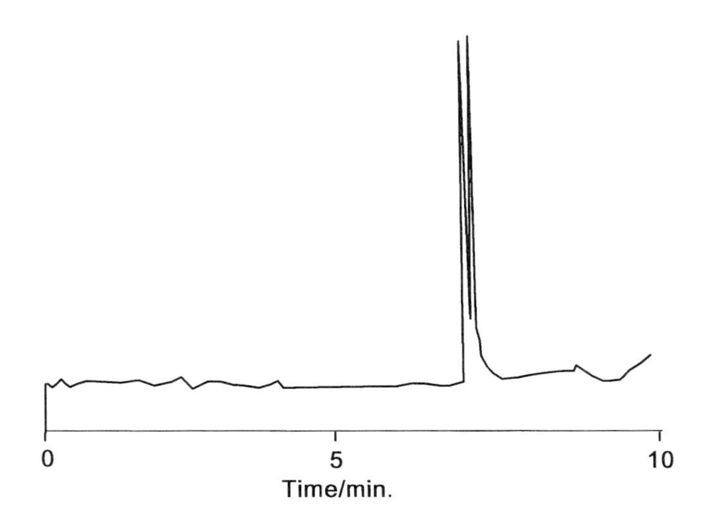

Figure 13.6. Optical resolution of RS-chlorpheniramine by CD-MECC. Separation solution: 50 mM of SDS, 100 mM of CD, 5 M of urea (pH 3.0). [Reprinted from (22) with permission].

compounds. The separation of complex mixtures is accomplished by the CD's ability to superimpose a chiral environment on the existing hydrophobic one. The resulting buffers are often quite complex, and their composition and development are not readily intuitive due to the lack of understanding of the mechanisms involved. A limitation that will more seriously impact the widespread use of this technique is the lack of availability of both CD enantiomers. This prevents migration-order reversal shown to be important in quantitation, particularly in instances where resolution is compromised (33).

13.3.2.2. Natural Product Chiral Micelles Formed from Bile Salt Surfactants

Another way to impart chirality to the CE buffer environment is to make the micelles themselves chiral. This was first reported in 1989 by Terabe, who used a bile salt (taurodeoxycholate) to separate dansylated amino acid enantiomers as shown in Figure 13.7 (34). Although the resolution in this first electropherogram is not complete, significant improvements have been made since.

Bile salts are surfactants that have a hydroxy-substituted steroid backbone. Examples are shown in Figure 13.8. In much the same way that SDS has a hydrophilic "head" and a hydrophobic "tail," bile salts have both hydrophilic and hydrophobic "faces" that also promote the formation of micelles (35). The result is a helical structure with the hydrophyllic region in the interior of the micelle (36). Their polar nature (relative to SDS) generally causes reduc-

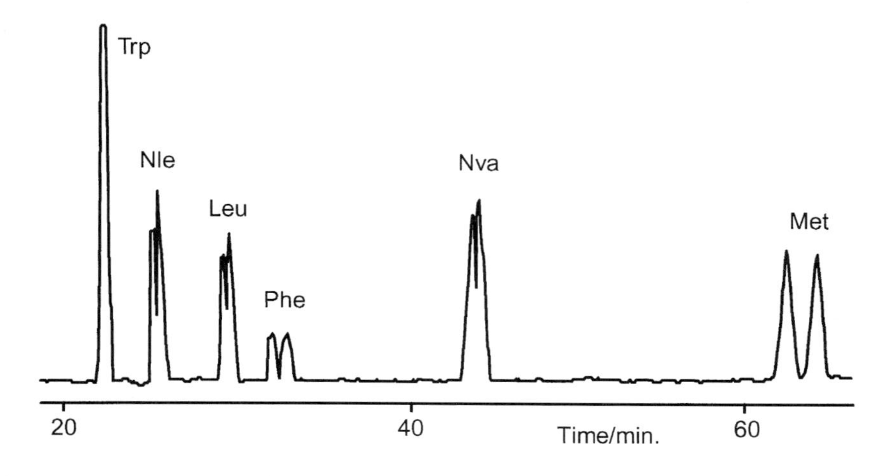

Figure 13.7. Chiral separation of DNS-AAs by MECC with 0.05 M of taurodeoxycholate at pH 3.0. [Reprinted from (34) with permission].

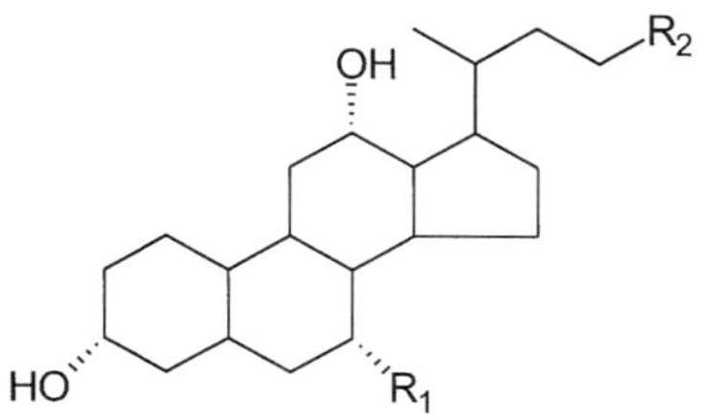

Compound	R_1	R_2
Sodium Deoxycholate	H	COONa
Sodium Taurodeoxycholate	H	$CONH(CH_2)_2SO_3H$
Sodium Cholate	OH	CooNa

Compound	CMC, mM	Aggregation #
Sodium Cholate	13	3
Sodium Deoxycholate	6	4
Sodium Taurodeoxycholate	9	11
Sodium Dodecyl Sulfate	8	63

Figure 13. 8. Common bile salt structures and associated data. Bile salt data from (36), SDS data from (37). CMC = critical micelle concentration.

tions in k' and expands the applicability of MECC to very hydrophobic compounds, since bile salt micelles have low aggregation numbers (Figure 13.8) (37). In addition, since the bile salt monomers are chiral, the separation of enantiomeric mixtures is also possible.

The use of bile salts for the separation of racemic amino acids (19, 32, 38), trimetoquinol and related compounds, and diltiazem (19, 39–41), as well as binaphthyl enantiomers (42), has also been reported. Applications to the enantiomeric purity testing of *S*-trimetoquinol to the 1% impurity level (*R*-form impurity) of five different batches of drug produced have also been observed (40). An example of the enantioselective separation of racemic trimetoquinol and some related compounds by this technique is presented in Figure 13.9.

The effects of various buffer parameters such as pH, type and concentration of bile salt, and temperature are all shown to be important for successful method development. Attempts at understanding the mechanisms have also presented (38, 40, 41). It is proposed that in these cases, resolution results from a negative elution window ($-t_0/t_{mc}$, Eq. 13.5). Separations using bile salts are often characterized by long analysis times as a consequence of working at low pH where electroosmotic flow is decreased or nonexistent. Long analysis

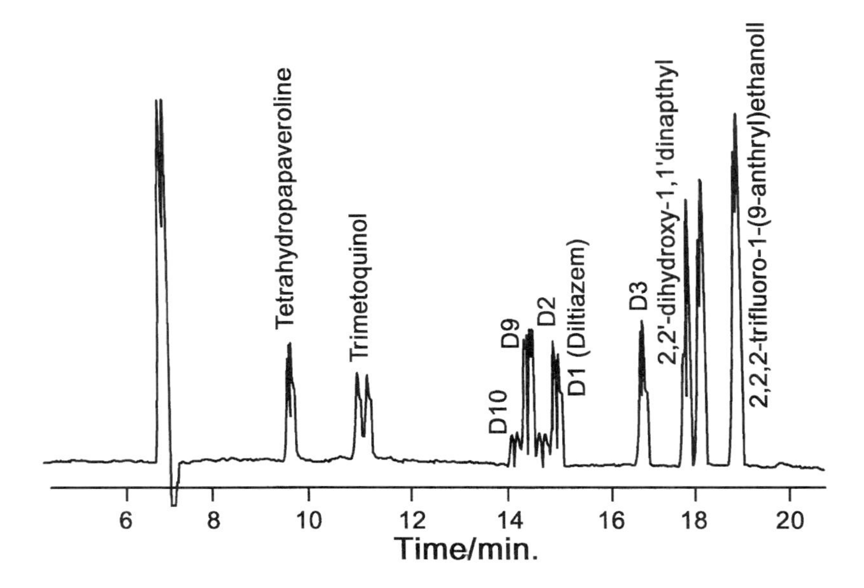

Figure 13.9. Chiral separation of trimetoquinol hydrochloride, tetrahydropapaveroline, five diltiazem-related compounds, 2,2′-dihydroxy-1,1′-dinapthyl and 2,2,2-trifluoro-1-(9-anthryl) ethanol using sodium taurodeoxycholate. [Reprinted from (40) with permission].

times are not a serious drawback if the separation cannot be achieved in any other manner. While bile salts have enjoyed success in this regard, the factor responsible for their success (polar nature resulting in the reduction in k') also limits their enantioselectivity. A more serious limitation, however, common to many natural product chiral selectands, is the unavailability of both bile salt enantiomers for migration-order reversal. Migration-order reversal is particularly important in this instance, where, as seen in Figures 13.7 and 13.8, baseline resolution is often not obtained.

13.3.2.3. Synthetic Chiral Surfactants

The second way to provide a chiral micelle is to utilize synthetic chiral surfactants. The use of synthetic surfactants, if properly designed, is potentially the most powerful enantioselective MECC technique, for a number of reasons. First, the molecule synthesized can still possess all the inherent advantages of MECC using "traditional" micelles, including high efficiencies, tolerance of complex sample matrices, use of small volumes of predominately aqueous solvents, and simple, straightforward methods development according to accepted theory. Second, both achiral and chiral separations can be performed. It should also be possible, through rational synthetic design and the systematic evaluation of structural perturbations, to tailor the enantio-

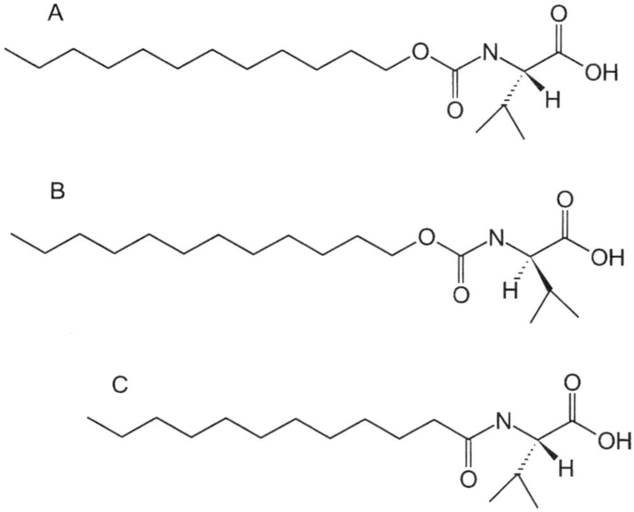

A
B
C

Figure 13.10. Example structures of chiral surfactants. A = (S)-N-dodecoxycarbonylvaline; B = (R)-N-dodecoxycarbonylvaline; C-(S)-N-dodecanoylvaline.

selectivity of the surfactant. And finally, it is possible to synthesize both enantiomers in high purity so that exact migration-order reversal can be obtained. Representative structures of some of the synthetic chiral surfactants used to date are displayed in Figure 13.10.

As illustrated in Figure 13.11, Dobashi et al. first described the use of synthetic chiral surfactants in MECC in 1989 (43, 44). With (*S*)-N-dodecanoylvaline used, neutral amino acid derivative enantiomers were separated. Otsuka and co-workers subsequently performed enantiomeric MECC separations with the same surfactant, again utilizing various amino acid derivative enantiomers (45–47). Otsuka found that adding methanol and/or urea to the buffer improved peak shape (46). A mixed micellar buffer with SDS was also employed to increase the elution window (t_0/t_{mc}) and k' (47). Since these early reports, work along these lines has been extended further, using (*S*)-N-dodecanoyl-l-glutamate, digitonin-sodium taurodeoxycholate (bile salt) mixed micelles, and (*S*)-N-dodecanoyl-*l*-serine for the separation of amino acids enantiomers, as well as a racemic mixture of the neutral compound benzoin (48, 49). In each instance, however, it was necessary to include several buffer additives in addition to the chiral surfactant, such as urea and methanol. In spite of these complex buffers, however, limited enantioselectivity and applicability were still obtained. Recently, the use of the synthetic chiral surfactant

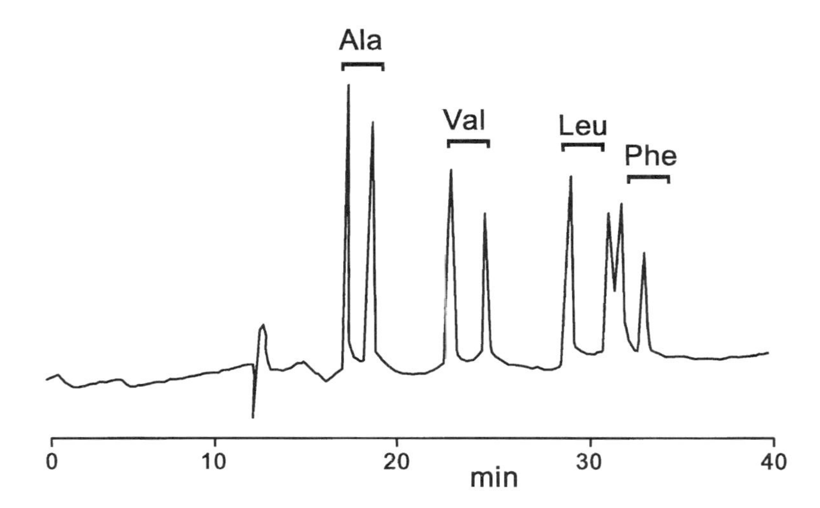

Figure 13.11. Optical resolution of a mixture containing four enantiomeric pairs of amino acids as their N-(3,5-dinitrobenzoyl)-O-isopropyl ester derivatives by MECC. Buffer consisted of using 0.025 M sodium N-dodecanoyl-l-valinate (SDVal) in 0.025 M-borate–0.025-M phosphate, pH 7.0. [Reprinted from (43) with permission].

Table 13.1. Comparison of Enantioselectivities (alpha) Obtained at pH 8.8 with 25 mm of (*S*)-N-Dodecoxycarbonylvaline versus 25 mm of (*S*)-N-Dodecanoylvaline

Analyte	(*S*)-N-Dodecanoylvaline Alpha	(*S*)-N-Dodecoxy-Carbonylvaline Alpha
Atenolol	1.00	1.04
Bupivacaine	1.06	1.05
Ephedrine	1.05	1.10
Homatotropine	1.02	1.03
Ketamine	1.05	1.01
Metoprolol	1.01	1.06
N-Methylpseudoephedrine	1.05	1.32
Norephedrine	1.04	1.10
Norphenylephrine	1.03	1.09
Octopamine	1.00	1.05
Pindolol	1.02	1.06
Terbutaline	1.00	1.01

Conditions: UV detection at 214 nm; 2-sec hydrostatic injection; 50 μm × 60 cm capillary; + 12 kV, pH 8.8 buffer: 25 mM of Na_2HPO_4/25mM of NaB_4O_7. [Reprinted from (12) with permission].

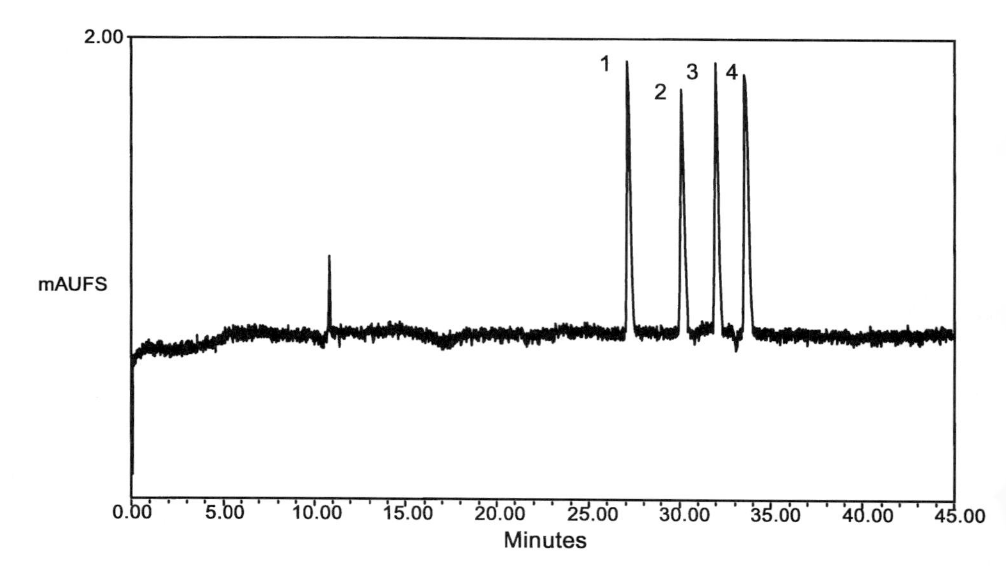

Figure 13.12. Chiral MECC separation of ephedrine and psuedoephedrine enantiomers. Buffer consisted of 25 mM Na_2HPO_4/NaB_4O_7, pH 8.8, and 50 mM (R)-N-dodecocycarbonylvaline. Sample concentration was 100 ug/mL each in buffer, and a 5-sec hydrostatic injection (10 cm) was used. Detection was at 214 nm. Capillary: 50 μm × 60 cm. Applied voltage: 16KV.

(S)-N-dodecoxycarbonylvaline was reported that provided increased enantioselectivity and applicability (12, 50, 51). The substitution of a carbamate group in (S)-N-dodecoxycarbonylvaline for the amide group in (S)-N-dodecanoylvaline resulted in a lower detector background, and increased enantioselectivities for 10 out of the 12 pharmaceutical amine compounds studied, as presented in Table 13.1 (12). These amines all have pK_a's in the 8–9 range and become more positively charged as pH decreases. Partitioning increases due to charge attraction to the negatively charged micelle, resulting in increased enantioselective interactions, or higher alpha values. An example of the separation of an enantiomeric mixture of two basic compounds, ephedrine and psuedoephedrine, utilizing (S)-N-dodecoxycarbonylvaline is shown in Figure 13.12. In addition to pH, another way to increase partitioning is to increase the hydrophobicity of the buffer by increasing surfactant concentration. Higher concentrations of surfactant were necessary to accomplish the separation of the acidic derivatized amino acid enantiomers shown in Figure 13.13. These amino acids are negatively charged (acidic) over the pH range 7–10, therefore, pH does not play a role in their separation. In order to resolve the enantiomers, it was necessary to increase the surfactant

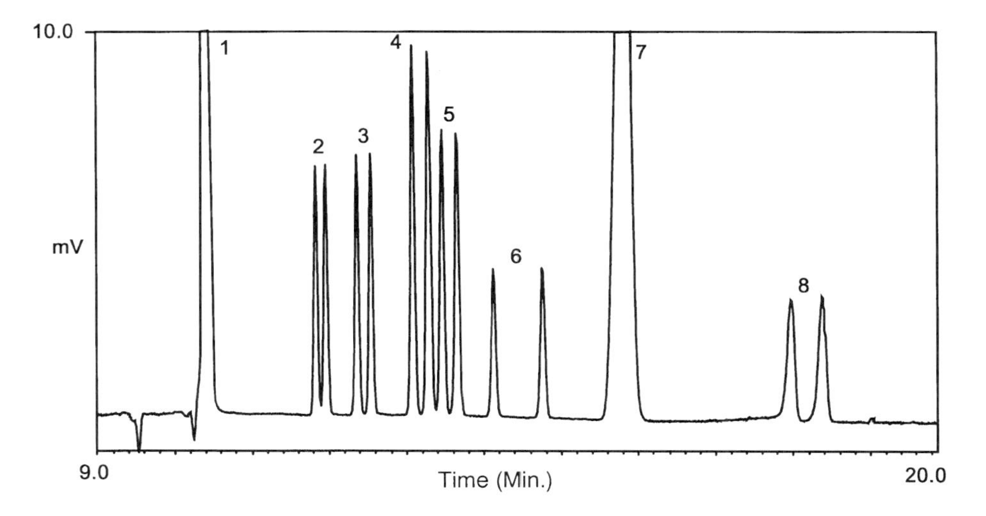

Figure 13.13. Chiral MECC enantioselective separation of six AQC-tagged amino acids. Buffer consisted of 25 mM Na_2HPO_4/NaB_4O_7 pH 9.0, and 100 mM (R)-N-dodecocycarbonylvaline. Sample concentration was 25 ug/mL each in buffer, and a 20-sec hydrostatic injection was used. Detection was at 254 nm. Capillary: 50 μm × 60 cm. Applied voltage: 16 KV. Peaks: 1 = AQC reagent blank; 2 = *d*, *l*-ile; 3 = *d*, *l*-leu; 4 = *d*, *l*-orn; 5 = *d*, *l*-lys; 6 = *d*, *l*-trp; 7 = AQC by-product; 8 = *d*, *l*-arg. [Reprinted from (50) with permission].

concentration until hydrophobicity overcame charge repulsion. The increased hydrophobic nature of the buffer resulted in increased partitioning and higher alpha values.

In recognition of success in this area, additional synthetic chiral surfactants have been developed and utilized for various other applications, such as dodecyl-D-glucopyranoside (52).

With the advent of these synthetic chiral surfactants, the advantages of simplified buffers, simultaneous achiral and chiral separations, and fast, efficient separations were finally realized. Figure 13.14 again illustrates the separation of ephedrine enantiomers, this time from a complex electropherogram resulting from the direct injection of a spiked, filtered, urine sample. Separations such as this highlight the ability of the chiral MECC technique for the separation of complex mixtures of both chiral and achiral components. In Figure 13.15, the separation of N-methylpseudoephedrine enantiomers that is complete in under 90 sec is presented. In addition, the ability to reverse migration order for quantitative purposes made possible by the

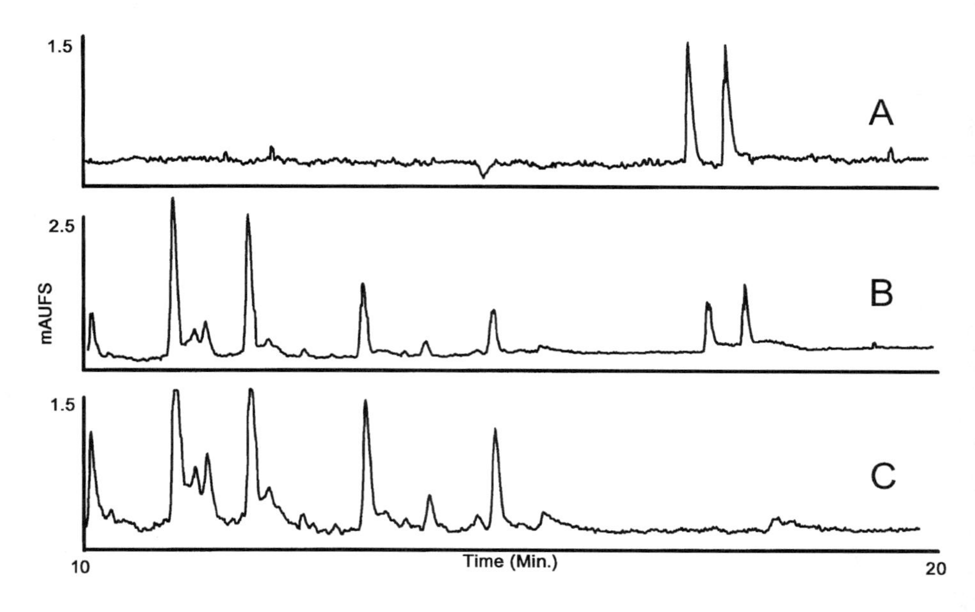

Figure 13.14. Separation of ephedrine enantiomers spiked to urine. Buffer consisted of 25 mM Na_2HPO_4/NaB_4O_7, pH 9.0, and 50 mM (*R*)-N-dodecocycarbonylvaline. Sample concentration was 100 ug/mL in buffer, and a 20-sec hydrostatic injection was used (10 cm). Detection was at 214 nm. Capillary: 50 μm × 60 cm. Applied voltage: 16 KV. [Reprinted from (12) with permission]. A, Ephedrine standard; B, spiked urine; C, urine blank.

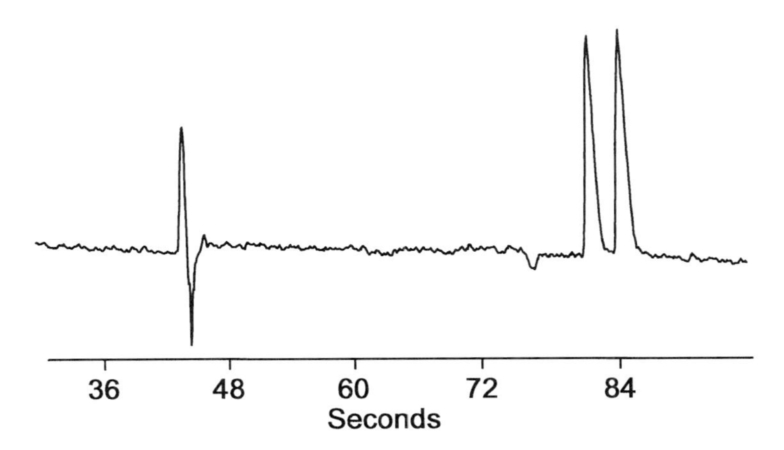

Figure 13.15. Fast separation of N-methylpsuedoephedrine enantiomers. Buffer consisted of 50 mM CHES, pH 8.8, and 25 mM (*S*)-N-dodecocycarbonylvaline. Sample concentration was 100 ug/mL in buffer, and a 5-sec hydrostatic injection was used (10 cm). Detection was at 214 nm. Capillary: 50 μm × 35 cm. Applied voltage: 30KV. [Reprinted from (12) with permission].

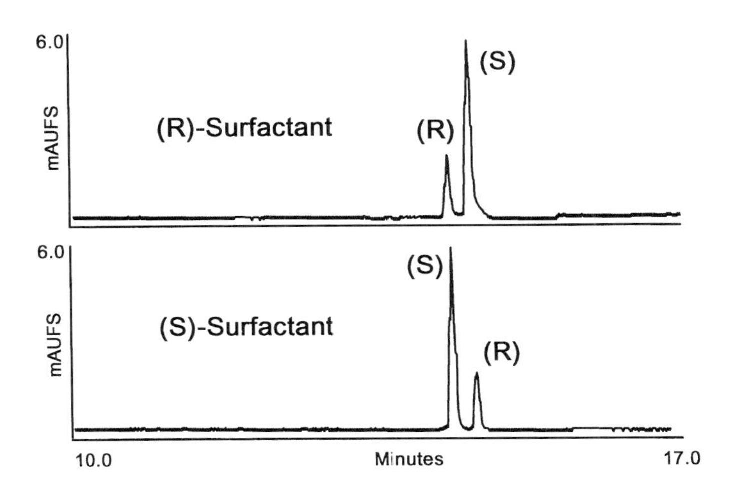

Figure 13.16. Migration order reversal of benzoin enantiomers. Buffer consisted of 25 mM Na_2HPO_4/NaB_4O_7, pH 8.8, and 25 mM (*R*)- or (*S*)-N-dodecocycarbonylvaline. Sample consisted of a 3:1 ratio of (*S*)- to (*R*)-benzoin dissolved in buffer. Detection was at 214 nm. Capillary: 50 μm × 60 cm. Applied voltage: 16KV. [Reprinted from (12) with permission].

availability of both enantiomers of the synthetic surfactants is shown in Figure 13.16. Exact reversals such as these can also be used to identify enantiomers when they are present in complex mixtures, for example, that shown in Figure 13.14.

13.3.2.4. Additional Chiral Micelle Applications

With any system of classification, there are always a few items that defy definition. Ironically, the very first report of the use of a chiral micelle that clearly showed the validity of the approach is the only report of its type in the literature (53). In this report from 1987, Cohen et al. employed didecyl-L-alanine in combination with SDS for the separation of six dansylated amino acid enantiomers.

More recently, separations employing new natural products have appeared using glycyrrhizic acid and β-escin in combination with SDS for the separation of dansylated and phenylthiohydantoin derivatized *d,l*-amino acids (54). The use of these natural products has shown low enantioselectivity, and again, only one enantiomer is available, limiting their quantitative ability as outlined above. An approach applying microemulsions of (2*R*, 3*R*)-di-*n*-butyl tartrate in combination with SDS (55), and polymerized chiral micelles of poly(sodium N-undecylenyl-*l*-valinate (56) has also been presented for a limited number of solutes. Although not reported, both enantiomers of this novel synthetic micelle could be synthesized.

13.4. CONCLUSION

MECC using chiral surfactants or additives shows great promise as a solution to many chiral compound separation problems. Although capillary electrophoresis has long been thought of as complementary to other analytical techniques, the separation of enantiomers is one area where CE can be seen to have significant advantages. High efficiencies allow the separation of a wider range of compounds given identical alpha, and lead to the likelihood that a smaller number of chiral selectands need to be evaluated during method development. With the novel synthetic surfactants currently in development, the increased resolution and the ability to reverse migration order aid in the quantitation and identification of enantiomeric impurities. Improvements in enantioselectivity and decreased detector background resulting in increased sensitivity and linearity/dynamic range have also been made. It is only a matter of time before these improvements ultimately result in the evolution of CE techniques from a research mode to the transfer of validated methods in routine use.

REFERENCES

1. Stinson, S. C. Sept., 1993. Chiral drugs. *C&EN*, 28.
2. *Chirality* (1992). **4**, 338.
3. Zief, M. and Crane, L. J. (ed.) (1988). *Chromatographic Chiral Separations*. New York: Marcel Dekker.
4. Jorgenson, J. W. and Lukacs, K. D. (1981). *Anal. Chem.*, **53**, 1298.
5. Jorgenson, J. W. and Lukacs, K. D. (1981). *J. Chromatogr.*, **218**, 209.
6. Jorgenson, J. W. and Lukacs, K. D. (1983). *Science*, **222**, 266.
7. Terabe, S., Otsuka, K., and Ando, T. (1985). *Anal. Chem.*, **57**, 834.
8. Weinberger, R. (1983). *Practical Capillary Electrophoresis* Bostan, M.A.: Academic Press.
9. Terabe, S., Chen, N., and Otsuka, K. (1994). In *Advances in Electrophoresis*, Vol. 7, A. Chrambach, M. J. Dunn, and B. J. Radola, (eds.), New York: VCH Publishers, Chap. 2, pp. 89–153.
10. Nielson, K. R. and Foley, J. P. (1993). In P. Camiller (ed.), *Capillary Electrophoresis, Theory and Practice*, Boca Raton, FL: CRC Press. Chap. 4, pp. 117–161.
11. Khaledi, M. (1994). In J. P. Landers, (ed.), *Handbook of Capillary Electrophoresis*, Boca Raton, FL: CRC Press, Chap. 30, pp. 43–93.
12. Mazzeo, J. R., Grover, E. R. Swartz, M. E., and Petersen, J. S., (1994). *J. Chromatogr. A*, **680**, 125.
13. Foley, J. (1990). *Anal. Chem.*, **62**, 1302.
14. Gal, J. (1987). *LC/GC*, **5**, 106.
15. Kang, L. and Buck, R. H. (1992). *Amino Acids*, **2**, 103.
16. Nishi, H., Fukuyama, T., and Matsuo, M. (1990). *J. Microcol. Sep.*, **2**, 234.
17. Tran, A. D., Blanc, T., and Leopold, E. J. (1990). *J. Chromatogr.*, **516**, 241.
18. Lurie, I. S. (1992). *J. Chromatogr.*, **605**, 269.
19. Nishi, H. and Terabe, S. (1990). *Electrophoresis*, **11**, 691.
20. Nishi, H., Fukuyama, T., and Terabe, S. (1991). *J. Chromatogr.*, **553**, 503.
21. Francotte, E., Cherkaoui, S., and Faupel, M. (1993). *Chirality*, **5**, 516.
22. Otsuka, K. and Terabe, S. (1993). *J. Liq. Chromatogr.*, **16**, 945.
23. Aumatell, A. and Wells, R. J. (1993). *J. Chrom. Sci.*, **31**, 502.
24. Siren, H., Jumppanen, J., Manninen, K., and Riekkola, M. (1994). *Electrophoresis*, **15**, 779.
25. Desiderio, C., Fanali, S. Kupfer, A., and Thormann, W. (1994). *Electrophoresis*, **15**, 87.
26. Furuta, R. and Doi, T. (1994). *Electrophoresis*, **15**, 1322.
27. Furuta, R. and Doi, T. (1994). *J. Chromatogr. A*, **676**, 431.

28. Ueda, T., Kitamura, F., Mitchell, R., Metcalf, T., Kuwana, T., and Nakamoto, A. (1991). *Anal. Chem.*, **63**, 2979.
29. Terabe, S., Miyashita, Y., Ishihama, Y., and Shibata, O. (1993). *J. Chromatogr.*, **636**, 47.
30. Noroski, J. E., Mayo, D. J., and Moran, M. (1995). *J. Pharm. Biomed. Anal.*, **13**, 45.
31. Lin, M., Wu, N., Barker, G. E., Sun, P., Huie, C. W., and Hrtwick, R. A. (1993). *J. Liq. Chrom.*, **16**, 3667.
32. Okafo, G. N., Bintz, C., Clarke, S. E., and Camilleri, P. (1992). *J. Chem. Soc. Chem. Comm.*, 1189.
33. Perryt, J. A., Rateiko, J. D., and Szczerbo, T. J. (1987). *J. Chromatogr.*, **389**, 57.
34. Terabe, S. (1989). *TRAC*, **8**, 129.
35. Small, D. M., Nair, P. P., and Kritchevsky (ed.). *The Bile Acids*, Vol. 1, New York: Marcel Dekker, p. 119.
36. Pavel, N. V., Giglio, E., Eposito, G., and Zanabi, A. (1987). *J. Phys. Chem.*, **91**, 356.
37. Attwood, D. and Florence, A. T., (1983). In *Surfactant Systems* London: Chapman and Hall, pp. 185–188.
38. Terabe, S., Shibata, M., and Miyashita, Y. (1989). *J. Chromatogr.*, **480**, 403.
39. Nishi, H., Fukuyama, T., Matsuo, M., and Terabe, S. (1990). *Anal. Chem. Act.*, **236**, 281.
40. Nishi, H., Fukuyama, T., Matsuo, M., and Terabe, S. (1990). *J. Chromatogr.*, **515**, 233.
41. Nishi, H., Fukuyama, T., Matsuo, M., and Terabe, S. (1989). *J. Microcol. Sep.*, **1**, 234.
42. Cole, R. O., Sepaniak, M. J., and Hinze, W. L. (1990). *JHRCCC*, **13**, 579.
43. Dobashi, A., Ono, T., Hara, S., and Yamaguchi, J. (1989). *J. Chromatogr.*, **480**, 413.
44. Dobashi, A., Ono, T., Hara, S., and Yamaguchi, J. (1984). *Anal. Chem.*, **61**, 1984.
45. Otsuka, K. and Terabe, S. (1990). *Electrophoresis*, **11**, 982.
46. Otsuka, K. and Terabe, S. (1990). *J. Chromatogr.*, **515**, 221.
47. Otsuka, K., Kawahara, J., Tatekawa, K., and Terabe, S. (1991). *J. Chromatogr.*, **559**, 209.
48. Otsuka, K., Kashihara, M., Kawaguchi, Y., Koike, R., and Terabe, S. (1993). *J. Chromatogr. A*, **652**, 253.
49. Otsuka, K., Karuhaka, K., Higashimori, M., and Terabe, S. (1994). *J. Chromatogr. A*, **680**, 317.
50. Swartz, M. E., Mazzeo, J. R., Grover, E. R., and Brown, P. R. (1995). *Anal. Biochem.*, **231**, 65–71.
51. Swartz, M. E., Mazzeo, J. R., Grover, E. R., Brown, P. R., and Aboul-Enein, H. Y. (1996). *J. Chromatogr. A.*, **724**, 307–316.

52. Tickle, D. C., Okafo, G. N., Camilleri, P., Jones, R. F. D., and Kirby, A. J. (1994). *Anal. Chem.*, **66**, 4121.
53. Cohen, A. S., Paulus, A., and Karger, B. L. (1987). *Chromatographia*, **24**, 15.
54. Ishihama, Y. and Terabe, S. (1993). *J. Liq. Chrom.*, **16**, 933.
55. Aiken J. H. and Huie, C. W. (1993). *Chromatographia*, **35**, 448.
56. Wang, J. and Warner, I. H., (1994). *Anal. Chem.*, **66**, 3773.

CHAPTER

14

UNIFIED ENANTIOSELECTIVE CHROMATOGRAPHY INVOLVING CHIRASIL-DEX

VOLKER SCHURIG, SABINE MAYER, MARTIN JUNG, MARKUS FLUCK, HANSJÖRG JAKUBETZ, ALEXANDRA GLAUSCH, and SIMONA NEGURA

Institut für Organische Chemie der Universität
D-72076 Tübingen, Germany

14.1. INTRODUCTION

In the chiral polymer Chirasil-Dex (Scheme 1), permethylated β-cyclodextrin is linked via an octamethylene bridge to polydimethylsiloxane (1). The polymer can be immobilized thermally to an extend of $\approx$ 85% onto the inner surface of fused silica capillaries or on silica particles via crosslinking and/or surface bonding (2). This feature enabled us to probe the use of Chirasil-Dex for a unified enantioselective chromatographic approach utilizing the common methods of GLC, SFC, LC (open tubular versus packed), and CEC for chiral separation.

14.2. EXPERIMENTAL

Chirasil-Dex has been prepared as previously described in detail (1–3). In our work, it typically contained 24% (w/w) permethylated cyclodextrin (CD, m = 0.22 mol/kg). Thus, statistically one out of 60 silicon atoms in the polymer chain carries a CD moiety. Fused silica columns were coated by the static method (3). GLC measurements were performed with a Carlo Erba HRGC 5300 Mega Series, and SFC instrumentation consisted of a computer-controlled SFC 3000 system (Fisons, Mainz-Castel, Germany). A "Kapillar-Elektrophorese System 100" (Grom Herrenberg, Germany) was employed for CEC. Open tubular LC was performed in CE apparatus with a pressure-

The Impact of Stereochemistry on Drug Development and Use, Edited by Hassan Y. Aboul-Enein and Irving W. Wainer. Chemical Analysis Series, Vol. 142.
ISBN 0-471-59644-2

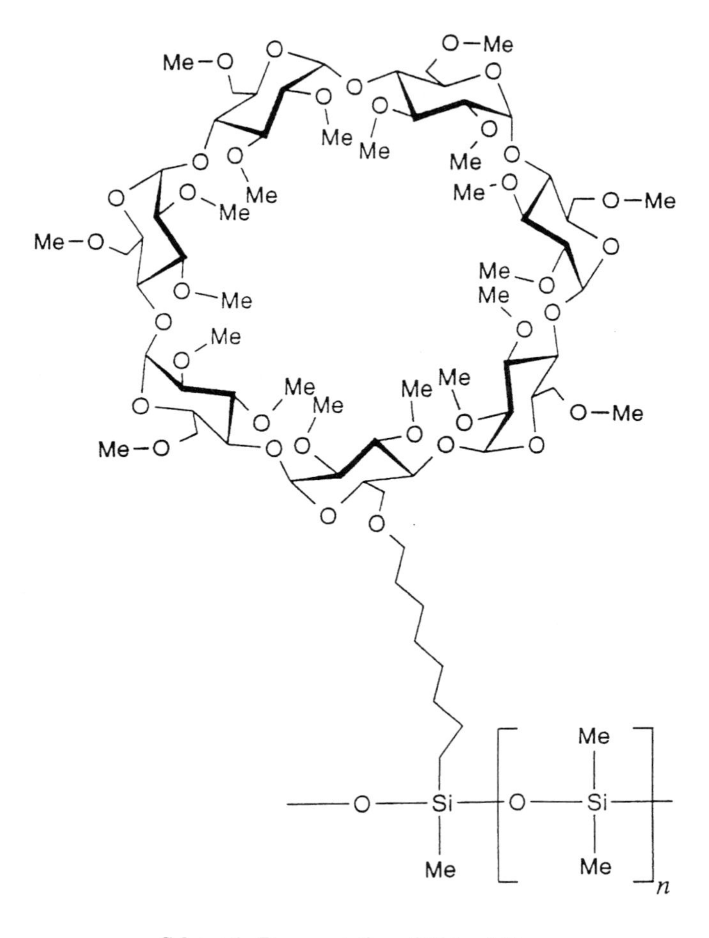

Scheme 1. Representation of Chirasil-Dex.

driven PRINCE injection system (Bischoff, Leonberg, Germany). Further details are given in the figure captions.

14.3. RESULTS

14.3.1. Unified Enantioselective Chromatography

The GOLAY equation predicts that at a given inner diameter and retention factor k, the optimum efficiency (H_{min}) of a capillary column is *independent of*

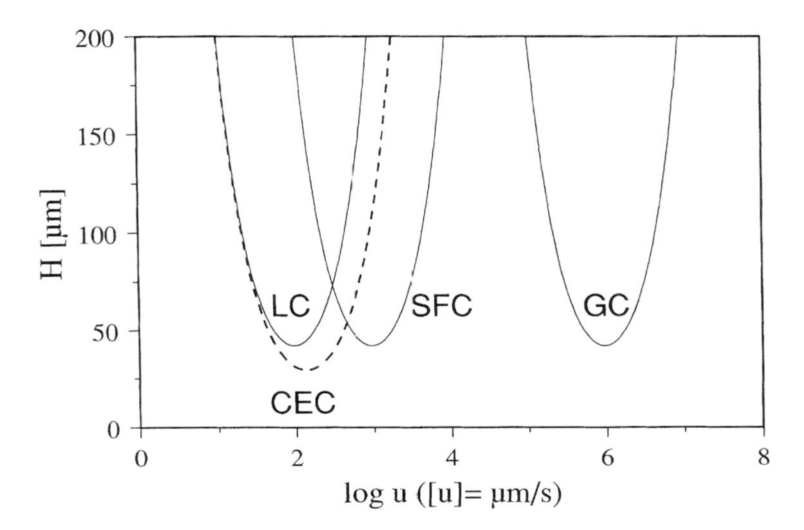

Figure 14.1. Height equivalent to a theoretical plate H versus *log* of mobile phase velocity u (neglecting C_s) for a 0.05-mm (i.d.) open tubular column [k = 5, D_m ($\mu m^2\ sec^{-1}$) = 10^7(GC), 10^4(SFC), 10^3(LC,CEC), $f(k)_{parabol} = (1 + 6k + 11k^2)/(96(1 + k)^2$] [Note that in CEC, H_{min} and u_{opt} are different due to the nearly flat peak profile (dotted curve), $f(k)_{flat} = k^2/(16(1 + k)^2$.]

the nature of the mobile phase, whereas the optimum velocity (u_{opt}) depends only on the diffusion coefficient (GC >> SFC > LC ~ CEC) (Figure 14.1).

Thus, toward the goal of *unified enantioselective capillary chromatography*, it has been shown that, for example, hexobarbital can be separated into enantiomers employing the same 85 cm × 0.05 mm (i.d.) fused silica capillary column coated with immobilized Chirasil-Dex **1** (film thickness 0.15 µm) by four independent methods, that is, by GC, SFC, OT-LC, and CEC (Figure 14.2) (4, 5).

As the result of the different diffusion coefficient D_m in gases and liquids (see Figure 14.1, caption), analysis times at optimum efficiency in capillary LC and CEC are longer by four orders of magnitude as compared to GC (see Figure 14.1). However, H_{min} and u_{opt} on the one hand, and total analysis time t on the other, depend also on the retention factor k. Fortunately, it has been observed (see Figure 14.2) that k is large in GC and intermediate in SFC, but very small indeed ($k < 1$) in LC and CEC for the first eluted enantiomer.

For hexobarbital, inspection of Figure 14.2 shows that CEC (capillary electrochromatography) is superior to GC and SFC in respect to all parameters important in enantiomer separation under the given experimental conditions (5).

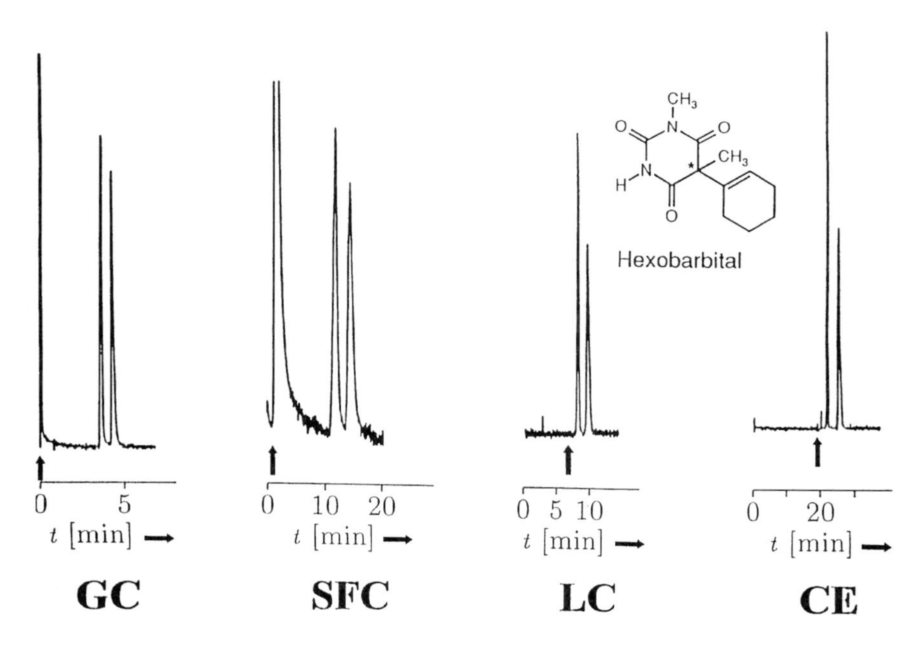

Figure 14.2. Enantiomer separation of hexobarbital by GC, SFC, OL-LC, and CEC. Column: 85 cm × 0.05 mm (i.d.) fused silica column coated and thermally immobilized (85%) with Chirasil-Dex (film thickness of 0.15 μm) (2, 3). The arrows in the chromatograms mark the hold-up times t_M.

Chiral separation factor α:	CEC > LC > SFC > GC
Peak resolution R_s:	CEC > LC ≈ GC > SFC
Efficiency N (first peak):	CEC > LC ≈ GC > SFC

Enantiomer separation by electrochromatography, however, is the slowest method due to reduced electroosmotic flow in the coated capillary (6–8). The chiral separation factor $\alpha = 3$ (Figure 14.2, right) is the highest value ever observed for Chirasil-Dex.

Until this point, the unified approach has employed a single capillary column coated with a chiral stationary phase via different chromatographic instrumentation. Current research is focused on enantioselective separations through the use of unified equipment. To the best of our knowledge, open tubular LC (OT-LC) has been used for enantiomer separation on chiral stationary phases for the first time. A CE apparatus equipped with a pressure-driven PRINCE injection system has been employed for OT-LC. The online coupling of OT-LC to electrospray mass spectrometry in enantiomer separation is demonstrated in Figure 14.3.

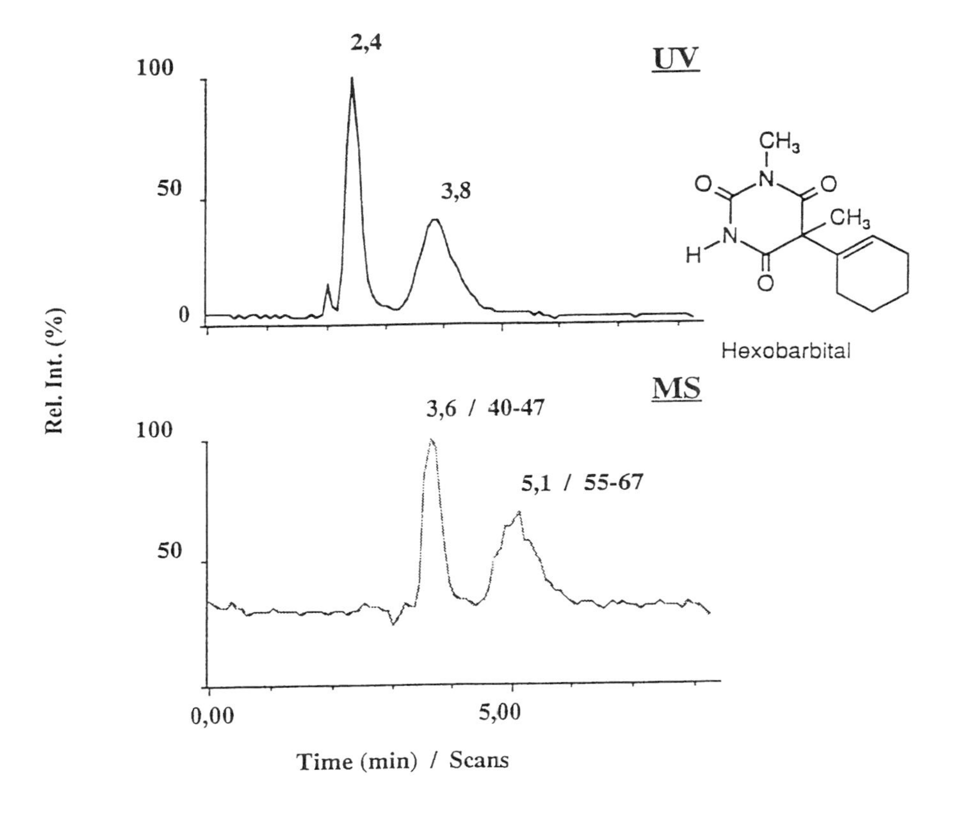

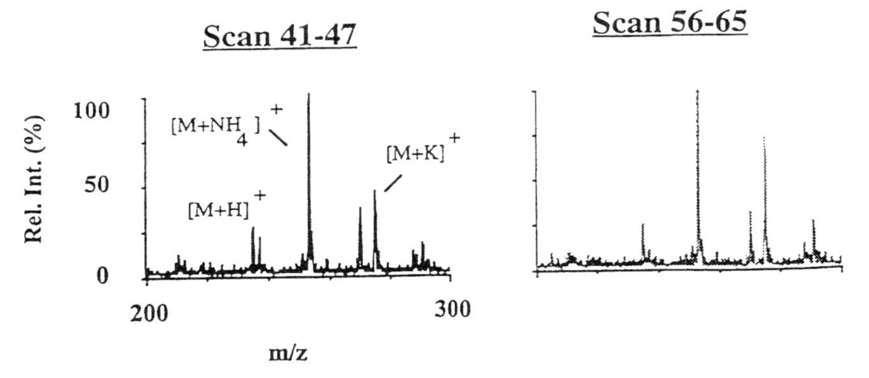

Figure 14.3. Enantiomer separation of hexobarbital by online OT-LC–electrospray mass spectrometry coupling. Column: 100 cm × 0.05 mm (i.d.) fused silica coated and thermally immobilized (85%) with Chirasil-Dex (film thickness 0.2 μm). Flow rate: 1.5 μl/min methanol/water (10:90).

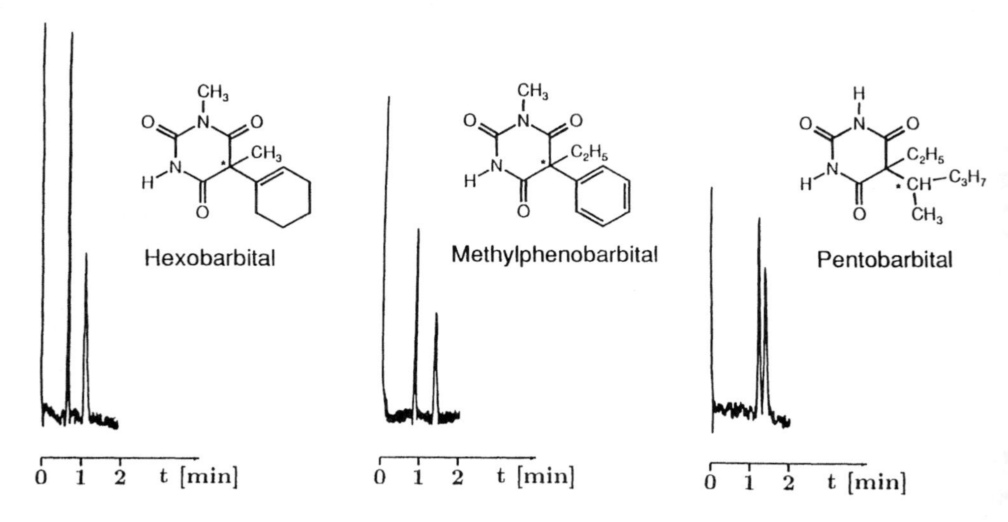

Figure 14.4. Rapid enantiomer separation of barbiturates by GLC. Column: 35 cm × 0.05 mm (i.d.) fused silica coated and thermally immobilized (85%) with Chirasil-Dex (film thickness 0.06 μm). Conditions: $T = 130°C$, $p = 1$ bar (hydrogen).

14.3.1.1. Miniaturization

Supported by theoretical considerations (9), it is anticipated that CEC employing chiral stationary phases will further benefit from smaller column diameters. Although the merit of miniaturization in the GLC mode is already evident in Figure 14.2 (left), a further reduction in the column length and film thickness of Chirasil-Dex allows enantiomer separations in less than 100 sec for, for example, barbiturates (Figure 14.2). The elution temperature of enantiomers can be reduced with miniaturized columns, whereby the chiral separation factor α is increased. The decrease of sample capacity (injection) is often outweighed by an increase in the signal-to-noise ratio (detection).

14.3.1.2. Effect of Cyclodextrin (CD) Dilution

In Chirasil-Dex, the cyclodextrin selector is diluted by the apolar environment of the polysiloxane matrix, thus decreasing analysis times for very polar compounds. Contrary to undiluted CD stationary phases, the chiral separation factor α_{dil} is concentration-dependent in Chirasil-Dex coated columns according to Eq. (14.1) (10):

$$\alpha_{dil} = \frac{K_R m + 1}{K_S m + 1} = \frac{R_R + 1}{R_S + 1} \quad \text{and} \quad R' = \frac{r}{r_0} - 1 \qquad (1)$$

with K = association constant, m = molality of the CD moiety in the polysiloxane matrix, R' = retention increase or chemical capacity factor, r = relative retention on CD column, r_0 = relative retention on pure polysiloxane.

According to Eq. (14.1), the α versus CD concentration curves level off at higher weight percentages. The optimum is already reached at low CD concentrations when the chemical association is strong (i.e., large K). Plots of α versus m (mol/kg) are shown in Figure 14.5 along with a practical example (10). The relationship between α and the weight of the CD in the polysiloxane matrix implies that any improvement in enantioselectivity would only be gained in the case of weakly associated analytes.

The concentration dependence of α should be taken into account when thermodynamic measurements are performed. The general relation $-\Delta_{R,S}\Delta G^o = RT \ln \alpha$ is not valid in systems where the cyclodextrin moiety is diluted by a polysiloxane matrix.

14.3.1.3. Multidimensional Techniques

The prerequisite of Chirasil-Dex in unified enantioselective chromatography is the immobilizability on the fused silica surface. This feature is also important when utilizing highly sensitive detection devices [GC-MS(SIM), ECD]. Here, the multidimensional GLC analysis of chiral atropisomeric polychlorinated biphenyls (PCBs) is demonstrated as an example (11). In Figure 14.6, "heart-cutting" of the congeners PCB 132 and 153 from a commercial Clophen A 60 mixture and subsequent separation on Chirasil-Dex is shown. This method allows the screening of chiral PCBs in biological matrices for expected enantiomeric bias (13). For PCB 132, an interesting deviation of the racemic composition has been detected in human milk samples in our laboratory.

14.3.1.4. Enantiomerization

Interestingly, the PCB 132 atropisomers are stable toward enantiomerization (11) via rotation about the central biphenyl bond at 170°C since no peak distortion is observed (Figure 14.6, right). Only at temperatures above 300°C, can enantiomerization be detected by the stopped-flow technique (12). On the other hand, *o,o'*-diisopropylbiphenyl undergoes an inversion of conformation already at 100°C, causing an interconversion profile characterized by plateau formation (14). Using a computer program featuring enantiomerization

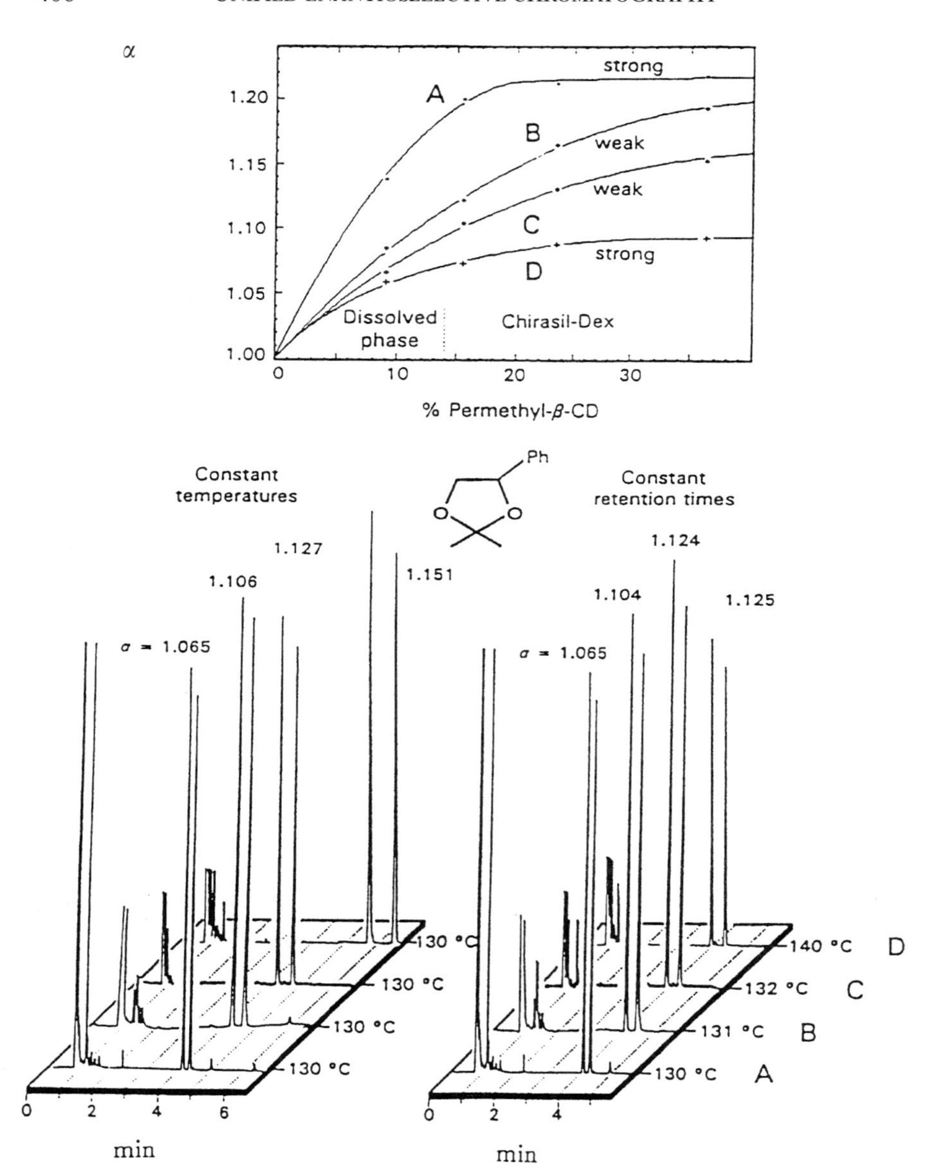

Figure 14.5. Top: Plots of α versus CD weight percentage for four racemates exhibiting strong or weak association K (A = 1-methylcyclohex-1-en-3-ol; B = *trans*-2,5-diethoxy-tetrahydrofuran; C = 2,2-dimethyl-4-phenyl-1,3-dioxolane, D = *cis*-pinane). Bottom: Dependence of the separation factor α of 2,2-dimethyl-4-phenyl-1,3-dioxolane on the weight percentages of the CD in Chirasil-Dex **1** at constant temperatures (left) and constant retention times (right). (A, B, C, D = 9, 16, 24, and 36%). Column: 25 m × 0.25 mm (i.d.) fused silica capillary, coated with Chirasil-Dex (film thickness 0.25 μm) (1).

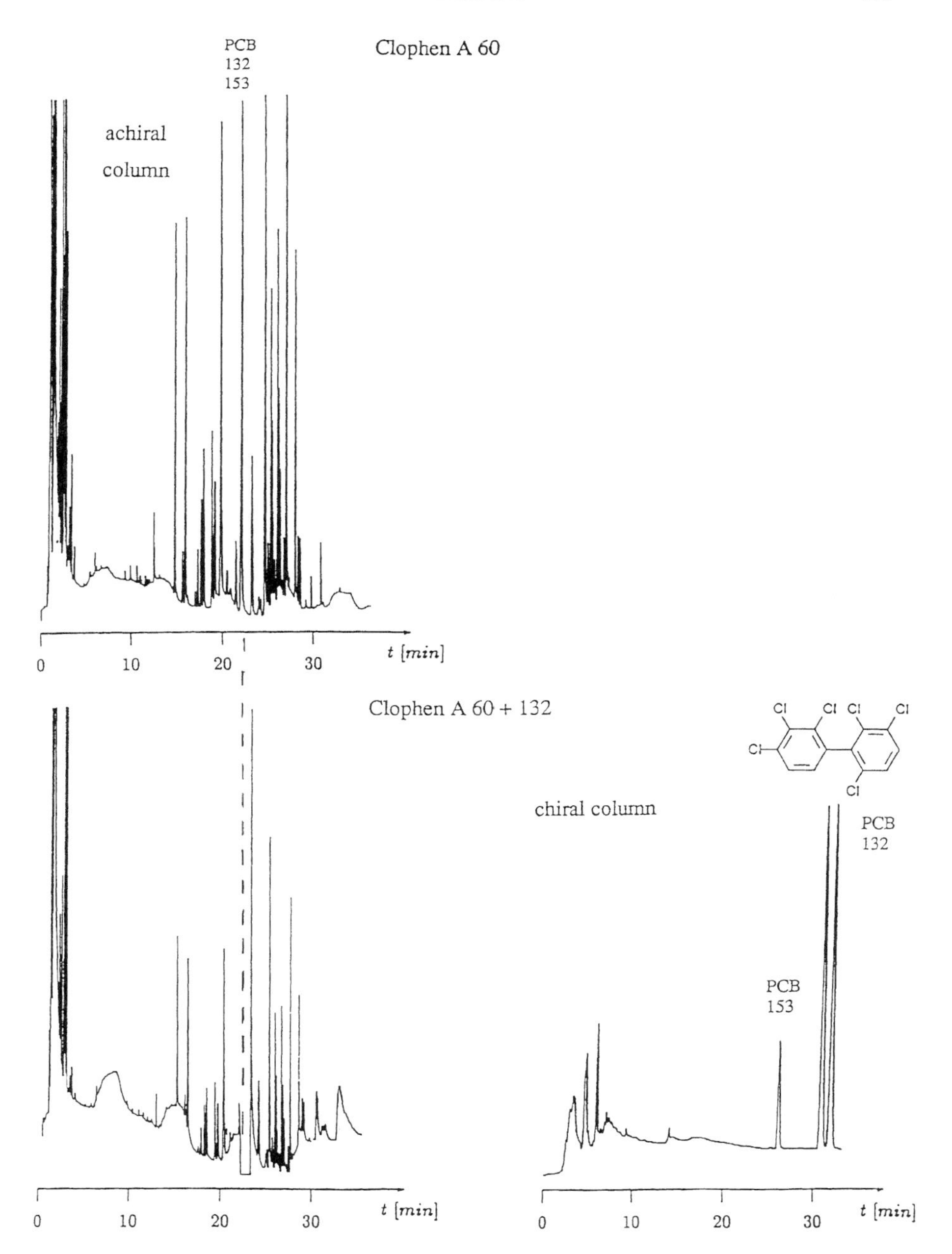

Figure 14.6. MDGC of a PCB mixture and "heart-cutting" of congeners 132 (enantiomers) and 153. Achiral column (left): 30 m × 0.22 mm (i.d.) fused silica coated with DB5. Chiral column (right): 10 m × 0.25 mm (i.d.) fused silica coated with thermally immobilized Chirasil-Dex (film thickness 0.2 μm). Temperature programmed; instrument: Siemens SiChromat-2.

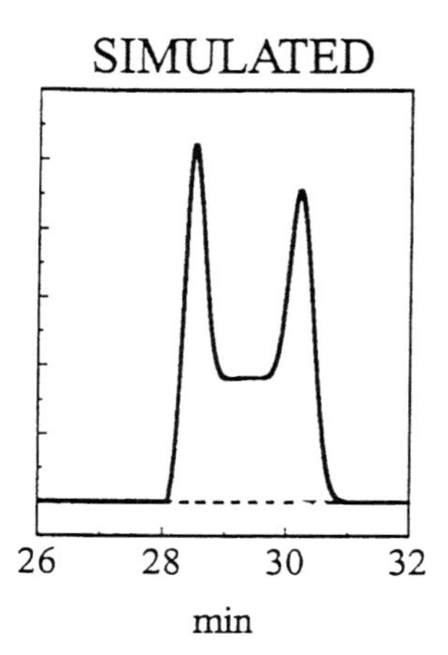

Figure 14.7. Simulated chromatogram featuring enantiomerization of *o,o'*-diisopropylbiphenyl (14). (Simulation: $N = 27\,000$, $k = 0.022\ \text{min}^{-1}$.)

kinetics (15), we have simulated the elution profile (Figure 14.7) and calculated an inversion barrier of $\Delta G^{\neq} = 115.1$ kJ/mol (100°C) (16). The simulated chromatogram is in excellent agreement with the experimental chromatogram published in the literature (14).

14.3.2. High-Pressure Liquid Chromatography

The use of Chirasil-Dex in high-pressure liquid chromatography (HPLC) with conventional and micro-packed columns is also possible [15]. The coating of the surface by a polysiloxane-anchored cyclodextrin moiety represents a new development involving the effective blocking of polar sites at the support.

A comparison of the enantiomer separation of hexobarbital and norgestrel with a conventional packed and micro-packed HPLC column containing immobilized (85%) Chirasil-Dex coated on Nucleosil (300–5) silica is shown in Figure 14.8. The merits of miniaturization in HPLC are low consumption of solvents, reduced amount of stationary phase, and easy coupling to mass spectrometry. The enantiomer separation of PCB 132 can also be performed by analytical HPLC on Chirasil-Dex in a semipreparative scale (see Figure 14.9).

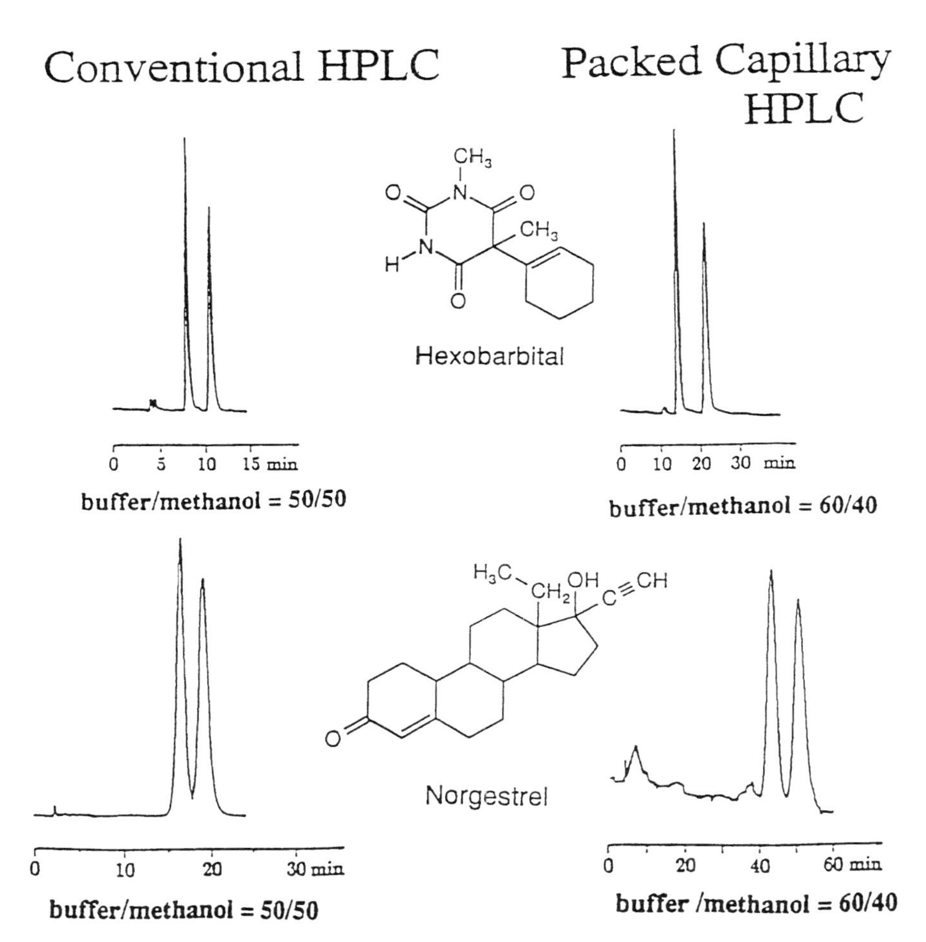

Figure 14.8. Comparison of the enantiomer separation of hexobarbital and norgestrel on a conventional packed HPLC column and a micro-packed capillary column containing Chirasil-Dex coated and thermally immobilized (85%) on Nucleosil (300–5). HPLC column: 200 mm × 4.6 mm (i.d.) stainless steel; capillary column: 150 mm × 0.25 mm (i.d.) fused silica. UV detection: 240 nm (hexobarbital), 254 nm (norgestrel).

14.4. CONCLUSION

Immobilized Chirasil-Dex can be employed as universal stationary phases in chromatography. The chiral polymer shows extended lifetimes and is configu-

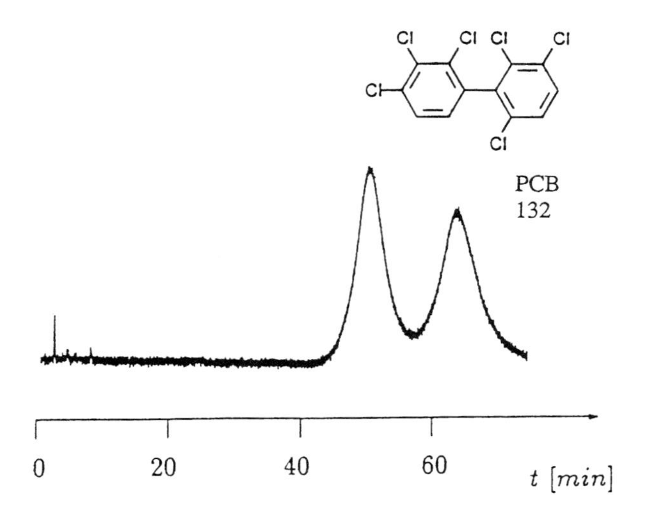

Figure 14.9. Semipreparative enantiomer separation of PCB 132 by HPLC. Column: 200 mm × 4.6 mm (i.d.) containing Chirasil-Dex coated and thermally immobilized (85%) on Nucleosil (300–5). Flow rate 0.7 μl/min methanol/water (70:30); UV detection: 278 nm.

rationally and thermally stable at least up to 250°C. In addition, it may be used as chiral polymer coating for nonchromatographic sensor and transducer devices.

ACKNOWLEDGMENT

This work was supported by Deutsche Forschungsgemeinschaft and Fonds der chemischen Industrie. The authors also wish to thank C. Kempter, University of Tübingen, for LC-electrospray-MS measurements.

REFERENCES

1. Schurig, V., Schmalzing, D., and Schleimer, M. (1991). *Angew. Chem. Int. Ed. Engl.*, **30**, 987.
2. Schmalzing, D., Nicholson, G. J., Jung, M., and Schurig V. (1992). *J. Microcol. Sep.*, **4**, 23.
3. Jung, M. and Schurig, V. (1993). *J. Microcol Sep.*, **5**, 11.
4. Jung, M., Mayer, S., and Schurig, V. (1994). *LC·GC Intern.*, **7**: 6, 340.
5. Schurig, V., Jung, M., Mayer, S., Negura, S., Fluck, M., and Jakubetz, H. (1994). *Angew. Chem. Int. Ed. Engl.*, **33**, 2222.

6. Mayer, S. and Schurig, V. (1992). *J. High Res. Chromatogr.*, **15**, 129.
7. Mayer, S. and Schurig, V. (1993). *J. Liq. Chromatogr.*, **16**, 915.
8. Mayer, S., Schleimer, M., and Schurig, V. (1994). *J. Microcol. Sep.*, **6**, 43.
9. Vindevogel, J. and Sandra, P. (1994). *Electrophoresis*, **15**, 842.
10. Jung, M., Schmalzing, D., and Schurig, V. (1991). *J. Chromatogr.*, **552**, 43.
11. Schurig, V. and Glausch, A. (1993). *Naturwissenschaften*, **80**, 468.
12. Schurig, V., Glausch, A., and Fluck, M. (1995). *Tetrahedron: Asymm.*, **6**, 2161.
13. Glausch, A., Nicholson, G. J., Fluck, M., and Schurig, V. (1994). *J. High Res. Chromatogr.*, **17**, 347.
14. König, W. A., Gehrcke, B., Runge, T., and Wolf, C. (1993). *J. High Res. Chromatogr.*, **16**, 376.
15. Jung, M. and Schurig, V. (1992). *J. Am. Chem. Soc.*, **114**, 529.
16. Jung, M., Fluck, M., and Schurig, V. (1994). *Chirality*, **6**, 510.
17. Schurig, V. Schleimer, M., and Mayer, S. (1994). German Patent Application, Deutsche Offenlegungsschrift DE 4324636, 11.5.1994.

CHAPTER

15

DERIVATIZED CYCLODEXTRINS AS CHIRAL GAS CHROMATOGRAPHIC STATIONARY PHASES AND THEIR POTENTIAL APPLICATIONS IN THE PHARMACEUTICAL INDUSTRY

WEIYONG LI and THOMAS M. ROSSI

The R. W. Johnson Pharmaceutical Research Institute
Spring House, Pennsylvania 19477

15.1. INTRODUCTION

In recent years, there has been an increased trend in the development of single-isomer chiral drugs in the pharmaceutical industry (1). This trend may be attributed to two main reasons. First, there is a greater awareness today among scientists, R&D decision makers, and regulators of the fact that stereoisomers may have distinguishable biological activities and toxicological properties. Second, specialized chiral technologies are currently available to provide not only stereospecific analytical methods for the analysis of chiral drugs, but also the ability to provide single enantiomers on a commercial scale.

Gas chromatographic (GC) separation of enantiomers using cyclodextrin (CD)-based chiral stationary phases (CSPs) is one of the modern analytical tools available for enantioselective analysis. The combination of the highly versatile enantioselectivity of CSPs with the high efficiency of the capillary GC system has resulted in the resolution of many enantiomers that are nonaromatic, have little organic functionality, and are very difficult to resolve using any other method. The new GC chiral separation methods are potentially very important and powerful complementry approaches to liquid chromatographic (LC) methods and may find applications throughout the process of discovery/development of chiral drugs. For example, enantioselective assay methods may be needed for the analysis of optically active starting materials or intermediates; in developing asymmetric synthesis procedures; for the

The Impact of Stereochemistry on Drug Development and Use, Edited by Hassan Y. Aboul-Enein and Irving W. Wainer. Chemical Analysis Series, Vol. 142.
ISBN 0-471-59644-2

analysis of the finished chiral drugs to assure their identity, quality, purity, and strength; and in stability testing of the chiral drugs. Furthermore, enantioselective assay methods may be used to study the pharmacokinetic properties (absorption, distribution, biotransformation, and excretion), and pharmacologic or toxicologic effects of chiral drugs.

Although the first enantiomeric separation using GC on a CSP was reported in 1966 (2), very few chiral GC stationary phases reached wide industrial applications in the last 20 years. In contrast, a variety of chiral LC columns became commercially available. The GC approach has some disadvantages compared with its LC counterpart. First, GC separation usually is performed at elevated temperatures. High column temperature may cause racemization of both the chiral stationary phase and analyte. Second, enantioselectivity of a CSP is temperature-dependent. Usually the enantioselectivity decreases with increased column temperature. However, GC methods may offer some unique advantages. For example, the GC separation process is less perturbed by the mobile phase, which may be advantageous in chiral separation. Furthermore, the very high efficiency of capillary GC system implies that a slight difference in intermolecular interactions between a pair of enantiomers and the chiral stationary phase may result in enantiomeric separation. Therefore, the search for new chiral GC stationary phases continued.

One area of investigation in the development of CSPs for GC has been the exploration of using derivatized cyclodextrins. Cyclodextrins are cyclic oligsaccharides composed of D(+)-glucopyranose units by α-(1,4) linkage (Figure 15.1). They can be obtained by the action of the enzyme cyclodextrin transglycolase on starch. α-, β- and γ-cyclodextrins which contain six, seven, and eight glucose units, respectively, are the main products of the fermentation process (3). The glucose units assume the chair conformation. Each glucose unit has three hydroxyl groups designated 2-OH, 3-OH, and 6-OH. Most derivatizations of the CDs take place via these groups. The 6-OH groups are primary hydroxyls and are located at the more narrow, "bottom" end of the CD torus. The 2-OH and 3-OH groups are secondary hydroxyl groups and located at the wide end or "mouth" of the CD molecule. Generally, it is thought that the 2-OH and 6-OH groups are most reactive for derivatization (the former being more acidic, the latter more accessible sterically), whereas the 3-OH groups are less reactive. The internal diameters of the cavities for the α-, β-, and γ-isomers range between 5–9 Å. The primary hydroxyl groups (6-OH) at the narrower side of the CD can rotate and thus partially block the cavity, whereas the secondary hydroxyl groups cannot rotate (4). The inside of the cavity contains only hydrogen atoms and glucoside oxygen atoms and is therefore relatively apolar compared with the outside. Nonpolar species or nonpolar moieties of compounds can enter into the hydrophobic cavity when in an aqueous environment to form so-called "inclusion complexes." Because

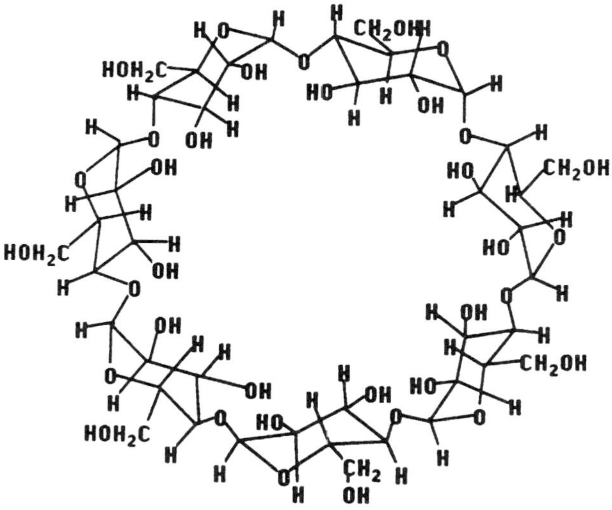

Figure 15.1. Structure of β-cyclodextrin.

all the CD molecules are chiral, they have the potential of forming diastereoisomeric complexes with enantiomers, which may result in the resolution of racemates.

Literature reporting the use of derivatized cyclodextrins in gas–liquid chromatography appeared in the early 1960s. The initial use of cyclodextrins in gas–liquid chromatography was not targeted toward developing enantioselective separation applications. Sand et al. utilized peracylated cyclodextrins such as cyclodextrin acetate, propionate, and butyrate as GC stationary phases for the separation of structural isomers (5, 6). These CD derivatives have high molecular weights, high melting points (e.g., β-CD acetate, MW=2018, m.p. 199–201°C), and high thermal stability (220–236°C). They were coated on traditional supporting materials such as Chromasorb R to prepare packed columns that were used at column temperatures above the melting points of these derivatives. Permethylated cyclodextrins were used as GC stationary phases by Casu and co-workers to study the inclusion phenomenon of CDs (7). The permethylated CDs were dissolved in a solvent stationary phase such as silicon oil to prepare a packed column. Recently, permethylated CD/silicon oil mixtures have been used to coat capillary columns for the separation of enantiomers (8).

Native cyclodextrins and some other derivatives also were used as stationary phases in gas–solid chromatography. Smolková et al. studied the behavior of α- and β-cyclodextrins as stationary phases in gas-solid chromatography (9, 10). Native cyclodextrins were deposited on Chromasorb W (10% w/w) and used at temperatures 50–80 °C. Possible inclusion complex formation between these stationary phases and solutes such as hydrocarbons, halohydrocarbons, and alcohols was investigated. Mizobuchi et al. synthesized cyclodextrin–polyurethane resins by the copolymerization of cyclodextrin with diisocynate in pyridine (11). The products were further granulated and silanized to form 177–250 μm packing materials. The resins had very strong interactions with molecules containing π-electrons or heteroatoms and could be used to separate xylene isomers and pyridine derivatives.

The first successful separation of enantiomers using CDs was reported by Sybilska and co-workers (12, 13). α-Cyclodextrin dissolved in formamide was used as a liquid phase and coated on celite to prepare a packed column. Racemic α- and β-pinene were resolved on the packed column. The development of a new generation of derivatized cyclodextrins appeared in the literature in the late 1980s (14–20). Remarkable enantiomeric separations have been achieved using these new CD-based chiral stationary phases.

In this chapter, we will review the use of the novel derivatized cyclodextrins as capillary GC stationary phases in enantiomeric separation. The discussion includes type of columns prepared from derivatized CDs that are commercially available, the physical properties of these derivatized CDs as well as their enantioselectivities, possible separation mechanisms, and the regulatory perspective when quality control methods are to be developed using these CSPs in the pharmaceutical industry.

15.2. DIFFERENCES BETWEEN EARLIER AND RECENTLY DEVELOPED CD DERIVATIVES

Currently, a majority of commercially available derivatized CD-based columns are wall-coated capillary columns. To understand the difference between the earlier and recently developed CD derivatives, it is helpful to compare them with the conventional GC liquid stationary phases. Nowadays, the most widely used achiral stationary phases in capillary GC are polymers such as methyl- and/or phenyl-substituted polysiloxanes. These materials have high average molecular weights, high thermostability, low melting points, and relatively uniform molecular-weight distribution. The surface tension of these phases matches that of the inside wall surface of capillary tubings, which ensure that the materials can be coated on the inside wall of a capillary tubing as a thin layer of uniform liquid film. Also, they retain high viscosities even at very high temperatures, which ensures that the integrity of the film will be maintained in the operating temperature range.

The earlier developed CD derivatives lack these important physical properties. They are either crystalline or amorphous solids at low temperatures, which cannot be used effectively for coating a capillary tubing. The new CD derivatives have very similar physical properties to those of achiral stationary phases, that is, they are liquids at room temperature and have the desired physical properties. There are several factors that control the physical state of CD derivatives. One such factor is the heterogeneity of these derivatives. This has been verified experimentally. For example, mass spectroscopic data confirmed that permethyl derivatives of O-(S)-2-hydroxypropyl cyclodextrins are composed of a series of analogs and isomers (18). HPLC analysis of the di-O-n-pentyl derivative of β-CD indicates that a mixture of compounds is present. The crude β-CD derivative is a viscous liquid, whereas extensive purification of this derivative yields a solid of melting point 90–92°C (19). Another factor is the size of substituent groups. Permethylated CDs are crystalline solids, whereas perpentylated CDs are liquids at room temperature. It was found that the perhexadecylated β-CD is an amorphows solid. Therefore, only moderate side-chain lengths generate liquid CD derivatives. The presence of hydroxyl groups also can affect the physical state of the CD derivatives. For example, the O-(*S*)-2-hydroxypropyl cyclodextrins are solids at room temperature. They become liquids after permethylation (18).

Because of their unique physical properties, many CD derivatives have been coated on fused silica or glass capillary tubings to prepare highly efficient capillary columns. The most widely used method to prepared a capillary column coated with derivatized CD is called the static method (21) that was successfully used to prepare glass (14–17) and fused silica capillary columns (18–20).

15.3. METHODS DEVELOPMENT USING COMMERCIALLY AVAILABLE CD COLUMNS

There are three types of CD-based GC columns commercially available, including (1) columns coated with liquid CD derivatives, (2) columns coated with CD derivative/silicone oil mixed phases, and (3) columns with immobilized CD derivatives. Table 15.1 lists the column types, CD derivatives used, column suppliers and trade names, classes of compounds resolved, and references.

Type A: Columns coated with liquid CD derivatives. This type of column is coated with one of the following CD derivatives: (1) fully pentylated CDs (phases I and II); (2) partially pentylated and partially acylated CDs (phases III, IV, VII, and VIII); (3) partially pentylated CD (phase

Table 15.1. Commercially Available CD CSPs and Their Applications

Phase	CD Derivative	Supplier[a]/ Trade Name	Class of Compound Resolved	Reference
Type A				
I	Hexakis (2,3,6-tri-O-*n*-pentyl)-α-cycoldextrin	A/Lipodex A	Alkyl alcohols, diols, triols, expoxy alcohols, glycerol derivatives, carbohydrates, ketones, alkyl halides, spiracetals, N-alkylated barbiturates, hydroxycarboxylic acid esters	(14)
II	Heptakis (2,3,6-tri-O-*n*-pentyl)-β-cycodextrin	A/Lipodex C	Alkyl alcohols, cyanohydrins, olefins, hydroxy acid esters, alkyl halides	(17)
III	Hexakis (3-O-acetyl-2,6-di-O-*n*-pentyl)-α-cyclodextrin	A/Lipodex B	Alkyl alcohols, diols, amino alcohols, hydroxycarboxylic acid esters, glycerol derivatives, carbohydrates, acyclic olifins, cyclic olifins, dienes, alkyl halides, lactones	(16)
IV	Heptakis (3-O-acetyl-2,6-di-O-*n*-pentyl) -β-cycodextrin	A/Lipodex D	Alkyl amines, amino alcohols, cyclic 1,2- and 1,3-diols, amino acid esters, pharmaceuticals	(15)
V	Heptakis (2,6-di-O-*n*-pentyl)-β-cycodextrin	B/Chiraldex B-DA	Aromatic alcohols, amino alcohols, aromatic amines, carbohydrates, nicotine analogs, barbitals	(19)
VI	Heptakis (O-(*S*)-2-hydroxypropyl-per-O-methyl)-β-cycolodextrin	B/Chiraldex B-PH	Cyclic alcohols and diols, aromatic alcohols, epoxides, aromatic hydrocarbons, carbohydrates, amines, glycerol derivatives	(18)

VII	Heptakis (2,6-di-O-*n*-pentyl1-3-O-trifluoroacetyl)-β-cyclodextrin	B/Chiraldex B-TA	Alkyl alcohols, diols, polyols, alkyl amines, bicyclic amines, amino alcohols, α-halo carboxylic acid esters, alkyl halides, pyran and furan derivatives, epoxides	(20)
VIII	Octakis (2,6-di-O-*n*-pentyl-3-O-trifluoroacetyl) γ-cyclodextrin	B/Chiraldex G-TA	Alkyl alcohols, diols, polyols, cyclic saturated hydrocarbons, lactones, amino alcohols, sugars, epoxides, alkyl halides, glyceril derivatives, halo carboxylic acid esters, haloepihydrins	(20)
Type B				
IX	Heptakis (2,3,6-tri-O-methyl)-β-cyclodextrin in OV-1701	C/DEX110	Alkyl alcohols, aromatic alcohols terpene alcohols, diols, furan derivatives, lactones, ketones, terpene ketones, halocarboxylic acid esters, saturated monocyclic hydrocarbons	(8)
Type C				
X	Copolymer of permethylated allyl-substitud-β-cyclodextrin and orano hydrosiloxane	B/Chiradex B-PA	Cyclic ketone, cyclic diol	(22)

[a]A = Macherey-Nagel GmbH & Co, Duren, Germany.
B = Astec, Whipanny, New Jersey.
C = Supelco, Inc., Bellefont, Pennsylvania.

V); (4) permethylated O-(*S*)-2-hydroxypropyl derivative of CD (phase VI). The fully pentylated CDs are relatively nonpolar and the O-(*S*)-2-hydroxypropyl derivatives are relatively polar. It is expected that the per-pentylated, partially pentylated, and O-(*S*)-2-hydroxypropyl derivatives are more stable because the substituted groups have C-O-C linkage, is more stable than the ester linkage. Actually, even the columns coated with the 2,6-di-O-pentyl-3-O-trifluoroacetyl CD derivatives had reasonable thermostability (18). However, as mentioned earlier, many of these liquid CD derivatives may be mixtures of analogs and homologies. The hetrogeneity may be desirable in order to obtain the proper physical properties. But this will affect the batch-to-batch reproducibility of the columns for the separation of certain types of chiral compounds. One has to take this into consideration, especially when a regulatory control method needs to be developed. On the other hand, this type of stationary phase has demonstrated very versatile enantioselectivity. Most chiral separations reported have been on this type of CSPs.

Type B: Columns coated with derivatized CDs dissolved in achiral stationary phases. The "solution stationary phase" approach allows the use of CD derivatives (e.g., permethylated CDs, phase IX) that do not have the desired physical properties to coat a capillary column. This also implies that one can use highly purified and well-characterized CD derivatives for column preparation to achieve better batch-to-batch reproducibility. This is important when a regulatory control method needs to be developed and will be discussed in more detail in the "Regulatory Perspective" section. Another advantage associated with this approach is that the overall polarity of the stationary phase can be adjusted by using solvent stationary phases with different polarities.

Type C: Columns with immobilized CD stationary phases. Astec (Whippany New Jersey) introduced a bonded phase that is a copolymer of permethylated allyl-substituted cyclodextrin and organohydroxsiloxane (phase X). This phase was developed by Armstrong and co-workers (22). The column also can be used in supercritical fluid chromatography (SFC) and capillary electrophoresis (CE). So far, only limited applications have been reported with the immobilized CD stationary phases in GC.

Choosing the right column probably is the most difficult and critical step in developing an enantioselective assay. The enantioselectivity of the phases in Table 15.1 is affected by many factors including size of the cyclodextrin cavity, type of substituent functionalities on the CD, degree of substitution, and so on. The enantioselectivity also is affected by structural changes of analytes.

Some racemates can be separated on many phases. For example, alkyl amines (N-TFA derivatives) were separated on phases IV, V, VI, VII, and VIII. Other compounds can only be separated on one particular phase. For instance, some racemic nicotine analogs were only separated on phase V. Therefore, usually it is difficult to predict the separation of a pair of enantiomers on a particular stationary phase. The information provided in Table 15.1 certainly is not enough as far as choosing a proper column is concerned. More detailed information for many compounds and structural types is given in Table 15.2, which

Table 15.2. Homologous Series of Enantiomers Separated on Commercially Available CD Stationary Phases

Compound Class	Structure	Stationary Phase	Reference
Alcohol	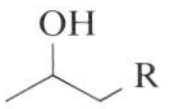 $R = C_1 - C_8$	VII, VIII	(20)
Alkyl halide	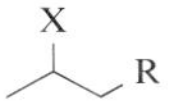 $X = Cl, Br, I$ $R = C_1 - C_4$	II, VII, VIII	(17, 20)
Alkyl halide	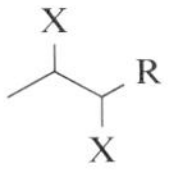 $X = Cl, Br$ $R = H, C_1, C_2$	II, VII, VIII	(17, 20)
Amine	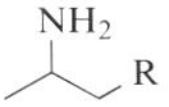 $R = C_1 - C_7$	IV, VIII	(15, 20)
Amino alcohol	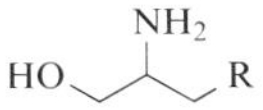 $R = H, C_1 - C_6$	VII, VIII	(20)
Cyanohydrin	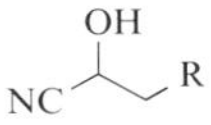 $R = C_1 - C_6$	II	(17)
Diol	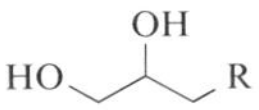 $R = H, C_1 - C_6$	I, III, VII, VIII	(14,16, 20)
Epoxide	 $R = C_1 - C_{12}$	VIII	(20)
Epihalohydrin	 $X = F, Cl, Br$	VIII	(20)
Cyclohexane	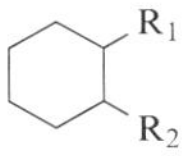 $R_1 = C_1$ $R_2 = C_2, C_3$	IX	(24)

(*continued*)

Table 15.2. *(contd.)*

Compound Class	Structure	Stationary Phase	Reference
Hydroxy acid ester	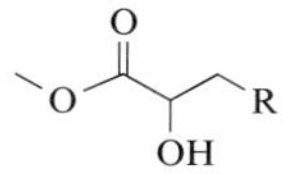 R = $C_1 - C_3$, phenyl	I, VII, VIII	(14, 20)
Cyclic diol	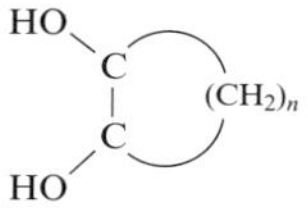 n = 4–8	IV, VI	(15, 18)
Bicyclic	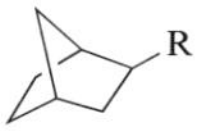 R = NH_2 ,OH, Cl, Br	VII, VIII	(20)
Glycerol derivative	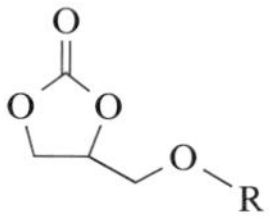 R = $C_1 - C_5$	III	(16)
Cyclic ether	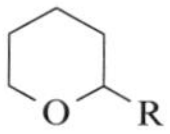 R = OCH_3, OC_2H_5 CH_2Cl, CH_2 Br	VI, VII, VIII	(18, 20)
Cyclic ether	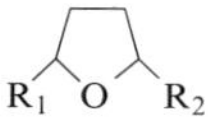 R_1 = H, C_1, OCH_3 R_2 = C_1, OCH_3	VI, VII, VIII	(18, 20)
Nicotine analogs	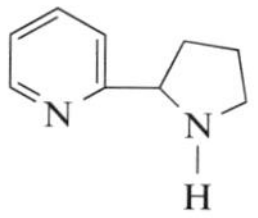	V	(19)
Lactone	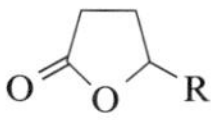 R = $C_2 - C_8$	III	(16)
Carboxylic acid ester	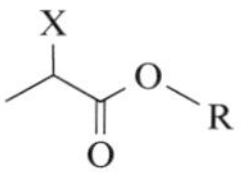 R = $C_1 - C_8$ X = Br, Cl	VII, VIII	(20)
Glycerol derivative	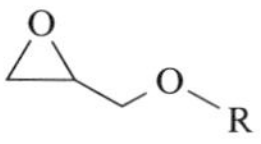 R = $C_1 - C_4$	VII, VIII	(20)

(continued)

Table 15.2. *(contd.)*

Compound Class		Structure	Stationary Phase	Reference
Aromatic hydrocarbon	[structure: benzene ring fused to ring with R and $(CH_2)_n$]	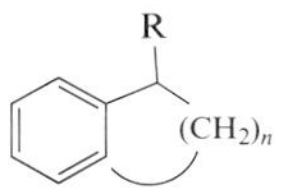 $R = -OH, -NH_2, C_1$ $n = 2,3$	VI	(18)
Carbohydrate	$C_n(H_2O)_n$	$n = 4–6$	I, V, VI	(14,18,19)
Terpene alcohol	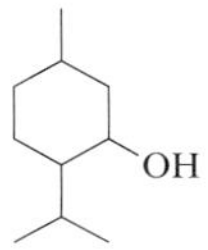 [structure: cyclohexane ring with OH]		IX	(8)
1,3-Dioxolane	[structure: dioxolane ring with R]	$R = CH_2Cl$, phenyl C_1, C_2	IX	(8)

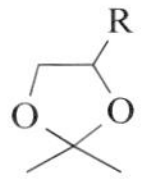

presents the separation of homolog series that have been reported in the literature. For these homologous series, the functional group(s) play a major role in chiral recognition. Changes in side-chain structure or length do not significantly affect enantioselectivity (23). Usually, most members of these homologous series are separated on the same phase. If one has a compound that has a structure similar to one of those listed in Table 15.2, it is probable that the compound can be separated using one of the phases listed for that homologous series, that is, provided that the compound can be eluted at reasonably low column temperatures. For compounds and structural types that have not been analyzed before, a trial-and-error approach is still inevitable. The readers are also referred to the references (14–20, 24, 25) and column manufacturer's brochures.

Another important factor in improving enantioselective separation is the nonstereospecific derivatization of analytes prior to GC analysis. In the GC separation of enantiomers, the purposes of solute derivatization are (1) to increase volatility, (2) to reduce polarity, (3) to improve enantioselectivity (26), and (4) to increase detectability. Some solutes (e.g., sugars) must be converted to more volatile, less polar derivatives to allow their analysis by GC. On many CD CSPs, amines, alcohols, and carboxylic acids must be converted to trifluoroacetamides or esters to achieve enantiomeric separation. Compounds become detectable with very high sensitivity after the introduction of electron-capturing groups or other detector-oriented substituents. Derivatization approaches in the GLC analysis of drugs were reviewed by VandenHeuvel and Zacchei (27).

15.4. TEMPERATURE VERSUS ENANTIOSELECTIVITY AND SEPARATION MECHANISMS

The GC separation of enantiomers is a thermodynamically controlled process. When a separation is taking place in a column with an optically active stationary phase, diastereomeric association complexes are formed:

$$\mathrm{CSP} + A_R \underset{}{\overset{K_R}{\rightleftharpoons}} \mathrm{CSP} - A_R \tag{15.1}$$

$$\mathrm{CSP} + A_S \underset{}{\overset{K_S}{\rightleftharpoons}} \mathrm{CSP} - A_S \tag{15.2}$$

where CSP is a chiral stationary phase and A_R and A_S are analytes in R and S configurations. If enantiomeric resolution is observed, the association constants K_R and K_S have different values. $\Delta(\Delta G)$ can be calculated from the ratio of K_R to K_S:

$$-\Delta(\Delta G) = RT \ln \frac{K_R}{K_S} \tag{15.3}$$

where R is the gas constant and T the temperature in Kelvins (K). Unfortunately, the measurements of K_R and K_S are not feasible in most cases. To a first approximation, $\Delta(\Delta G)$ can be calculated from the separation factor (α):

$$-\Delta(\Delta G) = RT \ln \alpha \tag{15.4}$$

It must be pointed out, however, that the retention of an enantiomer is determined by both gas-liquid equilibrium (achiral contribution) and diastereomeric association complexation (chiral contribution). Because the achiral contribution to retention cannot be eliminated, usually $\Delta(\Delta G)$ is underestimated using Eq. (15.4).

Based on the basic thermodynamic relationships, the corresponding $\Delta(\Delta H)$ and $\Delta(\Delta S)$ can be obtained by measuring α values at different temperatures and plotting $R \ln \alpha$ versus $1/T$.

$$R \ln \alpha = -\frac{\Delta(\Delta H)}{T} + \Delta(\Delta S) \tag{15.5}$$

If $\Delta(\Delta H)$ is a constant within a certain temperature range, a straight line should be obtained. The slope is $\Delta(\Delta H)$ and the intercept $\Delta(\Delta S)$. Thermodynamic parameters have been measured for a number of chiral compounds separated on derivatized CD phases (23, 24). For most compounds, $\Delta(\Delta H)$ and $\Delta(\Delta S)$ have the same sign (they both are negative). This means there is an isoenantioselective temperature (T_{iso}) for these compounds. According to Eq. (15.5), when $\alpha = 1$, we have the following:

$$T_{iso} = \frac{\Delta(\Delta H)}{\Delta(\Delta S)} \tag{15.6}$$

At temperature higher than T_{iso}, the enantiomer elution order should be reversed. However, for most compounds T_{iso} is much higher than the working temperature range and the retention time for these compounds is equal to t_0 (dead time) at T_{iso}. Table 15.3 presents the $\Delta(\Delta H)$ and $\Delta(\Delta S)$ values for four pairs of enantiomers resolved on phase VIII. These four chiral compounds have similar T_{iso} and the α value-temperature relationships for these compounds are plotted in Figure 15.2. From these plots, enantioselectivity is highly temperature-dependent for compounds that have high absolute $\Delta(\Delta H)$ values. For example, for *trans*-2,4,dimethoxytetrahydrofuran, lowering the column temperature 10°C will increase the α value by a factor of 10%. For compounds of low absolute $\Delta(\Delta H)$ values, however, enantioselectivity is almost temperature-independent. For *n*-pentyl 2-bromopropanoate, the same column temperature change will only increase the α value by a factor of 1%.

Based on the above discussion, for compounds that have moderate to high absolute $\Delta(\Delta H)$ values, it is advantageous to perform the separation at lower column temperature by using a short column. For compounds that have low absolute $\Delta(\Delta H)$ values, the analyst may have to use a long and highly efficient column or try a different stationary phase to improve enantiomeric separation.

Table 15.3. Thermodynamic Parameters Measured by GC on a Capillary Column Coated with Octakis (2,6-Di-O-Pentyl-3-O-Triflouroacetyl)-γ-Cyclodextrin (Phase VIII)

Compound	$-\Delta(\Delta H^0)$ (kcal/mol)	$-\Delta(\Delta S^0)$ (cal/mol·K)	T_{iso} (°C)
trans-2,5-Dimethoxytetrahydrofuran	3.1	8.0	110
2-Methylcyclohaxanone	0.7	1.8	120
2-Octanol (O-TFA derivative)	0.5	1.3	110
n-Pentyl 2-bromopropanoate	0.15	0.4	100

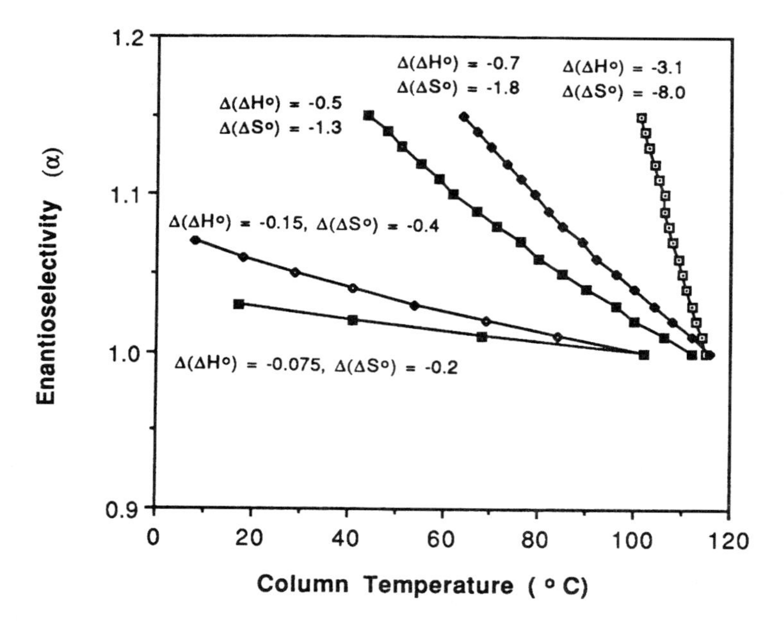

Figure 15.2. Plots showing the effect of temperature on enantioselectivity for compounds with different $\Delta(\Delta H)$ (kcal/mol) and $\Delta(\Delta S)$ (cal/mol·K) values. Stationary phase, phase VIII. Chiral compounds plotted are ( ) *trans*-2,5-dimethoxytetrahydrofuran, (♦) 2-methylcyclohexanone, ( ) 2-octanol (O-TFA derivative), (◆) n-pentyl 2-bromopropamoate and (■) 1,5-dimethylhexylamine (N-TFA derivative).

In 1952, Dalgliesh proposed a simplified chiral recognition model based on a "three-point-interaction" mechanism between a solute and the chiral moiety of a stationary phase. The model requires at least three simultaneous interactions between the solute and stationary phase and at least one of these interactions must be stereochemically dependent to distinguish a pair of enantiomers. That is, only one of a pair of enantiomers can better fulfill the three-point attachment (28). In 1985, Armstrong et al. proposed a modified model to explain the chiral recognition interactions between cyclodextrin LC stationary phases and solutes (29). According to this model, the nonpolar moiety (preferably aromatic moiety) of solute molecules is included in the CD cavity. Two polar substituents of the solute interact with the secondary hydroxyls at the wider opening of the cyclodextrin torus through hydrogen bonding.

It is well established that cyclodextrins can form inclusion complexes with certain inorganic or organic molecules in aqueous solution. Inclusion complexation is considered a main driving force in chiral recognition with cyclodextrin-bonded LC stationary phases. The following intermolecular interactions may be responsible for the chiral recognition in LC: hydrogen bonding, electrostatic interactions, dipole–dipole interactions, hydrophobic interactions, van der Waals interactions, size and shape selectivity of the cyclodextrin cavity, release of "high enthalpy" water molecules from the cyclodextrin cavity, and relief of the ring strain of the macrocycle.

For the GC separation of enantiomers on CD CSPs, Dalgliesh's principle of three-point interaction must hold true. However, it is very difficult to identify where these stereospecific three-point interactions may take place. Armstrong et al. performed mechanistic studies on phases VII and VIII and suggested the presence of at least two different enantioselective retention mechanisms. One mechanism may involve a classic inclusion complex formation. Only a small percentage of compounds resolved fall in this category (the "strong-interaction" group). For these compounds, the interactions may occur inside the cavity (inclusion complex). The main supporting evidence was the measurement of high absolute $\Delta(\Delta H)$ and $\Delta(\Delta S)$ values. For example, a $\Delta(\Delta H)$ of -3.1 kcal/mol and $\Delta(\Delta S)$ of -8 cal/mol·K were measured for *trans*-2,5-dimethyltetrahydrofuran on phase VIII, although no hydrogen bonding interactions are expected between the solute and CD stationary phase. The absolute $\Delta(\Delta H)$ and $\Delta(\Delta S)$ values for this group of compounds are much higher than those obtained for chiral separations of amino alcohols on Chirasil-Val (30). If we consider that enantiomeric separation on Chirasil-Val occurs because of hydrogen bonding, which usually is the strongest noncovalent intermolecular interaction, inclusion complex formation may be responsible for such a high absolute $\Delta(\Delta H)$ value. In addition, size and shape selectivities observed for this group of compounds also support an inclusion complex mechanism.

The separation mechanism for compounds that have low absolute $\Delta(\Delta H)$ and $\Delta(\Delta S)$ values (the "weak-interaction" group) is more ambiguous. Circumstantial evidence suggests that these solutes may form a loose, probably external (outside the CD cavity), multiple association (guest/host ratio >1) with the stationary phase. For example, an enthalpy–entropy compensation study suggests that within the weak-interaction group all compounds may follow the same separation mechanism on both phases VII and VIII. The entropy–entropy compensation relationship was not observed within the strong-interaction group compounds or between the two groups. In studies on the effect of sample size on column efficiency, weak-interaction group compounds show higher column capacity than strong-interaction group compounds, indicating a different ratio of association between solute and stationary phase (23).

15.5. REGULATORY PERSPECTIVE

The application of enantioselective analyses to pharmaceutical compounds has grown dramatically in the past five years (31, 32). This is a direct result of the increased development of single enantiomers to maximize therapeutic index and also of an increasingly stringent regulatory climate. In the chemistry, manufacturing, and control arena, the regulatory mandate for performing enantioselective assays results from several factors, including a need to demonstrate that the drug substance synthesis method consistently produces material of the desired enantiomeric composition, the need to control the inactive enantiomer as a potential impurity in the drug substance, and the need to demostrate that the drug substance and drug product do not racemize during the recommended shelf life of the drug. Even for drugs developed as racemates, the analysis of enantiomeric composition becomes an important regulatory point in explaining the metabolism of each enantiomer. The remainder of this discussion will focus on enantiomeric controls mandated by good manufacturing practices (GMPs), which govern the synthesis and manufacture of drug substances and products (33–35).

In order to satisfy a regulatory need in the pharmaceutical industry, enantioselective methods must meet three general criteria. First, the measurement must be performed at a logical control point in the manufacture of the drug. Second, the method must be capable of meeting pharmaceutical industry standards for validation. Third, the technology must be readily available or transferrable to quality control laboratories.

15.5.1. Enantiomeric Control Points

Control of the enantiomeric composition of a drug substance starts with an understanding of the synthesis of the material. In general, chirality is either carried through the synthesis from one of the starting materials or key reagents, or introduced in the synthesis via the reaction of two achiral materials (36, 37). Alternatively, achiral synthesis generating a mixture of enantiomers may be used, with the desired enantiomeric composition being introduced by a sample purification step (e.g., enantioselective crystallization or preparative LC). In either case, the synthesis method must be shown to produce a consistent enantiomeric composition in the drug sunstance. So, whether the drug substance is a racemic mixture of enantiomers, or a single enantiomer, the synthesis process must be proven to have a consistent outcome.

A generalized synthesis process is diagrammatically illustrated in Figure 15.3. In this example, the enantiomeric composition of the drug substance is governed by the composition of the starting material A. Thus, a logical enan-

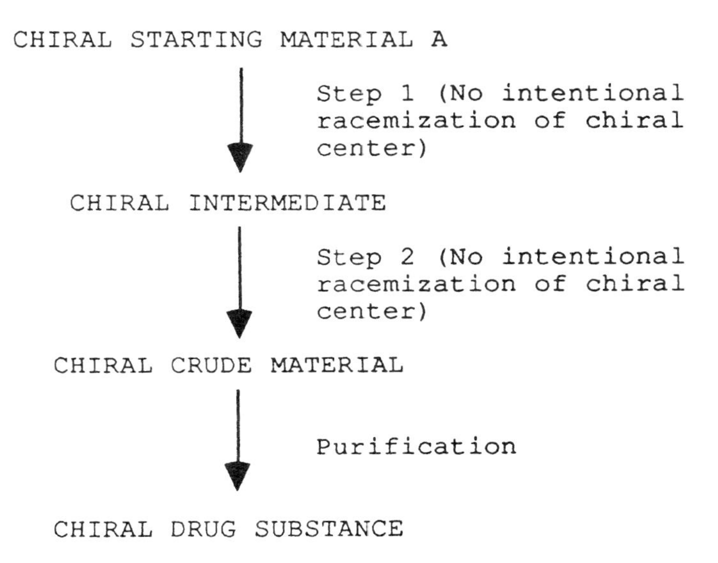

Figure 15.3. Example synthesis of an enantiomeric drug.

tiomeric control strategy for this synthesis would require that a regulatory method be developed for the determination of an enantiomeric composition of each batch of starting material A. Furthermore, a sound control strategy will be based on an understanding of the possible influence of each synthesis step on the enantiomeric composition of the drug substance. For example, if the starting material is enantiomerically pure, it is possible that epimerization may occur in one or more of the synthesis steps. Or, if on the other hand, the starting material is a racemate, it is possible that the crystallization and purification conditions may lead to enantiomeric enrichment. This may be desired or undesired, but must be studied and proven to be consistent. Finally, an enantiomeric control is likely to be required for the drug substance itself. This example is for a case where the chirality is introduced from the starting material. However, the same general logic holds true for other methods of introducing chirality. The control strategy will generally have to be designed to monitor enantiomeric composition at the point of introduction of chirality, to monitor enantiomeric composition of intermediates after synthesis steps, which are shown to have the potential to induce epimerization or enantiomeric enrichment, and to prove that the final drug substance meets appropriate specifications for enantiomeric composition. This logic defines three control points for drug substance enantiomeric composition: analysis of raw materials and key reagents; analysis of intermediates; analysis of drug substance.

A fourth point of control develops out of a need to determine whether epimerization occurs upon storage of the drug substance and formulated drug product. This is generally of concern only for drug products that are developed as a single enantiomer. In these cases, enantioselective assays become a necessary component of the stability-indicating assay strategy.

15.5.2. Validation of Enantioselective Assays

Validation standards in the pharmaceutical industry dictate that the analytical performance characteristics of regulatory assays as well studied and documented. For chromatographic methods of analysis, including enantioselective assays, the criteria for evaluations are linearity, selectivity, accuracy, precision, stability of analytical solutions, sensitivity, and robustness. These criteria are not unique for enantioselective GC separations. Some of these requirements deserve further discussion because they present unique or unexplored challenges to this approach to enantioselective separations.

Selectivity for enantioselective assays includes the obvious requirement that the assay method must be able to separate the enantiomers of interest. However, this problem is complicated by the fact that most assays must be performed in the presence of multiple potential interfrences that are not enantiomers of the compound interest. For example, synthesis intermediates may contain complex mixtures of reagents, solvents, and side products at the point of enantiomeric control. Many drugs have multiple asymmetric centers, leading to the possibility of many diastereromers and enantiomeric pairs being present in the intermediates and finished drug substance. In drug products, the enantiomers must be separated from pharmaceutical excipients and degradation products. Since achieving separations from these potential interferents is heavily dependent on achiral contributions to retention and selectivity, there is a high likelihood that these conditions will pose considerable challenges for CD-based GC columns.

Accuracy must be demonstrated for each enantiomer. In the drug product, this includes a demonstration that each enantiomer is completely extracted from the matrix of excipients. It is important to study both enantiomers since many pharmaceutical excipients are chiral, and thus may interact selectively with one of the enantiomers.

Validation of robustness requires that the assay be characterized as to its failure conditions. The purpose of this requirement is to ensure that the assay conditions in the method leave enough room for varying analytical conditions and so as to enable the method to be reproduced easily by other analysts on different equipment and in different laboratories. Generally, this is proven by varying the recommended operating conditions of the assay to explore the latitude available for change in each important parameter. For enantioselective assays performed on cyclodextrin-based stationary phases in GC, the following parameters should be explored: range in column temperature that will still allow sufficient α; multiple lots of columns and, if possible, multiple suppliers of columns; variation in derivatization conditions where appropriate; variation in flow rate; multiple instruments; multiple analysts; multiple laboratories. The limited number of commercial sources for CD-based GC columns is a liability in method robustness. This is because in cases where the stationary phase of choice is available from only a single source, the long-term use of the method to provide enantiomeric control over a 10- to 15-year horizon is dependent on the supplier's continued, consistent manufacture of the column.

Sensitivity is another key point in achieving a validated method. The use of the term "sensitivity" in the context of assay validation is somewhat of a misnomer. In general, this term is used as a catch-all synonym for the limit of detection, limit of quantitation, and dynamic range. In most cases for regulatory control methods, the amount of material available for analysis is plentiful. Therefore, sample concentrations and column loadings can frequently be adjusted freely to yield the desired sensitivity. The exception to this rule is cases in which the solubility of the analyte is severely limited. However, the dynamic range of the assay procedure may prove to be limiting, particularly in cases where the drug is being developed as a single enantiomer. In such cases, the inactive or undesired enatiomer will generally be treated as an impurity in the drug substance or product. Current standards for impurity controls are tending toward limiting individual impurities to approximately 1 ppt relative to the active ingredient. Therefore, enantioselective assays must be capable of quantitating both enantiomers with linear response over three orders of magnitude.

Based on these considerations, the primary concerns for the validation of regulatory control methods based on derivatized CD GC columns are robustness, selectivity and sensitivity. Robustness primarily because the long-term stability of columns and consistent supply of columns are yet to be demonstrated, selectivity because the complexity of the problem requires both chiral and achiral contributions to retention and selectivity, and sensitivity because the appropriate dynamic range has yet to be demonstrated.

15.5.3. Operability in the Quality Control Environment

In order for CD-based GC columns to meet with acceptance, there is an additional area of concern related to the regulatory aspects already discussed. That is, separations based on this technology must be easily and inexpensively reproduced in the quality control laboratory. This is an extension of the concept of method robustness discussed above.

Because GC enantiomeric separations are performed on commonly available instrumentation, there is an advantage to this approach relative to some other technologies, such as enantioselective capillary electrophoresis, for which equipment is not yet widely available in quality control laboratories. In order for a separation to be considered operable in quality control laboratories, it must meet several basic criteria: All components must be commercially available on an international scale; the transfer of methodology from research to Quality Control must be validated; and the long-term dependability of the method must be relatively certain. It is the first and latter points that pose a challenge for enantioselective GC assay methods. The first point dictates that this technology will find applicability in regulatory methods only in cases where the few commercially available stationary phases work. Custom column packing would be a liability, and it is very likely that only commercially prepared columns would find widespread acceptance. The last point brings us back to the earlier discussion on robustness. The stability of column suppliers and reproducibility of their product has yet to be demonstrated over a time horizon spanning the lifetime of a typical drug product (10–20 years).

Cost containment is a major concern at present in the pharmaceutical industry and every analytical test performed on incoming raw materials, intermediates, or finished products adds cost to the product. Maintaining infrequently used columns in inventory and changeover among many different columns add cost to testing. So, another critical aspect of utility in the pharmaceutical industry for CD-based GC columns is that the applications must be numerous enough to justify the use of this technology on a relatively frequent basis. Column cost and lifetime are also concerns in this regard.

15.6. CONCLUSION

The regulatory and cost containment pressures in the pharmaceutical industry impose exacting criteria for the introduction of new technologies for regulatory control methods. The relatively recent development of CD-based GC enantioselective separations leaves several open questions as to the eventual use of this particular technology to address regulatory control. It has yet to be demonstrated that this technology can provide cost-effective solutions to separation problems encountered in developing a complete enantiomeric control strategy. The major unknowns at this time center around the robustness of columns and their long-term availability, as well as further demonstrations of the ability of this technique to provide enantioselective separations in complex mixtures of potential interferences. However, these perceived limitations are based primarily on a lack of data, and there does not seem to be any indication that methods based on this technology cannot overcome these obstacles.

REFERENCES

1. Millership, J. S. and Fitzpatrick, A. (1993). *Chirality*, **5**, 573.
2. Gil-Av E., Feibush, B., and Charles-Sigler, R. (1967). In A. B. littlewood (ed.). *Gas Chromatography 1966*, London: Institute of Petroleum, p. 227.
3. Saenger, W. (1980). *Angew. Chem. Int. Ed. Engl*, **19**, 344. (and references citied there).
4. Hybl, A., Rundle, R. E., and Williams. D. E. (1965). *J. Am. Chem. Sco.*, **87**, 2779.
5. Sand, D. M. and Schlenk, H. (1961). *Anal. Chem.*, **33**, 1624.
6. Schlenk, H., Gellerman, J. L., and Sand, D. M. (1962). *Anal. Chem.*, **34**, 1529.
7. Casu, B., Reggani, M., and Sanderon, C. R. (1975). *Carbohydr. Res.*, **76**, 59.
8. Schurig, V., and Nowotny, H.-P. (1987). In A. Zlatkis (ed.) *Proceedings of Advances in Chromatography* 1987. Berlin: p. 8: In (1988) *J. Chromatogr.*, **441**, 155.
9. Smolková, E., Kralova, H., Krysl, S., and Feltl, L. (1983). *J. Chromatogr.*, **241**, 3.
10. Smolková-Keulemansová, E., Neumannová, E., and Feltl, L. (1986). *J. Chromatogr.*, **365**, 279, 289.
11. Mizobuchi, Y., Tanaka, M., and Shono, T. (1980). *J. Chromatogr.*, **194**, 153.
12. Koscielski, T., Sybilska, D., and Jurczak, J. (1983). *J. Chromatogr.*, **280**, 131.
13. Koscielski, T., Sybilska, D., Belniak, S., and Jurczak, J. (1984). *Chromatographia*, **19**, 292.
14. König, W. A., Lutz, S., Mischnick-Lubbecke, P., Brassat, B., and Wenz, G. (1988). *J. Chromatogr.*, **447**, 193.
15. König, W. A., Lutz, S., Wenz, G., and von der Bey, E. (1988). *J. High Res. Chromatogr. Chromatogr. Comm.*, **11**, 506.
16. König, W. A., Lutz, S., Colberg, C., Schmidt, N., Wenz, G., von der Bey E., Mosandl, E. A., Gunther, C., and Kustermann, A. (1988). *J. High Res. Chromatogr. Chromatogr. Comm.*, **11**, 621.
17. König, W. A., Lutz, S., Hagen, M., Krebber, R., Wenz, G., Baldenius, K., Ehlers, K. J., and Tom Dieck, H. (1989). *J. High Res. Chromatogr.*, **12**, 35.
18. Armstrong, D. W., Li W., Chang, C. D. and Pitha, J. (1990). *Anal. Chem.*, **62**, 914.
19. Armstrong, D. W., Li W., Stalcup, A. M., Secor, H. V., Izac, R. R., and Seeman, J. I. (1990). *Anal. Chim. Acta*, **234**, 365.
20. Li, W., Jin, H. L., and Armstrong, D. W. (1990). *J. Chromatogr.*, **509**, 303.
21. Bouche, J. and Verzele, M. (1968). *J. Gas Chromatogr.*, **6**, 501.
22. Armstrong, D. W., Tang, Y. B., Ward, T., and Nichols, M., (1993). *Anal. Chem.*, **65**, 1114.
23. Berthod, A., Li, W., and Armstrong, D. W., (1992). *Anal. Chem.*, **64**, 873.
24. Schurig, V. and Nowotny, H.-P. (1990). *Angew. Chem. Int. Ed. Engl.*, **29**, 939.
25. Li, W. (1990). Ph.D. dissertation, University of Missouri-Rolla.
26. Armstrong, D. W. and Jin, H. L. (1990). *J. Chromatogr.*, **502**, 154.

27. VandenHeuvel, W. J. A. and Zacchei, A. G. (1976). In J. C. Jiddings, E. Grushka, J. Cazes, and P. R. Brown (eds.), *Advances in Chromatography*. New York: Marcel Dekker, p. 199.
28. Dalgliesh, C. E. (1952). *J. Chem. Soc.*, **3**, 3940.
29. Hinze, W. L., Riehl, T. E., Armstrong, D. W., Demond, W., Alak, A., Ward, T. (1985). *Anal. Chem.*, **57**, 237.
30. Koppenhoefer, B. and Byer, E. (1984). *Chromatographia*, **19**, 123.
31. Subert, J. (1994). *Pharmazie*, **49**, 3.
32. Wainer, I. W. (1993). *Trends Anal. Chem.*, **12**, 153.
33. Tomaszewski, J. and Rumore, M. M. (1994). *Drug Dev. Ind. Pharm.*, **20**, 1119.
34. De Camp, W. H. (1993). *J. Pharm. Biomed. Anal.*, **11**, 1167.
35. Rauws, A. G. and Groen, K. (1994). *Chirality*, **6**, 72.
36. Seebach, D. (1990). *Angew Chem. Int. Ed. Engl.*, **23**, 1320.
37. Nugent, W. A., RajanBabu, T. V., and Burk, M. J. (1993). *Science*, **259**, 479.

CHAPTER

16

CHIRAL DERIVATIZATION REAGENTS IN THE BIOANALYSIS OF OPTICALLY ACTIVE DRUGS WITH CHROMOPHORE-BASED DETECTION

RALF BÜSCHGES

Boehringer Ingelheim KG
Pharmaceutics Department
D-55216 Ingelheim
Germany

ERIC MARTIN

Department of Pharmacology
Johann Wolfgang Goethe-University
D-60053 Frankfurt
Germany

HASSAN Y. ABOUL-ENEIN

Bioanalytical and Drug Development Laboratory
Biological and Medical Research
King Faisal Specialist Hospital and Research Centre
Riyadh 11211, Kingdom of Saudi Arabia

PETER LANGGUTH

Astra Hässle AB,
Kärragatan 5,S-41383 Mölndal
Sweden

and

HILDEGARD SPAHN-LANGGUTH

Department of Pharmacy
Martin-Luther-University Halle-Wittenberg
Wolfgang-Langenbeck-Strasse
D-06120 Halle/Saale
Germany

The Impact of Stereochemistry on Drug Development and Use, Edited by Hassan Y. Aboul-Enein and Irving W. Wainer. Chemical Analysis Series, Vol. 142.
ISBN 0-471-59644-2

16.1. INTRODUCTION

The growing attention to the pharmacokinetics and pharmacodynamics of stereoisomeric drugs—combined with an increased awareness of the drug regulatory authorities (Nitchuk, 1992; Batra et al., 1993; De Camp, 1993; Fassihi, 1993; Marzo, 1994; Tomaszewski and Rumore, 1994)—requires the development of enantiospecific analytical methods permitting the quantitation of trace amounts in biological matrices.

High-performance liquid chromatography (HPLC) has become a favorite tool in the quantitation of drugs and their metabolites in body fluids and tissue. Although the number of chiral stationary phases (CSPs) is rapidly growing—more than 400 were known in 1993 with some 80 commercially available (Koppenhoefer et al., 1993)—the interest in chiral derivatizing agents is still unbroken.

The reasons are obvious: Chiral derivatization to diastereomers followed by HPLC analysis of the derivatives on conventional normal- or reversed-phase columns ("indirect" method) is a versatile and comparatively low-cost technique to determine xenobiotics in biological fluids. Since a large number of reagents is commercially available or can be prepared by simple preparative steps, a variety of options concerning the physico-chemical nature of the diastereomeric derivatives are at hand.

Derivatization not only changes the physico-chemical attributes of the analyte and may provide improved chromatographic properties of the diastereomeric derivatives, but also selectivity may be enhanced. Despite the large number of available HPLC-CSPs, many separations only have been accomplished via derivatization to diastereomers. Chiral stationary phases often show appropriate selectivity to separate parent compound and metabolite enantiomers, but not enough selectivity to separate parent compounds and metabolites from each other.

Furthermore, the application of gradient elution is very restricted with most CSPs but not with conventional reversed-phase columns—a factor that may limit the determination of a drug plus its metabolites or several drugs in a single analytical run at CSPs. Finally, a low limit of quantitation is a very important factor in the performance of pharmacokinetic studies, which may be considerably improved via derivatization.

However, there are certain limitations with respect to the application of chiral derivatizing agents in precolumn derivatization. First, the substrate must have a reactive moiety that can be derivatized at all. The reagent must be of high and known enantiomeric purity. Reagents and derivatives have to be stable, especially with regard to isomerization during storage and use. The reaction should be selective to the functional moieties within the analyte gaining well-defined products. Derivatives of both enantiomers should be formed to the same extent and at the same rate (i.e., absence of so-called "kinetic resolution"). The detector response to both diastereomers should be almost identical. Finally, excess reagent and by-products—the cause of additional chromatographic peaks—should not interfere with the assay. All these criteria have to be considered in the development and validation of bioanalytical HPLC methods, a sometimes unpleasant duty.

An introduction to the basics of detection-oriented derivatization in liquid chromatography is provided by Lingeman and Underberg (1990). Recently, a special issue of the *Journal of Chromatography, Biomedical Applications* was dedicated to derivatization in liquid chromatography (special issue 1994). The topic of chiral derivatization in gas and liquid chromatography has been reviewed by Skidmore (1993) and Gal (1993). Görög and Gazdag (1994) reported on the biomedical applications of chiral derivatization, achiral derivatization prior to separation on CSPs, and chiral mobile phase additives. Publications focused on the enantiomeric separation of β-adrenoceptor antagonists (Egginger et al., 1993; Jira and Breyer, 1993; Olsen et al., 1993) and amino acids (Bhushan and Joshi, 1993) have been published recently.

The present chapter provides a survey of chiral derivatization in the HPLC analysis of drug enantiomers and other xenobiotic compounds from biogenic sources. Special emphasis is given to precolumn derivatization techniques and the detectability of derivatives with standard liquid chromatographic—that is, UV and fluorescence—detectors.

In this context, a closer look at three different detection modes may be helpful (Schemes 1 and 2):

- *Detection at the substrate wavelengths* (*substrate-oriented detection*). An easily detected substrate is coupled to a reagent lacking strong chromophores. This method of detection is especially useful for well-detectable substrates (e.g., fluorescent 2-arylpropionic acids like naproxen or carprofen). Interferences from matrix compounds are rare.
- *Detection at the reagent wavelengths* (*reagent-oriented detection*). Strong chromophoric or fluorophoric labels are attached to hardly detectable substrates. Derivatives are monitored at their absorption or fluorescence optima that are usually near the respective maxima of the reagent. Detectability may be enhanced by several orders of magnitude, but excess reagent and by-products are sometimes difficult to remove.
- *Detection at the wavelengths of a chromophore generated in the course of reaction.* Neither substrate nor reagent are well-detectable. The reaction is chromo- or fluorogenic and the chromophoric properties are characteristic for the derivatives [e.g., fluorescent isoindoles resulting from the reaction of primary amines with o-phthaldialdehyde and a chiral thiol

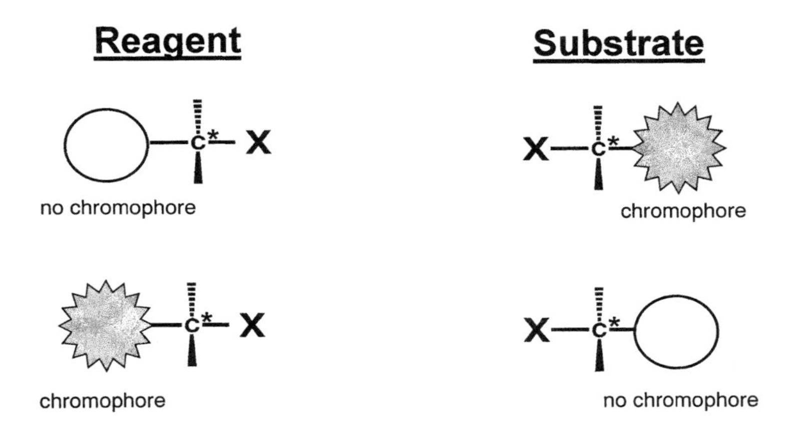

Scheme 1. Easily detected (e.g., fluorescent) derivatization products may result from either the reaction of a nonchromophoric substrate with a chromophoric CDA or the reaction of a chromophoric substrate with a nonchromophoric CDA. The use of chromophoric CDAs for chromophoric substrates is mostly avoided in LC since baseline noise increases.

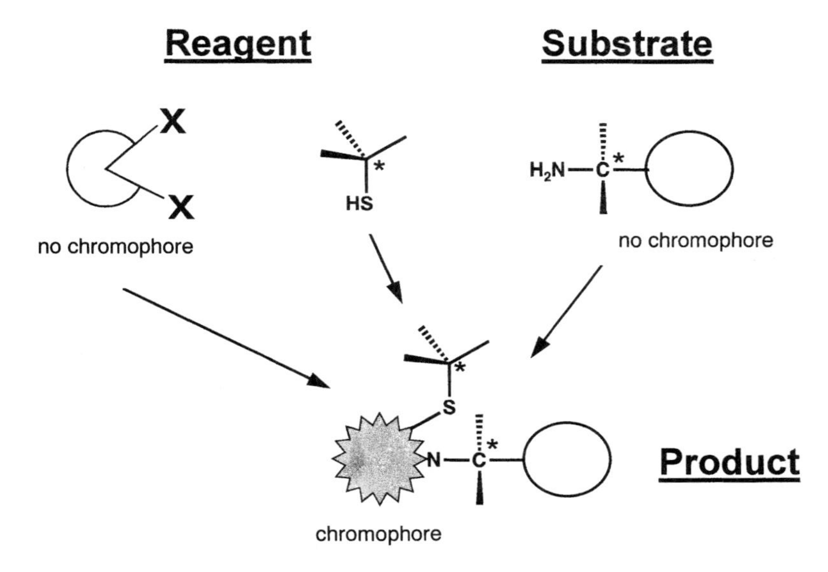

Scheme 2. Reagents with poor chromophoric properties (OPA, SH-donor) and a nonfluorescent substrate (primary amine) are reacting to form a strongly fluorescent product.

(Nimura and Kinoshita 1986)]. Removal of excess reagent is less important than with reagent-oriented detection, although highly fluorescent derivatives are selectively generated from primary amines.

The following sections give an overview of the chiral derivatizing agents (CDAs) presently available, their advantages and disadvantages, and their application to enantioselective bioanalytical methods.

16.2. ACTIVATED CARBOXYLIC ACIDS

Chiral carboxylic acids (see Figure 16.1) are very popular derivatizing reagents in liquid chromatography, since they are readily available or easily prepared from compounds of the chiral pool (e.g., amino acids, menthol). However, carboxylic acids have to be activated prior to derivatization with alcohols or amines to allow rapid reaction under acceptably mild conditions. The topics of esterification with alcoholic reagents and acylation with carboxylic acid derivatives—reaction mechanisms, catalysts, and applications—were reviewed by Blau and Darbre (Blau, 1993; Blau and Darbre, 1993). (A detailed description of carboxylic acid activation is provided in Section 16.7 on amine reagents.) Principally, three types of acylating reagents may be distinguished (Blau, 1993): acid halides, acid anhydrides, and other reactive acyl derivatives (e.g., imidazolides).

Acyl halides are highly reactive but rather unselective. They are preferentially used in the derivatization of sterically hindered molecules and weak nucleophiles—for example, hydroxyl groups. As with other acylating reagents, the derivatization of amino compounds, amides, and thiols is feasible as well. However, if milder reaction conditions are required, derivatization with anhydrides or acyl imidazoles will be more suitable.

Acylating reagents have been used frequently in gas chromatography. O-methylmandelyl chloride (2-methoxy-2-phenylacetyl chloride) is one of the first CDAs introduced to HPLC (Helmchen and Strubert, 1974). The reagent has been used for the determination of tocainide enantiomers in plasma (Hoffmann et al., 1984). 2-Methoxy-2-phenylacetic acid may also be applied as the succinimidylester (SMPA) (Husain et al., 1989, 1993).

A related and more popular CDA, Mosher's acid (2-methoxy-2-(trifluoromethyl)phenylacetic acid, MTPA), was applied first for chromatographic purposes in GC applications (Dale et al., 1969). However, HPLC separations of diastereomeric MTPA esters have been reported as well. This CDA has been applied to drug metabolism studies, preferentially for the analysis of epoxides and diol metabolites of polycyclic aromatic hydrocarbons (PAH)—some of them highly mutagenic and carcinogenic (Yagi et al., 1977, 1982;

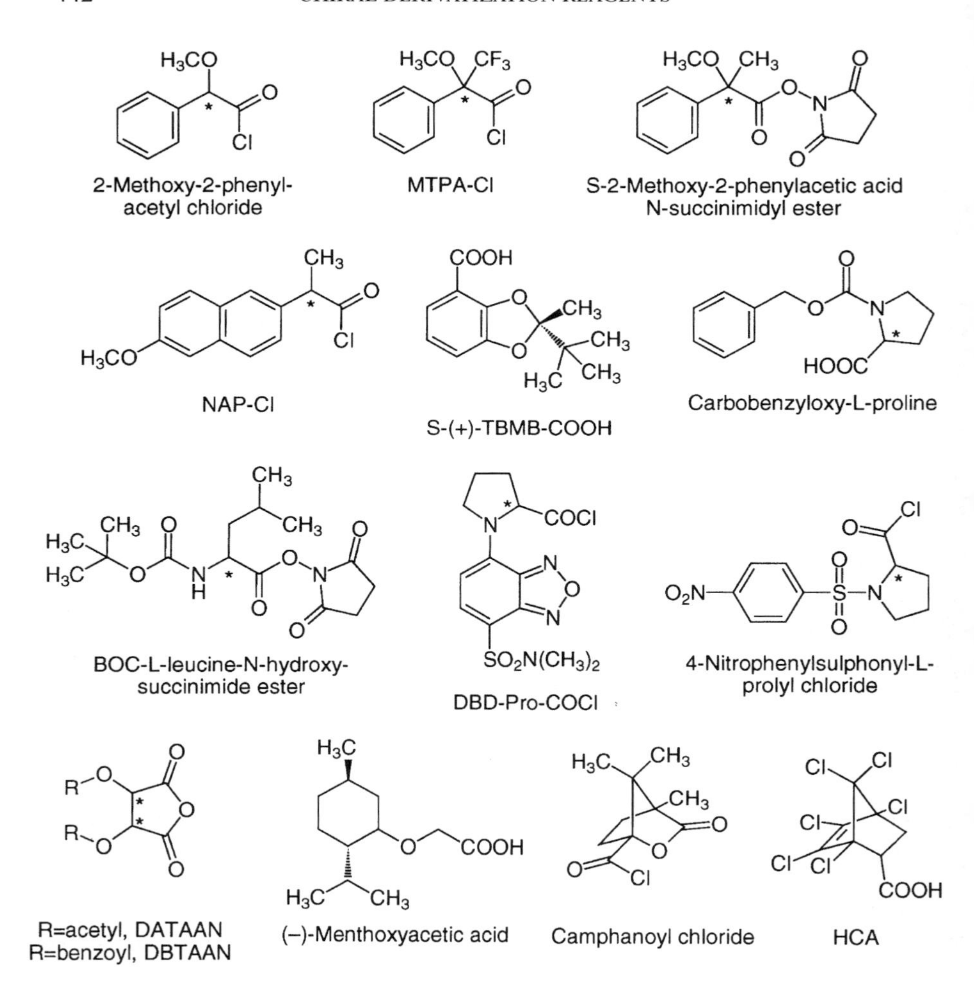

Figure 16.1. Selected carboxylic acid derivatives.

Armstrong et al., 1981; Vyas et al., 1982; Lehr et al., 1985; van Bladeren et al., 1985; Balani et al. 1987). HPLC applications for studies in biological matrices are rare. The configurationally stable reagent has been used instead for the determination of absolute configuration (Blaschke and Walther, 1985; Davis et al., 1985) than for bioanalytical purposes.

(−)-2-Methoxy-2-methyl-2-(1-naphthyl)acetic acid and its 2-naphthyl analog were introduced by Goto et al. (1978) for the separation of amino acids. The succinimidyl ester of the more promising 1-naphthyl derivative was used for the determination of 2,5-dimethoxy-4-methylamphetamine enantiomers in plasma (Goto et al., 1979). Shimizu et al. (1982), used (+)-2-(2-naphthyl) propionyl chloride for the separation of diltiazem isomers.

Acid chlorides of the fluorescent, chiral nonsteroidal antiinflammatory drugs benoxaprofen (BOP-CL) (Weber et al., 1984), naproxen (NAP-Cl) (Spahn, 1988c), and flunoxaprofen (FLOP-Cl) (Spahn, 1988b)—both 2-arylpropionic acids—were used for the enantiospecific separation of alcohols and amines. The separation of tocainide recovered from plasma and tranylcypromine is described. HPLC separations of different compounds are compared to those obtained from silica-gel TLC. NAP-Cl was used for the analysis of cyclophosphamide in plasma after amidoalkylation of the compound (Reid et al., 1989), diacylglycerols in phospholipids from brain tissue (Rastegar et al., 1990), baclofen (Spahn et al., 1988), and carvedilol in plasma (Spahn et al., 1990), and the antidepressant venlafaxine in biological specimens (Wang et al., 1992). FLOP-Cl was applied to the enantioselective determination of esterified amino acids and small peptides (Langguth et al., 1990) and to the derivatization of propranolol (Spahn-Langguth et al., 1991). The applicability of other 2-arylpropionic acid chlorides as CDAs was examined as well, but they were not used in a bioanalytical assay (Spahn, 1989).

Amino acid derivatives have been used frequently as CDAs in chromatographic applications (Skidmore, 1993). Since investigations with N-trifluoroacetyl-L-prolyl chloride (Hermansson and von Bahr, 1980; Silber and Riegelman, 1980) showed significant recemization of the CDA, more stable reagents of different structures had to be developed.

Optically pure amino acids may be coupled to chiral amino compounds when activated and N-protected. The application of *tert*-butyloxycarbonyl-L-leucine (BOC-L-Leu) to the chiral derivatization of amino compounds—adapting reactions from peptide chemistry—is feasible as either the N-hydroxysuccinimide ester (BOC-L-Leu-SU) (Mitchell et al., 1978) or the symmetrical anhydride (Hermansson, 1982). Hermansson synthesized the anhydrides of BOC-L-Leu and its alanine analog and applied both to the derivatization of racemic propranolol. The protecting groups were removed with trifluoroacetic acid prior to reversed-phase ion-pair chromatography. Separation was better for L-leucine derivatives than for the L-alanine analogs. The limit of detection was 1 ng/mL of enantiomeric propranolol in plasma using fluorescence detection at the substrate wavelengths. Comparable assays were established for alprenolol and metoprolol, with limits of detection as low as 500 pg of enantiomer/mL of plasma (Hermansson and von Bahr, 1982). The addition of an internal standard, cyclopentyldeisopropylpropranolol, and reduction of the sample volume were reported for the analysis of propranolol from rat blood (Guttendorf et al., 1989). Furthermore, analysis of penicillamine enantiomers (Nachtmann, 1980) and iodinated thyronines (Lankmayr et al., 1980) after derivatization with BOC-L-Leucine-N-hydroxysuccinimide ester has been reported.

Banfield and Rowland (1983) introduced carbobenzyloxy-L-proline to the derivatization of warfarin followed by HPLC separation of the diastereomeric derivatives. An improved and—due to postcolumn aminolysis and fluorescence detection—more sensitive assay (Banfield and Rowland, 1984) helped to determine warfarin and its metabolites in plasma and urine after low single oral doses of the anticoagulant.

Shimizu et. al. (1986) described the preparation of *S*(–)-N-1-(2-naphthylsulfonyl)-2-pyrrolidinecarbonyl chloride. The reagent derived from L-proline helped to resolve the four stereoisomers of diltiazem, a benzothiazepinone with two chiral centers. The application to the enantioselective determination of amino acids has been described (Nishi et al., 1989). Tosylated *S*-proline was used for the stereosisomeric separation of 2-[[[2-(2,6-dimethoxyphenoxy)-ethyl]amino]methyl]-1,4-benzodioxan, a potent α_1-adrenoceptor antagonist (Andrisano et al., 1992). Further CDAs derived from the amino acid proline were evaluated by Clark and Barksdale (1984), who applied 4-nitrophenylsulfonyl-*S*-prolyl chloride to the determination of amphetamine and derivatives (Barksdale and Clark, 1985). Interesting and highly fluorescent reagents, the enantiomers of 4-(N,N-dimethylaminosulfonyl)-7-(2-chloroformylpyrrolidin-1-yl)-2,1,3-benzoxadiazole (DBD-Pro-COCl) were recently synthesized and tested by Toyo'oka et al. (1993). The fluorescence wavelengths were λ_{ex} = 450 nm and λ_{ex} = 560 nm, respectively. Thus, interferences from underivatized matrix compounds can be avoided. Furthermore, the reagent seems highly suitable for laser-induced fluorescence detection.

R,*R*-O,O-diacetyltartaric acid anhydride (DATAAN), a reagent for substrate-oriented detection, and related CDAs with more prominent UV absorption were used for the derivatization of enantiomeric alkanolamines (β-adrenoceptor antagonists) (Lindner et al., 1984). The tartaric acid monoesters were well resolved on reversed-phase HPLC columns. Tartaric anhydride CDAs were used in the analysis of β-adrenoceptor antagonists in biological fluids (Lindner et al., 1989; Vermeulen et al., 1992). The procedure was adapted to preparative optical resolution of hydroxychloroquine—an antimalarial agent—and its determination in different biogenic matrices (Brocks et al., 1992) and the bioanalysis of delmopinol (Egginger and Lindner, 1993). Schützner et al. (1993) were able to separate amino acid derivatives of DATAAN employing capillary electrophoresis with polyvinylpyrrolidone as a buffer additive. Derivatization with DATAAN usually yields well-separated diastereomeric ester or amides, due to the formation of intramolecular hydrogen bonds and the resulting cyclic arrangement of the derivative (Lindner et al., 1984).

Activated (–)-menthoxyacetic acid, a CDA for substrate-oriented detection, was primarily applied to the derivatization of alcohols and diols. Most investigations were centered on the metabolism of polycyclic aromatic hydrocarbons (PAHs) (Harvey and Cho, 1977; Yang et al., 1977; Armstrong et al., 1981; Vyas et al., 1983; Thakker et al., 1984, 1986; Yang and Fu, 1984).

Examples of the application of (–)-camphanic acid or the respective acyl chloride for the liquid chromatographic analysis of xenobiotics are rare. The CDA has been applied to the chromatographic resolution of proxyphylline enantiomers (Ruud-Christensen and Salvesen, 1984) and β-adrenoceptor antagonists (Olsen et al., 1993).

Duke and Holder (1988) used (–)-(1*S*,2*R*,4*R*)-endo-1,4,5,6,7,7-hexachlorobicyclo[2.2.1]hept-5-ene-2-carboxylic acid (HCA) for the derivatization of PAH hydroxymetabolites. HCA was superior to menthoxyacetic acid, MTPA, and several chiral stationary phases with respect to the resolution of stereoisomeric di- and tetrahydrodiols of substituted benzacridines, which served as model compounds. HCA was further applied to the analysis of warfarin in plasma (Carter et al., 1992).

Carboxylic acid derivatives are well-established derivatizing agents in the enantioselective analysis in biogenic matrices for various reasons:

- A wide variety of carboxylic acid CDAs are available.
- Activated carboxylic acid derivatives are especially suitable in the analysis of less reactive analytes (e.g., aliphatic alcohols).
- The type of derivative (e.g., acid chloride vs. imidazolide) is highly influential on the reactivity of the reagent.

16.3. CHLOROFORMATES

Up to now, chiral chloroformates have found minor attention in the analysis of biogenic compounds and xenobiotics compared with different acylating agents, isocyanates, and isothiocyanates. They are readily prepared from a chiral alcohol and phosgene or one of its less toxic and less volatile substitutes [e.g., triphosgene (Nekrassow and Melnikow, 1930; Eckert and Forster, 1987)]. In general, chloroformates are versatile CDAs, applicable to the derivatization of primary and secondary amines (Westley and Halpern, 1968; Seeman et al., 1985; Einarsson et al., 1987; Turgeon et al., 1990; Witte et al., 1993; Büschges et al., 1995, 1996), amino acids (Einarsson et al., 1987), and aminoalcohols (Einarsson et al., 1987; Mehvar, 1989; Prakash et al., 1989; Rosseel et al., 1991; Roux et al., 1991; Miller and Guertin, 1992; Fiset et al.,

1993; Lai et al., 1993; Büschges et al., 1996). Reaction with tertiary amines in aprotic solvents and elevated temperatures leads to either carbamates via dealkylation or deaminated products (Hartvig et al., 1976; Gübitz et al., 1981; Wallace et al., 1981; Prakash et al., 1989a; Sintov et al., 1989; Witte et al., 1990; Kristensen and Angelo, 1992). Derivatization of alcohols—yielding carbonates—is feasible as well (Westley and Halpern, 1968; Annett and Stumpf, 1972; Prelusky et al., 1982; Jeyaraj and Porter, 1984; Baertschi et al., 1988; Meese et al., 1988).

Although a variety of functional groups may be subject to chloroformate derivatization, in many cases the choice of reaction conditions allows the selective and rapid formation of carbamates from primary and secondary amines. However, a certain cross-reactivity with phenolic compounds has to be considered. Unlike many acylating reagents, chloroformates are applicable to derivatizations in aqueous media, that is, basic borate buffers. Due to the rapid reaction kinetics of 1-(9-fluorenyl)ethyl chloroformate (FLEC), Lai and co-workers (1993) were able to develop an on-line precolumn derivatization method for the β-adrenoceptor antagonist propranolol.

The application of chiral chloroformates to the liquid chromatographic separation of drugs and other amino compounds of interest, especially from biological fluids, has been limited to basically two reagents: (–)-menthyl chloroformate (MCF) (Carpino, 1966; Westley and Halpern, 1968) and 1-(9-fluorenyl)ethyl chloroformate (FLEC) (Einarsson et al., 1987) (see Figure 16.2). Both reagents are commercially available.

Ahnoff et al. (1990) synthesized 10 different chloroformate reagents and compared their applicability to the HPLC analysis of metoprolol enantiomers with substrate-oriented detection. Recently, two more reagents have been added. Büschges et al. described two compounds related to either naproxen (NAP-C) or nabumetone (NAB-C) (Büschges, 1996), both nonsteroidal anti-inflammatory agents with extraordinary UV-absorbance and an additional intrinsic fluorescence.

Menthyl chloroformate (MCF) does exhibit neither relevant UV-absorbance nor any fluorescence, a CDA highly suitable for substrate-oriented detection. In contrast, 9-(1-fluorenyl)ethyl chloroformate (FLEC), closely related to the more frequently used but achiral 9-(1-fluorenyl)methyl chloroformate (FMOC-Cl) (Carpino and Han, 1972; Moye and Boning, 1979; Einarsson et al., 1983), is highly fluorescent (excitation at approximately 265 nm, emission at approximately 345 nm).

MCF has been applied to the preparative liquid chromatographic separation of warfarin enantiomers (Jeyaraj and Porter, 1984) and nornicotine (Seeman et al., 1985). Mehvar (1989) describes reversed-phase separations of various β-adrenoceptor agonists and antagonists. The precolumn derivatization of β-adrenoceptor antagonists and related compounds after extraction

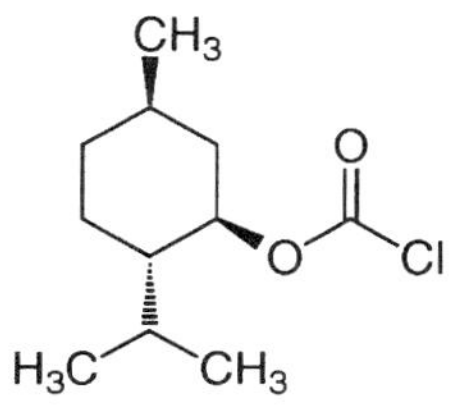

(−)-MCF 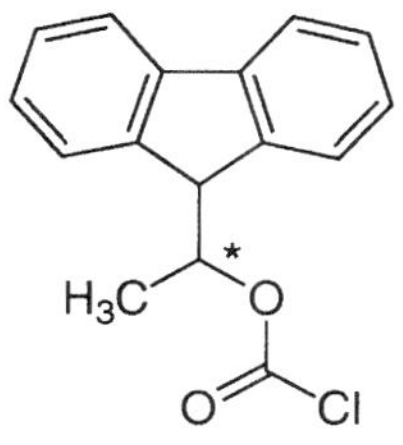 FLEC

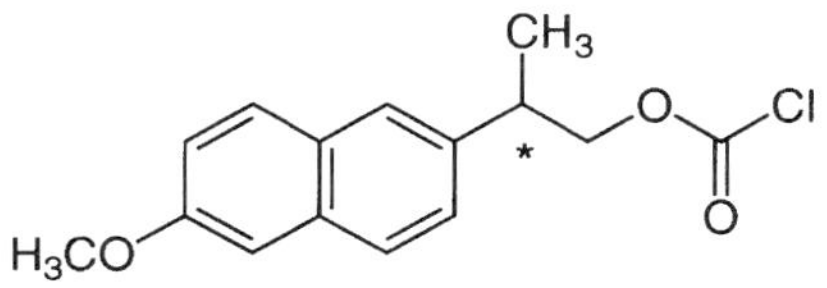

NAP-C 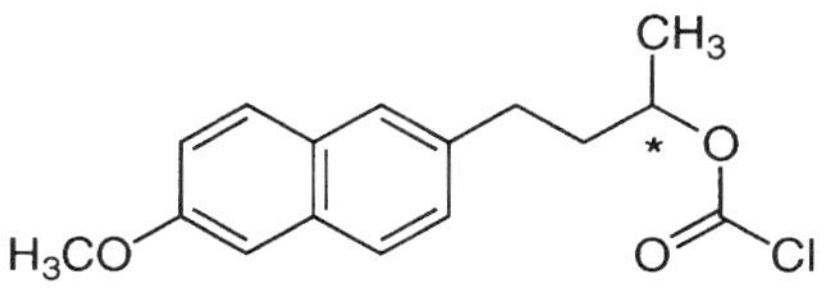 NAB-C

Figure 16.2. Selected chloroformate reagents.

from biological fluids has been described by different authors (Prakash et al., 1989b; Miller and Guertin, 1992; Fiset et al., 1993). Prakash et al. (1989a) and Turgeon et al. (1990) investigated the application of MCF to the bioanalysis of the antiarrhythmic agents encainide—a tertiary amine—and its metabolites and flecainide, respectively. The analysis of a second tertiary amine, promethazine, was reported by Witte et al. (1990).

FLEC was introduced by Einarsson et al. (1987) for the derivatization of amino acids, imino acids, primary and secondary amines, as well as β-blocking agents. They reported the reversed-phase separation of 17 racemic primary amino acids into their diastereomeric derivatives in a single run. Since FLEC is highly fluorescent and excess reagent may interfere with the analytes, sample clean-up is recommended. Either extraction with a lipophilic solvent (e.g., *n*-pentane) (Einarsson et al., 1987; Rosseel et al., 1991) or solid-phase extraction (Roux et al., 1991) is feasible. The addition of a hydrophilic compound (e.g., hydroxyproline) after derivatization of the analyte may also be used to remove interfering peaks (Einarsson et al., 1987).

Rosseel et al., (1991) reported on the analysis of atenolol enantiomers from rat plasma. Roux et al., (1991) and Lai et al., (1993) concentrated on the enantiomeric analysis of propranolol. Boursier-Neyret et al. (1993) employed FLEC derivatization for the analysis of a new drug with cognitive enhancing properties, (*R*,*S*)-1-methyl-8-[(morpholin-2-yl)methoxy]-1,2,3,4-tetrahydro-quinoline (S 12024) after extraction from plasma. Shimada et al. (1993) compared FLEC to a variety of other chiral derivatizing agents in the analysis of baclofen enantiomers from plasma. Interestingly, they employed cyclodextrin as a chiral mobile-phase additive. Witte et al. (1993) described the separation of the enantiomers N-6-(endo-2-norbornyl)-9-methyladenine. Frigerio et al. (1994) reported on the derivatization of the antidepressant reboxetine after extraction from plasma. Recently, derivatization with FLEC was applied to the analysis of amphetamine derivatives in forensic samples (Chen et al., 1994).

The following features of chiral chloroformate reagents have to be emphasized:

- Chloroformates are applicable to derivatization in aqueous reaction media.
- They allow selective coupling to amino compounds in the absence of phenols.
- They may facilitate on-line derivatizations.

16.4. ISOCYANATES

Chiral isocyanate reagents (Figure 16.3) have been applied frequently to the derivatization of amines, amino acids, and alcohols. Application to chromatographic separation was reported first by Pereira et al. (1970). They introduced *R*(+)-1-phenylethyl isocyanate [*R*(+)-PEIC] to the derivatization of chiral alcohols and subsequent GLC analysis. Pirkle and Hoekstra (1974) described the preparation of *R*(–)-1-(1-naphthyl)ethyl isocyanate [*R*(–)-NEIC] from the corresponding amine and its application to the preparative resolution of chiral alcohols. Pirkle and Hauske (1977) investigated the resolution of different carbamates on a silica-gel stationary phase.

Isocyanate reagents are moisture-sensitive and therefore kept and handled under anhydrous conditions, mostly in aprotic solvents. Application in aqueous solutions may be feasible as well (Dunlop and Neidle, 1987), although a large excess of reagent is required and precipitating by-products have to be removed prior to HPLC analysis. Compared to their sulfur analogs, isocyanates tend to be more reactive toward alcohol compounds (March, 1968). However, selectivity can be achieved by the choice of reaction conditions (e.g., solvent polarity, or selection of catalyst) that have to be monitored carefully. Figure 16.4. provides the normal-phase separation of alprenolol derivatives with *R*-1-(6-methoxy-2-naphthyl)ethyl isocyanate (NAPIC). The pair of unidentified peaks presumably originates from additional carbamate forma-

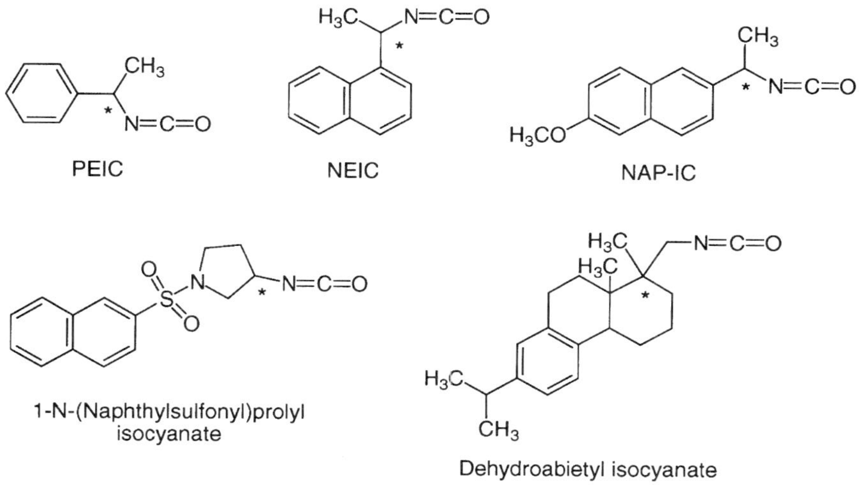

PEIC NEIC NAP-IC

1-N-(Naphthylsulfonyl)prolyl isocyanate

Dehydroabietyl isocyanate

Figure 16.3. Selected isocyanates.

tion at the hydroxyl moiety of the aminoalcohol substrate due to inappropiate reaction conditions.

Indeed, most of the isocyanate derivatizations have been performed with merely two reagents: 1-phenylethyl isocyanate (PEIC) and its 1-naphthyl analog (NEIC). Other reagents like dehydroabietyl isocyanate (Falck et al., 1984), 1-N-(naphthylsulfonyl)prolyl isocyanate and its phenyl analog (Zhou et al., 1994)—though exhibiting favorable chromatographic properties—have not yet been applied to the determination of drug enantiomers from biogenic fluids. Martin et al. (1989) introduced two isocyanate reagents of interest, NAPIC and FLOPIC, compounds derived from the 2-arylpropionic acids *S*(+)-naproxen and *S*(+)-flunoxaprofen, where as in the work of Herber et al. (1994), carprofen and N-methylcarprofen enantiomers were used as the starting material. FLOPIC especially shows excellent fluorescence properties due to the substituted benzoxazole moiety. The derivatives of different β-adrenoceptor antagonists, antiarrhythmic agents, and calcium channel blockers were well separated on normal- or reversed-phase columns. The reagents have been applied to pharmacokinetic studies, with propranolol providing a detection limit of 1 ng of enantiomer/mL for the derivatization with both reagents.

Misonidazole [(±)-1-(2-nitro-1-imidazole)-3-methoxy-2-propanol], a chiral alcohol, was tested as an adjuvant in the radiation therapy of tumor cells. In order to determine its pharmacokinetic behavior, diastereomeric carbamates were formed with *R*(+)-PEIC heating to 95°C for 5 h and analyzed on

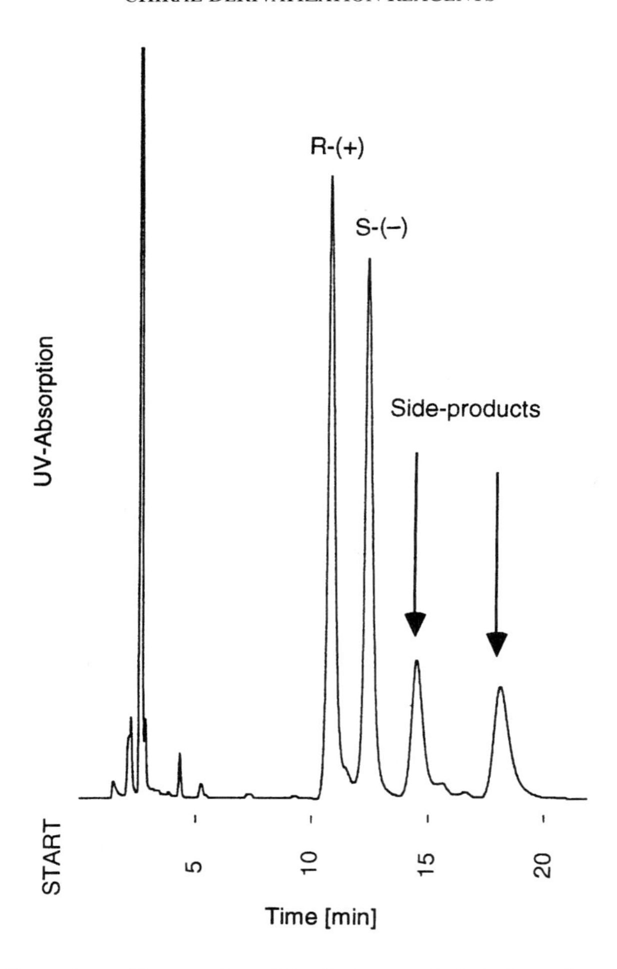

Figure 16.4. Chromatographic separation of the diastereomeric derivatives of racemic alprenolol after derivatization with the isocyanate derived from *S*-naproxen (NAP-IC). Arrows depict the di-NAP-IC derivatives. Derivatization conditions: solvent, toluene-triethylamine 100:1 by volume; reagent, 10 mM in toluene, 20-fold excess; 1 h at ambient temperature; chromatographic conditions: column, Zorbax Sil 5 μm, 250 × 4 mm mobile phase, *n*-hexane-2-propanol 100:4 by volume; flow, 1.0 mL/min; detection, UV 230 nm.

an ODS column to determine the plasma levels of the compound (Williams, 1984). Pirkle et al. (1984) reported on the derivatization of chiral lactams derived from primary amines. The compounds readily reacted with NEIC and PEIC to form diastereomeric ureides that were separated on a silica-gel stationary phase. An interesting example for the derivatization of tertiary amines. although not from biological matrices, was proposed by Maibaum (1988). Racemic phenothiazines were heated with vinyl chloroformate and thus deaminated, yielding secondary amines. Subsequent reaction with *R*(+)-PEIC led to diastereomeric urea, separated with normal-phase HPLC.

Miller et al. (1984) investigated the reversed-phase separation of amphetamine derivatives with 2,3,4,6-tetra-O-acetyl-β-D-glucopyranosyl isothiocyanate (GITC), 2,3.4-tri-O-acetyl-α-D-arabinopyranosyl isothiocyanate (AITC), 1-methoxy-1-(trifluoromethyl)phenylacetic acid (MTPA), and PEIC. Urea derivatives were in many cases less resolved than the thiourea and amide diastereomers. Mathys and Brenneisen (1992) determined the *S*(–)-cathinone and its metabolites *R*,*S* (–)-norephedrine and *R*,*R* (–)-norpseudoephedrine in urine after solid-phase extraction and reaction to the respective phenylethylamides. Under the employed conditions, PEIC derivatives were stable for at most 1 h. A reliable enantiospecific quantitation in urine was feasible for cathinone and norephedrine, not for the low amounts of norpseudoephedrine.

Chiral β-adrenoceptor antagonists have been the object of intensive investigation. Thompson et al. (1982) described the reversed-phase liquid chromatographic separation of diastereomeric urea derivatives of propranolol after reaction with PEIC. Wilson and Walle (1984) analyzed propranolol and its 4-hydroxy-metabolites in urine with a normal-phase assay. Langaniere et al. (1989) reported on an assay for the determination of propranolol in serum, applying fluorescence detection.

A reversed-phase assay for propranolol and 4-hydroxypropranolol in plasma was developed by Schaefer et al. (1990). The analytical method for propranolol was improved by Spahn-Langguth et al. (1991) with pronethalol added as an internal standard. Plasma samples were extracted and solvents evaporated to dryness. Residues were recovered in methanol before adding a solution of *R*(+)-PEIC. Thus, derivatization yields were significantly increased.

Dieterle and Faigle (1983) used *S*(–)-PEIC to determine oxprenolol in blood, plasma, and urine as a part of a multiple inverse isotope dilution assay. Gulaid et al. (1985) elaborated on an assay for acebutolol and its metabolite diacetolol in plasma. Pindolol was determined after derivatization with *S*(–)-PEIC (Hsyu and Giacomini, 1986) using reversed-phase HPLC and fluorescence detection at the substrate wavelengths. Pflugmann et al. (1987) reported on the analysis of metoprolol in plasma and urine. Ethanolamine was added to remove excess reagent. A correlation of data from HPLC analysis to a metoprolol radioreceptor assay (RRA) was presented by Spahn et al. (1989) and indicated that in plasma samples no detectable amounts of active metabolites are present. Sotalol—both enantiomers show additional class III antiarrhythmic activity—were recovered from plasma, derivatized with *S*(–)-PEIC and analyzed by HPLC. Atenolol served as an internal standard. Carvedilol enantiomers were monitored in a rat model for liver cirrhosis by Stahl et al.

(1993). Carvedilol was extracted from plasma, urine, and bile, derivatized with *R*(+)-PEIC, and analyzed with fluorescence detection.

Jira et al. (1991) compared the reversed-phase separation of the diastereomeric derivatives of selected β-adrenoceptor antagonists and α-adrenoceptor agonists obtained from reaction with PEIC, NEIC, and (1*S*)(+)-camphor-10-sulfonyl chloride. Olsen et al. (1993) presented a survey of the normal-phase separations of 18 β-adrenoceptor antagonists after derivatization with PEIC, GITC, and camphanoyl chloride.

Dunlop and Neidle (1987) proposed the derivatization of amino acid enantiomers with NEIC. The derivatives were resolved by reversed-phase HPLC with fluorescence detection (λ_{ex} = 230 nm and λ_{em} = 320 nm cutoff). The authors applied aqueous borate buffer, pH 9, to dilute the samples before NEIC in acetone was added. Gietl et al. (1988) employed NEIC to the determination of the calcium antagonist prenylamine—a secondary amine with unfavorable chromophoric properties—from plasma and urine. Matuszewski et al. (1990) offered an assay for a chiral sulfonamide—thieno [2,3b]thiopyran-2-sulfonamide—with a secondary amine moiety, which proved to be applicable to routine analysis in pharmacokinetic studies. The chromatographic separation of diastereomers usually diminishes, with an increasing distance between the chiral centers (Rose et al., 1996). However, baseline separation was obtained from the enantiomers of 9-amino-20(*S*)-camptothecin after derivatization with NEIC (Kagel et al., 1993), although the chiral centers are separated by 12 bonds.

An enantioselective assay for the antiarrhythmic agent mexiletine and its two major metabolites—all primary amines—in microbial fermentation media is described (Freitag et al., 1993). Substrates were allowed to react for 10 sec, the solvents were evaporated, and the residue was reconstituted in a solution containing *n*-butylamine. Thus, reaction with the hydroxyl groups of the metabolites was prevented. The antiarrhythmic agents tocainide (Carr et al., 1991b) and propafenone (Mehvar, 1990) were determined in plasma using naphthylethyl isocyanate. The prevention of late eluting peaks by a solid-phase clean-up of the reagent solution is described (Mehvar, 1990).

The serotonin reuptake inhibitor and antidepressant fluoxetine, (±)-N-methyl-3-phenyl-3-(4-trifluoromethyl-1-phenoxy)propylamine, and its desmethyl metabolite were determined in plasma or tissue after derivatization with *R*(–)-NEIC (Peyton et al., 1991; Potts and Parli, 1992). Diastereomers were separated on normal-phase HPLC columns and monitored with fluorescence detection. Detection wavelengths were either λ_{ex} = 285 nm, λ_{em} = 313 nm (Peyton et al., 1991) with a limit of quantitation (LOQ) of 1 ng/mL or λ_{ex} = 218 nm, λ_{em} = 333 nm (Potts and Parli, 1992) (LOQ = 5 ng/mL).

Several authors reported on the enantioselective analysis of β-adrenoceptor antagonists with NEIC as a chiral derivatizing agent (Darmon and Thenot,

1986; Piquette-Miller and Foster, 1990; Piquette-Miller et al., 1990; Carr et al., 1991a, 1992; Jira et al., 1991; Lave et al., 1991; Stagni et al., 1991; Bhatti and Foster, 1992; Laethem et al., 1993; Piquette-Miller and Jamali, 1993). Due to the fluorescent properties of the reagent, detection limits were lower than for comparable derivatives with the nonfluorescent phenylethyl isocyanate.

Elevated temperatures are necessary for the carbamate formation from alcohols and isocyanates. Methocarbamol, a muscle relaxant, is a secondary alcohol. Since other CDAs (acyl chlorides, chloroformates, anhydrides) failed to react with the hydroxyl group, the compound was coupled with *S*(+)-NEIC (Alessi-Severini et al., 1992). Diastereomeric derivatives were well resolved on a silica-gel stationary phase.

Isocyanate reagents are especially helpful when highly reactive CDAs are required:

- Slow-reacting or sterically hindered substrates (e.g., aliphatic alcohols) can be analyzed.
- Short derivatization times may help to save time with respect to large sample numbers.
- Isocyanates are rather unspecific reagents.
- Therefore, side effects due to increased reactivity may occur and have to be carefully avoided.

16.5. ISOTHIOCYANATES

Nimura et al. (1980) first applied 2,3,4,6-tetra-O-acetyl-β-D-glucopyranosyl isothiocyanate (GITC) to the HPLC analysis of amino acids, initiating the breakthrough of isothiocyanate CDAs in liquid chromatography. GITC, and other isothiocyanate reagents, readily reacts with primary and secondary amino compounds, yielding diastereomeric thiourea derivatives that are usually well separated on reversed-phase HPLC systems. Figure 16.5 provides selected chiral isothiocyanates that have been applied successfully in high-performance liquid chromatography.

Isothiocyanates tend to be less reactive than their O-analogs and, therefore, do not couple to alcohols under the applied reaction conditions. Consequently, they are more selective toward amino groups, which is highly appreciated in the derivatization of aminoalcohols (i.e., epinephrine, ephedrine-derivatives, β-adrenoceptor antagonists) (Nimura et al., 1981; Sedman and Gal, 1983; Gal and Sedman, 1984; Allgire et al., 1985; Noggle et al., 1986; Zhong-Yuan and Ru-Zheng, 1989; Stockley et al., 1991).

Normally, isothiocyanates are applied in polar reaction media (e.g., acetonitrile, dimethylformamide, dimethylsulfoxide) or even aqueous solvents.

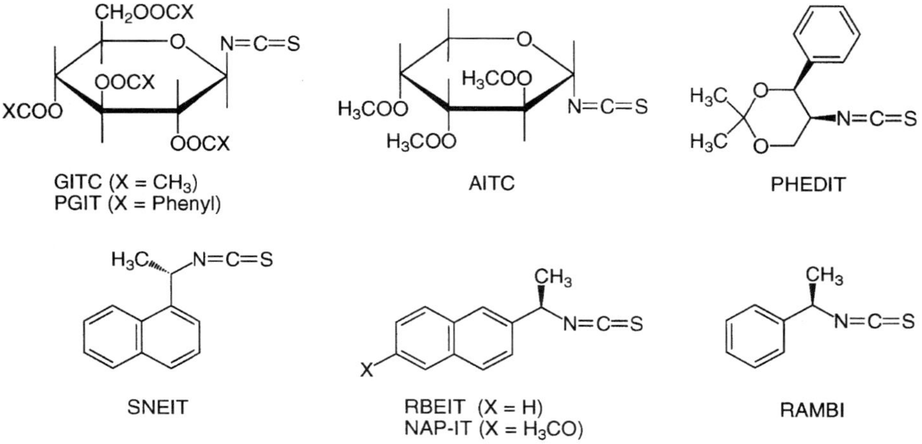
CH2OOCX
N=C=S
OOCX
XCOO
OOCX
GITC (X = CH3)
PGIT (X = Phenyl)
H3COO
H3COO
H3COO
N=C=S
AITC
H3C
H3C
N=C=S
PHEDIT
H3C
N=C=S
SNEIT
CH3
N=C=S
X
RBEIT (X = H)
NAP-IT (X = H3CO)
CH3
N=C=S
RAMBI

Figure 16.5. Selected isothiocyanates.

The hydrolytic stability of these reagents facilitates the derivatization of hydrophilic compounds (e.g., amino acids) and the analysis of substrates from biological matrices, where traces of water may interfere with isocyanate reagents. A basic catalyst, such as triethylamine, may reduce derivatization time but often is not essential.

In the last 15 years, chiral isothiocyanates have found widespread application to the analysis of physiological as well as xenobiotic compounds. Several authors (Nimura et al., 1980; Kinoshita et al., 1981; Gal and Sedman, 1984; Nimura et al., 1984; Lobell and Schneider, 1993) reported on the analysis of free or esterified amino acids with different isothiocyanate reagents. Recently, Miyazawa et al. (1993) focused on the separation of β-amino acids and cyclic imino acids. The derivatization of β-adrenoceptor antagonists or their phase I metabolites after extraction from biological samples (e.g., plasma, urine, tissue homogenates) has been described by various authors (Christ and Walle, 1985; Schuster et al., 1988; Webb et al., 1988; Eisenberg et al., 1989; Hasegawa et al., 1989; Fujimaki et al., 1990; Zhou and Sun, 1990; Rose and Randinitis, 1991; Miller, 1992). Walle and co-workers introduced GITC to the direct analysis of sulfate conjugates of 4-hydroxypropranolol, prenalterol, and terbutaline (Walle et al., 1985; Walle and Walle, 1989). Antiarrhythmic agents, such as mexiletine, flecainide, tocainide, and propafenone, have been investigated by Grech-Belanger et al. (1985) and Gal et al. (1990). Synthetic by-products of chloramphenicol and thiamphenicol, both broad-spectrum antibiotics, were quantified after derivatization with different chiral isothiocyanates (Gal and Meyer-Lenhert, 1988).

However, epoxides (oxiranes) also can be substrates to isothiocyanate derivatization after conversion to their corresponding aminoalcohols (Gal, 1985; Geerlof et al., 1993; Lobell and Schneider, 1993). Ito et al. (1992) proved the applicability of GITC to the derivatization of various thiol compounds, yielding dithiocarbamates under mild conditions.

Presently, most investigations with isothiocyanate reagents are based on the application of 2,3,4,6-tetra-O-acetyl-β-D-glucopyranosyl isothiocyanate (GITC) (Nimura et al., 1980). The reagent is commercially available or can be prepared easily from acetobromo-α-D-glucose. However, Ahnoff et al. (1992) described the presence of reactive impurities even in specimens from commercial suppliers. Side effects were diminished by adding a limited amount of another amine to the reagent solution prior to derivatization.

The derivatives from GITC and the respective substrate are usually analyzed by reversed-phase high-performance liquid chromatography with UV detection. Yet, the separation of the diastereomeric thioureas from β-adrenoceptor antagonists after derivatization with GITC also may be accomplished by normal-phase HPLC (Olsen et al., 1993). A remarkable approach has been made by Nishi et al. (1990), who separated the thiourea derivatives of amino acids, applying micellar electrokinetic chromatography (MEKC). Recently, Lurie (1992) employed MEKC for the separation of amphetamine and its precursors after derivatization with GITC. Figure 16.6 displays the reversed-phase separation of carvedilol enantiomers and three phase I metabolites after extraction from human plasma, derivatization with GITC, and fluorescence detection (A. Kurz, H. Spahn-Langguth, and E. Mutschler, unpublished data, 1994)—a vivid example for the extraordinary separability of the thioureas derived from 2,3,4,6-tetra-O-acetyl-β-D-glucopyranosyl isothiocyanate.

GITC itself does not exhibit any outstanding chromophoric properties. Nevertheless, the UV absorption of its thiourea derivatives (λ_{max} = 250 nm) is increased about one order of magnitude compared to the reagent (Nimura et al., 1980; Kinoshita et al., 1981). Therefore, GITC may be regarded as suitable for both: either reagent-oriented or substrate-oriented detection. Examples of fluorescence detection at substrate wavelengths have been given by various authors (Ahnoff et al., 1992; He and Stewart, 1992; Ito et al., 1992; Miller, 1992; Kitani et al., 1993). Excess reagent does not interfere with the fluorescence peaks of the investigated products. Thus, a highly selective detection mode is at hand—a favorable feature in the analysis of biological samples.

2,2,4-Tri-O-acetyl-α-D-arabinopyranosyl isothiocyanate (AITC) was introduced by Kinoshita and co-workers (1981), but has found less attention than its glucose analog. Still, good resolutions of diastereomeric amino acid

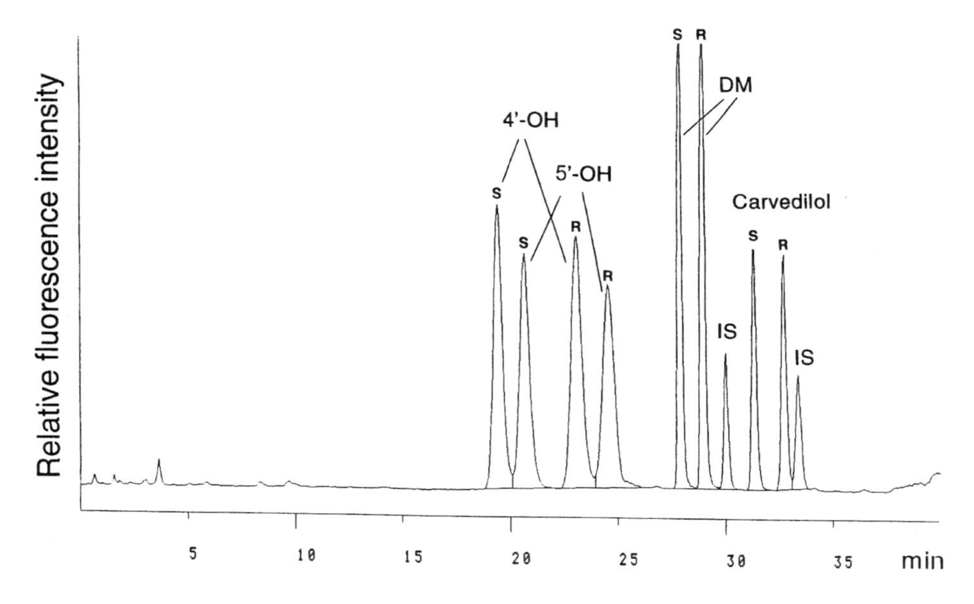

Figure 16.6. Reversed-phase separation of three of its carvedilol and phase I metabolites (4′-hydroxy- and 5′-hydroxy carvedilol as well as desmethyl carvedilol; internal standard = propranolol) after derivatization with GITC in a plasma/acetonitrile mixture, solid-phase extraction (ODS), and reversed-phase chromatography. (Chromatographic conditions: column, Sherisorb ODS II 3 μm, 125 mm × 4 mm mobile-phase gradient step I (0–36 min), acetontrile/0.1% aqueous triethylamine solution (adjusted to pH 2.5 with phosphoric acid) 37.63 (by volume); mobile = phase gradient step 2 (36–50 min), acetonitrile/0.1% triethylamine solution, pH 2.5; flow, 1 mL/min; detection, fluorescence at 285–355 nm (A. Kurz, H. Spahn-Langguth, and E. Mutschler, unpublished results 1994).

derivatives have been reported. Their elution order was reversed compared to GITC derivatives. AITC has been used for the analysis of β-blocking agents, amphetamines and epinephrine (Nimura et al., 1981; Sedman and Gal, 1983; Miller et al., 1984).

Lobell and Schneider (1993) investigated the influence of bulkiness on the separability of various amino compounds after derivatization with 2,3,4,6-tetra-O-benzoyl-β-D-glucopyranosyl isothiocyanate (BGIT) and 2,3,4,6-tetra-O-pivaloyl-β-D-glucopyranosyl isothiocyanate (PGIT) compared to GITC. The best resolutions of different racemic β-aminoalcohols were achieved with BGIT. The pivaloyl residues in the case of PGIT, although even bulkier than the benzoyl groups in BGIT, could not lead to enhanced separation. Besides, the application of 2,3,4,6-tetra-O-benzoyl-β-D-glucopyranosyl isothiocyanate (BGIT) instead of GITC increases maximum UV absorption of the diastereomeric thiourea derivatives (λ_{max} = 230 nm) by a factor of 5 and, thus, may help to lower detection limits.

R(+)-phenylethyl isothiocyanate [*R*(+)-α-methylbenzyl isothiocyanate, RAMBI] was used as an alternative to GITC in the derivatization of different substrates (Gal and Sedman, 1984; Gal and Meyer-Lehnert, 1988). *S*-1-(1-naphthyl)ethyl isothiocyanate (SNEIT) (Pianezzola et al., 1989) and *R*-1-(2-naphthyl)ethyl isothiocyanate (RBEIT) (Gal et al., 1990) are related CDAs with a higher UV absorption than RAMBI and GITC. Gal et al. employed all three reagents in the analysis of different antiarrhythmic agents and compared their resolution ability to GITC. In this respect, the glucose-derived CDA turned out to be superior to all other reagents. A reagent comparable to RBEIT has been prepared from *S*(+)-naproxen (Büschges et al., 1996). 1-(6-Methoxy-2-naphthyl)ethylisothiocyanate (NAP-IT) exhibits considerably high UV absorbance as well as moderate fluorescence.

Since all available isothiocyanate reagents failed to resolve the four stereoisomers of labetalol, Desai and Gal (1992) proposed a CDA with different structure. 4*S*-*cis*-2,2-Dimethyl-5-isothiocyanato-4-phenyl-1,3-dioxane (PHEDIT) was able to provide nearly baseline resolution for both diastereomeric pairs of the α- and β-adrenoceptor antagonistic substance within 30 min.

Coming to a conclusion, we must acknowledge that chiral isothiocyanate reagents have become an important element in the bioanalysis of primary and secondary amino compounds. The present isothiocyanate CDAs exhibit the following advantages:

- They are more selective toward amino compounds compared to isocyanates and different acylating reagents.
- They are applicable in the presence of water.
- The diastereomeric thioureas of many compounds are excellently resolved on reversed-phase HPLC columns.

However, the addition of some highly fluorogenic reagents—not available at present—may enlarge the applicability of chiral isothiocyanates to substrates with unfavorable chromophoric properties and would be highly appreciated.

16.6. ORTHO-PHTHALDIALDEHYDE

The chiral derivatization procedure with o-phthaldialdehyde (OPA) and a chiral SH-group donor appears to be a most elegant method, which is specific for primary amines. The significant advantage of the procedure—as opposed to

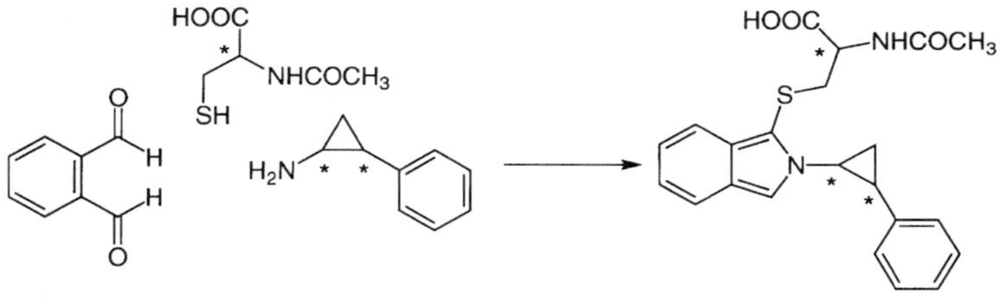

Figure 16.7. Derivatization reaction of tranylcypromine enantiomers with o-phthaldialdehyde (OPA) and N-acetylcysteine as a chiral SH-donor, yielding fluorescent isoindoles.

the application of reagent-oriented derivatization and detection—is the fact that the procedure is fluorogenic, that is, the resulting isoindole derivatives are intensely fluorescent (Figure 16.7), where as the substrate as well as the reagents are not. Alternatives to the fluorogenic o-phthaldialdehyde were proposed by Beale and co-workers (1990), who demonstrated the applicability of 3-(2-furoyl)quinoline-2-carbaldehyde for the high-sensitivity analysis of amino acids by laser-induced fluorescence.

Originally, the OPA method was applied to the achiral pre- and postcolumn derivatization of amino acids, their derivatives, as well as peptides (Magat et al., 1982; Meyer et al., 1991; Fermo et al., 1992; Min et al., 1992; Langguth et al., 1994), but also for xenobiotics, such as oxiracetam (Boppana et al., 1992) and aminoglycosides (Caturla et al., 1992; Sar et al., 1992). For these methods, mercaptoethanol and mercaptopropionic acid were used as thiols, where as more recent developments include chiral SH-group donors for the analysis of amino acids and peptides as well as catecholamines (e.g., Saito et al., 1992; Desai and Gal, 1993; Skidmore, 1993, and references therein) in addition to the application of the OPA method to enantiospecific drug analysis (Desai and Gal, 1993), also with respect to bioanalytical methods, for example, for baclofen, norephedrine, and tranylcypromine (Wuis et al., 1987; Hahn et al., 1992; Weber-Grandke et al., 1993). In general, the procedure includes derivatization in pH 9–11 buffer media at ambient temperature. The reaction is rapidly completed, as for example, demonstrated for tranylcypromine (TCP) enantiomers (Spahn-Langguth et al., 1992). The fluorescent derivatives are not necessarily stable (Duchateau et al., 1992), which usually requires rapid analysis of the reaction mixture. Separation is performed via high-performance liquid chromatography on reversed-phase stationary phases (mainly ODS). For automatization in routine analysis, on-line derivatization is necessary. An interesting application of the OPA procedure was introduced into stereospecific analysis by Duchateau and co-workers in 1993, who converted chiral epoxides into the corresponding thioglycol derivatives by reaction with sodium sulfide. Subsequently, the products were reacted with OPA and an optically pure amino acid and the mixture was analyzed on a reversed-phase system.

Several investigators described the chromatographic behavior and resolution of various amino acid and amine drug derivatives when different chiral thiols were employed (Buck and Krummen, 1984, 1987; Euerby et al., 1989; Duchateau et al., 1992; Desai and Gal, 1993). Whereas in previous studies, N-acetyl-L-cysteine (NAC), (Lam, 1986; Nimura et al., 1987; Maurs et al., 1988), thiosugars, N-acetyl-D-penicillamine (NAP), and N-*tert*-butyloxycarbonyl-L-cysteine (BocC) as well as captopril had been used as chiral thiols (as reviewed by Skidmore, 1993; Görog and Gazdag, 1994, and in the introductory statement of Duchateau et al., 1992), the 1989 work of Euerby et al. describes the synthesis and comparative testing of alternative N-substituted L-cysteine derivatives in addition to NAC and BocC. In their studies, the resulting OPA amino thiol adducts showed enhanced resolution and better fluorescent properties. Good resolvability and high sensitivity were also found by Duchateau et al. (1992) for the derivatives of primary amines, when D-3-mercapto-2-methylpropionic acid was used as chiral thiol. With respect to the applicability of the stereospecific OPA method to biological material, very few articles have been published, although the number is increasing. Hashimoto and co-workers (1992) discovered amino acids in rat brain and serum following their reaction with BOC-L-cysteine and OPA and detected large amounts of free D-serine in the brain. In the field of drug analysis Desai and Gal (1993) reported on the derivatization of a number of xenobiotics [amphetamine, p-chloroamphetamine, hydroxyamphetamine, tranylcypromine, 3-amino-1-phenylbutane, 3-amino-1-(4-hydroxyphenyl)butane, tocainide, mexiletine, and rimantidine] with different commercially available enantiomerically pure thiols (NAC, NAP, 1-thio-β-D-glucose, and 2,3,4,6-tetra-O-acetyl-β-D-glucopyranoside). They found that the resolution of most of the tested amines could be optimized by varying the chiral thiol in the derivatization step. The systemic screening of the chromatographic behavior of derivatives as performed by these authors may be regarded as an appropriate basis for the future development of bioanalytical methods. The application of the OPA/NAC method to the determination of tranylcypromine (Figure 16.8) as well as norephedrine enantiomers (Figure 16.9) in plasma and urine was described by Hahn et al. (1992) and Spahn-Langguth and co-workers (1992). Both bioanalytical methods were fully validated and applied in extensive pharmacokinetic studies, in which a significant first-pass effect was detected for perorally dosed tranylcypromine, whereas no relevant difference between enantiomers was found in the pharmacokinetic characteristics of norephedrine.

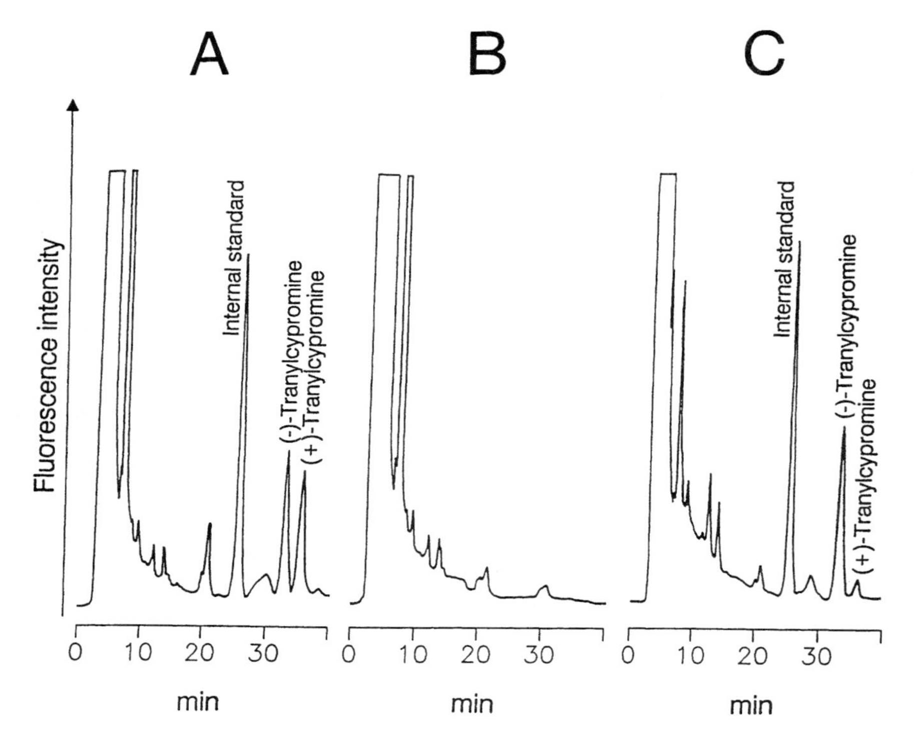

Figure 16.8. HPLC separation of diastereomeric isoindoles of tranylcypromine enantiomers and of *S*(+)-amhetamine after reaction from the reaction with OPA/NAC, following extraction from plasma at alkaline pH using diethylether (containing 1.5% ethanol). Chromatographic conditions: column, Zorbax ODS; mobile phase, phosphate buffer, pH 6.5 (0.05 M)/methanol/tetrahydrofuran, 50:60: (by volume); flow, 1.5 mL/min; detection, fluorescence at 344–442 nm (Spahn-Langguth et al., 1992). (*a*) Blank plasma containing 25 μg/mL of each tranylcypromine enantiomer; (*b*) blank plasma; (*c*) plasma sample 4 h after the p.o. dosage of 20 mg of *rac*-tranylcypromine sulfate.

Summing up, we note that the following features characterize derivatizations with OPA and a chiral thiol:

- Derivatization with OPA and a chiral thiol is specific for primary amines.
- OPA derivatives are usually largely fluorescent, although the reagents are not, thus allowing selective detection of the derivatives.
- Derivatization kinetics allow on-line derivatization, but derivatives are often unstable.

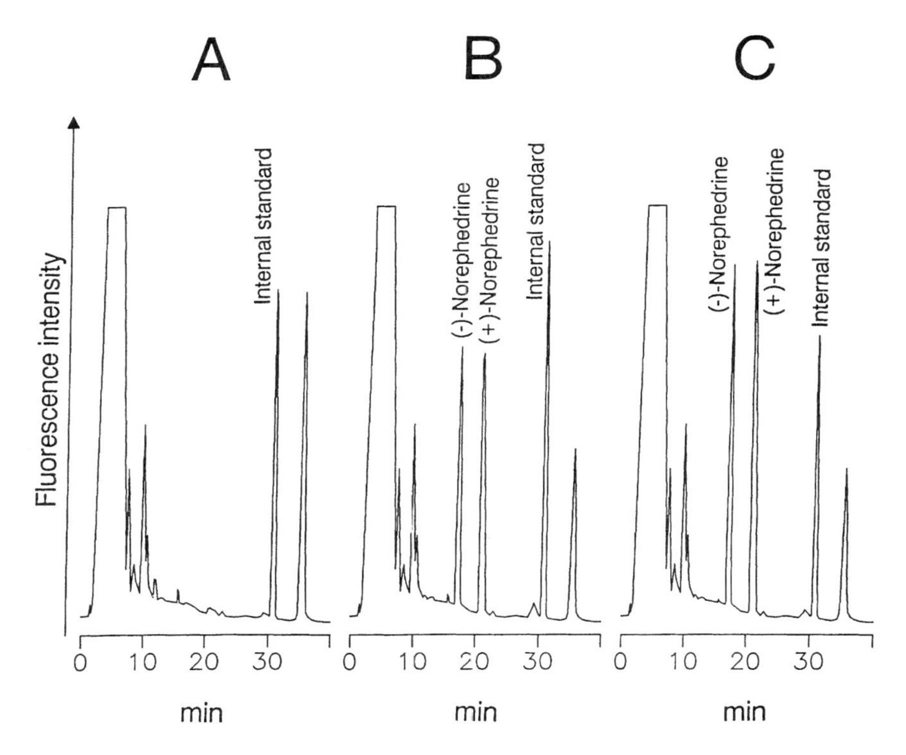

Figure 16.9. HPLC separation of diastereomeric isoindoles of norephedrine with levonorpseudoephedrine as the internal standard, following extraction from plasma at alkaline pH using diethylether (cont. 1.5% ethanol) and derivatization with OPA/NAC. Chromatographic conditions: column, Zorbax C-8; mobile phase, phopshate buffer, ph 6.5/methanol/acetonitrile, 100:40:30 (by volume); flow, 1 mL/min; detection, fluorescence at 344/442 nm (Hahn, G. et al., 1992). (*a*) Blank plasma; (*b*) Blank plasma containing 50 ng/mL of each norephedrine enantiomer; (*c*) plasma sample 3 h after the dose of 50 mg of *rac*-norephedrine hydrochloride

16.7. AMINES

Chiral amine reagents have been used primarily for the derivatization of carboxylic acid substrates. Since the chromatographic behavior of diastereomeric amides is often superior to that of respective esters and since amides are readily formed under mild reaction conditions, amine reagents are the preferred CDAs in the quantitation of carboxylic acid enantiomers. Additionally, chiral amines may be used for the derivatization of aldehydes and ketones, yielding Schiff base adducts (Sampath and Balaram, 1986a,b). However, amine reagents have not found broad application in the liquid chromatographic bioanalysis of enantiomeric carbonyl compounds.

Prior to derivatization with amine CDAs, activation of the carboxylic acid substrates is obligatory. Several methods have been proposed:

- Reaction with acid chlorides [e.g., thionyl chloride (McKay et al., 1979), oxalyl chloride (Valentine et al., 1976)]
- Formation of mixed anhydrides with aliphatic chloroformates (e.g., ethyl chloroformate), adding a basic amine as a proton scavenger (Björkman, 1985)
- Reaction with carbodiimides, in many cases adding 1-hydroxybenzotriazole (1-HOBT) (Goto et al., 1978, 1982a)
- Formation of imidazolide intermediates with 1,1′-carbonyl diimidazole (Kemmerer et al., 1979; Maitre et al., 1984)
- Coupling with 2,2′-dipyridyl disulfide/triphenylphosphine (Toyo'oka et al., 1992a).

Although activation with thionyl chloride is a very simple way of preparing reactive intermediates, harsh reaction conditions may lead to unwanted aromatic chlorination, racemization, or even decomposition of the substrate molecule. Acid chlorides are highly reactive and, therefore, subject to various side effects limiting their application to stable carboxylic acid molecules. Though carbodiimide activation and the preparation of imidazolides are relatively mild procedures taking place at ambient or subambient temperatures, they are mostly applied in nonwater miscible solvent systems that may complicate derivatization procedures. The 2,2′-dipyridyl disulfide/triphenylphosphine method (Toyo'oka et al., 1992a) as well as the formation of mixed anhydrides with aliphatic chloroformates (Björkman, 1985) may be employed in acetonitrile: a highly favorable solvent for the reversed-phase HPLC analysis of chiral compounds in biological matrices. The derivatization scheme developed by Björkman (1985) consisting of ethyl chloroformate activation and subsequent derivatization with L-leucinamide allows complete reaction within 3 min.

However, Wright and Jamali (1993) reported that a stereochemical conversion may be occurring during the derivatization of 2-arylpropionic acids with ethyl chloroformate and *R*(+)-1-phenylethylamine or L-leucinamide, especially at low concentrations of the chloroformate reagent. Consequently, the quantitation of small amounts of one enantiomer in the presence of large amounts of the optical antipode is limited.

Mild reaction conditions combined with simple "single-pot" chemistry, rapid and complete conversion to the activated intermediate, are, of course,

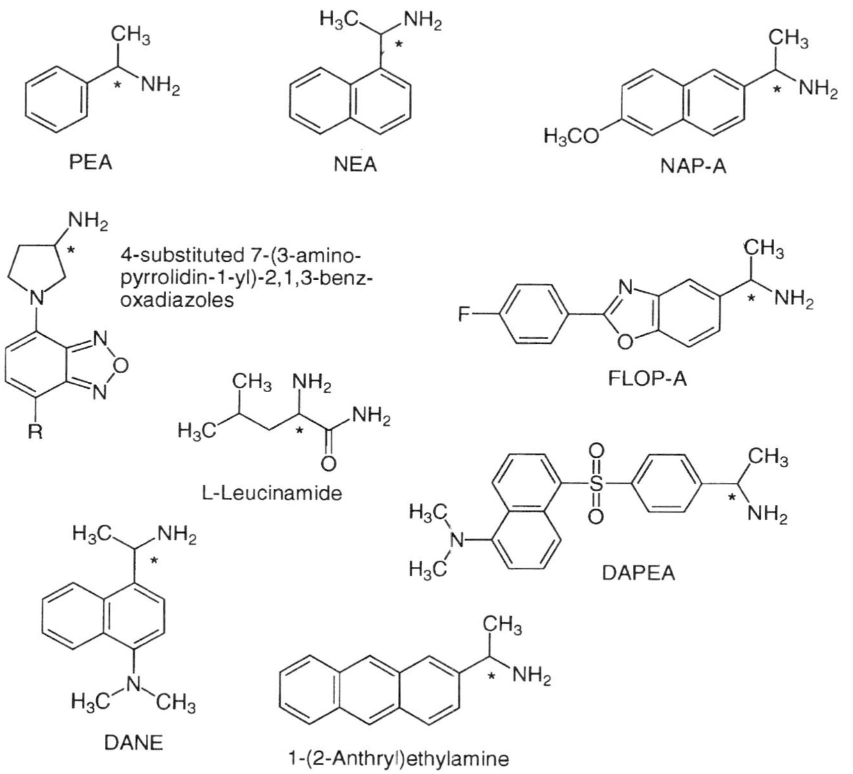

Figure 16.10. Selected amines used as CDAs in high-performance liquid chromatography. *R*,1-piperazinyl (ABD-PZ), 5-aminopentylamino (ABD-AP), or 2-aminoethylamino (ABD-AE)

favorable. Nevertheless, activation procedures have to be optimized for every single substrate considering the nature of the carboxylic acid as well as the requirements of the chromatographic system.

The HPLC separation of the diastereomeric amide derivatives of chiral carboxylic acids has been systematically investigated by Helmchen et al. (1977). They employed 1-phenylethylamine (PEA) and 1-(1-naphthyl)ethylamine (NEA) as chiral derivatizing agents. Figure 16.10 provides the molecular formula of selected amines applied to the analysis of carboxylic acid enantiomers. Included are some frequently used CDAs as well as compounds with outstanding chromophoric properties.

Up to now, numerous investigations have been performed with 1-phenylethylamine (α-methylbenzylamine, PEA) (Helmchen et al., 1977).

Consequently, only selected bioanalytical applications may be discussed in this context. *S*(–)-1-phenylethylamine was used by Maitre et al. (1984) for the derivatization of 2-arylpropionic acids followed by normal-phase HPLC separation of the derivatives. The method was applicable to metabolism studies in rat liver preparations. Pedrazzini et al. (1987) describe the reversed-phase separation of flunoxaprofen enantiomers after extraction from plasma and urine. The 2-arylpropionic acid was activated with 1,1′-carbonyldiimidazole prior to derivatization with *S*(–)-PEA. Carprofen enantiomers were measured in plasma and urine by, for example, Iwakawa et al. (1989) They derivatized with *S*(–)-PEA prior to reversed-phase HPLC with fluorescence detection at the substrate wavelengths. Hayball et al. (1991) reported on the enantiospecific analysis of ketoprofen in plasma. The phenylethylamides of drug and the internal standard, *S*(+)-naproxen, were baseline-resolved within 8 min, improving the method of Sallustio et al. (1986). A related method is described for the 2-arylpropionic acid furprofen (Carlucci et al., 1992). Ibuprofen enantiomers were analyzed by Wright et al. (1992) using *R*(+)-PEA and activation with ethyl chloroformate. A fully validated assay is reported for ketorolac (Hayball et al., 1993), including reaction with thionyl chloride/*S*(–)-PEA and reversed-phase HPLC analysis.

S(–)-1-(1-Naphthyl)ethylamine (NEA) presents significantly higher UV absorbance than its phenyl analog. The reagent was used by Hutt et al. (1986) in the derivatization of different 2-arylpropionic acids and in the analysis of 2-phenylpropionic acid in urine. All derivatives were baseline-resolved on a silica-gel radial compression column with *n*-hexane/ethyl acetate as a mobile phase. The reagent was used for enantiospecific pharmacokinetic studies after the repeated oral administration of ibuprofen (Avgerinos and Hutt, 1987) and for monitoring the metabolites of 2-[4-(2-oxocyclohexylidene methyl)phenyl]propionic acid in human and rat plasma (Takasaki et al., 1993).

L-Leucinamide was introduced by Björkman (1985) for the enantioselective analysis of indoprofen in plasma. The carboxylic acid substrate was activated with ethyl chloroformate beforehand. L-Leucinamide derivatives of 2-arylpropionic acids are usually well separated on reversed-phase columns. The absence of a strong chromophore within the reagent is ideal for substrate-oriented detection. Several authors (Björkman, 1987; Foster and Jamali, 1987; Spahn, 1987; Jamali et al., 1989; Palylyk and Jamali, 1991; Liang et al., 1992; Strolin Benedetti et al., 1992) have employed L-leucinamide to the separation of 2-arylpropionic acids. Lehr and Damm (1988) described the quantitation of ofloxacin in plasma and urine activating the gyrase inhibitor substrate with diphenylphosphinyl chloride prior to derivatization with L-leucinamide. Compared to direct separation on a BSA column, the diastereomeric derivatives of ofloxacin were much better resolved. A separation factor α of 1.5 for the diastereomers seems quite remarkable, since both chiral centers are six bonds apart!

1-(4-Dimethylamino-1-naphthyl)ethylamine (DANE) is a fluorescent CDA introduced by Goto et al. (1980). It has been used for the enantioselective determination of naproxen in human serum (Goto et al., 1982a)—with an detection limit of 100 pg on-column—and loxoprofen as well as its metabolites (Nagashima et al., 1984, 1985) in rat plasma and human urine.

Many investigations have been centered on 2-arylpropionic acids—chiral nonsteroidal antiinflammatory agents—their pharmacokinetics, and the metabolic chiral inversion, which part of them undergo *in vivo* (Hutt and Caldwell, 1983). The fluorescent properties found with several of these compounds (Spahn, 1989) made them an interesting starting material for the preparation of CDAs for reagent-oriented detection.

Reagents derived from the 2-arylpropionic acids $S(+)$-naproxen and $S(+)$-flunoxaprofen were presented by Spahn and Langguth (1990). Especially FLOP-A, the amine derived from flunoxaprofen, is a highly fluorescent benzoxazole derivative. FLOP-A has been applied to pharmacokinetic investigations of beclobric acid (Mayer et al., 1991) (= the active metabolite of the chiral lipid-regulating agent beclobrate) and α-phenylcyclopentylacetic acid (a metabolite of the antimuscarinic drug ciclotropium) (Liebmann et al., 1992). Even the low amounts of beclobric acid covalently bound to plasma proteins could be determined in the course of these studies (Mayer et al., 1993). CDAs derived from cicloprofen, flurbiprofen, indoprofen, and ketoprofen were evaluated in the derivatization of different 2-arylpropionic acids, showing excellent resolution of the diastereomers on normal- and reversed-phase chromatographic systems (Mayer, 1992; Büschges, 1996).

The enantiomers of 1-(1-anthryl)-and 1-(2-anthryl)ethylamine were synthesized by Goto et al. (1986). A detection limit equivalent to 100 fmol naproxen was achieved by fluorescence detection after normal-phase separation of the derivatives. Recently, Toyo'oka et al. (1992a) prepared some highly fluorescent 4-substituted-7-(3-aminopyrrolidin-1-yl)-2,1,3-benzoxadiazoles. They were tested with laser-induced fluorescence detection (Toyo'oka et al., 1992b) and peroxyoxalate chemiluminescence detection (Toyo'oka et al., 1992c) pushing down the detection limits of naproxen enantiomers even to attomole levels.

Recently, Iwaki et al. (1994) synthesized a CDA carrying the dansyl (N-dimethylaminonaphthalene-5-sulfonyl) moiety as a fluorophore: The enantiomers of 1-(4-dansylphenyl)ethylamine (DAPEA) were tested in the derivatization of 2-arylpropionic acids, with a limit of detection of 170 fmol for the derivative with 2-phenylpropionic acid.

A variety of well-detectable amines is available for the derivatization of carboxylic acid enantiomers. These groups of CDAs present the following features:

- Chiral amines are the preferred reagents for the derivatization of 2-arylpropionic and other carboxylic acids prior to HPLC analysis.
- Carboxylic acid substrates have to be activated prior to derivatization ("reversed labeling") with a variety of activation procedures at hand.
- The diastereomers are usually well separated by either normal- or reversed-phase HPLC.

16.8. MISCELLANEOUS CDAs

16.8.1. Solid-Phase Reagents

The use of covalently bonded chiral derivatizing agents for off-line and on-line derivatization in high-performance liquid chromatography is a relatively new approach to the analysis of enantiomeric drugs. (Chou et al., 1989) designed fluorenylmethyl-L-proline (FMOC-L-proline) and fluorenylmethyl-L-alanine (FMOC-L-alanine) reagents bonded to a polymeric support for the derivatization of chiral amines, for example, phenylethylamine. The fluorescent tag was coupled to a polystyrene matrix via a 4-hydroxy-3-nitrobenzophenone intermediate. The use of a more reactive benzotriazole spacer is described by Zhang et al. (1992). An example for the application of polymeric FMOC-L-proline is the direct analysis of amphetamine in plasma (Zhou and Krull, 1993). Plasma samples were injected onto a small derivatization column packed with the polymeric reagent. A column switching system was used for back-flushing the derivatized analytes onto an ODS stationary phase. Mixtures of acetonitrile/water containing sodium dodecylsulfate (SDS) as an ion-pairing reagent and surfactant served as a mobile phase.

The scopes and limits of solid-phase reagents have been reviewed recently by Krull et al. (1994a,b). This type of CDAs offers some interesting features:

- Feasibility of on-line derivatizations.
- Combined solid-phase sample preconcentration and derivatization in a single step, that is, direct injection of biogenic samples may be suitable.
- Reduced background noise due to excess reagent compared to conventional derivatization techniques.

16.8.2. Diastereomeric Derivatives via a Metabolic Conjugation Step

For several chiral compounds, phase II metabolites are generated *in vivo*, which are diastereomers, because the conjugation partner has one or several centers of chirality. Such conjugates may be directly resolved on achiral stationary phases. Potential *in vivo* coupling components are glutathione, D-glucuronic acid, and glucose as well as chiral amino acids and their derivatives. Chromatographic systems—usually ODS stationary phases and mixtures of acetonitrile and water with the addition of pH modifiers or ion-pairing reagents—were described for various phase II conjugates including the glucuronides of propranolol (Thompson et al., 1981), oxazepam (Ruelius et al., 1979), amobarbital (Soine et al., 1986), oxaprotilin (Dieterle et al., 1984), pentobarbital (Soine et al., 1994), bromoisovaleranylurea as the model substrate for glutathione conjugation (Te Koppele et al., 1987), as well as the glucuronides of various 2-arylpropionic acids (Spahn, 1988a). The applicability of such biotransformation reactions as chiral derivatization methods *in vitro* (with substrate-oriented detection) was studied with respect to phenolic glucuronides (Gerding et al., 1989) as well as labile ester glucuronides (Spahn, 1989; Spahn-Langguth and Benet, 1993) formed by incubation with microsomal protein (containing uridinediphospho glucuronosyltransferases) and uridinediphospho glucuronic acid (UDPGA) in the presence of detergents and inhibitors of hydrolytic enzymes.

However, if the reaction is to be used for chiral derivatization, this is only possible under certain conditions. Important prerequisites are complete conversion into the product, dependence of the yield on the substrate concentration only, and no competition between the enantiomers (or other compounds) at the enzyme. Yields lower than 100% are tolerable. However, additional competition of the enantiomers at the enzyme may be a factor that significantly complicates the quantification of enantiomer concentrations, because it will be necessary to establish several calibration curves based on the initial rates of formation: The two optical antipodes will have to be derivatized separately, and an additional competition curve will be necessary. Although the method described by Gerding et al. (1989) appeared to yield reasonable results, the studies on acyl glucuronidation as an analytical derivatization method performed by our group indicated that the problem was more complicated than initially anticipated.

In the course of studies, it has been shown that even with optimized conditions—extended incubation time, elevated UDPGT levels due to enzyme induction, sheep vs. rat liver microsomes (*R*/*S* ratio of V_{max} approximately 1.0 vs. 0.4), substrate concentration lower than substrate inhibitory level—the yield never approached 100%. Furthermore, the observed competitive inhibition at the enzyme with typical inhibition curves clearly demonstrated the necessity to determine the inhibitory constant IC_{50} and to include the equations valid for competitive inhibition, in order to be able to calculate unknown concentrations from the observed peak areas and *R*/*S* ratios (Spahn, 1989). Studies with flunoxaprofen enantiomers and sheep liver microsomes, however, revealed that the enantiomers may be treated as independent and the unknown concentrations can be calculated merely from the substrate/product-concentration relationship, when the concentrations of "inibitor" are significantly below the IC_{50} value. In addition to these aspects, other substrates for UDPGTs also may reduce the yields by competitive inhibition.

16.8.3. Other Derivatizing Agents

Although it was not our intention to cover all CDAs ever applied to the liquid chromatographic separation of xenobiotics in the present review, there are still some reagents left that could not be included in the previous sections. However, these reagents offer some valuable features and should at least be mentioned briefly.

Chiral alcohols like 2-octanol have been used for the esterification of carboxylic acid substrates (Johnson et al., 1979; Lee et al., 1984, 1985; MacGregor et al., 1992). However, to-date alcohols have lost some of their importance since chiral carboxylic acids are more easily derivatized with amine reagents and the resulting diastereomeric amides are often better separated than their ester equivalents.

Not only carboxylic acid derivatives may be used as CDAs, sulfonic acid chlorides may be feasible as well. Enantiomerically pure 10-camphorsulfonyl chloride has been applied to the derivatization of amino acids, amines, and aminoalcohols (β-adrenoceptor antagonists) (Furukawa et al., 1975; Souter, 1976; Jira and Breyer, 1993). Etilefrine enantiomers were separated via derivatization with *R*(+)-tetrahydrofurfuryl-(1*S*)-camphor-10-sulfonate (Knorr et al., 1984). Reactive derivatives of other organic acids (e.g., carbamoyl chlorides) may be feasible as well, but may not be discussed in detail.

Goto et al. (1982b) introduced a highly fluorescent cyanide to the derivatization of alcohols under mild conditions. Interestingly, the molecule does not exhibit a chiral carbon atom but a chiral axis due to its substitued 1,1′-binaphthalene skeleton (see Figure 16.11).

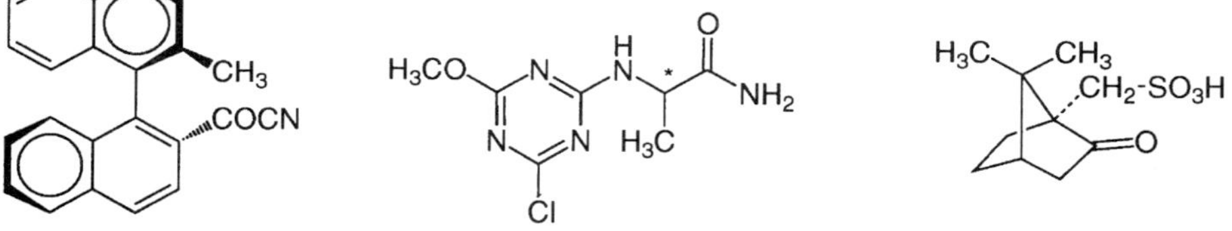

1,1'-binaphthalene-2'-methyl-2-carbonyl nitrile

N²-[2-(4-chloro-6-methoxy-1,3,5-triazine)yl]-L-alanine amide

(+)-Camphor-10-sulfonic acid

Figure 16.11. Miscellaneous reagents.

Finally, amines and amino acids can be derivatized with chiral aromatic halides like N^2-(5-fluoro-2,4-dinitrophenyl)-L-alanine amide (Marfey's reagent) (Marfey, 1984) or related chiral fluorodinitrobenzenes (Brückner and Gah, 1991) or chiral monohalo-s-triazine reagents (Brückner and Strecker, 1992a,b).

16.9. CONCLUSION

A large number of reagents are presently available for chiral derivatization in high-performance liquid chromatography. Included are CDAs with low UV-absorbance as well as highly fluorogenic compounds. No matter, which reactive moiety characterizes the analyte or which detection mode is chosen, the researcher may select from a wide variety of analytical tools. Still, many of the suggested CDAs have not yet been applied to the enantioselective *bioanalysis* of xenobiotics.

There may be a number of reasons preventing their use: Some of the reagents are not yet commercially available. So, many laboratories may not have the time or the equipment for synthetic experiments. However, method development for the quantitation of chiral compounds at trace levels is far more difficult when associated with complex matrices (i.e., body fluids). Here, the limit of quantitation is predominantly set by matrix effects.

However, various additional aspects have to be considered when dealing with "true" samples, that is blood, urine, tissue. And although for the development of bioanalytical assays for chiral drugs, the availability of a separation system is of crucial importance, the whole procedure starts with sample pretreatment, where additional sources of error reside. The stability of drugs and metabolites needs to be assured to avoid enantiospecific, yet false data for a particular compound. For example with chiral carboxylic acids out of the group of NSAIDs, it was found that epimeric acyl glucuronide metabolites occur in blood and urine samples, which degrade at different velocities and under regeneration of the aglycone (Benet et al., 1993; Spahn-Langguth and Benet, 1993). This is of particular relevance when the glucuronide/aglycone ratios in the samples as well as the degradation rates are high.

Recent studies performed by our group with (chiral) adrenoceptor antagonists—for example, the highly liver-extracted and highly metabolized carvedilol—applying a combination of enantiospecific high-performance liquid chromatographic and radioreceptor assays nicely illustrated the relevance of (chiral) oxidative metabolites with respect to the overall therapeutic effect and emphasizes the importance of including such metabolites in the assay. However, with such compounds sophisticated chromatographic systems may be necessary, in order to be able to include all relevant components in one assay as mentioned above.

In summary, the development of enantiospecific assay methods for an administered compound with the aim of measuring therapeutically relevant concentrations needs to compile various aspects: the stability of the drug and metabolites during sample processing, the resolvability of enantiomers, and the occurrence of additional, possibly active and chiral moieties.

Moreover, for the evaluation of the assay quality, information is necessary about the expected concentration range, that is the kinetic behavior of the compounds (drug and/or metabolites) in the population under investigation. Concentrations of the two enantiomers obtained from *in vivo* studies in healthy volunteers may differ by a factor significantly higher than 2 (particularly with high-extraction ratio drugs following oral dosage, e.g., verapamil or tranylcypromine), whereas the concentrations may be considerably higher and the differences smaller in a group of individuals with a lower hepatic extraction ratio (e.g., patients with liver cirrhosis). The reverse may be true for the metabolites formed in the liver.

This illustrates the complexity of aspects to be considered when evaluating the suitability and validity of an enantiospecific bioassay and emphasizes that the kinetic behavior of the compounds under investigation represents an important basis for method optimization.

With respect to the derivatization step itself, in the future, major progress may be achieved by more selective detection modes or derivatization reactions of higher specificity, but—what seems more substantial—improved separation techniques and postderivatization sample clean-up. Excess reagent and reaction by-products have to be removed as well as derivatized matrix compounds. A first step into this direction has been made by the elaboration of solid-phase derivatization techniques, a topic recently reviewed by Krull and co-workers (1994b).

Progress also has been made in the miniaturization of analytical methods, with a subsequent reduction of sample volume. Solvent-saving microbore HPLC combined with conventional fluorescence or laser-induced fluorescence (LIF) detectors can help to reduce detection limits by enhanced separation and sensitive detection. Sample amounts are even smaller in the rapidly developing techniques of micellar electrokinetic capillary chromatography (MEKC) and capillary electrophoresis (CE). Only nanoliter volumes are required for electrophoretic separations.

With the development of "nano-scale" techniques, the significance of highly fluorescent or at least strong UV-absorbing reagents gains additional importance. Few reports on chiral derivatization combined with MEKC or CE are presently available (Nishi et al., 1990; Tran et al., 1990; Ueda et al., 1991; Lurie, 1992; Houben et al., 1993; Schützner et al., 1993). Most authors have concentrated on direct separations so far. Still, there will be a demand for simple and effective chiral derivatizing agents.

ACKNOWLEDGMENTS

Part of the included studies performed at the Department of Pharmacology at the Johann Wolfgang Goethe-University in Frankfurt/M. was supported by the Deusche Forschungsgemeinschaft.

REFERENCES

Ahnoff, M., Chen, S., Green, A., and Grundeurik, I. (1990). Chiral chloroformates as transparent reagents for the resolution of metoprolol enantiomers by reversed-phase liquid chromatography. *J. Chromatogr.*, **506**, 593–599.

Ahnoff, M., Balmér, K., and Lindman, Y. (1992). By-products in the derivatization of amines with the chiral reagent 2,3,4,6-tetra-O-acetyl-β-D-glucopyranosyl isothiocyanate and their elimination. *J. Chromatogr.*, **592**, 323–329.

Alessi-Severini, S., Coutts, R. T., Jamali, F., and Pasutto, F. M. (1992). High-performance liquid chromatographic analysis of methocarbamol enantiomers in biological fluids. *J. Chromatogr.*, **582**, 173–179.

Allgire, J. F., Juenge, E. C., Damo, C. P., Sullivan, G. M., and Kirchhoefer, R. D. (1985). High-performance liquid chromatographic determination of d-/l-epinephrine enantiomer ratio in lidocaine-epinephrine local anesthetics. *J. Chromatogr.*, **325**, 249–254.

Andrisano, V., Marucci, G., Melchiorre, C., and Tumiatti, V. (1992). Stereoselectivity at α-adrenoreceptor subtypes: Observations with the enantiomers of WB 4101 separated through their amides of N-tosyl-(S)-proline. *Chirality*, **4**, 16–20.

Annett, R. G. and Stumpf, P. K. (1992). L-(–)-menthyloxycarbonyl derivatization of hydroxy acid methyl esters. *Anal. Biochem.*, **47**, 638–640.

Armstrong, R. N., Kedzierski, B., Levin, W., and Jerina, D. M. (1981). Enantioselectivity of microsomal epoxide hydrolase toward arene oxide substrates. *J. Biol. Chem.*, **256**, 10, 4726–4733.

Avgerinos, A. and Hutt, A. J. (1987). Determination of the enantiomeric composition of ibuprofen in human plasma by high-performance liquid chromatography. *J. Chromatogr.*, **415**, 75–83.

Baertschi, S. W., Ingram, C. D., Harris, T. M., and Brash, A. R. (1988). Absolute configuration of cis-12-oxophytodienoic acid of flaxseed: Implications for the mechanism of biosynthesis from the 13(*S*)-hydroperoxide of linoleic acid. *Biochemistry*, **27**, 18–24.

Balani, S. K., van Bladeren, P. J., Cassidy, E. S., Boyd, D. R., and Jerina, D. M. (1987). Synthesis of the enantiomeric K-region arene 5,6-oxides derived from chrysene, 7,12-dimethyl-benz[a]anthracene, and benzo[c]phenanthrene. *J. Org. Chem.*, **52**, 137–144.

Banfield, C. and Rowland, M. (1983). Stereospecific high-performance liquid chromatographic analysis of warfarin in plasma. *J. Pharm. Sci.*, **72**, 8, 921–924.

Banfield, C. and Rowland, M. (1984). Stereospecific fluorescence high-performance liquid chromatographic analysis of warfarin and its metabolites in plasma and urine. *J. Pharm. Sci.*, **73**, 10, 1392–1396.

Barksdale, J. M. and Clark, C. R. (1985). Liquid chromatographic determination of the enantiomeric composition of amphetamine and related drugs by diastereomeric derivatization. *J. Chromatogr. Sci.*, **23**, 176–180.

Batra, S., Seth, M., and Bhaduri, A. P. (1993). Chirality and future drug design. In *Progress in Drug Research.* Basel: edited by Tucker E. Birkhauser Verlag, pp. 191–248.

Beale, S. C., Hsieh, Y.-Z., Wiesler, D., and Novotny, M. (1990). Application of 3-(2-furoyl)quinoline-2-carbaldehyde as a fluorogenic reagent for the analysis of primary amines by liquid chromatography with laser-induced fluorescence detection. *J. Chromatogr.*, **499**, 579–587.

Benet, L. Z., Spahn-Langguth, H., Iwakawa, S., Volland, C., Mizuma, T., Mayer, S., Mutschler, E., and Lin, E. T. (1993). Prediction of covalent binding of acidic drugs in man. *Life Sci.*, **53**, PL 141–146.

Bhatti, M. M. and Foster, R. T. (1992). Stereospecific high-performance liquid chromatographic assay of metoprolol. *J. Chromatogr.*, **579**, 361–365.

Bhushan, R., and Joshi, S. (1993). Resolution of enantiomers of amino acids by HPLC. *Biomed. Chromatogr.*, **7**, 5, 235–250.

Björkman, S. (1985). Determination of the enantiomers of indoprofen in blood plasma by high-performance liquid chromatography after rapid derivatization by means of ethyl chloroformate. *J. Chromatogr.*, **339**, 339–346.

Björkman, S. (1987). Determination of the enantiomers of ketoprofen in blood plasma by ion-pair extraction and high-performance liquid chromatography of leucinamide derivatives. *J. Chromatogr.*, **414**, 465–471.

Blaschke, G. and Walther, B. (1985). Konfiguration der Praziquantel-Enantiomeren. *Chem. Ber.*, **118**, 4620–4622.

Blau, K. (1993). Acylation. In Blau and J. M. Halket (eds.), *Handbook of Derivatives for Chromatography*, 2nd ed. New York: Wiley, pp. 31–50.

Blau, K. and Darbre, A. (1993). Esterification. In K. Blau and J. M. Halket (eds.), *Handbook of Derivatives for Chromatography*, 2nd ed. New York: Wiley, pp. 11–30.

Boppana, V. K., Simpson, R. C., Anderson, K., Miller-Stein, C., Blake, T. J. A., Hwang, B. Y.-H., and Rhodes, G. R. (1992). High-performance liquid chromatographic determination of monohydroxy compounds by a combination of pre-column derivatization and post-column reaction detection. *J. Chromatogr.*, **593**, 29–36.

Boursier-Neyret, C., Baune, A., Klippert, P., Castagne, I., and Sauveur, C. (1993). Determination of S 12024 enantiomers in human plasma by liquid chromatography after chiral precolumn derivatization. *J. Pharm. Biomed. Anal.*, **11**, 11–12., 1161–1166.

Brocks, D. R., Pasutto, F. M., and Jamali, F. (1992). Analytical and semi-preparative high-performance liquid chromatographic separation and assay of hydroxychloroquine enantiomers. *J. Chromatogr.*, **581**, 83–92.

Brückner, H. and Gah, C. (1991). High-performance liquid chromatographic separation of DL-amino acids derivatized with chiral variants of Sanger's reagent. *J. Chromatogr.*, **555**, 81–95.

Brückner, H. and Strecker, B. (1992a). Chiral monochloro-s-triazines as derivatizing reagents for resolving DL-amino acids by HPLC. *Chromatographia*, **33**, 11–12, 586–587.

Brückner, H. and Strecker, B. (1992b). Use of chiral monohalo-s-triazine reagents for the liquid chromatographic resolution of DL-amino acids. *J. Chromatogr.*, **627**, 97–105.

Buck, R. H. and Krummen, K. (1984). Resolution of amino acid enantiomers by high-performance liquid chromatography using automated pre-column derivatisation with a chiral reagent. *J. Chromatogr.*, **315**, 279–285.

Buck, R. H. and Krummen, K. (1987). High-performance liquid chromatographic determination of enantiomeric amino acids and amino alcohols after derivatization with o-phthaldialdehyde and various chiral mercaptans. *J. Chromatogr.*, **387**, 255–265.

Büschges, R. (1995). *Chiral derivatizing agents derived from 2-arylpropionic acids and related compounds: Synthesis and application to the analysis of xenobiotics*. Ph.D. thesis, Johann Wolfgang Goethe-University, Frankfurt/M.

Büschges, R., Linde, H., Mutschler, E., and Spahn-Langguth, H. (1996). Chloroformates and isothiocyanates derived from 2-arylpropionic acids as chiral reagents: Synthetic routes and chromatographic behaviour of the derivatives. *J. Chromatogr.*, A **725**, 323–334.

Büschges, R., Devaut, R., Mutschler, E., and Spahn-Langguth, H. (1996). 4-(6-Methoxy-2-naphthyl)-2-butyl chloroformate enantiomers: new reagents for the enantiospecific analysis of amino compounds in biogenic matrices. *J. Pharm. Biomed. Anal.*, **15**, 201–220.

Carlucci, G., Mazzeo, P., and Palumbo, G. (1992). Indirect stereoselective determination of enantiomers of furprofen in human plasma by high-performance liquid chromatography. *Chromatographia*, **34**, 11–12, 618–620.

Carpino, L. A. (1966). Easily cleaved, optically active protective groups of the urethane type. *J. Chem. Soc., Chem. Commun.*, **23**, 858–859.

Carpino, L. A. and Han, G. Y. (1972). The 9-fluorenylmethoxycarbonyl amino-protecting group. *J. Org. Chem.*, **37**, 22, 3404–3409.

Carr, R. A., Foster, R. T., and Bhanji, N. H. (1991a). Stereospecific high-performance liquid chromatographic assay of sotalol in plasma. *Pharm. Res.*, **8**, 9, 1195–1198.

Carr, R. A., Foster, R. T., Freitag, D., and Pasutto, F. M. (1991b). Stereospecific high-performance liquid chromatographic determination of tocainide. *J. Chromatogr.*, **566**, 155–162.

Carr, R. A., Foster, R. T., Lewanczuk, R. Z., and Hamilton, P. G. (1992). Pharmacokinetics of sotalol enantiomers in humans. *J. Clin. Pharmacol.*, **32**, 12, 1105–1109.

Carter, S. R., Duke, C. C., Cutler, D. J., and Holder, G. M. (1992). Sensitive stereospecific assay of warfarin in plasma: Reversed-phase high-performance liquid chromatographic separation using diastereoisomeric esters of (–)-(1S, 2R, 4R)-endo-1,4,5,6,7,7-hexachlorobicyclo[2.2.1]-hept-5-ene-2-carboxylic acid. *J. Chromatogr.*, **574**, 77–83.

Caturla, M. C., Cusido, E., and Westerlund, D. (1992). High-performance liquid chromatography method for the determination of aminoglycosides based on automated pre-column derivatization with o-phthalaldehyde. *J. Chromatogr.*, **593**, 69–72.

Chen, Y.-P., Hsu, M.-C., and Chien, C. S. (1994). Analysis of forensic samples using precolumn derivatization with (+)-1-(9-fluorenyl)ethyl chloroformate and liquid chromatography with fluorimetric detection. *J. Chromatogr., A*, **672**, 135–140.

Chou, T.-Y., Gao, C.-X., Grinberg, N., and Krull, I. S. (1989). Chiral polymeric reagents for off-line and on-line derivatizations of enantiomers in high-performance liquid chromatography with ultraviolet and fluorescence detection: An enantiomer recognition approach. *Anal. Chem.*, **61**, 1548–1558.

Christ, D. D. and Walle, T. (1985). Stereoselective sulfate conjugation of 4-hydroxypropranolol in vitro by different species. *Drug Metab. Dispos.*, **13**, 3, 380–381.

Clark, C. R. and Barksdale, J. M. (1984). Synthesis and liquid chromatographic evaluation of some chiral derivatizing agents for resolution of amine enantiomers. *Anal. Chem.*, **56**, 6, 958–962.

Dale, J. A., Dull, D. L., and Mosher, H. S. (1969). α-Methoxy-α-trifluoromethylphenylacetic acid, a versatile reagent for the determination of enantiomeric composition of alcohols and amines. *J. Org. Chem.*, **34**, 2543–2549.

Darmon, A. and Thenot, J. P. (1986). Determination of betaxolol enantiomers by high-performance liquid chromatography. Application to pharmacokinetic studies. *J. Chromatogr.*, **374**, 321–328.

Davis, P. J., Yang, S.-K., and Smith, R. V. (1985). Microbial models of mammalian metabolism: Stereospecificity of ketone reduction with pentoxifylline. *Xenobiotica*, **15**, 12, 1001–1010.

De Camp, W. H. (1993). Chiral drugs: The FDA perspective on manufacturing and control. *J. Pharm. Biomed. Anal.*, **11**, 11–12, 1167–1172.

Desai, D. M. and Gal, J. (1992). Reversed-phase high-performance liquid chromatographic separation of the stereoisomers of labetalol via derivatization with chiral and non-chiral isothiocyanate reagents. *J. Chromatogr.*, **579**, 165–171.

Desai, D. M. and Gal, J. (1993). Enantiospecific drug analysis via the ortho-phthalaldehyde/homochiral thiol derivatization method. *J. Chromatogr.*, **629**, 215–228.

Dieterle, W. and Faigle, J. W. (1983). Multiple inverse isotope dilution assay for the stereospecific determination of R-(+)- and S-(–)-oxprenolol in biological fluids. *J. Chromatogr.*, **259**, 311–318.

Dieterle, W., Faigle, J. W., Kriemler, H. P., and Winkler, T. (1984). The metabolic fate of ^{14}C-oxaprotiline HCl in man. II. Isolation and identification of metabolites. *Xenobiotica*, **14**, 311–319.

Duchateau, A. L. L., Knuts, H., Boesten, J. M. M., and Guns, J. J. (1992).

Enantioseparation of amino compounds by derivatization with o-phthaldialdehyde and D-3-mercapto-2-methylpropionic acid. *J. Chromatogr.*, **623**, 237–245.

Duchateau, A. L. L., Jacquemin, N. M. J., Straatman, H., and Noorduin, A. J. (1993). Liquid chromatographic determination of chiral epoxides by derivatization with sodium sulfide, o-phthalaldehyde and an amino acid. *J. Chromatogr.*, **637**, 29–34.

Duke, C. C. and Holder, G. M. (1988). Endo-1,4,5,6,7,7-Hexachlorobicyclo[2.2.1]-hept-5-ene-2-carboxylic acid, a superior resolving agent for the high-performance liquid chromatographic separation of enantiomers of hydroxylated derivatives of two azaaromatic hydrocarbons. *J. Chromatogr.*, **430**, 53–64.

Dunlop, D. S. and Neidle, A., The separation of D/L amino acid pairs by high-performance liquid chromatography after precolumn derivatization with optically active naphthylethyl isocyanate. *Anal. Biochem.*, **165**, 38–44.

Eckert, H. and Forster, B. (1987). Triphosgen, ein kristalliner Phosgen-Ersatz. *Angew, Chem.*, **99**, 9, 922–923.

Egginger, G. and Lindner, W. (1993). Development of a stereoselective bioassay for delmopinol in HPLC. *Chirality*, **5**, 400–402.

Egginger, G., Lindner, W., Vandenbosch, C., and Massart, D. L. (1993). Enantioselective bioanalysis of β-blocking agents: Focus on atenolol, betaxolol, carvedilol, metoprolol, pindolol, propranolol and sotalol. *Biomed. Chromatogr.*, **7**, 6, 277–295.

Einarsson, S., Josefsson, B., and Lagerkvist, S., (1983). Determination of amino acids with 9-fluorenylmethyl chloroformate and reversed-phase high-performance liquid chromatography. *J. Chromatogr.*, **282**, 609–618.

Einarsson, S., Josefsson, B., Möller, P, and Sanchez, D. (1987). Separation of amino acid enantiomers and chiral amines using precolumn derivatization with (+)-1-(9-fluorenyl)ethyl chloroformate and reversed-phase liquid chromatography. *Anal. Chem.*, **59**, 1191–1195.

Eisenberg, E. J., Patterson, W. R., and Kahn, G. C. (1989). High-performance liquid chromatographic method for the simultaneous determination of the enantiomers of carvedilol and its O-desmethyl metabolite in human plasma after chiral derivatization. *J. Chromatogr.*, **493**, 105–115.

Euerby, M. R., Partridge, L. Z., and Gibbons, W. A. (1989). Study of the chromatographic behaviour and resolution of α-amino acid enantiomers by high-performance liquid chromatography utilizing pre-column derivatization with o-phthaldialdehyde and new chiral thiols. *J. Chromatogr.*, **483**, 239–252.

Falck, J. R., Manna, S., Jacobson, H. R., Estabrook, R. W., Chacos, N., and Capdevilla, J. (1984). Absolute configuration of epoxyeicosatrienoic acids (EETs) formed during catalytic oxygenation of arachidonic acid by purified rat liver microsomal cytochrome P-450. *J. Am. Chem. Soc.*, **106**, 3334–3336.

Fassihi, A. R. (1993). Racemates and enantiomers in drug development. *Int. J. Pharm.*, **92**, 1–3, 1–14.

Fermo, I., Arcelloni, C., De Vecchi, E., Vigano, S., and Paroni, R. (1992). High-performance liquid chromatographic method with fluorescence detection for the determination of total homocyst(e)ine in plasma. *J. Chromatogr.*, **593**, 171–176.

Fiset, C., Phillipon, F., Gilbert, M., and Turgeon, J. (1993). Stereoselective high-performance liquid chromatographic assay for the determination of sotalol enantiomers in biological fluids. *J. Chromatogr.*, **612**, 231–237.

Foster, R. T. and Jamali, F. (1987). High-performance liquid chromatographic assay of ketoprofen enantiomers in human plasma and urine. *J. Chromatogr.*, **416**, 388–393.

Freitag, D. G., Foster, R. T., Coutts, R. T., and Pasutto, F. M. (1993). High-performance liquid chromatographic method for resolving the enantiomers of mexiletine and two major metabolites isolated from microbial fermentation media. *J. Chromatogr.*, **616**, 253–259.

Frigerio, E., Pianezzola, E., and Strolin Benedetti, M. (1994). Sensitive procedure for the determination of reboxetine enantiomers in human plasma by reversed-phase high-performance liquid chromatography with fluorimetric detection after chiral derivatization with (+)-1-(9-fluorenyl)ethyl chloroformate. *J. Chromatogr.*, **660**, 1–2, 351–358.

Fujimaki, M., Murakoshi, Y., and Hakusui, H. (1990). Assay and disposition of carvedilol enantiomers in humans and monkeys: Evidence of stereoselective presystemic metabolism. *J. Pharm. Sci.*, **79**, 7, 568–572.

Furukawa, H., Sakakibara, S., Kamei, A., and Ito, K. (1975). Separation of L- and D-amino acids as diastereomeric derivatives by high-performance liquid chromatography. *Chem. Pharm. Bull.*, **23**, 1625–1626.

Gal, J. (1985) Determination of the enantiomeric composition of chiral epoxides using chiral derivatization and liquid chromatography. *J. Chromatogr.*, **331**, 349–357.

Gal, J., (1993). Indirect methods for the chromatographic resolution of drug enantiomers. In Wainer, I. W. (ed.), *Drug Stereochemistry. Analytical Methods and Pharmacology*, 2nd ed. New York: Marcel Dekker, pp. 65–106.

Gal, J. and Meyer-Lehnert, S. (1988). Reversed-phase liquid chromatographic separation of enantiomeric and diastereomeric bases related to chloramphenicol and thiamphenicol. *J. Pharm. Sci.*, **77**, 12, 1062–1065.

Gal, J. and Sedman, A. J. (1984). R-α-Methylbenzyl isothiocyante, a new and convenient chiral derivatizing agent for the separation of enantiomeric amino compounds by high-performance liquid chromatography. *J. Chromatogr.*, **314**, 275–281.

Gal, J., Desai, D. M., and Meyer-Lehnert, S. (1990). Reversed-phase LC resolutions of chiral antiarrhythmic agents via derivatization with homochiral isothiocyanates. *Chirality*, **2**, 43–51.

Geerlof, A., van Tol, J. B. A., Jongejan, J. A., and Duine, J. A. (1993). Methods for the determination of the enantiomeric purity of the C_3-synthons glycidol (2,3-epoxy-1-propanol) and solketal [2,2-dimethyl-4-(hydroxymethyl)-1,3-dioxolane]. *J. Chromatogr.*, **648**, 1, 119–129.

Gerding, T. K., Drenth, B. F. H., Van de Grampel, V. J. M., Niemeijer, N. R., De Zeeuw, R. A., Tepper, P. G., and Horn, A. S. (1989). Determination of enantiomeric purity of the new D_2-dopamine agonist 2-(N-propyl-N-2-thienylethylamino)-5-hydroxytetralin (N-0437) by reversed-phase high-performance liquid chromatography after pre-column derivatization with D(+)-glucuronic acid. *J. Chromatogr.*, **487**, 125–134.

Gietl, Y., Spahn, H., and Mutschler, E. (1988). Simultaneous determination of R- and S-prenylamine in plasma and urine by reversed-phase high-performance liquid chromatography. *J. Chromatogr.*, **426**, 305–314.

Görög, S. and Gazdag, M. (1994). Enantiomeric derivatization for biomedical chromatography. *J. Chromatogr. B*, **659**, 51–84.

Goto, J., Hasegawa, M., Nakamura, S., Shimada, K., and Nambara, T. (1978). New derivatization reagents for the resolution of amino acid enantiomers by high-performance liquid chromatography. *J. Chromatogr.*, **152**, 413–419.

Goto, J., Goto, N., Hikichi, A., and Nambara, T. (1979). Separation and determination of 2,5-dimethoxy-4-methylamphetamine enantiomers in plasma by high-performance liquid chromatography. *J. Liq. Chromatogr.*, **2**, 8, 1179–1190.

Goto, J., Goto, N., Hikichi, A., Nishimaki, T., and Nambara, T. (1980). Sensitive derivatization reagents for the resolution of carboxylic acid enantiomers by high-performance liquid chromatography. *Anal. Chim. Acta*, **120**, 187–192.

Goto, J., Goto, N., and Nambara, T. (1982a). Separation and determination of naproxen enantiomers in serum by high-performance liquid chromatography. *J. Chromatogr.*, **239**, 559–564.

Goto, J., Goto, N., and Nambara, T. (1982b). New type of derivatization reagents for liquid chromatographic resolution of enantiomeric hydroxyl groups. *Chem. Pharm. Bull.*, **30**, 12, 4597–4599.

Goto, J., Ito, M., Katsuki, S., Saito, N., and Nambara, T. (1986). Sensitive derivatization reagents for optical resolution of carboxylic acids by high-performance liquid chromatography with fluorescence detection. *J. Liq. Chromatogr.*, **9**, 2–3, 683–694.

Grech-Bélanger, O., Turgeon, J., and Gilbert, M. (1985). High-performance liquid chromatographic assay for mexiletine enantiomers in human plasma. *J. Chromatogr.*, **337**, 172–177.

Gübitz, G., Wintersteiger, R., and Hartinger, A. (1981). Fluorescence derivatization of tertiary amines with 2-naphthyl chloroformate. *J. Chromatogr.*, **218**, 51–56.

Gulaid, A. A., Houghton, G. W., and Boobis, A. R. (1985). Separation of acebutolol and diacetolol diastereomers by reversed-phase high-performance liquid chromatography. *J. Chromatogr.*, **318**, 393–397.

Guttendorf, R. J., Kostenbauder, H. B., and Wedlund, P. J. (1989). Quantification of propranolol enantiomers in small blood samples from rats by reversed-phase high-performance liquid chromatography after chiral derivatization. *J. Chromatogr.*, **489**, 333–343.

Hahn, G., Langguth, P., and Möhrke, W. (1992). Enantiospecific fluorescence assay for norephedrine (phenylpropanolamine) and its applicability in pharmacokinetic studies. *Naunyn Schmiedeberg's Arch. Pharmacol.* Suppl., R9.

Hartvig, P., Ahnfelt, N. O., and Karlsson, K. E., Electron capture gas chromatography of nanogram amounts of tertiary amines as trichloroethyl carbamates. *Acta Pharm. Suec.*, **13**, 181–189.

Harvey, R. G. and Cho, H. (1977). Efficient resolution of the dihydrodiol derivatives of benzo[a]pyrene by high-pressure liquid chromatography of the related (–)-dimenthoxyacetates. *Anal. Biochem.*, **80**, 540–546.

Hasegawa, R., Murai-Kushiya, M.,Komura, T., and Kimura, T. (1989). Stereoselective determination of plasma pindolol in endotoxin-pretreated rats by high-performance liquid chromatography. *J. Chromatogr.*, **494**, 381–388.

Hashimoto, A., Nishikawa, T., Oka, T., Takahashi, K., and Hayashi, T. (1992). Determination of free amino acid enantiomers in rat brain and serum by high-performance liquid chromatography after derivatization with N-tert.-butyloxycarbonyl-L-cysteine and o-phthaldialdehyde. *J. Chromatogr.*, **582**, 41–48.

Hayball, P. J., Nation, R. L., Bochner, F., and Le Leu, R. K. (1991). Enantiospecific analysis of ketoprofen in plasma by high-performance liquid chromatography. *J. Chromatogr.*, **570**, 446–452.

Hayball, P. J., Tamblyn, J. G., Holden, Y., and Wrobel, J. (1993). Stereoselective analysis of ketorolac in human plasma by high-performance liquid chromatography. *Chirality*, **5**, 31–35.

He, L. and Stewart, J. T. (1992). A high performance liquid chromatographic method for the determination of albuterol enantiomers in human serum using solid phase extraction and chemical derivatization. *Biomed. Chromatogr.*, **6**, 291–294.

Helmchen, G. and Strubert, W. (1974). Determination of optical purity by high-performance liquid chromatography. *Chromatographia*, **7**, 12, 713–715.

Helmchen, G., Völter, H., and Schuhle, W. (1977). Directed resolution of enantiomers via liquid chromatography of diastereomeric derivatives. III. A convenient method to determine the absolute configuration of carboxylic acids R_1R_2 HCCOOH. *Tetrahedron Lett.*, **16**, 1417–1420.

Herber, B., Büschges, R., and Spahn-Langguth, H. (1994). Fluorescent carbazoles as chiral derivatizing agents in the analysis of drug enantiomers: Isocyanates deriving from carprofen and N-methylcarprofen. *Eur. J. Pharm. Sci.*, **2**, 155.

Hermansson, J. (1982). Separation and quantitation of (R)- and (S)-propranolol as their diastereomeric derivatives in human plasma by reversed-phase ion-pair chromatography. *Acta Pharm. Suec.*, **19**, 11–24.

Hermansson, J. and von Bahr, C. (1980). Simultaneous determination of d- and L-propranolol in human plasma by high-performance liquid chromatography. *J. Chromatogr.*, **221**, 109–117.

Hermansson, J. and von Bahr, C. (1982). Determination of (R)- and (S)-alprenolol and (R)- and (S)-metoprolol as their diastereomeric derivatives in human plasma by reversed-phase liquid chromatography. *J. Chromatogr.*, **227**, 113–127.

Hoffmann, K.-J. Renberg, L., and Bäärnhielm, C. (1984). Stereoselective disposition of RS-tocainide in man. *Eur. J. Drug Met. Pharmacokinet.*, **9**, 3, 215–222.

Houben, R. J. H., Gielen, H., and van der Wal, S. (1993). Automated preseparation derivatization on a capillary electrophoresis instrument. *J. Chromatogr.*, **634**, 317–322.

Hsyu, P.-H., and Giacomini, K. M. (1986). High performance liquid chromatographic determination of the enantiomers of β-blocking agents in biological fluids. I. Studies with pindolol. *J. Pharm. Sci.*, **75**, 6, 601–605.

Husain, P. A., Colbert, J. E., Sirimanne, S. R., VanDerveer, D. G., Herman, H. H., and

May, S. W. (1989). N-succinimidyl methoxyphenylacetic acid ester, an amine-directed chiral derivatizing reagent suitable for enzymatic scale resolutions. *Anal. Biochem.*, **178**, 177–183.

Husain, P. A., Debnath, J., and May, S. W. (1993). HPLC-based method for determination of absolute configuration of α-chiral amines. *Anal. Chem.*, **65**, 1456–1461.

Hutt, A. J. and Caldwell, J. (1983). The metabolic chiral inversion of 2-arylpropionic acids—a novel route with pharmacological consequences. *J. Pharm. Pharmacol.*, **35**, 693–704.

Hutt, A. J., Fournel, S., and Caldwell, J. (1986). Application of a radial compression column to the high-performance liquid chromatographic separation of the enantiomers of some 2-arylpropionic acids as their diastereoisomeric S-(–)-1-(naphthen-1-yl)ethylamides. *J. Chromatogr.*, **378**, 409–418.

Ito, S., Ota, A., Yamamoto, K., and Kawashima, Y. (1992). Resolution of the enantiomers of thiol compounds by reversed-phase liquid chromatography using chiral derivatization with 2,3,4,6-tetra-O-acetyl-β-D-glucopyranosyl isothiocyanate. *J. Chromatogr.*, **626**, 187–196.

Iwakawa, S., Suganuma, T., Lee, S.-F., Spahn, H., Benet, L. Z., and Lin, E. T. (1989). Direct determination of diastereomeric carprofen glucuronides in human plasma and urine and preliminary measurements of stereoselective metabolic and renal elimination after oral administration of carprofen in man. *Drug Metab. Dispos.*, 17, **4**, 414–419.

Iwaki, K., Bunrin, T., Kameda, Y., and Yamazaki, M. (1994). Resolution and sensitive detection of carboxylic acid enantiomers using fluorescent chiral derivatization reagents by high-performance liquid chromatography. *J. Chromatogr. A*, **662**, 1, 87–93.

Jamali, F., Pasutto, F. M., and Lemko, C. (1989). HPLC of ketorolac enantiomers and application to pharmacokinetics in the rat. *J. Liq. Chromatogr.*, **12**, 10, 1835–1850.

Jeyaraj, G. L. and Porter, W. R. (1984). New method for the resolution of racemic warfarin and its analogues using low-pressure liquid chromatography. *J. Chromatogr.*, **315**, 378–383.

Jira, T. and Breyer, C. (1993). HPLC analysis of chiral β-adrenergic active agents. *Pharmazie*, **48**, 11, 829–833.

Jira, T., Toll, C., Vogt, C., and Beyrich, T. (1991). Zur Trennung einiger racemischer β-Blocker und α-Sympathikomimetika durch HPLC nach Derivatisierung. *Pharmazie*, **46**, 6, 432–434.

Johnson, D. M., Reuter, A., Collins, J. M., and Thompson, G. F. (1979). Enantiomeric purity of naproxen by liquid chromatographic analysis of its diastereomeric octyl esters. *J. Pharm. Sci.*, **68, 1**, 112–114.

Kagel, J. R., Stella, V. J., and Riley, C. M. (1993). A liquid chromatographic method for the determination of the enantiomeric purity of the anticancer drug, 9-amino-20(S)-camptothecin. *J. Pharm. Biomed. Anal.*, **11**, 9, 793–802.

Kemmerer, J. M., Rubio, F. A., McClain, R. M., and Koechlin, B. A. (1979). Stereospecific assay and stereospecific disposition of racemic carprofen in rats. *J. Pharm. Sci.*, **68**, 10, 1274–1280.

Kinoshita, T., Kasahara, Y., and Nimura, N. (1981). Reversed-phase high-performance liquid chromatographic resolution of non-esterified enantiomeric amino acids by derivatization with 2,3,4,6-tetra-O-acetyl-β-D-glucopyranosyl isothiocyanate and 2,3,4-tri-O-acetyl-α-D-arabinopyranosyl isothiocyanate. *J. Chromatogr.*, **210**, 77–81.

Kitani, M., Miyamoto, G., and Odomi, M. (1993). Stereoselective high-performance liquid chromatographic assay for the determination of OPC-18790 enantiomers in human plasma and urine. *J. Chromatogr.*, **620**, 1, 97–104.

Knorr, H., Reichl, R., Traunecker, W., Knappen, F., and Brandt, K. (1984). Asymmetrische Synthese des (R)-(–)- und (S)-(+)-Etilefrins durch enantioselektive Hydrierung und die Wirkungen der Enantiomere im Tierexperiment. *Arzneim.-Forsch./Drug Res.*, **34(II)**, 12, 1709–1713.

Koppenhoefer, B., Nothdurft, A., Pierrot-Sanders, J., Piras, P., Popescu, C., Roussel, C., Stiebler, M., and Trettin, U. (1993). CHIRBASE, a graphical molecular database on the separation of enantiomers by liquid-, supercritical fluid-, and gas chromatography. *Chirality*, **5**, 213–219.

Kristensen, K. and Angelo, H. R. (1992). Stereospecific gas chromatographic method for determination of methadone in serum. *Chirality*, **4**, 263–267.

Krull, I. S., Deyl, Z., and Lingeman, H. (1994a). General strategies and selection of derivatization reactions for liquid chromatography and capillary electrophoresis. *J. Chromatogr. B*, **659**, 1–17.

Krull, I. S., Szulc, M. E., Bourque, A. J., Zhou, F.-X., J., and Strong, R. (1994b). Solid-phase derivatization reactions for biomedical liquid chromatography. *J. Chromatogr. B*, **659**, 19–50.

Kurz, A., Spahn-Langguth, H., and Mutschler, E. (1994). (unpublished data).

Laethem, M. E., Rosseel, M. T., Wijnant, P., and Belpaire, F. M. (1993). Chiral high-performance liquid chromatographic determination of oxprenolol in plasma. *J. Chromatogr.*, **621**, 2, 225–229.

Lai, F., Mayer, A., and Sheehan, T. (1993). Chiral separation and detection enhancement of propranolol using automated pre-column derivatization. *J. Pharm. Biomed. Anal.*, **11**, 2, 117–120.

Lam, S. (1986). Resolution of D- and L-amino acids after precolumn derivatization with o-phthalaldehyde by mixed chelation with Cu(II)-L-proline. *J. Chromatogr.*, **402**, 387–391.

Langaniere, S., Kwong, E., and Shen, D. D. (1989). Stereoselective high-performance liquid chromatographic assay for propranolol enantiomers in serum. *J. Chromatogr.*, **488**, 407–416.

Langguth, P., Spahn, H., and Merkle, H.-P. (1990). Fluorescence assay for small peptides and amino acids: high-performance liquid chromatographic determination of selected substrates using activated S-flunoxaprofen as a chiral derivatizing agent. *J. Chromatogr.*, **528**, 55–64.

Langguth, P., Merkle, H. P., and Amidon, G. L. (1994). The effect of absorption site and enzyme inhibition on the systemic availability of metkephamid. *Pharm. Res.*, **11**, 528–535.

Lankmayr, E. P., Budna, W., and Nachtmann, F. (1980). Separation of enantiomeric iodinated thyronines by liquid chromatography of diastereomers. *J. Chromatogr.*, **198**, 471–479.

Lave, T., Efthymiopoulos, C., Koffel, J. C., and Jung, L. (1991). Determination of tertatolol enantiomers in biological fluids by high-performance liquid chromatography. *J. Chromatogr.*, **572**, 203–210.

Lee, E. J., Williams, K. M., Graham, G. G., Day, R. O., and Champion, G. D. (1984). Liquid chromatographic determination of plasma concentration profile of optical isomers of ibuprofen in humans. *J. Pharm. Sci.*, **73**, 11, 1542–1544.

Lee, E. J., Williams, K., Day, R., Graham, G., and Champion, D. (1985). Stereoselective disposition of ibuprofen enantiomers in man. *Br. J. Clin. Pharmacol.*, **19**, 669–674.

Lehr, K.-H. and Damm, P. (1988). Quantification of the enantiomers of ofloxacin in biological fluids by high-performance liquid chromatography. *J. Chromatogr.*, **425**, 153–161.

Lehr, R. E., Kumar, S., Shirai, N., and Jerina, D. M. (1985). Synthesis of enantiomerically pure bay-region 3,4-diol 1,2-epoxide diastereomers and other derivatives of the potent carcinogen dibenz[c,h]acridine. *J. Org. Chem.*, **50**, 98–107.

Liang, W. T. C., Brocks, D. R., and Jamali, F. (1992). Stereospecific high-performance liquid chromatographic assay of pirprofen enantiomers in rat plasma and urine. *J. Chromatogr.*, **577**, 317–324.

Liebmann, B., Mayer, S., Mutschler, E., and Spahn-Langguth, H. (1992). Studies on the metabolic clearance of ciclotropium to α-phenylciclopentylacetic acid using a new enantiospecific metabolite assay. *Arzneim.-Forsch./Drug Res.*, **42 (II)**, 11, 1354–1358.

Lindner, W., Leitner, C., and Uray, G. Liquid chromatographic separation of enantiomeric alkanolamines via diastereomeric tartaric acid monoesters. *J. Chromatogr.*, **316**, 605–616.

Lindner, W., Rath, M., Stoschitzky, K., and Uray, G. (1989). Enantioselective drug monitoring of (R)- and (S)-propranolol in human plasma via derivatization with optically active (R,R)-O,O-diacetyl tartaric acid anhydride. *J. Chromatogr.*, **487**, 375–383.

Lingeman, H. and Underberg, W. J. M. (eds.) (1990). *Detection-Oriented Derivatization Techniques in Liquid Chromatography*. New York: Marcel Dekker.

Lobell, M. and Schneider, M. P. (1993). 2,3,4,6-Tetra-O-benzoyl-β-D-glucopyranosyl isothiocyanate: An efficient reagent for the determination of enantiomeric purities of amino acids, β-adrenergic blockers and alkyloxiranes by high-performance liquid chromatography using standard reversed-phase columns. *J. Chromatogr.*, **633**, 287–294.

Lurie, I. S. (1992). Micellar electrokinetic capillary chromatography of the enantiomers of amphetamine, methamphetamine and their hydroxyphenethylamine precursors. *J. Chromatogr.*, **605**, 269–275.

MacGregor, R. R., Fowler, J. S., and Wolf, A. P. (1992). Determination of the enantiomeric composition of samples of cocaine by normal-phase high-performance liquid chromatography with UV detection. *J. Chromatogr.*, **590**, 354–358.

Magat, H., Kucera, P., and Wen, H. D. (1982). Total amino acid analysis using pre-column fluorescence derivatization. *J. Chromatogr.*, **239**, 463–474.

Maibaum, J. (1988). Indirect high-performance liquid chromatograpic resolution of racemic tertiary amines as their diastereomeric urea derivatives after N-dealkylation. *J. Chromatogr.*, **436**, 269–278.

Maitre, J.-M., Boss, G., and Testa, B. (1984). High-performance liquid chromatographic separation of the enantiomers of anti-inflammatory 2-arylpropionates: suitability of the method for in vitro metabolic studies. *J. Chromatogr.*, **299**, 397–403.

March, J. (1968). *Advanced Organic Chemistry: Reactions, Mechanisms, and Structure.* New York: McGraw-Hill.

Marfey, P. (1984). Determination of D-amino acids. II. Use of a bifunctional reagent, 1,5-difluoro-2,4-dinitrobenzene. *Carlsberg Res. Comm.*, **49**, 591–596.

Martin, E., Quinke, K., Spahn, H., and Mutschler, E. (1989). (–)-(S)-Flunoxaprofen and (–)-(S)-naproxen isocyanate: Two new fluorescent chiral derivatizing agents for an enantiospecific determination of primary and secondary amines. *Chirality*, **1**, 223–234.

Marzo, A. (1994). Incoming guidelines on chirality. A challenge for pharmacokinetics in drug development. *Arzneim.-Forsch./Drug Res.*, **44**, 6, 791–793.

Mathys, K. and Brenneisen, R. (1992). Determination of (S)-(–)-cathinone and its metabolites (R,S)-(–)-norephedrine and (R,R)-(–)-norpseudoephedrine in urine by high-performance liquid chromatography with photodiode-array detection. *J. Chromatogr.*, **593**, 79–85.

Matuszewski, B. K., Constanzer, M. L., Hessey, G. A., and Bayne, W. F. (1990). Indirect stereoselective determination of the enantiomers of a thieno[2,3-b]thiopyran-2-sulfonamide in biological fluids. *Anal. Chem.*, **62**, 1308–1315.

Maurs, M., Trigalo, F., and Azerad, R. (1988). Resolution of α-substituted amino acid enantiomers by high-performance liquid chromatography after derivatization with a chiral adduct of o-phthalaldehyde. Application to glutamic acid analogues. *J. Chromatogr.*, **440**, 209–215.

Mayer, S. (1992). *Analytik und Pharmakokinetik des chiralen Lipidsenkers Beclobrat. Ph. D. Thesis*, Department of Pharmacology, Johann Wolfgang Goethe-University, Frankfurt/Main.

Mayer, S., Mutschler, E., and Spahn-Langguth, H. (1991). Pharmacokinetic studies with the lipid-regulating agent beclobrate: Enantiospecific assay for beclobric acid using a new fluorescent chiral coupling component (S-FLOPA). *Chirality*, **3**, 35–42.

Mayer, S., Mutschler, E., Benet, L. Z., and Spahn-Langguth, H. (1993). In vitro and in vivo irreversible plasma protein binding of beclobric acid enantiomers. *Chirality*, **5**, 120–125.

McKay, S. W., Mallen, D. N. B., Shrubsall, P. R., Swann, B. P., and Williamson, W. R. N. (1979). Analysis of benoxaprofen and other α-methylarylacetic acids using high performance liquid chromatography. *J. Chromatogr.*, **170**, 482–485.

Meese, C. W., Fischer, C., and Eichelbaum, M. (1988). Stereoselectivity of the 4-hydroxylation of debrisoquine in man, detected by gas chromatography/mass spectrometry. *Biomed. Environ. Mass Spectrom.*, **15**, 63–66.

Mehvar, R. (1989). Stereospecific liquid chromatographic analysis of racemic adrenergic drugs utilizing precolumn derivatization with (–)-menthyl chloroformate. *J. Chromatogr.*, **493**, 402–408.

Mehvar, R. (1990). Liquid chromatographic analysis of propafenone enantiomers in human plasma. *J. Chromatogr.*, **527**, 79–89.

Meyer, M. W., Meyer, V. R., and Ramseyer, S. (1991). The kinetics of diastereomeric amino acids with o-phthaldialdehyde. *Chirality*, **3**, 471–475.

Miller, K. J., Gal, J., and Ames, M. M. (1984). High-performance liquid chromatographic resolution of enantiomers of 1-phenyl-2-aminopropanes (amphetamines) with four chiral reagents. *J. Chromatogr.*, **307**, 335–342.

Miller, R. B. (1992). High-performance liquid chromatographic assay for the derivatized enantiomers of propranolol and 4-hydroxypropranolol in human plasma. *J. Pharm. Biomed. Anal.*, **9**, 10–12, 953–958.

Miller, R. B., and Guertin, Y. (1992). High-performance liquid chromatographic assay for the derivatized enantiomers of atenolol in whole blood. *J. Liq. Chromatogr.*, **15**, 8, 1289–1302.

Min, S., Yisheng, Y., and Lu, Y. (1992). Determination of urinary 3-methylhistidine by high-performance liquid chromatography with o-phthaldialdehyde precolumn derivatization. *J. Chromatogr.*, **581**, 272–276.

Mitchell, A. R., Kent, S. B. H., Chu, I. C., and Merrifield, R. B. (1978). Quantitative determination of D- and L-amino acids by reaction with tert-butyloxycarbonyl-L-leucine N-hydroxysuccinimide ester and chromatographic separation as L,D and L,L dipeptides. *Anal. Chem.*, **50**, 4, 637.

Miyazawa, T., Iwanaga, H., Yamada, T., and Kuwata, S. (1993). Resolution of cyclic imino acid and β-amino acid enantiomers by derivatization with 2,3,4,6-tetra-O-acetyl-β-D-glucopyranosyl isothiocyanate followed by reversed-phase HPLC analysis. *Anal. Lett.*, **26**, 2, 367–378.

Moye, H. A. and Boning, A. J., Jr. (1979). A versatile fluorogenic labelling reagent for primary and secondary amines: 9-Fluorenylmethyl chloroformate. *Anal. Lett.*, **12**, 25–35.

Nachtmann, F. (1980). Determination of the L-isomer in D-penicillamine by derivatization liquid chromatography. *Int. J. Pharm.*, **4**, 337–345.

Nagashima, H., Tanaka, Y., Watanabe, H., Hayashi, R., and Kawada, K. (1984). Optical inversion of (2R)- to (2S)-isomers of 2-[4-(2-oxocyclopentylmethyl)-phenyl]propionic acid (loxoprofen), a new anti-inflammtory agent, and its monohydroxy metabolites in the rat. *Chem. Pharm. Bull.*, **32**, 1, 251–257.

Nagashima, H., Tanaka, Y., and Hayashi, R. (1985). Column liquid chromatography for the simultaneous determination of the enantiomers of loxoprofen sodium and its metabolites in human urine. *J. Chromatogr.*, **345**, 373–379.

Nekrassow, W. and Melnikow, N. (1930). Über die Ester der halogenierten Alkohole.

II. Mitteilung: Über die Reaktionen der Ester, die die Trichlormethoxylgruppe enthalten. *J. Prakt. Chem.*, **126**, 81–96.

Nimura, N. and Kinoshita, T. (1984). o-Phthalaldehyde-N-acetyl-L-cysteine as a chiral derivatization reagent for liquid chromatographic optical resolution of amino acid enantiomers and its application to conventional amino acid analysis. *J. Chromatogr.*, **352**, 169–177.

Nimura, N., Ogura, H., and Kinoshita, T. (1980). Reversed-phase liquid chromatographic resolution of amino acid enantiomers by derivatization with 2,3,4,6-tetra-O-acetyl-β-D-glucopyranosyl isothiocyanate. *J. Chromatogr.*, **202**, 375–379.

Nimura, N., Kasahara, Y., and Kinoshita, T. (1981). Resolution of enantiomers of norepinephrine and epinephrine by reversed-phase high-performance liquid chromatography. *J. Chromatogr.*, **213**, 327–330.

Nimura, N., Toyama, A., and Kinoshita, T. (1984). Optical resolution of amino acid enantiomers by high-performance liquid chromatography. *J. Chromatogr.*, **316**, 547–552.

Nimura, N., Iwaki, K., Kinoshita, T. (1987). Separation of norepinephrine enantiomers, DOPA enantiomers and dopamine derivatized with o-phthalaldehyde-N-acetyl-L-cysteine by high-performance liquid chromatography. *J. Chromatogr.*, **402**, 387–391.

Nishi, H., Ishii, K., Taku, K., Shimizu, R., and Tsumagari, N. (1989). New chiral derivatization reagent for the resolution of amino acids as diastereomers by TLC and HPLC. *Chromatographia*, **27**, 7–8, 301–305.

Nishi, H., Fukuyama, T., and Matsuo, M. (1990). Resolution of optical isomers of 2,3,4,6-tetra-O-acetyl-β-D-glucopyranosyl isothiocyanate (GITC)—Derivatized DL-amino acids by micellar electrokinetic chromatography. *J. Microcol. Sep.*, **2**, 5, 234–240.

Nitchuk, W. M. (1992). Regulatory requirements for generic chiral drugs. *J. Clin. Pharmacol.*, **32**, 953–954.

Noggle, T. T., Jr., DeRuiter, J., and Clark, C. R. (1986). Liquid chromatographic determination of the enantiomeric composition of methamphetamine prepared from ephedrine and pseudoephedrine. *Anal. Chem.*, **58**, 1643–1648.

Olsen, L., Bronnum-Hansen, K., Helboe, P., Herlev-Jorgensen, G., and Kryger, S. (1993). Chiral separations of β-blocking drug substances using derivatization with chiral reagents and normal phase high-performance liquid chromatography. *J. Chromatogr.*, **636**, 231–241.

Palylyk, E. L. and Jamali, F. (1991). Simultaneous determination of ketoprofen enantiomers and probenecid in plasma and urine by high-performance liquid chromatography. *J. Chromatogr.*, **568**, 187–196.

Pedrazzini, S., Zanoboni-Muciaccia, W., Sacchi, C., and Forgione, A. (1987). Determination of flunoxaprofen enantiomers in biological fluids by high-performance liquid chromatography. *J. Chromatogr.*, **415**, 214–220.

Pereira, W., Bacon, V. A., Patton, W., Halpern, B., and Pollok, G. E. (1970). The use of R(+)-1-phenylethylisocyanate in the optical analysis of asymmetric secondary alcohols by gas chromatography. *Anal. Lett.*, **3**, 23–28.

Peyton, A. L., Carpenter, R., and Rutkowski, K. (1991). The enantiospecific determination of fluoxetine and norfluoxetine in human plasma by high-pressure liquid chromatography (HPLC) with fluorescence detection. *Pharm. Res.*, **8**, 12, 1528–1532.

Pflugmann, G., Spahn, H., and Mutschler, E. (1987). Determination of metoprolol enantiomers in plasma and urine using (S)-(–)-phenylethyl isocyanate as a chiral reagent. *J. Chromatogr.*, **421** 161–164.

Pianezzola, E., Belloti, V., Fontana, E., Moro, E., Gal, J., and Desai, D. M. (1989). Determination of the enantiomeric composition of salsolinol in biological samples by high-performance liquid chromatography with electrochemical detection. *J. Chromatogr.*, **495**, 205–214.

Piquette-Miller, M. and Foster, R. T. (1990). High-performance liquid chromatographic determination of diacetolol enantiomers. *J. Chromatogr.*, **533**, 300–303.

Piquette-Miller, M. and Jamali, F. (1993). Selective effect of adjuvant arthritis on the disposition of propranolol enantiomers in rats detected using a stereospecific HPLC assay. *Pharm. Res.*, **10**, 2, 294–299.

Piquette-Miller, M., Foster R. T., Pasutto, F. M., and Jamali, F. (1990). Stereospecific high-performance liquid chromatographic assay of acebutolol in human plasma and urine. *J. Chromatogr.*, **526**, 129–137.

Pirkle, W. H. and Hauske, J. R. (1977). Broad spectrum methods for the resolution of optical isomers. A discussion of the reasons underlying the chromatographic separability of some diastereomeric carbamates. *J. Org. Chem.*, **42**, 11, 1839–1844.

Pirkle, W. H. and Hoekstra, M. S. (1974). An example of automated liquid chromatography. Synthesis of a broad-spectrum resolving agent and resolution of 1-(1-naphthyl)-2,2,2-trifluoroethanol. *J. Org. Chem.*, **39**, 26, 3904–3906.

Pirkle, W. H., Robertson, M. R., and Hyun, M. H. (1984). A liquid chromatographic method for resolving chiral lactams as their diastereomeric ureide derivatives. *J. Org. Chem.*, **49**, 2433–2437.

Potts, B. D. and Parli, C. J. (1992). Analysis of the enantiomers of fluoxetine and norfluoxetine in plasma and tissue using chiral derivatization and normal-phase liquid chromatography. *J. Liq. Chromatogr.*, **15**, 4, 665–681.

Prakash, C., Jajoo, H. K., Blair, I. A., and Mayol, R. F. (1989a). Resolution of enantiomers of the antiarrhythmic drug encainide and its major metabolites by chiral derivatization and high-performance liquid chromatography. *J. Chromatogr.*, **493**, 325–335.

Prakash, C., Koshakji, R. P., Wood, A. J. J., and Blair, I. A. (1989). Simultaneous determination of propranolol enantiomers in plasma by high-performance liquid chromatography with fluorescence detection. *J. Pharm. Sci.*, **78**, 9, 771–775.

Prelusky, D. B., Coutts, R. T., and Pasutto, F. M. (1982). Stereospecific metabolic reduction of ketones. *J. Pharm. Sci.*, **71**, 12, 1390–1393.

Rastegar, A., Pelletier, A., Duportail, G., Freysz, L., and Leray, C. (1990). Sensitive analysis of phospholipid molecular species by high-performance liquid chromatography using fluorescent naproxen derivatives of diacylglycerols. *J. Chromatogr.*, **518**, 157–165.

Reid, J. M., Stobaugh, J. F., and Sternson, L. A. (1989). Liquid chromatographic determination of cyclophosphamide enantiomers in plasma by precolumn chiral derivatization. *Anal. Chem.*, **61**, 441–446.

Rose, H. C., Stern, R. L., and Karger, B. L. (1996). Studies on the mechanism of separation of diastereoisomeric esters by gas-liquid chromatography. Effect of bulk dissymmetry and distance between optical centers. *Anal. Chem.*, **38**, 3, 469–472.

Rose, S. E. and Randinitis, E. J. (1991). A high-performance liquid chromatographic assay for the enantiomers of bevantolol in human plasma. *Pharm. Res.*, **8**, 6, 758–762.

Rosseel, M. T., Vermeulen, A. M., and Belpaire, F. M. (1991). Reversed-phase high-performance liquid chromatographic analysis of atenolol enantiomers in plasma after chiral derivatization with (+)-1-(9-fluorenyl)ethyl chloroformate. *J. Chromatogr.*, **568**, 239–245.

Roux, A., Blanchot, G., Baglin, A., and Flouvat, B. (1991). Liquid chromatographic analysis of propranolol enantiomers in human blood using precolumn derivatization with (+)-1-(9-fluorenyl)ethyl chloroformate. *J. Chromatogr.*, **570**, 453–461.

Ruelius, H. W., Tio, C. O., Knowles, J. H., McHugh, S. L., Schillings, R. T., and Sisenwine, S. F. (1979). Diastereomeric glucuronides of oxazepam: Isolation and stereoselective enzymic hydrolysis. *Drug Metab. Dispos.*, **7**, 40–43.

Ruud-Christensen, M. and Salvesen, B. Separation of (R)- and (S)-proxyphylline as diastereoisomeric camphanates by reversed-phase liquid chromatography. *J. Chromatogr.*, **303**, 433–435.

Saito, K., Horie, M., Nose, N., Nakagomi, K., and Nakazawa, H. (1992). High-performance liquid chromatography of histamine and 1-methylhistamine with on-column fluorescence derivatization. *J. Chromatogr.*, **595**, 163–168.

Sallustio, B. C., Abas, A., Hayball, P. J., Purdie, Y. J., and Meffin, P. J. (1986). Enantiospecific high-performance liquid chromatographic analysis of 2-phenylpropionic acid, ketoprofen and fenoprofen. *J. Chromatogr.*, **374**, 329–337.

Sampath, D. S. and Balaram, P. (1986a). A rapid procedure for the resolution of racemic gossypol. *J. Chem. Soc., Chem. Commun.*, 649–650.

Sampath, D. S. and Balaram, P. (1986b). Resolution of racemic gossypol and interaction of individual enantiomers with serum albumins and model peptides. *Biochim. Biophys. Acta*, **882**, 183–186.

Sar, F., Leroy, P., Nicolas, A., Archimbault, P., and Ambroggi, G. (1992). Determination of amikacin in dog plasma by reversed-phase ion-pairing liquid chromatography with post-column derivatization. *Anal. Lett.*, **25**, 7, 1235–1250.

Schaefer, H. G., Spahn, H., Lopez, L. M., and Derendorf, H. (1990). Simultaneous determination of propranolol and 4-hydroxypropranolol enantiomers after chiral derivatization using reversed-phase high-performance liquid chromatography. *J. Chromatogr.*, **527**, 351–359.

Schuster, D., Modi, M. W., Lalka, D., and Gengo, F. M. (1988). Reversed-phase high-performance liquid chromatographic assay to quantitate diastereomeric derivatives of metoprolol enantiomers in plasma. *J. Chromatogr.*, **433**, 318–325.

Schützner, W., Fanali, S., Rizzi, A., and Kenndler, E. (1993). Separation of diastereomeric derivatives of enantiomers by capillary zone electrophoresis with a polymer network: Use of polyvinylpyrrolidone as buffer-additive. *J. Chromatogr.*, **639**, 375–378.

Sedman, A. J. and Gal, J. (1983). Resolution of the enantiomers of propranolol and other beta-adrenergic antagonists by high-performance liquid chromatography. *J. Chromatogr.*, **278**, 199–203.

Seeman, J. I., Chavdarian, C. G., and Secor, H. V. (1985). Synthesis of the enantiomers of nornicotine. *J. Org. Chem.*, **50**, 5419–5421.

Shimada, K., Mitamura, K., Morita, M., and Hirakata, K. (1993). Separation of the diastereomers of baclofen by high-performance liquid chromatography using cyclodextrin as a mobile phase additive. *J. Liq. Chromatogr.*, **16**, 15, 3311–3320.

Shimizu, R., Ishii, K., Tsumagari, N., Tanigawa, M., Matsumoto, M., and Harrison, I. T. (1982). Determination of optical isomers in diltiazem hydrochloride by high-performance liquid chromatography. *J. Chromatogr.*, **253**, 101–108.

Shimizu, R., Kakimoto, T., Ishii, K., Fujimoto, Y., Nishi, H., and Tsumagari, N. (1986). New derivatization reagent for the resolution of optical isomers in diltiazem hydrochloride by high-performance liquid chromatography. *J. Chromatogr.*, **357**, 119–125.

Silber, B. and Riegelman, S. (1980). Stereospecific assay for (–)-and (+)-propranolol in human and dog plasma. *J. Pharmacol. Exp. Ther.*, **215**, 3, 643–649.

Sintov, A., Siden, R., and Levy, R. J. (1989). Sensitive high-performance liquid chromatographic assay using 9-fluorenylmethylchloroformate for monitoring controlled-release lidocaine in plasma. *J. Chromatogr.*, **496**, 335–344.

Skidmore, M. W. (1993). Derivatization for chromatographic resolution of optically active compounds. 2nd ed. In Blau, K., and Halket, J. N. (eds.), *Handbook of Derivatives for Chromatography*, 2nd ed. Chichester: Wiley, pp. 215–252.

Soine, W. H., Soine, P. J., Overton, B. W., and Garrettson, L. K. (1986). Product enantioselectivity in the N-glycosylation of amobarbital. *Drug Metab. Dispos.*, **14**, 619–621.

Soine, W. H., Soine, P. J., England, T. M., Graham, R. M., and Capps, G. (1994). Identification of the diastereomers of pentobarbital N-glycosides excreted in human urine. *Pharm. Res.*, **11**, 1535–1539.

Souter, R. W. (1976). High-performance liquid chromatographic separation of enantiomeric amines. *Chromatographia*, **9**, 12, 635–637.

Spahn, H. (1987). Formation of diastereomeric derivatives of 2-arylpropionic acids using L-leucinamide. *J. Chromatogr.*, **423**, 334–339.

Spahn, H. (1988a). Assay method for the product formation in in-vitro enzyme kinetic studies of uridinediphospho glucuronyltransferases: 2-Arylpropionic acid enantiomers. *J. Chromatogr.*, **426**, 305–314.

Spahn, H. (1988b). S-(+)-Flunoxaprofen chloride as chiral fluorescent reagent. *J. Chromatogr.*, **427**, 131–137.

Spahn, H. (1988c). S-(+)-Naproxen chloride as acylating agent for separating the enantiomers of chiral amines and alcohols. *Arch. Pharm.* (*Weinheim, Ger.*), **321**, 847–850.

Spahn, H. (1989). *Characterization of stereoselective processes in drug metabolism and pharmacokinetics*. Habilitation thesis, Johann Wolfgang Goethe-University, Frankfurt/Main.

Spahn, H., and Langguth, P. (1990). Chiral amines derived from 2-arylpropionic acids: Novel reagents for the liquid chromatographic: (LC) fluorescence assay of optically active carboxylic acid xenobiotics. *Pharm. Res.*, **7**, 12, 1262–1268.

Spahn, H., Krauss, D., and Mutschler, E. (1988). Enantiospecific determination of baclofen and its fluoro analogue in biological material. *Pharm. Res.*, **5**, 107–112.

Spahn, H., Wellstein, A., Pflugmann, G., Mutschler, E., and Palm, D. (1989). Radioreceptor assay of metoprolol in human plasma: Comparison with an enantiospecific high-performance liquid chromatographic (HPLC) procedure. *Pharm. Res.*, **6**, 2, 152–155.

Spahn, H., Henke, W., Langguth, P., Schloos, J., and Mutschler, E. (1990). Measurement of carvedilol enantiomers in human plasma and urine using S-naproxen chloride for chiral derivatization. *Arch. Pharm.* (*Weinheim, Ger.*), **323**, 465–469.

Spahn-Langguth, H. and Benet, L. Z. (1993). Microsomal acyl glucuronidation: Enzyme kinetic studies with labile glucuronides. *Pharmacology*, **46**, 268–273.

Spahn-Langguth, H., Podkowik, B., Stahl, E., Martin, E., and Mutschler, E. (1991). Improved enantiospecific RP-HPLC assays for propranolol in plasma and urine with pronethalol as internal standard. *J. Anal. Toxicol.*, **15**, 4, 209–213.

Spahn-Langguth, H., Hahn, G., Mutschler, E., Möhrke, W., and Langguth, P. (1992). Enantiospecific fluorescence assay for the monoamine oxidase inhibitor tranylcypromine and its applicability to pharmacokinetic studies. *J. Chromatogr.*, **584**, 221–228.

Special issue. (1994). General strategies and selection of derivatization reactions for LC *J. Chromatogr. B*, **659**, 1–273.

Stagni, G., Davis, P. J., and Ludden, T. M. (1991). Human pharmacokinetics of betaxolol enantiomers. *J. Pharm. Sci.*, **80**, 4, 321–324.

Stahl, E., Baumgartner, U., Henke, D., Schölmerich, J., Mutschler, E., and Spahn-Langguth, H. (1993). Rats with portacaval shunt as a potential experimental pharmacokinetic model for liver cirrhosis: Application to carvedilol stereopharmacokinetics. *Chirality*, **5**, 1–7.

Stockley, C. S., Wing, L. M. H., and Miners, J. O. (1991). Stereospecific high-performance liquid chromatographic assay for the enantiomers of phenylpropanolamine in human plasma. *Ther. Drug Monit.* **13**, 332–338.

Strolin Benedetti, M., Frigerio, E., Tamassi, V., Noseda, G., and Caldwell, J. (1992). The dispositional enantioselectivity of indobufen in man. *Biochem. Pharmacol.*, **43**, 9, 2032–2034.

Takasaki, W., Asami, M., Muramatsu, S., Hayashi, R., Tanaka, Y., Kawabata, K., and Hoshiyama, K. (1993). Stereoselective determination of the active metabolites of a new anti-inflammatory agent (CS-670) in human and rat plasma using antibody-mediated extraction and high-performance liquid chromatography. *J. Chromatogr.*, **613**, 67–77.

Te Koppele, J. M., van der Mark, E. J., Boerrigter, J. C. O., Brusse, J., van der Green, A., van der Greef, A. and Mulder, G. J. (1987). α-Bromovaleranylurea as model substrate for studies on pharmacokinetics of glutathione conjugation in the rat. I. (Bio)Synthesis, analysis and identification of diastereomeric glutathione conjugates and mercapturates. *J. Pharmacol. Exp. Ther.*, **239**, 898–904.

Thakker, D. R., Yagi, H., Sayer, J. M., Kapur, U., Levin, W., Chang, R. L., Wood, A. W., Conney, A. H., and Jerina, D. M. (1984). Effects of a 6-fluoro substituent on the metabolism of benzo[a]pyrene 7,8-dihydrodiol to bay-region diol epoxides by rat liver enzymes. *J. Biol. Chem.*, **259**, 18, 11249–11256.

Thakker, D. R., Levin, W., Yagi, H., Yeh, H. J. C., Ryan, D. E., Thomas, P. E., Conney, A. H., and Jerina, D. M. (1986). Stereoselective metabolism of the (+)-(S,S)- and (–)-(R,R)-enantiomers of trans-3,4-dihydroxy-3,4-dihydrobenzo[c]-phenanthrene by rat and mouse liver microsomes and by a purified and reconstituted cytochrome P-450 system. *J. Biol. Chem.*, **261**. 12, 5404–5413.

Thompson, J. A., Hull, J. E., and Norris, K. J. (1981). Glucuronidation of propranolol and 4′-hydroxypropranolol. Substrate specificity and stereoselectivity of rat liver microsomal glururonyltransferases. *Drug Metab. Dispos*, **9**, 5, 466–471.

Thompson, J. A., Holtzman, J. L., Tsuru, M., Lerman, C. L., and Holtzman, J. L. (1982). Procedure for the chiral derivatization and chromatographic resolution of R-(+)- and S-(–)-propranolol. *J. Chromatogr.*, **238**, 470–475.

Tomaszewski, J. and Rumore, M. M. (1994). Stereoisomeric drugs: FDA's policy statement and the impact on drug development. *Drug Dev. Ind. Pharm.*, **20**, 2, 119–139.

Toyo'oka, T., Ishibashi, M., and Terao, T. (1992a). Fluorescent chiral derivatization reagents for carboxylic acid enantiomers in high-performance liquid chromatography. *Analyst* (*London*), **117**, 4, 727–733.

Toyo'oka, T., Ishibashi, M., and Terao, T. (1992b). Resolution of carboxylic acid enantiomers by high-performance liquid chromatography with highly sensitive laser-induced fluorescence detection. *J. Chromatogr.*, **625**, 357–361.

Toyo'oka, T., Ishibashi, M., and Terao, T. (1992c). Resolution of carboxylic acid enantiomers by high-performance liquid chromatography with peroxyoxalate chemiluminescence detection. *J. Chromatogr.*, **627**, 75–86.

Toyo'oka, T., Ishibashi, M., Terao, T., and Imai, K. (1993). 4-(N,N-Dimethylaminosulfonyl)-7-(2-chloroformylpyrrolidin-1-yl)-2,1,3-benzoxadiazole: Novel fluorescent chiral derivatization reagents for the resolution of alcohol enantiomers by high-performance liquid chromatography. *Analyst* (*London*), **118**, 7, 759–763.

Tran, A. D., Blank, T., and Leopold, E. J. (1990). Free solution capillary electrophoresis and micellar electrokinetic resolution of amino acid enantiomers and peptide isomers with L-and D-Marfey's reagent. *J. Chromatogr.*, **516**, 241–249.

Turgeon, J., Kroemer, H., Prakash, C., Blair, I. A., and Roden, D. M. (1990).

Stereoselective determination of flecainide in human plasma by high-performance liquid chromatography with fluorescence detection. *J. Pharm. Sci.*, **79**, 2, 91–95.

Ueda, T., Kitamura, F., Mitchell, R., Metcalf, T., Kuwana, T., and Nakamoto, A. (1991). Chiral separation of naphthalene-2,3-dicarboxaldehyde-labeled amino acid enantiomers by cyclodextrin-modified micellar electrokinetic chromatography with laser-induced fluorescence detection. *Anal. Chem.*, **63**, 2979–2981.

Valentine, D., Chan, K. K., Scott, C. G., Johnson, K. K., Toth, K., and Saucy, G. (1976). Direct determinations of R/S enantiomer ratios of citronellic acid and related substances by nuclear magnetic resonance spectroscopy and high pressure liquid chromatography. *J. Org. Chem.*, **41**, 62–65.

van Bladeren, P. J., Sayer, J. M., Ryan, D. E., Thomas, P. E., Levin, W., and Jerina, D. M. (1985). Differential stereoselectivity of cytochromes P-450b and P-450c in the formation of naphthalene and anthracene 1,2-oxides. *J. Biol. Chem.*, **260**, 18, 10226–10235.

Vermeulen, A. M., Belpaire, F. M., Moerman, E., De Smet, F., and Bogaert, M. G. (1992). The influence of aging on the stereoselective pharmacokinetics of propranolol in the rat. *Chirality*, **4**, 73–79.

Vyas, K. P., Thakker, D. R., Levin, W., Yagi, H., Conney, A. H., and Jerina, D. M. (1982). Stereoselective metabolism of the optical isomers of trans-1,2-dihydroxy-1,2-dihydrophenanthrene to bay-region diol epoxides by rat liver microsomes. *Chem.-Biol. Interact.*, **38**, 203–213.

Vyas, K. P., van Bladeren, P. J., Thakker, D. R., Yagi, H., Sayer, J. M., Levin, W., and Jerina, D. M. (1983). Regioselectivity and stereoselectivity in the metabolism of trans-1,2-dihydroxy-1,2-dihydrobenz[a]anthracene by rat liver microsomes. *Mol. Pharmacol.*, **24**, 115–123.

Wallace, J. E., Shimek, E. L., Jr., Harris, S. C., and Stavchansky, S. (1981). Determination of promethazine in serum by liquid chromatography. *Clin. Chem. (Winston-Salem, N. C.)*, **27**, 2, 253–255.

Walle, T., Christ, D. D., Walle, U. K., and Wilson, M. J. (1985). Separation of the enantiomers of intact sulfate conjugates of adrenergic drugs by high-performance liquid chromatography after chiral derivatization. *J. Chromatogr.*, **341**, 213–216.

Walle, U. K., and Walle, T. (1989). Stereoselective sulfatation of terbutaline by the rat liver cytosol: Evaluation of experimental approaches. *Chirality*, **1**, 121–126.

Wang, C. P., Howell, S. R., Scatina, J., and Sisenwine, S. F. (1992). The disposition of venlafaxine enantiomers in dogs, rats, and humans receiving venlafaxine. *Chirality*, **4**, 84–90.

Webb, J. G., Street., J. A., Bagwell, E. E., Walle, T., and Gafney, T. E. (1988). Stereoselective secretion of atenolol from PC12 cells. *J. Pharmacol. Exp. Ther.*, **247**, 958–965.

Weber, H., Spahn, H., Mutschler, E., and Möhrke, W. (1984). Activated α-alkyl-α-arylacetic acid enantiomers for the stereoselective thin-layer chromatographic and high-performance liquid chromatographic determination of chiral amines. *J. Chromatogr.*, **307**, 145–153.

Weber-Grandke, H., Hahn, G., Mutschler, E., Möhrke, W., Langguth, P., and Spahn-Langguth, H. (1993). Pharmacokinetics of tranylcypromine enantiomers in healthy volunteers after p.o. dosage of racemic drug and the single enantiomers. *Br. J. Clin. Pharm.*, **36**, 363–365.

Westley, J. W. and Halpern, B. (1968). The use of (–)-menthyl chloroformate in the optical analysis of asymmetric amino and hydroxyl compounds by gas chromatography. *J. Org. Chem.*, **33**, 10, 3978–3980.

Williams, K. M. (1984). Kinetics of misonidazole enantiomers. *Clin. Pharmacol. Ther.*, **36**, 817–823.

Wilson, M. J. and Walle, T. (1984). Silica gel high-performance liquid chromatography for the simultaneous determination of propranolol and 4-hydroxypropranolol enantiomers after chiral derivatization. *J. Chromatogr.*, **310**, 424–430.

Witte, D. T., de Zeeuw, R. A., and Drenth, B. F. H. (1990). Chiral derivatization of promethazine with (–)-menthylchloroformate for enantiomeric separation by RP-HPLC. *J. High Resolut. J. Chromatogr.*, **13**, 569–571.

Witte, D. T., Ahnoff, M., Karlsson, K.-E., Franke, J.-P., and de Zeeuw, R. A. (1993). Liquid chromatographic chiral separations of the N-6-(endo-2-norbornyl)-9-methyladenine enantiomers. *J. Chromatogr.*, **641**, 39–48.

Wright, M. R. and Jamali, F. (1993). Limited extent of stereochemical conversion of chiral nonsteroidal anti-inflammatory drugs induced by derivatization methods employing ethyl chloroformate. *J. Chromatogr.*, **616**, 59–65.

Wright, M. R., Sattari, S., Brocks, D. R., and Jamali, F. (1992). Improved high-performance liquid chromatographic assay method for the enantiomers of ibuprofen. *J. Chromatogr.*, **583**, 259–265.

Wuis, E. W., Beneken Kolmer, E. W. J., van Beijsterveldt, L. E. C., Burgers, R. C. M., Vree, T. B., and van der Kleyn, E. (1987). Enantioselective high-performance liquid chromatographic determination of baclofen after derivatization with a chiral adduct of o-phthaldialdehyde. *J. Chromatogr.*, **415**, 419–422.

Yagi, H., Akagi, H., Thakker, D. R., Mah, H. D., Koreeda, M., and Jerina, D. M. (1977). Absolute Stereochemistry of the highly mutagenic 7,8-diol 9,10-epoxides derived from the potent carcinogen trans-7,8-dihydroxy-7,8-dihydrobenzo-[a]pyrene. *J. Am. Chem. Soc.*, **99**, 7, 2358–2359.

Yagi, H., Vyas, K. P., Tada, M., Thakker, D. R., and Jerina, D. M. (1982). Synthesis of the enantiomeric bay-region diol epoxides of benz[a]anthracene and chrysene. *J. Org. Chem.*, **47**, 1110–1117.

Yang, S. K. and Fu, P. P. (1984). Stereoselective metabolism of 7-methylbenz[a]anthracene: Absolute configuration of five dihydrodiol metabolites and the effect of dihydrodiol conformation on circular dichroism spectra. *Chem.-Biol. Interact.*, **49**, 71–88.

Yang, S. K., Gelboin, H. V., Weber, J. D., Sankaran, V., Fischer, D. L., and Engel, J. F. (1977). Resolution of optical isomers by high-pressure liquid chromatography. The separation of benzo[a]pyrene trans-diol derivatives. *Anal. Biochem.*, **78**, 520–526.

Zhang, Z., Malikin, G., and Lam, S. (1992). Novel polymeric reagent for synthesizing 9-fluorenylmethoxycarbonyl L-prolinyl derivatives for chiral high-performance liquid chromatography of amino acids. *J. Chromatogr.*, **603**, 279–284.

Zhong-Yuan, Y. and Ru-Zheng, X. (1989). Investigation of the enantiomeric impurity of epinephrine hydrochloride injections. *Chirality*, **1**, 92–93.

Zhou, F.-X. and Krull, I. S. (1993). Direct enantiomeric analysis of amphetamine in plasma by simultaneous solid phase extraction and chiral derivatization. *Chromatographia*, **35**, 3–4, 153–159.

Zhou, Y. and Sun, Z. P. (1990). Resolution of the enantiomers of some β-aminoalcohols using chiral derivatization and reversed-phase liquid chromatography. *Acta Pharm. Sin.*, **25**, 4, 311–314.

Zhou, Y., Luan, P., Liu, L., and Sun, Z. P. (1994). Chiral derivatizing reagents for drug enantiomers bearing hydroxyl groups. *J. Chromatogr. B*, **659**, 109–126.

CHAPTER

17

CIRCULAR DICHROISM SPECTROSCOPY IN THE ANALYSIS OF CHIRAL DRUGS

PIERO SALVADORI, CARLO BERTUCCI

Centro Studio CNR Macromolecole Stereordinate Otticamente Attive
Dipartimento di Chimica e Chimica Industriale
Università di Pisa
56126, Italy

CARLO ROSINI

Dipartimento di Chimica,
Università della Basilicata a Potenza, via Nazario Sauro,
85, 85100 Potenta, Italy

17.1. INTRODUCTION

The aim of the present chapter is to discuss the principles of circular dichroism (CD) spectroscopy and its main applications to the determination of the molecular absolute stereochemistry of organic molecules, with particular reference to those having pharmaceutical activity. Taking into account the complexity of the problem, we cannot provide here an exhaustive review, but rather, we wish to recall the basic principles of optical activity and to illustrate the use of CD for stereochemical and analytical determinations in the area of chiral drugs, by making use of some selected examples.

17.2. BASIC DEFINITIONS*

Optical rotation (OR) and circular dichroism (CD) are different aspects of the same physical phenomenon: the interaction of plane-polarized electromagnetic radiation with a collection of chiral molecules. Right- and left-

*See (1–12).

The Impact of Stereochemistry on Drug Development and Use, Edited by Hassan Y. Aboul-Enein and Irving W. Wainer. Chemical Analysis Series, Vol. 142.
ISBN 0-471-59644-2

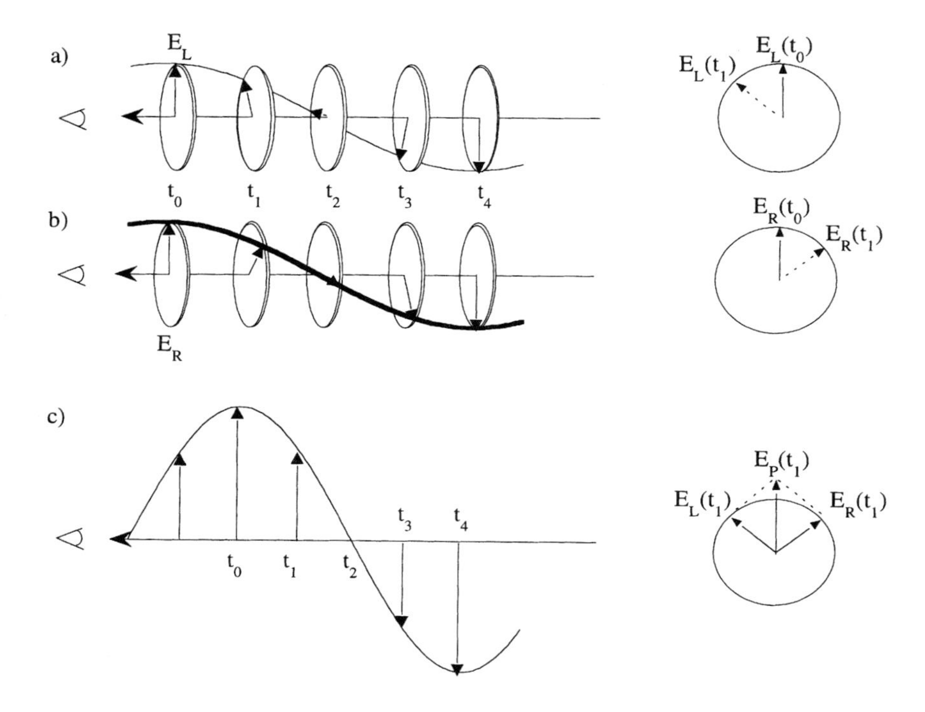

Figure 17.1. Left-circularly (*a*) and right- circularly (*b*) polarized radiation, with their corresponding views by an observer looking toward the source: Their result, a plane-polarized radiation beam, is represented in (*c*), with its projection.

circularly polarized radiation beams having the same frequency, intensity, and phase (Figures 17.1*a* and 17.1*b*) can combine to afford plane-polarized radiation (Figure 17.1*c*) (for simplicity, only the electric vector is represented). The left- and right-circularly polarized components pass through the optically active medium with different speeds, owing to their different refractive indices, that is, n_L and n_R. This difference in velocity leaving the chiral medium results in a difference in the phase of the two circularly polarized components and then a rotation of the plane of polarization. Figure 17.2 illustrates the electric vector of a plane-polarized (p_0) beam, with its right-($^RE_1$) and left-($^LE_1$) circularly polarized components. Passing through an optically active medium, where, for instance, $n_L < n_R$ and then $v_R < v_L$, the arc $^RE_1 - {}^RE_2$ becomes minor than the corresponding arc $^LE_1 - {}^LE_2$. Leaving the optically active medium, the combination of the two circularly polarized beams ($^RE_3$ and $^LE_3$) gives rise to radiation polarized in the plane p'_0, which is rotated through α negative angle

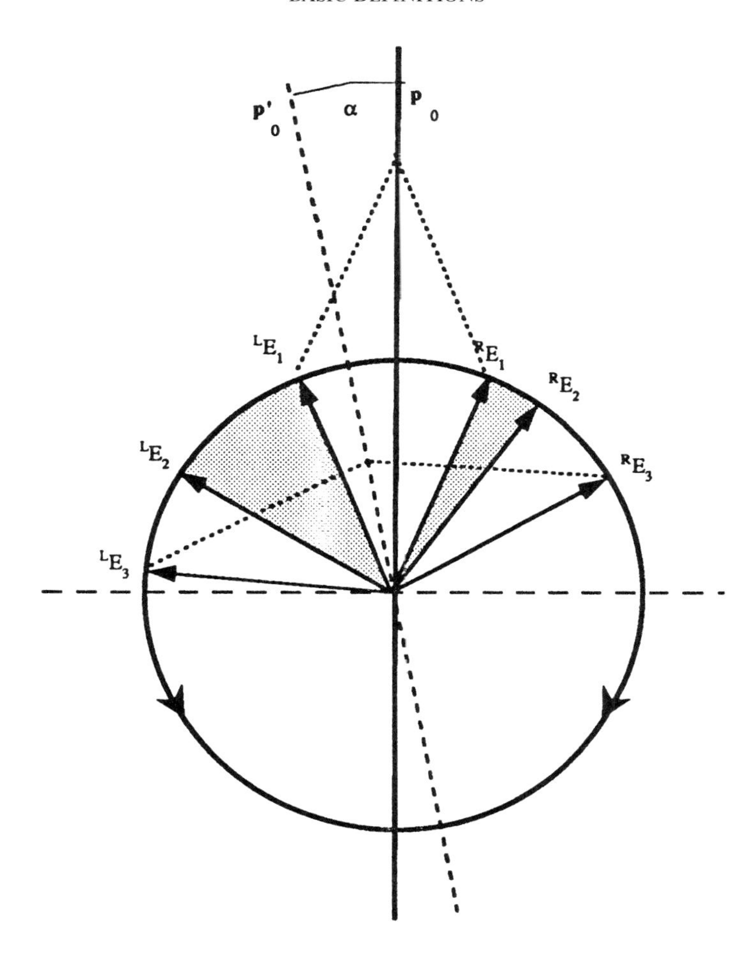

Figure 17.2. The optical rotation α results from the different speed of the circularly polarized components in a chiral medium. ^{L}E and ^{R}E are the left- and right-circularly polarized components, respectively, of the plane-polarized incident radiation p_0. When leaving such a medium, the two radiations have passed arcs $(^{R}E_2 - {}^{R}E_1)$ and $(^{L}E_2 - {}^{L}E_1)$ having a different length.

a with respect to that of the incident beam p_0. The specific optical rotation can be defined as:

$$[\alpha]_{\lambda}^{T} = \frac{\alpha 100}{l.\mathrm{c}}$$

where α is the measured angle of optical rotation in degrees, l the pathlength in

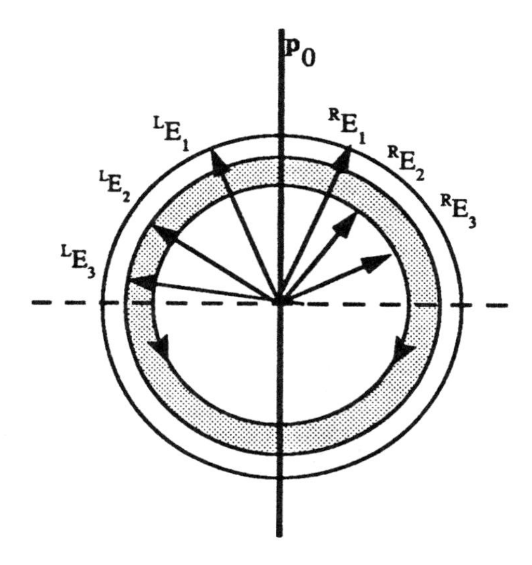

Figure 17.3. The two circularly polarized components (^{L}E and ^{R}E) are differentially absorbed by the optically active medium. This leads to a different molar extinction coefficient for the two circularly polarized radiation beams (i.e., $\varepsilon_L < \varepsilon_R$). The value $\Delta\varepsilon = \varepsilon_L - \varepsilon_R$ constitutes the circular dichroism.

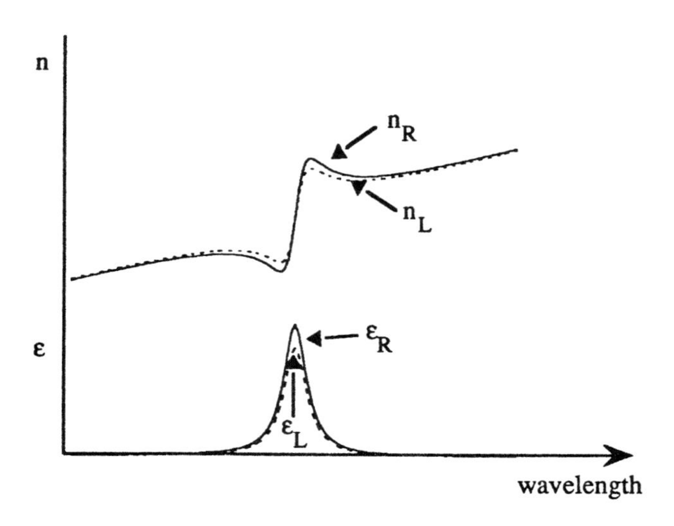

Figure 17.4. n and ε as functions of the wavelength.

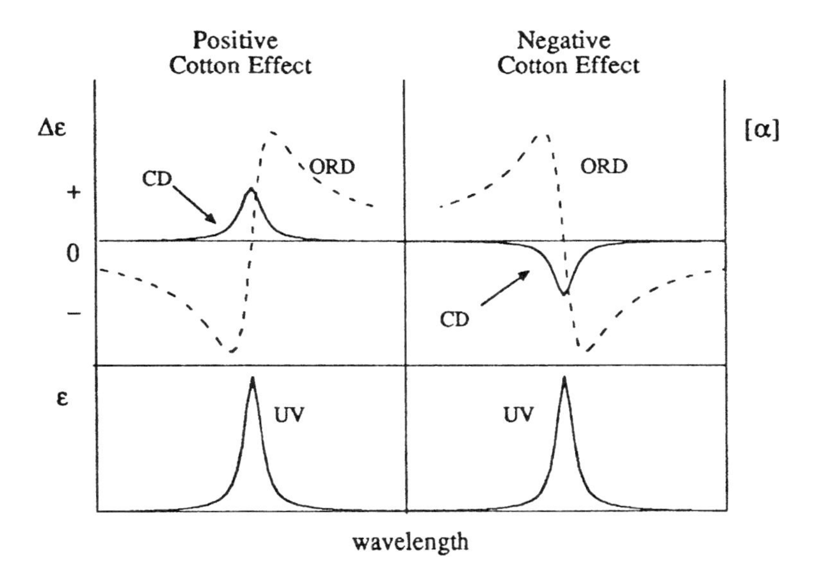

Figure 17.5. Absorption (UV), optical rotatory dispersion (ORD), and circular dichroism (CD) corresponding to the same electronic transition for the two antipodes of the same chiral molecule.

decimeters, c the concentration in grams of solute in 100 cc of solution, T the temperature, and λ the wavelength of the incident radiation in nanometer. The two circularly polarized components are also differentially absorbed by the optically active medium (Figure 17.3); the moduli of the two electric vectors of the right- and left-circularly polarized beams are reduced to a different extent, so two different molar extinction coefficients ε_L and ε_R originate. The difference $\Delta\varepsilon = \varepsilon_L - \varepsilon_R$ constitutes the circular dichroism CD. Both $n_L - n_R$, which determines $[\alpha]$, and $\varepsilon_L - \varepsilon_R$ depend (1) on λ (Figure 17.4). Figure 17.5 reports the shape of $[\alpha]$ and $\Delta\varepsilon$, corresponding to the same electronic transition (Cotton effect), for the two antipodes of a hypothetic chiral molecule, as well as the allied absorption curve (UV). A plot of $[\alpha]$ versus λ constitutes an ORD spectrum, whereas a plot of $\Delta\varepsilon$ versus λ provides a CD spectrum. ORD and CD give equivalent information; however, since the CD can be measured only in correspondence of an absorption band, the interpretation of CD data is generally easier than that of data from its dispersive counterpart, ORD, where it is more difficult to single out the contribution of each Cotton effect (6). Therefore, we will consider only CD spectroscopy in order to keep our treatment simpler.

17.3. NATURE OF THE ELECTRONIC TRANSITIONS

When a photon is absorbed by a molecule, passage from the ground (S_0) to an excited state (S_i) occurs. At a very simple level, we can talk about electronic transitions from an occupied (σ, π) or a nonbonding (n) MO to an empty (π^*,σ^*) MO. Therefore, we have $n \rightarrow \pi^*$, $\pi \rightarrow \pi^*$ transitions, and so on. The distribution of electronic charge is different in the two states and, during the transition, a redistribution of electronic charge occurs, giving rise to a transient moment. An electric dipole transition moment is generated if such redistribution corresponds to a linear displacement of charge, whereas a magnetic transition moment is generated by a rotation of charge. The transition charge density can be defined as follows:

$$\rho_{0i}(x, y, z) = -e\Psi_0(x, y, z)\Psi_i(x, y, z)$$

where Ψ_0 is the starting MO and Ψ_i the final MO. That is, it can be calculated by picking a point in space and determining the signed product of the two MO values for that point. If ρ is an electric dipole moment, we have an allowed transition and this dipole moment has a well-defined orientation within the chromophore. Let us consider the case of the $\pi \rightarrow \pi^*$ transition (13, 14) of the olefin chromophore, as presented in Figure 17.6. The transition charge density is given by the product $-e\pi\pi^*$ and it corresponds to an electric transition dipole moment $\vec{\mu}_{\pi\pi^*}$, directed from the atom with the defect of negative

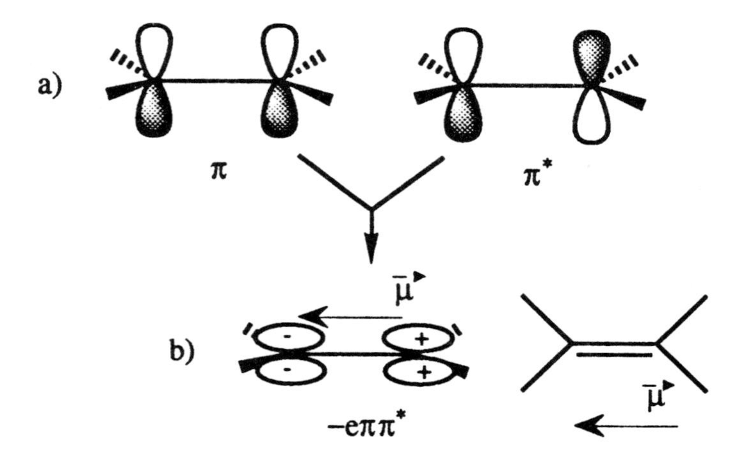

Figure 17.6. Origin of the electric dipole transition moment for the $\pi - \pi^*$ transition of the ethylene chromophore. (*a*) The two MOs involved and (*b*) the product-$e\pi\pi^*$, which gives rise to the dipole μ.

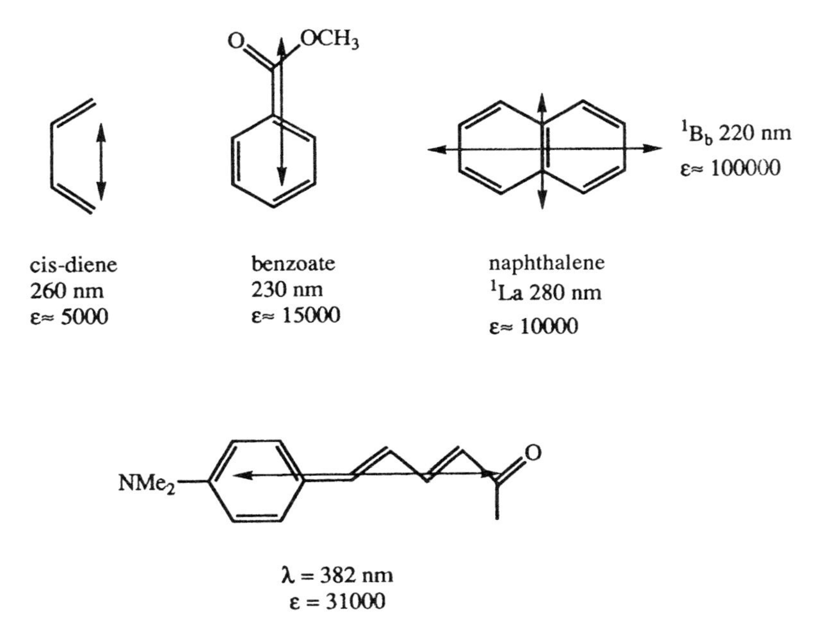

Figure 17.7. Polarization directions of the allowed transitions in some common organic chromophores.

charge to the atom with an excess of negative charge. This transition is dipole-allowed and then shows a high extinction coefficient (ε 10,000) (ca). Figure 17.7 illustrates other electrically allowed transitions present in the most common organic chromophores, with their polarization directions, which, in principle, can be established theoretically (13) (MO calculations) or experimentally (13) (linear dichroism measurements).

The dipolar strength of a given electronic transition $0 \rightarrow i$ is defined as

$$D_{0i} = \mu_{0_i} \mu_{0_i}$$

Experimentally, this quantity is given (2) by the area of the corresponding absorption band and can be approximated as follows (in c.g.s. units):

$$D_{0_i} = 92 * 10^{-40} \varepsilon_{\max} \Delta\tilde{\nu}/\Delta\tilde{\nu}_{\max}$$

where $\Delta\tilde{\nu}$ is the width of the band at half the maximum. In addition to the electrically allowed transitions, which dominate the isotropic absorption spectrum, there is another type of electronic transition that is very important

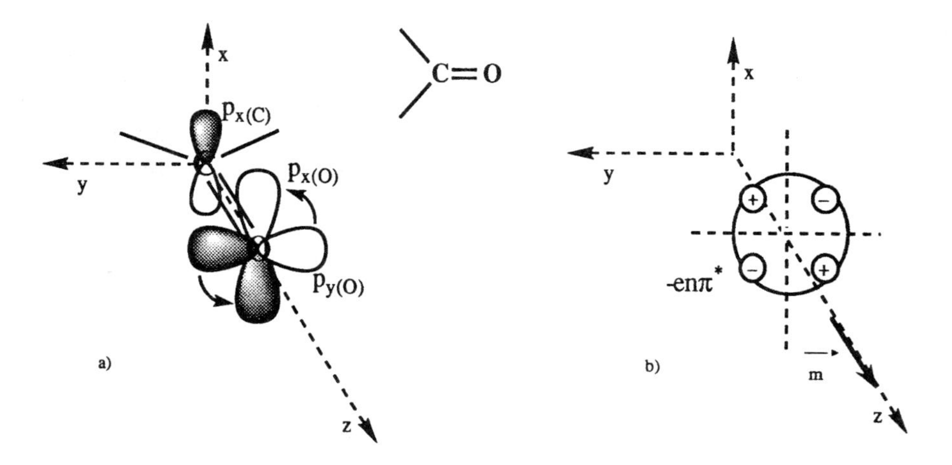

Figure 17.8. The molecular orbitals n and π^* (a), their product $-en\pi^*$ (an electric quadrupole), and the magnetic transition moment (b) of the carbonyl chromophore $n-\pi^*$ transition.

from the optical activity point of view. Let us consider the $n \rightarrow \pi^*$ transition of the carbonyl chromophore (9, 13, 14). With the choice of axis completed in Figure 17.8a, the n orbital is a pure p_y orbital localized on the oxygen atom, whereas the π^* MO is a linear combination of $p_x(C)$ and $p_x(O)$ with suitable coefficients c' and c'' for the p atomic orbitals on O and C, respectively. Therefore,

$$\rho_{n\rightarrow\pi^*}(x, y, z) = -en\pi^* = -ep_y(O)[c'p_x(O) - c''p_x(C)] = -c'ep_y(O)p_x(O)$$

Indeed, if we take into account that $p_y(O)$ is different from zero only in the proximity of the oxygen atom, the transition charge density will be different from zero only on the oxygen atom. This product will have the shape reported in Figure 17.8b: The transition charge density possesses only an electric quadrupole moment and then it will be not allowed. This transition corresponds to a $p_y \rightarrow p_x$ excitation on the oxygen atom. The charge rotates around the z axis and a magnetic transition moment m is developed along the positive z axis (right-hand rule). This transition is then electrically forbidden, magnetically allowed. In order for the electronic transition $S_0 \rightarrow S_i$ to be optically active, that is, to give rise to a Cotton effect, it must possess an electric and a magnetic transition moment and they must not to be orthogonal each other. In fact, the rotational strength R_{0_i}, allied to the transition $S_0 \rightarrow S_i$, is defined (1, 2) as

$$R_{0_i} = \mu_{0_i} m_{0_i} = /\mu_{0i}//m_{0_i}/\cos\theta$$

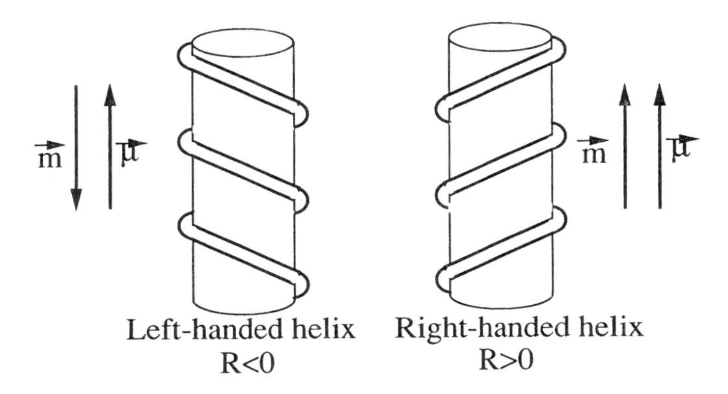

Figure 17.9. A left-handed motion of charge gives rise to a negative rotational strength (the electric and magnetic transition dipoles are antiparallel), whereas a right-handed charge motion gives rise to a positive rotational strength (the same vectors are parallel).

where θ is the angle between the two transition dipoles. Experimentally, this quantity is given (2) by

$$R_{0_i} = 23 * 10^{-40}\ \Delta\varepsilon_{max} \Delta\tilde{\nu} / \tilde{\nu}_{max}$$

The simultaneous presence of the electric and magnetic transition moments, not mutually perpendicular, requires (2) a transitional charge displacement along a helix: A right-handed helix will give a positive rotational strength, whereas a left-handed helix will yield a negative rotational strength (Figure 17.9). The electronic transition of a given chromophore becomes optically

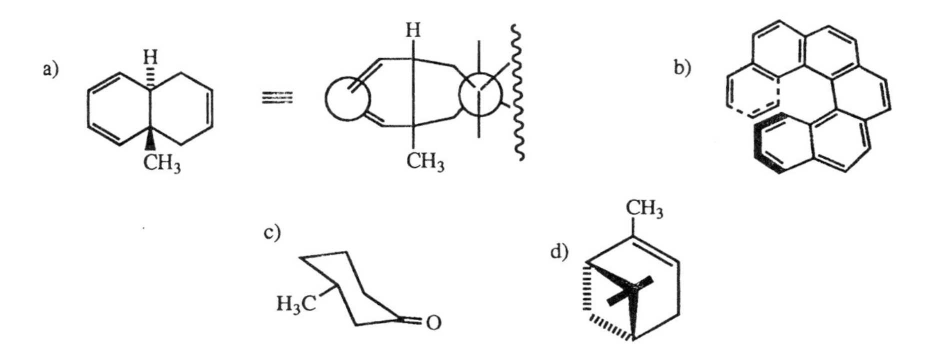

Figure 17.10. Two examples of intrinsically dissymmetric chromophores (*a*) and (*b*) and two examples of dissymmetrically perturbed chromophores (*c*) and (*d*).

active, according to Moscowitz (15), because it can occur in an intrinsically chiral chromophore [e.g., a distorted diene in (+)-*trans*-9-methyl-1,4,9,10-tetrahydronaphthalene or the extended aromatic chromophore of (*M*)-hexahelicene, Figures 17.10*a* and 17.10*b* respectively] or in a chromophore having local symmetry, but dissymmetrically perturbed [e.g., the carbonyl group in (*R*)-3-methylcyclohexanone or the olefinic double bond of (–)-α-pinene, Figures 17.10*c* and 17.10*d*, respectively]. In the case of intrinsically chiral chromophores, each transition generates collinear electric and magnetic dipole moments, and then a nonvanishing rotational strength is originated. In the case of dissymmetrically perturbed chromophores, each transition possesses only the electric or magnetic moment: The lacking moment (magnetic or electric) is induced in the molecule by the presence of the chirally disposed perturber.

17.4. THE INTERPRETATION OF CD DATA

The energy position, sign, and intensity of single Cotton effects depend on overall chemical structure and hence CD and more general chiroptical data are potential reservoirs of structural information. However, the analysis of a CD spectrum must be done with care and the extraction of geometrical information from CD data is not always immediate. The CD data can be interpreted by one of the following ways (4, 5): empirical, semiempirical, and nonempirical.

17.4.1. Empirical Analysis

Empirical analysis of the CD data implies building a spectra-structure relationship with reference to those of a molecule of known structure. For instance, in Figure 17.11, the CD spectra (16) of several enantiomeric 1,4-benzodiazepin-2-ones are reported. From the sign of the CD bands observed, it is possible to establish the absolute configuration. In fact, it is known (17) that the (*S*)-enantiomer of 7-chloro-1,3-dihydro-3-i-propyl-5-phenyl-2H-1,4-benzodiazepin-2-one shows positive ellipticity between 250 and 260 nm (Figure 17.11). Since the compounds examined are structurally very close to the reference substance, and they all show negative CD in the same spectral range, it is concluded that the *R* absolute configuration can be assigned to the benzodiazepines under investigation.

The chromophore acts as a probe of the chirality around it. A very instructive example has been reported by Mislow (18): Two ketones having a different chemical structure show mirror-image ORD curves in correspondence to the $n \rightarrow \pi^*$ transition, because the chromophore senses the chirality of its immediate environment, which is opposite in both cases. This fact can be exploited for configurational assignments (19). The enantiomers of terfenadine,

$$HO-C-\quad N-(CH_2)_3-\overset{OH}{\underset{H}{C}}-\quad -Bu^t$$

a nonsedating antihistamine and one of the most frequently prescribed antihistamines in the United States, have been separated by HPLC using a chiral stationary phase obtained by covalently bonding an ovomucoid protein to silica. The CD spectrum of the less strongly retained enantiomer shows a weak, but well-defined, negative Cotton effect at about 225 nm, reasonably due to an optically active $\pi \rightarrow \pi^*$ transition of the benzene chromophore. Taking into

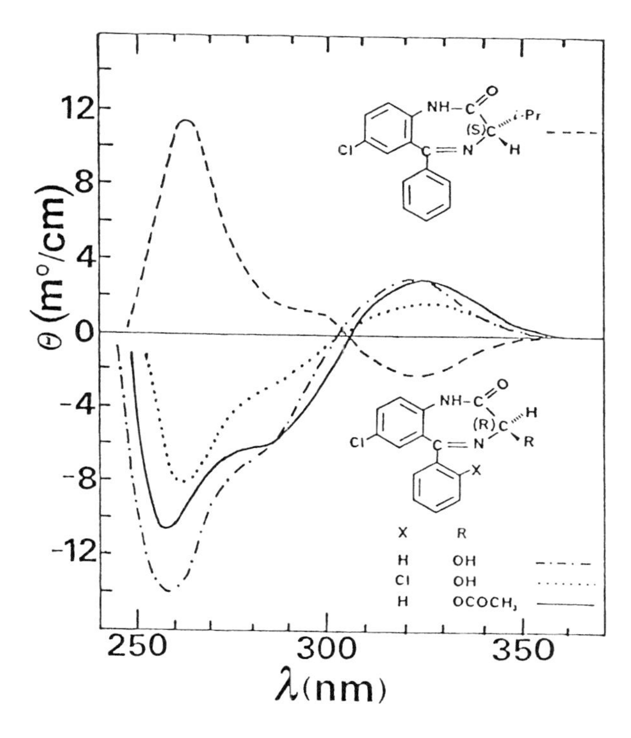

Figure 17.11. The comparison of the CD spectra of some optically active 3-substituted 1,4-benzodiazepin-2-ones with that of a structurally related compound having a known absolute configuration.

account the chemical structure of terfenadine, one immediately realizes that the optically active chromophore is a chiral benzyl alcohol. Then an external reference is provided by the series of alkylphenylcarbinols, whose absolute configuration has been safely established. The CD spectrum of (*S*)-1-phenyl-1-butanol shows a negative Cotton effect at 205 nm: The difference in wavelength between the two compounds can be explained by taking into account the presence of the *para* ter-butyl group in the terfenadine molecule. This group may exert a red-shift effect on the benzene transitions. The sign of this Cotton effect, which is the same as observed for the enantiomer of terfenadine studied, allows us to assign the (*S*) absolute configuration to the less retained antipode upon the ovomucoid CSP (19). It is important to note that such configurational assignment has been done because, even if the structure of the two compounds is quite different, they contain the same chromophore in surroundings that are reasonably comparable. However, making empirical correlations, one has to be very prudent: One uses them only if any other kind of aid is absent. Correlations of this kind are not possible between molecules having similar structures but different chromophores. An instructive example can be found in the field of natural product stereochemistry. Colchicine **1** and analogous alkaloids cannot be correlated (20) to other simple biphenyl systems (e.g., glaucine **2**):

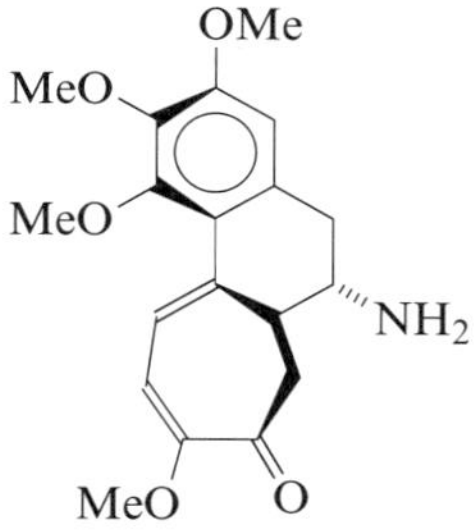

OMe
MeO
MeO
NH_2
MeO O
1

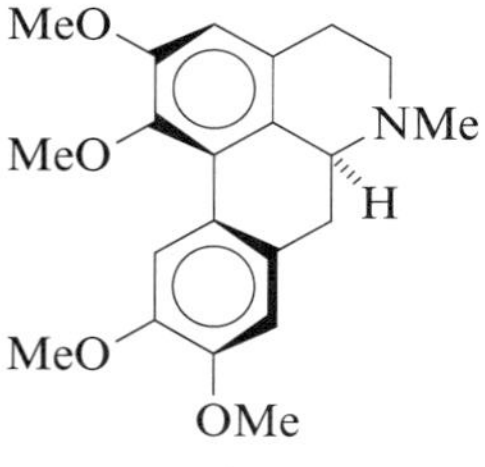

MeO
MeO
NMe
H
MeO
OMe
2

because the presence of the tropolone chromophore introduces deep differences in the CD spectra, making difficult and unsafe any configurational correlation. It is interesting to note that this configurational correlation can be actually established by making use of the liquid crystal-induced optical activity, which is more sensitive to the overall molecular shape than to the nature of the chromophores (21).

17.4.2. Semiempirical Approach

Since the beginning of the 1960s, attempts have been made to formulate rules (mainly on empirical grounds, even if often a theoretical basis has been found for some of them) to build spectrum-structure relationships. We can classify these rules as follows:

1. Helicity rules, which correlate the sense of helicity of an intrinsically dissymmetric chromophore with the CD sign of a given Cotton effect. Examples are provided by the *cis*-diene rule (22) and the disulfide chromophore rule (23). Other rules concern the α,β-unsaturated ketone chromophore (24) and the biphenyl chromophore (25). The latter is very useful for determining the absolute configuration of metabolites of polycyclic aromatic hydrocarbons (26) as well as in the field of natural product chemistry (e.g., aporphine alkaloids) (25).
2. Sector rules, which provide the sign of the contribution to a given Cotton effect due to a substituent, changing the position of the substituent with respect to the chromophore itself. The oldest and most reliable sector rule is the octant rule (27), for the saturated carbonyl chromophore, formulated in 1961. In more recent years, sector rules have been proposed for the oxirane (28) and benzene (29) chromophores. We shall discuss here, as an example, the determination of the absolute configuration of compound **3** (30).

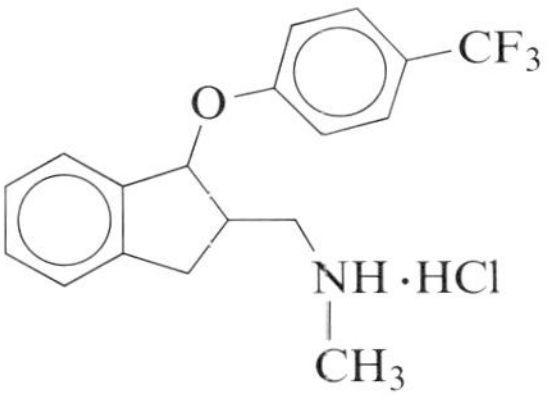

CF_3
O
NH·HCl
CH_3

3

Compound **3** is a new serotonin uptake inhibitor, and studies have shown that the dextrorotatory antipode is at least 10 times more active *in vitro* and *in vivo* than its levorotatory enantiomer. H.E. Smith et al. have formulated a sector rule that correlates the sign of the lowest-energy Cotton effect (^1L_b transition) of several compounds containing the benzene chromophore such as phenyl carbinols, carbinamines, as well as indane derivatives. For the last substances, the rule is as follows:

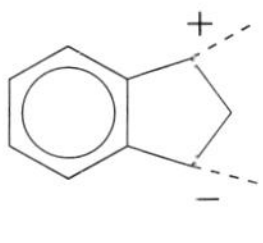

+
−

that is, a substituent located in the near sectors (over the benzene ring) contributes to CD as described in the above picture. Compound **3** can be treated as a disubstituted indane compound. On the reasonable assumption that the influence of the stereocenter at C-2 will have only a minor effect on the CD of ^1L_b transition, then the enantiomer of **3** that shows positive CD will have, at C-1, the absolute configuration shown below, that is, *S*.

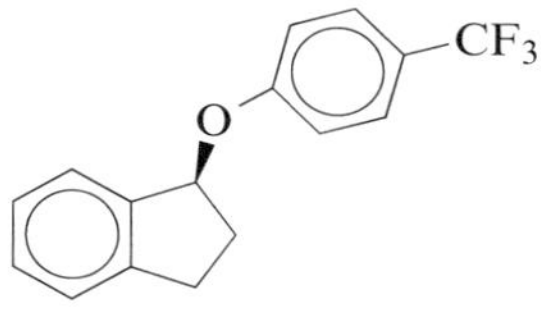

CF_3

O

Since the relative configuration of the two stereocenters has been already established as *cis*, the absolute stereochemistry of the dextrorotatory antipode of **3** is 1*S*, 2*S*.

17.4.3. Nonempirical Approach

Methods are also available that allow spectra–structure relationships to be established without any comparison to a reference molecule (nonempirical methods). It is in fact possible, once a given molecular geometry has been assumed, to calculate, for instance, the sign of the rotational strength of a chosen transition employing two different groups of methods: the molecular orbital (MO) approach (9,10,12,31) and the independent-system (IS) approach (9,10,12). The former makes use of overall molecular wavefunctions to provide theoretical rotatory strengths to be compared with the experimental ones. The reliability of this method depends mainly on the quality of the wavefunction of the ground and excited states. This aspect can constitute a very difficult task in particular for molecules having complex structures, such as several pharmaceutical compounds, and represents a limitation for the general applicability of this technique. The latter approach is based on the so-called independent-system methods: i.e., in a chiral molecule, one (or more) chomophore, where the electronic transition under examination is localized, and one (or more) perturber which "modifies" the above transition are separated. Optical activity arises from the interaction of the chromophore(s) and perturber(s). Among the independent-system techniques, those based on coupled oscillators or polarizability models have received considerable attention (9,10,12,32). Such models have been quite successful for molecules where two identical and interacting chromophores can be identified ("exciton model"). This nonempirical model for interpreting CD data in order to obtain stereochemical information was first applied (33) by Mason to determine the absolute configuration of the alkaloid calycanthine. The basic principle of the exciton model can be described as follows, by exploiting the example of the chiral dibenzoate rule due to Harada and Nakanishi (32). The *p*-bromobenzoate chromophore presents an electrically allowed transition at about 244 nm (ε 19,000), which is polarized along the axis joining the bromine atom and the carbonylic carbon atom.

Let us now consider a simple chiral structure where two *p*-bromobenzoate chromophores have been introduced, such as the *bis*-chromophoric derivative of (1*R*, 2*R*)-cyclohexanediol, (Figure 17.12*a*). In this situation, the two benzoate chromophores are near each other, interacting through space; then the excited states split into two levels, giving rise to red- and blue-shifted maxima (with respect to the wavelength position of the single monomer, Figure 17.12*b*). In absorption spectroscopy, the only observable phenomenon is the doubling of the molar extinction coefficient. On the contrary, in CD spectroscopy the two transitions give rise to Cotton effects of an opposite sign, leading to a bisignate curve ("exciton couplet," Figure 17.12*c*). In the case of the *p*-bromobenzoate chromophore, the 244-nm band gives rise to a couplet with extrema at 236 nm and 256 nm. The important observation, due to Harada and Nakanishi (32), is that the sign of the first Cotton effect at a longer wavelength reflects the chirality of the system: In the case reported in Figure 17.12, the positive first Cotton effect shows that the chirality defined by the transition dipole moments is positive. Since all esters have an *s-trans* conformation around the O—CO bond, the direction of the benzoate long axis approximately represents the direction of the alcoholic C—O bond. Thus, the chirality of the two axes gives the absolute sense of the twist of the two C—O bonds, and this establishes the molecular absolute configuration.

The applications of these considerations to structural determination are numerous and concern natural product stereochemistry (34); there have been stereochemical studies in the field of metabolism of polycyclic aromatic hydrocarbons (35) and of important drugs such as carbamazepine (36). We shall describe in the following the application (37) of the nonempirical approach to the determination of the absolute stereochemistry of *cis*-(+)-2-hydroxy-2-phenylcyclohexanecarboxylic acid (cicloxilic acid):

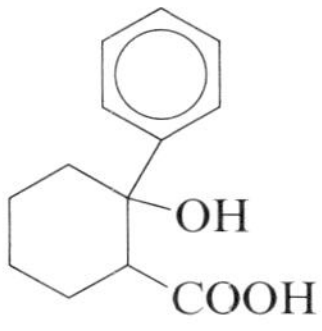

OH
COOH

an important choleretic agent, where a second suitable chromophore is introduced in order to obtain a CD exciton couplet and assign the absolute stero-

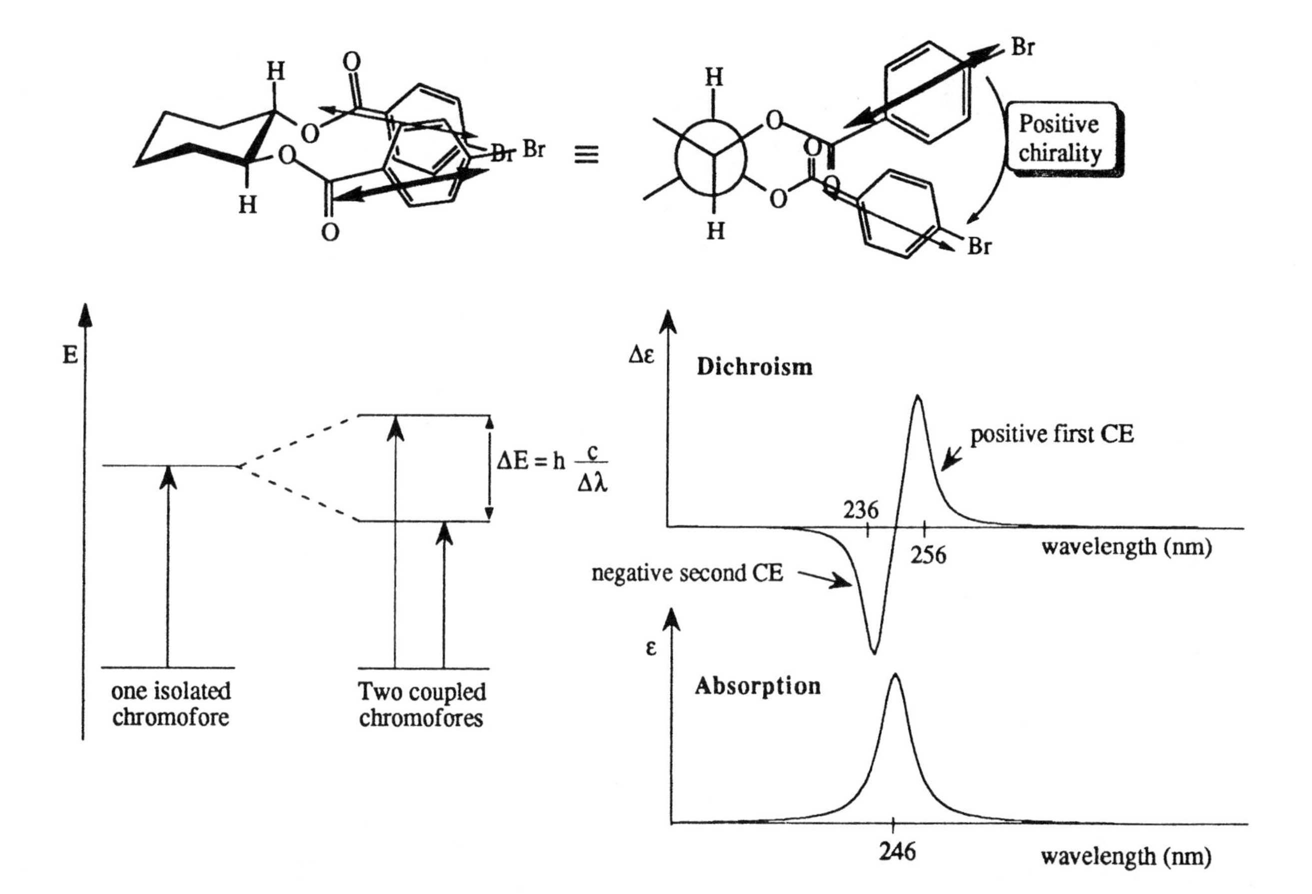

Figure 17.12. (*a*) The structure of the most stable conformer of (*R*, *R*)-cyclohexanediol, *bis*(*p*-bromo)benzoate, and its Newman projection along the C1—C2 bond. (*b*) The two excited states separated by the energy ΔE, derived from the coupling of the chromophores. (*c*) Related absorption and CD spectra of the dimer.

chemistry. This molecule possesses a chromophore, the alkyl-substituted benzene, the electronic transitions of which are well characterized. In particular, it shows an intense absorption band at about 180 nm assigned to the electrically allowed degenerate $^1A_{1g} \rightarrow {}^1E_{1u}$ transition, the two components of which are polarized as described below:

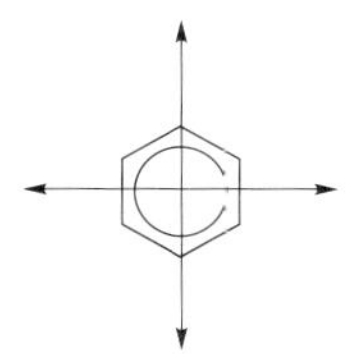

By introducing a second chromophore having a well-defined electrically allowed transition, exciton coupling effects can be observed in the CD spectrum, allowing the assignment of the absolute configuration. The transformation of COOH to NH_2 (Hofmann degradation of amides) affords the corresponding aminoalcohol, which, in turn, can be transformed in the corresponding *p*-chlorobenzamide, having the same absolute stereochemistry of the starting acid, since the Hofmann reaction takes place with a complete retention of configuration. The *p*-chlorobenzamide chromophore presents an allowed transition at 235 nm, polarized along the C*-N axis. Then the coupling between this amide transition and the aforementioned allowed transitions of the alkylbenzene chromophore could afford the CD spectral feature suitable for the stereochemical assignment. As a matter of fact, the absorption spectrum of the benzamide exhibits an absorption band centered at 235 nm, with an ε_{max} of 12,900 assignable to the *para*-chlorobenzamide chromophore. In the corresponding region, the CD spectrum shows (Figure 17.13) a negative Cotton effect ($\Delta\varepsilon = -4.9$); this effect is attributable to the exciton coupling between the benzamide transition and the allowed transitions of the alkylbenzene chromophore at C2. In order to build a spectrum–structure relationship, we assumed an *R*, *R* absolute configuration for the two stereogenic centers. If we take into account that the cyclohexane ring adopts the most stable chair conformation with the bulkiest substituents (phenyl and benzamide groups) both equatorial, then the relative position of the two chromophores can be reasonably represented by the Newman projection of Figure 17.13. The transoid disposition of the H—C—N—H fragment is indicated in the ^{1}H-NMR spectrum by the high value of the vicinal coupling constant H—C1—N—H (approximately 12 Hz), which corresponds to a dihedral angle of about 180°. Thus, since the phenyl group can rapidly rotate around the C2-phenyl bond (as determined by the analysis of ^{13}C-NMR relaxation times), then if we aver-

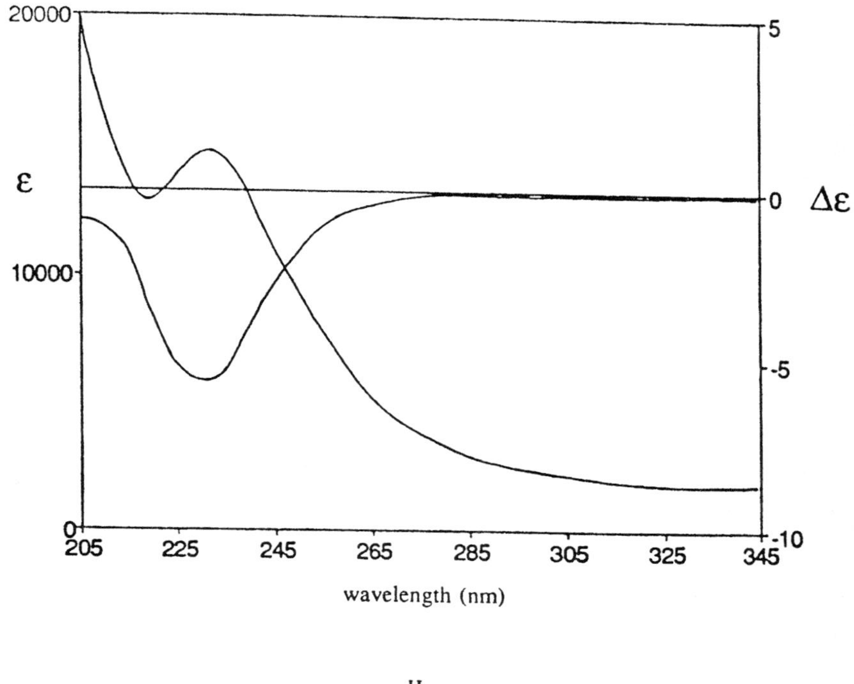

Figure 17.13. (*a*) The absorption and CD spectra of N-*cis*-(2-phenyl-2-hydroxy-cyclohexyl)-4-chlorobenzamide. (*b*) The exciton chirality defined by the transition dipoles of the benzamide and phenyl chromophores.

age to zero any contribution from the short-axis polarized transition, the only possible exciton coupling is that between the long-axis polarized transitions of the phenyl group at 190 nm and the transition of the *para*-chlorobenzamide chromophore at 230 nm. The two transition moments constitute a negative exciton chirality system that agrees with the negative sign of the observed CD Cotton effect at 230 nm. Therefore, the absolute configuration at the two stereogenic centers is 1*R*, 2*R*; then the 1*R*, 2*S* configuration has to be assigned to the dextrorotatory antipode of cycloxilic acid (37).

Coupled oscillators or polarizability methods can also be applied at a quantitative level to reproduce the intensity of the Cotton effects observed in a CD spectrum. From this point of view, the polarizability model due to DeVoe (38) looks particularly useful. In this model, a molecule is considered to be composed of a set of substituents, the chromophores: They are polarized by the external radiation field and coupled together by their own dipolar oscillating fields. The optical properties (absorption, refraction, optical rotatory dispersion, and circular dichroism) of the molecule under study can be calculated taking into account the above interaction among the subsystems. Each subsystem is represented in terms of one (or more) classical oscillator(s): Each oscillator represents an electric dipole-allowed transition, defined by the polarization direction and complex polarizability $\alpha(\tilde{\nu}) = R(\tilde{\nu}) + I(\tilde{\nu})$. $I(\tilde{\nu})$ is obtainable from the experiment, that is, from the absorption spectra of compounds that can be considered good models of the subsystems. $R(\tilde{\nu})$ can be

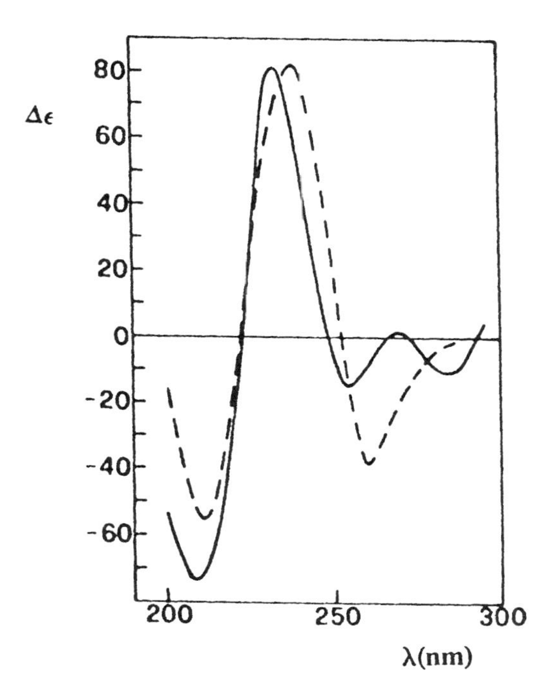

Figure 17.14. Experimental (—) and calculated (--) CD spectra of 8-O-methylrifamicin SV.

calculated from $I(\tilde{\nu})$ by means of the Kronig–Kramers transforms. DeVoe calculations of CD spectra have been applied (39,40) to the compounds of the rifamicyn family, a well-known class of semisynthetic antibiotics that have acquired a great importance in therapy. These molecules are constituted by a naphthoquinone (rifamicyn S) or hydronaphthoquinone (rifamicyn SV) ring spanned by an aliphatic bridge called "ansa," containing a dieneamide chromophore. The CD spectrum of rifamycin S, in the range between 300 and 190 nm, shows (39) two intense coupletlike features. This observation could be interpreted in terms of exciton interaction among allowed transitions localized on the aromatic ring and allowed transitions localized on the ansa bridge. The comparison of the absorption spectra of rifamycin S with those of ansa-hydrogenated derivatives, which show very low CD in this region without any coupletlike features, strongly supports this interpretation. DeVoe calculations then have been carried out, allowing interaction among oscillators localized on the aromatic chromophore and those localized on the ansa bridge. The CD spectrum can be reproduced by assuming a molecular geometry, where the aromatic ring is only partially conjugated with the diene amide chromophore. The CD spectrum of 8-O-methylrifamycin SV has been treated analogously (40). The intense excitonlike couplet present between 250 and 190 nm can be reproduced in an excellent way (Figure 17.14) using the same chromophores as before, but assuming the structure found in the solid state for rifamycin B *p*-iodoanilide.

17.5. RECENT APPLICATIONS

The above treatments of CD data have been proved useful in determining the absolute stereochemistry of biologically active organic molecules. This application led to new horizons for CD spectroscopy. The well-documented relationship between stereochemistry and biological activity makes any method leading to a full stereochemical characterization important. However, a major limitation to the use of CD spectroscopy is connected to the fact that the compound of interest has to be available in a relative large amount as an optically enriched form and with high chemical purity. Unfortunately, in dealing with biologically active compounds, this often represents a limitation. This problem can be overcome by coupling the CD technique with the HPLC resolution method. Many chiral stationary phases are now commercially available, allowing one to efficiently separate various classes of racemates, at the analytical and preparative scale (41, 42). When the separation has been obtained, the CD detection system can be reliably used to determine the stereochemistry of the single eluates, either working at a single wavelength to follow the chromatographic separation as well as to record the CD spectrum of the

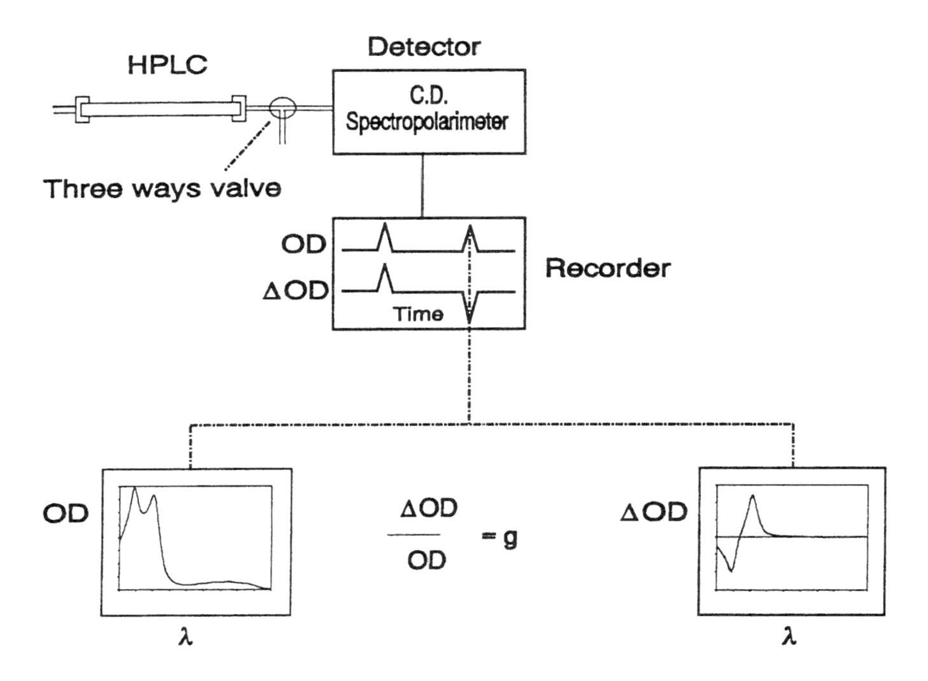

Figure 17.15. The block diagram of an HPLC-CD system.

enantiomeric eluates. In practice, this detection system allows the simultaneous measurement of circular dichroism, absorbance, and their ratio, the anisotropy factor $g(g = \Delta A/A)$ (43,44) (Figure 17.15). Then spectroscopic and stereochemical studies of substances that are difficult to resolve or to purify by traditional methods or even of species that are optically unstable can be performed without enantiomeric enrichment of the sample and even in complex mixtures (45). This fact greatly extends the number of products that can be investigated by CD and thus increases applications of this spectroscopic technique.

The peculiar advantages in the use of this detection system have been reviewed by several authors (45–47). Here, some selected examples of its application to the chiral analysis of biologically active compounds are reported.

17.5.1. Determination of the Dissymmetry Factor

The dissymmetry factor $g(g = \Delta A/A)$ is independent of the concentration and it is linearly related to the enantiomeric excess (43,44) [e.e.= $(g_{exp}/g_{max}) \times 100$].

This makes its application useful in the optimization of the fraction collection in performing preparative HPLC resolutions. As an example (44), the preparative resolution of (*R*)-(*S*)-7-chloro-1,3-dihydro-3-methyl-5-phenyl-2H-1,4-benzodiazepin-2-one is reported in Figure 17.16. A partial overlap of the peaks is observed because of the low efficiency in the resolution. However, the chromatographic profile obtained by the measurement of the dissymmetry factor allows one to optimize the fraction collection by measuring directly the e.e. The usefulness arises from the fact that preparative chromatography is routinely used to obtain relatively small amounts of the single enantiomers of a chiral drug to establish their pharmacological and toxicological properties. The measurement of the anisotropy factor permits a further and peculiar application, that is, the determination of the e.e. by HPLC, upon nonchiral stationary phases (48). As an example of such determination, the chromatographic analysis of an enriched sample of (+)(*S*)-3-methyl-N1-desmethyl-diazepam on a Silica diol stationary phase, followed at 254 nm, can be discussed. The concentration of the eluate is obtained by the value of the absorbance at the selected wavelength, whereas the sign and value of the anisotropy factor are related to the absolute configuration of the most abun-

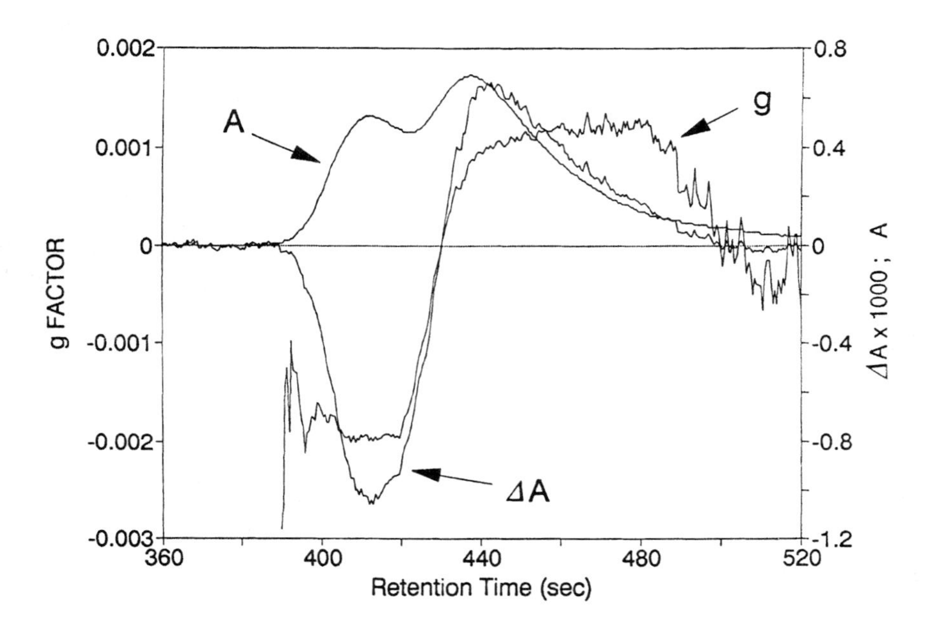

Figure 17.16. Chromatographic resolution of (*R*)-(*S*)-7-chloro-1,3-dihydro-3-methyl-5-phenyl-2H-1,4-benzodiazepin-2-one upon SiSquinmei CSP. Absorption (A), CD (ΔA) and dissymmetry factor (g) traces.

dant antipode and to the e.e., respectively. Obviously, the value of g_{max} has to be determined at the same wavelength selected for monitoring and under the same experimental conditions. The use of nonchiral stationary phases makes this method of potential wide use for the quality control in the pharmaceutical production and in the study of enantioselective biotransformation.

17.5.2. Circular Dichroism on Line

The method consists of the simultaneous measurements of the absorbance and CD of a single enantiomer obtained by chromatographic resolution upon a chiral stationary phase. The spectra can be obtained over the entire wavelength range covered by the spectrometers used, as mentioned above, if mixtures of transparent solvents are used as the mobile phase (45). Thus, stereochemical and spectroscopic determination can be quickly carried out, on any separable racemic mixture, avoiding expensive and time-consuming procedures for preparing pure antipodes, even if the antipodes are stereochemically labile (oxazepam, Figures 17.11 and 17.17; lorazepam, Figure 17.17) (16). The absolute configuration of the single eluates can be established on the basis of the sign of the observed CD. This is possible by comparing the

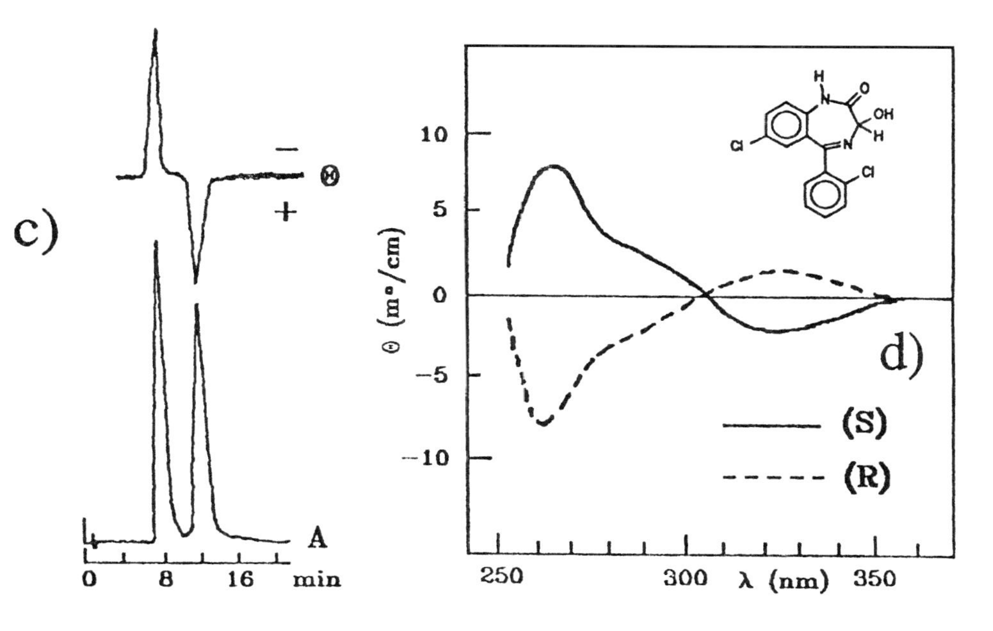

Figure 17.17. The resolution of lorazepam upon SiSQuinmei CSP and the on-line CD spectra of the two enantiomers.

obtained results with the CD of the structurally related compound for which the relationship between the absolute configuration and sign of the CD signal at the selected wavelength has been determined, as previously shown (Figure 17.11). This technique has also been applied to conformational analysis when the pure compound is not available in a suitable amount. As an example, Kurosu et al. (49) obtained data on the α-helix content of proteins analyzed by the CD technique after chromatographic separation. A further development has been recently introduced (50). The spectrometer acquires the CD signal simultaneously at all wavelengths in a definite spectral range ($\Delta\lambda = 60\,\text{nm}$).

17.5.3. Enantioselective Biotransformation

CD coupled with HPLC is very useful for monitoring the enantioselective biotransformation of chiral drugs. This method has been applied to check the presence of esterases in the binding tests of oxazepam and lorazepam esters to the receptor at the central nervous system (51). The presence of hydrolysis products, for example, oxazepam, strongly influences the results of the binding experiments because of its affinity to the receptor. The HPLC analysis of the mixtures extracted from the synaptosomal fractions after the incubation of oxazepam hemisuccinate allows one to determine the extent of the hydrolysis reaction by the area of the peaks. Further CD monitoring yields evidence that the biotransformation is enanantioselective, the *S*-enanatiomer being preferentially hydrolyzed by the enzyme (51).

REFERENCES

1. Mathieu, J. P. (1946). *Les theories moleculaires du pouvoir rotatoire naturel*. Paris: Gauthier-Villars.
2. Mason, S. F. (1963). Optical rotatory power. *Quart. Rev.*, **17**, 20–66.
3. Snatzke, G. (ed.) 1967. *Optical Rotatory Dispersion and Circular Dichroism in Organic Chemistry*. London: Heyden and Son Ltd.
4. Snatzke, G. (1968). Circular dichroism and optical rotatory dispersion—Principles and applications to the investigation of the stereochemistry of natural products. *Angew. Chem. Int. Ed. Engl.*, **7**, 14–25.
5. Barrett, G. C. (1972). Applications of optical rotation and circular dichroism. *Tech. Chem.*, **4**, 515–610.
6. *Fundamental aspects and recent developments of optical rotatory dispersion and circular dichroism*. London: Ciardelli, F. and Salvadori, P. (Eds.) (1973). Heyden and Son Ltd.
7. Legrand, M. and Rougier, M. (1977).In Kagan H. B., (ed.), *Stereochemistry: Fundamentals and Methods*. Thieme Publ., G., Stuttgart: Vol. 2, p. 33.

8. Mason, S. F. (ed.) (1978). *Optical Activity and Chiral Discrimination*. Dordrecht, Holland: D. Reidel Publishing Co.
9. Charney, E. (1979). *The Molecular Basis of Optical Activity–Optical Rotatory Dispersion and Circular Dichroism*. New York: J. Wiley and Sons.
10. Mason, S. F. (1982). *Molecular Optical Activity and Chiral Discrimination*. Cambridge: Cambridge University Press.
11. Drake, A.F. (1989). *Physical Methods of Chemistry*. B. W. Rossiter, and J. F. Hamilton (Eds.), New York: J. Wiley and Sons, 2nd ed., Vol. IIIB, pp.1–44.
12. Berova, N., Woody, R. W., and Nakanishi, K. (eds.)(1994). *Circular dichroism: application and interpretation*, New York: VCH Publ. *Analytical Applications of Circular Dichroism*, Brittain, H.G. and Purdie, N. (eds.), (1994). Elsevier Science B.V.; Salvadori, P. (1994). Circular dichroism spectroscopy in the recognition of molecular chirality. In *Seminars in Organic Synthesis*. Divisione di Chimica Organica, Società, Chimica Italiana Milano, Italy, pp. 25–52.
13. Michl, J. and Thulstrup, E. W. (1986). *Spectroscopy with Polarized Light*. New York: VCH Publ.
14. Snatzke, G. (1979). Circular dichroism and absolute conformation: Application of qualitative MO theory to chiroptical phenomena. *Angew. Chem. Int. Ed. Engl.*, **18**, 363–377.
15. Moscowitz, A. (1961). Some applications of the Kronig–Kramers theorem to optical activity, *Tetrahedron*, **13**, 48.
16. Bertucci, C., Rosini, C., Pini, D., and Salvadori, P. (1990). Sterochemical characterization of drugs by chiral chromatography and circular dichroism. In B. Holmstedt, H. Frank., and B. Testa (eds.), *Chirality and Biological Activity*, New York, pp. 71–74.
17. Corbella, A., Gariboldi, P., Iommi, G., Forgione, A., Marcucci, F., Martelli, P., Mussini, E., and Mauri, F. (1973). Stereochemistry of the enzymatic 3-hydroxilation of 1,3-dihydro-2H-1,4-benzodiazepinones. *J. Chem. Soc. Chem. Comm.*, 721–722.
18. Mislow, K. (1965). *Introduction to Stereochemistry*. New York: W. A. Benjamin, Inc.
19. Zamani, K., Conner, D. P., Weems, H. B., Yang, S. K., and Cantilena, L. R. (1991). Enantiomeric analysis of terfenadine in rat plasma by HPLC. *Chirality*, **3**, 467–470.
20. Hrbek, J., Hruben, L., Simanek, W., Santavy, F., Snatzke, G., and Yemul, S. (1982). Circular dichroism of alkaloids of colchicine type and their derivatives. *Collect. Czech. Chem. Comm.*, **47**, 2258–2279.
21. Gottarelli, G. and Spada, G. P. (1994). In Berova, N., Woody, R, W., and Nakanishi, K. (eds.), *Circular Dichroism: Application and Interpretation*. New York: VCH Publ., Chap. 5, pp. 105–119.
22. Moscowitz, A., Charney, E., Weiss, U., Ziffer, H. (1961). Optical activity in skewed dienes. *J. Am. Chem. Soc.*, **83**, 4661–4663.
23. Neubert, L. A. and Carmack, M. (1974). Circular dicroism of disulfides with dihedral angles of 0, 30, 60° in the 400–185 nm spectral range. *J. Am. Chem. Soc.*, **96**, 943–944.

24. Gawronski, J. K. (1982). Circular dichroism and stereochemistry of chiral conjugated cyclohexenones. *Tetrahedron*, **38**, 3–26.
25. Ringdahl, B., Chan, R. P. K., Cymerman Craig, J., Cava, M. P., and Shamma, M. (1981). Circular dichroism in aporphines. *J. Nat. Prod.*, **44**, 80–85.
26. Balani, S. K., van Bladeren, P. J., Sally Cassidy, E., Boyd, D. R., and Jerina, D. M. (1987). Synthesis of the enantiomeric K-region arene 5,6-oxides derived from chrisene, 7, 12-dimethylbenz[*a*]anthracene, and benzo[*c*]phenantrene. *J. Org. Chem.*, **52**, 137–144.
27. Moffit, W., Woodward, R. B., Moscowitz, A., Klyne, W., and Djerassi, C. (1961). Structure and optical rotatory dispersion of saturated ketones. *J. Am. Chem. Soc.*, **83**, 4013–4018.
28. Basil, A., Ben-Tazur, S., Gedanken, A., Rodger, A. (1991). An extension of the quadrant rule in oxiranes to non-alkyl substituents: The CD of R(–)- and S(+)-epichlorohydrin. *Chem. Phys. Lett.*, **180**, 482–485.
29. Lorentzen, R. J., Brewster, J. H., and Smith, H. E. (1992). Application of the benzene sector and the benzene chirality rules to perhydrobenzocycloalkenes and related compounds. *J. Am. Chem. Soc.* **114**, 2181–2187 (and references there).
30. Michals, D. R. and Smith, H. E. (1993). Absolute configuration of (+)-*cis*-2,3-dihydro-2-[(methylamino)methyl]-1-[4-(trifluoromethyl)phenoxy]-1-*H*-indene hydrochloride, a chiral serotonin uptake inhibitor. *Chirality*, **5**, 20–23.
31. Hansen, A. E. and Bouman, T. D. (1989). *Ab initio* calculations and mechanistic analysis of optical activity of organic molecules with extended chromophores. *Croat. Chem. Acta*, **62**, 227–243; Hansen, A. E. and Bouman, T. D. (1980). Natural chiroptical spectroscopy: Theory and computations. *Adv. Chem. Phys.*, **44**, 545–644.
32. Harada, N. and Nakanishi, K. (1983). *Circular Dichroic Spectroscopy: Exciton Coupling in Organic Stereochemistry*. Mill Valley, CA: University Science Books.
33. Mason, S. F. (1962). The absolute configuration of calycanthine. *Proc. Chem. Soc.*, 362.
34. Cai, G., Bozhkova, N., Odingo, J., Berova, N., and Nakanishi, K. (1993). *J. Am. Chem. Soc.*, **115**, 7192–7198. (and references there).
35. Yang, S. K., Weems, H. B., and Mushtaq, M. (1990). Chiral stationary phase HPLC separation of enantiomeric oxygenated derivatives of polycyclic aromatic hydrocarbons and application to metabolism studies. In Holmstedt, B., Frank, H., and Testa, B., (eds.), *Chirality and Biological Activity*. New York: Alan R. Liss, pp. 81–109.
36. Bellucci, G., Berti, G, Chiappe, C., Lippi, A., and Marioni, F. (1987). The metabolism of carbamazepine in humans: Steric course of the enzymatic hydrolysis of the 10,11-epoxide. *J. Med. Chem.*, **30**, 768–773.
37. Pini, D., Petri, A., Rosini, C., Salvadori, P., Giorgi, R., Di Bugno, C., Turbanti, L., and Marchetti, F. (1994). The absolute stereochemistry at C1 and C2 of *cis*-(+)-2-hydroxy-2-phenylcyclohexanecarboxylic acid (cicloxilic acid). *Tetrahedron*, **50**, 205–216.

38. DeVoe, H. (1964). Optical properties of molecular aggregates. II. Classical theory of refraction, absorption, and optical activity of solutions and crystals. *J.Chem.Phys.*, **43**, 3199–3208; for some applications in structural organic chemistry, see Rosini, C., Zandomeneghi, M., and Salvadori, P. (1993). Coupled oscillator calculations of circular dichroism intensities: Structural applications in organic chemistry. *Tetrahedron: Asym.*, **4**, 545–554.
39. Salvadori, P., Bertucci, C., Rosini, C., Zandomeneghi, M., Gallo, G. G., Martinelli, E., and Ferrari, P., (1981). Circular dichroism of rifamycin S. *J. Am. Chem. Soc.*, **103**, 5553–5557.
40. Rosini, C., Bertucci, C., Salvadori, P., and Zandomeneghi, M. (1985). Circular dichroism of rifamycin antibiotics. Circular dichroism and stereochemistry in solution of 8-*O*-methylrifamycin SV. *J. Am. Chem. Soc.*, **107**,17–19.
41. Taylor, D. R. and Maher, K. (1992). Chiral separations by high-performance liquid chromatography, *J. Chromatographic Sci.*, **30**, 67–85 (and references there).
42. Wainer, I. W. (ed.) (1993). *Drug Stereochemistry: Analytical Methods and Pharmacology*, 2nd ed., New York: Marcel Dekker.
43. Drake, A. F., Gould, J. M., and Mason, S. F. (1980). Simultaneous monitoring of light absorption and optical activity in the liquid chromatography of chiral substances. *J.Chromatogr.*, **204**, 103–107.
44. Bertucci, C., Domenici, E., and Salvadori, P. (1990). Circular dichroism detection in high-performance liquid chromatography: Evaluation of the anisotropy factor, *J. Pharm. Biomed. Anal.*, **8**, 843–846.
45. Salvadori, P., Bertucci, C., and Rosini, C. (1991). Circular dichroism detection in HPLC, *Chirality*, **3**, 376–385.
46. Mannschreck, A. (1992). Chiroptical detection during liquid chromatography: Applications to stereoanalysis and stereodynamics, *Chirality*, **4**, 163–169.
47. Gergely, A. (1994). The use of circular dichroism as liquid chromatographic detector. In Berova, N., Wood, R. W., and Natianishi K. (eds.), *Circular Dichroism: Application and Interpretation*. New York: VCH Publ., Chap. 9, pp. 279–292.
48. Bertucci, C., Salvadori, P., and Lopes Guimaraes, L. F. (1994). Determination of optical purity by high-performance liquid chromatography upon nonchiral stationary phases with dual circular dichroism/absorption detection. *J. Chromatogr. A*, **666**, 535–539.
49. Kurosu, Y., Sasaki, T., Takakuwa, T., Sakanayagy, N., Hibi, K., and Senda, M. (1990). Analysis of proteins by high-performance liquid chromatography with circular dichroism spectrophotometric detection. *J. Chromatogr.*, **515**, 407–414.
50. Brandl, G., Kastner, F., Fritsch, R., Zinner, H., and Mannschreck, A. (1992) Chiroptical detection during liquid chromatography. Part 5: On-line measurement of circular dichroism spectra $\Delta\varepsilon(\lambda)$ during stops of chromatographic flows. *Monatsh. Chemie*, **123**, 1059–1069.
51. Salvadori, P., Bertucci, C., Domenici, E., and Giannaccini, G. (1989). Chiral 1,4-benzodiazepin-2-ones: Relationship between stereochemistry and pharmacological activity. *J. Pharm. Biomed. Anal.*, **7**, 1735–1742.

CHAPTER

18

CIRCULAR DICHROISM IN THE STUDY OF STEREOSELECTIVE BINDING OF DRUGS TO SERUM PROTEINS

CARLO BERTUCCI, PIERO SALVADORI

Centro Studio CNR Macromolecole Stereordinate Otticamente Attive
Dipartimento di Chimica e Chimica Industriale
Università di Pisa
56126 Pisa, Italy

ENRICO DOMENICI

Department of Microbiology
Glaxo Wellcome SpA
Medicine Research Centre, 37135 Verona, Italy

18.1. INTRODUCTION

The investigation of drug–protein interactions and the determination of binding parameters are of fundamental importance to establish the mechanism of action of a drug and the extent of its activity at the target site. In the case of plasma proteins, which act as drug carriers in the body, binding interactions have a profound effect on overall drug activity (1). It has become evident from protein binding investigations that only unbound drug is available for either activity or tissue distribution. Tightly bound drugs are distributed in a smaller volume and tend to have delayed elimination as a result of glomerular filtration and hepatic uptake proportional to free drug in serum (2). It has been widely documented how drug-binding capacity affects the pharmacodynamic and pharmacokinetic properties of a drug and contributes to determining its toxic side-effects (3–5).

Enantioselective binding occurs if the drug is chiral, because of the different stability of the two diastereomeric complexes formed between the single enantiomers and the biomacromolecule. The single enantiomers act then as

The Impact of Stereochemistry on Drug Development and Use, Edited by Hassan Y. Aboul-Enein and Irving W. Wainer. Chemical Analysis Series, Vol. 142.
ISBN 0-471-59644-2

distinct compounds, their metabolism, distribution, and elimination being, in principle, different (6). If we consider that protein binding is not a simple transport mechanism, due to the occurrence of drug–drug (and even enantiomer–enantiomer) interactions, the impact of stereoselectivity on biological activity becomes evident (6–11). This justifies the increasing interest in the identification of the molecular mechanisms involved in stereodiscrimination by biological macromolecules.

18.1.1. Stereoselective Serum Protein Binding

Human serum albumin (HSA) and α_1-acid-glycoprotein (α-AGP) are the most abundant proteins in the plasma and they are also believed to be responsible for the plasma binding of most drugs. HSA binds predominantly acidic drugs, whereas α-AGP is responsible for the binding of the principle basic drugs (12, 13). The binding of acidic drugs, whether on albumin or to human plasma proteins, is uniformly stereoselective, whereas the binding of basic drugs, whether to α-AGP or other plasma protein, is relatively nonstereoselective (14).

HSA is known to possess the highest stereoselectivity among plasma proteins, as it is documented by the 35-fold difference in the binding affinity displayed for *S*- and *R*- 7-chloro- 1,3-dihydro-3-succinyloxy-5-phenyl-2H-1,4-benzodiazepin-2-one, that is, oxazepam hemisuccinate (15). Because of this rather exceptional enantioselectivity, comparable to that displayed by a specific drug target, HSA also has been regarded as a "silent receptor" (16).

Two relatively specific drug-binding sites have been identified on the HSA molecule: site I (also called the warfarin-azapropazone binding area) and site II (diazepam or tryptophan binding site) according to the definition of Sudlow et al. (17). The protein also displays affinity for a variety of other drugs, the binding of which cannot be ascribed to the two regions previously mentioned. For this reason, it also has been postulated that accessory drug-binding sites exist (18). Moreover, HSA displays affinity toward some endogenous molecules like bilirubin and fatty acids, and often bilirubin and fatty acid binding sites have been invoked, even if displacement experiments have shown that the binding of these molecules occurs in a region that is shared by the two main drug-binding sites (18).

The three-dimensional structure of human serum albumin has been determined recently by X-ray crystallography (19, 20). The crystal structure of different ligand-HSA complexes indicates the existence of two principal binding cavities in domains IIA and IIIA of the polypeptide. These data would support the model described by Sudlow et al. (17), Kragh-Hansen (18), and Fehske et al. (21), where two major, preformed, stable, high-affinity sites accounting for the binding of most drugs are present on the protein. However, the site-oriented approach leaves open many questions and does not fully explain all the data obtained by independent techniques.

The alternative model proposed by Honoré and Brodersen (22) suggests that many regions of the protein are highly flexible and conformationally labile. In practice, the binding sites may be induced in the presence of the drug. Recently, Rosen and co-workers (23) examined the ability of certain drugs to induce conformational changes in HSA by differential optical rotatory measurements at 233 nm. The results have been interpreted to suggest the existence of at least one stable, preformed, high-affinity site for the binding of drugs to HSA. It was also stressed that the relative flexibility of HSA could be important in the creation of the low-affinity binding sites that enable the protein to perform its physiological role as a nonsaturable carrier. At the same time, evidence was obtained that the albumin is a flexible macromolecule and even the high-affinity binding sites are conformationally labile. It appears that the site-oriented approach cannot be considered as a classic *lock-and-key* scheme, that is, a static discrimination mechanism. A more realistic picture of the above interaction can be obtained if we assume a dynamic discrimination mechanism. In order to gain molecular insight into the drug-binding phenomena, spectroscopies are receiving increasing attention for the characterization of ligand-protein complexes. Among them, circular dichroism (CD) represents the spectroscopic technique of choice when stereoselective binding has to be investigated. This is fundamental if we take into account that stereoselection is generally linked to the high-affinity (or primary) binding site on the protein and at therapeutic concentrations, it is the primary site that binds the drug and regulates its transport.

18.1.2. Circular Dichroism Approach

Classical methods, like ultrafiltration, ultracentrifugation, and equilibrium dialysis, are the most widely used for determining the binding of ligands to proteins and displacement phenomena (24). These techniques are based on the separation of the free drug from the bound one after the equilibrium is reached. Among these, equilibrium, dialysis is the most accurate one to determine binding parameters because it allows the determination of the free ligand concentration without disrupting equilibrium. However, the method is time-consuming and it cannot be applied when either the protein or the ligand is unstable under the experimental conditions, due to the time required to reach equilibrium. Another widely used technique is that of ultrafiltration, where the separation between the free and bound drug is achieved by centrifuging the protein-drug mixture through ultrafilters. The centrifugation step, however, should be as short as possible in order to prevent any perturbation of the equilibrium due to the increasing protein concentration in the retenate. Both techniques have some limitations due to the possible nonspecific absorption of the drug to the dialysis membrane or ultrafilters, and both cannot be applied for tightly bound drugs for which the amount of free drug in the dialysate or the ultrafiltrate approaches the detection limit of the drug itself. Other techniques based on the separation of the free drug from the bound one have been applied to define in a qualitative and quantitative manner the interaction of a drug with plasma proteins. We refer to some recent reviews that cover these aspects (25, 26).

A different approach involves the use of spectroscopic techniques, such as fluorescence, UV, CD, and resonance spectroscopies, which allow one to monitor directly in solution the binding interaction by the selective measurements of a signal allied to the drug protein complex. We would like here to focus our attention on the application of one of these methodologies, circular dichroism, to the study of the drug-plasma protein interactions of achiral and chiral drugs and to the elucidation of the stereoselection mechanism.

CD has been widely used to follow the binding of small organic ligands to serum proteins, by virtue of the contribution that can be generated upon binding (27–56). This induced CD can provide information on the binding affinity, the conformation of the bound ligand and possible protein conformational changes. In addition, this technique can be used to detect interaction between ligands in their binding to the protein. The induced CD can be observed, either for achiral or chiral drugs, at the wavelengths where the electronic transitions of the drug chromophores occur. In the case of an achiral drug, the contribution to the CD may arise from the stabilization of one of the chiral conformers in which the drug is in equilibrium, or in general from an asymmetric perturbation of the chromophores by the electrostatic field of the nearby amino acid residues. In the case of a chiral drug, the extrinsic Cotton effect will be added to the intrinsic one and the interpretation of the resulting induced ellipticity becomes more difficult. In fact, the extrinsic Cotton effect may arise from either the induction of a different conformation in the bound drug or a different polarity of the protein binding site that results in different solvation of the drug chromophores when bound to the protein and, therefore, in a shift of the absorption and CD bands observed.

CD has also been used to describe, either in qualitative or quantitative terms, the binding of racemic ligand drugs, such as (*R*,*S*)-warfarin (46) or racemic benzodiazepines (36), to serum proteins. However, it should be stressed that in both cases, the analysis of the data can be misleading since the signal observed arises from two distinct interactions. The CD observed upon binding to the protein of the racemic drug, which has a null CD in the absence of the protein, is a sum of two different induced signals due to the two interacting enantiomers.

As we will describe below, CD has been used extensively to study the interaction of chiral and achiral drugs with HSA and allowed not only the determination of binding parameters and the detection of displacement phenomena, but also the identification of structural and conformational features important in determining the stereoselectivity of binding.

In particular, applications of circular dichroism are reported that illustrate its contribution to the analysis of the following: conformation of the ligand; binding parameters; interactions between drugs; and conformation of the protein.

18.2. CONFORMATION OF THE LIGAND

Without any doubt, the determination of the ligand conformation represents a natural application of CD, this technique being the most widely used to determine the stereochemistry in solution of organic molecules (57). This methodology also has been employed to characterize the stereochemistry of small organic molecules when they are bound to a biomacromolecule (54). In this way, it is possible to gain information on the structural features relevant to the determination of the stereoselective binding. As an example, we illustrate the application of CD spectroscopy for the elucidation of the ligand conformational features of 1,4-benzodiazepine drugs when bound to serum proteins.

The binding of benzodiazepines to HSA has been studied extensively by CD (31, 33, 36, 39, 40, 41, 50). Diazepam, the most widely used benzodiazepine in therapy, is an achiral molecule present in solution in two mirror-image conformations, *M* and *P* (Scheme 1), in fast exchange between each other. When diazepam binds to HSA, an induced CD spectrum is observed closely resembling that of a chiral benzodiazepine, with an *S* absolute configuration at C3 of the heptaatomic ring (Figure 18.1). The (*S*)-enantiomers of such 1,4-benzodiazepines are known to exist in solution mainly in the *M* conformation, because the bulky substituent at C3 is allowed to stay in a pseudoequatorial position (58). If we take into account these considerations, it can be

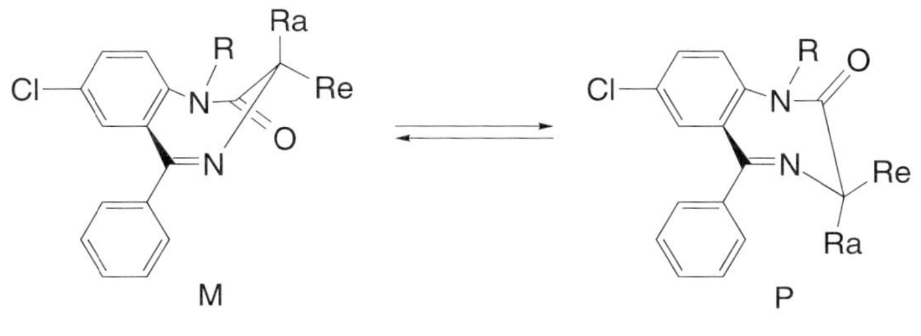

Scheme 18.1. The *M* and *P* conformers of benzodiazepines. Ra = quasiaxial substituent; Re = quasiequatorial substituent.

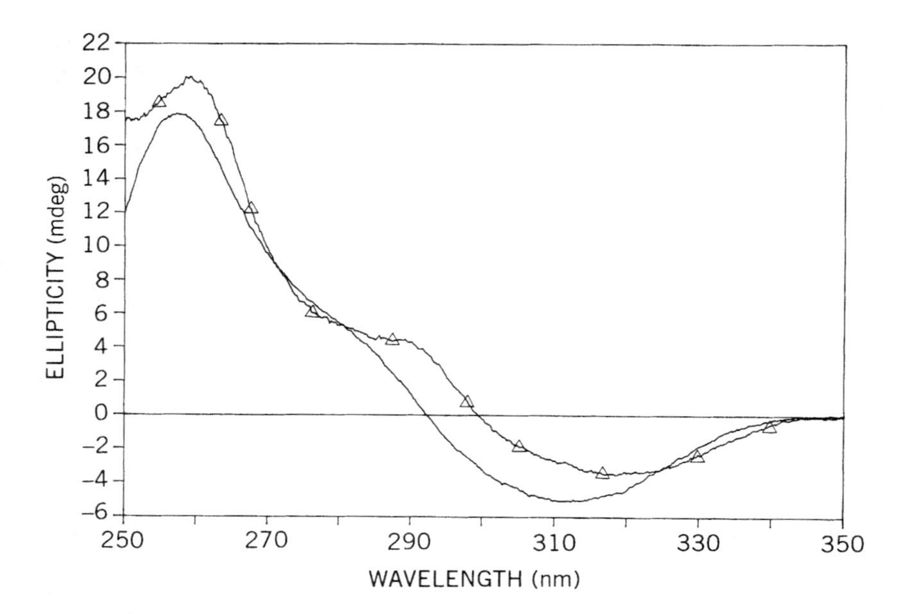

Figure 18.1. CD of (*S*)-3-methyl-N1-desmethyl-diazepam, 2×10^{-5}M (–), and of diazepam/HSA (2/1 complex), HSA 2×10^{-5}M (-Δ-Δ-Δ-). Phosphate buffer, pH 7.4, 1-cm cell.

reasonably assumed that the induced CD observed for diazepam reflects the shift of the $M \rightleftarrows P$ equilibrium toward the *M* conformer in the binding to the protein (40, 50, 58). It is noteworthy that the binding to α-AGP results in an induced CD having the opposite sign (48), suggesting the inverse conformational selectivity by a binding site present on this protein.

In the case of chiral drugs, the study of differential circular dichroism is more complicated because of the contribution of the intrinsic Cotton effects of the drug itself. This contribution can be subtracted (50) and the analysis of the resulting ΔCD becomes easier with respect to the direct analysis of the CD spectra of the complexes (40, 41). As an example, ΔCD spectra of the single enantiomers of 3-methyl-N1-desmethyl diazepam complexed to HSA are shown in Figure 18.2. These spectra have been obtained by subtracting the intrinsic contribution of the drug and of the protein from those of the complexes, carried out at a different [drug]/[HSA] molar ratio. In the case of the (*S*)-enantiomer, a significant change in the difference spectrum was obtained on varying the concentration of the drug. The observed induced CD in the complexation of the (*S*)-enantiomer of benzodiazepine to HSA can be explained by the red shift of the electronic transitions of the ligand when it binds to the protein (50). This hypothesis has been supported by the differences observed

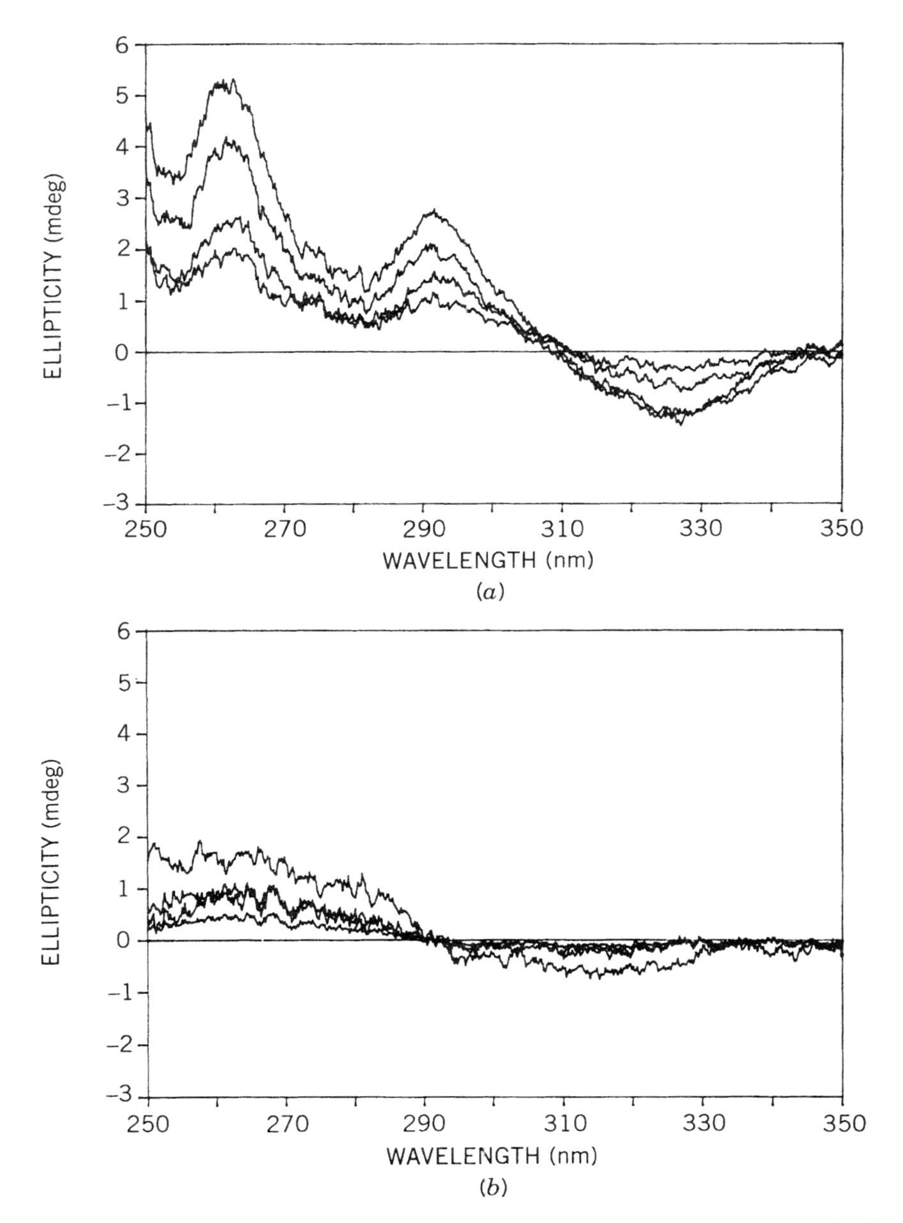

Figure 18.2. CD difference spectra (phosphate buffer solutions, pH 7.4, 2-cm cell). (*a*) (*S*)-3-methyl-N1-desmethyl-diazepam/HSA complexes: [ligand] = 0.25; 0.5; 1; 2×10^{-5} M; [HSA] = 1×10^{-5} M. The CD contributions of both the ligand and protein were subtracted. (*b*) (*R*)-3-methyl-N1-desmethyl-diazepam/HSA complexes: [ligand] = 0.5; 1; 2; 4×10^{-5} M; [HSA] = 1×10^{-5} M. The CD contributions of both the ligand and protein were subtracted.

between the absorption spectra of the complexes and the sum of the uncomplexed compounds. The differences arise from the red shift of the electronic transitions and thus of the absorption maxima of benzodiazepine when bound to the protein: This is determined by a more hydrophobic environment to the ligand with respect to its water solution. As a matter of fact, by subtracting the CD spectrum of (*S*)-3-methyl-desmethyldiazepam in buffer from that in methanol, which is red-shifted in this region, one can closely reproduce the experimental difference CD spectra obtained in the presence of HSA (50).

On the contrary, HSA does not significantly affect the CD spectra of the (*R*)-enantiomer. At higher drug concentrations, small differences appear in the region between 250 and 280 nm that could be due to both small quantities of complexed benzodiazepine or likely the perturbation of the aromatic residues of the protein by the ligand binding (50). The (*R*)-enantiomer is known to bind to HSA with a lower affinity than the (*S*) one. However, the (*R*)-enantiomer also should be complexed at the relatively high molar ratio used. Thus, the different behavior observed in the complexation of the single enantiomers to HSA can be rationalized, suggesting a different primary affinity binding site for the two enantiomers, with only those of the (*S*)-enantiomer being able to make a significant contribution to the CD observed (50). This interpretation is supported by HPLC experiments upon a HSA-based chiral stationary phase (59) and by CD competition experiments carried out on the same system in solution (55). Both approaches show that the (*S*)-enantiomer is displaced by the presence of ibuprofen, known to bind to site II on HSA, whereas the (*R*)-enantiomer is not affected, even at a high [competitor]/[drug] molar ratio.

18.3. BINDING PARAMETERS

Irrespective of the origin of the induced CD, the observed signal is proportional to the amount of bound ligand, and this property has been used since 1970 to determine drug-protein binding parameters. One of the first reports is that by Rosen (28) who determined the association constant of phenylbutazone and oxyphenylbutazone to HSA. The ΔCD of the drug-HSA complex was measured at increasing drug concentrations. A graphical method was used to determine the amount of free and bound drug from the plot of the induced CD versus the molar ratio of the complex. The tangent to the plot of induced ellipticity is drawn, providing an estimate of the free and bound drug concentration (28). This method is suitable for the determination of the binding constant if a single site is involved in the interaction, if we assume a proportionality between induced ellipticity and the amount bound. Such an assumption is valid when all sites contribute equally to the induced ellipticity because different classes of binding sites usually show a different contribution to the induced CD; therefore, proportionality is not maintained in the presence of additional sites. The dissymmetry induced in a drug by binding to a secondary site may differ, resulting in a different intensity of the induced CD signal or, as in the case of oxyphenylbutazone, an induced CD having the opposite sign (28). In an extreme case, the secondary sites may not produce any detectable induced Cotton effect, as observed in the case of the binding of diazepam to HSA (see below).

In 1973, Perrin and Nelson recognized that only the primary site of serum albumin was able to induce optical activity on sulfaethidole (32) by comparing the K_{aff} determined by equilibrium dialysis with that obtained by the CD method of Rosen (28). Years later, a method for the determination of binding constants from continuous CD titration data by numerical analysis was reported by Roodsorp and Sjöholm (37), with different programs based on the possibility of one-site, two-site, or three-site models used for the numerical calculations, which rely on the estimation of the ΔCD_{max} values for each complex. They investigated the binding of several benzodiazepines to HSA and found the analysis with one site more appropriate to describe the interaction for most derivatives.

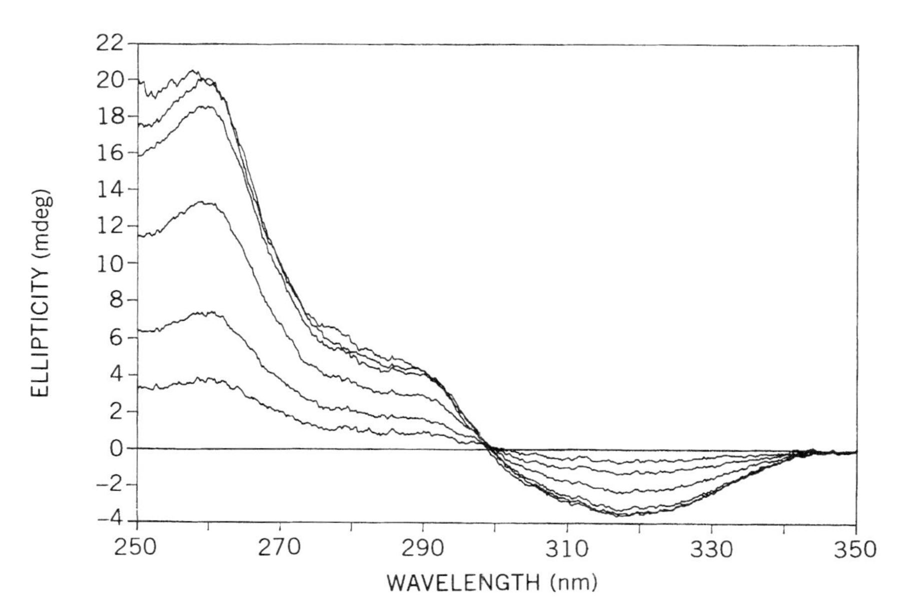

Figure 18.3. CD difference spectra of diazepam/HSA complexes in phosphate buffer solutions, pH 7.4; [HSA] = 1×10^{-5}, 2-cm cell. The spectrum of the protein was subtracted.

At least from a qualitative point of view, a correct interpretation can be done if one carries out parallel CD and UV difference studies on the protein binding of the drug at a different molar ratio. As an example, the difference spectra obtained by the subtraction of the CD spectra of HSA from those of the diazepam/HSA complexes at different molar ratios are shown in Figure 18.3. A two-fold molar excess of diazepam is enough to saturate the HSA binding site responsible for the observed CD. This can be observed also in Figure 18.4*a*, where the difference CD is followed at a fixed wavelength, 260 nm, upon the [diazepam]/[HSA] molar ratio. If one measures the ΔAbs, it can be observed that binding to the HSA still occurs after the molar ratio 1:1, even though the stereospecific binding site has been saturated (50), that is, HSA is no longer able to induce a preferential conformation to the diazepam ring (Figure 18.4*b*).

Thus, the quantitative analysis of the CD data, in terms of the dissociation constant, is relatively easy when we are dealing with an achiral drug (L) and a single-site interaction on the protein (P) responsible for the induced CD. In this case, the CD signal undoubtedly arises from the complexed drug (LP) and the result obtained reflects selectively the binding site producing the CD signal when the drug is complexed. The dissociation constant, related to the stereoselective binding site, can then be determined by measuring the lowest-energy CD-induced band of the complexed drug at different dilutions of the [HSA]/[drug] adduct (LP) in a 1:1 molar ratio (Figure 18.5) (55). In practice, the equilibrium constant of the complex:

$$K = \frac{[\text{HSA/drug}]}{[\text{HSA}][\text{drug}]} = \frac{[LP]}{[P][L]}$$

assumed with 1:1 stoichiometry and the Lambert and Beer law:

$$\text{CD} = A_L - A_R = [LP]l\Delta\varepsilon$$

can be combined, according to the Benesi–Hildebrand treatment [60], to give the following linear equation:

$$\frac{P}{\sqrt{\text{CD}/l}} = \frac{1}{\Delta\varepsilon}\sqrt{\text{CD}/l} + \frac{1}{\sqrt{K\Delta\varepsilon}}$$

p being the initial concentration of both the drug and protein: $p = [P] + [LP] = [L] + [LP]$.

This allows the determination of the $\Delta\varepsilon_{max}$ of the complex and its dissociation constant. The two values can be obtained by applying the above equation and plotting $[\text{HSA}]/(\text{CD}/l)^{1/2}$ against $(\text{CD}/l)^{1/2}$ (Figure 18.6). The method previously reported by Rosen (28) suffered from a huge inaccuracy, being based

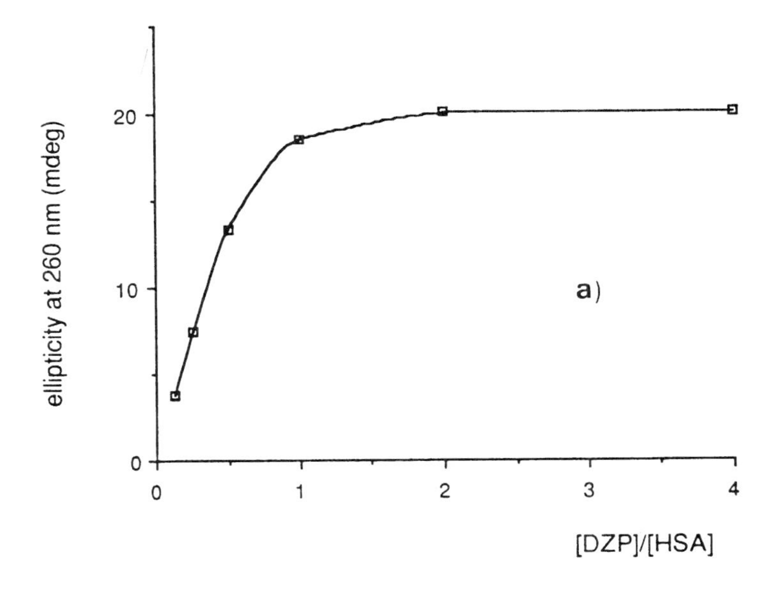

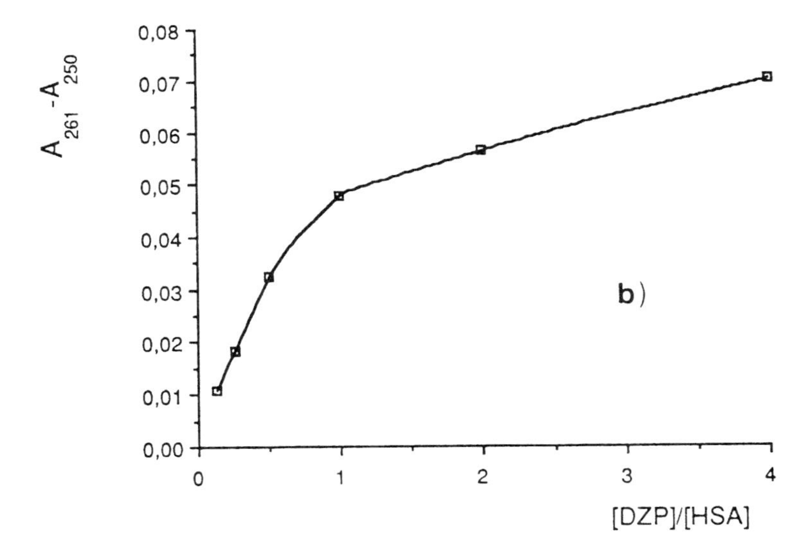

Figure 18.4. (*a*) Ellipticity and absorbance (*b*) from the difference CD and UV spectra of diazepam/HSA complexes. Phosphate buffer solutions, pH 7.4; [HSA] = 1×10^{-5} M, 2-cm cell. The spectra of the components were subtracted from those of the complexes.

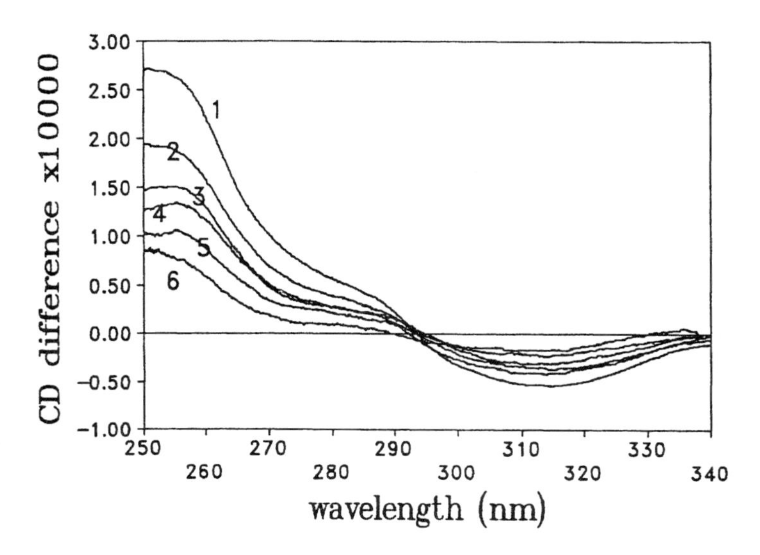

Figure 18.5. CD difference spectra of diazepam/HSA (1/1) complexes. 1 [HSA] 1-mM, 0.01-cm cell; 2 c 0.2-mM, 0.05-cm cell; 3 c 0.05-mM, 0.2-cm cell; 4 c 0.005-mM, 2-cm cell; 5 c 0.002-mM, 5-cm cell; 6 c 0.001-mM, 10-cm cell.

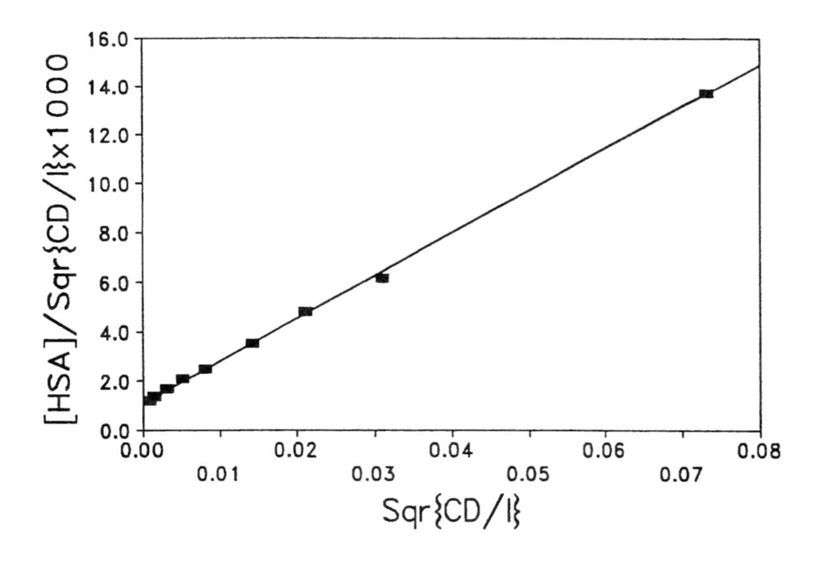

Figure 18.6. Linear fitting for 1:1 diazepam/HSA complexes, 315 nm.

on drawing a tangent to the origin of the experimental curve. Further, and even more important, the assumption of binding-site homogeneity was not fully justified because in that case, and also in the Roodsorp treatment, the variable parameter was the drug/HSA ratio and not the total concentration. Thus, the binding of the drug to low-affinity binding sites has to be taken into account as well as the possibility of the conformational change of the protein when a high concentration of the drug is present. It has to be noted that the method discussed above allows a definite characterization of the stereospecific binding that induces a preferential enantiomeric conformation in the achiral ligand (55). The obtained values of the dissociation constants can be used, as reported later on, in an experiment based on selective competition at those sites. Compared to other widely used methods, it allows the study of a stereospecific binding site, disregarding secondary or non-sterospecific binding on HSA, and has no need of separating the mixture. Indeed, only the bound drug, when it is achiral, is monitored.

18.4. INTERACTIONS BETWEEN DRUGS

Competition between two drugs for the same plasma protein or the same binding site may strongly affect the drug disposition of both drugs, resulting in serious physiological consequences (61, 62). As an example, the concurrent administration of oral anticoagulants with other agents may result in an enhanced risk of hemorrhage in patients due to the decreased metabolism and/or displacement of the drug from its binding sites on plasma proteins (63). When two drugs interact or bind to the same protein, in the simplest case the binding of one drug is not affected by the presence of the second drug. This can be interpreted as if the two drugs bind to distinct sites on the protein, or (less likely) they can accommodate in the same binding area without affecting each other. In any of the above cases, the binding of the first drug does not modify the binding capacity of the protein for the second drug.

The second possibility is the observation of a variation in the binding capacity or affinity of one or both drugs, which can be explained by assuming a simple competition phenomena, or by the existence of a cooperative or anticooperative interaction. The observation of competitive binding of two drugs to the same protein does not necessarily mean that they bind to the same site. In fact, mutual displacement may be observed even in the case where the two drugs bind to separate sites, whenever the binding of the first drug to its site results in conformational changes of the protein secondary structure that affect the conformation of the second site (anticooperative interaction). By virtue of the allosteric effect, it is possible to also have cooperative binding of two drugs to the same protein (64).

Classical biochemical methodologies were successfully applied to determine the type of mutual interaction of two drugs in the presence of HSA. However, a step of separation between macromolecule and ligand is necessary for determining the binding parameters. Moreover, these methods are extremely time-consuming when more than one ligand is present in solution. Honoré (64) and Kragh-Hansen (12) proposed graphical and mathematical treatments to obtain reproducible data analysis for competition experiments in solution. These experiments can be carried out also by HPLC upon a HSA chiral stationary phase, a technique that does not require separation between free and bound molecules (59, 65). Recently, Noctor et al. adapted a mathematical method to analyze affinity chromatography data, but this treatment does not markedly discriminate between binding to specific and nonspecific areas of the protein [66].

Another widely used method is that of the fluorescent probe (17, 67), in which the drugs are studied for their ability in displacing fluorescent marker ligands known to bind to specific sites on HSA. In this way, competition between two drugs may be determined indirectly by determining their capacity to displace the fluorescent marker from the protein.

CD has been used also to carry out competition experiments for obtaining information, selectively, on the high-affinity binding sites, where usually stereo-

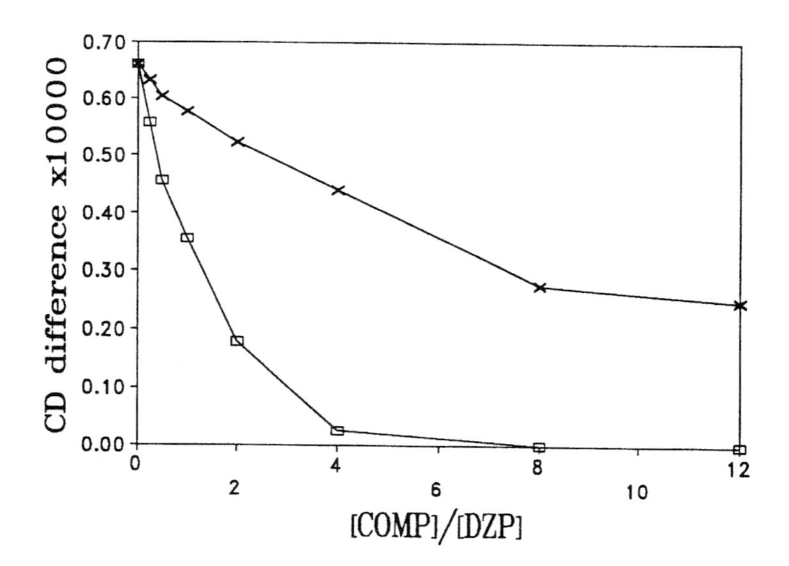

Figure 18.7. CD difference at 315 nm from the spectra of 1:1 diazepam/HSA complexes versus [(*R*)-(*S*)-ibuprofen]/[diazepam] (□-□-□-□) and [salicylate]/[diazepam] (x-x-x-x) molar ratios. 1.2-mM HSA, 0.01-cm cell.

selective binding occurs. In 1973, Perrin and Nelson in studying the induced CD of sulfaethidole have analyzed the binding to serum albumin of drugs that interact with the same primary site as sulfaethidole, but do not show a significant CD at the wavelengths observed. They were able to determine by competitive studies monitored by CD the binding constants of such antagonists (32). In 1975, Perrin et al. identified by using CD methods two distinct binding sites for dicoumarol on HSA, responsible for induced ellipticity of different intensity, and they measured displacement by specific competitors for the two classes of sites (34). In practice, these experiments are conducted by measuring the induced CD of the drug/HSA complex in the presence of an increasing amount of the competitor. The value of the CD signal allows the determination of the concentration of the bound drug and then the ratio of the affinity constants of the marker, that is, the drug bound to the protein, and that of the competitor, that is, the drug acting as a displacer. The phenomenon is strictly limited to the binding site responsible for the induced CD signal.

An example that illustrates competitive binding is the displacement of diazepam by ibuprofen, as reported in Figure 18.7 (55). The two drugs are known to bind to the same binding area, in particular site II, and indeed diazepam is completely displaced from its stereospecific binding site by a fourfold molar excess of the competitor. The ratio of the affinity constants (K_m and K_c) of the marker (m, diazepam) and of the competitor (c, ibuprofen) can be simply evaluated by applying the following equation:

$$\frac{K_m}{K_c} = \frac{[MP]([MP] + c - p)}{(m - [MP])(p - [MP])}$$

Interaction between drugs can occur even when they bind to different binding sites and a decreased affinity may arise from anticooperative binding if negative allosteric interactions take place. In Figure 18.7 we have an example of anticooperative binding, that is, an indirect competition between diazepam and salicylate, known to bind at a distinct primary site, that is, sites II and I, respectively. In this cross-competition experiment, only 42% of the diazepam stereospecific (CD-inducing) binding is influenced by a large excess of displacer. The calculation of the affinity constant ratio deserves the application of other equations, formulated for indirect competition:

$$\frac{K_m}{K_c} = \frac{[MP]([MP] + c - p)}{(m - [MP] - [MZ](p - [MP])} \qquad \text{if } \frac{c}{p} > 1$$

$$\frac{K_m}{K_c} = \frac{[MP]\left([MP] + c - p\dfrac{[MP]}{[MP] + [MZ]}\right)}{(m - [MP] - [MZ](p - [MP])} \qquad \text{if } \frac{c}{p} < 1$$

Actually, indirect competition could result in diminishing the induced CD due to the binding, but not the effective binding. As far as stereoselective sites are concerned, the two possibilities are not formally distinguishable [55]. Thus, data obtained by independent techniques like biochromatography or dialysis are needed to confirm the displacement of the marker from its site on HSA.

18.5. CONFORMATION OF THE PROTEIN

The degree of dissymmetry of a protein-bound ligand is strongly affected by any variation in the protein secondary or tertiary structure due to changes in the buffer conditions (ionic strength or pH) or to the binding of other ligands. The induced Cotton effect of a bound ligand can then be used as a sensitive tool to detect even subtle changes in the protein conformation resulting from a perturbation of the system. As an example, the study of the extrinsic CD spectrum of bilirubin bound to the protein (30, 35, 44) allowed the detection of conformational changes undergone by the serum albumin molecule upon variation of the pH. In fact, by going from pH 4–10, we may observe an inversion of the sign of the Cotton effect of the twisted albumin-bound bilirubin that has been ascribed to a conformational change called N-B transition. The N and B conformations of HSA (with the N predominant at pH 6 and the B at pH 9) have the same α-helix content, but differ in the relative positions of the half-domains in the N-terminal part of the molecule. The N-B transition, in turn, can strongly affect the binding of drugs to HSA. As an example, warfarin preferentially binds to the B conformation of HSA and not to the N one, the two conformations being in equilibrium at physiological pH (43). Wilting and co-workers have also shown, by using a combined CD-equilibrium dialysis approach, that the pH dependence of the diazepam binding to HSA can be fully explained by assuming two conformational states for the protein that display different binding affinities for the drug molecule (42).

Protein conformational changes can also arise from the binding of drugs or endogenous factors. This phenomenon is important because it may affect the protein biochemical properties and, in particular, its binding capacity for other ligands. A flexible HSA molecule is actually demanding in order to explain the existence of the cooperative and anticooperative binding of two ligands to HSA (64). It is worth mentioning that even direct competition may arise from the induction of a protein conformational change by one of the ligands and not necessarily from the binding of both ligands to the same site.

CD has been used successfully to study these mechanisms by detecting changes in the protein conformation by looking at the spectra of the first ligand, after the binding of the second one. An example of this application was reported by Maruyama et al. (68). These authors examined the effects of benzodiazepines on the binding of bilirubin to HSA by a combined CD and equilibrium dialysis study. They have shown that the bilirubin site is independent of the diazepam site, but the diazepam site acts cooperatively on the bilirubin site with a site-to-site effect. The addition of benzodiazepine to HSA caused an increase in the amount of bound bilirubin and a strong enhancement of the induced CD of the complex. On the other hand, the addition of site II binding drugs (such as mefenamic acid and flufenamic acid) had little or no effect on the bilirubin binding to HSA. These findings together indicate that the benzodiazepine binding site and site II may not be identical even though they overlap each other, and that the binding of benzodiazepine to HSA causes a conformational change around the bilirubin site as the basis of the cooperative mechanism (68).

Circular dichroism can be profitably used also to characterize the binding properties of chemically modified albumins. Amino acid-specific chemical modifications have been performed in order to identify specific binding areas on HSA (12). Some of these modifications also occur in physiological conditions, as in the case of the acetylation of Lys_{199} by aspirin (69, 70), or the glycosylation of Lys_{525} by D-glucose (71, 73). The significant role that similar modifications can have *in vivo* (69, 73) makes the elucidation of the molecular aspects that are involved of interest. The analysis of the induced CD permits the detection of changes in the ligand binding stereoselectivity (28, 56). An irreversible change in the protein conformation due to a stable chemical

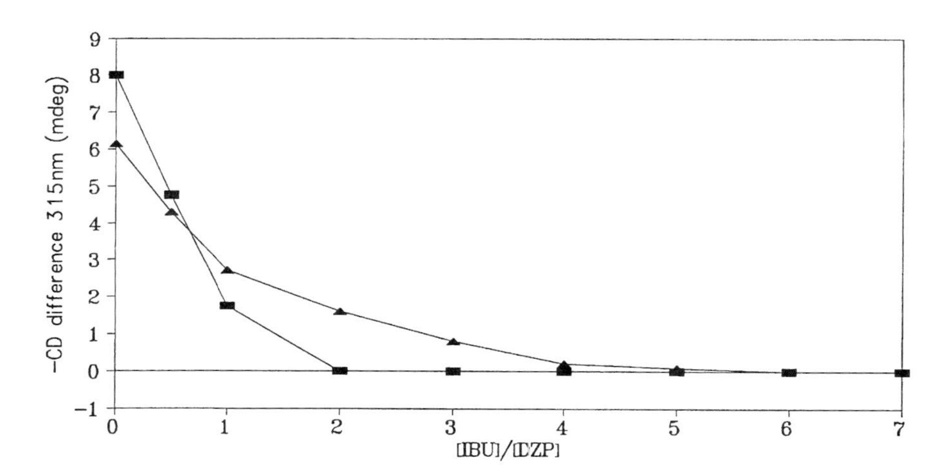

Figure 18.8. Ellipticity at 315 nm from the CD difference spectra of the 1:1 diazepam/HSA complex versus [(*R*)-(*S*)-ibuprofen]/[diazepam] molar ratio. [HSA] = 0.03 mM, phosphate buffer, pH 7.4, 1-cm cell.

modification may result in a different conformation of the bound ligand and in a variation of the bound fraction.

As an example, CD has been applied to study the binding of drugs to acetylated HSA (HSAAc) as well as the effects of the interaction between drugs on their binding to the protein when compared to the native one (56). In Figure 18.8, the induced CD for 1:1 [diazepam]/[albumin] has been measured in the presence of an increasing amount of a competitor, (*R*)–(*S*)-ibuprofen, known to bind to the same binding site. The ΔCD signal of the diazepam/protein complex collapses after the addition of the competitor. The CD value is close to zero for a [competitor]/[diazepam] ratio of 2 or 6 in the case of native or acetylated albumins, respectively. This behavior suggests a reduction in the effects of the interactions between drugs on their binding to the protein, or at least in their stereoselective binding, when the albumin is modified (57).

18.6. CONCLUSION

Difference circular dichroism technique can be used to determine binding parameters. This method, when compared with other widely used methods, allows the selective investigation of the stereospecific binding, disregarding secondary or nonstereospecific sites on HSA. The induced CD reveals only the bound fraction of the drug, making the method more simple (the separation of the drug–protein complex is not required) and more selective. This becomes important in the study of drugs having high-affinity constants, for which the determination of the free fraction would be problematical. In addition, CD gives information on the interactions between drugs for their binding to the protein, in particular at the primary/stereospecific binding sites.

The more peculiar feature of CD is the possibility to gain information on the conformation of the ligand as well the protein. This really gives a picture of the complex, at the molecular level, that more reliably accounts for the specificity of the binding process. Thus, CD studies should shed light on the relationship between drug structure and affinity to the stereoselective/primary binding site on HSA, usually the primary binding sites. CD studies also can be focused on the structural characterization of the proteins (74), particularly when they have been obtained by biotechnological techniques or chemical modifications. As in the case of recombinant HSA or HSA modified in physiological conditions, their conformational properties and binding capacity can be determined and compared with those of the native protein.

In conclusion, a parallel application of circular dichroism to classical protein binding studies provides a fundamental contribution to the identification of the structural features that determine the molecular mechanisms of the binding and the stereodiscrimination of drug enantiomers by biological macromolecules.

REFERENCES

1. Anton, A. H. and Solomon, H. M (eds.) (1973). Drug-protein binding. *Annals of the New York Academy of Science*, Vol. 226.
2. Vallner, J. J. (1977). Binding of drugs by albumin and plasma proteins. *J. Pharm. Sci.*, **66**: 447–65.
3. Tillement, J. P., Houin, G., Zini, R., Urien, S., Albengres, E., Barré, J., Lecompte, M., D'Athis, P., and Sebille, B. (1984). The binding of drugs to blood plasma macromolecules: Recent advances and therapeutic significance. In B. Testa (ed.), *Advances in Drug Research*, Vol. 13. London: Academic Press, pp. 59–94.
4. Levy, R. and Shand, D. (eds.) (1984). Clinical implications of drug-protein binding. *Clin. Pharmacokin.*, **9** (suppl. 1),1–104.
5. Fichtl, B., Nieciecki, A., and Walter, K. (1991). Tissue binding of drugs: General principles and pharmacokinetic consequences. In B. Testa (ed.), *Advances in Drug Research*, Vol. 20. London: Academic Press, pp. 118–166.
6. *Stereochemistry and Biological Activity of Drugs. Ariens.* E. J., Soudijn, W., and Timmermans, P. B. M. W. M. (eds.) (1983). London: Blackwell.
7. Simonyi, M. (1984). On chiral drug action. *Med. Res. Rev.*, **4**, 359–413.
8. Wainer, I. W. (Ed.). 2nd ed. (1993). *Drug Stereochemistry: Analytical Methods and Pharmacology*. New York: Marcel Dekker.
9. Jamali, J., Mehvar, R., and Pasutto, F. M. (1989). Enantioselective aspects of drugs action and disposition: Therapeutic pitfalls. *J. Pharm. Sci.*, **7**, 695–715.
10. Holmstedt, B., Frank, A., and Testa, B. (eds.) (1990). *Chirality and Biological Activity*. New York: Alan R. Liss.
11. Caldwell, J. (1992). The importance of stereochemistry in drug action and disposition. *J. Clin. Pharmacol.*, **32**, 925–929.
12. Kragh-Hansen, U. (1981) Molecular aspects of ligand binding to serum albumin. *Pharmacol. Rev.*, **33**, 17–53.
13. Kremer, J. M. H., Wilting, J., and Janssen, L. H. M. (1988). Drug binding to human alpha-1-acid glycoprotein in health and disease. *Pharmacol. Rev.*, **40**, 1–47.
14. Noctor, T. A. G. (1993). In Wainer I. W. (ed.), *Drug Stereochemistry: Analytical Methods and Pharmacology*, 2nd ed. New York: Marcel Deker, pp. 337–364.
15. Müller, W. E. and Wollert, U. (1975). High stereospecificity of the benzodiazepine binding site on human serum albumin. *Mol. Pharmacol.*, **11**, 52–60.
16. Müller, W. E. and Wollert, U. (1979). Human serum albumin as a silent receptor for drugs and endogeneous substances. *Pharmacology*, **19**, 59–68.
17. Sudlow, G., Birkett, D. J., and Wade, D. N. (1975). The characterisation of two drug specific binding sites on human serum albumin. *Mol. Pharmacol.*, **11**, 824–832.

18. Kragh-Hansen, U. (1990). Structure and ligand binding properties of human serum albumin. *Danish Med. Bull.*, **37**, 57–84.
19. He, X. M. and Carter, D. C. (1992). Atomic structure and chemistry of human serum albumin. *Nature*, **358**, 209–215.
20. Carter, D. C. and Ho, J. X. (1994). Structure of serum albumin. *Adv. Protein. Chem.*, **45**, 153–203.
21. Fehske, K. J., Müller, W. E., and Wollert, U. (1981). The location of drug binding sites in human serum albumin. *Biochem. Pharmacol.*, **30**, 687–692.
22. Honorè, B. and Brodersen, R. (1984). Albumin binding of anti-inflammatory drugs. Utility of a site-oriented versus a stoichiometric analysis. *Mol. Pharmacol.*, **35**, 137–150.
23. Walji, R., Rosen, A., and Hider, R. C. (1993). The existence of conformationally labile (preformed) drug binding sites in human serum albumin as evidenced by optical rotation measurements. *J. Pharm. Pharmacol.*, **45**, 551–558.
24. Sebille, B. (1990). Methods of drug protein binding determination. *Fund. Clin. Pharmacol.*, **4** (suppl.) **2**, 151–161.
25. Pacifici, G. M. and Viani, A. (1992). Methods of determining plasma and tissue binding of drugs. *Clin. Pharmacokinet.*, **23**, 449–468.
26. Sebille, B., Zini, R, Madjar, C. V., Thuaud, N., and Tillement, J. P. (1990). Separation procedures used to reveal and follow drug protein binding. *J. Chromatogr.*, **531**, 51–77.
27. Chignell, C. F. (1968). Circular dichroism studies of drug-protein complexes. *Life Sci.*, **7**, 1181–1186.
28. Rosen, A. (1970). The measurement of binding constants using circular dichroism. Binding of phenylbutazone and oxyphenbutazone. *Biochem. Pharmacol.*, **19**, 2075–2081.
29. Sjöholm, I. and Sjödin, T. (1972). Binding of drugs to human serum albumin. I. Circular dichroism studies on the binding of some analgesics, sedatives and antidepressive agents. *Biochem.Pharmacol.*, **21**, 3041–3052.
30. Blauer, G., Harmatz, D., and Snir, J. (1972). Optical properties of bilirubin-serum albumin complexes in aqueous solution. I. Dependence on pH. *Biochim. Biophys. Acta*, **278**, 68–88.
31. Müller, W. E. and Wollert, U. (1973). Interaction of benzodiazepines with human serum albumin. Circular dichroism studies. *Naunyn-Schmiedeberg's Arch. Pharmacol.*, **278**, 301–312.
32. Perrin, J. H. and Nelson, D. A. (1973). Competitive binding of two drugs for a single binding site on albumin: a circular dichroic study. *J. Pharm. Pharmacol.*, **25**, 125–130.
33. Müller, W. E. and Wollert, U. (1974). Influence of pH on the benzodiazepine-human serum albumin complex. Circular dichroism study. *Naunyn Schmiedeberg's Arch. Pharm.*, **283**, 67–82.
34. Perrin, J. H., Vallner, J. J., and Nelson, D. A. (1975). Some quantitative investiga-

tions of the binding to and the displacement of bishydroxycoumarin from human serum albumin. *Biochem. Pharmacol.*, **24**, 769–774.

35. Blauer, G. and Wagnière, G. (1975). Conformation of bilirubin and biliverdin in their complexes with serum albumin. *J. Am. Chem. Soc.*, **97**, 1949–1954.
36. Sjödin, T., Roosdorp, N., and Sjöholm, I. (1976). Studies on the binding of benzodiazepines to human serum albumin by circular dichroism measurements. *Biochem. Pharmacol.*, **25**, 2131–2140.
37. Roodsorp, N. and Sjöholm, I. (1976). Determination of binding constants from continuous circular dichroism titration data by numerical analysis. *Biochem. Pharmacol.*, **25**, 2141–2145.
38. Brown, N. A., Jähnchen, E., Müller, W. E., and Wollert, U. (1977). Optical studies on the mechanism of the interaction of the enantiomers of the anticoagulant drugs phenprocoumon and warfarin to human serum albumin. *Mol. Pharmacol.*, **13**, 70–79.
39. Brodersen, R., Sjödin, T., and Sjöholm, I. (1977). Independent binding of ligands to human serum albumin. *J. Biol. Chem.*, **252**, 5067–5072.
40. Alebic-Kolbah, T., Kajfez, F., Rendic, S., Sunjic, V., Konowal, A., and Snatzke, G. (1979). Circular dichroism and gel filtration studies of binding of prochiral and chiral 1,4-benzodiazepin-2-ones to human serum albumin. *Biochem. Pharmacol.*, **28**, 2457–2464.
41. Konowal, A., Snatzke, G., Alebic-Kolbah, T., Kajfez, F., Rendic, S., Sunjic, V. (1979). General approach to chiroptical characterization of binding of prochiral and chiral 1,4-benzodiazepin-2-ones to human serum albumin. *Biochem. Pharmacol.*, **28**, 3109–3113.
42. Wilting, J. T., Hart, B. J., and De Gier, J. J. (1980). The role of albumin conformation in the binding of diazepam to human serum albumin. *Biochim. Biophys. Acta*, **626**, 291–298.
43. Wilting, J., van der Giesen, W. F., Janssen, L. H. M. Weideman, M. M., Otagiri, M., and Perrin, J. H. (1980). The effect of albumin conformation on the binding of warfarin to human serum albumin. *J. Biol. Chem.*, **255**, 3032–3037.
44. Honoré, B. and Frandsen, P. C. (1986). Conformational changes in the bilirubin-serum albumin complex at extreme alkaline pH. *Biochem. J.*, **236**, 365–369.
45. Fitos, I., Simonyi, M., Tegyey, Z., Kajtar, M., and Ötvös, L. (1986) Binding of 3-alkyl-1,4-benzodiazepin-2-one stereoisomers to human serum albumin. *Arch. Pharm.*, (Weinheim) **319**, 744–749.
46. Otagiri, M., Maruyama, T., Imai, T., Suenaga, A., and Imamura, Y. A. (1987). A comparative study of the interaction of warfarin with human α1-acid glycoprotein and human albumin. *J. Pharm. Pharmacol.*, **39**, 416–420.
47. Honorè, B., Sato, H., and Brodersen, R. (1988). Cobinding of bilirubin and laurate to human serum albumin: spectroscopic characterization of stoichiometric complexes. *Arch. Biochem. Biophys.*, **266**, 189–196.
48. Otagiri, M., Yamamichi, R., Maruyama, T., Imai, T., Suenaga, A., Imamura, Y.,

and Kimachi, K. (1989). Drug binding to α1-acid glycoprotein studied by circular dichroism. *Pharm. Res.*, **6**, 156–159.

49. Fitos, I., Visy, J., Magyar, A., Kajtár, J., and Simonyi, M. (1989). Inverse stereoselectivity in the binding of acenocoumarol to human serum albumin and to α_1-acid glycoprotein. *Biochem. Pharmacol.*, **38**, 2259–2262.
50. Bertucci, C., Domenici, E., and Salvadori, P. (1990). Stereochemical features of 1,4-benzodiazepin-2-ones bound to human serum albumin: difference CD and UV studies. *Chirality* **2**, 167–174.
51. Oravcová, J., Mlynárik, V., Bystrický, S., Šoltès, L., Szalay, P., Bohácik, L., and Trnovec, T. (1991). Interaction of pirprofen enantiomers with human serum albumin. *Chirality*, **3**, 412–417.
52. Rahman, M. H., Maruyama, T., Okada, T., Yamasaki, K., and Otagiri, M. (1993). Study of interaction of carprofen and its enantiomers with human serum albumin. I. Mechanism of binding studied by dialysis and spectroscopic methods. *Biochem. Pharmacol.*, **46**, 1721–1731.
53. Rahman, M. H., Maruyama, T., Okada, T., Imai, T., and Otagiri, M. (1993). Study of interaction of carprofen and its enantiomers with human serum albumin. II. Stereoselective site-to-site displacement of carprofen by ibuprofen. *Biochem. Pharmacol.*, **46**, 1733–1740.
54. Brittain, H. G. and Purdie, N. (ed.) (1994). *Analytical Application of Circular Dichroism.*, New York: Elsevier Science B.V.
55. Ascoli, G., Bertucci, C., and Salvadori, P. (1995). Stereospecific and competitive binding of drugs to human serum albumin: A difference circular dichroism approach. *J. Pharm. Sci.*, **84**, 737–741.
56. Bertucci, C., Viegi, A., Ascoli, G., Salvadori, P. (1995). Protein binding investigation by difference circular dichroism: Native and acetylated human serum albumins. *Chirality*, **7**, 57–61.
57. Nakanishi, K., Berova, N., and Woody, R. W. (eds.) (1994). *Circular Dichroism, Principles and Applications*. New York: VCH.
58. Sunjic, V., Lisini, A., Sega, A, Kovac, T., and Kajfez, F. (1979). Conformation of 7-chloro-5-phenyl-3(S)-methyl-dihydro-1,4-benzodiazepin-2-one in solution. *J. Heterocyclic Chem.*, **16**, 757–761.
59. Domenici, E., Bertucci, C., Salvadori, P., Motellier, S., and Wainer, I. W. (1990). Immobilized serum albumin: Rapid HPLC probe of stereoselective protein binding interactions. *Chirality*, **2**, 167–174.
60. Benesi, H. and Hildebrand, J. H. (1949). A spectrophotometric investigation of iodine with aromatic hydrocarbons. *J. Amer. Chem. Soc.*, **71**, 2703–2707.
61. MacKichan, J. J. (1984). Pharmacokinetic consequences of drug displacement from blood and tissue proteins. *Clin. Phamacokin.*, **9** (suppl. 1), 32–41.
62. McElnay, D. C. and DeArcy, P. F. (1983). Protein binding displacement interactions and their clinical importance. *Drugs*, **25**, 495–513.
63. Goodman Gilman, A., Rall, T. W., Nies, A. S., and Taylor, P. (eds.) (1991). *The Pharmacological Basis of Therapeutics.*, New York: Pergamon, Chap. **55**, p. 1320.

64. Honoré, B. (1990). Conformational changes in human serum albumin induced by ligand binding. *Pharmacol. Toxicol.*, **66** (suppl. II), 1–26.

65. Wainer, I. W. (1994). Enantioselective high performance liquid affinity chromatography as a probe of ligand-biopolymer interactions: An overview of a different use for high performance liquid chromatographic chiral stationary phases. *J. Chromatogr.*, **666**, 221–234.

66. Noctor, T. A. G., Pham, C. D., Kaliszan, R., and Wainer, I. W. (1992). Stereochemical aspects of benzodiazepine binding to human serum albumin. I. Enantioselective high performance liquid affinity chromatography examination of chiral and achiral binding interactions between 1,4-benzodiazepines and human serum albumin. *Mol. Pharmacol.*, **42**, 506–511.

67. Sudlow, G., Birkett, D. J., and Wade, N. (1976). Further characterization of specific drug binding sites on human serum albumin. *Mol. Pharmacol.*, **12**, 1052–1061.

68. Maruyama, K., Nishigori, H., and Iwatsuru, M. (1985). Characterization of the benzodiazepine binding site (diazepam site) on human serum albumin. *Chem. Pharm. Bull.*, **33**, 5002–5012.

69. Farr, R. S., Reid, R. T., and Minden, P. (1966). Spontaneous and induced alterations in the anion binding properties of human albumin. *J. Clin. Invest.*, **45**, 1006–1012.

70. Walker, J. E. (1976). Lysine residue 199 of human serum albumin is modified by acetylsalicylic acid. *FEBS Lett.*, **66**, 173–175.

71. Shaklay, N., Garlick, R. L., and Bunn, H. F. (1984). Non-enzymatic glycosylation of human serum albumin alters its conformation and function. *J. Biol. Chem.*, **259**, 3812–3817.

72. Iberg, N. and Flückiger, R. (1986). Nonenzymatic glycosylation of albumin in vivo. Identification of multiple glycosylated sites. *J. Biol. Chem.*, **261**, 13542–13545.

73. Okabe, N. and Hashizume, N. (1994). Drug binding properties of glycosylated human serum albumin as measured by fluorescence and circular dichroism. *Biol. Pharm. Bull.*, **17**, 16–21.

74. Woody, R. W. (1994). In Nakanishi, K., Berova, N., and Woody, R. W. (eds.) *Circular Dichroism, Principles and Applications*. New York: VCH Publ., pp. 473–496 (and references there).

CHAPTER

19

CURRENT REGULATORY GUIDELINES OF STEREOISOMERIC DRUGS: NORTH AMERICAN, EUROPEAN, AND JAPANESE POINT OF VIEW

SYLVIE LAGANIÈRE*

Drugs Directorate,
Health Protection Branch, Canada
Ottawa, Ontario, Canada, K1A-OL2

19.1. INTRODUCTION

In recent years, the issues of stereochemistry in drug development and its implications in drug regulation have been raised at international meetings. In the literature, there are many examples illustrating the substantial differences between stereoisomers, enantiomers, as well as diastereomers, with regard to biological disposition, pharmacodynamic/therapeutic activity, or toxicological properties (William and Lee, 1985; Drayer, 1986; Jamali et al., 1989). *In vivo* and *in vitro* stereochemical inversion of a compound can occur (Kaiser et al., 1976; Nakamura, et al., 1980) and stereoselective drug interactions (Lewis, et al., 1974; Zhou, et al., 1990; Smith, et al., 1991) and enantiomer–enantiomer interactions have also been described (George, et al., 1972; Giacomini, et al., 1986; Kunze, et al., 1991; Mast, et al., 1992; Gibaldi, 1993; Koremer, 1994). Therefore, from a scientific point of view, stereoisomers should be treated as distinct substances and their combination as a mixture.

Although neither the scientific nor the regulatory issues are fundamentally new, the availability of new technologies, developed over the last 15 years, have overcome in a significant way limitations associated with obtaining large quantities of a single stereoisomer and have facilitated stereoselective analysis. These developments include novel asymmetric synthetic approaches, enantiomerically pure reagents for use in enantioselective synthesis or in

*The views expressed in this article are those of the author and not those of any regulatory agency.

The Impact of Stereochemistry on Drug Development and Use, Edited by Hassan Y. Aboul-Enein and Irving W. Wainer. Chemical Analysis Series, Vol. 142.
ISBN 0-471-59644-2

preparative scale separations of enantiomers, and enantioselective HPLC and GC methods of analysis. These recent developments have made the development of single enantiomeric drugs more cost-effective. Beyond the aspect of new technology, one should ask in drug development if the drug treatment would be improved using a single stereoisomer. Although for some drugs, treatment would be improved using a single stereoisomer, in other cases, it may offer little advantage over the mixture and may even be detrimental in some other cases (Lennard, 1991; Scott, 1991; Abernethy, and Andrawis, 1993).

Requirements for stereoisomeric drugs have been reevaluated by regulatory agencies around the world and guidelines were developed. There was a concern in the pharmaceutical industry on the type of stereoisomeric drug products that would be acceptable for submission. Were regulatory agencies reluctant about the marketing of a racemate? What would be the future requirements for a racemate? Also, in the context of the worldwide harmonization of regulatory requirements, how do the regulatory requirements differ among countries?

It appears that there was a need to clarify the regulatory position and to provide sponsors of New Drug Submissions (NDS) with guidance on specific areas to be addressed during the development of stereoisomeric drugs for new chemical entities and for second market entry products. As a result of the worldwide marketing of drug products, several manufacturers and authorities alike have shown interest in the harmonization of requirements for registration. This chapter intends to describe the current regulatory requirements of several countries for the assessment of the quality, safety, and efficacy of stereoisomeric drug products, to highlight the differences and similarities between major regulatory agencies and present unresolved preclinical and clinical issues related to the chemical, pharmaceutical, pharmacodynamic, and toxicological characteristics. Requirements for second and subsequent market entries for stereoisomeric products will also be presented.

19.2. REGULATORY GUIDELINES

In the United States, the 1987 New Drug Application (NDA) documentation guidelines concerning the manufacture of a drug substance (FDA, 1987) suggest that when developing drugs that can exist in multiple stereoisomeric forms, the sponsor should prepare the individual stereoisomers and study their physical, chemical, pharmacological, toxicological, and clinical properties. In January 1989, the Center for Drug Evaluation and Research (CDER) established a Stereoisomeric Committee charged with determining what requirements if any should be imposed on a sponsor developing a stereoisomeric compound. In 1992, the Food and Drug Administration (FDA) issued a policy statement for the development of new stereoisomeric drugs (FDA, 1992).

In the European Union (EU), the Committee for Proprietary Medicinal Products (CPMP) working parties on the quality, safety, and efficacy of medicinal products included in the document entitled "The Rules Governing Medicinal Products in the European Community" (CPMP, Vol. II, 1989), a section on stereoisomerism in the chapter on Expert Reports. This is now superseded by the note for guidance on the investigations of chiral active substances (CPMP, 1994). The CPMP's Notice to Applicants provides all the basic procedural information needed by any firm making national or multistate, applications for marketing authorizations for medicinal products for human use (CPMP, 1988; C. 1989; 1990). The Expert Report, which should be part of a submission, is intended to "consist of a critical evaluation of the quality of the product and the investigations carried out in animals and human."

In Canada, a working group from the Drugs Directorate, Health Protection Branch (HPB) was formed in August 1990 to develop guidelines on the issue of stereochemistry. The initial draft Canadian guidelines was distributed in May 1994 to major health and pharmaceutical organizations for comments prior to final approval. The last version of the guidelines on Stereochemical Issues in Chiral Drug Development was updated in July 1996.

The Japanese Ministry of Health and Welfare (MHW) has no plan to publish specific guidelines for the development of chiral drugs. In 1985, the MHW added a statement to the publication "Drug Approval and Licensing Procedures" specific to racemates (Japan, 1989). A statement for racemates was also included in the new Japanese guidelines for nonclinical pharmacokinetics published in 1991. In addition, the Central Pharmaceutical Affairs Council (CPAC), an advisory body to the Pharmaceutical Affairs Bureau, Ministry of Health and Welfare, had made public a series of inquiries regarding the submission of racemates. These CPAC inquiries with an explanation by MHW provide more detailed guidance for the final requirements for a racemate (Kawahara, 1990). Referral should be made to individual cases of CPAC's treatment of applications. The present status of the Japanese administration on chiral drugs was summarized by Dr. H. Kubota, Deputy Director, New Drug Division, Ministry of Health and Welfare, Japanese Government, at a workshop on chiral drugs in Paris, France in 1992 (Gross, et al., 1993). The Japanese previously were thought to be "anti-racemate." They now state they have no problems with racemates, provided there are no safety issues or the safety issues have been resolved. In the same workshop, the harmonization of regulatory requirements was discussed by representatives from academia, industries, and regulatory agencies from the European Union, United States,

Canada, Australia, Japan, and Sweden. Their viewpoints on the issue of chirality were summarized and a consensus report was written. This meeting constituted an important step toward a better understanding of issues to be addressed and the harmonization of regulatory requirements. The topic of stereochemistry was not addressed, however, at the International Committee on Harmonization (ICH) held in Japan in 1995.

Australia and the most recent members of the European Union, including Sweden and Finland, have adopted EU guidelines for the registration of drugs. In all regulatory agencies, these documents complement their respective guidelines including chemistry, manufacturing, preclinical, and clinical.

19.3. GENERAL POLICY

On the market today, single enantiomers are obtained mainly by biochemical processes such as fermentation or by chemical synthesis. Results from a recent study based on data available in *Therapeutic Drugs* (Millership and Fitzpatrick, 1993), a reference book of drugs giving an overview of drugs in use in 1991, indicated that the development of single-enantiomeric drugs increased in the period 1982–1991. This appears to be related to a large increase in the development of single-enantiomeric drugs derived from the synthetic processes discovered in recent years.

Although the regulatory guidelines address issues related to stereoisomers with one or more asymmetric centers, they usually focus primarily on the requirements for chemical entities containing a single stereogenic center (enantiomers and their mixtures) as they often require more specific and unconventional methods of analysis. The separation of geometric isomers and diastereomers can usually be readily accomplished without chiral techniques as these moieties are physically and chemically distinct. In general, therefore, geometric isomers and diastereomers should be treated as distinct drugs and developed accordingly.

In the United States and Canada, the pair of enantiomers forming a racemate is considered a drug product. There is no reason to consider developing mixtures of geometric isomers and diastereomers unless the mixture fortuitously represents a reasonable fixed-dose combination, *in vivo* interconversion of the isomers occurs, or it is impossible or too expensive to separate them. In the EU and Canada, the guidelines consider active substances with one or more stereogenic centers displaying chirality but not the *cis*/*trans* isomerism. In general, geometric isomers have been developed as single isomers.

Now, in all major jurisdictions, promising active substances should be examined for the presence of elements of chirality to assess the relevance of stereoisomerism for activity and disposition *in vivo*. As a result of these guidelines, there will be a trend to require more detailed information on the differences in pharmacodynamic and pharmacokinetic profiles between the individual enantiomers of a racemate.

The decision whether to develop a single enantiomer or a racemic mixture rests with the drug company and should be based on scientific data relating to quality, safety, and efficacy, and ultimately the risk/benefit assessment of the drug under the conditions of proposed use. The major regulatory agencies do not require single isomers to be separated or manufactured and enantiomers to be marketed. In Japan, sufficient justification must be provided for the development of a drug as a racemate to ensure especially the safety of the inactive enantiomer.

19.4. CHEMICAL AND PHARMACEUTICAL DOCUMENTATION REQUIRED FOR CHIRAL DRUGS

Among the major regulatory agencies (Japan, the United States, EU, Canada), there appear to be very similar requirements for the chemical and pharmaceutical documentation of drugs with one stereogenic center (enantiomer, racemates). All information related to the development of the chiral drug product must be submitted. Therefore, if technical problems, such as synthesis, isolation, or scale-up of the preferred enantiomer, pose a major obstacle to its development, the submission should contain a detailed description of such problems.

Appropriate manufacturing and control procedures should be used to characterize and ensure the stereoisomeric composition of a drug substance and the final drug product with respect to identity, strength, quality, and purity during and subsequent to the manufacture of the dosage form. Adequate control procedures and tests must be used to ensure that the enantiomeric composition of drug substances used for animal and human trials represents that proposed for marketing and is consistent with the given specifications.

19.4.1. Drug Substance

Guidelines require the determination of the absolute configuration when possible as part of structural elucidation studies, and the possibility of polymorphism should be investigated.

The physicochemical properties also should be determined. The EC document emphasizes that special attention must be given to the physicochemical properties of the active substance (enantiomer, racemate): crystallinity, polymorphism, rate of dissolution etc. If a racemate is developed, it should be determined whether it is a true racemate (homogeneous solid phase) or a conglomerate of enantiomers (e.g., a mixture of two crystalline enantiomers) by investigating its melting point, solutibility, crystal properties.

Relative solubilities or dissolution rates could be an important factor in deciding to develop a racemate over the enantiomer for a particular dosage form such as oral versus i.v. or aerosol (Jacques, et al., 1981).

For a single enantiomer, the submission must include a full description of the synthesis or isolation procedure used to obtain the individual enantiomer. More specifically, stereochemical identity and purity should be established for the drug substance. Information on the stereochemical identity and purity for chiral starting materials and reagents should also be provided according to the Canadian guideline. In-process testing for identity and purity should be stereoselective for key intermediates in which chirality has been introduced. In the case of an enantiomeric pure drug, the other enantiomer should be considered an impurity.

When a new active substance contains one or more chiral centers (racemate, diastereomer), in addition to the determination of the stereochemical configuration, the specifications for the drug substance should include validated stereospecific methods for identity and purity. The tests used will depend on the nature of the drug substance. The applicant should choose the methods that will adequately demonstrate these characteristics. A description of advantages and limitations associated with tests used to establish the identity and purity of a stereochemical drug was discussed by De Camp (1993). A stereoselective test and/or assay for identity is required in the Canadian, U.S. and European guidelines for the enantomeric drug substance and for the racemate to ensure that partial spontaneous resolution has not occurred during the purification of the bulk drug.

The Canadian guidelines define the requirements for different types of nonracemic mixture: (1) an enantiomerically enriched drug substance resulting from incomplete resolution or partial enantioselective synthesis; (2) a mixture of enantiomers in specified proportions. Enantioselective tests for identity and composition are required in the specifications for mixtures 1 and 2. Optical rotation may be suitable for identity, but a more specific and sensitive method is required for the determination of enantiomeric purity. Pure reference standards should be prepared for all components in a mixture. For a nonracemic mixture of enantiomers with one chiral center, an enantioselective method is required to determine the ratio of enantiomers and the limits for the ratio.

19.4.1.1. Purity

Enantiomeric purity should be established by an independent, suitably validated method. A reference standard of high stereochemical purity should be available for a specific enantiomer and, ideally, reference standards should be available for the antipode. A limit should be specified for the antipode and this limit should be consistent with the levels found in batches used in preclinical and clinical studies. According to the latest draft document of the ICH on chemistry and manufacturing, in general, it is expected that any impurity present in a quantity greater than 0.1% be isolated and identified. According to the Canadian experience, limits of not more than 1% on the antipode have been achievable; this corresponds to an enantiomeric excess of 98%, which is the usual norm for a "pure enantiomer." The U.S. policy document indicates that the maximum allowable level of impurity in a stereoisomeric product employed in clinical trials should not exceed that present in the material evaluated in nonclinical toxicity studies. In the United States, the enantiomeric purity of material used for the clinical studies should be defined.

Optical rotation that may be used as a stereoselective test for identity may not be sufficient for the determination of enantiomeric purity depending on the nature of the drug substance. A more specific and sensitive method is required for the determination of enantiomeric purity as optical rotation lacks sensitivity. In the case of a stereoisomeric drug marketed as a single enantiomer, the USP presently sets requirements for specific rotation. Results presented by P. Lacroix (Lacroix, 1994) at the 4th International Symposium on Chirality in Montreal in 1993 showed that for timolol, specific rotation was not specific and sensitive enought to detect the low levels of the other enantiomer. In this example, all timolol maleate raw materials met the USP requirements for specific rotation (–11.7°–12.5°). However, using a specific HPLC method, the percentage of *R*-enantiomer measured in the samples ranged from 0.1–4.1%. The NMR method provided confirmation of these results. Traditional methods of stereochemical analysis contained in the U.S. Pharmacopeia may not be sufficiently accurate to determine the percentage of impurity in the drug product. Purity requirements in pharmacopeial monographs are set on a case-by-case basis. it is foreseeable that in the future, as a result of technical advances in chiral technology, new standards may be applied for setting Pharmacopoeial limits for purity. The FDA's policy states that manufacturers should notify compendia of the specifications and tests to identify the strength, quality, and purity of stereoisomeric products.

The chemistry section of the Japanese document recommends that chromatographic tests be performed in addition to optical rotatory tests. The EU document lists method and test procedures that may be used in identity and stereoisomeric purity tests for an active substance; they range from more accessible methods such as optical rotation, melting point, liquid chromatography with a chiral stationary phase, and so on, to more sophisticated methods such as optical rotary dispersion, circular dichroism, NMR using chiral shift reagents, and so on. The EU document states that

> depending upon the intended use of the active substance and upon the specification which define the quality, it is for the applicant to choose those methods which will adequately demonstrate the stereoisomeric purity of his active substance.

The purity issue for chiral drugs is not only important in terms of the chemistry and toxicological aspects, but also with regard to the pharmacological effect. The implication of enantiomeric purity must be considered in safety assessment. The degree of enantiomeric impurity may significantly affect the eudismic ratio for a particular pharmacological effect. The eudismic ratio reflects the potency ratio of two enantiomers for a specified pharmacological effect. This factor is even more important if the true eudismic ratio is high (Barlow, et al., 1972). The larger the potency ratio, the more significant the presence of the other enantiomer on the overall pharmacological effect. But the higher the eudismic ratio, the more difficult it is to determine the true ratio, as a minute amount of the eutomer may be difficult to detect by a particular analytical method. During the assessment of the pharmacodynamic effect of each enantiomer, if large doses of the nonactive enantiomer are used in animals to examine adverse effects, it is important to determine the degree of contamination of this isomer with the active isomer, as the effects observed may be due to trace amounts of the active isomer rather than the effect of the inactive isomer. An example could be the presence of L-thyroxine in D-thyroxine.

19.4.1.2. Stability

When appropriate, examination of variation in stereoisomeric ratio between batches and the stability of the drug substance towards racemization under stress conditions and on long term storage should be investigated. Once it has been demonstrated that stereochemical inversion (*in vitro*) does not occur under appropriate conditions, routine stereoselective tests may not be needed.

19.4.2. Drug Product

The maintenance of the stereochemical integrity and physicochemical stability of the drug substance during and subsequent to manufacture of the dosage form are key considerations. In addition to the usual considerations when developing a dosage form, stereochemical aspects may be important when the drug is an enantiomer or a racemate. Solubility differences and the possibility of polymorphism were already stated previously. These may be important considerations for a particular dosage form. A stereoselective interaction between an active ingredient and a pharmaceutical excipient that is often chiral (e.g., cyclodextrin, cellulose derivatives) is also a possibility that could have implications on the dissolution rate of individual enantiomers from the formulation and possibly on the bioavailability of a solid dosage form. When appropriate, the possibility of racemization should also be investigated during the manufacture of the dosage form and during long-term storage.

For a single enantiomer, stereochemical identity and purity should be established for the finished drug substance. An investigation to ensure the maintenance of the crystal form is considered to be important for dosage forms where stability or bioavailability may be affected. Particularly for suspension aerosol products, the effect of grinding and micronization on crystallinity might be important. The shelf-life stability protocol should contain a test to monitor for enantiomeric purity.

The Canadian and EU guidelines indicate that for nonracemic mixtures, a test with limits on the composition of a dosage form is required in the finished product specifications and the relative proportions of the components should be monitored in the shelf-life stability studies.

19.4.3. Nomenclature

The issue of inconsistency in the way stereoisomers are named has been raised repeatedly in recent years (Gal, 1988; De Camp, 1989; Simonyi, et al., 1989). To avoid ambiguity in the dossier regarding nomenclature used to describe the chemical name of stereoisomers, the EU, United States, and Canada recognized that chemical names should be expressed using the appropriate stereochemical descriptors in accordance with the rules of an international organization such as the International Union of Pure and Applied Chemistry (IUPAC 1970). It is preferred that both descriptors (optical rotation and absolute configuration) be used.

It is recognized that most generic names do not indicate anything about stereoisomerism. Although different conventions are used, such as the International Nonproprietary Name (INN) or the U.S. Adopted Names (USAN), it would be desirable to adopt a single convention for standardization purposes that would reflect this characteristic. Simonyi et al. (1989) proposed the SIGNS nomenclature (stereochemically informative generic name system). In this system, seven prefixes are offered to describe the stereochemical nature of any drugs. The appropriate prefix would be attached to the generic name. Examples are rac-ibuprofen, levo-dilevalol, and levo-timolol.

Overall, the chemistry and pharmaceutical requirements are very similar between countries. Minor noted differences could probably be resolved easily in harmonious international document.

19.5. PRECLINICAL AND CLINICAL REQUIREMENTS

19.5.1. Stereoselective Assay

A validated stereoselective assay of enantiomers in biological fluids should be developed at an early stage of drug development and used whenever relevant, unless it has been clearly demonstrated that a nonenantioselective assay provides results equivalent to those obtained with the enantioselective assay in a targeted species and under certain conditions.

These methods must be validated for specificity, precision, accuracy, and sensitivity in the range of concentrations and ratios of enantiomers expected. Diastereomeric derivatives formed prior to GC and HPLC analysis have often been used. The purity of the resolving agents and of the reference standards should be determined carefully to ensure proper evaluation of the method.

19.5.2. Preclinical Pharmacology/Toxicology Requirements

19.5.2.1. Development of a Single Enantiomer of a New Active Substance

The documentation required for a single enantiomer of a nonmarketed drug is the same as that for any new active substance. The enantiomeric purity of the active substance used in preclinical and clinical studies should be defined using a sensitive method. The stability of the enantiomer must be established by *in vivo* studies. If the antipode is formed *in vivo*, it should be considered to be a metabolite, and should be addressed as such in accordance with normal drug development practices. The rate of occurrence of such an outcome would determine the need for any enantiospecific biological assay. Otherwise, the disposition of the enantiomer could be based on nonenantioselective methods. These requirements are basically the same between major countries.

19.5.2.2. Racemate of a New Active Substance

In all major regulatory agencies, it is possible to make a successful application for the marketing of a racemate. However, companies are being asked for extended experimental data on the individual enantiomers, including, as a minimum, information on the activity of single enantiomers and the use of enantioselective methods of analysis in animal pharmacology/toxicology and human pharmacokinetic and metabolic studies. In the EU, companies need to justify in an Expert Report their decision to market a racemate instead of a single enantiomer (CPMP, 1989).

A drug containing a pair of enantiomers is usually marketed in a 1:1 ratio. For purposes of new drug application approval, the regulatory agencies do not currently consider racemates to be fixed-combination drugs unless specific claims for the actions of individual enantiomers are made. If a drug product contains enantiomers in different ratios for which a specific claim is made, this product will be assessed as a drug combination. The regulations governing the approval of fixed combinations usually require that the contribution made by each active ingredient to the overall effectiveness of a fixed combination be defined and that each active ingredient independently be shown to be effective and safe. These requirements have not been applied as yet to racemates because historically enantiomers were not easily separable or resolvable.

The documentation required for a new racemate is the same as that for any new active substance, plus generally the following for the major regulatory agencies.

19.5.2.2.1. Pharmacodynamics. Primary pharmacodynamic effects related to the therapeutic effect and other important pharmacodynamic effects of each enantiomer administered individually and as a racemate should be evaluated in animals and *in vitro* systems where appropriate. These requirements are common to most regulatory agencies to define the profile of pharmacodynamic effects for each enantiomer. However, if the racemate undergoes rapid interconversion *in vivo*, then this requirement would not be necessary.

In humans, the safety and efficacy studies should be examined with the racemate intended for marketing, and plasma concentrations of individual isomers after administration of the racemate will be required.

19.5.2.2.2. Toxicology. Major differences are observed between the European, North American, and Japanese regulatory agencies with regard to the information required for a racemate. In Japan, the minimum preclinical requirements with respect to stereoisomers are a comparison of the acute toxicology of each enantiomer and of the racemate in animals (Hutt, 1991; Shindo, 1991).

In North America and Europe, it would usually suffice to carry out toxicity studies on the racemate only. The disposition of each enantiomer should be followed after administration of the racemate, when applicable, in *in vitro* systems and in animals an humans. If the toxicology data of the racemic/non-racemic mixture indicate toxicity findings other than those that might be a natural extension of the pharmacological effects of the drug, especially if they are unusual and occur at doses closely associated with effective concentrations in several animal species or close to the proposed human exposure or adverse reactions that might prevent or delay the product to enter in the clinical trial, additional toxicology data will be necessary to examine the relative contribution of each enantiomer to the overall toxicity.

19.5.2.2.3. Pharmacokinetics. The pharmacodynamic activity of each enantiomer established *in vitro* cannot be readily extrapolated to the *in vivo* situation without taking into account the stereoselective disposition of the drug. For enantiomers that differ significantly in their pharmacological effect, the importance of the stereoselective disposition of a drug could significantly impact on the expected response.

For a racemate, the concentration-time profile of each isomer should be examined in animals and humans after administration of the racemate to compare the pharmacokinetic disposition of the enantiomers within and between species and to validate the results of acute and multiple-dose toxicity studies performed with the racemate. The potential for interconversion should also be evaluated. Pharmacokinetic studies should be performed with the same doses/routes/animal species as the studies in animals and at therapeutic doses in humans. Studies in humans (healthy volunteers and patients) should be carried out using stereoselective methods unless it is shown that there is no qualitative and quantitative difference in the fate of both enantiomers in the target population.

The information required for the racemate by the Japanese regulatory agency is more extensive compared with the European and North American regulatory agencies. As reported at the Paris conference, the Japanese authorities recommend investigation of the absorption, distribution, metabolism, and excretion of each of the enantiomers separately and of the racemate, including the possible occurrence of inversion *in vivo* if a racemate is considered for marketing. For a mixture of isomers, in particular, it is necessary to investigate how each isomer is metabolized and disposed of and how each isomer contributes to efficacy.

19.5.2.2.4. Labelling. This issue related to the labelling of stereoisomeric drug products was discussed at several meetings, but, with the exception of the United States, it was not incorporated as a specific item in any of the regulatory guidelines for the assessment of stereoisomeric drugs: Should the warnings section include information relating to which stereoisomer is responsible for an adverse reaction, and should the individual half-life be reported for each enantiomers? This aspect is likely to be raised during discussion for the harmonization of regulatory requirements. The FDA's policy states that the labelling should include a unique established name and a chemical name with the appropriate stereochemical descriptors. Up to now, a general practitioner would find it difficult to know from a monograph whether the prescribed drug is a racemate or a single enantiomer.

19.5.3. Racemate–Enantiomer Switches

In principle, the development of a compound with a stereogenic center, in a stereoisomeric form different from the one that has already received approval from regulatory agency, requires a completely new application. The concept of bridging studies was brought forward in the context of new drug submissions, in particular for the development of a single stereoisomer after the racemate has been approved and in the case of drug development when the racemate was initially investigated preclinically before the decision was made to develop the enantiomer for marketing (CPMP, 1994). The intention is to make use of prior knowledge on the racemate. Bridging studies may be useful in avoiding the replication of studies.

This concept was recognized in the consensus report from Paris, which indicated that

> for a racemate-to-enantiomer switch, certain studies that would normally be required may not need to be repeated with the selected enantiomer when relevant studies have previously been conducted with the racemate and the data from these studies are available to the applicant of the enantiomer (Gross et al., 1993).

Bridging studies, for example, could be relevant pharmacologic/toxicologic studies that would allow the use of existing information on the racemate to be applied to the enantiomer for the purpose of submitting an abbreviated drug submission for only the single enantiomer.

Most guidelines refer with caution to this concept of bridging studies. Although the EU and U.S. documents cite the possibility of bridging studies for preclinical and clinical studies, Canada proposes this option only in the context of pharmacological/toxicological evluation. The EU is probably the most explicit, providing also guidelines for the development of a new enantiomer from an approved racemate, the development of a racemate from an approved single enantiomer, and the development of a nonracemic mixture from an approved racemate or single enantiomer.

The U.S. document refers to the case where

> a sponsor wishes to develop a single stereoisomer from a mixture that has already been studied non-clinically, indicated that an abbreviated appropriate pharmacology/toxicology evaluation could be conducted to allow the existing knowledge of the racemate to the sponsor to be applied to the pure stereoisomer.

If a chiral drug is submitted to the FDA as a racemate, studies with the individual enantiomers can be carried out under the existing Notice of Claimed Investigational Exemption for a New Drug (IND) as long as appropriate chemistry, manufacturing, and control data are submitted for the individual enantiomers. In this situation, bridging studies may be performed. For racemates already on the market, the FDA requires a new drug application for the corresponding single enantiomer. If a racemate has been marketed and the sponsor wishes to develop the single enantiomer, evaluation should include a determination of whether there is significant conversion to the other isomer, and whether the pharmacokinetics of the single isomer are the same as they were for that isomer as part of the racemate. The FDA's policy statement does not define the types of bridging studies that would be acceptable for clinical assessment. Clinical trials must be carried out under a new IND for that particular enantiomer.

Examples of preclinical toxicology bridging studies, as cited in the Canadian document, could consist of repeat-dose comparisons of the selected enantiomer and the racemate in the most appropriate species. The duration of the repeat-dose studies would depend on the proposed duration of use in man and, in general, should not be shorter than 3 months in Canada with exceptions (e.g., single-use drugs such as neuromuscular relaxants) and up to 3 months with United States and the EU. In addition, the reproductive toxicology study should be completed in the most sensitive and relevant species, using the single enantiomer. The pharmacodynamic and pharmacokinetic profile of the selected enantiomer should be compared with that of the racemate. The approach to bridging studies and the requirements for this example are similar in the United States, the EU and Canada.

The pharmacokinetic profile of the enantiomer following administration of the racemate could be used for comparison with the pharmacokinetics of the enantiomer proposed for development to identify possible interaction and evaluate its implications for dynamic interaction. The usual studies required to demonstrate the safety and efficacy of the single enantiomer proposed should be fully documented as well as any claims for the superiority of the single enantiomer over the racemic mixture.

Japan did not address at all the concept of bridging studies in official documents.

The scientific literature reports numerous examples of enantiomers that differ substantially with respect to their pharmacokinetics, pharmacodynamics, toxicity, protein binding, and so on. Two recent examples of enantioselective toxicity were raised by Shah (1993). Racemic terolidine was marketed clinically for the treatment of urinary incontinence. It was withdrawn from the market worldwide in September 1991 following the report of serious ventricular arrhythmias, with a fatal outcome in some cases (Anonymous, 1991). *S*(–)-

terolidine is almost 10 times more potent than its antipode as a calcium antagonist, whereas *R*(+)-terolidine is almost 10 times more potent than *S*(–)-terolidine in its anticholinergic activity (Larrsson-Backstrom, et al., 1985). Therapeutic efficacy is related to the *R*-enantiomer, whereas cardiotoxicity is associated with the *S*-enantiomer. Dilevalol is the R,R-diastereomer of labetolol. This isomer was selected for development as an antihypertensive agent because of four-fold β-blocking, seven-fold vasodilatory, and 1/3 α-blocking activities compared to labetolol. Following its approval, the toxicity of dilevalol was revealed as reports showed with use an increase in the incidence of hepatotoxicity compared with the mixture labetolol (Shah, 1993).

It is therefore essential that, because of the distinct characteristics of stereoisomers in a chiral environment such as the body and the limited regulatory experience, the nature of the bridging studies must be defined on a case-by-case basis and justified by the applicant. It is recommended that advice be sought from the regulatory agency on the necessary requirements for a particular compound prior to filing the submission. Because of the disadvantage of a case-by-case where several authorities may decide differently, this aspect will need to be addressed at a future ICH meeting on stereoisomeric drugs.

19.6. SECOND AND SUBSEQUENT MARKET ENTRIES OF SOLID ORAL DOSAGE FORMS FOR CHIRAL PRODUCTS

Generic drugs are usually approved on the basis of adequate chemistry, manufacturing, quality control procedures, and comparative pharmacokinetics (bioequivalence). An abbreviated New Drug Application can be submitted for generic products.

The requirements for second and subsequent market entries of solid oral dosage forms for chiral products were discussed in the EU and Canadian documents only. The Center for Drug Evaluation and Research (CDER) at the FDA has presented and discussed this issue at several symposia, but it was not included in the policy statement of the FDA, which examined only new stereoisomeric drugs. Japan does not address the requirements of stereoisomeric generic products in official documents.

Consensus at the Paris meeting recommended that generic drug products containing a chiral drug substance be held to the same manufacturing and control standards that were originally applied in the approval of the innovator product. However, although it might be claimed that a generic drug product should be required to undergo the same chemical and pharmaceutical evaluation as did the innovator, for example, when stereoselective assays were not previously available, there may be instances when the approval of a generic product requires measurement of individual enantiomers of a chiral drug.

Generic formulations must be both pharmaceutically and therapeutically equivalent to the innovator or first market entry. The comparative bioavailability of these products should be assessed against an acceptable reference. Comparison should be made between the pharmaceutical equivalent and pharmaceutical alternative products (the United States, EU, Canada), which also includes both qualitative and quantitative equivalence in stereoisomeric composition. The pharmaceutical equivalent and/or alternative between a generic drug and its innovator is defined as (1) the same active moiety, including the same salt or ester; (2) the same strength; (3) intended for the same route of administration; (4) the same dosage form (e.g., capsule or tablet); and (5) similar or identical labeling. The Office of Generic Drugs does not require comparability in the chiral composition of excipients, provided the *in vivo* or *in vitro* performance of the drug product remains comparable between the innovator and generic product (Shriniwas, 1992).

Therapeutic equivalent products are pharmaceutically equivalent products that have been shown to be bioequivalent through either an *in vivo* or *in vitro* bioequivalence study. In Canada, the assessment of therapeutic equivalence falls under provincial jurisdiction, but the generic product should meet bioequivalence standards.

Japan did not discuss the issue of stereochemistry in relation to generic products.

It is recognized that enantiomers may have a different pharmacokinetic disposition and pharmacodynamic effect *in vivo* and, therefore, stereoselective bioassay would be necessary to determine the characteristic of each enantiomer when a racemate is administered to humans. However, the characteristics of a chiral drug will be important in assessing the bioequivalence of a racemate only if it is associated with the absorption phase. The requirements for a stereoselective assay will need to be met in the case of nonlinear active absorption, where the enantiomeric ratio changes with the change in input rate. The same rationale will apply to chiral inversion, which will be important in bioequivalence if it influenced one enantiomer preferentially before systemic absorption. If the rate of entry of a drug in the systemic circulation affects the enantiomeric ratio, it might be necessary to meet requirements for bioequivalence on individual enantiomers unless pharmacodynamic/clinical data are provided.

The Office of Generic Drugs indicated (Williams, 1993) that in the case of drugs exhibiting passive absorption, passive and linear active absorption, poorly extracted drugs exhibiting nonlinearity after absorption, enantiomer-specific genetic polymorphism or absorption, the measurement of the racemate as opposed to individual enantiomers in a bioequivalence study is acceptable because differences in the rate and extent of absorption from the gut to the systemic circulation should affect both enantiomers equally. For a genetic polymorphism, measurement of the racemate in a bioequivalence study is still acceptable under the assumption that the individual enantiomer clearances remain constant in phenotypically distinct individuals during both treatments of a two-treatment cross-over bioequivalence study.

The EU document indicates that bioequivalence studies supporting generic applications of chiral medicinal products should be based on enantiospecific bioanalytical methods, unless (1) both products contain the same, stable single enantiomer as the active substance , or (2) both products contain the racemate and both enantiomers show linear pharmacokinetics.

The Canadian guideline states that specific bioequivalence requirements may have to be met on individual enantiomer(s) of drugs for which it is known that the rate of dissolution and/or absorption in the systemic circulation significantly affects the *in vivo* enantiomeric (e.g., drugs with enantioselective non-linear first pass metabolism) ratio. In general, for conventional solid oral dosage forms, for which the rate of release of the drug from the formulation does not influence the relative concentration ratio of the enantiomers, the parameters to be determined and the standards to be met will be the same as those specified for total drug.

The rate of input (release and/or absorption) and in turn the ratio of individual enantiomers in the systemic circulation is more likely to be different for modified release dosage forms than conventional dosage forms. Therefore, in the case of these formulations generally, comparative bioavailability requirements will have to be met for both enantiomers unless it can be clearly demonstrated that they possess similar pharmacokinetic and pharmacological properties.

Enantioselective assay may be required for modified release products containing only a single enantiomer, unless it is known that no interconversion of the enantiomer occurs *in vivo*.

Therefore, it appears that in general regulatory agencies will not impose a requirement if there is no compelling reason to do so. For uncomplicated conventional solid oral dosage forms, for which the rate of release of the stereoisomers from the formulation does not influence their relative concentration ratio, the parameters to be determined for bioequivalence will be the same as those described in the EU, U.S., or Canadian guideline for bioequivalence. These guidelines essentially require the superimposability of total (non-stereoselective) drug concentration-time profiles within statistical limits.

19.7. CONCLUSIONS

Regulatory requirements are set to ensure that the quality, safety, and efficacy of drug products are properly evaluated and modified when necessary to improve the benefit to the population. Recently available guidelines for the development of stereoisomeric drugs should help to clarify a regulatory position with regard to the stereochemistry issues, to provide sponsors of New Drug Submission (NDS) with guidance on specific areas to be addressed during the development of stereoisomeric drugs, and to provide a basis document for the future harmonization of regulatory requirements between countries. Although the requirements on the chemistry and manufacturing, and pharmacodynamic and pharmacokinetic aspects are very similar between major countries, some discussion will be necessary before harmonization can be achieved on the toxicological issues and bridging studies.

ACKNOWLEDGMENTS

I would like to thank the members of the Working Group on stereochemistry and Dr. McGilveray for their contribution to the development of the Canadian guidelines. I would also like to thank Dr. G. Rauws, the Netherlands, for his review of this manuscript.

REFERENCES

Abernethy, D. R. and Andrawis, N. S. (1993). In I. W. Wainer (ed.), *Stereoisomeric Drugs in Therapeutics in Drug Stereochemistry*, 2nd ed. New York: Marcel Dekker, pp. 385–397.

Anonymous (1991). Terodiline withdrawn worldwide. Scrip. no. 1653 20, Sept., p. 19.

Barlow, R. B., Frank, F. M., and Pearson, J. D. M. (1972). The relation between biological activity and the degree of optical resolution of optical isomers. *J. Pharm. Pharmacol.*, **24**, 753–761.

CPMP (1988). Notice to Applicants for Marketing Authorization for Proprietary Medicinal Products in the Member States of the European Union. Brussels Commission of the European Union III/118/87, rev., 53–54.

CPMP (1989). The Rules Governing Medicinal Products in the European Union. Vol. II, Notice to Applicants for Marketing Authorizations for Medicinal Products for Human Use in the Member States of the European Union, Brussels. Commission of the European Union, cat. no. CD-55-89-293 EN-C.

CPMP (1990). The Rules Governing Medicinal Products in the European Union. Vol. III, Guidelines on the Quality, Safety and Efficacy of Medicinal Products for Human Use, Brussels. Commission of the European Union, cat. no. CB-55-89-843-EN-C.

CPMP (1994). Note for Guidance. Investigation of Chiral Active Substance. Brussels: Commission of the European Union, III/3501/91 final.

De Camp, W. H. (1989). Letter to the editor. *Chirality*, **1**, 97.

De Camp, W. H. (1993). FDA perspective on the development of new stereoisomeric drugs: Chemistry, manufacturing and control issues. In I. W. Wainer (ed.), *Drug Stereochemistry*. New York: Marcel Dekker, Chap. 13, pp. 365–374.

Drayer, D. E. (1986). Pharmacodynamic and pharmacokinetic differences between drug enantiomers in man—An overview. *Clin. Pharmacol. Ther.*, **40**, 125–133.

FDA (1992). FDA's Policy Statement for the Development of New Stereoisomeric Drugs. *Chirality*, **4**, 338–340

FDA (1987). Guideline for Submitting Documentation in Drug Applications for the Manufacture of Drug Substance. Rockville, MD.

Gal, J. (1988). Stereoisomerism and drug nomenclature. *CPT*, **3**, 251–253.

George, C. F., Fenyvsi, C., Conolly, M. E., and Dollery, C. T. (1972). Pharmacokinetics of dextro, laevo, and racemic propranolol in man. *Eur. J. Clin. Pharmacol.*, **56**, 485–490.

Giacomini, K. M., Nelson, W. L., Pershe, R. A., Valdivieso, L., Turner-Tamiyasu, K., and Blaschke, T. R. (1986). In vivo interaction of the enantiomers of disopyramide in human subjects. *J. Pharmacokin. Biopharm.*, **14**, 335–356.

Gibaldi, M. (1993). Stereoselective and isozyme-selective drug interactions. *Chirality*, **5**, 407–413.

Gross, M., Cartwright, A., Campbell, B., Bolton, R., Holmes, K., Kirkland, K., Salmonson, T., and Roberts, J.-L. (1993). Regulatory requirements for chiral drugs. *Drug Inform. J.*, **27**, 453–457.

Gross, M. (1991). Development of chiral drugs in an evolving regulatory environment. *Reg. Affairs*, **3**, 483–494.

Hutt, A. J. (1991). Drug chirality: Impact on pharmaceutical regulation. *Chirality*, **3**, 161–164.

IUPAC, (1970). *J. Organic Chem.*, **35**, 2849–2867.

Jacques, J., Collet, A. and Wilen, S. H. (1981). *Enantiomers, Racemates and Resolutions*. Wiley, New York.

Jamali, F., Mehvar, R., and Pasutto, F. M. (1989). Enantioselective aspects of drug action and disposition: Therapeutic pitfalls. *J. Pharm. Sci.*, **78**: 9, 695–715.

Japan, (1989). Drug Approval and Licensing Procedures. Yakugyo Jiho Co. Ltd. Tokyo, Japan.

Kaiser, D. G., van Geissen, G. J., Reisher, R. J., and Wechter, W. J. (1976). Isomeric inversion of ibuprofen R enantiomer in humans. *J. Pharm. Sci.*, **65**, 269.

Kawahara, A. (1990). Present and future aspects of the Japanese Pharmaceutical Affairs Administration. *Drug Inform. J.*, **24**, 153–167.

Kroemer, H. K., Fromm, M. F., Buhl, K., Terefe, H., Blaschke, G., and Eichelbaum, M. (1994). An enantiomer-enantiomer interaction of (S)-propafenone and (R)-propafenone modifies the effect of racemic drug-therapy. *Circulation*, **89**: 5, 2396–2400.

Kunze, K. L., Eddy, A. C., Gibaldi, M., and Trager, W. F. (1991). Metabolic enantiomeric interactions: the Inhibition of human (S)-warfarin-7-hdroxylase by (R)-warfarin. *Chirality*, **3**, 24–29.

Lacrois, P. M., Dawson, B. A., Sears, R. W., and Black, D. B. (1994). HPLC and NMR methods for the quantitation of the (R)-enantiomer in (–)(S)-timolol maleate drug raw materials. *Chirality*, **6**, 484–491.

Larsson-Backstrom, C., Arhenius, E., and Sagge, K. (1985). Comparison of the calcium antagonistic effects of terolidine, nifedipine and verapamil. *Acta Pharmacologica et Toxicologia*, **57**, 8–17.

Lennard, M. S. (1991). Clinical pharmacology through the looking glass: Reflections on the racemate and enantiomer debate. *Br. J. Clin. Pharmacol.*, **31**, 623–625.

Lewis, R. J., Trager, W. F., Chan, K. K., Brechenridge, A., Orme, M., Roland, M., and Schary, W. (1974). Warfarin-stereochemical aspects of its metabolism and the interaction with phenylbutazone. *J. Clin. Invest.*, **53**, 1607.

Mast, V., Fischer, C., Mikus, G., and Eichelbaum, M. (1992). Use of pseudoracemic nitrendipine to elucidate the metabolic steps for stereoselective disposition of nitrendipine enantiomers. *Br. J. Clin. Pharmacol.*, **33**, 51–59.

Millership, J. S. and Fitzpatrick, A. (1993). Commonly used chiral drugs: A survey. *Chirality*, **5**, 573–57.

Nakamura, Y., Yamaguchi, T., Takahashi, S., Hashimoto, S., Iwatani, K., and Natiagawa, Y. (1980). *Proc. Symp. Drug Metab. Action*, **12**, S1.

Scott, A. K. (1991). Stereoisomers in clinical pharmacology. *DN & P*, **48**, 476–482.

Shah, R. R. (1993). Clinical pharmacokinetics: Current requirements and future perspectives from a regulatory point of view. *Xenobiotica*, **23**:11, 1159–1193.

Shindo, H. and Caldwell, J. (1991). Regulatory aspects of the development of chiral drugs in Japan. *Chirality*, **3**, 91–93.

Shriniwas, G. N., Dighe, S. V., and Williams, R. L. (1992). Bioequivalence of racemic drugs. *J. Clin. Pharmacol.*, **32**, 935–943.

Simonyi, M., Gal, J., and Testa, B. (1989). Signs of the times: The need for a stereochemically informative generic name system. *TiP S*, **10**, 349–354.

Smith, D. A., Chandler, M. H. H., Shedlofsky, S. I., Wedlund, P. J., and Blouin, R. A. (1991). Age-dependent stereoselective increase in the oral clearance of hexobarbitone isomers caused by rifampicin. *Br. J. Clin. Pharmacol.*, **32**, 735–739.

William, K. and Lee, E. (1985). Importance of drug enantiomers in clinical pharmacology. *Drugs*, **30**, 333–354.

Williams, R. L. CDER, FDA. (1993). International harmonization of stereochemical drug regulation. Paper presented at 4th International Symposium on Chiral Discrimination, Montréal, 19–22.

Zhou, H.-H., Anthony, L. B., Roden, D. M., and Wood, A. J. J. (1990). Quinidine reduces clearanceof (+)-proparanolol more than (–)-propranolol through marked reduction in 4-hydroxylation. *Clin. Pharmacol. Ther.*, **47**, 686–693.

CHAPTER

20

ENANTIOSELECTIVE ANALYSIS AND THE REGULATION OF CHIRAL DRUGS

MICHAEL GROSS

Chiros International
Ho-Ho-Kus, New Jersey 07423

20.1. INTRODUCTION

The importance of stereochemistry in drug action has been appreciated for almost three quarters of a century (1). However, the development of chiral drugs has been significantly influenced by inadequacies in synthetic and analytical technologies. As a result, about one-fourth of the drugs marketed today are registered in racemic form (2). Current interest in applying stereochemical principles in drug development reveals a renewed appreciation for the enantioselective nature of the interactions between small molecules and endogenous macromolecules in man. This is the direct result of technical developments in the field of enantioselective separations and analysis.

The question of the continuing acceptability of racemic mixtures as new drugs is only one aspect of the *racemic drugs issue* (3). The racemic drugs issue concerns defining the appropriate level of chemical, biological, and clinical information necessary to assure the safety, efficacy, consistency of manufacture, and adequacy of labeling of a chiral drug regardless of its stereochemical form. Regulatory agencies are now requesting chemical, biological, and clinical data to support the choice of a specific stereochemical form of a drug. Over the past 10 years, these requests have evolved from a diffuse concern about stereochemistry to the issuance of regulatory guidelines and policy statements. The introduction of stereochemical awareness into the registration process has produced significant changes in the way chiral drugs are developed. Increased stereochemical awareness on the part of both sponsors and regulators is providing a new driving force for the development of drugs with increased specificity and reduced toxicity. Modern enantioselective syn-

The Impact of Stereochemistry on Drug Development and Use, Edited by Hassan Y. Aboul-Enein and Irving W. Wainer. Chemical Analysis Series, Vol. 142.
ISBN 0-471-59644-2

thetic and analytical technologies provide powerful tools to help in the pursuit of this ideal.

The milestone that marks the change in thinking about the complexities of developing racemic drugs was an infamous 1984 paper entitled, "Stereochemistry, a Basis for Sophisticated Nonsence in Pharmacokinetics and Clinical Pharmacology" (4). This and a number of related articles criticized the widespread practice of collecting pharmacokinetic and pharmacodynamic data with racemic drugs while ignoring the separate contributions of the enantiomers (2, 5–13). These seminal publications have served to stimulate scientific and regulatory discussions involving industry, academia, and government concerning the significance of applying stereochemical principles in drug development (14–17). These discussions recognize that individual enantiomers may contribute differently to the pharmacokinetic/pharmacodynamic, toxicokinetic/toxicodynamic, and clinical effects of a drug.

It is now recognized that at best, a racemic drug is composed of enantiomers that may contribute in a favorable way to the clinical effects of a drug, or that one of the enantiomers may be inert with respect to both the pharmacologic and toxicologic effects and represents only, *isomeric ballast*. At worst, a racemate may contain enantiomers, one of which may contribute in a beneficial way and the other in a deleterious way to the clinical effects of a drug.

In 1990, a survey of industry practice was conducted to assess the impact of this regulatory emphasis on the U.S. pharmaceutical industry (18). Twenty-six companies were surveyed and 17 responded to a questionnaire. The majority of those responding indicated that no formal corporate policy had been established within their organizations concerning the development of chiral drugs. The responses indicated, however, that when faced with a choice of stereochemical form, there was a tendency within the industry to develop a single enantiomer rather than a racemic mixture. Although the question was not specifically asked in the survey, it is inferred from the responses that, at the time, the use of enantioselective methods of analysis in chiral drug development was not widely adopted within the pharmaceutical industry.

In 1992, an international group of representatives from the pharmaceutical industry, academia, and government participated in a workshop aimed at producing recommendations for harmonizing preclinical, clinical, and quality requirements for chiral drugs from which a basis for harmonized requirements could be developed. Their report emphasizes the use of enantioselective methods of analysis throughout chiral drug development, regardless of which stereochemical form is chosen (17). It suggests that when single enantiomers are developed, the optical antipode will often be an impurity and should be measured. When a racemate is to be developed the report also recommends defining the separate contributions of the constituent enantiomers to the biological profile of the drug.

20.2. FDA POLICY ON CHIRAL DRUGS

The 1962 Kefauver–Harris Amendments to the Food, Drug and Cosmetic Act, which followed the thalidomide tragedy, strengthened the act by tightening safety requirements for investigational and marketed drugs. Although drug stereochemistry was not specifically adressed in the Amendments, they require the inclusion in a new drug application (NDA) of a full description of the production methods including tests for identity, strength, quality, and purity. FDA regulations also do not specifically mention requirements concerning drug stereochemistry. The first mention of stereochemical requirements to be applied in drug development is found in three separate sections of a 1987 NDA guideline concerning manufacturing documentation (19). Regarding the manufacture of drug substances, it suggests that the FDA may consider enantiomers to be impurities if they are represented in small quantities in the drug substance or even when they represent as much as 50% of its composition (i.e., in a racemic mixture). Concerning drug substance manufacturing controls, the guideline indicates that for a chiral drug substance, regardless of its stereochemical form, the specifications and tests applied should assure that a material with requisite properties for therapeutic activity is produced. Concerning the physical and chemical characterization of a new drug substance (NDS) entering clinical investigations, the guideline states,

> When a NDS is asymmetric (e.g., contains one or more chiral centers, or has cis-trans or other types of isomers), the sponsors should ideally (and prior to the submission of an IND) have either separated the various potential stereoisomers of the NDS or synthesized them independently. Physical/chemical information about each stereoisomer should be provided (in detail), or may be requested. Individual stereoisomers may need to be studied for pharmacological and toxicological properties (and/or safety and efficacy).

The wording of the FDA guideline revealed that sponsors could be required by the FDA to provide, *prior to clinical investigations*, physical, chemical, and even biological data on the individual enantiomers of a racemic mixture. This raised concerns within the pharmaceutical industry about how the FDA might regulate stereochemistry and this led to the formation of the Pharmaceutical Manufacturers Association (PMA) Ad Hoc Committee of Racemic Mixtures. The PMA has published its position describing points to consider in the development and registration of chiral drugs with the intent of providing an industrial perspective on regulatory requirements (20). It provides examples where the registration of a racemate might be justified and emphasizes that racemates should remain viable drug development candidates, with each candidate being considered on a case-by-case basis by its sponsor. The PMA position also describes how the registration of an enantiomerically pure drug derived from a previously approved racemic mixture (i.e., a racemate-to-enantiomer switch) might be based, in part, on pharmacokinetic *bridging* comparisons. The position paper also raised the possibility that early studies of a chiral drug might be conducted with a racemic mixture, while the later studies might be conducted with an enantiomerically pure form that could ultimately undergo registration on the basis of combined data.

In 1992, the FDA issued a policy statement that focused on regulatory issues in the development and registration of drugs based on either single enantiomers or racemates (21). The policy statement provides insight into FDA general philosophy concerning the development and registration of chiral drugs. The policy cites examples in which a racemate has been registered, where each enantiomer exhibits unique biological characteristics (e.g., pharmacology, toxicology, disposition), and states that in such cases the optimal ratio of the enantiomers would not be expected to be unity. However, the document stops short of considering a racemate to be fixed-combination drug product. The policy recognizes that the practice of developing and registering racemates has not been problematic and further states that the registration of racemates may continue to be acceptable. Concerning the registration of a mixture of diastereoisomers, the document suggests that there should be no generalized rationale for developing such mixtures and these could be regulated as fixed-combination products.

The FDA's policy statement stresses the appropriate use of enantioselective methods of analysis in drug development regardless of the stereochemical form of a chiral drug. For pharmacokinetic (ADME), pharmacologic, toxicologic, and clinical studies, defining the contributions of individual enantiomeric species is recommended. For pharmacokinetics, only in subsequent, more routine, pharmacokinetic evaluations might a nonenantioselective assay be accepted and only after the pharmacokinetics of the individual enantiomers are established and shown to be predictable. It recommends the investigation of possible stereogenic interconversion (i.e., unidirectional inversion or racemization). It generally allows the study of a racemic mixture as a surrogate for studies of the pure enantiomer that may ultimately undergo registration when the toxicology model has been appropriately exposed to meaningful levels of the enantiomers. If an unexpected toxicity is observed, this may require attribution to an individual enantiomer and this could be a reason for the development of a single enantiomer. More specific requirements with regard to chemistry, manufacture, and controls (CMC) issues are currently in draft at the FDA and the release of further regulatory guidance is anticipated.

Note added in proof: In 1996 an expanded policy statement issued which provides additional guidance on CMC, pharmacology–toxicology, and labelling issues (24).

The FDA policy concerning drug stereochemistry recognizes that it may be appropriate to pursue a truncated development strategy for developing a pure enantiomer from a previously registered racemic mixture when such a program includes appropriately designed bridging studies. In the case where the pharmacology and toxicology of the enantiomers of racemic mixture are equivalent and little difference is observed in the disposition of the enantiomers, or when one enantiomer is inert, a racemate may be an appropriate development choice. However, when one enantiomer is more toxic or exhibits an undesirable pharmacologic effect or when one enantiomer is less potent or less pharmacologically specific than the other, it may be more appropriate to develop a single enantiomer. In the unique case where both enantiomers exhibit different but desirable properties, the policy suggests that it might be appropriate to regulate the racemate as a fixed-combination drug product.

Concerning chemistry issues, the FDA policy requires that stereochemistry should be appropriately considered when establishing specifications to assure the identity, purity, strength, and quality of the drug substance and drug product. This would include stereochemically specific identity tests and, at least initially, the use of enantioselective assays for stability and content assays.

20.3. EUROPEAN COMMUNITY GUIDANCE ON CHIRAL DRUGS

A European Community (EC) "Notice to Applicants" issued in 1988 requires for drugs containing chiral centers that the stereochemical form used in preclinical and clinical studies as well as the steriochemical form intended for marketing be specified in the registration dossier and that the methods of preparation of the drug substance be fully described (22). The Expert Report section of the dossier must contain discussions of the batch-to-batch consistency of enantiomeric ratios as well as discussions of relevant preclinical and clinical pharmacologic, toxicologic, and pharmacokinetic considerations as they relate to the stereochemistry of the drug undergoing registration. Results of preclinical and clinical studies should be contrasted considering possible species differences in ADME. The notice requires a complete registration dossier when a single enantiomer is developed from a previously approved racemate.

In October 1993, the CPMP finalized a guidance on chiral drugs that became effective in May 1994 (23). It recognizes that a new active substance application for the marketing of a single enantiomer should be treated in the same way as an application for any new active substance. It also suggests that the pharmacodynamic profiles of the individual enantiomers and the racemate as well as the pharmacokinetics of the individual enantiomers after administration of the racemate be established to support the registration of a racemic mixture. As with the FDA policy, the CPMP guidance allows the use of toxicology data obtained with a racemic mixture to support the registration of a single enantiomer, provided that in the preclinical toxicology model, proper exposure to the enantiomers is achieved. In the case of either a racemate or single enantiomer, the possibility of unidirectional *in vivo* stereogenic inversion or racemization should be assessed. When a new single enantiomer is to be developed from a previously approved racemate, the choice must be justified in terms of pharmacotherapeutic advantage. It is recognized that it may be possible through the use of suitable pharmacodynamic/pharmacokinetic and toxicologic bridging studies to link previously developed racemate data to the marketing application for a single enantiomer. In such a case, the pharmacokinetic profile of the selected enantiomer would be compared with that of the racemate. Similarly, the results of animal and human pharmacokinetic studies should be compared. For the development of a racemate where a previously approved single enantiomer is approved, the guidance requires a new application although enantiomer data may be used to supplement the application. Mixtures of enantiomers, other than racemic mixtures, are considered by the guidance to be fixed-combination products.

For drugs that contain chiral centers, the synthetic step in which the center is formed must be described in full detail in the marketing application, including in process and final testing and the stability methods used to assess, and precautions taken to maintain, stereochemical integrity. The resulting product must be fully described with respect to its stereochemical characteristics. Although it is the applicant's responsibility to choose the appropriate analytical methods to be used to assess stereochemical purity at appropriate points in the production process, the methods must be validated and be capable of appropriately controlling essential stereochemical purity. If a racemic mixture is to be marketed as the result of the sponsor's inability to produce or separate pure enantiomers, these failed efforts should be described in the application. In the case where a single enantiomer is to be marketed, its optical antipode is considered to be an impurity.

20.4. CONCLUSIONS

In spite of regulatory emphasis on drug stereochemistry, racemic drugs may continue to be viable drug development candidates when justified in terms of risk–benefit. Problems in production scale-up may be appropriate to justify the registration of a racemate where drug saftey or effectiveness would be only marginally improved by the registration of a pure enantiomer. When one enantiomer is relatively inert in both a pharmacologic or toxicologic sense, then less justification may be acceptable. When both enantiomers of a racemate contribute in a beneficial way to the therapeutic and safety profile of a chiral drug, this is an acceptable reason to register a racemic mixture.

The concept that data obtained with a racemic mixture may be useful to understanding the effects of its constituent enantiomers seems to be generally accepted. Such an approach may be used as a strategy in which drug development is initiated with a racemic mixture while the intent is to register a single enantiomer. The development of a single enantiomer from a previously approved racemate by way of a truncated development strategy also seems to be acceptable from a regulatory point of view. Bridging studies play a central role in both of these development strategies and these studies will frequently rely on enantioselective pharmacokinetic analysis. The use of enantioselective methods of analysis is equally important in process control, content, and stability testing of both chiral drug substances and drug products containing chiral drug substances and in the assessment of the stereochemical purity of test articles used in pharmacologic and toxicologic assessments. Their use is essential to understanding the pharmacokinetics of individual enantiomers comprising a racemic mixture and in assessing the possibility of stereochemical biotranformations that may occur in either animals or humans.

Risk–benefit considerations play an important role in decision making in drug development. To assess risk–benefit, information must be available concerning the pharmacology, toxicology, pharmacokinetics, pharmacodynamics, toxicokinetics, and toxicodynamics of drug constituents and potential significant impurities. The use of enantioselective methods of analysis is therefore essential, regardless of which stereochemical form of a chiral drug is developed. The application of these methods is rapidly becoming the norm in the development of chiral drugs. Enantioselective analysis serves as the keystone in the development of chiral drugs.

REFERENCES

1. Cushny, A. R. (1926). In *Biological Relations of Optical Isomeric Substances*. London: Balliere, Tyndal and Cox.
2. Ariens, E. J. and Wuis, E. (1987). *Clin Pharmacol. Ther.*, **42**, 361.
3. Gross, M. (1991). *Reg. Affairs*, **3**, 483.
4. Ariens, E. J. (1984). *Eur. J. Pharmacol.*, **26**, 663.
5. Ariens, E. J., Wuis, E. W., and Varinga, W. J. (1986). *Biochem. Pharmacol.*, **37**, 9.
6. Ariens, E. J. (1986). *Trends Pharmacol Sci.*, **7**, 200.
7. Ariens, E. J. (1987). *Drug Intel. Clin. Pharm.*, **21**, 627.
8. Ariens, E. J. (1987). *Clin. Pharmacol. Ther.*, **42**, 361.
9. Ariens, E. J. (1986). *Med. Res. Rev.*, **6**, 451.

10. Ariens, E. J. (1987). *Med. Res. Rev.*, **7**, 367.
11. Ariens, E. J. (1988). *Med. Res. Rev.*, **8**, 309.
12. Ariens, E. J. (1988). *Eur. J. Drug Metabol. Pharmacokin.*, **13**, 307.
13. Ariens, E. J. (1989). *Pharmacol. Toxicol.*, **64**, 319.
14. DeCamp, (1989). *Chirality*, **1**, 2.
15. Weissenger, J. (1989). *Drug Inform. J.*, **23**, 663.
16. Gross, M. (1990). *Annual Repts. Med. Chem.*, **25**, 323.
17. Gross M., Cartwright, A., Campbell, B., Bolton, R., Holmes, K., Kirkland, K., Salmonson, T., and Robert, J. (1993). *DIA J.*, **27**, 453.
18. Cayen, M. (1991). *Chirality*, **3**, 94–98.
19. FDA (1987). Guideline for Submitting Supporting Documentation in New Drug Applications for the Manufacture of Drug Substances. Rockville, MD: FDA.
20. Holmes, K., Jr., Baum, R., Brenner, G., Eaton, C., Gross, M., Grundfest, C., Margerson, R., Morton, D., Murphy, P., Palling, D., Repic, O., Simon, R., and Stoll, R. (1990). *Pharmaceut. Technol.*, **14**, 46.
21. FDA (1992). Policy Statement for the Development of New Stereoisomeric Drugs. Rockville, MD: FDA.
22. CPMP (1986). Notice to Applicants: The Rules Governing Medicinal Products in the European Community, Vol. II. Commission of the European Communities.
23. CPMP (1993). Note for Guidance: Investigation of Chiral Active Substances. Commission of the European Communities. Brussels.
24. FDA (1996). Policy Statement for the Development of New Stereoisomeric Drugs. Rockville, MD: FDA.

CHAPTER

21

FIRST-PASS PHENOMENA: SOURCES OF STEREOSELECTIVITIES AND VARIABILITIES OF CONCENTRATION-TIME PROFILES AFTER ORAL DOSAGE

HILDEGARD SPAHN-LANGGUTH

Department of Pharmacy
Martin-Luther-University Halle-Wittenberg
D-06120 Halle/Saale

LESLIE Z. BENET

Department of Pharmacy
University of California San Francisco
San Francisco, California 94143-0446

WERNER MÖHRKE

Procter & Gamble Pharmaceuticals
Intl. Inc., D-64331 Weiterstadt, Germany

and

PETER LANGGUTH

Astra Hässle AB, Kärragatan 5, S-43183 Mölndal, Sweden

21.1. INTRODUCTION

Any interaction of drug enantiomers with endogenous chiral molecules may result in concentration differences between enantiomers, either in the effect compartment or in the sampling compartments for pharmacokinetic studies. Thus, apart from the pharmacologic receptor itself, drug input and disposi-

Dedicated to Prof. Dr. Ernst Mutschler on the occasion of his 65th birthday

The Impact of Stereochemistry on Drug Development and Use, Edited by Hassan Y. Aboul-Enein and Irving W. Wainer. Chemical Analysis Series, Vol. 142.
ISBN 0-471-59644-2

tion may also play important roles, actually define the amount of active agent present at the pharmacologic receptor, or create other active or toxic species. "Chiral discriminators" of relevance in pharmacokinetics are transport proteins, binding proteins in blood and tissues, and metabolizing enzymes. Depending on the extent of discrimination and the relevance of the respective processes in the overall kinetic profile of the enantiomers, the pathway may or may not lead to significantly stereoselective kinetics. While distribution processes and carrier-mediated intestinal, biliary, and renal transports via multisubstrate transporters appear to be of minor relevance, the most significant differences between enantiomers usually result from highly stereoselective metabolic clearances. Stereoselective drug metabolism may involve substrate and product stereoselectivity as well as metabolic chiral inversion. Apart from these processes, the nonenzymatic racemization of configurationally unstable molecules may occur. When one or both enantiomers undergo intestinal secretion (with the concentration-dependent transport-to-antitransport ratio as the relevant measure), various metabolic processes in the gut, or efficient extraction by the liver with subsequent biotransformation or secretion/excretion into bile, then considerable differences may be observed, particularly following peroral administration where the total absorbed fraction of the dose passes through eliminating organs, before reaching the systemic circulation. From various *in vitro* and *in vivo* studies, we know that metabolic clearance represents the most important source of marked blood concentration differences between enantiomers.

Today, research in drug-metabolizing enzymes has to face an emerging base of knowledge, which enables us to understand, from a molecular biological viewpoint, the differences between the various multigene superfamilies of enzymes that are responsible for the disposition and elimination of most drugs. We know a great deal about the cytochromes P-450 and the various isozymes in both humans and animals; we also know quite a bit about the acetyltransferases, one of the first groups of enzymes at which we looked for a genetic difference (between slow acetylators and fast acetylators). We are beginning to identify in detail the various glutathione-*S*-transferases, their subfamilies, and—more difficult to deal with—some of the phase II enzymes, the glucuronosyltransferases and sulfotransferases. This is where the cutting edge science in drug metabolism with the application of molecular biology affects pharmacokinetics because a lot of the differences that we are observing between subjects, genders, and races can be explained on differences, potentially predictable differences, in drug-metabolizing enzymes. We know that the individual isoforms have specific but potentially overlapping substrate specificities, and there can be very large differences in the regulation of each of these enzymes. There can be a genetic basis, there can be a development basis (fetus, neonate adult, elderly), there can be differences in various tissues in the body responsible for eliminating the drug, there can be *inhibition* of these particular enzymes (most likely when more than one drug is a substrate for the enzyme), and there can be *induction* of these enzymes. And a lot of the interindividual variability in catalytic activity is explained by these differences in the drug-metabolizing enzymes.

The organism's ability to biotransform a drug and the potential pharmacodynamic response mechanisms are going to very much depend on these enzymes and the various factors that affect the enzymes: the genetic and also the environmental basis, nutrition, the meals, potential use of cigarettes, alcohol, or drugs, pollutants, and an existing disease. All have the potentials to change these enzymes from their basic characteristics in a healthy volunteer.

The purpose of this chapter is not to provide a complete overview of the literature published about the first-pass extraction of racemically administered drugs, but to define and illustrate particular phenomena and their stereoselectivity as observed under different physiological conditions.

21.2. CLEARANCE AND ITS RELEVANCE WITH RESPECT TO THE PLASMA CONCENTRATIONS OF STEREOISOMERS AFTER p.o. DOSING

21.2.1. Stereoselectivities in Concentration-Time Profiles

When a racemate is dosed perorally, various processes may affect its concentration-time profile in the systemic circulation or at the site of action. (Ariens, 1984; Fassihi: 1993; Jamali et al., 1989; Levy and Boddy, 1991). These include gastrointestinal transit processes delivering the drug to its absorption site, processes occurring inside the gastrointestinal lumen, during the drug's passage through the wall of the gastrointestinal tract and in the blood vessels, as well as the transport and metabolic processes into and out of the hepatocyte, the smallest functional unit of the liver.

The analysis of concentration-time profiles regarding stereochemical aspects was mostly empirical until recently. However, today enzymatic as well as molecular biology techniques are providing keys for a mechanistic approach to stereoselective pharmacokinetics, may explain observations made in previous *in vivo* studies, and permit predictions.

Drugs that are characterized by a high clearance based on efficient presystemic extraction are usually undergoing a significant first-pass effect following absorption from the G.I. tract. However, in most cases, no information is available yet about the mechanisms leading to the concentration differences in the systemic circulation, since, as opposed to hepatic processes, processes such as gut wall metabolism and intestinal secretion have largely been neglected until today (Klaassen, 1988).

21.2.2. General Pharmacokinetic Principles as Applied to *in vivo* Studies: Drug Clearance, First-Pass Effect, and Chirality

21.2.2.1. Clearance Definition

When a racemate is dosed and both enantiomers are measured, knowledge of general pharmacokinetic principles, particularly with respect to drug clearance, is a prerequisite for data interpretation.

Drug clearance principles are similar to the clearance concepts known from renal physiology, where we determine, for example, creatinine clearance from the rate of elimination of the compound in the urine relative to the plasma drug concentration. The total clearance CL [blood clearance (CL_b), plasma clearance (CL_p), clearance based on unbound drug concentration (CL_u)] is the rate of elimination by all routes relative to the drug concentration in the respective biological fluid and includes the excretion of unchanged compound as well as biotransformation. The total clearance of a compound represents the sum of all clearance processes, that is clearances of different organs (e.g., gut, liver, lung, and kidneys) are additive. When blood flow is used to define clearance, the maximum clearance possible is equal to the sum of blood flows to the various organs of elimination. For a drug solely eliminated by the liver, blood clearance is limited by the flow of blood to that organ (approximately 1.5 L/min).

Generally, organ blood clearance cannot exceed blood flow in the particular organ, and total clearance for a compound cannot be higher than cardiac output (except when clearance also occurs in the vascular bed).

21.2.2.2. Clearance, Extraction Ratio, and First-Pass Effect

Clearance appears to be constant for most drugs under normal clinical settings, that is the rate of drug elimination is directly proportionate to concentration. The clearance by a specific organ (CL_{organ}) is dependent on the respective blood flow Q, as well as the extraction ratio ER, which equals the difference between the drug concentration entering the organ (C_{in}) and that exiting from the organ (C_{out}) normalized by C_{in}. According to this relationship, an ER value of 1 can be obtained when the drug is efficiently extracted from the perfusate by the organ, whereas the ER approaches 0 when the drug is not extracted by the organ. When a drug is highly extracted by the liver, blood clearance is limited by hepatic blood flow and not intrahepatic processes. When a drug is completely absorbed from the gastrointestinal tract and presystemic elimination occurs only in the liver, the systemic availability of the compound (F) is diminished by hepatic extraction. When the metabolizing or biliary excretion capacity of the liver is great, the (reducing) effect on the extent of availability will be substantial (*first-pass effect*). In the case of a first-pass effect, some of the active drug absorbed from the G.I. tract is inactivated by hepatic processes before the drug can reach the general circulation and be distributed to its sites of action. The maximum bioavailability (F_{max}) equals $1 - ER$, and the decrease in availability is a function of the physiologic site from which absorption takes place. Another aspect that may have to be taken into consideration when a significant fraction of the dose is first-pass extracted is the occurrence of active or toxic metabolities, which may be generated to a significantly higher extent than that observed after i.v. dosage. In general, drugs with high hepatic extraction ratios show marked intersubject variabilities in bioavailability because of variations in hepatic function or blood flow or both (Benet, 1989). For such drugs, disease-induced reduction of the hepatic extraction ratio from 0.95 to 0.90 would lead to a doubling of bioavailability, as may be calculated from the relationship given above.

21.2.2.3. Low vs. High Extraction Ratio Drugs

Generally, we can differentiate between compounds with a high extraction ratio in a particular organ and compounds that are poorly extracted when passing through an organ. While for low extraction drugs, protein binding and biotransformation capacity play a considerable role, the perfusion of the organ limits the clearance of high extraction ratio drugs, as may be derived from the conditions assumed in the well-stirred model (e.g., Rowland and Tozer, 1989). This model states that $CL_H = (Q_H \times CL_{int} \times f_u)/(Q_H + CL_{int} \times f_u)$, where CL_H represents the hepatic clearance, Q_H is the hepatic blood flow, CL_{int} is the intrinsic clearance that relates the rate of metabolism at steady state to the unbound concentration at the enzyme site, and f_u is the fraction unbound. The hepatic extraction ratio, $CL_H/Q_H = E_H$, is equal to $(CL_{int} \times f_u)/(Q_H + CL_{int} \times f_u)$. When E_H is very high and approaches 1, the clearance approaches Q_H and even large, changes in the intrinsic clearance and the unbound fractions will not affect E_H considerably. When the extraction ratio is small, it depends on all three factors, whereas CL depends only on $CL_H = f_u \times CL_{int}$.

21.2.2.4. Chirality and Clearance

When a drug is chiral and the racemate is dosed, the extraction ratio may be different, resulting in different organ clearances for the two enantiomers (E1, E2). As illustrated in Figure 21.1, concentration differences may be magnified following p.o. rather than after i.v. dosage, since here the total absorbed fraction has to pass through the liver prior to reaching the systemic circulation and distributing into other tissues, whereas only 20% of the cardiac output reaches the liver after i.v. administration. Verapamil represents a well-known

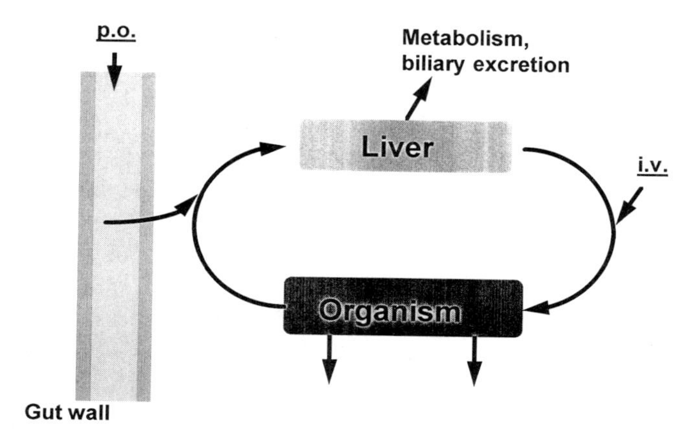

Figure 21.1. *Schematic illustration of hepatic first-pass effect for oral as opposed to intravenous dosing*. The clearance of a drug by the liver is separated from the remaining systemic clearance to illustrate hepatic first-pass extraction following oral dosing, due to which some of the absorbed drug may not be available to the remainder of the body.

example for a more pronounced concentration difference after p.o. dosing [AUCs, *S*(–)/*R*(+)-ratios: i.v., 0.56; p.o., 0.23] (Vogelgesang et al., 1984; Walle and Walle, 1986), although almost similar terminal half-lives were detected for the two enantiomers irrespective of the administration route. Similar phenomena were observed with carvedilol (Spahn et al., 1990) and propranolol (e.g., Spahn, 1989) as well as dl-threo-methylphenidate (Nugehally et al., 1991 and 1993).

As for verapamil enantiomers, similar stereoisomer $t_{1/2}$ values may be observed in spite of different hepatic extraction ratios and different plasma or blood clearances; differences in the terminal half-lives with constantly changing E1/E2 ratios may occur as well. [This may be understood on the basis of the determinants of $t_{1/2}$, which is directly proportional to the distribution volume and indirectly proportional to clearance ($t_{1/2} = 0.693$ V/CL)].

When the presystemic (e.g., hepatic) clearance for one of the enantiomers (E1) is larger than for the other one (E2), that is, when ER_{E1} exceeds ER_{E2}, then the systemic availability of E2 is higher when the racemate is given by a route that yields first-pass metabolic effects.

In general, at therapeutically administered doses, clearance is constant, that is independent of dose or the blood or plasma concentrations. However, for some compounds, dose dependence is observed, which again, may be different for the two enantiomers of a racemate. Another aspect that has not yet been sufficiently included studies on stereopharmacokinetics is the potential interaction between the two enantiomers at carriers or metabolizing enzymes.

21.2.3. Enantiomer Differences in Their Concentration-Time Profiles Following Oral Dosage, Potentially Through Stereoselective First-Pass Processes

The relative contributions of the enantioselectivities of various kinetic processes have to be considered when trying to explain the overall stereoselectivity of concentration-time profiles, with biotransformation and plasma protein binding being the most prominent factors. Tucker and Lennard (1990) summarized clearance characteristics of various compounds, including several high extraction drugs, with particular emphasis on the calculation of the stereoselectivity of unbound metabolic clearance. In this selected group of compounds, they calculated stereoselectivity ranges from virtually none to a factor of 14. The overall extent of stereoselectivity may represent the net balance of selectivity of different enzymes, since the drug may be metabolized by different pathways. Moreover, several isozymes with different steric preferences may be involved in the formation of the same product.

With respect to the β-adrenoceptor antagonist *propranolol*, the plasma concentrations of the *S*(–)-enantiomer (eutomer) are always higher, with larger differences observed after oral dosing (Walle et al., 1988; Lindner et al., 1989; Silber et al., 1971; Stahl, 1993). With this compound, various experimental studies have been carried out in different species. It was concluded that stereoselectivity in the systemic clearance of propranolol is the net result of stereoselectivities in the effects of the isomers on hepatic blood flow, differences in their plasma protein binding and enzymatic transformation. The measurement of partial metabolic clearances *in vivo* indicates that the higher oral clearance of *R*(+)-propranolol in man appears to be due to selective ring 4-hydroxylation. With respect to side-chain oxidation, *R*-propranolol appears to be favored as well, whereas glucuronidation was higher for the *S*-enantiomer (Ward et al., 1989). Furthermore, studies performed with human liver microsomes by Otton et al. (1989) indicate that 4-hydroxylation of propranolol is mediated by two different enzymes, one of which is stereoselective. These same authors also found that *metoprolol* is demethylated by two isozymes, which show opposite stereoselectivities (Otton et al., 1988).

The (–)-enantiomer of the calcium channel blocker *verapamil* is mainly responsible for the negative dromotropic activity on atrioventricular conduction. After a racemic i.v. dose, systemic plasma clearance was found to be much higher for the (–)-enantiomer, yielding higher concentrations of the (+)-enantiomer (Vogelgesang et al., 1984; Harder et al., 1992). The difference is much more pronounced following an oral dose, indicating stereoselective presystemic elimination. The total systemic availability was 20% and significantly different between the isomers.

Longstreth (1993) reviewed the kinetic and dynamic characteristics of verapamil and described the dependence of the verapamil *R*/*S* ratio on the input rate following oral administration. Also, Karim and Piergies (1995) studied the dependence of the enantiomeric ratios in plasma on peak concentrations, oral input rate, or both and detected marked differences in the ratio between different input rates.

In a combined single- and multiple-dose study in 12 patients (160 mg of immediate-release and 240 mg of slow-release verapamil as cross-over), the *S*/*R* ratios of concentrations of verapamil and norverapamil were 0.27 and 0.45, respectively, and independent of the respective plasma concentration or time after dosage, but significantly smaller at steady state (Rosenthal et al., 1992).

In the vasodilating β-blocker *rac-carvedilol*, a compound with α- and β-antagonistic properties in two structural elements of the molecule, only the *S*(–)-enantiomer exhibits both a significant β-blockade as well as a vasodilating effect, and is therefore the actual hybrid component, whereas the R(+)-enantiomer is almost exclusively an α-antagonist.

Moreover, the compound undergoes significant metabolic clearance with the formation of (in part β-antagonistic) phase I and (nonactive) phase II products as well as subsequent phase I and phase II metabolic steps (Neugebauer and Neubert, 1991). Following a 1-h i.v. infusion of 12.5 mg of *rac-carvedilol* in a clinical study, the total clearance was found to be higher for the *S*-enantiomer, reaching 541 mL/min for *R*- and 718 mL/min for *S*-carvedilol (Henke et al., 1985; Spahn et al., 1990; Neugebauer et al., 1990). Following a 50 mg oral dose, considerably higher clearance values were obtained (5500 mL/min for *S* and 1382 mL/min for *R*), which resulted in lower *S*/*R* ratios after oral rather than i.v. administration and indicated that high and stereoselective first-pass extraction occurs after the oral dose. From rat data (Stahl et al., 1993a,b), absolute bioavailabilities of 19% for *S*- and 32% for *R*-carvedilol were estimated, indicating that rat and man are comparable with respect to the extent and stereoselectivity of the first-pass effect.

Following racemate dosage, distinct differences were found in the various kinetic parameters of *prenylamine* enantiomers, an orally administered WHO class V calcium antagonist (Gietl et al., 1988, 1989, 1990). The maximum plasma concentrations of the *R*-enantiomer exceeded those of the *S*-enantiomer five-fold. The CL/F of *S*-prenylamine was five times and the (low) renal clearance three times higher than for the *R*-form. Acid-catalyzed hydrolysis of urine samples released more *S*-prenylamine, indicating stereoselective glucuronidation. The hypothesis of a stereoselective first-pass extraction was confirmed by Paar et al. (1990) who compared a conventional oral dosage form and a solution containing the deuterated compound.

Table 21.1. Pharmacokinetic Parameters for Tranylcypromine (TCP) Enantiomers in Man (p.o. dose: 20 mg of TCP sulfate or 10 mg of the respective enantiomers) (arithmetical means ± SD)

	(–)-TCP		(+)-TCP	
Adminstration of	Enantiomer	Racemate	Enantiomer	Racemate
CL/F (mL/min)	1058 ± 796	597 ± 224	5490 ± 4126	13,797 ± 18,629
AUC (ng mL^{-1} h)	130 ± 62	197 ± 71	29 ± 19	26 ± 24
C_{max} (ng/mL)	49 ± 26	56 ± 25	19 ± 4	14 ± 17
t_{max} (h)	1.0 ± 0.6	1.2 ± 0.5	0.8 ± 0.5	0.8 ± 0.6
$t_{1/2}$ (h)	1.6 ± 0.5	1.7 ± 0.3	0.75 ± 0.4	1.2 ± 0.6
MRT (h)	2.6 ± 1.1	3.3 ± 0.3	1.6 ± 0.8	2.0 ± 0.8
Ae_{inf} (ng)	62 ± 40	210 ± 178	25 ± 17	51 ± 51
CL_R (mL/min)	8.1 ± 4.4	15.3 ± 9.0	17.7 ± 12.5	24.9 ± 17.9

CL/F = apparent oral clearance.
AUC = area under the concentration-time curve.
C_{max} = maximal plasma concentration.
t_{max} = time for C_{max}.
$t_{1/2}$ = terminal half-life.
MRT = total mean residence time.
Ae_{inf} = amount excreted into urine until infinite time.
CL_R = renal clearance.

rac-Tranylcypromine represents another example of a compound exhibiting stereoselectivities in its concentration-time profiles. Human data obtained more recently with p.o. administration of the racemic drug as well as the single enantiomers showed that the concentrations of the (–)-enantiomer always exceed those of the (+)-enantiomer, with C_{max} being 5-10 times higher for (–)- than (+)-tranylcypromine (Weber-Grandke et al., 1993; Spahn-Langguth et al., 1994 a,b) (Figure 21.2). From the respective pharmacokinetic data (Table 21.1) with low renal clearances and apparently high metabolic clearances, it was hypothesized that tranylcypromine may undergo a significant and stereoselective first-pass extraction upon p.o. dosage in humans. However, this hypothesis was not proven because of the lack of a parenteral dosage form approved for use in clinical studies. Absolute bioavailability was determined in a rat study (Langguth et al., 1996) at a dose level anticipated to yield linear pharmacokinetics. When compared to the data obtained in man, kinetics in the rat were characterized by higher (–)- than (+)-tranylcypromine concentrations, with a difference much smaller than detected for humans. The calculated hepatic extraction ratio was also smaller than expected, yielding a maximum first-pass extraction of 57% for (+)- and 34% for (–)-tranylcypromine.

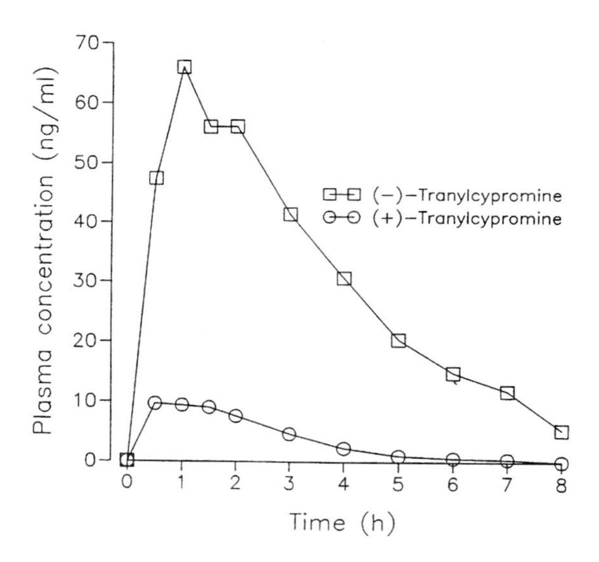

Figure 21.2. *Plasma concentration-time curves of (−)- and (+)-tranylcypromine (calculated as tranylcypromine sulfate) obtained following peroral administration of 20 mg of rac-tranylcypromine sulfate (arithmetical means of six healthy volunteers).* From the large difference between the two enantiomers and the high CL/F values, it was hypothesized that tranylcypromine undergoes a high and stereoselective first-pass. However, absolute bioavailability was not estimated in man due to the lack of an i.v. dosage form (Spahn-Langguth et al., 1992).

21.2.4. Kinetic Behavior of Enantiomers at Higher Dose Levels

In order to study the saturability of the first-pass effect observed with carvedilol, the plasma concentration-time profiles were analyzed following different i.v. and p.o. doses of carvedilol in rats. Both in humans and rats, carvedilol appears to be subject to a marked stereoselective first-pass effect with preferential extraction of the *S*(−)-enantiomer, particularly when dosed orally. While in rats, no dose dependence of clearance and *S*/*R* ratios was observed for the maximum tolerable i.v. dose of 10 mg/kg (AUC_S/AUC_R: approximately 0.6), a significant increase was detected from the lowest p.o. dose of *rac-carvedilol* (5 mg/kg: AUC_S/AUC_R 0.4) to the highest dose (30 mg/kg: AUC_S/AUC_R 0.9). The plasma concentrations of the respective glucuronides in rats were considerably lower (1/50 to 1/100) than those of the aglycones, whereas their *S*/*R* ratios were comparable for i.v. and higher for p.o. dosage (due to a smaller increase of *R*- than *S*-glucuronide with increasing dose) (Stahl et al., 1993a).

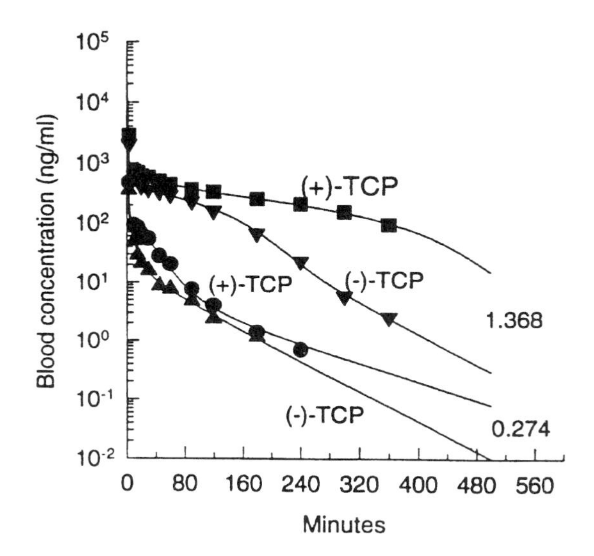

Figure 21.3. *Blood concentration-time profiles of (+)- and (−)-tranylcypromine in rats.* The racemic compound was applied intravenously (low dose: 0.274 mg/kg; high dose: 1.368 mg/kg of each enantiomer, administered as *rac*-tranylcypromine) and blood samples were drawn under urethane anesthesia. Each point represents the arithmetical mean of three to four rats (Langguth et al., 1996).

In a systematic study of the kinetic behavior of tranylcypromine enantiomers following racemate dosage in rats (including absolute bioavailability through different administration sites), rather complex dose-dependent disposition characteristics were detected, when doses comparable with those in humans and significantly higher were applied. The average concentration-time profiles for tranylcypromine enantiomers obtained after two of the four different doses are depicted in Figure 21.3 showing fairly complex kinetic characteristics, which we may observe when saturable processes are involved. Here, the E1/E2 ratios were shown to be a function of the dose and time. However, if we try to extrapolate the findings from rats to man, problems may arise. The enantiomer ratios obtained in rats are much closer to unity than those in man, demonstrating that the elimination pathways and/or their stereoselectivities are different between species.

21.2.5. Prehepatic Clearance Processes: An Overview

Any enzyme-catalyzed or carrier-mediated process on a chiral drug's way into the systemic circulation is potentially stereoselective. This includes gut-related

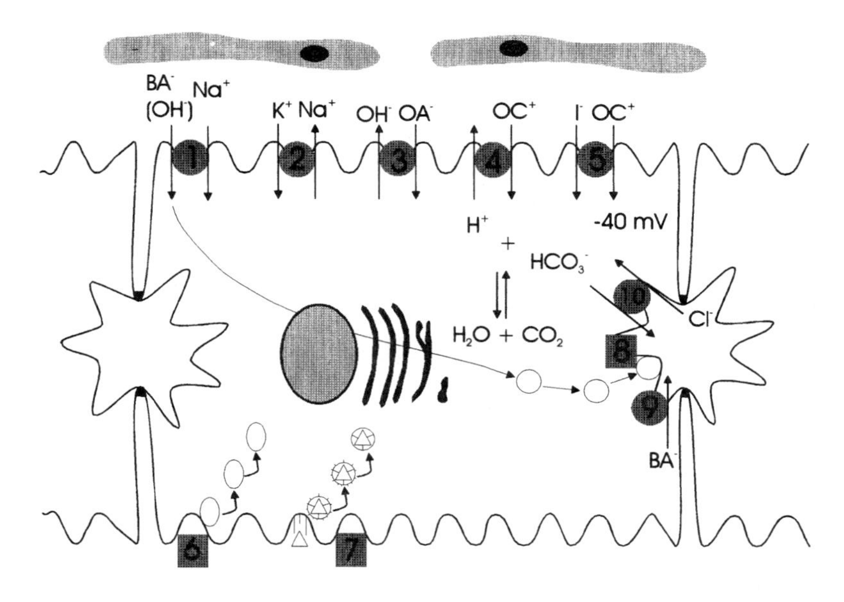

Figure 21.4. *Transport into the hepatocyte and subsequent processes, potential rate-limiting steps in hepatobiliary transport.* Hepatic uptake rate (= uptake into the hepatocyte, i.e., into the most important type of liver cells, occurring besides endothelial cells, Kupffer cells, and lipocytes) can be influenced by hepatic blood flow, protein binding, and net sinusoidal membrane transports [primary sinusoidal uptake and sinusoidal efflux (secretion)]. Elimination from the cell occurs via metabolism and canalicular excretion (net canalicular excretion consists of membrane transport into the canalicular lumen and counter-transport processes), with its rate depending on intracellular binding to macromolecules and sequestration to organelles.

Functional experiments performed by various groups yielded different uptake mechanisms in the hepatocyte: (1) the sodium-dependent uptake of organic anions (sodium-bile acid cotransporter), which is maintained by (2) a potassium-sodium ATPase; (3) the sodium-independent uptake of organic anions via an anion antiport system (bilirubin uptake); (4,5) uptake of organic cations via proton antiport (4) for low-molecular-weight compounds and via ion pair-transport (5) for compounds with higher molecular weight and—most probably—also for uncharged organic molecules; (6) fluid endocytosis; (7) receptor-mediated endocytosis, which appears to explain the uptake of asialoglycoproteins, a mechanism that is of interest with respect to drug targeting to the liver. Secretion into the canalicular lumen may occur via exocytosis (8) or as a carrier-mediated process (9). Secretion of bicarbonate is based on chloride antiport (10) (Steen and Meijer, 1991; Meijer and Groothuis, 1991).

Molecular biology techniques permitted cloning of some carrier proteins. Moreover, when applying such techniques, structural similarities between carrier in different tissues were revealed and multidrug resistance (MDR)-related carriers detected, such as mdr2 and mdr3 in the canalicular membrane (Roepe, 1995).

1. *Canalicular transport*: An ATP-dependent multiorganic anion transport system (MOAT) was cloned, for which for example, bilirubin diglucuronide and glutathione conjugates are substrates. In addition, a separate bile acid carrier (CBST) was detected, which has not yet been cloned.

(*continued*)

and hepatic processes, such as intestinal secretion and metabolism, hepatic uptake and intracellular transport, and metabolism as well as secretion into bile canalicula (Meijer and van der Sluijs, 1989; Oude Elferink et al., 1995). Possible pathways into, inside, and out of the hepatocyte are depicted in Figure 21.4.

Potential intestinal secretion as a gut-related presystemic clearance or sustaining process is mediated by P-glycoprotein (P-gp), a multisubstrate carrier (Endicott and Ling, 1989; Gottesman, 1993; Gottesman and Pastan, 1993). Another important factor is the presence of drug-metabolizing enzymes in the gut. For example, Kolars et al. (1992) detected, localized, and characterized P450IIIA enzymes (CYP3A, see below) in the digestive tract of male and female rats. P450IIIA-related proteins were detected in the mature epithelia throughout the gut; however, the specific proteins expressed appeared to differ among digestive organs and between genders.

In order to study the extrahepatic localization of the CYP3A isoenzymes, an immunohistochemical method (based on an avidin-biotin complex) was developed using polyclonal and monoclonal anti-CYP3A. This assay was applied to 158 routinely fixed and paraffin-embedded human tissues. The different antibodies used showed identical results. In addition to the liver, CYP3A was detected in various extrahepatic tissues (such as the stomach, duodenum, and pancreas). High abundance was observed in the gastrointestinal tract, with staining intensity decreasing from the stomach to the rectum (Fritz et al., 1993).

Apparently, also phase II metabolic pathways and stereoinversion (Caldwell et al., 1988) occur in the gut to a significant extent (Jamali et al., 1992; Berry and Jamali, 1991).

21.2.6. Oxidative Metabolism in Different Tissues: A Major Factor Leading to Stereoselectivities of First-Pass Effect of High Extraction Ratio Drugs

21.2.6.1. Cytochromes P450

For many chiral and achiral drugs, biotransformation represents the major route of drug elimination. Here, the membrane-bound polysubstrate monooxygenases (cytochromes P-450) play a major role. The products result-

---

Figure 21.4. (*contd.*)

2. *Uptake into the hepatocytes*: A sodium-bile acid cotransporter (51-KDa protein) was found in liver as well as kidney. Because of the lack of multispecificity of the Ntcp, it was hypothesized that amphiphatic molecules are transported via a different carrier, the organic anion-transporting polypeptide (Oatp), a sodium-independent approximately 80-KDa transport protein with 12 transmembrane domaines. This transport system was found to occur in various tissues (including rat brain) and to transport anionic, cationic, and neutral compounds.

ing from this process, the oxidative phase I metabolites of the substrates, are usually less lipophilic than their precursors. Furthermore, such metabolites are usually assumed to have no or less activity than the parent drug, although we know today that the hydrophilic metabolites may contribute to the effect, even when they occur in low concentrations, since they are mostly less protein-bound and the fraction that has access to the pharmacologic receptor(s) may be considerably higher than for the parent drug (e.g., Brockmeier et al., 1988). There are also examples where a biotransformation step leads to compounds, which are as effective or much more effective than the parent drug (Mayer et al., 1993). Another aspect to be discussed with respect to the introduction of polar groups into the molecule is the facilitation of excretion or subsequent phase II metabolic steps such as glucuronidation followed by renal or biliary (or potentially intestinal) excretion via passive diffusion or multisubstrate carriers.

With respect to cytochrome P450 (CYP), it is known that a so-called super-gene family of P-450 enzymes exists and many different enzymes of this family are expressed simultaneously, yet under at least partially independent regulation. A comprehensive recent update about the nomenclature, sequencing, and gene mapping was provided by Nelson et al. (1993). For human enzymes, the abbreviation **CYP** denotes the gene encoding for cytochrome P450, the following numeral (e.g., CYP**1**) characterizes the family, and the subsequent character (e.g., CYP1**A**) the subfamily, whereas the final character (e.g., CYP1A**2**) defines the single cytochrome P450 gene. Usually, the gene nomenclature is transferred to the respective enzymes. From the various CYP families, only three appear to be relevant with respect to human xenobiotic phase I metabolism: CYP1, CYP2, and CYP3, for which 1, 8, and 2 subfamilies and 2, 57, and 10 different forms, respectively, are known (Gonzalez, 1992). Knowing the contribution of a particular isozyme permits the prediction of potential drug–drug interactions and changes of hepatic clearance with altered liver function. Increased or decreased availability of the respective enzyme, for example, in the liver, may significantly affect drug elimination. The enzymes may be selectively or specifically inhibited or induced by various chemicals as summarized by Parke et al. (1990), Gonzalez (1992), and Murray and Reidy (1990). According to these data, verapamil is a substrate for the 3-methylcholanthrene- and aroclor1254-inducible CYP1A2, for CYP2C8 and CYP2C9, as well as for the phenobarbital-, triacetyloleandomycin-, and dexamethasone-inducible CYP3A4.

Well-known examples of cytochrome P450 isoenzymes and their respective substrate, characterized by different substrate specificities, are CYP1A2 with for examples, phenacetin, verapamil, caffeine, theophylline, and triamterene as substrates; CYP2B6 (cyclophosphamide), CYP2C8 (tolbutamide, verapamil), and CYP2C9 (warfarin, tolbutamide, verapamil), CYP2C19 with mephenytoin and omeprazole as substrates, the sparteine- and debrisoquin-hydroxylating CYP2D6 with, for example, metoprolol, propranolol, and propafenone as substrates; and CYP3A3/4, mainly responsible for the oxidative biotransformation of, for example, nifedipine, verapamil, and cyclosporine as well as numerous other compounds (Eichelbaum et al., 1992; Kober et al., 1996).

From these enzymes, those showing polymorphic drug oxidation are of particular interest, also with respect to enantioselectivities in drug metabolism. A recent review of the geographical and interracial differences in polymorphic drug oxidation, as far as CYP2D6 and CYP2C19 are concerned, was provided by Bertilsson (1995), as mentioned below.

21.2.6.2. Available in vitro Systems to Evaluate the Contribution of the Various CYP Isoenzymes to Metabolic Drug Clearance

When comparing the relevance of possible *in vitro* systems (isolated-perfused liver, liver slices, liver homogenates, cell cultures, subcellular fractions, expressed enzymes), the prediction of the metabolic clearance to be expected in humans is closer to reality when intact tissues are used. In spite of higher predictive values, such systems are scarcely applied due to their limited availability. In commercially available cell-culture-expressed enzymes, background activity may be high and turnover rates low. Today, mostly microsomal systems are employed, where isoenzyme-specific inhibitors or antibodies are used to characterize the cytochrome P450 isoenzymes involved in the metabolic degradation of a compound to defined metabolites.

When using liver tissue microsomes for the prediction of *in vivo* metabolic clearances, the fractional distribution may have to be considered, since in uninduced livers the CYP3A content amounts to 30% and the CYP2D6 content to 1.5% of total CYPs in the liver. Moreover, as indicated above, most cytochromes P450 exhibit local multiplicity, that is they are expressed in various organs (liver, G.I. tract, lung, kidneys, CNS, skin). CYP3A but little CYP2D6 was detected in the gut, whereas CYP1A2 occurs mainly in the liver.

21.2.6.3. In vitro Data and Predictability of Clearance from in vitro Studies

The potential role of *in vitro* stereoselectivity studies in the metabolism of specific substrates and isozymes is discussed with respect to the possibility of distinguishing the isoforms of an enzyme involved in the metabolism *in vivo*. By monitoring several metabolic pathways as well as the ratio of enantiomeric and diastereomeric metabolites *in vitro* with purified enzyme systems, one can more specifically define the isoforms responsible for *in vivo* metabolism.

Moreover, this method for the identification of enzymes responsible for metabolism should provide valuable information in predicting potential drug interactions and assist in rational drug development.

From enzyme kinetic parameters (V_{max} and K_m) determined for a particular metabolic step, the intrinsic clearance (CL_{int}) is calculated as the V_{max}/K_m ratio. The respective liver clearance is calculated from CL_{int}, the microsomal protein concentration obtained from a defined tissue weight, and the liver weight. Total metabolic clerance of the liver represents the sum of all single clearances calculated for defined biotransformation steps.

In our metabolism studies with carvedilol and human liver microsomes, we defined the isozymes involved in carvedilol metabolism. The demethylation and 4′- and 5′-hydroxylation favored the *R*-enantiomer in these studies. The addition of inhibitors (e.g., furafylline, erythromycin, and ketoconazole, as well as quinidine) suggested that neither CYP1A2 nor CYP3A3/4 play an important role, but demonstrated that CYP2D6 is a major enzyme involved in the phase I biotransformation of carvedilol enantiomers (Kurz et al., 1996).

With carvedilol, higher systemic concentrations of the *R*-enantiomers of the major oxidative metabolites were detected, whereas in the *in vitro* studies, the investigated pathways exhibited different stereoselectivities. This *in vitro*/*in vivo* discrepancy is possibly explained by sequential metabolic steps, for example, further phase I or phase II metabolism with consecutive stereoselective biliary or renal excretion.

Very complex data were presented by Rettie et al. (1992) for the stereoselective metabolism of warfarin. The warfarin metabolite profiles were generated using the human cDNA expressed P-450's 1A2, 2C9, and 3A4. CYP1A2 favored the *R*-enantiomer of warfarin with the *R*-6-hydroxy metabolite as the major product, followed by the *R*-7-hydroxy and *R*-8-hydroxy compound. On the other hand, with CYP2C9, *S*-7-hydroxy warfarin was the major product, followed by *S*-6- and *S*-4′-hydroxy warfarin. No clear preference was found for CYP3A4, where the major products were *R*-10-hydroxy-, *S*-4′-hydroxy-, *S*-10-hydroxy-, and *R*-4′-hydroxy warfarin (Rettie et al., 1992).

Kroemer et al. (1992) studied the *in vitro* metabolism of verapamil and the respective enantioselectivity of different oxidative metabolic steps, including the O- and N-dealkylation steps (Table 21.2). Furthermore, they tried to correlate their *in vitro* data with that obtained from a clinical study to evaluate the potential for a prediction of *in vivo* drug disposition from *in vitro* studies with human liver microsomes to be applied in the early stages of drug development. *In vitro* intrinsic clearances showed the same quantitative order in the formation of metabolites from *S*- and *R*-verapamil. In general, O-dealkylation as the initial step was inferior to N-dealkylation as the initial step. Furthermore, both *in vitro* and *in vivo* data indicated a preferential metabolism of *S*-verapamil.

Table 21.2. ***In vivo*** **Intrinsic Clearances for Metabolite Formation after Administration of *S*- or *R*-Verapamil in Relation to *in vitro* Intrinsic Clearance (according to Kroemer et al., 1992) (Correlation of *in vivo* intrinsic clearance with metabolic clearance as determined *in vitro* shows a good correlation of the data; however, the correlation obtained for the *S*-enantiomer of verapamil yielded a more significant result)**

I: Kinetic Data from Incubations with Human Liver Microsomes (10 livers)

Metabolic Pathways		K_m (μM)	V_{max} (pmol/mg per min)	CL_{int} (mL/min per g protein)
A. N-Dealkylation (formation of metabolite D-617)				
	S	44.7	1068	27.1
	R	92.4	1158	19.6
B. N-Demethylation (= norverapamil formation)				
	S	52.8	809	18.4
	R	63.8	817	14.7
C and D. Two O-demethylation reactions				
C. Formation of D-702				
	S	159	112	1.02
	R	329	146	0.84
D. Formation of D-703				
	S	59.4	309	7.0
	R	40.9	174	5.4

II. Formation Intrinsic Clearance (mL/min) for the Different Metabolic Pathways of S- and R-Verapamil as Determined from the Total Amounts Excreted into Urine

	S	*R*
A. N-Desalkyl plus consecutive metabolites	1857	568
B. Norverapamil plus consecutive metabolities	851	213
C. D-702 (no formation detected)	—	—
D. D-703	531	9.6

D-702: O-demethylation at the phenyl next to the chiral carbon (the *para*-position); D-703: O-demethylation at the other phenyl (also in the *para*-position).

21.2.7. Overlapping Substrate Specificities and Tissue Distribution of CYP3A and P-Glycoprotein or Complementary Roles in Drug Elimination at the Cellular Level

In a recent article, Wacher et al. (1995) discussed the potential coordinate regulation of CYP3A and the drug efflux pump P-gp in tumor cells. CYP3A, like P-gp, is located primarily in hepatocytes and the biliary epithelial cells of the liver, as well as the enterocytes (columnar epithelial cells) of the jejunum. For multidrug resistance, the levels of CYP3A are increased in a variety of tumors (Wacher et al., 1995). A review of the literature reveals a striking overlap between the substrates for and inhibitors of CYP3A and P-gp. In some cases, a direct relationship has not been observed; rather, the parent compound is a CYP3A substrate, whereas the metabolite is transported by P-gp [e.g., digitoxin/digoxin, estradiol/17β-estradiol glucuronide (Eberhart et al., 1991; de Lannoy and Silverman, 1992; Waxman et al., 1992; Gosland et al., 1992)]. Supporting data were also obtained for the angiotensin-II antagonist losartan and its active phase I metabolite EXP 3174 (Soldner et al., 1996).

21.2.8. Gut and Liver Metabolism: How to Differentiate

When we want to define the extent of first-pass metabolism *in vivo*, particularly in humans, for completely absorbed drugs, we usually may accomplish this by measuring venous blood or plasma concentrations in the systemic circulation and calculating systemic availability as AUC ratio.

In physiological pharmacokinetics, the contributions of different oragans to total clearance as well as a prediction of the extent of first-pass effect is made from the intrinsic clearances within a particular eliminating organ and the respective blood flow. Following racemate dosage, the two enantiomers are usually treated as individual entities and we assume no mutual interaction with respect to any of the processes. Estimates of the extent of first-pass extraction from *in vivo* data have usually been based on the metabolic clearance of a drug, calculated from the dose fraction that is recovered as metabolite(s), and on the assumption that metabolic processes occur in the liver only and all metabolites are excreted into urine. In clinical studies, we usually use the fraction of the dose excreted into urine following biotransformation for this purpose. When an oral dose is administered, CL_o (CL/F) replaces CL and the assumption of complete absorption is made (Kroemer et al., 1992). When the unchanged renal clearance of a drug is known, the difference between total and renal clearance, the nonrenal clearance, may be a reasonable basis for an estimation of metabolic (= hepatic) clearance and the extent of first-pass effect.

Significant gut metabolism is to be expected when a drug is mainly and highly metabolized via CYP3A, since significant amounts of this class of cytochrome P450 isozymes are present in the gut. In general, studies of subcellular fractions of the respective tissues may be carried out for differentiation.

In order to characterize the fractional contribution of the metabolizing enzymes in gut vs. liver tissue in experimental animals, it is possible to measure

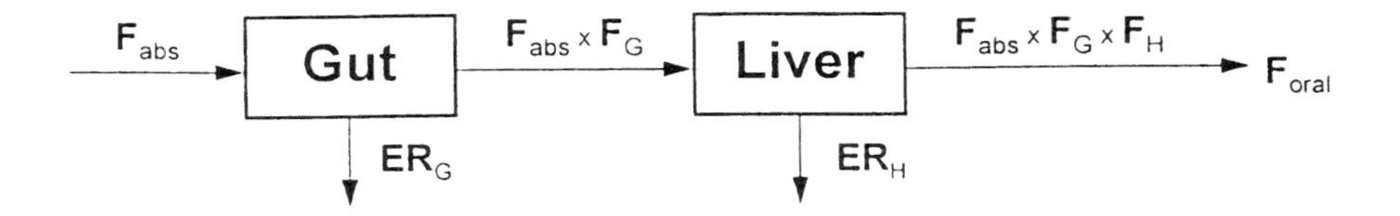

Figure 21.5. Schematic diagram depicting the effects of absorption and gut and hepatic first-pass extraction on drug/enantiomer oral bioavailability (Wu et al., 1995). (F_{abs} = fraction of the dose absorbed into and through the gastrointestinal membranes. ER_G = gut extraction ratio. F_G = fraction of the absorbed dose that passes through the gut into the hepatic portal blood unmetabolized. ER_H = hepatic extraction ratio. F_H = the hepatic first-pass availability. F_{oral} = oral bioavailability.)

blood in the mesenteric or portal vein(s) in addition to peripheral blood concentrations. However, such procedures cannot be readily carried out in humans because of technical and ethical limitations.

The problem of differentiation between gut and liver metabolism has already been addressed in a publication of Minchin and Ilett from 1982, who tried to quantify their relative contributions in the presystemic elimination of drugs from blood/plasma and urine data.

It may be expected that we will be able to largely predict the extent of gut wall and liver metabolism on the basis of our knowledge about the occurrence of certain metabolizing enzymes in the respective tissues and *in vitro* studies with for example, microsomal enzymes.

Some experimental and theoretical considerations were described in recent articles on cyclosporine systemic availability (Wu et al., 1995; Hebert et al.; 1992; Gupta et al., 1989; Gomez et al., 1995). As depicted in Figure 21.5, oral bioavailability (F_{oral}) is the product of the fraction of the dose absorbed into and through the gastrointestinal membranes (F_{abs}), the fraction of the absorbed dose that passes through the gut into the hepatic portal blood unmetabolized (F_G), and the hepatic first-pass availability (F_H):

$$F_{oral} = F_{abs} \times F_G \times F_H$$

Gut and hepatic availability may be defined as 1 minus the extraction ratio (ER) at each site:

$$F_{oral} = F_{abs} \times (1-ER_G) \times (1-ER_H)$$

When a drug is administered intravenously, it is possible to estimate the hepatic extraction ratio from the ratio of hepatic blood clearance (CL_H) and hepatic blood flow (Q_H):

$$ER_H = \frac{CL_H}{Q_H}$$

In this approach, metabolism in gut epithelial tissue that is negligible with respect to drug in the systemic circulation, linear pharmacokinetics, and negligible extrahepatic nonrenal clearance are assumed. Then it is possible to determine $F_{max} \times F_G$ [33% for cyclosporine according to Hebert et al. (1992)].

Three pharmacokinetic interaction studies were performed: one with rifampicin, an inducer of cyclosporine metabolism, and two other studies with erythromycin and ketoconazole, two inhibitors of cyclosporine metabolism (Hebert et al., 1992; Gupta et al., 1989; Gomez et al., 1995). Pharmacokinetic characteristics following i.v. and p.o. dosing were analyzed from blood samples before and after oral administration of the inducer/inhibitor of cyclosporine metabolism.

If we assume that the absorption of cyclosporine was unaffected by concomitant (not simultaneous) dosing, the equations can be modified for the administration of an interacting drug (Wu et al., 1995). This approach permits one to evaluate boundary conditions for F_{abs}. The lower boundary for F_{abs} would be the minimum availability obtained in a study in which metabolism is inhibited, as in the erythromycin and ketoconazole interaction studies. The analysis of data indicates that in the absence of intestinal metabolism, cyclosporine absorption averages at least 65% in healthy volunteers and 77% in kidney transplant patients. Moreover, the calculations suggest that the extraction ratio for cyclosporine in the gut is approximately twice the hepatic extraction and an average of 86% of the drug is absorbed intact from its commercially available dosage form in healthy volunteers. Hence, it is well absorbed, and the major reason for its poor bioavailability is oxidative metabolism mediated by CYP3A in the gut.

F_{abs} may be estimated on the basis of *in vitro* and *in vivo* permeability studies, such as in human epithelial cell cultures or perfusion experiments.

21.3. PHYSIOLOGICAL FACTORS AFFECTING THE EXTENT OF FIRST-PASS EXTRACTION AND ITS STEREOSELECTIVITY

21.3.1. Genetic Polymorphism in Oxidative Drug Metabolism

21.3.1.1. General Aspects

The phenomenon of polymorphism was detected initially for acetyltransferases involved in the elimination of isoniazide. Since the time when it became obvious that the phenomenon of "extensive" and "poor metabolizers" (EMs and PMs) exists with respect to metoprolol elimination, various compounds were found to undergo polymorphic oxidative metabolism. CYP2D6 (debrisoquine/sparteine hydroxylase) is involved in the stereospecific metabolism of several important groups of drugs, for example, β-adrenoceptor blockers, antiarrhythmics, antidepressants, and neuroleptics (e.g., metoprolol, propranolol, carvedilol, timolol, encainide, flecainide, propafenone, sparteine, etc.). This enzyme catalyzes the metabolism of lipophilic bases only. Approximately 7% of Caucasians but only 1% of Orientals are poor metabolizers of debrisoquine. Usually, the mutated allele CYP2D6B is detected in Caucasian PMs, which is almost absent in Oriental PMs. Furthermore, the enzyme activity detected with EMs is lower for Orientals than Caucasians, which is explained by the frequent occurrence (in 50% of EMs) of a partially deficient CYP2D6 allele, causing an amino acid exchange (Pro^{34} to Ser). As a consequence, lower doses of the respective drugs are used in Oriental populations.

Interestingly, the metabolic activity may also be different within the group of EMs. As reported by Dahl et al. (1995), approximately 1% of a Swedish Caucasian population has one or more extra active CYP2D6 gene(s), causing ultrarapid metabolism of the substrates of CYP2D6 and further contributing to the extent of variabilities to be observed with extensively metabolized substrates.

While CYP2D6 catalyzes the oxidation of lipophilic bases only, CYP2C19 is involved in the metabolism of acids (e.g., *S*-mephenytoin), bases (e.g., imipramine, propranolol, and omeprazole, which shows a 10-fold difference in oral clearance between the EMs and PMs of *S*-mephenytoin), and neutral drugs (e.g., diazepam). About 3% of Caucasians and 12–22% of Orientals were found to be PMs of *S*-mephenytoin (Bertilsson, 1995).

When analyzing the information available with respect to the various CYP isoforms and their substrates, it becomes obvious that a tremendous prevalence of only two isozymes exists in humans. It appears in the group of cytochromes P450 that approximately 50% of all human drugs are metabolized by one isozyme, CYP3A4. And about another 25% of human drugs are metabolized by the isozyme 2D6, the "debrisoquine" isozyme. 2D6-enzymes largely work on drugs that affect the heart and drugs that affect the brain. So, an understanding of two isozymes provides essential information about drugs in humans, and there are only a few others that one needs to understand to make predictions about metabolism in humans.

21.3.1.2. Examples

Although a number of polymorphic traits responsible for altered metabolism have been described, the best studied is probably the CYP2D6 enzyme, the importance of which may be derived from the large number of drugs metabolized via this enzyme, including a number of β-blockers such as alprenolol, bufuralol, betaxolol, metoprolol, propranolol, and timolol (Dayer et al., 1982; McGourty et al., 1985; Kirch et al., 1985; Lennard et al., 1983, 1986, 1989; Gut et al., 1984; Jonkers et al., 1993; Raghuram et al., 1984.)

Lennard et al. (1983) first detected a difference in enantiomer ratios following metoprolol racemate dosage between debrisoquine EMs and PMs, with higher apparent oral clearance of the *R*-enantiomer in EMs, yet similar clearances for both enantiomers or slightly higher values for *R*-metoprolol and higher plasma concentrations in PMs. In the EM group, metoprolol was a medium to high clearance drug and the systemic availability of the active *S*-enantiomer was 1.4 times that of the less active *R*-enantiomer. In the PM group, metoprolol was a low clearance drug, and the systemic availabilities of the enantiomers were equal.

Recently, Zhou and Wood (1995) were able to detect the dependence of the clearance of carvedilol enantiomers from the pharmacogenetic status of the volunteers, as predicted from the above mentioned *in vitro* data with human liver microsomes. They studied EMs of sparteine/ debrisoquine and mephenytoin, PMs of debrisoquine but EMs of mephenytoin, and PMs of mephenytoin but EMs of debrisoquine. In all groups, the clearance of carvedilol was significantly lower for *R*- than *S*-carvedilol. No effect of the mephenytoin phenotype on carvedilol kinetics was observed (CL/F: *R*, 119; *S*, 198 L/h); however, PMs of debrisoquine (CL/F, *R*, 39; *S*, 164 L/h) had a significantly lower clearance for *R*-carvedilol than EMs (CL/F, *R*, 119; *S*, 215 L/h). These authors found the partial metabolic clearance of carvedilol to the two ring-hydroxylated metabolites to be significantly reduced in PMs of debrisoquine and based their calculations on the renal excretion data for the unconjugated and conjugated oxidative metabolites.

The stereoselective disposition of the class IC antiarrhythmic agent flecainide in relation to the sparteine/debrisoquine metabolizer phenotype was presented by Gross et al. (1989). In their study, five EMs and five PMs received 50 mg of *rac*-flecainide under conditions of high urinary flow rate and acidic urinary pH. In the group of EMs, no significant enantiomer differences were observed in any of the kinetic parameters (e.g., CL/F: *S*, 793; *R*, 768 mL/min; CL_R: *S*, 290; *R*, 274 mL/min), whereas in PMs the apparent oral clearance was smaller for *R*- than *S*-flecainide (*R*, 467; *S*, 620 mL/min) and the half-lives longer for *R*-flecainide (12.9 vs 9.8 h). Differences in renal clearances were detected neither between the two enantiomers nor between EMs and PMs. However, the analysis of meta-O-dealkylated flecainide and the lactam yielded a significant difference between EMs and PMs with respect to amounts excreted into urine, which were higher for EMs (12.0 vs. 17.7%).

21.3.2. Other Genetic Factors: Gender Differences, Menopausal Status, and Age

In rats it was found that males have significantly higher levels of an isozyme of the 3A family than females (Perotti et al., 1994), whereas current data for humans suggest that CYP3A4 activity is higher in women than men. On the other hand, the activity of many other systems involved in drug metabolism may be higher in men than women. Female-specific issues such as pregnancy, menopause, oral contraceptives, and the menstrual cycle may also have profound effects on drug metabolism (Kato and Yamazone, 1992). A comprehensive survey about gender differences in drug metabolism, with an emphasis on cytochromes P450, was provided by Harris et al. (1995), who also emphasized that differences in gut enzymes may be of importance.

As discussed by Harris et al. (1995), the higher CYP3A4 levels in women appear to decrease in menopausal status, whereas similar age-related changes are not observed in men (Gustavson et al., 1986; Hunt et al., 1992; Gustavson and Benet, 1994).

A variety of changes in pharmacokinetically relevant processes may occur with higher age. With respect to processes relevant to first-pass elimination, the diminished activity of metabolizing enzymes and changes in blood flow have to be mentioned. It is known that cardiac output and liver blood flow are reduced in relation to aging (Vestal, 1989; Wynne et al., 1989).

The effect of *aging* (Wynne et al., 1989) on *verapamil* stereopharmacokinetics was investigated in two clinical studies performed by Schwartz et al. (1993, 1994). When the two enantiomers were administered on separate occasions, verapamil clearance was found to be decreased in an age-related stereoselective manner, with significant reductions in *S*(–)-verapamil clearance in older subjects but no age-related change in *R*(+)-verapamil clearance (Schwartz et al., 1993). In the second study, an age-dependent reduction in the clearance of both enantiomers was found when the racemate was dosed (Schwartz et al., 1994). However, the respective doses were administered intravenously; no p.o. data are available. From the clearance changes, we would expect a decrease in hepatic extraction and an increase in oral availability.

Chandler et al. (1988) evaluated *age*-associated alterations in *hexobarbital* metabolism. They found that the mean oral clearance of *d*-hexobarbital did not differ significantly between young and elderly subjects. However, *l*-hexobarbital mean oral clearance was approximately two-fold greater in the young than in the elderly subjects, yielding a significantly greater enantiomeric oral clearance ratio in the young when compared with the elderly subjects.

In a study performed in young healthy volunteers and multimorbid patients, we evaluated potential *age*-related changes in the kinetics of *propranolol* enantiomers following racemate dosage. In general, the concentrations of both propranolol enantiomers were two-fold higher in the patients, and the differences between enantiomers were less pronounced (CL/F: young volunteers, 2286 (*S*), 3962 mL/min (*R*); elderly patients, 1222 (*S*), 1722 mL/min (*R*)]. The concentrations of the glucuronides were at a similar level for both groups; however, the average *S*/*R* ratio increased slightly with age, yielding values of 2.4 for the C_{max} of the controls and 2.9 for the patients (Stahl, 1993).

In the repetitive dosage study with *rac-propranolol* on *age and gender* influences performed by Gilmore et al. (1992), the intrinsic hepatic clearance of the *S*-enantiomer of propranolol appeared to be 30% smaller for the elderly than young volunteers. Furthermore, only females displayed a stereoselective difference in the intrinsic hepatic clearance of the unbound enantiomers, with that of the *S*-enantiomer being about 40% greater than that for the *R*-enantiomer.

Hooper and Qing (1990) investigated the influence of *age and gender* on the stereoselective metabolism and pharmacokinetics of *mephobarbital* in humans (Lim and Hooper, 1989). In this study, the apparent total body clearance of *R*-mephobarbital, a substrate of CYP2C8 and CYP2C9, was much greater in young men than in any of the other three groups. This enantiomer displayed an age-dependent gender effect and a gender-dependent age effect in its metabolism. CL/F of the *S*-enantiomer was much lower than that of the *R*-enantiomer in all subjects with no difference between groups, although $t_{1/2}$ was slightly less in young males. Consequently, an apparently enhanced stereoselectivity in the metabolism of mephobarbital was detected in young men.

21.3.3. Diseased States

21.3.3.1. General Aspects

Pathological changes in an eliminating organ may lead to reductions in the extraction ratio in the respective organ (Krähenbühl and Reichen, 1988; Wosilait and Luecke, 1988). When the extraction ratios are different for the two enantiomers, the disease may either affect both enantiomers to the same extent or potentially—when the enantiomers are substrate for different elimination systems, which are differently affected by the disease—to different extents.

The influences of *gastrointestinal diseases* on the stereoselectivity of drug input have not yet been evaluated. But we may be able to predict that such changes will occur when we consider that the metabolic and transport systems located in the enterocytes are affected.

We are expecting the most prominent changes in enantiomer concentration-time profiles in *liver disease* when the compound is undergoing a considerable hepatic (metabolic) clearance. Changes in the half-lives of the enantiomers may result from disease-induced alterations (reductions) in protein binding, with increased distribution volumes due to reduced plasma protein concentrations. With respect to the clearance concept, we need to differentiate between low and high clearance drugs when evaluating the potential influence of liver disease. When the capability of the eliminating organ to metabolize the drug enantiomers is large in comparison to the rate of drug presentation to the organ ($f_u \times CL_{int}$ is much smaller than Q; high extraction ratio drug), then the organ clearance will approximate the organ blood flow (CL = Q), that is, drug enantiomer elimination is limited by the blood flow rate. On the other hand, when the metabolic capability is small in comparison to drug presentation (Q is much smaller than $f_u \times CL_{int}$), the clearance will be proportionate to the unbound fraction of the drug in blood and the intrinsic clearance ($CL = f_u \times CL_{int}$). From this concept, major clearance changes are expected for low clearance compounds only, disease-induced changes in CL_{int} should not be expected for high clearance compounds when Q remains unchanged. However, in liver cirrhosis different pathophysiological stages may be observed, where alterations of the hemodynamic characteristics (blood flow changes) as well as severe impairments of metabolic processes are possible. As a consequence, compounds, characterized as high extraction ratio drugs in healthy volunteers, also potentially exhibit deviating kinetic characteristics in severe liver disease.

21.3.3.2. Experimental and Clinical Data

With respect to kinetic data from *in vivo* studies, we usually have to deal with reduced stereoselectivity in concentration-time profiles. However, an increased stereoselectivity of clearance and first-pass effect is also feasible. The shape and stereoselectivity of the concentration-time curves are dependent on the relative contribution of all stereoselective clearance processes and the influence of the disease on the respective process.

The results obtained from a preclinical study with carvedilol (Stahl et al., 1990, 1993) were confirmed by a subsequently performed clinical study by Neugebauer et al. (1992). These authors found the plasma clearances to be close for the *R*- and the *S*-enantiomer, with a slight preference for *S*-carvedilol in cirrhotic patients (medians: *S*, 421, *R*, 321 mL/min). Following oral administration, the absolute bioavailabilities were 83.7% for *R*- and 71.3% for *S*-carvedilol (n.s.). In contrast to healthy subjects, the apparent volume of distribution was 90% greater for *S*- than *R*-carvedilol in these patients. The renal excretion of carvedilol and the respective glucuronide exhibited stereoselective behavior, but with opposite configurational preference. The authors concluded from their data that in patients with liver cirrhosis, stereoselective metabolism is still operative and, probably because of portacaval shunts, the hepatic first-pass extraction is reduced markedly, eliminating the difference in bioavailability between enantiomers.

21.3.4. Food Constituents and Food Effects

Food consitituents as well as pharmaceutical excipients may influence the amount of drug (enantiomer) entering the systemic circulation, due to physicochemical interactions inside the gut lumen, to the inhibition of transport processes or to an influence on the activity of metabolizing enzymes.

The "grapefruit juice interaction" represents one of the probably best-known drug interaction with food constituents. Already known for a considerable period of time (e.g., Buening et al., 1981), this type of interaction has recently been further investigated with respect to its mechanism and stereoselectivity (Ducharme et al., 1993). The specificity of the grapefruit effects suggested the role of the major flavonoid naringin. Naringenin, the naringin aglycone, was shown to inhibit the cytochrome P450 isoforms responsible for the metabolism of a number of drugs sensitive to this interaction (e.g., nifedipine, felodipine, nitrendipine, nisoldipine, nimodipine, verapamil, terfenadine, verapamil, caffeine, and cyclosporine), that is CYP3A and CYP1A2, respectively. In a study with human liver microsomes and verapamil as the substrate performed by Fuhr et al. (1993), naringin was found to specifically inhibit one metabolic pathway, whereas naringenin inhibited all tested pathways. When relative bioavailability was evaluated, the AUCs of both isomers of verapamil were elevated to a comparable extent (approximately 150%), with a slight preference for the *S*-enantiomer. Norverapamil AUCs were enhanced to 122% for *R*- and 160% for *S*-verapamil (Fuhr et al., 1994).

To better understand the mechanism of interaction, Fuhr and Kummert (1995) studied the fate of naringin in humans. The authors concluded that the data suggest that cleavage of the glycone—presumably by intestinal bacteria—is the first step in naringin metabolism. Apparently, the formation of narigenin, which undergoes rapid glucuronidation, is the crucial step. This hypothesis would provide an explanation for some apparently contradictory results from *in vitro* and *in vivo* drug interaction studies with grapefruit juice and naringin.

In general, other mechanisms for food-drug interactions are feasible as well, including the influence on blood perfusion of organs and G.I. motility in addition to interactions at the enzymes.

An overview about food effects was provided recently by Kuhlmann and Weihrauch (1995). However, stereochemical aspects have largely been neglected so far. This is also valid for studies performed with propranolol by Woodruff Modi et al. (1988), who tried to explain the increased AUC to be observed with protein-rich food. They came to the conclusion that the increased oral availability is not principally a result of a change in Q_H, but must be due to other mechanisms such as a decrease in apparent intrinsic metabolic clearance (through an inhibition of metabolizing enzymes), altered plasma protein binding in hepatic sinusoidal blood, etc.

21.3.5. Drug–Drug Interactions

Numerous drug–drug interaction studies have been performed with the largely nonsubtype-specific CYP inhibitor cimetidine. With respect to the interaction mechanism, however, effects on the systemic availability of drugs may also be mediated by the pharmacologic profile of the H_2-antagonist (alterations of gastrointestinal pH and motility as well as blood flow changes).

Stereochemical aspects were not always included in studies, particularly in the early 1980s. While, for example, Abernethy et al. (1985) were not able to detect any influence of cimetidine on verapamil kinetics in a study where verapamil was dosed intravenously as well as perorally to nine healthy volunteers, Mikus et al. (1990) detected a significant AUC increase for both verapamil enantiomers when cimetidine was coadministered to six healthy volunteers (average AUC increase: 150.3% for *S*; 117.8% for *R*; Table 21.3). In five of the six subjects, fractional metabolic clearance to two N-dealkylated metabolites was decreased for both enantiomers. Slight increases in the average renal clearance were detected for both enantiomers of verapamil and norverapamil. Tubular secretion of *S*-enantiomer of one of the metabolites was reduced by cimetidine, indicating stereoselective interaction at both the hepatic and renal level.

The AUC of *R*-metoprolol increased 84% with cimetidine coadministration, whereas that of the pharmacologically active *S*-enantiomer increased by only 40% (Toon et al., 1988).

When studying the carvedilol/cimetidine interaction in six healthy volunteers with a single 100-mg *rac*-carvedilol dose and 3 days of pretreatment with cimetidine (at a total daily dose of 1000-mg), a significant increase of AUC and reduction of CL/F were found for both carvedilol enantiomers, particularly for the *R*-enantiomer (Henke et al., 1992). In addition to the apparent oral clearances, renal clearances of the parent carvedilol enantiomers as well as of the respective conjugates were reduced under cimetidine coadministration (Table 21.3)

With respect to pindolol, only a minor prevalence of the *S*-enantiomer was found in six volunteers. Cimetidine coadministration increased slightly the maximum levels of both enantiomers (0–20% of the control) as well as the half-lives of both enantiomers (means: control, 3.54 h (*R*), 3.90 (*S*); cimetidine, 4.56 (*R*), 4.76 (*S*) (Spahn, 1989). With cimetidine, renal excretion of both oxprenolol enantiomers of the unconjugated as well as conjugated compound

Table 21.3. Results from Interaction Studies of Drugs with Cimetidine (A: verapamil/cimetidine) (verapamil 160 mg of as a single dose and 400 mg of cimetidine/bid); B: carvedilol/cimetidine (Carvedilol 100 mg of as a single dose and 1000 mg of cimetidine/die); parameters are given as arithmetical means ± SD

Period		Control Period	Cimetidine
A: Verapamil			
AUC (nmol mL^{-1} min)	*S*	29.2 ± 31.8	41.2 ± 33.7
	R	124.7 ± 112.2	156.8 ± 105.0
CL/F (mL/min/kg bw)			
	S	126.5 ± 74.1	74.1 ± 31.8
	R	26.0 ± 11.4	18.1 ± 7.4
B: Carvedilol			
AUC (ng mL^{-1} h)	*S*	198 ± 94	228 ± 81
	R	350 ± 180	442 ± 134
CL/F (mL/min)	*S*	3616 ± 2036	4297 ± 2351
	R	3187 ± 1999	2075 ± 707
CL_R (mL/min)	*S*	5.65 ± 0.55	4.21 ± 1.96
	R	8.56 ± 1.96	5.44 ± 2.03
$CL_{R,conj}$ (mL/min)	*S*	73.5 ± 51.2	51.1 ± 26.2
	R	33.8 ± 15.4	25.2 ± 12.4

AUC = area under the concentration-time curve.
CL/F = apparent oral clearance.
CL_R = renal clearance of parent drug enantiomers.
$CL_{R,conj}$ = renal clearance of the conjugates of the enantiomers.

were lower, with the highest reduction detected for *S*-oxprenolol glucuronide (Spahn, 1989), a finding comparable with that for carvedilol glucuronide.

Ranitidine and, for example, nizatidine are both H_2-receptor antagonists of the "second generation", that is, compounds superior to cimetidine with respect to potency/efficacy and side-effects and with less affinity to the cytochrome P450 system and thus less likely to cause pharmacokinetic interactions than cimetidine, as shown for metoprolol as the coadministered drug (Spahn, 1989). However, for most H_2-antagonists, a slight to significant reduction in liver blood flow was detected, which may be an additional source for pharmacokinetic interactions with respect to high clearance drugs.

Knowing the involved (iso)enzymes and carriers as well as a substrate's affinity to the respective macromolecule, we should be able to predict the possible types of interaction as well as their stereoselectivities.

21.4. CONCLUSIONS

Presystemic elimination, the prediction of which was previously based on hepatic metabolism only, is a fairly complex sum of transport and metabolic processes occurring in different organs with varying (and potentially inverse) stereoselectivities (Figure 21.6). We know today that the cause of previous false predictions may be the neglection of, for example, gut metabolism or intestinal secretion.

The availability of enzyme and carrier proteins has a genetic basis, but it is also influenced by various physiological and environmental factors as well as by other drugs and diseases. Prediction of the overall stereoselectivities and variabilities of the first-pass effect under various conditions requires detailed knowledge about the most relevant presystemic elimination processes.

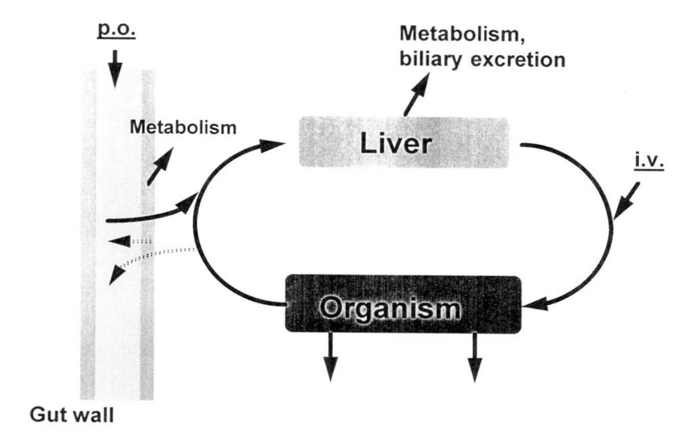

Figure 21.6. *Schematic illustration of first-pass processes possible with oral dosing.* In addition to hepatic first-pass extraction, also gut-related processes may affect considerably the amount of drug reaching the systemic circulation after p.o. dosing. We may have to expect metabolism in the gut wall, since various metabolic enzymes were detected in the gut enterocytes. Moreover, intestinal secretion via P-glycoprotein or other carrier mechanisms may affect drug input into the systemic circulation and, potentially, contribute to drug elimination from the organism (Wetterich et al., 1996). Both intestinal metabolism and secretion may be influenced by, for example, other drug products and food constituents. Furthermore, with respect to the dosage of racemic drugs, competition of the stereoisomers may occur at the different metabolizing enzymes and carriers.

ACKNOWLEDGMENTS

The authors wish to thank the Deutsche Forschungsgemeinschaft (Sp 290/2), the Fonds der Chemischen Industrie (grant to H.S.-L.), and the Dr. Robert-Pfleger-Stiftung (grant to H.S.-L.), as well as the Schweizer Nationalfonds (research grant to P.L.) and the NIH (GM 26691 to L.Z.B.), who supported part of the studies from our laboratories that were included in this overview.

REFERENCES

Abernethy, D. R., Schwartz, J. B., and Todd, E. L. (1985). Lack of interaction between verapamil and cimetidine. *Clin. Pharmacol. Ther.*, **38**, 342–349.

Ariens, E. J. (1984). Stereochemistry, a basis for sophisticated nonsense in pharmacokinetics and clinical pharmacology. *Eur. J. Clin . Pharmacol.*, **26**, 663–668.

Benet, L. Z. (1989). Pharmacokinetics: Absorption, distribution and elimination. In B.G. Katzung (ed.), *Basic and Clinical Pharmacology*. East Norwalk, Appleton & Lange.

Berry, B. W. and Jamali, F. (1991). Presystemic and systemic chiral inversion of R-(–)-fenoprofen in the rat. *J. Pharmacol. Exp. Ther.*, **258**, 695–701.

Bertilsson, L. (1995). Geographical/interracial differences in polymorphic drug oxidation: Current state of knowledge of cytochromes P450 (CYP) 2D6 and 2C19. *Clin. Pharmacokinet.*, **29**, 192–209.

Brockmeier, D., Hajdu, P., Henke, W., Mutschler, E., and Palm, D. (1988). Pharmacokinetics of penbutolol, its effect on exercise heart rate, and *in vitro* inhibition of radioligand binding using plasma samples. *Eur. J. Clin. Pharmacol.*, **35**, 613–623.

Buening, M. K., Chang, R. L, Huang, M. T., Fortner, J. G., Wood, A. W., and Conney, A. H. (1981). Activation and inhibition of benzo(a)pyrene and aflatoxin B1 metabolism in human liver microsomes by naturally occurring flavonoids. *Cancer Res.*, **41**, 67–72.

Caldwell, J. C., Hutt, A. J., and Fournel-Gigleux, S. (1988). The metabolic chiral inversion and dispositional enantioselectivity of the 2-arylpropionic acids and their biological consequences. *Biochem. Pharmacol.*, **37**, 105-114.

Chandler, M. H. H., Scott, S. R., and Blouin, R. A. (1988). Age-associated stereoselective alterations in hexobarbital metabolism. *Clin. Pharmacol. Ther.*, **43**, 436-441.

Dahl, M. L., Johansson, I., Bertilsson, L., et al. (1995). Ultrarapid hydroxylation of debrisoquine in a Swedish population: Analysis of molecular genetic basis. *J. Pharmacol. Exp. Ther.*, **274**, 516–520.

Dayer, P., Balant, L., Courvoisier, F. et al. (1982). The genetic control of bufuralol metabolism in man. *Eur. J. Drug Metab. Pharmacokin.*, **7**, 73–77.

de Lannoy, I.A. and Silverman, M. (1992). The MDR1 gene product, P-glycoprotein, mediates the transport of the cardiac glycoside, digoxin. *Biochem. Biophys. Res. Commun.*, **189**, 551–557.

Ducharme, M. P., Provenzano, R., Dehoorne-Smith, M., and Edwards, D. J., (1993). Trough concentrations of cyclosporine in blood following administration with grapefruit juice. *Br. J. Clin. Pharmacol.*, **36**, 457–459.

Eberhart, D. C., Gerrzik, B., Halvorson, M. R., and Parkinson, A. (1991). Species differences in the toxicity and cytochrome P450 III A-dependent metabolism of digitoxin. *Mol. Pharmacol.*, **40**, 859–867.

Eichelbaum, M., Gross, A. S., and Kroemer, H. K. (1992). Genetic polymorphism in drug metabolism and its role in patient medication. In D. A. Crommelin, and K. K. Midha, eds. '*Topics in Pharmaceutical Sciences 1991*.' Stuttgart: Medpharm Scientific Publishers.

Endicott, J. A. and Ling, V. (1989). The biochemistry of P-glycoprotein-mediated multidrug resistance. *Annu. Rev., Biochem.*, **58**, 137–171.

Fassihi, A.R. (1993). Racemates and enantiomers in drug development. *Int. J. Pharm.*, 92, 1–14.

Fritz, P., Behrle, E., Eichelbaum, M., Kroemer, H. K., and Beaune, P. (1993). Expression of cytochrome P4503A in various human tissues. *Nanunyn Schmiedeberg's Arch. Pharmacol.*, **347** (Suppl.), 121.

Fuhr, U. and Kummert, A. L. (1995). The fate of naringin in humans: A key to grapefruit juice-drug interactions? *Clin. Pharmacol. Ther.*, **58**, 365–373.

Fuhr, U., Kroemer, H. K., and Schymanski, P. (1993). Effects of naringenin and naringin on verapamil metabolism in human liver microsomes. *Naunyn Schmiedeberg's Arch. Pharmacol.*, **347** (Suppl.), R31.

Fuhr, U., Harder, S., Lopez-Rojas, P., Müller-Peltzer, H., Kern, P., and Staib, A. H. (1994). Increase verapamil concentrations in steady state by coadministration of grapefruit juice. *Naunyn Schmiedeberg's Arch. Pharmacol.*, **349** (Suppl.), R134.

Gietl, Y., Spahn, H., and Mutschler, E. (1988). Simultaneous determination of *R*- and *S*-prenylamine in plasma and urine by reversed-phase high-performance liquid chromatography. *J. Chromatogr. Biomed. Appl.*, **426**, 305–314.

Gietl, Y., Spahn, H., and Mutschler, E. (1989). Enantioselective disposition of *R*-(–)- and *S*-(+)-prenylamine in the rat. *Arzneim.-Forsch./Drug Res.*, **39**, 853–856.

Gietl, Y., Spahn, H., Knauf, H., and Mutschler, E. (1990). Single- and multiple-dose pharmacokinetics of *R*-(–)- and *S*-(+)-prenylamine in man. *Eur. J. Clin. Pharmacol.*, **38**, 587–593.

Gilmore, D. A., Gal, J., Gerber, J. G., and Nies, A. S. (1992). Age and gender influence the stereoselective pharmacokinetics of propranolol. *J. Pharmacol. Exp. Ther.*, **261**, 1181–1186.

Gomez, D. Y., Wacher, V. J., Tomlanovich, S. J., Hebert, M. F., and Benet, L. Z. (1995). The effects of ketoconazole on the intestinal metabolism and bioavailability of cyclosporine. *Clin. Pharmacol. Ther.*, **58**, 15–19.

Gonzalez, F. J. (1992). Human cytochrome P450: Problems and prospects. *Trends Pharmacol. Sci.*, **13**, 346–352.

Gosland, M., Tsuboi, C., Hoffman, T., Goodin, S. and Vore, M. (1992). 17-β-Estradiol glucuronide: An inducer of cholestasis and a physiological substrate for the multidrug resistance transporter. *Cancer Res.*, **53**, 5382–5385.

Gottesman, M. M. (1993). How cancer cells evade chemotherapy: Sixteenth Richard and Hinda Rosenthal Foundation Award Lecture. *Cancer Res.*, **53**, 747–754.

Gottesman, M. M. and Pastan, I. (1993). Biochemistry of multidrug resistance mediated by the multidrug transporter. *Ann. Rev. Biochem.*, **62**, 385–427.

Gross, A. S., Mikus, G., Fischer, C., Hertrampf, R., Gundert-Remy, U., and Eichelbaum, M. (1989). Stereoselective disposition of flecainide in relation to the sparteine/debrisoquine metabolizer phenotype. *Br. J. Clin. Pharmol.*, **28**, 555–566.

Gupta, S. K., Bakran, A., Johnson, R. W. G., Hebert, M. F., and Benet, L. Z. (1989). Cyclosporine-erythromycin interaction in renal transplant patients. *Br. J. Clin. Pharmacol.*, **27**, 475–481.

Gustavson, L. E. and Benet, L. Z. (1994). Menopause: Pharmacodynamics and pharmacokinetics. *Exp. Gerontol.*, **29**, 437–444.

Gustavson, L. E., Legler, U. F., and Benet, L. Z. (1986). Impairment of prednisolone disposition im women taking oral contraceptives or conjugated estrogens. *J. Clin. Endocrinol. Metab.*, **62**, 234–237.

Gut, J., Gasser, R., Dayer, P., Kronbach, T., Catin, T., and Meyer, U.A. (1984). Debrisoquine-type polymorphism in drug oxidation: Purification from human liver of a cytochrome P450 isozyme with high activity for bufuralol hydroxylation. *FEBS Lett.*, **173**, 287–290.

Harder, S., Thürmann, P., Siewert, M., Blume, H., Rietbrock, N., van der Kleijn, J., and Gierend, M. (1992). Concentration/effect relationship and enantioselective analysis of verapamil in hypertensive patients. *J. Cardiovasc. Pharmacol.*, **19**, 665–669.

Harris, R. Z., Benet, L. Z., and Schwartz, J. B. (1995). Gender effects in pharmacokinetics and phrmacodynamics. *Drugs*, **50**, 222–239.

Hebert, M. F., Roberts, J. P., Prueksaritanont, T., and Benet, L. Z. (1992). Bioavailability of cyclosporine with concomitant rifampicin administration is markedly less than predicted by hepatic enzyme induction. *Clin. Pharmacol. Ther.*, **52**, 453–455.

Henke, W., Spahn, H., von Möllendorf, E., and Mutschler, E. (1985). Pharmacokinetics of the enantiomers of the β-adrenoceptor blocking carvedilol. Paper presented at the annual meeting of the German Pharmaceutical Society, Braunschweig.

Henke, W., Henke, D., and Spahn-Langguth, H. (1992). The influence of cimetidine on the stereopharmacokinetics of carvedilol. *Naunyn Schmiedeberg's Arch. Pharmacol.*, **345** (Suppl.), R9.

Hooper, W. D. and Qing, M. S. (1990). The influence of age and gender on the stereoselective metabolism and pharmacokinetics of mephobarbital in humans. *Clin. Pharmacol. Ther.*, **48**, 633–640.

Hunt, C. M., Westerkam, W. R., and Stave, G. M. (1992). Effect of age and gender on the activity of human hepatic CYP3A. *Biochem. Pharmacol.*, **44**, 275–283.

Jamali, F., Mehvar, R., and Pasutto, F.M., (1989). Enantioselective aspects of drug action and disposition: Therapeutic pitfalls. *J. Pharm. Sci.*, **78**, 695–715.

Jamali, F., Mehvar, R., Russell, A. S., Sattari, S., Yakimets, W. W., and Koo, J. (1992). Human pharmacokinetics of ibuprofen following different doses and formulations: Intestinal chiral inversion. *J. Pharm. Sci.*, **81**, 221–225.

Jonkers, R. E., Koopmans, R. P., Portier, E. J. G., and van Boxtel, C. J. (1991). Debrisoquine phenotype and the pharmacokinetics and beta-2-receptor pharmacodynamics of metoprolol and its enantiomers. *J. Pharmacol. Exp. Ther.*, **256**, 959–966.

Karim, A. and Piergies, A. (1995). Verapamil steroisomerism: Enantiomeric ratios in plasma dependent on peak concentrations, oral input rate, or both. *Clin. Pharmacol. Ther.*, **58**, 174–184.

Kato, R. and Yamazone, Y. (1992). Sex-specific cytochrome P450 as a cause of sex- and species-related differences in drug toxicity. *Toxicol Lett.*, **64**, 661–667.

Kirch, W., Ohnhaus, E. E., Zekorn, C., Eichelbaum, M., Spahn, H., and Mutschler, E. (1985). Pharmacokinetics of metoprolol in relation to polymorphic sparteine oxidation. *Arch. Toxicol.*, **8** (Suppl.), 401–402.

Klaassen, C. D. (1988). Intestinal and hepatobiliary disposition of drugs. *Toxicol Pathol.*, **16**, 130–137.

Kober, S., Spahn-Langguth, H., Kurz, A., Zaigler, M., Fuhr, U., and Mutschler, E. (1996). Triamterene hydroxylation in human liver microsomes is mediated by the cytochrome P450 isoform CYP1A2. *Arch Pharmacol.*, **353** (Suppl.), R157.

Kolars, J. C., Schmiedlin-Ren, P., Dobbins, W.O., Schuetz, J., Wrighton, S. A., and Watkins, P. B. (1992). Heterogeneity of cytochrome P450IIIA expression in rat gut epithelia. *Gastroenterology*, **102**, 1186–1198.

Krähenbühl, S. and Reichen, J. (1988). Canalicular bile flow and bile salt secretion are maintained in rats with liver cirrhosis. Further evidence for the intact hepatocyte hypothesis. *J. Hepatol.*, **7**, 63–71.

Kroemer, H. K., Echizen, H., Heidemann, H., and Eichelbaum, M. (1992). Predictability of the *in vivo* metabolism of verapamil from *in vitro* data: Contribution of the individual metabolic pathways and stereoselective aspects. *J. Pharmacol. Exp. Ther.*, **260**, 1052–1057.

Kuhlmann, J. and Weihrauch, T. R. (1995). Food-drug interactions. *Clinical Pharmacology*, Vol. 12. W. Zuckschwedt Verlag, München., pp. 34–56.

Kurz, A., Jensen, C., Mutschler, E., Schloos, J., and Spahn-Langguth, H. (1994a). Stereoselectivities in carvedilol phase-I metabolism: Studies with human liver microsomes. *Naunyn Schmiedeberg's Arch. Pharmacol.*, **349** (Suppl.).

Kurz, A., Fuhr, U., Mutschler, E., Staib, A. H., and Spahn-Langguth, H. (1994b). Characterization of phase-I metabolism of carvedilol in human liver microsomes. *Eur. J. Pharm. Sci.*, **2**, 171.

Langguth, P., Möhrke, W., Möhler, H., and Spahn-Langguth, H. (1997). Drug absorption via intestinal vs. nasal mucosae: Dependence of rate and extent of systemic availability of tranylcypromine enantiomers upon the application site. *Biopharm. Drug Dispos.*, (submitted for publication).

Lennard, M. S., Tucker, G. T., Silas, J. H., Freestone, S., Ramsey, L. E., and Woods, H.

F. (1983). Differential stereoselective metabolism of metoprolol in extensive and poor debrisoquine metabolizers. *Clin. Pharmacol. Ther.*, **34**, 732–737.

Lennard, M. S., Tucker, G. T., and Woods, H. F. (1986). The polymorphic oxidation of β-adrenoceptor antagonists. *Clin. Pharmackinet.*, **11**, 1–17.

Lennard, M. S., Lewis, R. V., Brawn, L. A., Tucker, G. T., Ramsey, L. E., Jackson, P. R., and Woods, H. F. (1989). Timolol metabolism and debrisoquine oxidation polymorphism: A population study. *Br. J. Clin. Pharmacol.*, **27**, 429–434.

Levy, R. H. and Boddy, A. V. (1991). Stereoselectivity in pharmacokinetics: A general theory. *Pharm. Res.*, **8**, 551–556.

Lim, W. H., and Hooper, W. D. (1989). Stereoselective metabolism and pharmacokinetics of racemic methylphenobarbital in humans. *Drug Metab. Dispos.*, **17**, 212–217.

Lindner, M., Rath, M., Stoschitzky, and Semmelrock, H. J. (1989). Pharmacokinetic data of propranolol enantiomers in a comparative human study with (*S*)- and (*R,S*)-propranolol. *Chirality*, **1**, 10–13.

Longstreth, J. A. (1993). Verapamil. A chiral challenge to the pharmacokinetic and pharmacodynamic assessment of bioavailability and bioequivalence. In I.W. Wainer (ed.) *Drug Stereochemistry: Analytical Methods and Pharmacology*, 2nd ed. New York: Marcel Dekker.

Marzo, A., Cardace, G., and Arrigoni Martinelli, E. (1992). Enzymes in stereoselective pharmacokinetics of endogenous substances. *Chirality*, **4**, 247–251.

Mayer, S., Spahn-Langguth, H., Gikalov, I., and Mutschler, E. (1993). Pharmacokinetics of beclobric acid enantiomers and their glucuronides after single and multiple p.o. dosage of *rac*-beclobrate. *Arzneim-Forsch/Drug Res.*, **42**:2, 40–43.

McGourty, J. C., Silas, J. H., Lennard, M. S., Tucker, G. T., and Woods, J. W. (1985a). Metoprolol metabolism and debrisoquine oxidation polymorphism—population and family studies. *Br. J. Clin. Pharmacol.*, **20**, 555–566.

McGourty, J. C., Silas, J. H., Fleming, J. J., McBurney, A., and Ward, J. W. (1985). Pharmacokinetics and β-blocking effects of timolol in poor and extensive metabolizers of debrisoquine. *Clin. Pharmacol. Ther.*, **38**, 409–413.

Mehvar, R. (1992). Input rate-dependent stereoselective pharmacokinetics: enantiomeric oral availability and blood concentration ratios after constant oral input. *Biopharm. Drug. Dispos.*, **13**, 597–615.

Mehvar, R. (1994). Input rate-dependent stereoselective pharmacokinetics: Effect of pulsatile oral input. *Chirality*, **6**, 185–195.

Meijer, D. K. F. (1987). Current concepts on hepatic transport of drugs. *J. Hepatol.*, **4**, 259–268.

Meijer, D. K. F. and Groothuis, G. M. M. (1991). Hepatic transport of drugs and proteins. In *Oxford Textbook of Clinical Hepatology*, Vol. 1. N. McIntyre, J. P. Benhamour, J. Bircher, M. Rizzetto, J. Rodes (eds.), Oxford: Oxford University Press, pp. 40–78.

Meijer, D. K. F., van der Sluijs, P. (1989). Covalent and noncovalent protein binding of drugs: Implications for hepatic clearance, storage, and cell-specific drug delivery. *Pharm. Res.*, **6**, 105–118.

Mikus, G., Eichelbaum, M., Fischer, C., Gumulka, S., Klotz, U., and Kroemer, H. K. (1990). Interaction of verapamil and cimetidine: Stereochemical aspects of drug metabolism, drug disposition and drug action. *J. Pharmacol. Exp. Ther.*, **253**, 1042–1048.

Minchin, R. F. and Ilett, K. F. (1982). Presystemic elimination of drugs: Theoretical considerations for quantifying the relative contribution of gut and liver. *J. Pharm. Sci.*, **71**, 458–460.

Murray, M. and Reidy, G. F. (1990). Selectivity in the inhibition of mammalian cytochromes P-450 by chemical agents. *Pharm. Res.*, **42**, 85–101.

Nelson, D. R., Kamataki, T., Waxman, D. J., Guengerich, F. P., Estabrook, R. W., Feyereisen, R., Gonzalez, F. J., Coon, M. J., Gunsalus, I. W., Gotoh, O., Okuda, K., and Nebert, D. W. (1993). The P450 Superfamily: Update of new sequences, gene mapping, accession numbers, early trivial names of enzymes and nomenclature. *DNA Cell Biol.*, **12**, 1–51.

Neugebauer, G. and Neubert, P. (1991). Metabolism of carvedilol in man. *Eur. J. Drug. Metab. Dispos.*, **16**, 257–260.

Neugebauer, G., Akpan, W., Kaufmann, B., and Reiff, K. (1990). Stereoselective disposition of carvedilol in man after intravenous and oral administration of the racemic compound. *Eur. J. Clin. Pharmacol.*, **38**, S108–S111.

Neugebauer, G., Gabor, M., and Reiff, K. (1992). Disposition of carvedilol enantiomers in patients with liver cirrhosis: Evidence for disappearance of stereoselective first-pass extraction. *J. Cardiovasc. Pharmacol.*, **19** (Suppl. 1), S142–S146.

Nuggehally, R. S., Hubbard, J. W., Quinn, D., Korchinski, E. D., and Midha, K. K. (1991). Extensive and enantioselective presystemic metabolism of DL-threo-methylphenidate in humans. *Prog. Neuro-Psychopharmacol. Biol. Psychiatry*, **15**, 213–220.

Nuggehally, R. S., Hubbard, J. W., Korchinsky, E. D., and Midha, K. K. Enantioselective pharmacokinetics of dl-threo-methylphenidate in humans. *Pharm., Res.*, **10**, 14–21.

Otton, S. V., Crewe, H. K., Lennard, M. S., Tucker, G. T., and Woods, H. F. (1988). Use of quinidine inhibition to define the role of sparteine/debrisoquine cytochrome P-450 in metoprolol oxidation by human liver mnicrosomes. *J. Pharmacol. Exp. Ther.*, **247**, 242–247.

Otton, S. V., Lennard, M. S., Tucker, G. T., and Woods, H. F. (1989). Cumene hydroperoxide-supported oxidation of propranolol enantiomers by human liver microsomes. *Eur. J. Clin. Pharmacol.*, **36** (Suppl.), 66.

Oude Elferink, R. P., Meijer, D. K., Kuipers, F., Jansen, P. L., Groen, A. K., and Groothuis, G. M. (1995). Hepatobiliary secretion of organic compounds: Molecular mechanisms of membrane transport. *Biophys. Acta*, **1241**, 215–268.

Paar, W. D., Brockmeier, D., Hirzebruch, M., Schmidt, E. K., von Unruh, G. E., and Dengler, H. J. (1990). Pharmacokinetics of prenylamine racemate and enantiomers in man. *Arzneim-Forsch/Drug Res.*, **40** :1, 657–661.

Parke, P. V., loannides, C, and Lewis, D. F. V. (1990). The 1990 pharmaceutical manufacturers association of Canada keynote lecture. The role of the cytochrome P450 in the detoxication and activation of drugs and other chemicals. *J. Physiol. Pharmacol.*, **69**, 537–549.

Perotti, B. Y. T., Okudaira, N., Prueksaritanont, T., and Benet, L. Z. (1994). FK 506 metabolism in male and female rat liver microsomes. *Drug Metab. Dispos.*, **22**, 85–89.

Raghuram, R. C., Koshakji, R. P., Wilkinson, G. R., and Wood, A. J. J. (1984). Polymorphic ability to metabolize propranolol alters 4-hydroxypropranolol levels but not beta blockade. *Clin Pharmacol. Ther.*, **36**, 51–56.

Rettie, A. E., Korzekwa, K. R., Kunze, K. L., Lawrence, R. F., Eddy, A. C., Aoyama, T., Gelboin, H. V., Gonzalez, F. J., and Trager, W. F. (1992). Hydroxylation of warfarin by human cDNA-expressed cytochrome P450: A role for P-4502C9 in the ethiology of the (*S*)-warfarin drug interaction. *Chem. Res. Toxicol.*, **5**, 54–59.

Roepe, P. D. (1995). The role of the MDR protein in altered translocation across tumor cell membranes. *Biochim. Biophys. Acta*, **1241**, 385–406.

Rosenthal, J., Aschoff, A., Bühler, G., Hahn, G., Mutschler, E., Spahn-Langguth, H., Möhrke, W., and Völger, K. D. (1992): Pharmacokinetics and pharmacodynamics of verapamil isomers after administration of immediate- and sustained-release racemic verapamil dosage forms. *Clin. Pharmacol. Ther.*, **51**, 153.

Rowland, M. and Tozer, T. N. (1989). Clinical Pharmacokinetics. Lea & Febiger, Philadelphia, pp. 148–176.

Schwartz, J. B., Troconiz, I. F., Verotta, D., Liu, S., and Capili, H. (1993). Aging effects on stereoselective pharmacokinetics and pharmacodynamics of verapamil. *J. Pharmacol. Exp. Ther.*, **265**, 690–698.

Schwartz, J. B., Capili, H., and Wainer, I. W. (1994). Verapamil stereoisomers during racemic verapamil administration: Effects of aging and comparisons to administration of individual stereoisomers. *Clin. Pharmacol. Ther.*, **56**, 368–376.

Silber, B., Holford, N. H. G., and Riegelman, S. (1971). Stereoselective disposition and glucuronidation of propranolol in humans. *J. Pharm. Sci.*, **71**, 1043–1044.

Soldner, A., Spahn-Langguth, H., Voith, B., Stapff, M., and Mutschler, E. (1996). Losartan and EXP 3174 in rats: Concentrations in central and peripheral compartments as basis for PK/PD correlations. (Submitted for publication).

Spahn, H. (1989). Characterization of stereochemical processes in drug metabolism and pharmacokinetics. Habilitation thesis, Johann Wolfgang Goethe-University Frankfurt/Main.

Spahn, H., Henke, W., Langguth, P., Schloos, J., and Mutschler, E. (1990). Measurement of carvedilol enantiomers in human plasma and urine using *S*-naproxen for chiral derivatization. *Arch. Pharm.* (*Weinheim, Ger.*), **323**, 465–469.

Spahn-Langguth, H., Hahn, G., Mutschler, E., Möhrke, W., and Langguth, P. (1992). Enantiospecific high-performance liquid chromatographic assay with fluorescence detection for the monoamine oxidase inhibitor tranylcypromine and its applicability in pharmacokinetic studies. *J. Chromatogr.*, **584**, 229–237.

Spahn-Langguth, H., Hahn, G., Mutschler, E., Elze, M., Potthast, H., Blume, H., and Möhrke, W. (1994). Determination of the bioavailability of tranylcypromine with a stereospecific assay. *Eur. J. Pharm. Sci.*, **2**, 160.

Spahn-Langguth, H., Möhrke, W., and Langguth, P. (1994b). Variabilities and discontinuities in TCP input profiles following oral administration. *Eur. J. Pharm. Sci.*, **2**, 160.

Stahl, E., Baumgartner, U., Krauss, D. Spahn-Langguth, H., Mutschler, E., Schölmerich, J. (1990): Pharmacokinetics of carvedilol enantiomers in rats with portacaval shunt after racemate dosing. *Naunyn-Schmiedeberg's Arch Pharmacol.*, **341** (Suppl.), R4.

Stahl, E. (1993). Stereochemische Aspekte der Pharmakokinetik der lipophilen β-Adrenozeptor-Antagonisten Carvedilol und Propranolol. Ph.D. thesis, Department of Pharmacy, Johann Wolfgang Goethe-University, Frankfurt/Main.

Stahl, E., Henke, D., Mutschler, E., and Spahn-Langguth, H. (1993a). Saturable enantioselective first-pass effect for carvedilol after high racemate doses in rats. *Arch. Pharm.* (*Weinheim, Ger.*), **326**, 123–125.

Stahl, E., Baumgartner, U., Mutschler, E., and Spahn-Langguth, H. (1993b). Carvedilol stereopharmacokinetics in rats: Affinities to blood constituents and tissues. *Arch. Pharm.* (*Weinheim, Ger.*), **326**, 529–533.

Stahl, E., Baumgartner, U., Henker, D., Schölmerich, J., Mutschler, E., and Spahn-Langguth, H. (1993c). Rats with portocaval shunt as a potential pharmacokinetic model for liver cirrhosis. *Chirality*, **5**, 120–125.

Steen, H., and Meijer, D. K. F. (1991). Organic cations. *Prog. Pharmacol. Clin. Pharmacol.*, **8**(4), pp. 239–272.

Toon, S., Davidson, E. M., Garstang, F. M., Batra, H., Bowes, R. J., and Roland, M. (1988). The racemic metoprolol H_2-antagonist interaction. *Clin. Pharmacol. Ther.*, **43**, 283–289.

Tucker, G. T. and Lennard, M. S. (1990). Enantiomer specific pharmacokinetics. *Pharmacol. Ther.*, **45**, 309–329.

Vestal, R. E. (1989). Aging and determinants of hepatic drug clearance. *Hepatology*, **9**, 331–333.

Vogelgesang, B., Echizen, H., Schmidt, E., and Eichelbaum, M. (1984). Stereoselective first-pass metabolism of highly cleared drugs: Studies of the bioavailability of L- and D-verapamil examined with a stable isotope technique. *Br. J. Clin. Pharmacol.*, **18**, 733–740.

Wacher, V. J., Wu, C. Y., and Benet, L. Z. (1995). Overlapping substrate specificities and tissue distribution of cytochrome P450 3A and P-glycoprotein: Implications for drug delivery and activity in cancer chemotherapy. *Mol. Carcinog.*, **13**, 129–134.

Walle, T. and Walle, U. K. (1986). Pharmacokinetic parameters obtained with racemates. *Trends Pharmacol. Sci.*, **7**, 155–158.

Walle, T., Webb, J. G., Bagwell, E. E., Walle, U. K., Daniell, H. B., and Gaffney, T. E. (1988). Stereoselective delivery and actions of beta-receptor antagonists. *Biochem. Pharmacol.*, **37**, 115–124.

Ward, S. A., Walle, T., Walle, U. K., Wilkinson, G. R., and Branch, R. A. (1989). Propranolol's metabolism is determined by both mephenytoin and debrisoquine hydroxylase activities. *Clin. Pharmacol. Ther.*, **45**, 72–79.

Waxman, D. J., Lapenson, D. P., Aoyama, T., Gelboin, H. V., Gonzalez, F. J., and Korzekwa, K. (1992). Steroid hormone hydroxylase specificities of eleven cDNA-expressed cytochrome P450s. *Arch. Biochem. Biophys.*, **290**, 160–166.

Weber-Grandke, H., Hahn, G., Mutschler, E., Möhrke, W., Langguth, P., and Spahn-Langguth, H. (1993). The pharmacokinetics of tranycypromine enantiomers in healthy subjects after oral administration of racemic drug and the single enantiomers. *Br. J. Clin. Pharmacol.*, **36**, 363–365.

Wetterich, U., Spahn-Langguth, H., Mutschler, E., Terhaag, B., Rösch, W., Langguth, P. (1996). Evidence for intestinal secretion as an additional clearance pathway of talinolol enantiomers: Concentration- and dose-dependent absorption *in vitro* and *in vivo*. *Pharm. Res.*, **13**, 526–534.

Woodruff Modi, M., Hassett J. M., and Lalka, D. (1988). Influence of posture on hepatic perfusion and the presystemic biotransformation of propanolol. Simulation of the food effect. *Clin Pharmacol. Ther.*, **44**, 268–274.

Wosilait, W. D. and Lueck, R. H. (1988). Multifactorial modeling, drug interactions, liver damage and aging. *Gen. Pharmacol.*, **19**, 143–151.

Wu, C. Y., Benet, L. Z., Hebert, M. F., Gupta, S. K., Rowland, M., Gomez, D. Y., and Wacher V. J. (1995). Differentiation of absorption and first-pass gut and hepatic metabolism in humans: Studies with cyclosporine. *Clin Pharmacol. Ther.*, **58**, 492–497.

Wynne, H. A., Cope, L. H., and Mutch, E. (1989). The effect of age upon liver volume and apparent liver blood flow in healthy man. *Hepatology*, **9**, 297–301.

Zhou, H. H. and Wood, A. J. J. (1995). Stereoselective disposition of carvedilol is determined by CYP2D6. *Clin. Pharmacol. Ther.*, **57**, 518–520.

CHAPTER

22

GASTROINTESTINAL TRANSPORT PROCESSES: POTENTIALS FOR STEREOSELECTIVITIES AT SUBSTRATE-SPECIFIC AND NONSPECIFIC EPITHELIAL TRANSPORT SYSTEMS

PETER LANGGUTH

Astra Hässle AB, Kärragatan 5, S-43183 Mölndal, Sweden

GORDON L. AMIDON

College of Pharmacy
The University of Michigan
Ann Arbor, Michigan 48109-1065

ELKE LIPKA

TSRL Inc., Ann Arbor, MI 48108

and

HILDEGARD SPAHN-LANGGUTH

Department of Pharmacy
Martin-Luther-University Halle-Wittenberg
D-06120 Halle/Saale

22.1. INTRODUCTION

Generally, provided that the two enantiomers of a racemate differ in their effect, we may distinguish between the enantiomer with higher or the entire pharmacological activity, the eutomer, and the enantiomer with either lower or no activity or potentially undesired effects, the distomer. Enantioselective drug action of agonists and antagonists may be caused by differences in the affinity with a receptor population, characterized by different equilibrium dissociation constants, different intrinsic acitivities, or maximal effects inducible

The Impact of Stereochemistry on Drug Development and Use, Edited by Hassan Y. Aboul-Enein and Irving W. Wainer. Chemical Analysis Series, Vol. 142.
ISBN 0-471-59644-2

by the ligand, or interaction of the two enantiomers with different receptor types. Similar to chromatography, a three-point attachment at a receptor area may be hypothesized for the enantiomer exhibiting higher affinity. The enantiomer difference in affinity was frequently found to be greater the higher the affinity of the more active enantiomer was (the so-called "Pfeiffer's rule") (Pfeiffer, 1956; Lien et al., 1982). This is evident on the basis of the three-point attachment, since the loss of one binding point means the considerable loss of binding forces.

We may hypothesize that this relationship is also of importance when carrier-mediated epithelial transport processes are involved. As may be derived from the available kinetic data, lumen-to-blood directed transporters appear to exhibit high substrate selectivity as well as stereoselectivity, whereas for enterocyte-to-lumen directed ("outside"-directed) transport processes mediated by multidrug transporters [e.g., P-glycoprotein (P-gp), mdr1], stereochemical aspects appear to be of inferior relevance.

Preferentially, for most drugs, uptake into the enterocyte and the blood occurs via passive diffusion, the most prominent process in drug absorption.

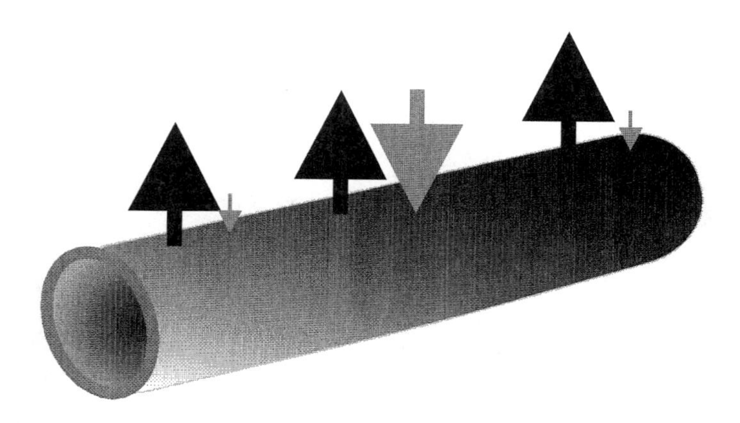

Figure 22.1. A drug passing through the intestinal lumen at variable transit-related velocities may be subject to various processes: uptake into the enterocytes and subsequent entrance into the blood vessels and exsorption (intestinal secretion) from the enterocyte back into the luminal space. For compounds exhibiting carrier-mediated uptake and/or secretion, it may be relevant that the amount of transporter can vary from site to site. For example, significant intestinal secretion from the enterocyte to the lumen may be the predominant process in one intestinal segment, while later the secreted (exsorbed) drug reenters the gut wall. As a consequence, discontinuous input profiles may result. For carrier-mediated transports, which are saturable, the concentration at the absorption site codetermines the relevance of the process. With respect to a racemically adminstered drug, its enantiomers may have different affinities to transporters. Moreover, intestinal content affects drug transport.

Dependent on the site-specific relative contribution of these processes, the net amount of drug reaching the systemic circulation is the sum of all involved "inside"- and "outside"-directed processes (Figure 22.1).

In addition, metabolic processes within the enterocyte (and/or the hepatocyte) may limit bioavailability. Moreover, phase I- or phase II metabolites may be substrates for the multidrug transporters present in, for example, the intestine, liver, and kidneys.

With respect to carrier-mediated processes, the interaction of eutomer and distomer is feasible. From the data obtained with talinolol, a substrate for intestinal secretion with a slight preference for the levorotatory eutomer, we may hypothesize that at low concentrations at the absorption site, the distomer enhances the amount of eutomer reaching the systemic circulation by the competitive inhibition of P-gp.

22.2. FACTORS THAT INFLUENCE THE EXTENT AND VELOCITY OF DRUG UPTAKE AFTER ORAL DOSING

When a drug racemate is adminstered, various processes within the gastrointestinal tract may affect the rate and extent of drug input into the systemic circulation, some with the potential for stereoselectivity, due to the involvement of chiral macromolecules, and some without this potential.

The motility-dependent transit through the G.I. tract should be similar for both enantiomers of the racemate and so should be processes based on the G.I. pH gradient across the stomach/gut wall. For various other factors, chiral partners are involved, such as excipients for the release from the formulation (Duddu et al., 1993), proteins, carbohydrates, etc., for the binding to food constituents (Schneider et al., 1996), (endogenous and bacterial) enzymes for intraluminal degradation, and mucines with respect to binding to the mucus layer. This is supplemented by uptake mechanisms into the epithelial cells (passage through the mucosal epithelium), where passive and vesicular transports are assumed to yield no stereoselectivities in spite of the chiral nature of the membrane constitutents (Fassihi, 1993), whereas carrier-mediated uptake may be stereospecific or stereoselective (Ott and Giacomini, 1993) (Figure 22.2).

The flow of intestinal content (including the dissolved drug or the respective formulation) through the gastrointestinal lumen determines the transport velocity to the absorption or uptake site as well as the dilution factor. A particular drug concentration is yielded, which may (co-)determine the velocity of drug (enantiomer) uptake from the respective intraluminal space. Within the enterocyte and, subsequently, in the major eliminating organs, stereoselective metabolism and excretion are feasible.

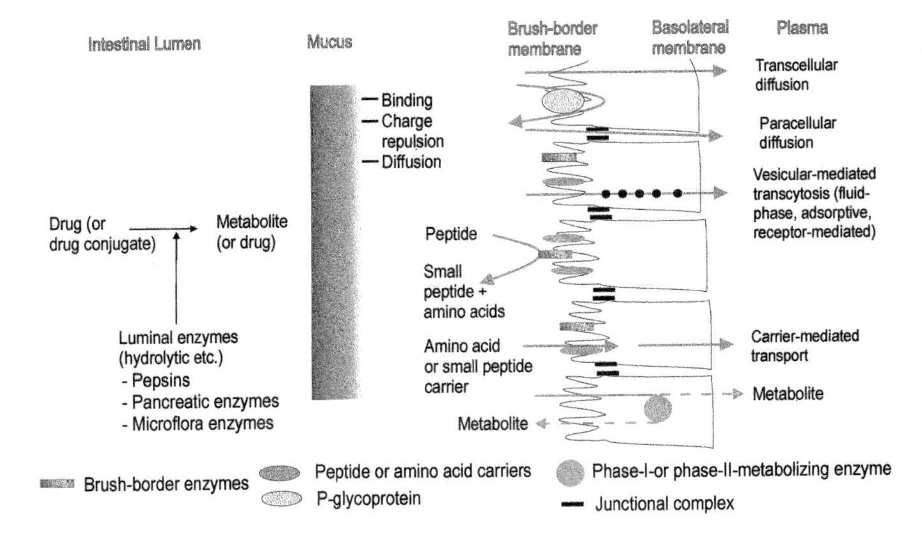

Figure 22.2. Uptake of drug enantiomers from the intestinal lumen to blood and processes potentially affecting the rate and extent of absorption: A drug (enantiomer) may be affected by luminal enzymes, has to pass through the mucus layer, and enters the systemic circulation via passive diffusion, carrier-mediated uptake, or other mechanisms. It may also undergo intestinal secretion (exsorption) through membrane-bound multidrug transporters and be metabolized via phase-I and phase-II pathways within the enterocyte. Metabolites may also enter the systemic circulation via diffusion or be substrates for the respective transporters. (An extended overview on carrier-mediated brush-border membrane and basolateral transport mechanisms is provided by Tsuji and Tamai (1996).)

22.3. KINETIC CHARACTERISTICS OF TRANSPORT AND METABOLIC TURNOVER PROCESSES

Although membranes need to be passed during the absorption process and membranes are composed of chiral elements, it is generally assumed that stereoselectivity "is not an issue with drugs that are passively absorbed by lipid diffusion" (Fassihi, 1993), whereas differences between enantiomers are feasible when active transport mechanisms are involved.

In most kinetic approaches, we assume that processes are not saturable in the therapeutic range. We know that this is a simplification, which is, however, permitted even when saturable processes occur, yet only as long as the saturable process is of inferior impact and, for example, the dose/AUC relationship remains linear.

With respect to the potential for transport enantioselectivities during the passage through the (gastro-)intestinal wall, hepatic uptake as well as other presystemic processes, secretion into the intestinal lumen (in addition to carrier-mediated absorption), carrier-mediated uptake into the hepatocytes, and biliary secretion represent processes where a chiral partner is involved. Such processes are saturable, because a limited number of binding sites, and, hence, a maximum turnover rate exist. Furthermore, the affinity of various ligands (drugs and metabolites) may be different. Usually, simple active transport and metabolic processes appear to be characterized by Michaelis–Menten-type kinetics. Hence, for each of the enantiomers, the rate of transport (t) or metabolism (m) follows the relationship:

$$\text{Rate } (t \text{ or } m) = (V_t \times C_u) / (K_t + C_u) \qquad \text{or} \qquad (V_m \times C_u) / (K_m + C_u)$$

in which C_u is the unbound drug concentration, V is the enzyme- or carrier-concentration-dependent maximum rate, and K represents the Michealis–Menten constant (C_u at which $V/2$ is observed), an inverse function of the affinity between the ligand (drug) and enzyme or transport protein. With respect to intestinal transports, the absorption rate for inside-directed transport processes adds to absorption via passive diffusion, whereas outside-directed transport (intestinal secretion) reduces the amount absorbed per unit of time. For both situations, a linear relationship between concentration (dose) and rate of absorption is obtained when the respective transporter operates at maximum velocity.

When a racemate is dosed, the two enantiomers may—in similarity to their interaction with receptors—exhibit similar or different affinities to the respective binding site(s). When both enantiomers are transported via the same macromolecule, they may compete at the binding site. For such compounds, a competition model may be included in data analysis, and potentially the competition characteristics may be determined experimentally.

22.4. GASTROINTESTINAL TRANSIT AND ABSORPTION OF DRUGS FROM THE GASTROINTESTINAL TRACT

22.4.1. G.I. Transit and Double Peaks

Several research articles on drug absorption have focused on the impact of gastrointestinal physiology on the absorption process and the overall kinetic profile. This mainly includes gastric emptying (Oberle and Amidon, 1987; Langguth et al., 1994) as well as the site dependence of absorption (Spahn-Langguth et al., 1993), complexation within the gut (Lennernäs and Regardh, 1993c), and secretion processes into the gut (e.g., Klaassen, 1988; Meijer and van der Sluijs, 1989; Langguth et al., 1990; Oude Elferink et al., 1995). An interesting observation made throughout the last decade is the phenomenon of discontinuous drug input with the occurrence of "unusual" input profiles and double peaks and shoulders in the concentration-time profiles, with the achiral cimetidine as a well-known example (Funaki et al., 1986; Langguth et al., 1994) and as reported for, for example, penicillamine (Bergström et al., 1981), pafenolol (Regardh et al., 1990; Lennernäs and Regardh, 1993a), or celiprolol (Hartmann et al., 1990). Potential explanations for such discontinuities are enterohepatic circulation, interruption of gastric emptying, pH-dependent solubilities, and site-dependent permeabilities.

The simulation studies as well as various experimental investigations—for example, the recent studies with the β-adrenoceptor antagonist celiprolol in dogs—demonstrated unambiguously that with p.o. dosing of celiprolol, the occurrence of double peaks and shoulders in concentration-time profiles is correlated with the phase of high motility (Langguth et al., 1994; Lipka et al., 1995). Of the generally feasible, potential causes to explain double peaks in profiles, those for which no stereoselectivity is anticipated due to their entirely physicochemical nature are pH-dependent solubility and different permeabilities in various regions of the G.I. tract.

Since these solubility- or transit-related processes occur prior to membrane passage, they usually are not assumed to affect the stereoselectivity of the blood concentration-time profiles. In dog studies with celiprolol, a compound with only minor differences between the concentration-time profiles of its enantiomers, there was no apparent relationship between input velocity (lag-time, input rate constant) and *S*/*R* ratios (Lipka et al., 1995). However, for talinolol there was a tendency for smaller *S*/*R* ratios at low plasma concentrations during the drug input period in a study in humans.

Discontinuous input profiles were also detected for tranylcypromine, a compound exhibiting very high metabolic clearance. Moreover, *rac*-tranylcypromine represents a compound showing significant stereoselectivities in its concentration-time profiles, which are hypothesized to be entirely of a metabolic nature (Spahn-Langguth et al,. 1994 a,b; Langguth et al., submitted, 1997). It is one of the few compounds in the group of antidepressive agents that act through an inhibition of monoamine oxidases (MAO). Since it has always been used as a racemate, different groups investigated possible pharmacological differences between the enantiomers in various species showing that the enantiomers have different pharmacological characteristics (Hampson et al., 1986), with the 1*S*,2*R*(+)-isomer being more potent at inhibiting MAO-A and MAO-B *in vivo* and at elevating brain levels of phenylethyl amine and serotonine, and the 1*R*,2*S*-(–)-isomer being more potent in inhibiting the uptake of noradrenaline, dopamine, and serotonine in brain slices. From the respective pharmacokinetic data (Weber-Grandke et al., 1993) that include low renal clearances and apparently high metabolic clearances, it was hypothesized that tranylcypromine may undergo a significant and stereoselective first-pass extraction upon p.o. dosage in humans. However, this hypothesis was not proven because of the lack of a parenteral dosage form approved for use in clinical studies.

Although we do not yet know for these compounds whether first-pass extraction occurs solely in the liver or in both gut wall as well as liver and whether in addition other factors contribute to the stereoselectivity, we may conclude that the respective extraction ratios are different for the two enantiomers. Hence, also different variabilities were to be expected, where the more efficiently extracted (+)-enantiomer should show a greater variability in bioavailability than its optical antipode. Another source of variability should be gastrointestinal motility, which may be the causative factor for the observed discontinuities in the profiles of tranylcypromine (Figure 22.3) (Spahn-Langguth et al., 1994b).

The relative bioavailability of tranylcypromine isomers following the oral dosage of three different formulations, one solution and two solid dosage forms, was estimated in a randomized cross-over trial in 18 volunteers. Although the input and disposition characteristics are enantioselective as shown by an almost ten-fold larger AUC of the levorotatory enantiomer, the bioavailability parameters (AUC and C_{max}) ratios were not different between enantiomers. Interestingly, the relative bioavailability of tranylcypromine was

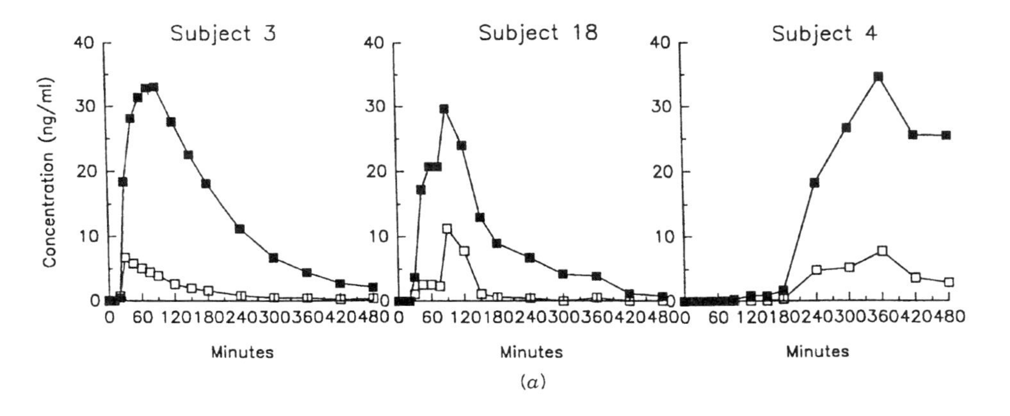

Figure 22.3(a). Typical plasma concentration-time profiles as observed in a clinical study following p.o. dosing of 20 mg of tranylcypromine sulfate (13.7-mg free base), with higher concentrations observed for the levorotatory enantiomer. Subject 1 shows a profile without discontinuities; the profile of subject 18 is characterized by two concentration maxima in the profiles, with approximately 50% of the dose absorbed as the first fraction, whereas subject 4 shows a delayed drug input with a late overall concentration maximum and a small fraction of the dose absorbed initially.

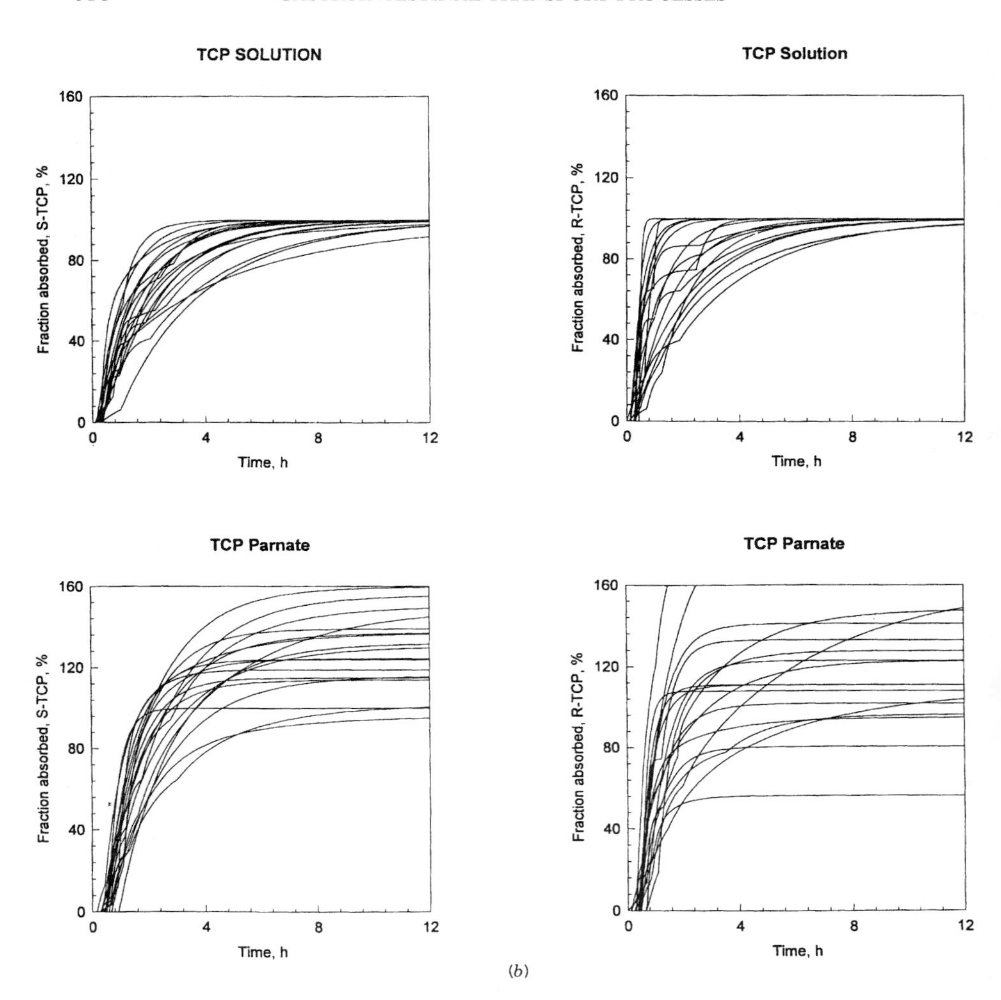

Figure 22.3(b). Input profiles (cumulative amounts) as observed for tranylcypromine enantiomers (S, left; R, right) following peroral administration of the solution and one of the solid dosage forms (parnate) (dose: 20 mg of the racemic compound). The solution serves as a reference (100%). The variability was considerably higher for the R-enantiomer, which was present in plasma in a lower concentration range than its optical antipode. Interestingly, the average relative bioavailability of the solid dosage form exceeded that of the solution.

higher for both solid dosage forms, yielding approximately 30% higher values. Concentration-dependent uptake into the systemic circulation is a possible explanation, since gut metabolism and intestinal secretion appear to be readily saturable processes.

22.4.2. Passive Diffusion and Carrier-Mediated Uptake Mechanisms

Differences in the rates and extents of absorption may be observed for compounds with affinity to amino acid or peptide carriers or other transport systems located in the gut wall. With respect to amino-acid- and peptide-type drugs, which are mostly used in a stereochemically pure form, stereoselective transport is assumed in many cases, yet not directly tested (Taylor and Amidon, 1995).

A well-known example for a drug undergoing stereoselective active transport in the gut is DOPA, where the L-enantiomer enters the systemic circulation via active transport and the respective D-enantiomer via passive diffusion with differences between enantiomers in the rate, yet not in the extent of absorption (Wade et al., 1971). Comparable behavior is assumed for the structurally related α-methyldopa, and its permeability in humans shows the expected pH and concentration dependence (Merfeld et al., 1986). Also 5-fluorouracil and methotrexate are structural analogs of endogenous compounds and are believed to be transported via the respective carrier systems (Rowland and Tozer, 1989; Hendel and Brodthagen, 1984). A carrier-mediated absorption process also has been observed for aminocephalosporins such as cefadroxil and cephradine (Sezaki and Kimura, 1983; Yuasa et al., 1993, 1994). These compounds showed saturable and inhibitable transport into the blood, which was also inhibited by dipeptides, suggesting that drugs may have some affinity for the dipeptide transport system in the small intestine. Furthermore, data obtained with ceftibuten (Yoshikawa et al., 1989) indicate that such carrier mechanisms show some limited stereoselectivity.

A comprehensive overview, including some stereochemical aspects, of the mechanisms involved in the intestinal absorption of different types of antibiotics and the respective experimental approaches, structure-absorption relationships, and interactions with other compounds was provided by Tsuji (1995), and in a summary of the oral absorption of ACE inhibitors and peptide prodrugs by Yee and Amidon (1995).

Extensive studies on the bioavailabilities of the enantiomers of baclofen as well as its fluoro analog were carried out in our laboratories. For baclofen, a centrally acting antispastic agent with a direct mimetic effect on $GABA_B$ receptors as the major mechanism of action, the *R*-enantiomer was found to be the eutomer with a eudismic ratio of approximately 100 (Olpe et al., 1982). Although no relevant enantiomer differences were detected for the disposition parameters (CL: 152 mL/min for *S*, 139 mL/min for *R*, CL_R: 113 for *S*, 112 for *R*, V_{ss}: 72 mL/min for *R* and *S*), significantly higher *R*-enantiomer levels were detected in plasma and urine during the initial period when baclofen was dosed orally. The individual *R*/*S*-ratios ranged from 1.6–2.2 for the 0 to 2-h collection period and from 1.1–1.3 for the 2 to 4-h fraction. Similar results were achieved for the respective fluoro analog. The renal clearances of both enantiomers approached the glomerular filtration rate, which indicates that neither active secretion nor significant tubular reabsorption occur and contribute to the overall stereoselectivity of the plasma concentration-time profile. For both compounds, the absolute bioavailability was fairly high and slightly stereoselective. Our studies with rats and with the rat everted-sac absorption model (Ader et al., in preparation) appear to confirm our hypothesis that the levorotatory *R*-enantiomer is more quickly absorbed from the G.I. tract, most probably via an active transport mechanism. The data published by Merino et al. (1989) also indicate that the mechanism of absorption of baclofen is not passive diffusion.

22.4.3. "Inverse" Transport Processes into the Gut, Intestinal Secretion

So far, no or little attention has been drawn to the relevance and stereoselectivity of inverse transport (counter-transport) mechanisms, the driving force of which may be a considerable pH gradient (with no expected stereoselectivities) or an energy-dependent carrier.

In the case of pafenolol (Lennernäs and Regardh, 1993b), intestinal secretion was found to account for approximately 25% of an i.v. dose in rat studies. However, except for talinolol, which is mentioned below, the stereochemical aspects of this "inverse" transport mechanism have not yet been investigated.

Although the mechanism of intestinal secretion has not been fully elucidated yet, more evidence is evolving that suggests the secretory process is mainly maintained by P-glycoprotein (P-gp), a 170-kD plasma membrane protein in the apical membranes of enterocytes (Hunter et al., 1993) (Figure 22.4). In tumor cells, P-gp functions as an ATP-dependent efflux pump for many anticancer drugs and other compounds that are hydrophobic cationic to neutral compounds with molecular weights of 400–1000 (Endicott and Ling, 1989; Gottesman, 1993; Gottesman and Pastan, 1993) and is known as the protein causing multidrug resistance in cancer chemotherapy. Recent observations on the intestinal absorption mechanism of drugs with peptide and nonpeptide structure suggest that the bioavailability of some drugs on peroral administration may be limited by the intestinal secretion process that is mediated by P-gp (Bohner et al., 1995), which may secrete drugs, during their absorption, back into the intestinal lumen at low drug concentrations. At higher doses (concentrations), the secretion may be saturated, thus leading to dose-dependent bioavailability upon oral administration. It is suspected today that P-gp is expressed in several epithelia, such as those of the gastrointestinal tract, liver, pancreas, kidney, reproductive organs, and in the endothelial cells of the brain, testes, and adrenal glands. Immunocytochemistry techniques with monoclonal antibodies to human P-gp have revealed its localization in the plasma membranes of several normal human and animal tissues as well, such

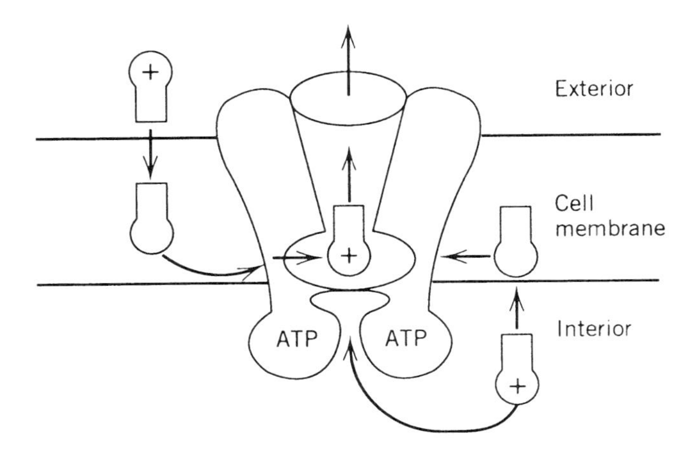

Figure 22.4. P-glycoprotein is a membrane-bound carrier-protein (functioning as a "lipophilic vacuum cleaner"). It consists of 1280 amino acids and is characterized by 12 transmembrane domains.

as the liver, the small and large intestine, and the brain capillaries (Thiebault et al., 1987, Wacher et al., 1995).

One reason for its existence in the intestine may be the protection against chemical carcinogens, which are abundant in plant and animal components of our diet. The interference of P-gp with the absorption of several drugs may hence seem preprogrammed, and further knowledge on this interference may help in explaining some of the phenomena and peculiarities seen after oral drug administration. With respect to research in overcoming multidrug resistance toward anticancer drugs, for example, quinidine and verapamil racemate and enantiomers have been investigated as transport inhibitors. All verapamil species were found to exhibit a comparable effect with respect to the inhibition of P-gp and to apparently lead to a similar increase of intracellular cytostatic drug concentrations, yet with fairly high verapamil doses (Plumb et al., 1990).

When studying P-gp-mediated intestinal transport processes, human colon carcinoma cells (Caco-2 cells), a well-established cell line for drug transport studies through the intestinal epithelium, may be used. The aim of recent investigations carried out by our group was to localize and quantify the expression of P-gp in Caco-2 cells. Confluently grown cell monolayers were trypsinized and suspended. P-gp was either directly labeled with an internal monoclonal antibody or indirectly with an external one in combination with FITC-marked antiserum. The expression of P-gp was quantified by FACScan sorting (Anderle et al., 1996). It was shown with both antibody types that P-gp was expressed in all examined passages of the Caco-2 cells. The incidence of the protein expression was higher at large passage numbers (58–80) than small ones (21–48). In addition, there was an influence of growth phase on the amount of P-gp. Furthermore, P-pg was detected in varying amounts and with different substrate specificities in different regions of the intestinum, an observation that may explain the presence of "absorption windows" in the G.I. tract.

22.4.4. Evidence for Intestinal Secretion with Moderate Stereoselectivities in the Case of Talinolol and Celiprolol

Celiprolol and talinolol, two compounds with discontinuous input profiles assumed to be, at least in part, due to different permeabilities in various gut regions (Gramatté, 1995), are suitable examples of drugs undergoing a different type of first-pass effect following p.o. administration, which is of neither metabolic nor biliary nature. There is evidence for both of these compounds, and also for various other chiral and achiral drugs, that active intestinal secretion occurs in different mammalian species (Karlsson et al., 1993; Kuo et al., 1994; Gramatté, 1995; Wetterich et al., 1996). For these compounds, dose-dependent absorption was found in *in vivo* studies.

The overall purpose of detailed studies with *talinolol* (Wetterich et al., 1996) was the evaluation of the *evidence for carrier-mediated intestinal secretion* of talinolol enantiomers *in vivo* in humans and *in vitro* and their relevance and included the quantification of accessible *total and partial clearances* (renal, biliary, residual), in order to elucidate to what extent intestinal secretion may contribute to the overall kinetic profiles of the two talinolol enantiomers. A potential *in vivo saturability* (= *concentration/dose dependence*) of the hypothesized intestinal secretion process should result in the nonlinearity of the relationship between dose normalized AUC as well as apparent oral clearance and the administered dose. Another part of the studies included the proof that in a cell culture model with (human) Caco-2 cells, *in vitro permeability* is also:

Concentration-dependent and saturable
Basolateral-to-apical directed
Inhibited by verapamil, a P-glycoprotein inhibitor

In the course of these studies, CL and CL/F were determined in a group of six cholecystectomized patients who received either 30 mg of *rac*-talinolol as an i.v. short infusion or 100 mg orally postsurgery. Total plasma clearance was always slightly higher for *S*(–)- than *R*(+)-talinolol (CL, *S*: 321, *R*: 269 mL/min, amounting to 44 and 31% of CL, respectively), with a tendency for

Table 22.1. Talinolol Enantiomer Clearances in Humans [total, apparent oral and partial clearances obtained for talinolol enantiomers after i.v. and p.o. dosing of 30 or 100 mg of *rac*-talinolol (15 and 50 mg of each enantiomer), respectively, in cholecystectomized patients ($n = 3$ for each group, arithmetical means ± SD)]

		i.v. Treatment	p.o. Treatment
CL, CL/F (min/min)	*S*(–)	321 ± 65	685 ± 221
	R(+)	269 ± 31	616 ± 181
CL_R (mL/min)	*S*(–)	153 ± 15	160 ± 11
	R(+)	163 ± 5	154 ± 14
CL_B, CL_{Bapp} (mL/min)	*S*(–)	28 ± 12	55 ± 30
	R(+)	23 ± 10	36 ± 16
$CL_{residual}$ (mL/min)	*S*(–)	140 ± 84	471 ± 207
	R(+)	84 ± 43	426 ± 182

CL = total plasma clearance.
CL/F = apparent oral clearance.
CL_R = renal clearance.
CL_B = biliary clearance.
$CL_{residual}$ = clearance not explained by renal and biliary excretion

the difference to be higher after the oral dose. Approximately one-third of the total clearance was not explained by renal and biliary excretion as summarized in Table 22.1. The slightly stereoselective residual clearances of 140 for *S*- and 84 for *R*-talinolol should not be explained by metabolic processes, since only 1% of the dose were found to undergo biotransformation. If the residual clearance is mainly explained by intestinal secretion, then it might be feasible due to the saturability of this transport mechanism to detect the dose dependence of the apparent oral clearance, CL/F, when increasing p.o. doses are administered. In fact, following oral dosage of 25–400 mg of *rac*-talinolol in a randomized cross-over trial, the dose-normalized AUC increased with increasing dose, whereas absorption rates as well as CL/F decreased. Average CL/F values amounted to 1062 (*S*) and 830 mL/min (*R*) for the lowest and 498 (*S*) and 452 mL/min (*R*) for the highest dose, with the respective average *S*/*R* ratios of 1.3:1.1 as depicted in Figure 22.5. This indicates a lack of stereoselectivity with higher doses. Permeation studies through Caco-2 cell monolayers supported the hypothesis of intestinal secretion, since lower apical-to-basolateral (a→b) transport direction, increased permeabilities, and transport inhibiton by verapamil were clearly shown. The addition of verapamil led to a significant increase in the amount of talinolol reaching the basolateral side and now negligible differences between a→b and b→a transport rates. *In vitro*, overall permeability was smaller for the *S*-enantiomer in

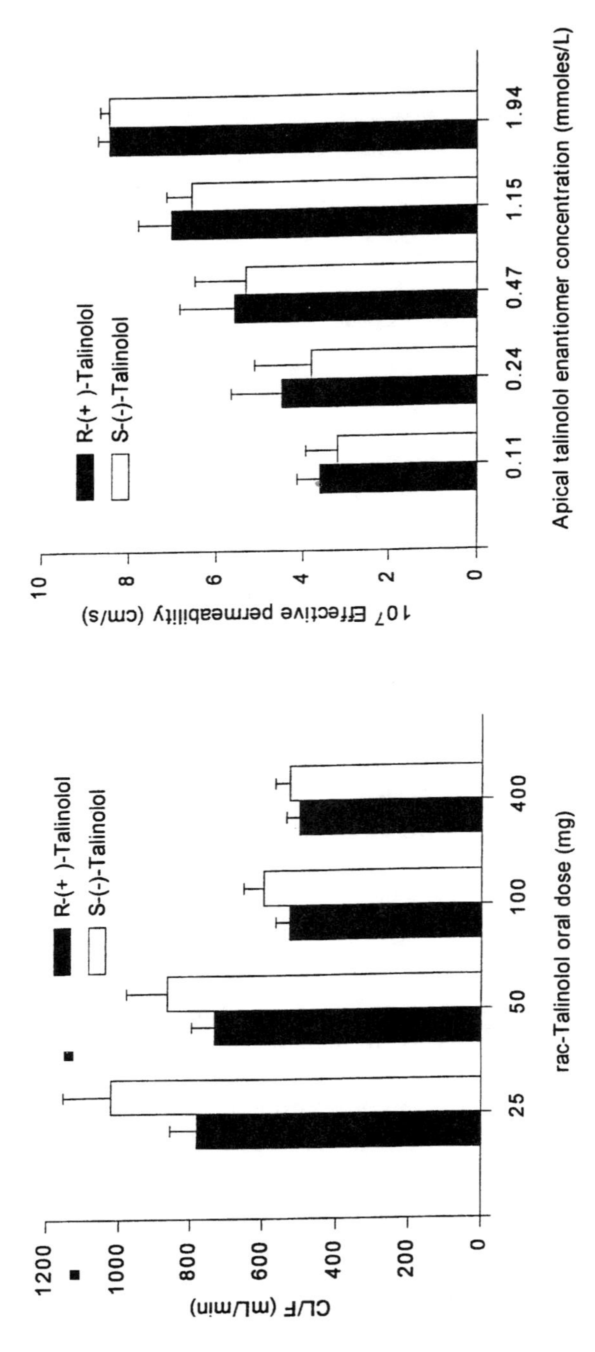

Figure 22.5. Dose-dependent oral clearance of talinolol enantiomers as obtained in a clinical study (Wetterich et al., 1996) and concentration-dependent effective permeability of talinolol enantiomers through Caco-2 cell monolayers. The *S*-enantiomer appears to exhibit slightly higher affinity to the transporter than the *R*-enantiomer (Wetterich et al., 1996).

the lower concentration range, whereas the relative difference became negligible for higher talinolol concentrations.

Both *in vivo* and *in vitro* data support the hypothesis that *S*-talinolol is a slightly better substrate for P-gp. Hence, net absorption into the systemic circulation is smaller, and plasma concentrations of the *R*-enantiomer are higher.

For the second example, celiprolol, no human i.v. data are available. However, the permeability data from Caco-2 cell studies of Karlsson et al. (1993), who did not differentiate between the two enantiomers, and the results obtained after different oral racemate doses in healthy volunteers and patients with reduced renal function, do not contradict the hypothesis that intestinal secretion plays a role in the kinetics of celiprolol. As for talinolol, a slight preference for the *R*-enantiomer was detected in plasma, yet the *S*/*R* ratios changed with time in most of the volunteers included in the studies. Usually, slightly higher *R*-celiprolol concentrations were also found in bioavailability and perfusion studies in dogs (Lipka et al., submitted for publication). Additionally, effective permeability was found to depend on the perfusate concentration and increase with higher doses (Table 22.2). At the same time, a tendency for a change in the *S*/*R* ratio from below unity at lower concentrations to above 1 for higher concentrations was noted. Double peaks and other

Table 22.2. Intestinal Perfusions with β-Adrenoceptor Antagonists with Different Lipophilicities [intestinal permeability data (effective permeabilities in 10^4 cm/sec; estimated employing a mixed tank model and given as arithmetical means ± SEM) obtained for the enantiomers of celiprolol, atenolol, and propranolol in dogs when perfusing intestinal segments with perfusate containing the respective racemic compound (Lipka et al., 1996)]

Perfusate concentration [mM] (Racemate concentration)		Effective Permeability * 10^4 (cm/sec)	
		S-Enantiomer	*R*-Enantiomer
Celiprolol	0.18	1.90 ± 0.30	1.48 ± 0.24
	0.36	2.10 ± 0.34	1.94 ± 0.35
	0.73	1.92 ± 0.22	1.79 ± 0.21
	1.45	3.24 ± 0.18	3.55 ± 0.36
Atenolol	0.19	0.59 ± 0.08	0.53 ± 0.08
	0.38	0.76 ± 0.17	0.72 ± 0.14
Propranolol	0.31	7.49 ± 1.14	7.82 ± 1.48
	0.62	13.82 ± 2.28	15.74 ± 3.24

The permeability differences between the various perfusate concentrations were statistically significant when comparing the respective lowest and highest concentration levels.

discontinuities in the input profiles were found to correlate with phases of fasted-state gastric motility in dogs (Lipka et al., 1995). However, when volunteers received the drug in a fasted versus fed state, the frequency of discontinuities was unaffected, indicating that "site-specific absorption" is also important. In addition to nonlinearities and discontinuities, which may be interpreted as related to intestinal secretion processes, another supporting observation was the higher t_{max} observed with lower celiprolol doses.

Current investigations performed with the largely unmetabolized talinolol and celiprolol as well as pindolol include the detection of drug enantiomers in the gut lumen following an i.v. dose and the influence of P-gp inhibitors on absolute bioavailability.

22.5. FORMATION OF P-GP SUBSTRATES VIA PHASE I- AND PHASE II METABOLIC STEPS

Interesting results were obtained in an extended preclinical study with the achiral angiotensin-II antagonist losartan and its active metabolite EXP 3174 in rats. The AUCs of both compounds were significantly lower in almost all tissues following a multiple dose than after a single dose, whereas the concentrations in the intestinal contents were higher. It may be hypothesized—and this currently is being investigated by our group—that this behavior is explained by decreased input rates possibly due to increased intestinal secretion of (drug and) metabolite—potentially generated through gut wall metabolism via CYP3A4—with chronic dosage (Yun et al., 1995; Soldner et al., 1996). Since the drug/metabolite AUC ratio remains unaffected or increases, the induction of metabolic enzymes in the gut wall does not appear to play a role in this particular case.

22.6. INTESTINAL VS. BILIARY SECRETION: SIMILAR SUBSTRATE AND ENANTIOMER SELECTIVITIES?

Noting the variety of carriers involved in blood-to-bile transport—including multidrug transporters for hepatocyte-to-bile transport—in addition to hypothesized passive diffusion, we see that it is possible the stereochemical aspects are of importance with respect to the biliary excretion of unmetabolized compounds or the unmetabolized residual. Data about the biliary excretion of unmetabolized drug enantiomers are scarce, whereas the biliary excretion of drug enantiomer or metabolite conjugates was investigated for several drugs. As summarized by Jamali et al. (1989), the stereoselectivity of biliary excretion may be explained mostly by stereoisomer differences in plasma or blood concentrations (e.g., acenocoumarol, various ester glucuronides of NSAIDs). This implies that biliary clearance is nonstereoselective for these compounds. We studied the biliary excretion of the β-adrenoceptor antagonists, celiprolol and talinolol, in humans (Wetterich et al., 1996) and propranolol and carvedilol in rats (Stahl, 1993; Stahl et al., 1993). The data showed considerable differences when comparing the highly metabolized carvedilol and propranolol with the poorly metabolized celiprolol and talinolol. At the therapeutic dose level, there was a slight preference for excretion of the *S*-enantiomer of these respective β-blockers with very low biliary clearances for parent carvedilol and propranolol, but a considerable contribution to total clearance for the largely unmetabolized talinolol and celiprolol.

The overall contribution of secretion into bile to stereoselectivities in blood concentration-time profiles appears to be of little relevance.

Interestingly, the unmetabolized celiprolol and talinolol are both substrates for carrier systems in the intestine, liver, and kidneys. However, the available data indicate that only little stereoselectivity occurs at intestinal, hepatic, and renal multidrug transporters, at least for these compounds.

22.7. COMPETITION OF DRUG ENANTIOMERS FOR INTESTINAL SECRETION: A BIOAVAILABILITY-ENHANCING FACTOR?

Whenever carrier-mediated processes are involved, the potential for interactions through competitive or noncompetitive inhibition exists. With respect to carrier-mediated absorption, competition would result in decreased bioavailablity for the actively transported molecule. That is, when the eutomer and distomer are both substrates for the carrier, the distomer would reduce eutomer transport when it is present as well.

Also, for a drug that is a substrate for intestinal secretion via P-gp, the potential for interactions with food constituents, excipients, and other drugs exists. An interaction will lead to increased blood concentrations of the drug and increased systemic availability, since extensive intestinal secretion might reduce the amount of drug reaching the systemic circulation and an inhibition of secretion would increase it.

The results obtained from the studies with talinolol yielded concentration-dependent absorption as well as an only slightly higher intestinal secretion and, hence, almost comparable affinity of the enantiomers to the transporter. On the basis of these data, it can be hypothesized that the *R*-enantiomer of talinolol, which does not contribute to the β-adrenoceptor blocking effect to a significant extent, enhances the absorbed fraction of the dose for the true β-antagonist *S*-talinolol, particularly in the concentration range where the “inverse” transport rate is not yet maximal.

Further investigations of substrate- and stereoselectivities at multidrug transporters in the G.I. tract will include an evaluation of the structure-affinity relationship, since it is assumed that higher stereoselectivities may occur for high-affinity P-gp substrates.

In general, potential implications of secretion via the enterocyte for the development of sustained-release as opposed to immediate-release products might be lower systemic availabilities with increasing drug release velocities for compounds undergoing saturable carrier-mediated uptake and higher systemic availabilities with increasing release velocities for compounds that are subject to intestinal secretion (Amidon et al., 1995).

ACKNOWLEDGMENTS

The preparation of this article and included experimental work that was performed in our own laboratories, were supported by grants of the Swiss National Science Foundation, the KTI (Commission of Technology and Innovation, Bern), Dr. Robert-Pfleger-Stiftung Bamberg, as well as the Fonds der Chemischen Industrie.

REFERENCES

Ader, P., Langguth, P., and Spahn-Langguth, H. (1997). Stereoselective uptake of baclofen enantiomers in the gut: Studies with the rat everted-sac model. *Arch. Pharm.* (*Weinheim*) (submitted).

Amidon, G., Lennernäs, H., Shah, V. P., and Crison, J. R. (1995). A theoretical basis for a biopharmaceutic drug classification: The correlation of in vitro drug product dissolution and in vivo bioavailability. *Pharm. Res.*, **12**, 413–420.

Anderle, P., Niederer, E., Spahn-Langguth, H., Merkle, H. P., and Langguth, P. (1996). Expression of bioavailability-limiting ATP-dependent efflux pump in intestinal epithelium. *Naunyn Schmiedeberg's Arch. Pharmacol.*, **353** (Suppl.), R153.

Bergström, R. F., Kay, D. R., and Wagner, J. G. (1981). The pharmacokinetics of penicillamine in a female mongrel dog. *J. Pharmacokinet. Biopharm.*, **9**, 603–621.

Bohner, V., Rutishauser. B., Pernard, J. C., Ottinger, C., Wunderli-Allenspach, H., Merkle, H. P., and Langguth, P. (1995). *Proc. Intern. Symp. Control Release Bioact. Mater.*, p. 540, Seattle.

Duddu, S. P., Vakilynejad, M., Jamali, F., and Grant, D. J. (1993). Stereoselective dissolution of propranolol hydrochloride from hydroxypropyl methylcellulose matrices. *Pharm. Res.*, **10**, 1648–1653.

Endicott, J. A. and Ling, V. (1989). The biochemistry of P-glycoprotein-mediated multidrug resistance. *Annu. Rev. Biochem.*, **58**, 137–171.

Fassihi, A. R. (1993). Racemates and enantiomers in drug development. *Int. J. Pharm.*, **92**, 1–14.

Funaki, T., Furuta, S., and Kaneniwa, N. (1986). Discontinuous absorption property of cimetidine. *Int. J. Pharm.*, **31**, 119–123.

Gottesman, M. M. (1993). How cancer cells evade chemotherapy: Sixteenth Richard and Hinda Rosenthal Foundation Award Lecture. *Cancer Res.*, **53**, 747–754.

Gottesman, M. M. and Pastan, I. (1993) Biochemistry of multidrug resistance mediated by the multidrug transporter. *Ann. Rev. Biochem.*, **62**, 385–427.

Gramatté, T. (1995). Small intestinal secretion of the β-blocker talinolol: Direct demonstration is possible by perfusion studies in man. *Naunyn Schmiedeberg's Arch. Pharmacol.*, **351** (Suppl.), R1.

Hampson, D. R., Baker, G. B., and Coutts, R. T. (1986). A comparison of the neurochemical properies of the stereoisomers of tranylcypromine in the central nervous system. *Cell. Mol. Biol.*, **32**, 593–599.

Hartmann, C., Frölich, M., Krauss, D., Spahn, H., Knauf, H., Mutschler, E. (1990). Comparative enantioselective pharmacokinetic studies of celiprolol in healthy volunteers and patients with renal insufficiency. *Eur. J. Clin. Pharmacol.*, **38**, 587–593.

Hendel, J. and Brodthagen, H. (1984). Enterohepatic cycling of methotrexate estimated by use of the D-isomer as a reference marker. *Eur. J. Clin. Pharmacol.*, **26**, 103–107.

Hunter, J., Jepson, M. A., Tsuruos, T., Simmons, N. L., and Hirst, B. H. (1993). Functional expression of P-glycoprotein in apical membranes of human intestinal Caco-2 cells. *J. Biol. Chem.*, **268**, 14991–14997.

Jamali, F., Mehvar, R., and Pasutto, F. M. (1989). Enantioselective aspects of drug action and disposition. *J. Pharm. Sci.*, **78**, 695–715.

Karlsson, J., Kuo, S. M., and Artursson, P. (1993). Transport of celiprolol across human intestinal epithelial (Caco-2) cells: Mediation of secretion by multiple transporters including P-glycoprotein. *Biochem. Pharmacol.*, **110**, 1009–1016.

Klaassen, C. D. (1988). Intestinal and hepatobiliary disposition of drugs. *Toxicol. Pathol.*, **16**, 130–137.

Kuo, S. M., Whitby, B. R., Artursson, P., and Ziemniak, J. A. (1994). The contribution of intestinal secretion to the dose-dependent absorption of celiprolol. *Pharm. Res.*, **11**, 648–653.

Langguth, P., Khan, P. J., and Garrett, E. R. (1990). Pharmacokinetics of morphine and its surrogates XI: Effect of simultaneously administered naltrexone and morphine on the pharmacokinetics and pharmacodynamics of each in the dog. *Biopharm. Drug Dispos.*, **11**, 419–444.

Langguth, P., Lee, K. M., Spahn-Langguth, H., and Amidon, G. L. (1994). Variable gastric emptying and discontinuities in drug absorption profiles: Dependence of rates and extent of cimetidine absorption on motility phase and pH. *Biopharm. Drug Dispos.*, **15**, 719–746.

Langguth, P., Möhrke, W., Möhler, H., and Spahn-Langguth, H. (1997). Drug absorption via intestinal vs. nasal mucosae: Dependence of rate and extent of systemic availability of tranylcypromine enantiomers upon the application site. *Biopharm. Drug Dispos.* (submitted for publication).

Lennernäs, H. and Regardh, C. G. (1993a). Regional gastrointestinal absorption of the beta-blocker pafenolol in the rat and intestinal transit rate determined by the movement of ^{14}C-polyethylene glycol (PEG) 4000. *Pharm. Res.*, **10**, 130–135.

Lennernäs, H. and Regardh, C. G. (1993b). Dose-dependent intestinal absorption and significant intestinal secretion (exsorption) of the beta-blocker pafenolol in the rat. *Pharm. Res.*, **10**, 727–731.

Lennernäs, H. and Regardh, C. G. (1993a). Evidence for an interaction between the beta-blocker pafenolol and bile salts in the intestinal lumen of the rat leading to dose-dependent oral absorption and double-peaks in the plasma concentration-time profile. *Pharm. Res.*, **10**, 879–883.

Lien, E. J., Rodriguez de Miranda, J. F., and Ariens, E. J. (1982). Quantitative structure-activity correlation of optical isomers: A molecular basis for Pfeiffer's rule. *Mol. Pharmacol.*, **12**, 598–604.

Lipka, E., Lee, I. D., Langguth, P., Spahn-Langguth, H., Mutschler, E., and Amidon, G. L. (1995). Celiprolol double peak occurrence and gastric motility: Nonlinear mixed effects modeling of bioavailability data obtained in dogs. *J. Pharmacokin. Biopharm.*, **23**, 267–286.

Lipka, E., Spahn-Langguth, H., Mutschler, E., and Amidon, G. L. (1996). The nonlinear intestinal permeability and pharmacokinetics of celiprolol in conscious dogs: Evidence for in-vivo intestinal secretion. *Eur. J. Pharm. Sci.* (submitted for publication).

Meijer, D. K. F. and van der Sluijs, P. (1989). Covalent and noncovalent protein binding of drugs: Implications for hepatic clearance, storage, and cell-specific drug delivery. *Pharm. Res.*, **6**, 105–118.

Merfeld, A. E., Mlodozeniec, A. R., Cortese, M. A., Rhodes, J. B., Dressman, J. B., and Amidon, G. L. (1986). The effect of pH and concentration on α-methyldopa absorption in man. *J. Pharm. Pharmacol.*, **38**, 815–822.

Merino, M., Peris-Ribera, J. E., Torres-Molina, F., and Sanchez-Pico, A. (1989). Evidence of a specialized transport mechanism for the intestinal absorption of baclofen. *Biopharm. Drug Dispos.*, **10**, 279–297.

Oberle, R. L. and Amidon, G. L. (1987). The influence of variable gastric emptying and intestinal transit rates on the plasma level curve of cimetidine, an explanation for the double peak phenomenon. *J. Pharmacokin. Biopharm.*, **15**, 529–544.

Olpe, H. R., Baudry, M., Fagni, L., and Lynch, G. (1982). The blocking action of baclofen on excitatory transmission in the rat hippocampal slice. *J. Neurosci.*, **2**, 698–703.

Ott, R. J. and Giacomini, K. M. (1993). Stereoselective transport of drugs across epithelia. In Wainer I.W. (ed.), *Drug Stereochemistry*, New York: Marcel Dekker, pp. 281–314.

Oude Elferink, R. P., Meijer, D. K., Kuipers, F., Jansen, P. L., Groen, A. K., and Groothuis, G. M. (1995). Hepatobiliary secretion of organic compounds: Molecular mechanisms of membrane transport. *Biophys. Acta*, **1241**, 215–268.

Pfeiffer, C. C. (1956). Optical isomerism and pharmacological action, a generalization. *Science*, **124**, 29–31.

Plumb, J. A., Milroy, R., Kaye, S. B. (1990). The activity of verapamil as a resistance modifier in vitro in drug resistant human tumour cell lines is not stereospecific. *Biochem. Pharmacol.*, **39**, 787–792.

Regardh, C. G., Lundborg, P., Gabrielsson, M., Heggelund, A., and Kylberg-Hanssen, K. (1990). Pharmacokinetics of a single intravenous dose of pafenolol—a beta$_1$-adrenoceptor antagonist with atypical absorption and disposition properties—in man. *Pharm. Res.*, **7**, 1222–1227.

Rowland, M. and Tozer, T. N. (1989). Clinical Pharmacokinetics: Concepts and Applications, 2nd Edition, Lea & Febiger, Philadelphia, P.A., pp. 148–173.

Schneider, E., Gleiter, C., Schug, B., Hermann, R., Gundert-Remy, U., Elze, M., and Blume, H. (1996). Influence of food intake on the bioavailability of α-lipoic acid. *Naunyn Schmiedeberg's Arch. Pharmacol.*, **353** (Suppl.), R153.

Sezaki, H. and Kimura, T. (1983). Carrier-mediated transport in drug absorption. In D. D. Breimer and P. Speiser (eds.), *Topics in Pharmaceutical Sciences*. Amsterdam: Elsevier, 133–142.

Soldner, A., Spahn-Langguth, H., Voith, B., Stapff, M. and Mutschler, E. (1996). Losartan and EXP 3174 in rats: Concentrations in central and peripheral compartments as basis for PK/PD correlations. (submitted for publication.)

Spahn-Langguth, H., Langguth, P., and Brockmeier, D. (1993). The pharmacokinetics of piretanide in humans: An update. In S. Johnson and F. N. Johnson (eds.), *The Diuretic Agents 2: Piretanide*. Carnforth, UK: Marius Press, 11–29.

Spahn-Langguth, H., Hahn, G., Mutschler, E., Elze, M., Potthast, H., Blume, H., and Möhrke, W. (1994). Determination of the bioavailability of tranylcypromine with a stereospecific assay. *Eur. J. Pharm. Sci.*, **2**, 160.

Spahn-Langguth, H., Möhrke, W. and Langguth, P. (1994). Variabilities in input profiles of tranylcypromine enantiomers following p.o.-administration. *Eur. J. Pharm. Sci.*, **2**, 160.

Stahl, E. (1993). Stereochemische Aspekte der Pharmakokinetik der lipophilen β-Adrenozeptor-Antagonisten Carvedilol und Propranolol" Ph.D. thesis, Department of Pharmacy, Johann Wolfgang Goethe-University, Frankfurt/Main.

Stahl, E., Baumgartner, U., Henke, D., Schölmerich, J., Mutschler, E., and Spahn-Langguth, H. (1993). Rats with portocaval shunt as a potential pharmacokinetic model for liver cirrhosis. *Chirality*, **5**, 120–125.

Taylor, M. D. and Amidon, G. L. (eds.) (1995). *Peptide-based drug design: Controlling transport and metabolism*. Washington, DC: ACS.

Thiebault, F., Tsuruo, T., Hamada, H., Gottesman, M. M., Pastan, I., and Willingham, M. C. (1987). Cellular localization of the multidrug resistance gene product P-glycoprotein in normal human tissues. *Proc. Natl. Acad. Sci. USA*, **84**, 7735–7738.

Tsuji, A. (1995). Intestinal absorption of β-lactam antibiotics. In M. D. Taylor and G. L. Amidon (eds.), (1995). *Peptide-Based Drug Design: Controlling Transport and Metabolism*. Washington, DC: ACS, pp. 101–134.

Tsuji, A. and Tamai, I. (1996). Carrier-mediated intestinal transport of drugs. *Pharm. Res.*, **13**, 963–977.

Wacher, V. J., Wu, C. Y., and Benet, L. Z. (1995). Overlapping substrate specificities and tissue distribution of cytochrome P450 3A and P-glycoprotein: Implications for drug delivery and activity in cancer chemotherapy. *Mol. Carcinog.*, **13**, 120–134.

Wade, D. N., Mearrick, P. T., and Morris, J. L. (1977). Active transport of L-dopa in the intestine. *Nature* (*London*), **242**, 463–465.

Weber-Grandke, H., Hahn, G., Mutschler, E., Möhrke, W., Langguth, P., and Spahn-Langguth, H. (1993). The pharmacokinetics of tranycypromine enantiomers in healthy subjects after oral administration of racemic drug and the single enantiomers. *Br. J. Clin. Pharmacol.*, **36**, 363–365.

Wetterich, U., Spahn-Langguth, H., Rösch, W., Mutschler, E., Terhaag, B., and Langguth, P. (1996). Evidence for intestinal secretion as additional clearance pathway of talinolol enantiomers. *Pharm. Res.*, **13**, 526–534.

Yee, S. and Amidon, G. L. (1995). Oral absorption of angiotensin-converting enzyme inhibitors and peptide prodrugs. In M. D. Taylor, G. L. Amidon (eds.), *Peptide-based drug design: Controlling transport and metabolism*. Washington, DC: ACS, pp. 135–147.

Yoshikawa, T., Muranishi, N., Yoshida, M., Oguma, T., Hirano, K., and Yamada, H. (1989). Transport characteristics of ceftibuten (7432-S), a new oral cepham, in rat intestinal brush-border membrane vesicles: Proton-coupled and stereoselective transport of ceftibuten. *Pharm. Res.*, **6**, 302–307.

Yuasa, H., Amidon, G. L., and Fleisher, D. L. (1993). Peptide carrier-mediated transport in intestinal brush border membrane vesicles of rats and rabbits: Cephradine uptake and inhibition. *Pharm. Res.*, **10**, 400–404.

Yuasa, H., Fleisher, D. L., and Amidon, G. L. (1994). Noncompetitive inhibition of cephradine uptake by enalapril in rabbit intestinal brush-border membrane vesicles: An enalapril specific inhibitory binding site on the peptide carrier. *J. Pharmacol. Exp. Ther.*, **269**, 1107–1111.

Yun, C. H., Lee, H. S., Rho, J. K., Jeong, H. G., and Guengerich, F. P. (1995). Oxidation of the angiotensin-II receptor antagonist losartan (DuP 755) in human liver microsomes. Role of cytochrome P4503A in formation of the active metabolite EXP 3174. *Drug Metab. Dispos.*, **23**, 285–289.

CHAPTER

23

PREPARATION OF DRUG ENANTIOMERS BY CHROMATOGRAPHIC RESOLUTION ON CHIRAL STATIONARY PHASES

ERIC FRANCOTTE

Pharmaceuticals Division
CIBA-GEIGY Limited
Basel, Switzerland

23.1. INTRODUCTION

The principle of chirality is the source of diverse phenomena on the macromolecular and molecular levels that govern our entire environment and the existence of all living organisms. The influence of this principle on most biological mechanisms and its significance have by now become generally accepted. Interest in the consequences of chirality on the physical and biological properties of molecules has created a rapidly growing demand for new general methods of obtaining optically pure compounds. The preparation of enantiomerically pure compounds is a topic of increasing importance in relation to the development of new biologically active substances containing one or several chiral centers, because it is now apparent that many chiral drugs and agrochemicals (herbicides, fungicides, insecticides) display different activity and toxicity profiles according to their absolute configuration (1–7). Potential methods of supplying optically pure isomers are being intensively pursued, and among the different techniques developed for this purpose, preparative chromatographic resolution on chiral stationary phases (CSPs) has been recognized as a useful approach (8).

There are two possible approaches to preparing enantiomerically pure compounds (Figure 23.1). The first consists of preparing the racemic compound, which is subsequently resolved into its antipodes, usually leading to the production of both enantiomers. The second is to design and elaborate a stereoselective synthesis, leading to the production of one enantiomer. In this

The Impact of Stereochemistry on Drug Development and Use, Edited by Hassan Y. Aboul-Enein and Irving W. Wainer. Chemical Analysis Series, Vol. 142.
ISBN 0-471-59644-2

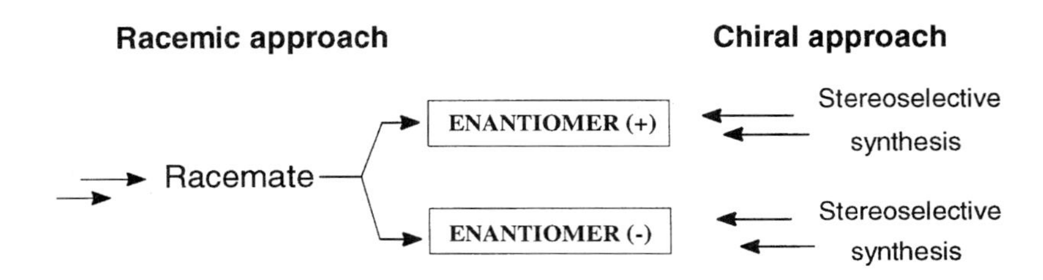

Figure 23.1. General approaches to optically active drugs.

latter approach, if both enantiomers are desired, a stereoselective synthesis has to be devised for each enantiomer. Both approaches have their advantages and disadvantages.

In the first step of the "racemic" approach, the substrate is prepared as the racemate by a reaction sequence that generally presents a much lower degree of difficulty than for the corresponding optically active forms. In the second step, the enantiomers of the racemate are separated via the formation of diastereoisomers, or by chromatography on chiral stationary phases. Although the separation by way of the formation of diastereoisomers is still often used, especially for compounds bearing acid or base functions, the direct separation of enantiomers by chromatography on chiral stationary phases is rapidly gaining favor (9). The main advantages of this latter technique are (a) applicability to a wide variety of racemic structures; (b) both enantiomers are usually obtained; (c) a high degree of optical purity of the isolated enantiomers; (d) rapid and easy achievement; (e) separation of the enantiomers of racemates with special features, such as compounds that cannot be derivatized (hydrocarbon), racemize easily, or show an unusual type of chirality (e.g., C_2 chirality).

This method is especially attractive for the pharmaceutical industry, because it greatly facilitates the provision of the desired compounds in amounts suitable for biological testing, toxicological studies, and even, at a later stage, clinical trials. Moreover, data on the activity and the toxicity profiles of the individual pure enantiomers are now routinely required by the authorities for all new chiral drugs submitted for registration. At least during the preliminary testing phase, chromatography can replace the often lengthy elaboration of an enantioselective synthesis to access the pure enantiomers needed for biological investigations. Even taking into account that only a limited number of drugs proceed to more intensive investigations, a systematic synthesis would occupy much time and manpower before the preliminary tests needed to decide on the further development of either a pure enantiomer or the racemate could be started. Additionally, if a drug has already been selected for further study, chromatography can ensure a continuous supply of

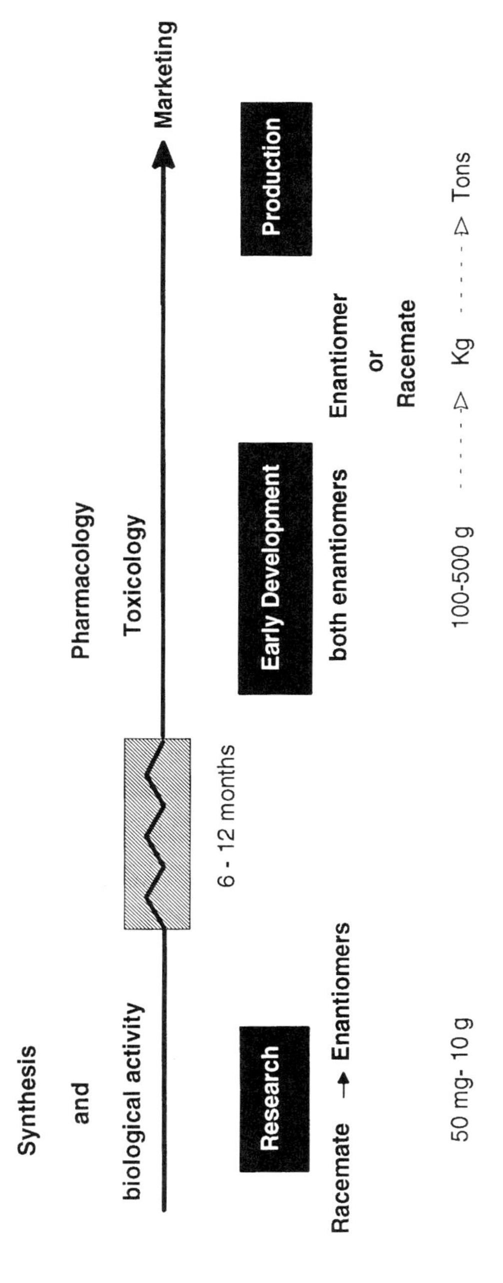

Figure 23.2. Chiral drug development.

further quantities of an enantiomerically pure substance for the biological investigations while an enantioselective synthesis is being developed. Furthermore, the chromatographic method can furnish both enantiomers, which obviously are required for comparative biological testing.

The preparation of enantiomers by chromatographic resolution on chiral stationary phases is already being more and more frequently used to obtain new, chiral, biologically active materials during the research phase (Figure 23.2). At this stage, a few milligrams to a few grams are generally enough to allow the performance of the biological tests and the full physicochemical characterization of the enantiomers necessary for a patent application, for example. The method is also increasingly being used to prepare enantiomers of new drugs in quantities of 100 g to a few kilograms. Chromatography on this scale permits pharmacological and toxicological studies to be carried out before an asymmetric synthesis has been developed, which often takes several months to a year, or more. This appreciably shortens the time needed to develop a chiral drug.

Although chromatography is generally considered an expensive technique, it is coming to be regarded as both technically and even economically attractive for the preparation of high-value compounds, or of optical isomers that are otherwise only accessible with difficulty. Amounts of up to several kilograms of optically pure isomers are already being produced using this technology, and it would seem likely to afford a useful approach to the manufacture of chiral drugs, especially in the case of very potent substances that need only be produced in relatively small amounts and that cannot easily be prepared by other methods. The combination of this application to chiral compounds with the simulated moving-bed chromatographic process, in particular, could considerably reduce production costs and save large amounts of mobile phase.

The utilization of chromatography on CSPs has permitted the preparation of optically active isomers of chiral drugs needed for the investigation of biological activities, for pharmacological or toxicological studies, and for clinical studies. These applications are reviewed in this chapter, emphasizing the versatility of the method.

23.2. CHIRAL STATIONARY PHASES

The chromatographic separation of enantiomers on chiral stationary phases (CSPs) has undergone spectacular development owing to the concomitant evolution of the chromatographic techniques and the design of numerous new chiral phases. Numerous analytical columns have become available and are now used routinely for the determination of the enantiomeric composition of mixtures of optical isomers from enantioselective syntheses, from biological investigations, or from pharmacokinetic or toxicological studies (10–12). The preparative potential of the method has also been recognized (8, 9), and the number of applications is rapidly growing, mainly because of the need to isolate the optically pure enantiomeric forms of new chiral biologically active compounds.

A fundamental distinction can be drawn between two kinds of CSPs (Figure 23.3), chiral polymers (type I) and achiral matrices (usually silica gel) modified with chiral moieties (type II).

In the first class (type I), due to the nature of the polymeric structure, the density of chiral information is generally high and the simultaneous participation of several chiral interaction sites or several polymeric chains is conceivable. The polymer can exist in pure form (type Ia) or in a diluted form when coated (type Ib) or grafted (type Ic). Oligo- and polysaccharides and their derivatives, polyacrylamides, polyacryl esters, and the protein-based phases, belong to this type of CSP, although the participation of several chiral interaction sites is unlikely for all kinds of polymers.

The second most commonly used approach to preparing chiral sorbents consists of attaching optically units to achiral carriers (mainly silica gel) by means of ionic or covalent bonds (type II). A wide range of optically active moieties have already been applied, including amino-acid derivatives, crown ethers, cinchona alkaloids, carbohydrates, amines, tartaric-acid derivatives, cyclodextrins, binaphthyl compounds, etc. Although the silica carrier is also a polymer, in this class of CSPs the chiral interaction sites distributed at the surface or in the network of the achiral support are relatively far away from each other and virtually only a "biomolecular" stereoselective interaction is possible between the chiral solute and chiral selector. Cyclodextrins as chiral selectors constitute an intermediate case, because the complexation with these macromolecules usually involves the simultaneous interaction with several glucose residues by inclusion into the cavity.

In both classes of CSPs, the classical interaction forces such as ionic, hydrophobic, dipolar, and π–π interactions, or hydrogen bonding can be involved. A recent review of the properties of the CSPs developed for preparative purposes includes a table listing those now available commercially (8), although new preparative CSPs have been introduced in the mean time. The CSPs used nowadays for practical preparative resolutions of chiral drugs are briefly presented below, and a list of the CSPs referring to reported applications is given in Table 23.1. The choice will be mainly governed by the selectivity obtained on the different CSPs, but other factors such as the solubility of the solute, and the availability and costs of the CSPs can also be determinants.

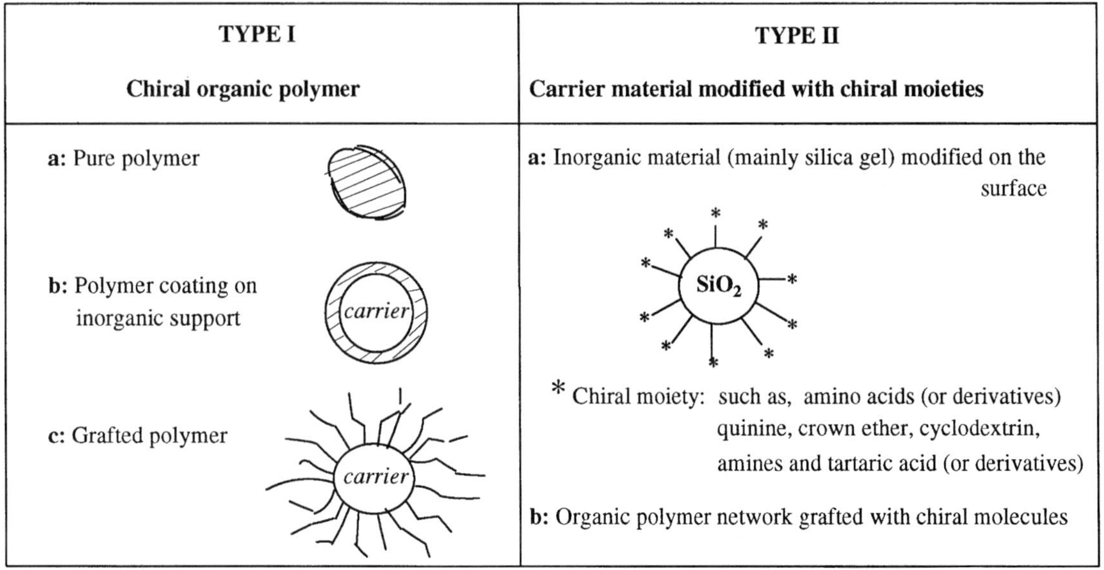

TYPE I **Chiral organic polymer**	**TYPE II** **Carrier material modified with chiral moieties**
a: Pure polymer **b:** Polymer coating on inorganic support *carrier* **c:** Grafted polymer *carrier*	**a:** Inorganic material (mainly silica gel) modified on the surface SiO_2 * Chiral moiety: such as, amino acids (or derivatives) quinine, crown ether, cyclodextrin, amines and tartaric acid (or derivatives) **b:** Organic polymer network grafted with chiral molecules

Figure 23.3. Classification of chiral stationary phases.

Table 23.1. Chiral Stationary Phases Used for Preparative Resolutions of Chiral Drugs

Packing Name	Chiral Selector	CSP Type	Supplier
Cellulose and Amylose Derivatives			
Chiralcel OA	Cellulose triacetate	Ib	Daicel
Chiralcel OB	Cellulose tribenzoate	Ib	Daicel
Chiralcel OC	Cellulose phenylcarbamate	Ib	Daicel
Chiralcel OD	Cellulose 3,5-dimethylphenylcarbamate	Ib	Daicel
Chiralcel OJ	Cellulose *para*-methylbenzoate	Ib	Daicel
Chiralpak AD	Amylose 3,5-dimethylphenylcarbamate	Ib	Daicel
CTA-I	Cellulose triacetate	Ia	Daicel, Merck
CTB	Cellulose tribenzoate	Ia	Riedel de Haen
MMBC	Cellulose *meta*-methylbenzoate beads	Ia	
PMBC	Cellulose *para*-methylbenzoate beads	Ia	
TBC	Cellulose tribenzoate beads	Ia	
Cyclodextrin			
CD-Poly	Cross-linked β-cyclodextrin polymer	IIb	
ChiraDex	β-cyclodextrin	IIa	Merck
Cyclobond I	β-cyclodextrin	IIa	Astec
π-Acidic and π-Basic CSPs			
DACH-DNB	*trans*-Cyclohexyldiamine 3,5-dinitrobenzamide	IIa	
DNBLeu	3,5-Dinitrobenzoylleucine	IIa	Regis
DNGPG-co	3,5-Dinitrobenzoylphenylglycine (covalent bonding)	IIa	Regis

(*continued*)

Table 23.1. (*contd.*)

Packing Name	Chiral Selector	CSP Type	Supplier
DNBPG-io	3,5-Dinitrobenzoylphenylglycine (ionic bonding)	IIa	Regis
ChyRoSine-A	3,5-Dinitrobenzoyltyrosine butylamide	IIa	Sedere
DNB-Tyr-E	3,5-Dinitrobenzoyltyrosine methylester	IIa	
NAP-A1	Naphthylalanine	IIa	Regis
WHELK-O1	3,5-Dinitrobenzoyl tetrahydrophenanthrene amine	IIa	Regis
Protein-Based CSPs			
Chiral-AGP	α-Acid glycoprotein	Ic	ChromTech
Ultron ES-OVM	Ovomucoid protein	Ic	RTI
Polyacrylamide CSPs			
ChiraSpher	Poly[(*S*)-N-acryloylphenylalanine ethyl ester]	Ic	Merk
Poly-CHMA	Poly[(*S*)-N-methacryloylcyclohexylethyl amine]	Ia	
Poly-MA	Poly[N-acryloylmenthylamine]	Ia	
Poly-PEA	Poly[(*S*)-N-acryloylphenylalanine ethylester]	Ia	

Astec = Advanced Separation Technologies (Whippany, New Jersey). **ChromTech** = ChromTech AB (Stockholm, Sweden). **Daicel** = Daicel Chemical Industries (Tokyo, Japan). **Merck** = E. Merck (Darmstadt, Germany). **Regis** = Regis Chemical Company (Morton Grove, Illinois). **Riedel-de Haen** (Seelze, Germany). **RTI** = Rockland Technologies (Newport, Delaware). **Sedere** (Alfortville, France).

23.2.1. Cellulose and Amylose Derivatives

A wide range of cellulose-based stationary phases (Figure 23.4) have been developed during the last 10 years (13). These polymeric materials are used as pure polymers in a form adequate for chromatographic purposes, or as a coating on an inert achiral support conferring mechanical stability. The high potential of these CSPs is clearly demonstrated by the great number of analytical pharmaceutical applications reported in the literature.

So far, the most widely used derivative for separations on a preparative scale is cellulose triacetate (CTA-I), introduced in its fully acetylated form by Hesse and Hagel in 1973 (14). This CSP can resolve a broad range of structurally different racemates (8). The high versatility and the high loading capacity, as well as the low preparation costs, have certainly contributed to the more extensive use of this sorbent, even granting that there are some practical

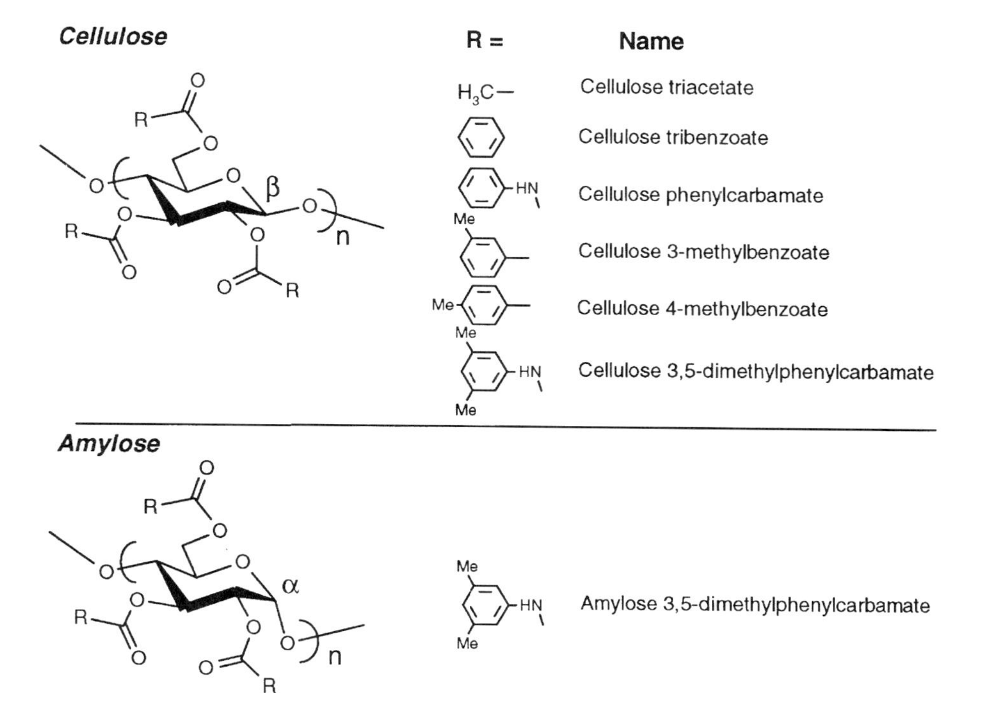

Figure 23.4. Structure of cellulose- and amylose-based CSPs.

limitations (8). Various benzoylcellulose derivatives were also developed in the pure polymeric form (15–17), or as a coating on silica gel (18). A wide range of racemic structures were resolved on an analytical scale on silica-coated phenylcarbamate derivatives of cellulose (19), especially on the 3,5-dimethylphenyl carbamoyl derivative. In spite of the high costs of these cellulose-based CSPs, the number of preparative applications is growing appreciably.

Some amylose derivatives (Figure 23.4) were also developed as a coating on silica and investigated as CSPs (20, 21). The 3,5-dimethylphenyl carbamate derivative seems to be particularly useful, and preparative resolutions have already been reported.

Except for microcrystalline cellulose triacetate (CTA I), which has been used mostly in the reversed-phase mode (usually methanol–or ethanol–water mixtures), the cellulose- and amylose-based CSPs were used under both normal- and reversed-phase conditions. Improvement of the selectivity and/or resolution is generally achieved by varying the composition of the mobile phase and by varying the temperature.

23.2.2. Cyclodextrin-Based Phases

Cyclodextrins are cyclic oligosaccharides (Figure 23.5) having the property of forming stable inclusion complexes in their highly hydrophobic cavity (chiral host) with a wide variety of molecules (22). Owing to the inherent chirality of their building units, namely, the α-1,4-linked glucose moiety, these cyclic

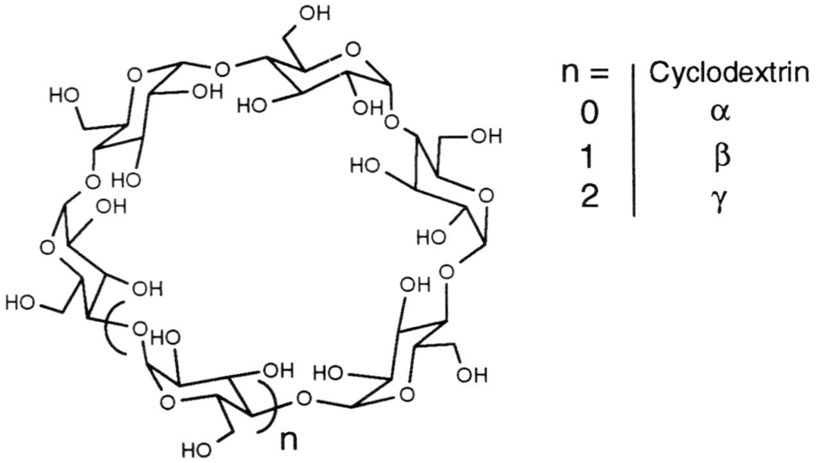

n =	Cyclodextrin
0	α
1	β
2	γ

Figure 23.5. Structure of cyclodextrins.

oligosaccharides are chiral and the formation of diastereomeric complexes with the two enantiomers of racemic compounds can be very stereoselective.

The size of the cavity, which differs for α-, β-, and γ-cyclodextrins and the substituent on cyclodextrin, plays a decisive role in determining the ability of these oligosaccharides to complex a defined molecule. Cyclodextrins CSPs were prepared by immobilizing CD in polymeric structures (23, 24), or on silica gel (25, 26), the latter CSPs showing good performance on an analytical scale. Preparative applications using cyclodextrins as chiral hosts were reported on polymers obtained by the cross-linking of cyclodextrin with ethylene glycol-bis(epoxypropyl) (24) and on the silica-modified materials Cyclobond I (27) and Chiral Dex (28).

Optimization of the enantioselectivity can be achieved by modifying different factors, such as the concentration and nature of organic modifiers, pH, and the temperature and buffer concentration. Although reversed-phase conditions are usually applied with the cyclodextrin-based CSPs, analytical applications in the normal mode also have been reported (29). In addition, further cyclodextrin derivatives have been introduced and turned out to exhibit better chiral recognition for some classes of racemates (30), but no preparative separation on these CSPs has been reported up to now.

23.2.3. Poly(meth)acrylamides

Cross-linked, optically active polyacrylamides and polymethacrylamides constitute another class of polymeric CSPs. These CSPs were introduced some years ago by Blaschke and co-workers, who rapidly demonstrated their usefulness for preparative applications (31). However, the gel structure of these cross-linked polymers prevents their utilization at high pressure. Improvement

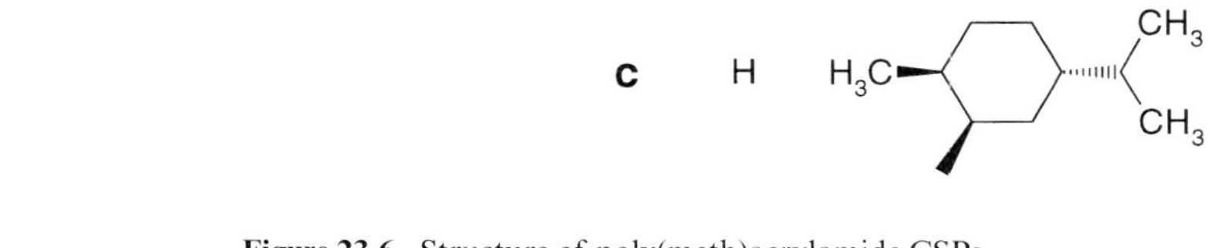

Figure 23.6. Structure of poly(meth)acrylamide CSPs.

of the mechanical performances of these CSPs was achieved by polymerization of the acrylic monomer on the surface of silica gel, giving a grafted polymer (32, 33).

The preparative separations reported in the literature have been carried out using (*S*)-phenylalanine ethyl ester (*a*), (*S*)-1-cyclohexylethylamine (*b*), and menthyl amine (*c*), (34, 35) as the chiral selector (Figure 23.6).

23.2.4. Protein-Based Phases

A number of CSPs have been developed by immobilizing proteins or enzymes such as bovine (36) or human (37) serum albumin, α_1-acid glycoprotein (38), α-chymotrypsin (39), cellulase (40), and ovomucoid (41, 42) on silica gel. These CSPs usually operate under reversed-phase conditions (phosphate buffers with the addition of organic modifiers) and exhibit a very high chiral-recognition ability toward various classes of racemates, including numerous important drugs. Enantioselectivity and retention can be regulated by changing the pH, the concentration or nature of the modifier, and the composition of the mobile phase (42–46). The presence of different types of highly stereoselective interaction sites renders these CSPs very versatile, but because the number of interaction sites is very limited, the loading capacity is low. This feature and the high costs of these CSPs are an obstacle to their use on a preparative scale. Nevertheless, a few semipreparative (0.1–3 mg scale) applications have been reported on the α_1-acid glycoprotein phase, demonstrating the feasibility of scaling-up (42, 47). Although this class of CSPs does not exhibit the features required for large-scale separations, it can be helpful for the resolution of racemic compounds for which only small amounts are required (reference, labeled compounds).

23.2.5. π-Acidic and π-Basic Phases

The best known CSPs of the π-acidic or π-basic type are the "Pirkle's phases," classified into π-acceptor and π-donor phases (48). The most frequently used π-acceptor phases are derived from the amino acids phenylglycine (DNBPG) or leucine (DNBLeu), covalently or ionically bonded to 3-aminopropyl silica gel (Figure 23.7*a*) (49, 50). These CSPs are commercially available for the analytical or preparative separation of enantiomers. Further CSPs based on other conventional chiral selectors, such as valine, phenylalanine, tyrosine (51), and 1,2-trans diaminocyclohexane, were also developed (Figures 23.7*b* and 23.7*c*).

The application of the reciprocality concept has led to the design of various phases of the π-donor type (52, 53) using, for instance, naphthylalanine (Figure 23.7*d*) or naphthylethylamine as the π-donor system. Recently, π-acidic CSPs of a new generation exhibiting high selectivities and broad applic-

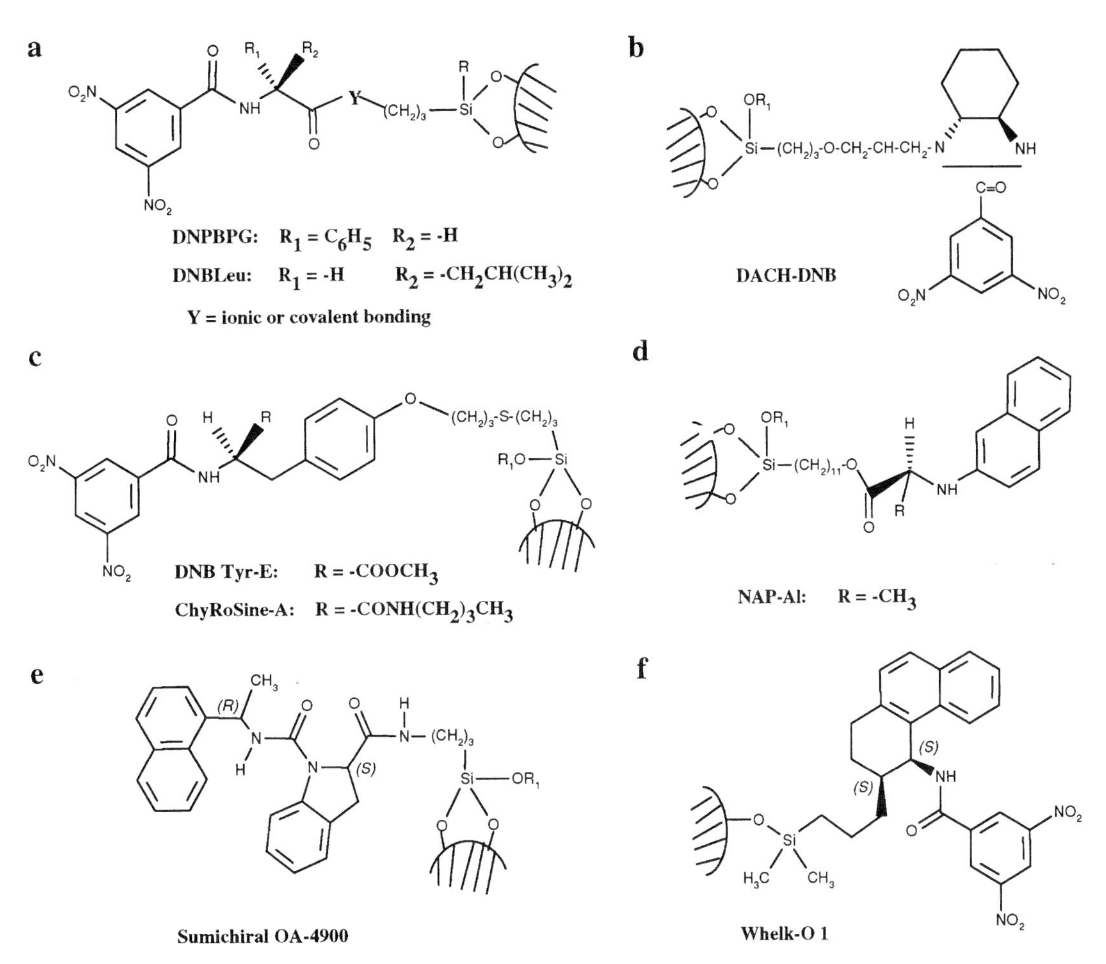

Figure 23.7. Structures of the π-acidic and π-basic phases.

ability were introduced by Pirkle and co-workers (54, 55). Preparative applications of chiral drugs such as some profens have already been reported on the Whelk-O 1 column (56). This CSP (Figure 23.7*f*) column was originally designed specifically for the resolution of the antiinflammatory agent naproxen. Further CSPs based on amino-acid derivatives or a combination of amino acids with chiral amines such as, for example, the (*S*)-indoline-2-carboxylic acid/(*R*)-1-(α-naphthyl)ethylamine CSP (Figure 23.7*e*) have been developed by Oi and co-workers following the same principle (57) and are available for preparative separations (8). However, to our knowledge, no preparative application has been reported.

These chemically modified silica gels are stable at high pressures and exhibit good chromatographic preformances. Usually, these CSPs were used under normal-phase conditions, but it has also been demonstrated that chiral resolutions of drugs could be achieved in the reversed-phase mode (58–60).

23.2.6. Chiral Phases for Ligand-Exchange Chromatography

Ligand-exchange chromatography (LEC) is based on the reversible formation of complexes between metal ions (usually Cu^{2+} or Ni^{2+}) and chiral complexing agents carrying functional groups capable of interacting as ligands. The most commonly used chiral complexing agents are α-amino acids chelating on the carboxylic and amino group (Figure 23.8).

CSPs of this type have been prepared by covalent bonding of the amino acid on organic polymeric and inorganic silica supports. Most of the initial work was performed with cross-linked polystyrene as an inert supporting

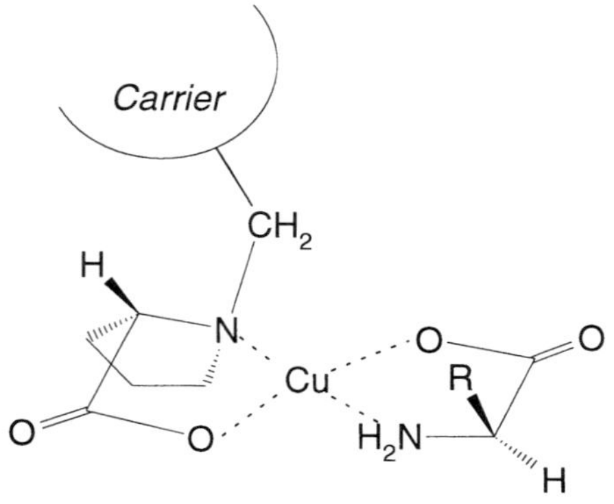

Figure 23.8. Schematic representation of the Cu-ligand complexation between a proline stationary phase and an amino acid.

phase. Further investigations in the field of LEC led to the introduction of an improved CSP using macronet isoporous styrene copolymers as supporting material (61). This sorbent achieves better chromatographic performances and has a higher exchange capacity than the previously developed CSPs based on cross-linked polystyrene. A wide range of racemic amino acids can be resolved analytically on these CSPs, and even preparative resolutions (up to 20 g of racemic proline) have been performed on an L-hydroxyproline CSP of this type. The importance of various parameters (mobile phase, metal ion, pH, temperature, supporting phase) in terms of the separation efficiency has been reviewed (62).

Recently, new ligand-exchanging resins based on highly cross-linked polyacrylamide have been introduced on the market and especially recommended for medium-pressure LC preparative resolutions of amino acids (63). Some preparative applications have been reported in the range of 200 mg to 10 g, but to our knowledge, no paper or text or practical application in the pharmaceutical field has been published.

Owing to the nature of the interactions, this type of CSP is particularly appropriate for the resolution of racemic compounds carrying chelating functionalities such as amino acids, hydroxy acids, and, in a few cases, amino alcohols. An advantage of the method is that the derivatization of the solute—even those as polar as amino acids—is usually not required prior to chromatography.

23.3. METHOD DEVELOPMENT FOR PREPARATIVE SEPARATIONS OF ENANTIOMERS

Having identified the appropriate CSP for the required enantiomeric resolution, the next step of method development consists in optimizing the separation to obtain a higher throughput. Therefore, different chromatographic aspects, such as the chromatographic mode, chromatographic running conditions, and fractionating conditions, can be varied.

23.3.1. Chromatographic Mode

First, the operator has to choose between various chromatographic modes for performing the preparative separation, depending on different factors, such as the available technique in his or her environment, the desired amount, the solubility of the solute, and the physical properties of the CSP. At least four different technologies, batch-mode, super(sub)critical fluid mode, simulated moving-bed process, and membrane process, have been applied to the preparative separation of enantiomers. Although most separations have been performed in the conventional batch-mode process, there is growing interest in simulated moving-bed (SMB) technology, because it permits large amounts of the mobile phase to be saved and productivity increased, thus reducing production costs (64–67). Moreover, this technique is particularly appropriate to the separation of binary mixtures, such as an enantiomeric mixture. However, setting up this technology, which is also fairly difficult to apply properly, entails comparatively large investments, and its practical uses probably will be limited to enantiomeric compounds that are needed in relatively large amounts. At present, the batch-mode process has the advantage of being a well-established and relatively simple technology that is applicable on any scale, according to the needs of the user. Recently, detailed studies have been carried out to compare the efficiency of the overloaded elution mode with that of the displacement mode, and the authors came to the conclusion that both modes give broadly similar purities and yields (68, 69).

Some preparative separations using supercritical or subcritical fluid chromatography (SFC or SubFC) (70–74) up to pilot scale with 6-cm (i.d.) axial compression columns have also been reported. Because of the low viscosity of the mobile phase (usually consisting mainly of carbon dioxide), SFC can be performed at high flow rates and generally is considered to be faster than HPLC (75, 76), thus permitting greater throughputs per unit of time. Moreover, a major advantage of SFC is that it can be operated with safer (nonflammable) and considerably cheaper mobile phases, two factors that are very important for large-scale separations. Nevertheless, the limited solubility of most drugs in carbon dioxide considerably reduces the applicability of this technique to the preparative separation of enantiomers. Recently an interesting analytical comparative study of enantiomeric separations under chiral HPLC versus chiral SFC on the Chiralcel OD CSP was carried out, showing that even if the selectivities are similar, SFC generally exhibits higher resolutions (77). Moreover, with some solutes such as amine derivatives, the interaction with carbon dioxide can alter considerably the resolution process (78, 79).

23.3.2. Strategy for Preparative Separations

The general strategy to be adopted for the performance of preparative chiral separations has been presented in detail in a previous review (8), the different steps can be summed up as follows (Figure 23.9): (a) selection of an appropriate CSP by screening analytical columns filled with chiral stationary phases also available in preparative amounts; (b) improvement of the separation and resolution factors by optimization of the various chromatographic parameters (mobile-phase composition, flow rate, temperature); (c) optimization of the chromatographic throughput (sample amount, column overloading); (d) transfer of the analytical separation to the preparative column, with a final

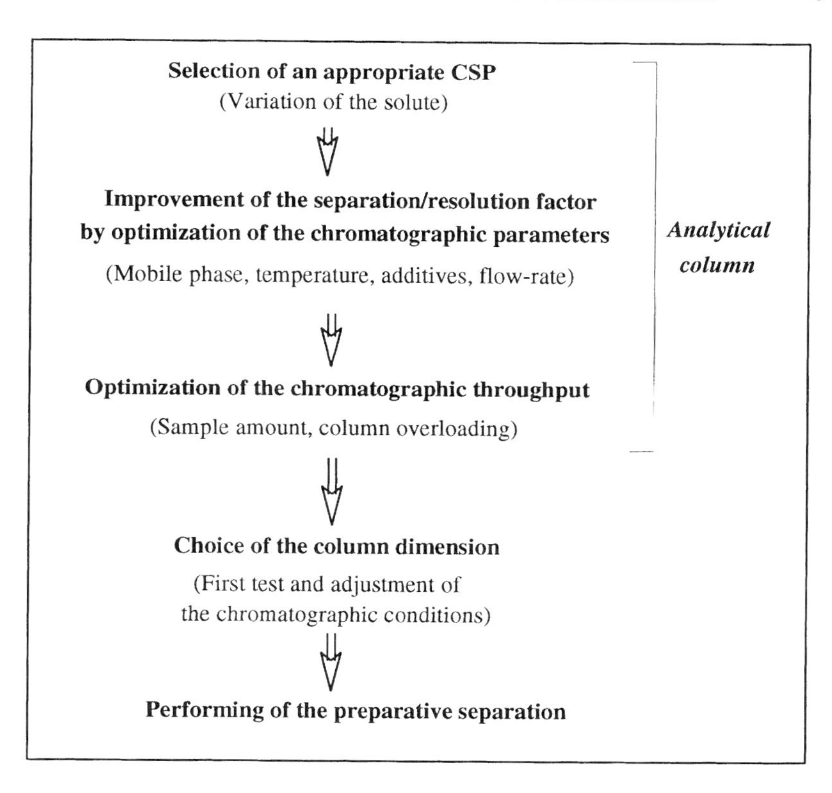

Figure 23.9. Strategy to perform preparative chiral separations.

adjustment of the chromatographic conditions; and (e) performance of the preparative separation and, if necessary, automation of the process.

The possibility of greatly improving the separation by achiral derivatization of the solute also should not be neglected. It has been shown that this strategy can lead to successful resolutions of racemates that otherwise could not be separated on a defined CSP (80–83). Finally, if the desired compound from a reaction sequence cannot be separated, it is often possible to achieve resolution on a precursor.

23.3.3. Isolation of the Solutes

Isolation of the resolved enantiomers is obviously a crucial step in preparative chromatographic separation. In the conventional batch-elution process, however, this step does not generally constitute a difficulty. Fractionation is usually achieved in a conventional way, that is, the fractions containing the solute at the desired optical purity are collected. To improve the separation and/or the throughput, recourse to such techniques as recycling and peak shaving is increasingly available. Recycling is especially useful for racemates that are poorly resolved, and consists in passing the solute through the column several times (9, 28, 84–88). Peak shaving usually is combined with recycling, allowing work to be performed under overload conditions and the throughput improved (86). In this case, the first and the last fractions containing the enantiomers at the desired optical purity are collected at each passage through the column, while the eluate containing a mixture is recycled.

23.4. APPLICATIONS TO CHIRAL DRUGS

The pharmaceutical field offers a rapidly expanding range of applications for chiral chromatographic separations. The utilization of CSPs as an analytical tool for the determination of the composition of enantiomeric mixtures in biological and pharmacological studies is now a well-established technique (10–12). However, the application of the method on a preparative scale for the production of optically active materials in amounts suitable for biological testing, toxicological studies, and even, at a later stage, clinical testing is gaining increasingly wide acceptance. In this respect, the chromatographic method affords the advantage of furnishing both enantiomers obviously required for preliminary comparative biological testing, and this approach is also widely applicable to elaborated drug structures that would be not easily accessible by a synthetic route. At least during the preliminary test phase of new chiral drugs, chromatography allows rapid access to the pure enantiomers and can replace advantageously the often lengthy elaboration of an enantioselective synthesis. A broad variety of racemic drugs already have been resolved on CSPs, covering different therapeutic classes of compounds. Most of the applications have been performed on CTA-I (8), but an increasing number of racemic drugs, or drug intermediates, were resolved on further cellulose, based CSPs such as cellulose tribenzoate, cellulose phenyl carbamate, and cellulose 3,5-dimethylphenyl carbamate. Polyacrylamide-type CSPs were also found to be useful for some drugs such as ifosfamide (chiral phosphorus atom), phthalimides (thalidomide, chlortalidone), and the diuretic classes of benzothiadiazines. The preparative resolution of a series of racemic benzodiazepinones could be achieved on various π-acid CSPs derived from phenylglycine and leucine.

For several drugs such as the hypnotic and anticonvulsive drug methaqualone (89, 90), the antiaromatase fadrozole (91), the antiinflammatory oxindanac (92), the diuretic chlortalidone (93), (94), the diuretic benzothiadiazines (95), the benzodiazepinone oxazepam (93), and the muscle relaxant chlormezanone (96, 97), the enantiomers were obtained for the first time by chromatography on CSPs. The chromatographic method also is applicable to the resolution of larger amounts of racemate, as demonstrated by the enantiomeric separation of the PAF antagonist WEB 2170 (98), of which not less than 1.3 kg of racemate have been chromatographed per run on a CTA-I column.

This approach is particularly suitable for the amounts usually required when the enantiomers of labeled compounds are desired. This strategy has been applied, for instance, for ^{14}C-zileuton (99), and ^{14}C-warfarin (100) on a milligram scale and for the ^{14}C- or deuterium-labeled anticancer drug fadrozole on a gram scale (91).

The growing number of applications in the pharmaceutical field attests to the usefulness of the method, which in many cases yields several grams of the pure enantiomers of chiral drugs in a simple way and in a relatively short time compared to other approaches. Moreover, considering that a great number of the analytical pharmaceutical applications reported up to now are theoretically transposable on a preparative scale, one can extrapolate the potential value of the method for generally supplying optically pure drugs. Chromatographic resolution on CSPs probably will become the method of choice for the preparation of enantiomeric drugs, at least in the first stage of development.

23.4.1. Cardiovascular System

Various classes of racemic compounds belonging to the drugs used for treatment of cardiovascular disorders have been resolved by preparative chromatographic resolution on chiral stationary phases (Table 23.2). In most instances, a particular CSP was found to be appropriate for resolving a broad range of racemic drugs belonging to the same structural classes. For β-adrenergic blocking agents, for instance, the 3,5-dimethylphenyl carbamate derivative of cellulose (Chiralcel OD) generally exhibits good chiral recognition on an analytical scale (101, 102), and the possibility of scaling up the chromatographic separation has been exemplified for propranolol, alprenolol (Figure 23.10), and oxprenolol (102, 103). Further reported examples applying the SMB technology (104), allow the production of 100 mg of pure enantiomer per hour using a battery of eight columns (25 cm × 1 cm) placed in series. The ability to resolve these arylpropoxy-1,2 amino alcohols as their corresponding oxazolidinone derivative on CTA-I also has been demonstrated (105). For large-scale separations, this way could be attractive, if we consider that the CSP costs are much lower for CTA-I. The cost factor also was decisive in choosing this CSP for the preparative resolution of a racemic intermediate of an angiotensin-converting-enzyme (ACE) inhibitor at a scale of 45 g per run

Table 23.2. Preparative Resolutions of Racemic Drugs for Cardiovascular Disorders

Structure	Name	Sample Amount	Column[a]	CSP	Reference
ACE Inhibitor					
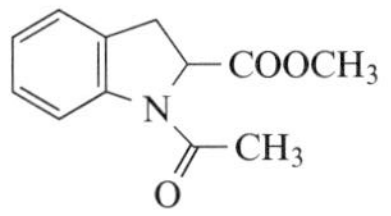	Intermediate	45 g	100 × 20	CTA-I	(106)
	β-*Adrenergic Blocking Agent*				
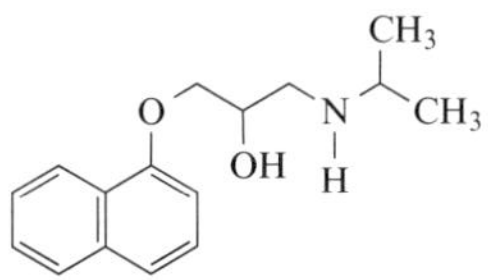	Propranolol	100 mg	50 × 2	Chiralcel OD	(102)
	Propranolol	SMB	(25 × 1)8	Chiralcel OD	(104)
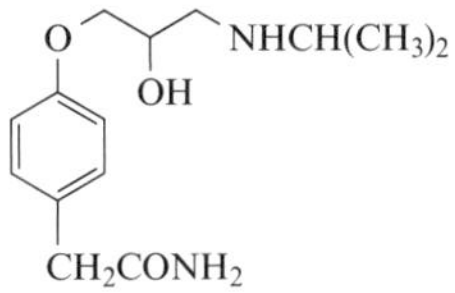	Atenolol	SMB	(25 × 1)8	Chiralcel OD	(104)
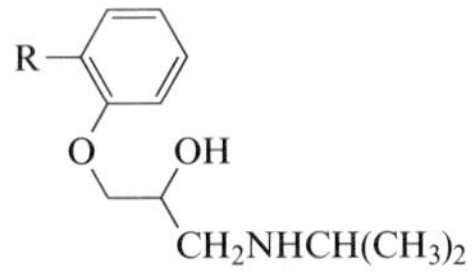	Alprenolol $R = -CH_2CH{=}CH_2$	150 mg	50 × 2	Chiralcel OD	(102, 103)
	Oxyprenolol $R = -O(CH_2)_2CH{=}CH_2$	400 mg	50 × 2	Chiralcel OD	(102)

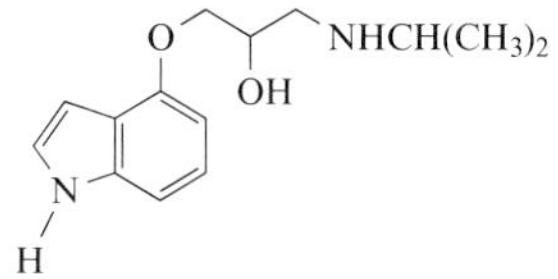 Pindolol	SMB	(25 × 1)8	Chiralcel OD	(104)
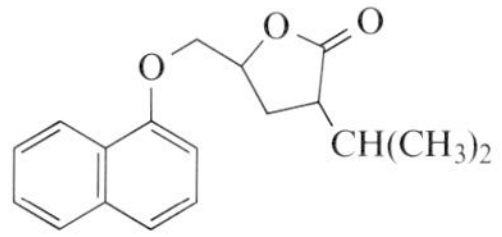 Propranolol oxazolidinone	10 mg	60 × 1	CTA-I	(105)
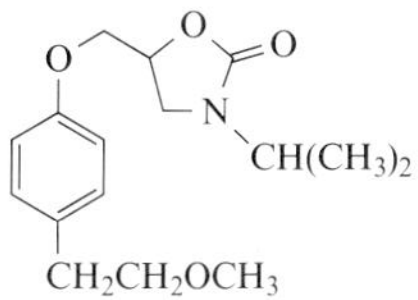 Metoprolol oxazolidinone	10 mg	60 × 1	CTA-I	(105)
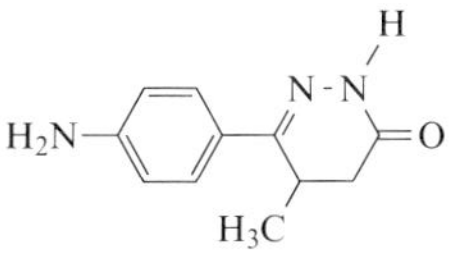 Intermediate	1 g	n.m.	DNBPG-io	(109)
Anticoagulant				
[14C]-warfarin	n.m.	25 × 0.46	Chiralcel OD	(100)
Warfarin	200 mg (SFC)	25 × 2.54	Whelk-O1	(74)

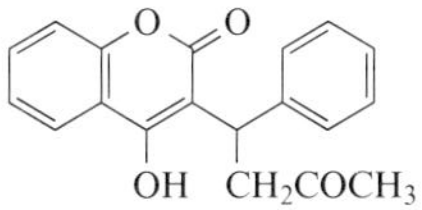

(continued)

Table 23.2. *(contd.)*

Structure	Name	Sample Amount	Column[a]	CSP	Reference
Antiarrhytmic					
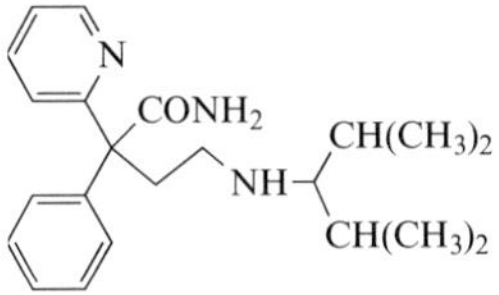	Disopyramide	0.8	15 × 1	Chiral-AGP	(47)
Ca Antagonist					
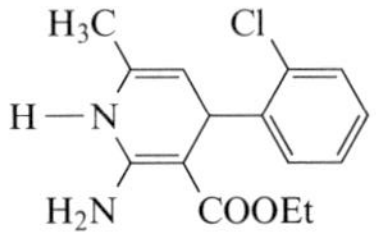	Dihydropyridine	n.m.	60 × 1	CTA-I	(107)
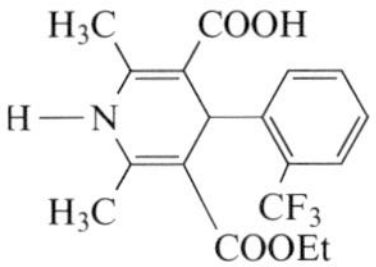	Dihydropyridine	50 g	n.m.	Poly-MA	(34)
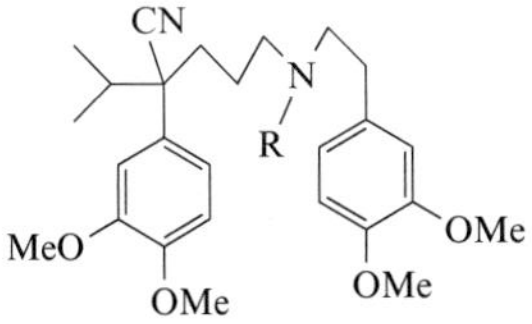	Verapamil R = —CH_3	25–100 mg	50 × 1	Chiralcel OD	(110)
	Norverapamil R = —H	25–100 mg	50 × 1	Chiralcel OD	(110)

Structure	Compound	Amount	Column	CSP	Ref.
K Channel Opener					
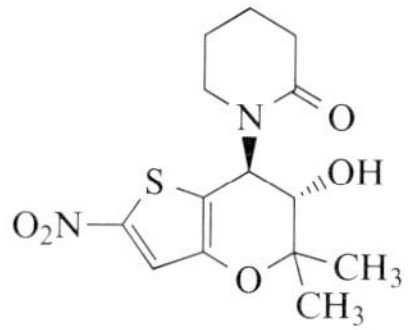	Thienopyran	30 mg	25 × 1	Cycobond I	(27)
Diuretic					
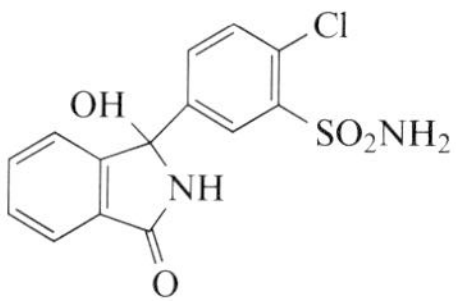	Chlortalidone	530 mg	36 × 3.2	Poly-PEA	(93, 94)
	Chlortalidone	250 mg	46 × 2.6	ChiraSpher	(108)
	Chlortalidone	530 mg	n.m.	Poly-MA	(35)
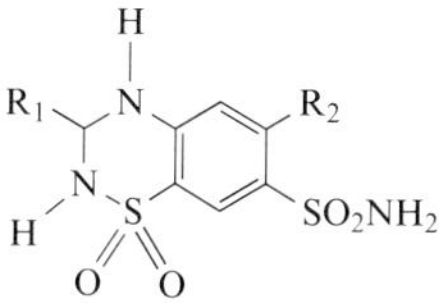 R_1= / R_2=					
$—(CH_2)CH_3$ / $—CF_3$	Penflutizide	210 mg	79 × 3.8	Poly-PE	(95)
$—CH_2C_6H_5$ / $—CF_3$	Bendroflumethiazide	210 mg	79 × 3.8	Poly-PE	(95)
$—CH_2C_6H_5$ / $—CF_3$	Bendroflumethiazide	2.32 mg	15 × 1	Chiral-AGP	(47)
$—CH_2CH(CH_3)_2$ / $—Cl$	Buthiazide	253 mg	79 × 3.8	Poly-PE	(95)

(continued)

Table 23.2. *(contd.)*

Structure	Name	Sample Amount	Column[a]	CSP	Reference
PAF Antogonist					
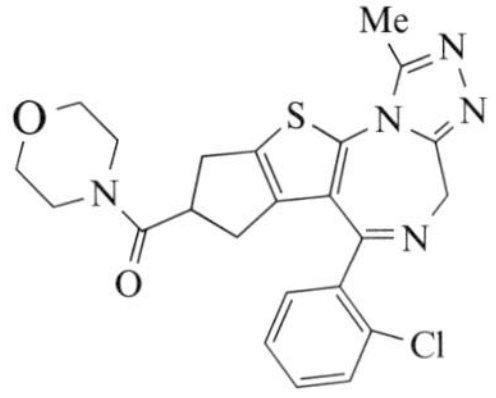	WEB 2170	1.3 kg	50 × 48	CTA-I	(98)
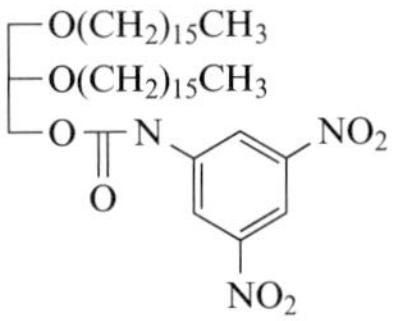	Intermediate	20 mg	(20 × 0.46)2	NAP-A1	(68)
Vasodilator					
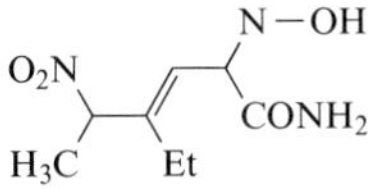	FK 409	676 mg	50 × 10	Chiralcel OJ	(112)

[a]Dimensions = length × i.d in centimeters.
n.m. = not mentioned
SMB = simulated moving bed.
SFC = supercritical or subcritical fluid chromatography.

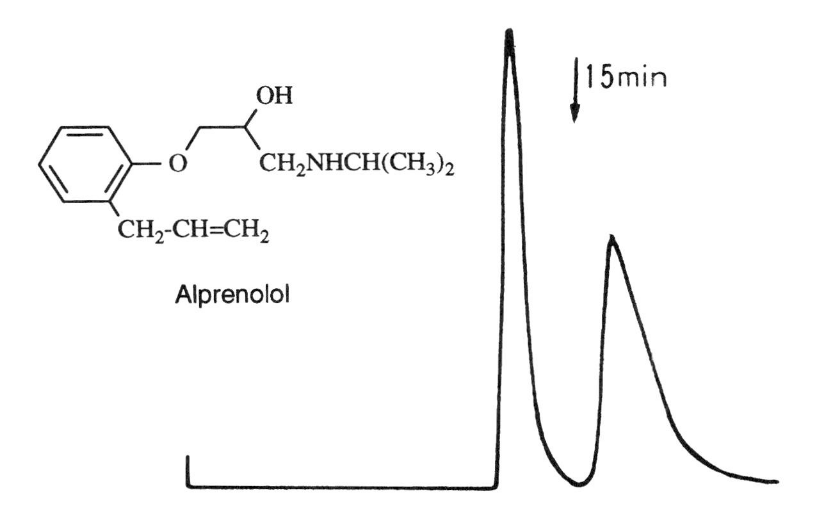

Figure 23.10. Preparative separation of the enantiomers of racemic alprenolol (150 mg) on Chiralcel OD (103).

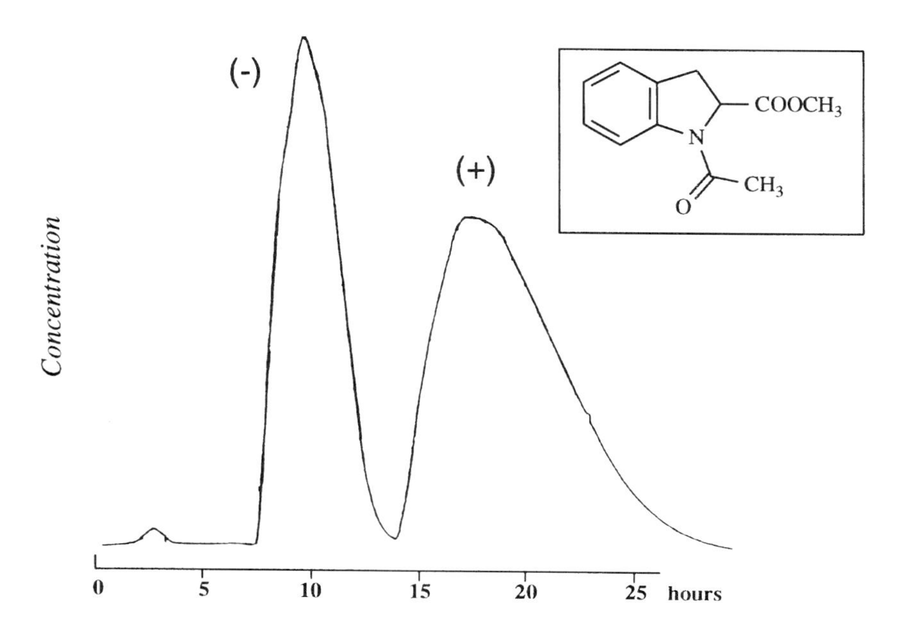

Figure 23.11. Preparative separation of the enantiomers of 45 g of racemic N-acetyl-indoline-2-carboxylic acid methylester on cellulose triacetate (CTA-I) (106). Column, 20 cm (i.d) × 100 cm (12 kg CTA-I); mobile phase, ethanol/water 94:6; flow rate, 5 L/h.

(Figure 23.11) (106) and of a PAF-antagonist drug at a scale of 1.3 kg per run (98). For calcium antagonist drugs, different kinds of CSPs have been used, comprising some cellulose derivatives and a polyacrylamide-type CSP, depending on the structural class of the racemate. Up to 50 g of a dihydropyridine-type calcium antagonist has been resolved in one run on the menthylacrylamide CSP Poly-MA (34). Chromatographic isolation of the pure enantiomers of the 1,4-dihydropyridine compound H 160/51 allowed the biological study to be performed, showing that the two enantiomers have opposing actions, (107). For the diuretics chlortalidone and other diuretics of the benzothiadiazine type, the polyacrylamide CSPs seems to be particularly appropriate (35, 93–95, 108). The enantiomers of all benzothiadiazine derivatives reported in Table 23.2 were isolated for the first time using this chromatographic technique. Small amounts of the enantiomers of C-14 labeled warfarin also have been separated for stereoselective binding studies on human serum albumin (100).

23.4.2. Central Nervous System

A wide variety of racemic structures belonging to the therapeutic area covering the central nervous system, such as anesthetic, anticholinergic, anticonvulsant, antidepressant, anxiolytic, hypnotic, and nootropic agents, have been resolved by chromatography on CSPs (Table 23.3). Depending on the structure of the racemate, successful separation of the enantiomers could be achieved on a cellulose-based CSP, π-acidic CSP, or polyacrylamide-type CSP. Bendiazepinone derivatives such as oxazepam and lorazepam have been resolved preparatively on the π-acidic phases (113–117), whereas Blaschke and co-workers have been able to resolve various barbiturates on CTA-I (93, 118). The hypnotic methaqualone also was resolved on CTA-I for the first time (89, 90). This compound has no center of chirality but possesses a C_2 symmetry (atropisomers), and the lack of functionalities prevents any resolution by the formation of diastereoisomers. Studies on the barrier to rotation of the phenyl-nitrogen bond have been carried out in order to determine the racemization rate of this molecule (90). Various cyclopyrrolone derivatives have been resolved on Chiralcel OC in amounts required for toxicological trials (111). The preparative resolution of racemic thalidomide (Figure 23.12) in 1979 (119, 120) on a polyacrylamide CSP developed by Blaschke remains a typical example of the application of chromatography on CSP, because it greatly facilitated access to the pure enatiomers needed for the biological study carried out to elucidate the controversial question of whether only the (*S*)(–)-enantiomer was responsible for the teratogenic activity (121). Chromatographic resolution was also the unique way to obtain the pure enantiomers of the muscle relaxant chlormezanone, which cannot be derivatized

Table 23.3. Preparative Resolutions of Racemic Drugs for Disorders of the Central Nervous System

Structure	Name	Sample Amount	Column[a]	CSP	Reference
Alkaloids					
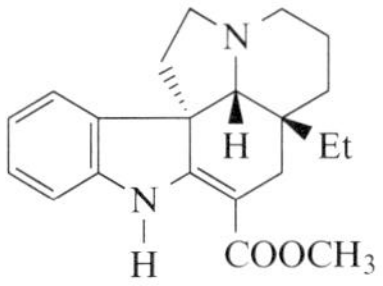	Vincadifformine	500 mg	95 × 5	CD-Poly	(24)
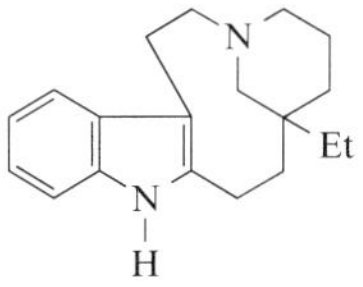	Quebrachine	200 mg	95 × 5	CD-Poly	(24)
Anesthetic					
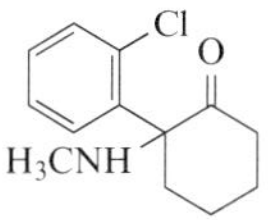	Ketamine	450 mg	70 × 3.8	CTA-I	(93)
Anticholinergic					
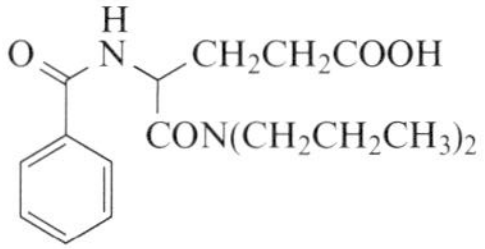	Proglumide	0.84 mg	15 × 1	Chiral-AGP	(47)

(continued)

Table 23.3. *(contd.)*

Structure	Name	Sample Amount	Column[a]	CSP	Reference
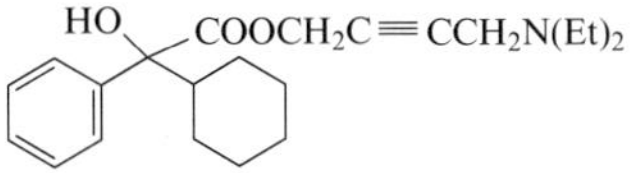 HO, $COOCH_2C{\equiv}CCH_2N(Et)_2$	Oxybutynin	0.5 mg	25 × 1	Chiralpak AD	(122)
Anticonvulsive					
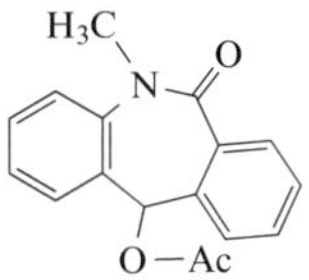 H_3C, N, O, O—Ac	Etazepine acetate	300 mg	100 × 5	CTA-I	(123)
Antidepressant					
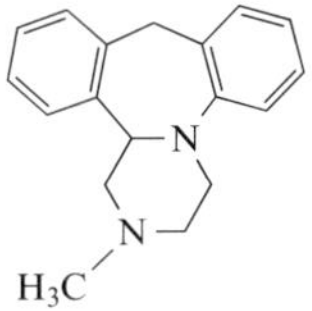 N, N, H_3C	Mianserin	8 mg	20 × 1.5	CTA-I	(93)
Muscle Relaxant					
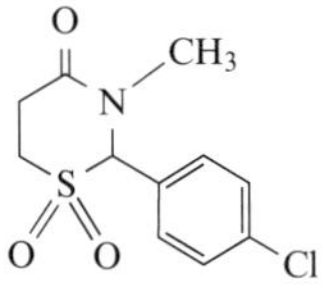 O, CH_3, N, S, O, O, Cl	Chlormezanone	250–700 mg	70 × 3.8	CTA-I	(96, 97)

Anxiolitic

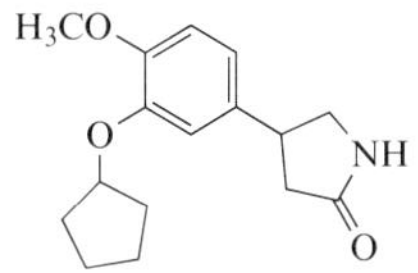	Rolipram	205 mg	85 × 2.4	CTA-I	(93)
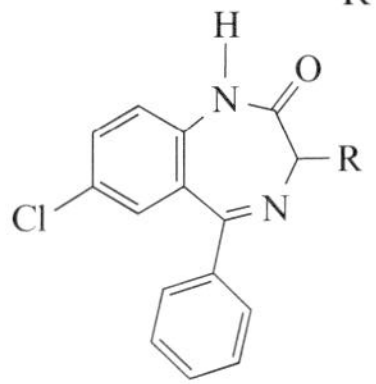 R = OH	Oxazepam	46 mg	76 × 3.2	Poly-PEA	(94)
OH	Oxazepam	5 mg	25 × 2	DNBPG-io	(113)
$OCOC(CH_3)_3$	Oxazepam ester	200 mg	15 × 4	DNBPG-co	(114)
CH_3	Benzodiazepinone	25–1000 mg	25 × 1	DNBPG-co	(114, 115)
$CH(CH_3)_2$	Benzodiazepinone	25–1000 mg	25 × 1	DNBPG-co	(114, 115)
$CH(CH_3)_2$	Benzodiazepinone	100 mg	25 × 1	DNBLeu	(116)
$CH_2CH_2SCH_3$	Benzodiazepinone	104 mg	25 × 1	DNBLeu	(115)
CH_2Ph	Benzodiazepinone	95 mg	25 × 1	DNBLeu	(117)
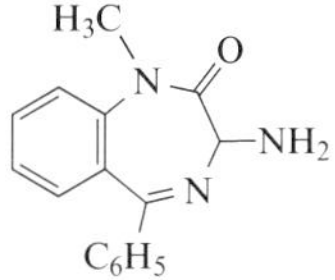	Intermediate	2 g	20 × 4	Chiralcel OC	(125)
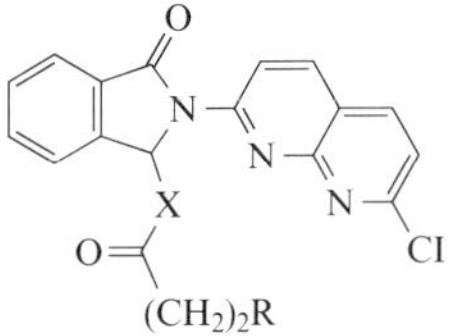	R=-$CH(CH_3)_2$	500 mg	22 × 7	Chiralcel OC	(111)
	R=-$CH_2NHCOCH_3$	2 g	22 × 7	Chiralcel OC	(111)
	R=-$CH_2NHCOEt$	4 g	22 × 7	Chiralcel OC	(111)

(continued)

Table 23.3. *(contd.)*

Structure	Name	Sample Amount	Column[a]	CSP	Reference
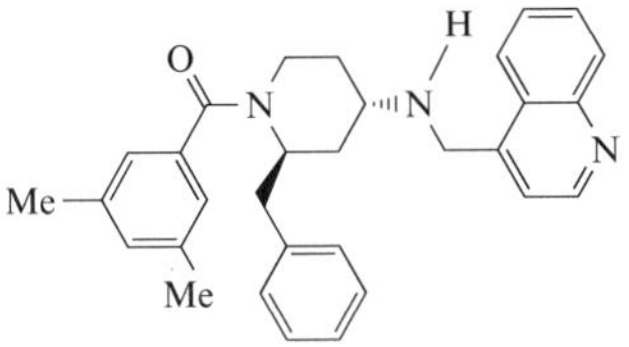	CGP 49823	1 g	50 × 5	Chiralcel OD	(91)
Nootropic					
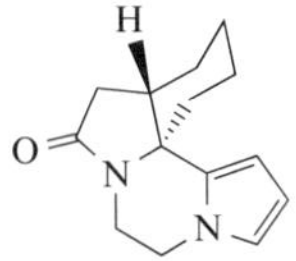	CGS 16920	2 g	45 × 5	TBC	(91)
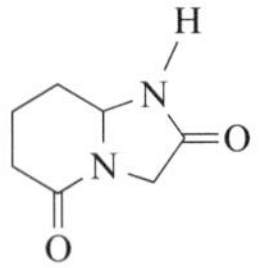	Dimiracetam	40 mg	25 × 1	Chiralcel OC	(124)

Hypnotic					
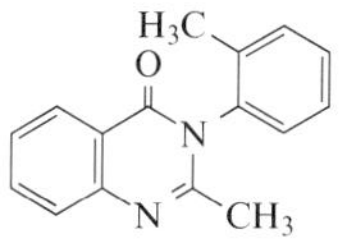	Methaqualone	50–300 mg	24 × 4	CTA-I	(89, 80)
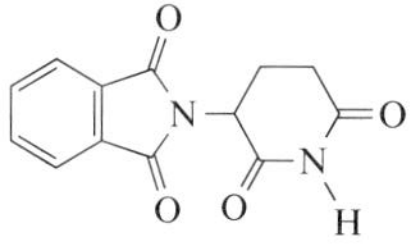	Thalidomide	505 mg	62 × 2.3	Poly-CHMA	(119, 120)
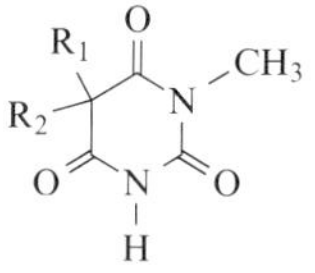	Barbiturates	140–257 mg	85 × 2.5	CTA-I	(93, 118)

R_1 =	R_2 =
Me, Et, n-Pr	C_6H_5
Me, Et	cyclohexene
Me, Et	cyclohexane
Et	cyclopentene

[a] Dimensions-length × i.d in centimeters.

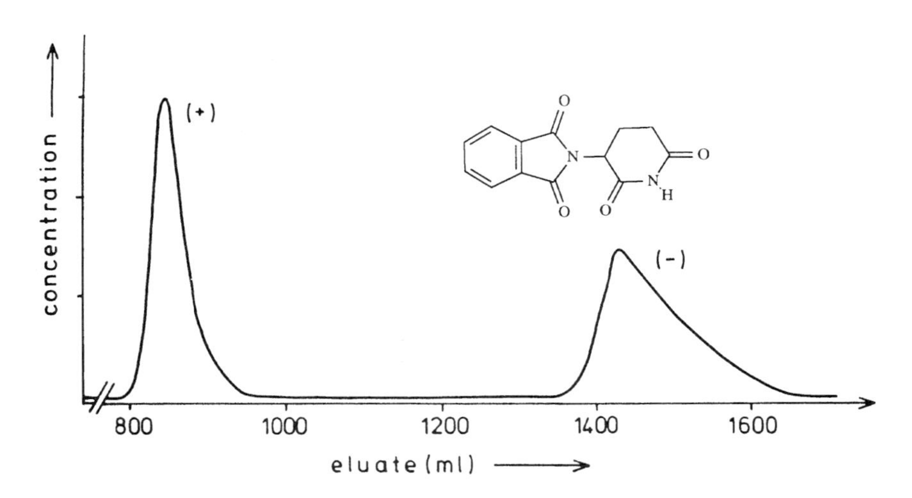

Figure 23.12. Chromatographic separation of the enantiomers of thalidomide (52 mg) on poly-methacrylamide (Poly-CHMA). Column, 80 × 2.3 cm; mobile phase, benzene/dioxane 4:1. [Reprinted with permission from G. Blaschke (1986). *J. Liq. Chromatogr.*, **9**, 341, courtesy of Marcel Dekker.]

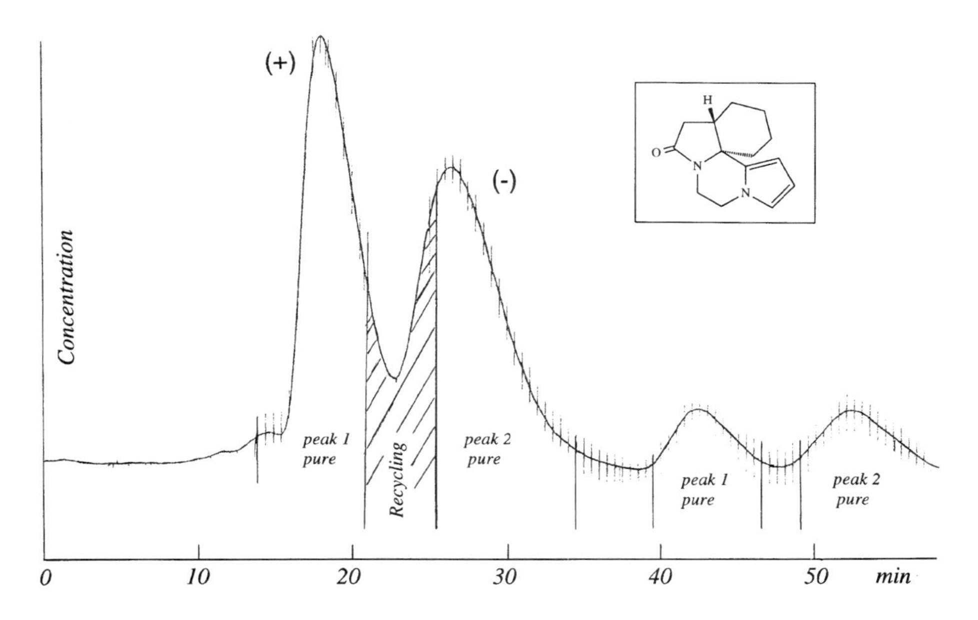

Figure 23.13. Chromatogram of the resolution of the racemic nootropic drug CGS 16920 (2 g) on tribenzoylcellulose beads (TBC) with recycling and peak shaving (91). Column, 5 cm × 45 cm; mobile phase, methanol 100%; flow rate, 60 mL/min.

(96, 97). For the nootropic CGS 16920, a tetracyclic compound, the enantiomers easily have been isolated by preparative resolution on tribenzoylcellulose beads, using the recycling/peak shaving technique (Figure 23.13) (91).

23.4.3. Tumor and Infection

Drugs used against tumors and infections include the important classes of antibiotics and anticancer agents (Table 23.4). Racemic lactone and sulfoxide derivatives used as intermediates for the preparation of β-lactam antibiotics were resolved into their corresponding enantiomers by chromatography on cellulose-based phases (126, 127). Multigram amounts of the γ-lactone were resolved in this way on cellulose triacetate (126).

The enantiomers of the anticancer drug fadrozole (91, 128) and antiaromatase agent CGP27 216 (123, 133) were isolated for the first time by chromatographic resolution on cellulose-based CSPs in our laboratories. For fadrozole, the chromatographic method also has been applied to the ^{14}C-labeled and deuterated compounds (Figure 23.14). For another anticancer

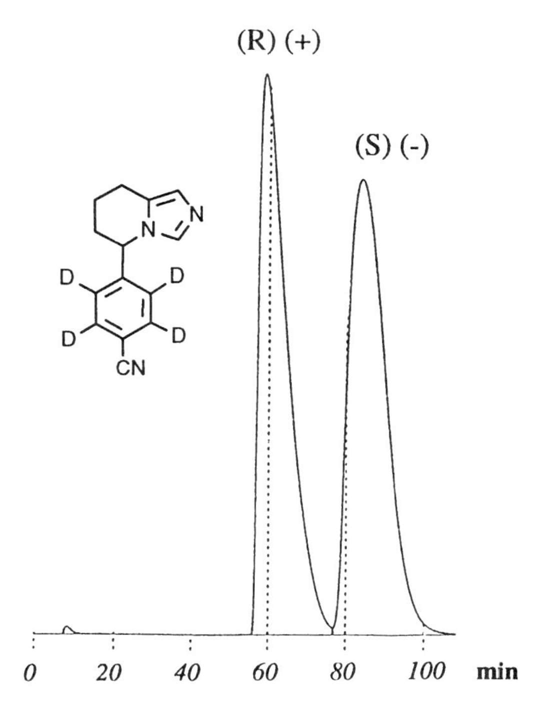

Figure 23.14. Separation of the enantiomers of deuterium labeled fadrozole (1 g) on Chiralcel OD (91). Column, 5 cm × 50 cm i.d.; mobile phase, hexane/2-propanol 9:1; flow rate, 100 mL/min.

Table 23.4. Preparative Resolutions of Racemic Drugs for Tumor and Infections

Structure	Name	Sample Amount	Column[a]	CSP	Reference
Antibiotic					
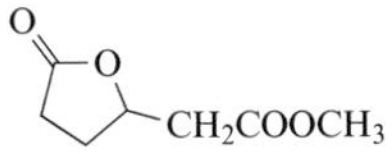	Intermediate	4.2 g	60 × 5	CTA-I	(126)
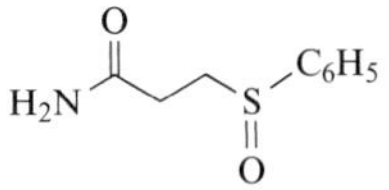	Intermediate	100 mg	50 × 2	Chiralcel OB	(127)
Anticancer					
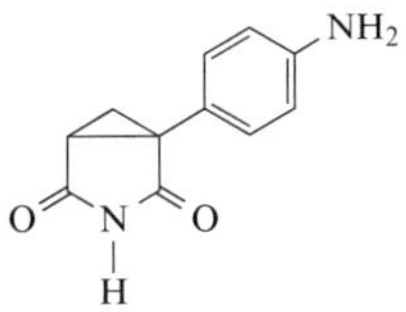	CGP 27216	300 mg	60 × 5	CTA-I	(123, 133)
	Fadrozole	1 g	75 × 5	TBC	(91, 128)
	[^{14}C]-Fadrozole	100 mg	75 × 5	TBC	(136)
	D_4-Fadrozole	1 g	50 × 5	Chiralcel OD	(91)

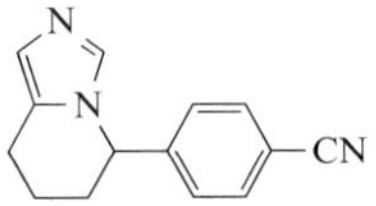

Structure	Compound	Amount	Column[a]	CSP	Ref.
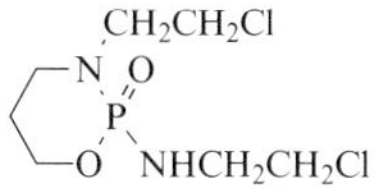 CH_2CH_2Cl, N, O, P, O, $NHCH_2CH_2Cl$	Ifosfamide Ifofamide	1.55 g 8 mg	155 × 5 25 × 1	Poly-PEA Chiralcel OD	(129, 130) (131)
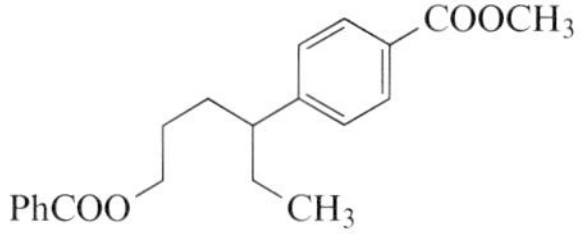 $COOCH_3$, PhCOO, CH_3	Edatrexate intermediate	4 g	70 × 5	TBC	(91)
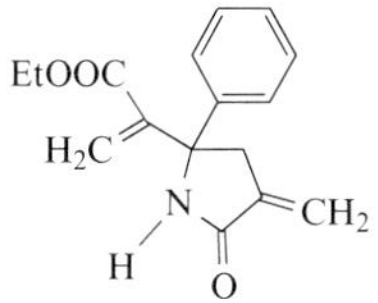 EtOOC, H_2C, N, H, O, CH_2	—	120 mg	40 × 2.6	DNB-Tyr-A	(132)
Infection					
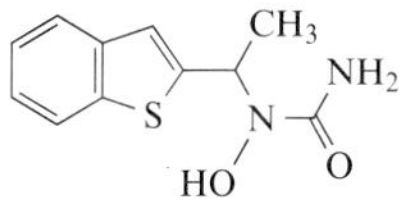 CH_3, NH_2, S, N, HO, O	[^{14}C]-Zileuton	7 mg	25 × 2	Chiralpak AD	(99)

[a]Dimensions-length × i.d in centimeters.

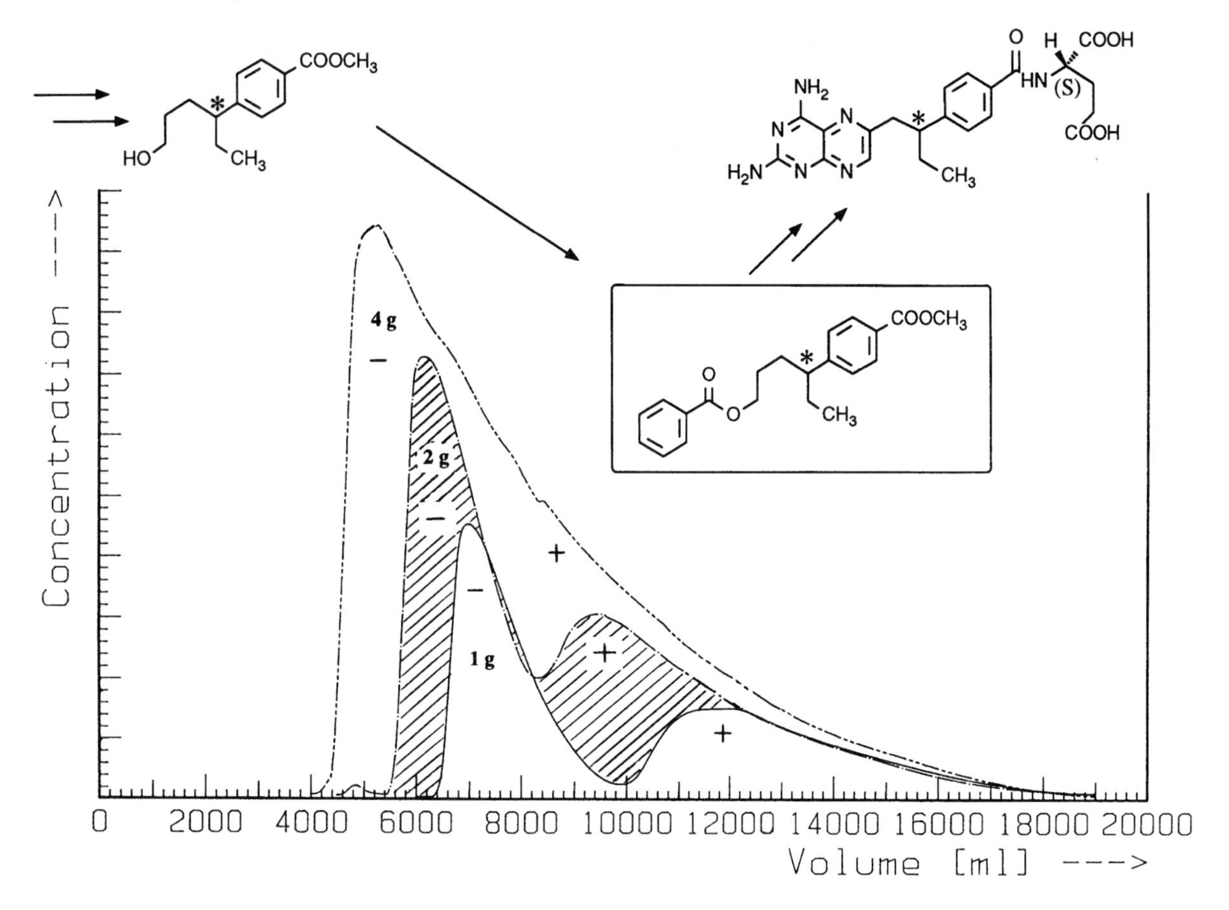

Figure 23.15. Preparative resolution of the benzoate derivative of a racemic intermediate used for the synthesis of the anticancer drug edatrexate on tribenzoylcellulose beads (CTB) (91). Influence of the sample size on the elution profile. Column, 70 cm × 5cm i.d.; mobile phase, hexane/2-butanol 9:1.

drug, edatrexate, the separation has been performed at the stage of the enantiomers of an alcohol intermediate (Figure 23.15) that is not separated on tribenzoylcellulose, but which after transformation into the corresponding benzoate is resolved completely on this CSP (91). This example illustrates the optimization strategy discussed above that consists of modifying the structure of the solute to obtain a separation on a defined CSP. Figure 23.14 shows the influence of the sample amount, varying in the range of 1–4 g per injection, on the elution profile. After several injections, 95 g of each pure enantiomer were isolated by chiral chromatography. The enantiomers of ifosfamide, an anticancer drug containing a chiral phosphorus atom, could be obtained by chromatography on two different CSPs (Chiralcel OD and the polyacrylamide Poly-PEA) and were used for clinical studies (129–131).

23.4.4. Antiinflammatory Agents and Analgesics

Most of the preparative chromatographic resolutions of racemic antiinflammatory and analgesic agents reported nowadays have been performed on CTA-I (Table 23.5) in amounts ranging between a few milligrams to several grams.

For the nonsteroidal antiinflammatory (NSAID) oxindanac, all the other methods of separation of the enantiomers failed because the center of chirality is very sensitive to racemization. For the preparative separation, the acid has been derivatized to an ester. The choice of the derivative also has been optimized to achieve a good separation, but at the same time assuring that the optical purity is preserved upon removal of the ester group. Various ester derivatives have been investigated (92), and the benzyl ester, which is smoothly cleaved by hydrogenolysis, was found to best fulfill the required conditions. For pharmacological studies, amounts up to 200 mg of racemate were resolved (92), on a 5 cm (i.d.) × 50 cm column, whereas for clinical trials, a larger column was used (106). Portions of 22–23 g of racemates were continuously injected on this large CTA column, and the waiting time between two injections has been reduced to the minimum in order to increase the throughput per unit of time (Figure 23.16). The same strategy has been applied by Miller et al. for the preparative resolution of SC-51146, an antiinflammatory agent, on the amylose-based CSP Chiralpak AD (134). For ibuprofen, the displacement chromatography has been applied using Cylobond I as the CSP and 4-*tert*-butylcyclohexanol as the displacer (69). Although this study affords useful insight into the application of displacement chromatography for preparative chiral resolutions, the relatively low loading capacity of the cyclodextrin-based CSP used in these investigations makes this application not really practical compared with other methods of resolving this drug. For instance, various 2-arylpropionic acid NSAID enantiomers have been

Table 23.5. Preparative Resolutions of Racemic Antiinflamatory and Analgesic Agents

Structure	Name	Sample Amount	Column[a]	CSP	Reference
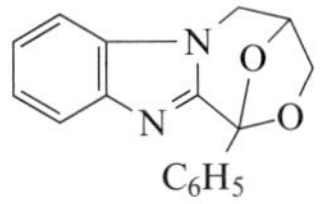	Oxapadol	2.1 g	70×3.8	CTA-I	(93)
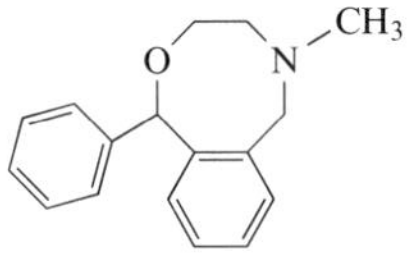	Nefopam	100 mg	75×5	CTA-I	(137)
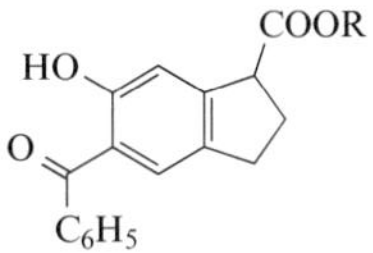	Oxindazac ester R=-$C(CH_3)_3$	100 mg	75×5	CTA-I	(92)
	Oxindazac ester R=-$CH_2C_6H_5$	20 g	100×20	CTA-I	(92, 106)
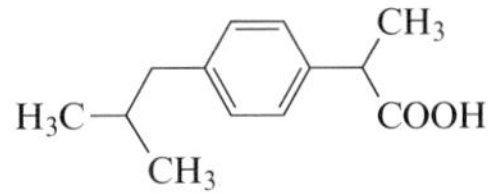	Ibuprofen	0.5 mg	(25×0.46) 2	Cyclobond I	(69)
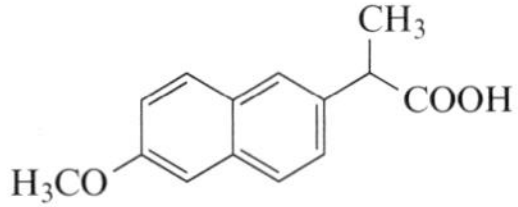	Naproxen	12 mg	25×0.46	Whelk-O 1	(56)

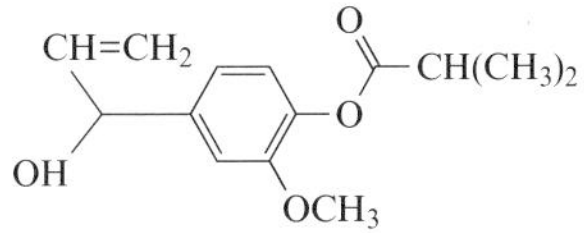	Hydroxyeugenol	20 mg	50 × 2	CTA-I	(138)
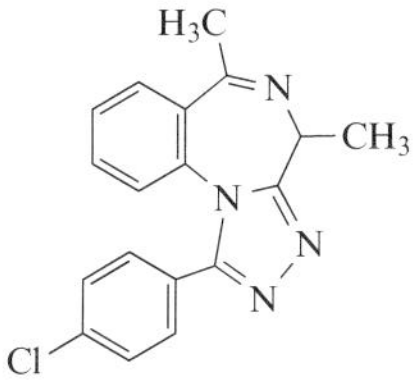	RU 48159	1 g	25 × 5	DNBPG-co	(135)
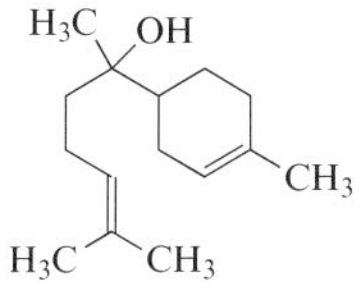	α-bisabolol	500 mg	20 × 10	CTB	(139)
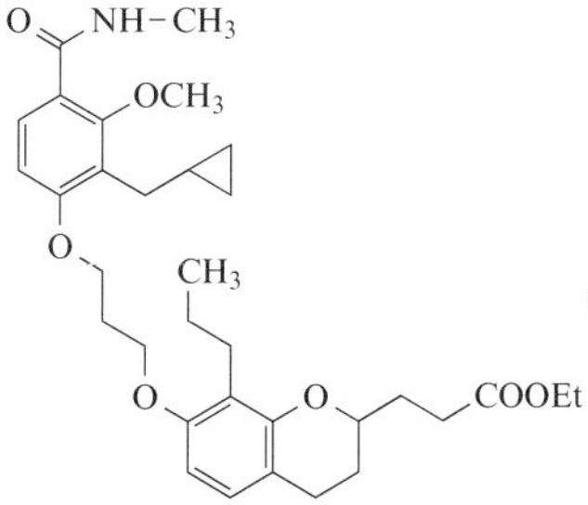	SC 51146	10 mg	50 × 1	Chiralpak AD	(134)

[a]Dimensions-length × in centimeter.

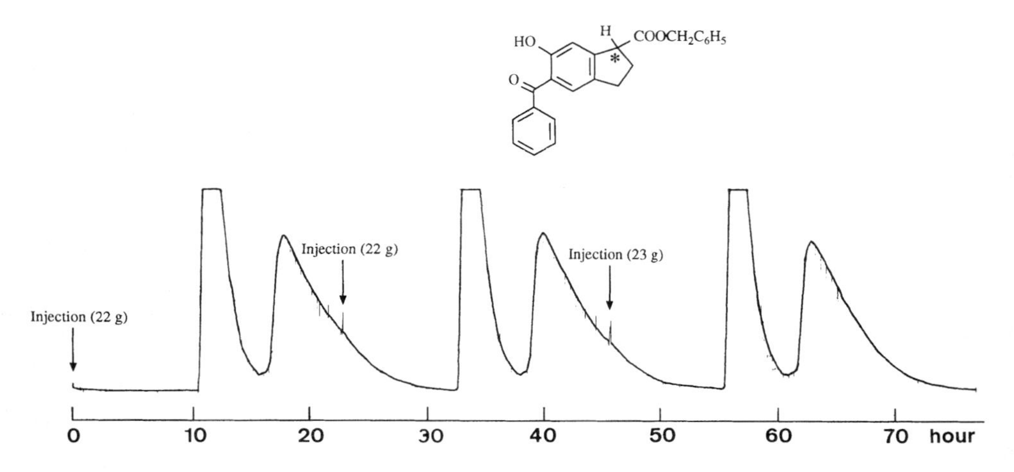

Figure 23.16. Continuous preparative chromatographic resolution of the benzylester of the racemic antiinflammatory agent oxindanac on cellulose triacetate (CTA-I) (106). Column, 20 cm (i.d.) × 100 cm (12 kg CTA-I); mobile phase, ethanol/water 94:6; flow rate, 5 L/h.

resolved on a preparative scale on the recently developed Whelk-O 1 CSP, which has good loading capacity (56). Sometimes, the solubility of the racemate can be a determining factor in choosing the CSP, as in the case of the analgesic RU 48159, for example; since this substance is only readily soluble in chlorinated alkanes, the authors opted preferentially for a Pirkle-type CSP, which is compatible with a larger number of mobile phases that the commonly cellulose-based CSPs (135).

23.4.5. Prostaglandins

Cyclopentenones constitute an important class of intermediates in the synthesis of prostaglandins. Various chiral intermediates of this type have been resolved into the corresponding enantiomers by preparative chromatography on different cellulose-based CSPs (Table 23.6). The phenylcarbamate derivative of cellulose (Chiracel OC) especially seems to be appropriate for resolving racemates of this class (140–142). Miller and Bush have reported a detailed optimization study for a cyclopentanone derivative of this type (140), showing that mobile-phase composition, flow rate, loading, and temperature can have a strong influence on the throughput of a preparative separation.

Table 23.6. Preparative Resolutions of Prostaglandine Precursors

Structure	Name	Sample Amount	Column[a]	CSP	Reference
$OSi(C_2H_5)_3$; O; $(CH_2)_2R$ R = $CH{=}CH(CH_2)_2COOMe$ $C{\equiv}C(CH_2)_2COOMe$ $(CH_2)_4CH_3$	Intermediate	500 mg	50×2	Chiralcel OC	(140, 141)
O; $OSiMe_2tBu$	Intermediate	10 mg	25×2	Chiralcel OC	(142)
O; OH	Intermediate	n.m.	55×7	CTA-I	(13)

[a]Dimensions-length × in centimeters.
n.m.:-not mentioned.

23.4.6. Miscellaneous Drugs

The enantiomers of miscellaneous drugs or drug intermediates used in various other therapeutic indications, such as the antiulcer agent omeprazole, the anthelmintic praziquantel, or the gastrointestinal agent ML-1035, also have been obtained by preparative chromatographic resolution (Table 23.7), and subjected to biological and pharmacological studies. For praziquantel (66) and an epoxide used as intermediate for drug synthesis (67), simulated moving-bed adsorption technology has been applied and shown to result in a marked improvement of the throughput. In particular, the amount of mobile phase needed to isolate a predetermined quantity of enantiomer can be reduced greatly (10–50 times) compared with that needed in the conventional batch process.

The SFC mode has been successfully applied to the preparative separation of the enantiomers of the racemic antitussive guaifenesine, allowing the resolution of 17 mg of racemate per hour per gram of CSP (79, 146). Up to 60 mg

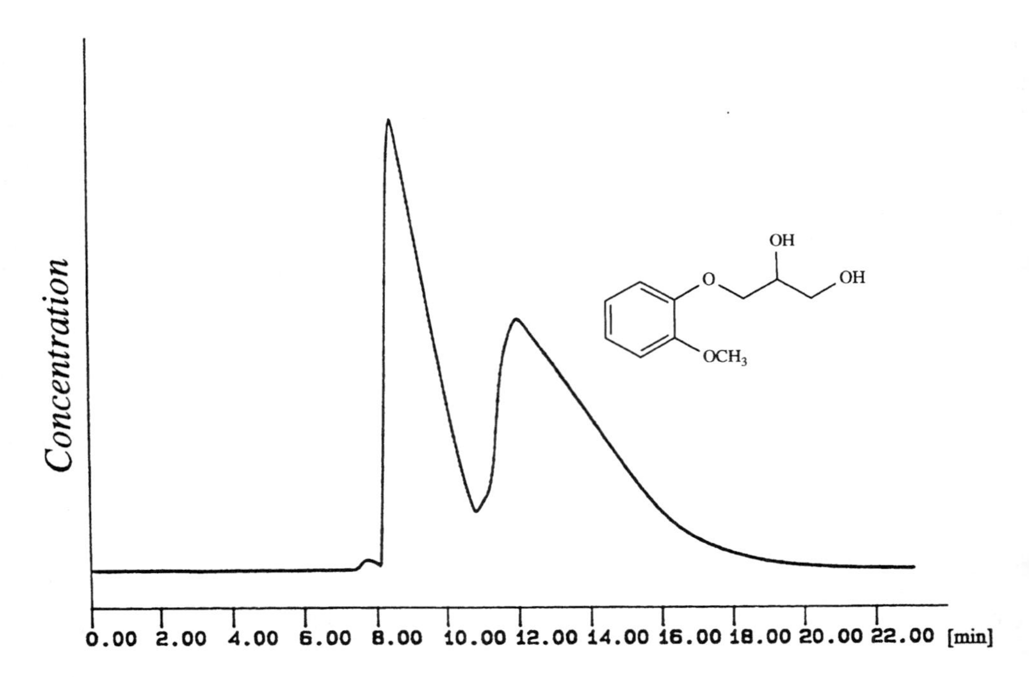

Figure 23.17. Chromatographic resolution of 60 mg of racemic guaifenesine on 3,5-dimethylphenylcarbamate of cellulose under SubFC conditions (carbon dioxide/2-propanol/methanol 86:7:7) (146). Column, 25 cm × 1 cm; 5 mL/min.

Table 23.7. Preparative Resolutions of Miscellaneous Chiral Drugs

Structure	Name	Sample Amount	Column[a]	CSP	Reference
Antiulcer					
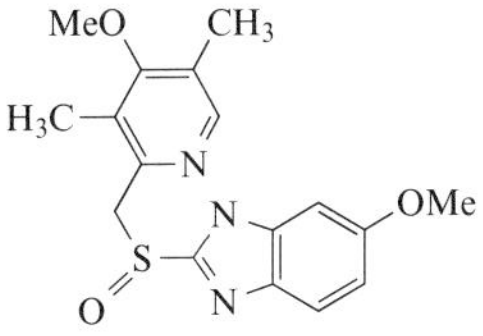	Omeprazole	5 mg	30×1	CTA-I	(143)
Gastrointestinal					
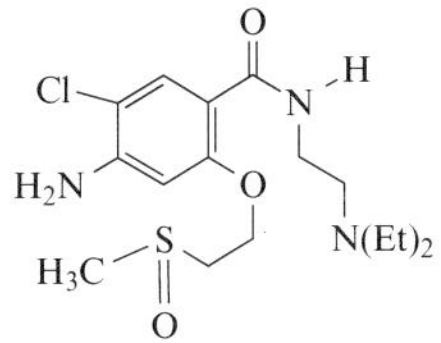	ML-1035	0.5 mg	25×0.46	Chiralcel OD	(144)
Hypocholesterolemic					
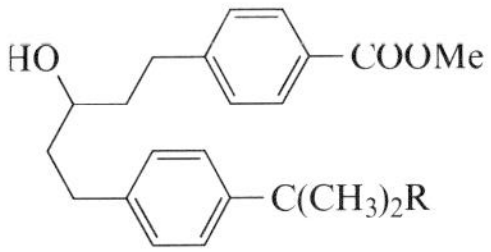 R = CH_2OH, —COOMe	Lifibrol metabolite	100–250 mg	50×2	Chiralcel OD	(145)

(*continued*)

Table 23.7. *(contd.)*

Structure	Name	Sample Amount	Column[a]	CSP	Reference
Antihelmintic					
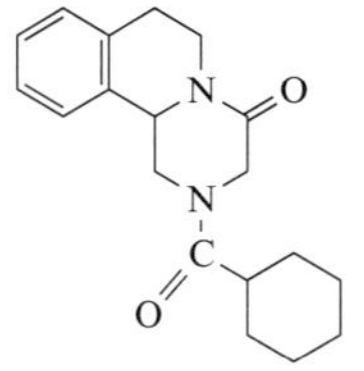	Praziquantel	1 g/h[(SMB)]	(44 × 1.25) 4	CTA-I	(66)
Antitussive					
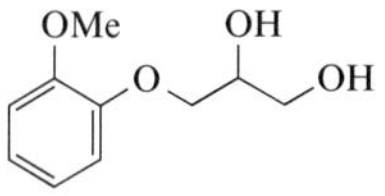	Guaifenesine	60 mg[(SFC)]	25 × 1	Chiralcel OD	(79, 146)
Various					
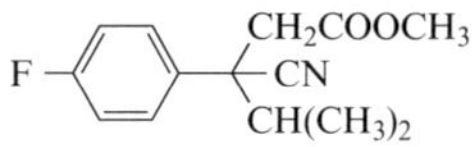	Intermediate	250 mg	50 × 1	Chiralpak AD	(134)
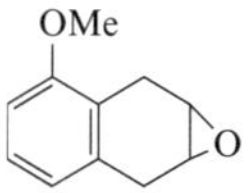	Intermediate	[(SMB)]	(11 × 2.6) 10	CTA-I	(67)

[a]Dimensions-length × i.d in centimeters
SMB=simulated moving bed.
SFC=supercritical or subcritical fluid chromatography.

of racemate could be injected per run on a 25 cm × 1 cm column containing 3,5-dimethylphenyl carbamate of cellulose as a CSP, and the separation took less than 20 min (Figure 23.17).

23.5. CONCLUSION

A variety of CSPs are now available for the preparative chromatographic resolution of racemic drugs. Among the different preparative techniques nowadays available, the batch-elution process has been the most commonly applied, but simulated moving-bed technology is arousing increasing interest. The chromatographic method has now been recognized as a powerful and widely applicable tool for preparing the enantiomers of potential new drugs. Numerous racemic molecules that could not be resolved by conventional methods, especially those lacking functional groups, were separated into their antipodes for the first time by chromatography on CSPs. Most of the preparative resolutions (about 70%) have been performed on cellulose-based CSPs. For large-scale separations, up to 1 Kg of racemate, CTA-I has been preferred, probably because of the relatively low costs of this CSP. Preparative applications on other cellulose derivatives have been reported, and the feasibility of gram-scale separation has been demonstrated. Some drugs, such as the diuretics chlortalidone and the benzothiadiazines, were particularly well resolved on the poly(meth)acrylamide types of CSP, whereas the benzodiazepines are generally well resolved on the π-acidic CSPs. All the examples reported in this review show that, even if new developments are to be expected in this field, the presently available CSPs already allow the isolation of the optically pure enantiomers of many racemic drugs, and that the chromatographic method has became an essential tool in the pharmaceutical industry for supplying the individual enantiomers needed to evaluate biological activity, at least during the early phase of development.

REFERENCES

1. Simonyi, M. (1984). *Med. Res. Rev.*, **4**, 359.
2. Williams, K. and Lee, E. (1985). *Drugs*, **30**, 333.
3. Tucker, G. T. and Lennard, M. S. (1990). *Pharm. Ther.*, **45**, 309.
4. Testa B. and Trager, W. F. (1990). *Chirality*, **2**, 129.
5. Crossley, R. (1992). *Tetrahedron*, **48**, 8155.
6. Stison, S. T. (1992). *Chem. & Eng. News. Sept.*, **28**, 46.
7. Ramos, G. M. and Bellus, D. (1991). *Angew. Chem. Int. Ed.*, **30**, 1193.
8. Francotte E. and Junker-Buchheit, A. (1992). *J. Chromatogr.*, **576**, 1.

9. Francotte, E. (1994). *J. Chromatogr. A*, **666**, 565.
10. Krstulovic, A. M. (ed.) (1989). *Application to Pharmaceutical Compounds. In Chiral Separation by HPLC*. Chichester: Ellis Horwood Ltd.
11. Zief, M. and Crane, L. J. (eds.) (1988). *Chromatographic chiral separations*. Chromatographic Science Series, Vol. 40. New York: Marcel Dekkar, p. 91.
12. Wainer, I. W. (ed.) (1993). *Drug Stereochemistry, Analytical Methods and Pharmacology*, 2nd ed. New York: Marcel Dekker, Chap. 6.
13. Shibata, T., Mori, K., and Okamoto, Y. (1989). In Krstulovic A. M. (ed.), *Chiral Separations by HPLC*. Chichester: Ellis Horwood Ltd., Chap. 13, p. 337.
14. Hesse G. and Hagel, R. (1973). *Chromatographia*, **6**, 277.
15. Rimböck, K.-H., Kastner, F., and Mannschreck, A. (1986). *J. Chromatogr.*, **351**, 346.
16. Francotte, E. and Wolf, R. M. (1991). *Chirality*, **3**, 43.
17. Francotte, E. and Wolf, R. M. (1992). *J. Chromatogr.*, **595**, 63.
18. Okamoto, Y., Aburatani, R., and Hatada, K. (1987). *J. Chromatogr.*, **389**, 95.
19. Okamoto, Y., Kawashima, M., and Hatada, K. (1986). *J. Chromatogr.*, **363**, 173.
20. Okamoto, Y., Aburatani, R., Fukumoto, T., and Hatada, K. (1987) *Chem. Lett.*, 1857.
21. Okamoto, Y., Kaida, Y., Hayashida, H., and Hatada, K. (1990). *Chem. Lett.*, 909.
22. Clarke, R. J., Coates, J. H., and Lincoln, S. F. (1988). *Adv. Carbohyd. Chem. Biochem.*, **46**, 205.
23. Harada, A., Furue, M., and Nozakura, S.-I. (1978). *J. Polym. Sci.*, **16**, 189.
24. Zsadon, B., Décsei, L., Szilasi, M., and Tüdos, F. (1983). *J. Chromatogr.*, **270**, 127.
25. Haan, S. M. and Armstrong, D. W. (1989). HPLC Separation of enantiomers and other isomers with cyclodextrin-bonded phases. In Krstulovic, A. M. (ed.), *Chiral Separations by HPLC*. Chichester: Ellis Horwood Ltd., Chap. 10, pp. 208–284.
26. Cabrera, K., Lubda, D., and Spezial, G. I. T. (1992). *Chromatographie*, **2**, 77.
27. Shaw, C. J., Sanfilippo, P. J., McNally, J. J., Park, S. A., and Press, J. B. (1993). *J. Chromatogr.*, **631**, 173.
28. Bruche, G., Mosland, A., and Kinkel, J. N. (1993). *JHRC*, **16**, 254.
29. Armstrong, D. W., Hilton, M., and Coffin, L. (1992). *LC-GC Int.*, **5**(1), 28.
30. Chang, S. C., Reid III, G. L., Cheng, S., Chang, C. D., and Armstrong, D. W. (1993). *Trends Anal. Chem.*, **12**, 144.
31. Schwanghart, A. -D., Backmann, W., and Blaschke, G. (1977). *Chem. Ber.*, **110**, 778.
32. Kinkel, J. N., Fraenkel, W., and Blaschke, G. (1987). *Kontakte (Darmstadt)*, 3.
33. Blaschke, G., Bröker, W., and Fraenkel, W. (1986). *Angew. Chem.*, **98**, 808.
34. Arlt, D., Bömer, B., Grosser, R., and Lange, W. (1991). Poster at 2nd International Symposium on Chiral Discrimination, Rome, May.

35. Arlt, D., Bömer, B., Grosser, R., and Lange, W. (1991). *Angew. Chem.*, **103**, 1685.
36. Allenmark, S. (1983). *J. Chromatogr.*, **264**, 63.
37. Noctor, T. A. G., Felix, G., and Wainer, I. W. (1991). *Chromatographia*, **31**, 55.
38. Hermansson J. and Eriksson, M. (1986). *J. Liq. Chromatogr.*, **9**, 621.
39. Jadaud, P., Thelohan, S., Schonbaum, G. R., and Wainer, I. W. (1989). *Chirality*, **1**, 38.
40. Marle, I., Jönsson, S., Isaksson, R., Pettersson, C., and Pettersson, G. (1993). *J. Chromatogr.*, **648**, 333.
41. Miwa, T., Miyakawa, T., Kayano, M., and Miyake, Y. (1987). *J. Chromatogr.*, **408**, 316.
42. Kirkland, K. M., Neilson, K. L., McCombs, D. A., and DeStefano, J. J. (1992). *LC-GC Sep. Sci.*, **10**, 322.
43. Schill, G., Wainer, I. W., and Barkan, S. A. (1986). *J. Chromatogr.*, **365**, 73.
44. Enquist, M. and Hermansson, J. (1990). *J. Chromatogr.*, **519**, 285.
45. Iredale, J., Aubry, A.-F., and Wainer, I. W. (1991). *Chromatographia*, **31**, 329.
46. Kirkland, K. M., Neilson, K. L., and McCombs, D. A. (1991). *J. Chromatogr.*, **545**, 43.
47. CHIRAL-AGP (ChromTech) (1993). Application prospect. Hägersten (Sweden).
48. Macaudière, P., Lienne, M., Tambutè, A., Caude, M. (1989). Pirkle-type and related chiral stationary phoses for enantiomeric resolution. In A. Krstulovic (ed.), *Chiral Separations by HPLC*. Chichester: Ellis Horwood Ltd. Ch. 14.
49. Pirkle, W. H., Finn, J. M., Hamper, B. C., Schreiner, J., and Pribish, J. A. (1982). In E. L. Eliel and S. Otsuka (eds.), *Asymmetric Reactions and Processes in Chemistry*, ACS Symposium Series, No. 185. Washington, DC: ACS, Chap. 18, pp. 245–260.
50. Pirkle, W. H. and Pochapsky T. C. (1987). *Adv. Chromatogr.*, **27**, 73.
51. Caude, M., Tambuté, A., and Siret, L. (1991). *J. Chromatogr.*, **550**, 357.
52. Pirkle, W. H., Hyun, M. H., and Bank, B. (1984). *J. Chromatogr.*, **316**, 585.
53. Pirkle, W. H. and Hyun, M. H. (1984). *J. Org. Chem.*, **49**, 3043.
54. Pirkle W. and Welch, C. J. (1992). *J. Liq. Chromatogr.*, **15**, 1947.
55. Pirkle W. H. and Burke III, J. A. (1991). *J. Chromatogr.*, **557**, 173.
56. Welch, C. J. (1993). *Chem. New Zealand*, 9.
57. Oi, N. and Kitahara, H. (1986). *J. Liq. Chromatogr.*, **9**, 511.
58. Pirkle, W. H. and Hyun, M. H. (1985). *J. Chromatogr.*, **322**, 287.
59. Pirkle, W. H. and Murray, P. G. (1990). *J. Liq. Chromatogr.*, **13**, 2123.
60. Pirkle, W. H. and Welch, C. J. (1991). *J. Liq. Chromatogr.*, **14**, 2027.
61. Davankov, V. A., Navratil, J. D., and Walton, H. F. (1988). *Ligand Exchange Chromatography*. Boca Raton, FL: CRC Press.
62. Davankov, V. A., Bochkov, A. S., Kurganov, A. A., Roumeliotis, P., and Unger, K. K. (1980). *Chromatographia*, **13**, 677.

63. Chimie, J. P. S. CH-2022 (1987). Application guide, Bevaix, Switzerland.
64. Negawa M. and Shoji, F. (1992). *J. Chromatogr.*, **590**, 113.
65. Nicoud, R. M. (1992). *LC-GC Intl.*, **5**: (5), 43.
66. Ching, C. B., Lim, B. G., Lee, E. J. D., and Ng, S. C. (1993). *J. Chromatogr.*, **634**, 215.
67. Nicoud, R.-M., Fuchs, G., Adam. P., Bailly, M., Küsters, E., Antia, F. D., Reuille, R., and Schmid, E. (1993). *Chirality*, **5**, 267.
68. Camacho-Torralba, P. L., Beeson, M. D., Vigh, G., and Thompson, D. H. (1993). *J. Chromatogr.*, **646**, 259.
69. Farkas, G., Irgens, L. H., Quintero, G., Beeson, M. D., Al-Saeed, A., and Vigh, G. (1993). *J. Chromatogr.*, **645**, 67.
70. Macaudiére, P., Caude, M., Rosset, R., and Tambuté, A. (1989). *J. Chromatogr. Sci.*, **27**, 383.
71. Fuchs, G. Doguet, L., Perrut, M., Tambuté, A., and Le Goff, P. (1991). Paper presented at 2nd Symposium on Supercritical Fluids, Boston, MA, May.
72. Berger, C. and Perrut, M. (1990). *J. Chromatogr.*, **505**, 37.
73. Fuchs, G., Doguet, L., Barth, D., and Perrut, M. (1992). *J. Chromatogr.*, **623**, 329.
74. Blum, A. M. and Kumar, M. L. (1993). Poster at 4th International Symposium on Chiral Discrimination, Montreal, Canada, Sept.
75. Macaudière, P., Caude, M., Rosset, R., and Tambuté, A. (1989). *J. Chromatogr. Sci.*, **27**, 583.
76. Perrut, M. (1994). *J. Chromatogr. A*, **658**, 293.
77. Lynam, K. G. and Nicolas, E. C. (1993). *J. Pharm. Biochem. Anal.*, **11**, 1197.
78. Siret, L., Bargmann, N., Tambuté, A., and Caude, M. (1992), *Chirality*, **4**, 252.
79. Anton, K., Eppinger, J. and Francotte, E. (1993). Poster at Supercritical Chemistry Symposium, Nottingham, UK, Sept.
80. Francotte, E. and Wolf, R. M. (1990). *Chirality*, **2**, 16.
81. Pirkle, W. H. and McCune, J. E. (1988). *J. Liq. Chromatogr.*, **11**, 2165.
82. Dyas, A. M., Robinson, M. L., and Fell, A. F. (1990). *Chromatographia*, **30**, 73.
83. Uzunov, D. and Stoev, G. (1993). *J. Chromatogr.*, **645**, 233.
84. Schlögl, K. and Widhalm, M. (1984). *Monatsh. Chem.*, **115**, 1113.
85. Werner, A. (1989). *Kontakte* (*Darmstadt*), 50.
86. Dingenen, J., Somers, I., and Mermans, R. (1991). Poster at 2nd International Symposium on Chiral Discrimination, Rome, May.
87. Francotte, E., Auer, K., Huxley, A., and Eckhardt, W. (1992). Paper presented at 3rd International Symposium on Chiral Discrimination, Tübingen, Germany, Oct.
88. Kinkel, J. N. and Dingenen, J. (1992). Poster at 3rd International Symposium on Chiral Discrimination, Tübingen, Germany, Oct.
89. Rimböck, K.-H., Kastner, F., and Mannschreck, A. (1985). *J. Chromatogr.*, **329**, 307.

90. Mannschreck, A., Koller, H., Stühler, G., Davies, M. A., and Traber, J. (1984). *Eur. J. Med. Chem.-Chim. Ther.*, **19**, 381.
91. Francotte, E. and Richert, P. (1992). CIBA, Basel, Switzerland (unpublished results).
92. Francotte, E., Stierlin, H., and Faigle, J. W. (1985). *J. Chromatogr.*, **346**, 321.
93. Blaschke, G. (1986). *J. Liq. Chromatogr.*, **9**, 341.
94. Blaschke, G. and Markgraf, H. (1980). *Chem. Ber.*, **113**, 2031.
95. Blaschke, G. and Maibaum, J. (1985). *J. Pharm. Sci.*, **74**, 438.
96. Allenmark, S. and Thompson, R. A. (1987). *Tetrahedron Lett.*, **28**, 3751.
97. Blaschke, G., Fraenkel, W., Fröhlingsdorf, B., and Marx, A. (1988). *Liebigs Ann. Chem.*, 753.
98. Brandt, K. H., Hampe, T. R., Nagel, J., and Schmitt, E. (1992). Poster at 3rd International Symposium on Chiral Discrimination, Tübingen, Germany, Oct.
99. Thomas, S. B. and Surber, B. W. (1992). *J. Chromatogr.*, **623**, 390.
100. Fitos, I., Visy, J., Magyar, A., Katjar, J., and Simonyi, M. (1990). *Chirality*, **2**, 161.
101. Vandenbosch, C., Massart, D. L., Egginger, G., and Lindner, W. (1993). *Trends Anal. Chem.*, **12**, 169.
102. Okamoto, Y., Kawashima, M., Aburatani, R., Hatada, K., Nishiyama, T., and Masuda, M. (1986). *Chem. Lett.*, 1237.
103. Daicel Chemical Industries Ltd. Tokyo. Application sheet on CHIRALCEL OD.
104. Ikeda, H., Negawa, M., Makino, N., and Murata, K. (1993). Paper Presented at 4th International Symposium on Chiral Discrimination, Montreal, Canada, Sept.
105. Isaksson R. and Lamm., B. (1986). *J. Chromatogr.*, **362**, 436.
106. Lohmann, D., Francotte, E., and Auer, K. (1986). CIBA, Basel, Switzerland (unpublished results).
107. Gjörstrup, P., Harding, H., Isaksson, R., and Westerlund, C. (1986). *Eur. J. Pharmacol.*, **122**, 357.
108. Jork, H., Ganz, J., and Junker-Buchheit, A. (1989). In Büchi Laboratoriums, Technik AG (ed.), Präparative Mitteldruck-flüssig Chromatographie, Teil 6, Anwendungsbeispiele, Flawil, Switzerland, pp. 3–8.
109. Howson, W., Kitteringham, J., Mistry, J., Mitchell, M. B., Novelli, R., Slater, R. A., and Swayne, G. T. G. (1988). *J. Med. Chem.*, **31**, 352.
110. Miller, L. and Bergeron, R. (1993). *J. Chromatogr.*, **648**, 381.
111. Mackenzie, R., Dutruc-Rosset, G., Beaudoin, F., Bourzat, J.-D., Chevé, M., and Manfré, F. (1992). Poster at 3rd International Symposium on Chiral Discrimination, Tübingen, Germany, Oct.
112. Kiriyama, K., Iwamoto, R., Morokoshi, N., Matsushita, J., Endoh, K., Nakashima, N., Koda, S., Yasuda, T., Sato, N., and Fukui, Y. (1993). Paper presented 4th International Symposium on Chiral Discrimination, Montreal, Canada, Sept.

113. Salvadori, P., Bertucci, C., Domenici, E., and Giannaccini, G. (1989). *J. Pharm. Biomed. Anal.*, **7**, 1735.
114. Pirkle, W. H., Tsipouras, A., and Sowin, T. J. (1985) *J. Chromatogr.*, **319**, 392.
115. Pirkle, W. H. and Tsipouras, A. (1984). *J. Chromatogr.*, **291**, 291.
116. Pirkle, W. H., Hyun, M. H., Tsipouras, A., Hamper, B. C., and Bank, B. (1984). *J. Pharm. & Biomed. Anal.*, **2**, 173.
117. Pirkle W. H. and Hamper, B. C. (1987). The direct preparative resolution of enantiomers by liquid chromatography on chiral stationary phases. In B. A. Bidlingmeyer (ed.), *Preparative Liquid Chromatography*. Amsterdam: Elsevier, Chap. 7, pp. 235–287.
118. Blaschke, G. and Markgraf, H. (1984). *Arch. Phram.*, **317**, 465.
119. Blaschke, G., Kraft, H.-P., and Markgraf, H. (1980). *Chem. Ber.*, **113**, 2318.
120. Blaschke, G., Kraft, H.-P., Fickentscher, K., and Köhler, F. (1979). *Arzneim.-Forsch./Drug Res.*, **29**, 1640.
121. Winter W. and Frankus, E. (1992). *Lancet*, **339**, 365.
122. Miyamoto, E., Demizu, Y., Murata, Y., Yamada, Y., Kawashima, S., Kontani, H., and Sakai, T. (1993). *J. Chromatogr. A*, **653**, 135.
123. Francotte, E. (1985). CIBA, Basel, Switzerland (unpublished results).
124. Camilleri, P., Eggleston, D., Farina, C., Murphy, J. A., Pfeiffer, U., Pinza, M., and Senior, L. A. (1993). *J. Chromatogr. A*, **654**, 207.
125. Bouchaudon, J., Dutruc-Rosset, G., Alasia, M., Beaudoin, F., Bourzat, J.-D., Chevé, M., Cotrel, C., and James, C. (1991). Poster at 2nd International Symposium on Chiral Discrimination, Rome, May.
126. Capraro, H.-G., Francotte, E., Kohler, B., Rihs, G., Schneider, P., Scartazzini, R., Zak, O., and Tosch, W. (1988). *J. Antibiotics*, **41**, 759.
127. Kaneko, T., Okamoto, Y., and Hatada, K. (1987). *J. Chem. Soc., Chem. Com.*, 1511.
128. Furet, P., Batzl, C., Bhatnagar, A., Francotte, E., Rihs, G., and Lang, M. (1993). *J. Med. Chem.*, **36**, 1393.
129. Blaschke, G., Hilgard, P., Maibaum, J., Niemeyer, U., and Pohl, J. (1986). *Arzneim.-Forsch.*, **36**, 1493.
130. Blaschke, G. and Maibaum, J. (1986). *J. Chromatogr.*, **366**, 329.
131. Masurel, D. and Wainer, I. W. (1989). *J. Chromatogr.*, **490**, 133.
132. Lienne, M., Macaudière, P., Caude, M., Rosset, R., and Tambuté, A. (1989). *Chirality*, **1**, 45.
133. Stanek, J., Alder, A., Bellus, D., Bathnagar, A., Häusler, A., and Schieweck, K. (1991). *J. Med. Chem.*, **34**, 1329.
134. Miller, L., Honda, D., Fronek, R., and Howe, K. (1994). *J. Chromatogr. A*, **658**, 429.
135. Droux, S., Gouraud, Y., Gigliotti, G., and Petit, F. (1993). Paper presented at 4th International Symposium on Chiral Discrimination, Montreal, Canada, Sept.

136. Francotte, E., Richert, P., and Wirz, B. (1991). CIBA, Basel, Switzerland (unpublished results).
137. Isaksson, R., Sandström, J., Eliaz, M., Israely, Z., and Agranat, I. (1988). *J. Pharm. Pharmacol.*, **40**, 48.
138. Herweck, U., Zimmermann, H., and Reichlin, J. (1992). *J. Chromatogr.*, **600**, 312.
139. Günther, K., Carle, R., Fleischhauer, I., and Merget, S. (1993). *Fresenius J. Anal. Chem.*, **345**, 787.
140. Miller, L. and Bush, H. (1989). *J. Chromatogr.*, **484**, 337.
141. Miller, L. and Weyker, C. (1993). *J. Chromatogr. A*, **653**, 219.
142. Okamoto, Y., Aburatani, R., Kawashima, M., Hatada, K., and Okamura, N. (1986). *Chem. Lett.*, 1767.
143. Erlandsson, P., Isaksson, R., Lorenton, P., and Lindberg, P. (1990). *J. Chromatogr.*, **532**, 305.
144. Butler, B. T., Silvey, G., Houston, D. M., Borcherding, D. R., Vaughn, V. L., McPhail, A. T., Radzik, D. M., Wynberg, H., Ten Hoeve, W., Van Echten, E., Ahmed, N. K., and Linnik, M. T. (1992). *Chirality*, **4**, 155.
145. Reiter, F., Pohl, G., and Grill, H. (1992). Poster at 3rd International Symposium on Chiral Discrimination, Tübingen, Germany, Oct.
146. Anton, K., Eppinger, J., Frederiksen, L., Francotte, E., Berger, T. A., and Wilson, W. H. (1994). *J. Chromatogr. A*, **666**, 395.

INDEX